MW01169551

Proceedings

MICROSCOPY AND MICROANALYSIS 2002

Microscopy Society of America
60th Annual Meeting

Microscopy Society of Canada /
Société de Microscopie du Canada
29th Annual Meeting

Microbeam Analysis Society
36th Annual Meeting

International Metallographic Society
35th Annual Meeting

Québec City, Québec, Canada

August 5–8, 2002

Edited by
E. Voelkl
D. Piston
R. Gauvin
A. J. Lockley
G. W. Bailey
S. McKernan

Microscopy and Microanalysis, Volume 8, Supplement 2, 2002

Published by the Press Syndicate of the University of Cambridge
The Pitt Building, Trumpington Street, Cambridge CB2 1RP, England
40 West 20th Street, New York, NY 10011, USA
10 Stamford Road, Oakleigh, Melbourne, Victoria 3166, Australia

First published 2002

Printed in the United States of America

Library of Congress Cataloging-in-Publication Data

CIP data pending

This publication constitutes Supplement 2 to Volume 8, 2002 of *Microscopy and Microanalysis*.

Microscopy AND Microanalysis

THE OFFICIAL JOURNAL OF

MICROSCOPY SOCIETY OF AMERICA
MICROBEAM ANALYSIS SOCIETY
MICROSCOPICAL SOCIETY OF CANADA / SOCIÉTÉ DE MICROSCOPIE DU CANADA
MEXICAN MICROSCOPY SOCIETY
BRAZILIAN SOCIETY FOR MICROSCOPY AND MICROANALYSIS
VENEZUELAN SOCIETY OF ELECTRON MICROSCOPY
EUROPEAN MICROBEAM ANALYSIS SOCIETY

PUBLISHED IN AFFILIATION WITH

ROYAL MICROSCOPICAL SOCIETY
GERMAN SOCIETY FOR ELECTRON MICROSCOPY
BELGIAN SOCIETY FOR MICROSCOPY
MICROSCOPY SOCIETY OF SOUTHERN AFRICA

Editorial Office

Charles E. Lyman, Editor in Chief, Department of Materials Science and Engineering, Lehigh University, 5 East Packer Avenue, Bethlehem, PA 18015, USA; Tel.: (610) 758-4249; Fax: (610) 758-4244; E-mail: charles.lyman@lehigh.edu.

Office of Publication

Cambridge University Press, 40 West 20th Street, New York, NY 10011-4211, USA, Tel: (212) 924-3900; Fax: (212) 645-5960.

Table of Contents

Titles of Sessions and Papers v
Society Presidents xlii
Foreword xliii
Letter from the Retiring Editor xliv
Microscopy Society of America xlv
Microbeam Analysis Society lii
Microscopial Society of Canada / Societé de Microscopie du Canada lvii
International Metallographic Society lix
2002 Program Committee Chairs and Co-Chairs and Local Arrangements Committee lxi
Abstracts of Invited Papers 2
Author Index 467
Subject Index 475

Note: *Invited papers appear in both the printed Proceedings and in the CD-ROM. The suffix "CD" following a page number indicates a submitted paper; these appear only on the CD-ROM.*

Aberration Correction in TEM and STEM and its Application to Real-World Materials – A Symposium Honoring the Contributions of Albert Crewe and Harald Rose

The Emergence of Aberration Correctors for Electron Lenses - J Silcox, Cornell University 2
Some Chicago Aberrations - A V Crewe, University of Chicago 4
Correction of Aberrations-Past, Present and Future - H Rose, Argonne National Laboratory 6
Application of Abberation-corrected Transmission Electron Microscopy to Materials Science - K Urban, M Lentzen, Forschungszentrum Jülich GmbH 8
Applications of a Cs Corrected HRTEM in Materials Science - J L Hutchison, J M Titchmarsh, D J H Cockayne, G Moebus, C J Hetherington, University of Oxford; R C Doole, University of Oxford ; F Hosokawa, JEOL Ltd; P Hartel, M Haider, CEOS GmbH 10
Benefits and Possibilities of Cc-Correction for TEM/STEM - H Müller, S Uhlemann, M Haider, CEOS GmbH 12
Sub-Angstrom Probe Size in HADF-STEM at 120KV - P Batson, IBM ; N Delby, O L Krivanek, Nion Co. ... 14
The Ultimate Resolution in Aberration-Corrected STEM - S J Pennycook, A R Lupini, Oak Ridge National Laboratory; P D Nellist, Nion Co 16
Enhancing the Resolution and Sensitivity of STEM by Aberration Correction - N D Browning, K Sun, R F Klie, University of Illinois; J Liu, Monsanto; M M Disko, ExxonMobil; P D Nellist, N Delby, O L Krivanek, Nion Co. 18
STEM Aberration Correction: Where Next? - O Krivanek, N Dellby, M Murfitt, P Nellist, Z Szilagyi, Nion Co. 20
Quantification of the Resolved Phase Change in Reconstructed Electron Exit Waves of Gold [110] in Different Electron Microscopes - JR Jinschek, C Kisielowski, Lawrence Berkeley National Laboratory; M Lentzen, K Urban, Forschungszentrum Juelich 466CD
Atomic imaging in aberration-corrected HRTEM with application to Al alloys - J H Chen, Delft University of Technology ; K Urban, B Kabius, M Lentzen, Research Center Juelich; J Jansen, H W Zandbergen, Delft University of Technology 468CD
The superSTEM: An Aberration Corrected Analytical Microscopy Facility - A Blochel, L M Brown, University of Liverpool; R Brydson, University of Leeds; A Craven, University of Glasgow; P Goodhew, C J Keily, University of Liverpool 470CD

Effects of Detector Black Level in ADF-STEM Imaging - Z Yu, Cornell University; P Batson, IBM; J Silcox, Cornell University . . . 472CD
Principles of Phase Reconstruction with Aberration Correction Using Three-Dimensional Fourier Filtering Method - T Kawasaki, Y Takai, Osaka University, Japan . . . 474CD
Initial Results from Aberration Correction in STEM - A R Lupini, S J Pennycook, Oak Ridge National Laboratory; O L Krivanek, N Dellby, P D Nellist, Nion Co . . . 476CD
A New Laboratory Designed to Provide an Optimum Environment for Aberration-Corrected Electron Microscopes - L Allard, D A Blom, T A Nolan, W H Sides, L J Degenhardt, J A Mayo, Oak Ridge National Laboratory; W Vogen, Vibration Engineering Associates; E St. Romain, Lord, Aeck & Sargent . . . 478CD
Estimation of the Electron Beam Energy Spread for TEM Information Limit - M O'Keefe, National Center for Electron Microscopy; P Tiemeijer, FEI; M Sidorov, Advanced Micro Devices . . . 480CD
Dynamic Observation of an Atom-sized Gold Wire by Real-time Defocus Image Modulation Processing Electron Microscope - Y Takai, T Kawasaki, Y Kimura, Osaka University, Japan . . . 482CD
New Approach for Ultra-Stable TEM-Column Support Frame - E Essers, G Benner, A Orchowski, LEO Elektronenmikroskopie GmbH; R Kappel, IDE Integrated Dynamics Engineering; M Trunz, IST Ingenieurbüro für Strukturmechanik Trunz, Germany . . . 484CD
Realization of a Field Emission Gun with Advanced Koehler Illumination - G Benner, G Lang, A Orchowski, W-D Rau, LEO Elektronenmikroskopie GmbH; M Haider, CEOS GmbH, Germany . . . 486CD

Image Simulation and Image Processing Techniques

Electron Holography - Where We are and Where to Go - H Lichte, Dresden University . . . 22
Effects of Fresnel Corrections for Phase-Shifting Electron Holography - K Yamamoto, Japan Science and Technology; T Hirayama, Japan Fine Ceramics Cente; T Tanji, Nagoya University; M Hibino, Aichi Institute of Technology, Japan . . . 24
Bright Electron Beams and Their Applications to Electron Phase Microscopy - A Tonomura, Hitachi, Ltd. . . . 26
Low Voltage Electron Holography - High Voltage Electron Holography - B Frost, A Thesen, D C Joy, University of Tennessee . . . 28
Holographic Observation of Magnetic Fine-Structures in New Magnetic Materials - T Tanji, S Hasebe, T Suzuki, Nagoya University, Japan . . . 30
Novel Techniques for Image Process in Electron Probe Microanalysis - N Mori, H Takahashi, M Takakura, C Nielsen, JEOL Ltd . . . 488CD
Image Simulation of Gold Cluster Detection in TEM and STEM - J Wall, Brookhaven National Laboratory . . . 490CD
Deconvolution Process of High-Resolution HAADF STEM Images - M Shiojiri, Kyoto Institute of Technology; K Watanabe, Tokyo Metropolitan Colleage of Technology; N Nakanishi, T Yamazaki, Tokyo University of Science; M Kawasaki, JEOL USA Inc.; A Recnik, M Ceh, Institute of Jozef Stefan . . . 492CD
High-Resolution HAADF STEM of Inversion Boundaries in Sb2O3-Doped Zinc Oxide - M Shiojiri, Kyoto Institute of Technology; A Recnik, Institute of Jozef Stefan; T Yamazaki, Tokyo University of Science; M Kawasaki, JEOL USA Inc.; M Ceh, Institute of Jozef Stefan; K Watanabe, Tokyo Metropolitan College of Technology . . . 494CD
Families of Particle-Like Fractals with Differing Shapes and Boundary Fractal Dimensions - D S Bright, National Institute of Standards and Technology . . . 496CD
Semi-Automatic Tools to Segment SEM images of Particles - D S Bright, National Institute of Standards and Technology . . . 498CD
Pit depth quantification of PM Aluminum Composites using Scanning Electron Microscopy and Image Analysis - N Martinez, D Busquets, V Amigó, M D Salvador, N Valero, Universidad Politécnica de Valencia, Spain . . . 500CD
Image Analysis in PM Aluminum Composites. Matrix/Reinforcement Characterization - N Martinez, V Amigó, D Busquets, M D Salvador, N Valero, Universidad Politécnica de Valencia, Spain . . . 502CD
HREM Study of Fullerenes Impact in a Metallic Matrix - J Pacaud, A Michel, C Jaouen, F Pailloux, Laboratoire de Metallurgie Physique; S Della Negra, Institut de Physique Nucleaire, France . . . 504CD
A New Computerised Method of Multiple Labelling Detection and Particle Evaluation. - L H Monteiro-Leal, State University of Rio de Janeiro; H Tröster, Biomedical Structure Analysis, Germany; L Campanati, Inst.de Biofísica Bl-G subsolo, Brazil; H Spring, M Trendelenburg, Biomedical Structure Analysis, Germany . . . 506CD

Electron Holography, Interference Phenomena and Related Techniques: A Symposium Honoring the Contributions of Hannes Lichte and Akira Tonomura

Fourier Methods for Field and Phase-shift Calculations of Long-range Electromagnetic Fields - G Pozzi, University of Bologna, Italy 32

Contributions of Elastically and Inelastically Scattered Electrons to High-Resolution Off-Axis Electron Holograms - a Quantitative Analysis - M Lehmann, H Lichte, University of Dresden 34

Mapping of Process Induced Dopant Redistributions by Electron Holography - W D Rau, LEO Elektronenmikroskopie GmbH 36

New Developments and Applications of Electron Holography - T Hirayama, Z Wang, Japan Fine Ceramics Center; K Yamamoto, Japan Science and Technology Corporation; T Kato, Japan Fine Ceramics Center; N Kato, ITES IBM Japan; K Sasaki, H Saka, Nagoya University 38

Electron Holographic Characterization of Nanoscale Magnetic and Electrostatic Fields - M McCartney, Arizona State University 40

The Determination and Interpretation of Electrically Active Charge Density Profiles at Reverse Biased p-n Junctions From Electron Holograms - R Dunin-Borkowski, A Twitchett, P Midgley, University of Cambridge 42

Confocal Holography - R A Herring, D Laurin, Canadian Space Agency 508CD

Phase Retrieval, Symmetrization Rule and Transport-of-Intensity Equation in Application to Induction Mapping of Magnetic Materials - V V Volkov, Y Zhu, Brookhaven National Laboratory 510CD

Image Interpretation of Magnetic Domains in Nd2Fe14B Hard Magnets - M Beleggia, M A Schofield, V V Volkov, Y Zhu, Brookhaven National Laboratory 512CD

Observation of Chain Structure of Superconducting Vortices by Lorentz Microscopy - T Matsuda, O Kamimura, H Kasai, K Harada, T Yoshida, T Akashi, A Tonomura, Hitachi Ltd.; Y Nakayama, J Shimoyama, K Kishio, T Hanaguri, K Kitazawa, University of Tokyo, Japan 514CD

Practical Considerations for Electron Holography on Doped Semiconductor Devices - A Thesen, B G Frost, University of Tennessee; D C Joy, University of Tennessee, Knoxville, Oak Ridge National Lab 516CD

Quantitative Examination of Reverse-Biased Semiconductor Devices using Off-axis Electron Holography - A C Twitchett, R Dunin-Borkowski, P Midgley, University of Cambridge 518CD

Low Voltage Nanotip Interferometry - A Thesen, B G Frost, D C Joy, University of Tennessee 520CD

Development of a Direction-Free Magnetic Field Application System - K Harada, J Endo, N Osakabe, A Tonomura, Hitachi Ltd.; K Kitazawa, University of Tokyo, Japan 522CD

Eleven Thousand Interference Fringes by 1-MV Field Emission Electron Microscope - T Akashi, K Harada, T Furutsu, N Moriya, T Matsuda, H Kasai, T Kawasaki, T Yoshida, A Tonomura, Hitachi Ltd.; K Kitazawa, H Koinuma, Japan Science and Technology Corporation 524CD

Observation of Vortices and Columnar Defects by Using Lorentz Microscopy - H Kasai, O Kamimura, T Matsuda, K Harada, A Tonomura, Hitachi Ltd.; S Okayasu, M Sasase, Japan Atomic Energy Research Institute; Y Nakayama, J Shimoyama, K Kishio, T Hanaguri, K Kitazawa, University of Tokyo, Japan 526CD

Phase Contrast Images of Superconducting Pancake Vortices - M Beleggia, Brookhaven National Laboratory; G Pozzi, University of Bologna; A Tonomura, Hitachi Ltd. 528CD

Vortex Modeling in High-Tc Anisotropic Materials - M Beleggia, Brookhaven National Laboratory; J Masuko, Tokyo Institute of Technology; N Osakabe, Hitachi Ltd; G Pozzi, University of Bologna; A Tonomura, Hitachi Ltd 530CD

Fast and Robust Phase Unwrapping Algorithm for Electron Holography - M A Schofield, Y Zhu, Brookhaven National Laboratory 532CD

Quantitative Phase Imaging and Differential Interference Contrast for Biological TEM - B E Allman, R R van Driel, IATIA Ltd, Australia; P J McMahon, E D Barone-Nugent, K A Nugent, University of Melbourne, Australia 534CD

Holographic Setup for 2D-Dopant Profiling using the Lorentz-lens - M Lehmann, K Brand, H Lichte, University of Dresden 536CD

Ferroelectric Electron Holography - H Lichte, M Reibold, K Brand, M Lehmann, Dresden University 538CD

Improved Information Recovery in Phase Contrast EM for non-Two-fold Symmetric Boersch Phase Plate Geometry - E Majorovits, R R Schroeder, Max-Planck-Institute for Medical Research 540CD

Applications and Developments of Focused Ion Beams

Comparison of FIB TEM Specimen Preparation Methods TEM - R M Anderson, Microscopy Today 44
Focused Ion Beam Based Sample Preparation Techniques - R Langford, A Petford-Long, University Of Oxford; P Gnauck, LEO Elektronenmikroskopie GmbH 46
A Newly Developed FIB System For TEM Specimen Preparation - T Kamino, T Yaguchi, Y Kuroda, T Hashimoto, T Ohnishi, T Ishitani, K Umemura, K Asayama, Hitachi Ltd. 48
FIB Damage in Silicon: Amorphization or Redeposition? - S Rajsiri, B Kempshall, S Schwarz, L Giannuzzi, University of Central Florida 50
Gallium Phase Formation in Cu and Other FCC Metals During Near-Normal Incidence Ga-FIB Milling and Techniques to Avoid this Phenomenon - M W Phaneuf, J Li, Fibics Incorporated; J D Casey Jr., FEI Co. ... 52
Focused Ion Beam: Much More Than a Sample Preparation Tool - P E Russell, North Carolina State University; T J Stark, J P Viterelli, Materials Analytical Services; A R Guchard, J Wang, K L Bunker, J C Gonzalez, D P Griffs, North Carolina State University 54
Dual-beam Focused Ion Beam: A Multifunctional Tool for Nanotechnology - N Yao, E Kung, S Allameh, W Soboyejo, Princeton University 56
The DualBeam FIB in a Materials Science Laboratory - H O Colijn, The Ohio State University 58
Advances in Dual Beam TEM Sample Preparation - M Moore, FEI Company 60
FIB Preparation of Mesa Structures for SIMS Analysis - J M McKinley, B B Rossie, M A Decker, Agere Systems; F A Stevie, North Carolina State University 542CD
EBSD Performed "In-situ" on a Dual-beam FIB - J Farrer, M Chipman, M Tiner, TSL/EDAX 544CD
Enhanced Site Specific Preparation of SEM Cross Sections and TEM Samples by Using CrossBeam Technology - P Gnauck, P Hoffrogge, M Schumann, G Bauhammer, Leo Elektronenmikroskopie GmbH 546CD
Identification of Cleavage Origins Using Focused Ion Beam (FIB) Sectioning - S Xu, R Bouchard, CANMET-Materials Technology Laboratory; J Li, Fibics Inc.; WR Tyson, CANMET-Materials Technology Laboratory 548CD
A Site- and Layer-Specific Sample Preparation Technique for Plan View TEM of Laser Diodes - J Tanimura, K Kawasaki, Y Yoshida, H Kurokawa, Mitsubishi Electric Corporation 550CD
3D Microscopy and Microanalysis of Heterogeneous SEM Samples by Broad Ion Beam Processing: Cutting - Etching - Coating - W Hauffe, D Gloess, Dresden University of Technology; R J Mitro, Gatan 552CD
Deformation of InxGa1-xAs Superlattices under Bending and Nanoindentation - S J Lloyd, University of Cambridge; K M Y P'ng, A J Bushby, D J Dunstan, University of London; P Kidd, Philips; W J Clegg, University of Cambridge 554CD
Auger Analysis of Focused Ion Beam Prepared Lift-Out Specimens - B B Rossie, R M Mills, S D Anderson, Agere Systems; M Antonell, RF Micro Devices; F A Stevie, North Carolina State University 556CD
Focused Ion Beam (FIB) Microscopy and Technology - P E Russell, F A Stevie, North Carolina State University 558CD
Marriage of Focused and Broad Ion Beam: Sample Preparation Optimized for High Performance Analytical (S)TEM - M Sidorov, Advanced Micro Devices 560CD
Application of FIB and TEM for the Characterization of Dewetting Behavior on Ceramics - S R Gilliss, N Ravishankar, University of Minnesota; P G Kotula, J R Michael, Sandia National Laboratories; C B Carter, University of Minnesota 562CD
Planarization Process for Pre-FIB Sample Preparation - T C Lee, J Y Huang, L C Chen, D H I Su, Taiwan Semiconductor Manufacturing Company, Ltd., Taiwan 564CD
"H-Bar Lift-Out" and "Plan-View Lift-Out": Robust, Re-thinnable FIB-TEM Preparation for Ex-Situ Cross-Sectional and Plan-View FIB Specimen Preparation - R J Patterson, D Mayer, L Weaver, M W Phaneuf, Fibics Incorporated 566CD
Applications (Fun and Practical) of FIB Nano-Deposition and Nano-Machining - M W Phaneuf, Fibics Incorporated 568CD

EELS and EFTEM Analysis

Probing the Electronic Structure of Transition Metal Oxides using EELS - D W McComb, University of Glasgow 62

Structures and Energetics of Interfaces in Materials -Ab-initio Local-Density-Functional Theory - C Elsaesser, Max-Planck-Institut fuer Metallforschung Stuttgart 64
Electron Energy-Loss Spectroscopy of Alternative Gate Dielectric Stacks - S Stemmer, D Klenov, Z Chen, Rice University; J-P Maria, A I Kingon, D Niu, G N Parsons, North Carolina State University 66
Development of An 0.2eV Energy Resolution Analytical Electron Microscope - M Tanaka, M Terauchi, K Tsuda, K Saitoh, Institute of Multidisciplinary Research for Advanced Materials; M Mukai, T Kaneyama, T Tomita, K Tsuno, JEOL LTD.; M Kersker, JEOL USA; M Naruse, T Honda, JEOL LTD. 68
Monochromized 200kV (S)TEM - P C Tiemeijer, J H A van Lin, B H Freitag, A de Jong, FEI Electron Optics 70
EFTEM at High Magnification: Principles and Practical Applications - W Grogger, Graz University of Technology; K M Krishnan, Lawrence Berkeley National Laboratory; F Hofer, Graz University of Technology, Austria 72
Understanding DNA Organization in the Nucleus By Fluorescence Microscopy and Energy Filtered Transmission Electron Microscopy - D P Bazett-Jones, The Hospital for Sick Children, Toronto, Canada 74
EELS Measurements on Wurtzite InN - A Mkhoyan, E S Alldredge, N W Ashcroft, J Silcox, Cornell University 570CD
The Electronic Structure of Threading Dislocations in GaN - I Arslan, N Browning, University of Illinois at Chicago 572CD
New Developments in EELS Applied to Interface Study in Magnetoresistive Tunnel Junctions With Manganites - D Imhoff, L Samet, M Tenc, A Gloter, Laboratoire Physique des Solides; F D R Pailloux, J L Maurice, CNRS, France; C Colliex, Laboratoire Physique des Solides, France 574CD
Artificial Charge Modulations in La-doped SrTiO3 Superlattices - D A Muller, A Ohtomo, J Grazul, H Y Hwang, Bell Labs Lucent Technologies 576CD
Atomic Scale Models for Grain Boundary Potentials in Perovskites - R Klie, J Buban, N Browning, University of Illinois 578CD
Probing the Metal-Insulator Transitions in Complex Oxides with EELS Near Edge Structures - G A Botton, A Safa-Sefat, J E Greedan, McMaster University 580CD
Analysis of 4d Transition Metal Oxides by EELS - Y Ito, R E Cook, Argonne National Laboratory; P W Klamut, B M Dabrowski, M Maxwell, Northern Illinois University 582CD
Experimental Set-Up of a Purely Electrostatic Monochromator for High Resolution and Analytical Purposes of a 200 KV TEM - S Uhlemann, M Haider, CEOS GmbH 584CD
Performance and First Results of a New 200 kV FEG-EFTEM - G Lang, G Benner, A Orchowski, W-D Rau, LEO Elektronenmikroskopie GmbH 586CD
A Simple Internal Active Field Compensator for Post Column Spectrometers - C Trevor, D A Ray, A Moonen, J A Hunt, H A Brink, Gatan 588CD
New Approach to Obtain Elemental Maps by Using Energy-Filterd Images - Y Taniguchi, K Kaji, Y Ueki, S Isakozawa, Hitachi Ltd. 590CD
Improvements in Spatially Resolved Characterization using EELS and EFTEM Spectrum-Image Datasets - P Thomas, C Trevor, R Harmon, M Kundmann, J Hunt, Gatan 592CD
Characterization of Exsolution Phenomena by Mapping EELS Fine Structure - U Golla-Schindler, G Lang, G Benner, LEO Elektronenmikroskopie GmbH 594CD
Combining EELS and Angle-Resolved AES to Measure Shallow Profiles of Thin Nitrided Oxide Films on SI - J Bruley, H Wildman, A Paterson, IBM 596CD
Background Removal and Data Analysis for Low-Loss Transmission Electron Energy-Loss Spectroscopy - B Reed, M Sarikaya, University of Washington 598CD
In Situ Determination of Local Ce Oxidation States During Redox Reactions - R Sharma, P Crozier, Arizona State University 600CD
Comparison of ELNES and NEXAFS of Vanadium Oxide V2O5 with Different Spectral Resolution - D S Su, M Hävecker, A Knop-Gericke, R Mayer, Fritz Haber Institute of the Max Planck Society; C Hebert, Technische Universität Wien, Austria; R Schlögl, Fritz Haber Institute of the Max Planck Society 602CD
Observation of Non-uniformities in Calcium Aluminosilicate Glass using EELS - Z Yu, J Silcox, Cornell University 604CD
Chemical Bonding Analysis of AlN Polytypes by ELNES - T Mizoguchi, M Kunisu, Kyoto University; M Yoshiya, Lawrence Berkeley National Laboratory; I Tanaka, H Adachi, Kyoto University; P Rulis, W Y Ching, University of Missouri 606CD
EXELFS and EXAFS: Complementary Probes Into the Structure of Metallic Glasses - F M Alamgir, H Jain, D B Williams, Lehigh University; G Hug, ONERA 608CD

Characterization of Boron Containing Graphite Using TEM and EELS - J S Kim, G H Kim, C H Chun, H H Koo, Agency for Defense Development, South Korea 610CD
Performance of Corrected Imaging Filters - E Essers, D Krahl, B Huber, A Orchowski, LEO Elektronenmikroskopie GmbH 612CD
A New High Stability, 4th Order Aberration Corrected Spectrometer and Imaging Filter for a Monochromated TEM - M Barfels, P Burgner, R Edwards, H Brink, Gatan Inc. 614CD
Parallel EELS Characterization of TiC Commercial Powder - J A Duarte Moller, C González-Valenzuela, Centro de Investigacion en Materiales Avanzados, S. C., México 616CD
AEM Investigation of Strontium Substituted La-Co-Perovskites - I Letofsky-Papst, W Grogger, I Rom, F Hofer, Graz University of Technology, Austria; E Bucher, W Sitte, University of Leoben, Austria 618CD
Optimizing EELS Data Sets Using 'a priori' Spectrum Simulation - N Menon, J A Hunt, Gatan Inc 620CD
Contributions of Inelastically Scattered Electrons to Defect Images - M A Kirk, Argonne National Laboratory; R Twesten, University of Illinois; S P Martin, C J D Hetherington, M L Jenkins, University of Oxford, UK ... 622CD
Electron Energy Loss Spectroscopy Study of NbC Precipitates in an Ultra-Low Carbon Hot-Rolled Microalloyed Steel - C A Hernandez Carreon, Centro de Investigacion en Materiales Avanzados, Mexico; L Bejar Gomez, Universidad Michoacana de San Nicolás de Hidalgo, Mexico; J E Mancilla Tolama, F Espinosa Magana, Centro de Investigacion en Materiales Avanzados, Mexico 624CD
Further Applications of Energy Filtered TEM in Semiconductor Devices - L Tsung, A Anciso, Texas Instruments; J Ringnalda, FEI co. 626CD
Electron Beam Induced Damage in Wurtzite InN - A Mkhoyan, J Silcox, Cornell University 628CD
A Next Generation Imaging Filter for High Voltage Electron Microscopy - D Moonen, C Trevor, D Chew, J A Hunt, P E Mooney, H A Brink, Gatan Inc 630CD
A Faster Approach to Spectrum Imaging and Elemental Mapping in STEM - A Fernandes, J Fung, P Ottensmeyer, University of Toronto 632CD
Chromophore Mapping via Low-Energy Loss Imaging in a Modified EM902: Formation of the Immune Synapse - C Hong, University of Toronto; A Fernandes, Ontario Cancer Institute; J Davis, J Penninger, P Ottensmeyer, University of Toronto, Canada 634CD
Localization of Low Energy Losses and the Mixed Dynamic Form Factor - P Schattschneider, C C Hébert, Vienna University of Technology, Austria; B Jouffrey, École Centrale Paris, France 636CD
An EFTEM Study of Polyimide Adhesion to Single-Walled Carbon Nanotube Bundles - N D Evans, Oak Ridge National Laboratory; C Park, ICASE; R Crooks, Lockheed Martin; E J Siochi, NASA; E A Kenik, Oak Ridge National Laboratory 638CD

Frontiers of X-ray Spectrometry

X-ray Spectrometry in the Fast Lane: An Introduction to High Speed Digital Processing Techniques and their Application to Emerging EDS Technologies - P Grudberg, W Warburton, J Harris, X-ray Instrumentation Associates 76
The Development of Microcalorimeter EDS Arrays - K Irwin, National Institute of Standards and Technology . 78
The Latest Experiences Using a Cryogen Free Microcalorimeter Energy Dispersive X-Ray Spectrometer - D Redfern, J Nicolosi, Edax; J Höhne, CSP Cryogenic Spectrometers; R Weiland, B Simmnacher, Infineon Technologies AG; C Hollerith, Technische Universität Müchen 80
A Well Dressed Microscope: Practical Experience with Microcalorimeter and Silicon Drift Detector Systems - J Small, D E Newbury, J H J Scott, L King, S W Nam, K Irwin, S Deiker, National Institute of Standards and Technology; S Barkan, J Iwanczyk, Photon Imaging Inc. 82
Real Time Color Scans with Scanning Electron Microscopes - A new Application of the XFlash(r) X-ray Detector Technology - G Mäurer, T Schülein, G Kommichau, RONTEC GmbH 84
Transition Edge Sensor Fabrication - K Nelms, D Liu, D McCammon, L Rocks, W Sanders, P Tan, J Vaillancourt, University of Wisconsin-Madison 640CD
Detectors Capable of High Countrates and High Speed Pulse Processing Electronics for X-Ray Spectroscopy - J Nicolosi, M Solazzi, D Redfern, EDAX Inc 642CD
Development of a High Energy-resolution Soft-X-ray Spectrometer for a Transmission Electron Microscope - M Terauci, M Kawana, Tohoku University, Japan 644CD

XML for Microanalysis: EMSA/MAS Spectrum File Format 2.0? - J H J Scott, S A Wight, B B Thorne, National Institute of Standards and Technology 646CD

Spectral Imaging: Getting the Most from All that Data

Spectral Image Analysis: Getting the Most from All that Data - P G Kotula, M R Keenan, Sandia National Laboratories 648CD

Electron Crystallography and Quantitative Electron Diffraction

Exploring the Valence Electron Distribution in High Temperature Superconductors with a Focused Electron Probe - J Tafto, University of Oslo; L Wu, Y Zhu, Brookhaven National Laboratory 86
Refinement of Crystal Structural Parameters and Charge Density using Convergent-Beam Electron Diffraction - K Tsuda, Y Ogata, M Tanaka, Tohoku University, Japan 88
Quantitative Convergent Beam Electron Diffraction - R Holmestad, J Friis, Norwegian University of Science and Technology 90
Bloch Wave Degeneracies and Critical Voltage Effect in CBED Patterns - H Matsuhata, National Institute of Advanced Industrial Scinece, Japan; J Gjønnes, University of Oslo, Norway 92
How to Select the Items for the Shopping List of Future High Resolution Electron Microscopists? - D Van Dyck, University of Antwerp; S Van Aert, A den Dekker, A van den Bos, Delft University of Technology 94
The 3D Structure of a Complex Quasicrystal Approximant Determined by Electron Crystallography - X D Zou, Z M Mo, S Hovmöller, Stockholm University; X Z Li, University of Nebraska-Lincoln; K H Kuo, Chinese Academy of Sciences 96
HRTEM Resolution Extension for Interface by Gerchberg-Saxton Algorithm with Supported Constraint - F R Chen, National Tsing Hua University, Taiwan; U Dahmen, Lawrence Berkeley National Laboratory; J J Kai, National Tsing Hua University, Taiwan 98
The collection of electron diffraction intensity data and their use in structure determination - J Gjonnes, University of Oslo; V Hansen, Stavanger University College; X Li, University of Nebraska Lincoln 100
The Accuracy of Crystal Structure Refinement from Electron Diffraction Data using Parallel Beam Illumination - J Jansen, Technische Universiteit Delft, The Netherlands 102
Single Crystal Electron Crystallography on Organic Molecules - U Kolb, Johannes Gutenberg Universität, Germany; G Matveeva, Russian Academy of Sciences 104
Structure Factor Phase and Amplitude Measurement in AlN by QCBED - B Jiang, J C H Spence, Arizona State University 650CD
Effect of Anisotropic Lattice Vibration in CBED Intensities and Detection of Local Change in Oxygen Deficiency of YBa2Cu3O7-x - Y Tomokiyo, Y Tanaka, D Koga, Kyushu University, Japan; Z Akase, Case Western Reserve University 652CD
A Study of Bonding in Copper by QCBED Measurements - J Friis, Norwegian University of Science and Technology; B Jiang, Arizona State University; R Holmestad, Norwegian University of Science and Technology ... 654CD
Automated Crystallography and Grain Mapping in the TEM - F B Clayton, Lake Highland Preparatory School; B Kempshall, S Schwarz, L Giannuzzi, University of Central Florida 656CD
Progress towards Quantitative Electron Nanodiffraction - J Zuo, R Twesten, B Q Li, J Tao, Y F Shi, J Bording, H Chen, I Petrov, University of Illinois 658CD
On the Amplitude Origin Problem in Dynamical Direct Methods - W Sinkler, UOP LLC; L D Marks, Northwestern University 660CD
Why does the hkl: h+k+l=4n+2 Reflections Reveal Intensity in Si [110]? - P Geuens, University of Antwerp, Belgium; C B Carter, University of Minnesota; D Van Dyck, University of Antwerp, Belgium 662CD
Crystallographic Analysis of Orientational Variants in PbZr0.52Ti0.48O3 Ferroelectric Perovskite - L Wu, Y Zhu, J Li, B Noheda, Brookhaven National Laboratory 664CD
Microstructure of Thick Polycrystalline Silicon Films for MEMS application - H Zhou, P Gouma, State University of New York; B G Kharas, Standard MEMS Inc. 666CD

Systematic Characterization of Reciprocal Space by SAED: Advantages of a Double-Tilt, Rotate Holder - S Turner, National Institute of Standards and Technology 668CD

Electron Backscatter Diffraction of Materials: Geology to Nanotechnology

Discussion of Ways to Energy-Filter the Electron Backscattering Pattern (EBSP) in the Scanning Electron Microscope (SEM) - O C Wells, IBM 106
EBSD Analysis Optimised for Twin-Related Boundaries - V Randle, University of Wales, UK 108
EBSD Spatial Resolution in the SEM when Analyzing Small Grains or Deformed Material - P Rolland, K G Dicks, Oxford Instruments Analytical; R Ravel-Chapuis, JEOL 670CD
A Comparison of Grain Size Measurements IN Al-Cu Thin Films: Imaging Verses Diffraction Techniques - L Gignac, C E Murray, K P Rodbell, M Gribelyuk, IBM 672CD
Microstructure of TiN Coatings by EBSD Techniques - B Y Jeong, S Yue, R Gauvin, R Drew, McGill University 674CD
Orientation Analysis of Ultra-fine Grained Bulk Materials Produced by Accumulative Roll-bonding (ARB) Process by the Use of EBSP Technique - R Yoda, H Haren, Kobelco Research Institute Inc., Japan; N Tsuji, R Ueji, T Toyoda, Y Minamino, Osaka University, Japan 676CD
Application of Orientation Imaging Microscopy in the TEM to Studies of Nano-crystalline Materials - D J Dingley, M J Tiner, S I Wright, TSL/EDAX 678CD
Determination of Activated Slip Systems in Experimentally Deformed Olivine-Orthopyroxene Polycrystals using EBSD - R De Kloe, EDAX; M Drury, Utrecht University; J Farrer, TSL/EDAX 680CD
Chemistry Assisted Phase Differentiation in Automated Electron Backscatter Diffraction - S I Wright, M M Nowell, EDAX 682CD
Misorientation Mapping for Visualization of Plastic Strain via Electron Back-Scattered Diffraction - L N Brewer, M A Othon, L M Young, T M Angeliu, General Electric Global Research Center 684CD
A New Method for Analyzing Electron Backscatter Diffraction (EBSD) Data for Texture using Inverse Pole Figures - C T Chou, P Rolland, K G Dicks, Oxford Instruments Analytical 686CD
The Random Orientation Probability in Consecutive Inverse Pole Figure Method for Texture Determination - C T Chou, P Rolland, K G Dicks, Oxford Instruments 688CD
Another Way to Implement Diffraction Contrast in SEM - X Tao, A Eades, Lehigh University 690CD
Alternatives to Image Quality (IQ) Mapping in EBSD - X Tao, A Eades, Lehigh University 692CD
Comparison of OIM and XRD Texture Determinations in a Deformation Processed B-Ti+Y Metal-Metal Composite - F C Laabs, A M Russell, Iowa State University 694CD
EBSD Characterization of Texture in Tungsten-Rhenium Foils - L N Brewer, B P Bewlay, General Electric Global Research Center 696CD
Electron Back-Scattered Diffraction Misorientation Mapping Applied to Stress Corrosion Cracking of Stainless Steels - M A Othon, L N Brewer, T M Angeliu, L M Young, General Electric Global Research Center 698CD
EBSD Study of Martensite in a Dual Phase Steel - B Y Jeong, R Gauvin, S Yue, McGill University 700CD

Current Topics in Low Voltage SEM

An Assessment of the Pros and Cons of Low Voltage X-ray Analysis in the SEM - E D Boyes, DuPont Company 110
About the Topographic Contrast in LVSEM - J Cazaux, DTI Faculty of Sciences Reims France 112
High Resolution Examination of Biological Samples Using Field Emission Scanning Electron Microscopy - S L Erlandsen, University of Minnesota; J Detry, Honeywell; C Ottenwaelter, C Frethem, University of Minnesota 114
Physics of Low Voltage Scanning Electron Microscopy - R Gauvin, McGill University 116
Advanced Instrumentation for Low Voltage Scanning Microscopy - D C Joy, University of Tennessee 118
Low Voltage Energy Dispersive Quantitative X-ray Microanalysis of Inorganic Light Elements in Bulk Frozen Hydrated Biological Specimens - P Echlin, Cambridge Analytical Microscopy 120

New FESEM Design for 1nm at 1kV Imaging, EDS and BSE Nanoanalysis, and a Discussion of Diffraction Limits, Depth of Field and the Future - E D Boyes, DuPont Company 122
Development of a Simulation Tool for Real World SEM Applications - D Drouin, ARA Couture, Universite de Sherbrooke 702CD
Electron Range at Low Energy (Eo < 10 KeV): Atomic Number Dependant? - P Hovington, M Lagacé, Hydro-Quebec Research Institut; D Drouin, Université de Sherbrooke, Canada; R Gauvin, McGill University, Canada 704CD
Ultra-Low-Voltage Scanning Electron Microscopy In The FEG-SEM - J Liu, Monsanto Company 706CD
Utilisation of Field Emission Scanning Electron Microscopy to Characterise an Active Brazing Alloy (ABA) Produced by an Electroless Plating Technique - M Brochu, D Knuutila, McGill University, Canada; M D Pugh, Concordia University, Canada; R Gauvin, R A L Drew, McGill University, Canada 708CD
FE-SEM Examination of Microalloyed Hypereutectoid Steels - A M Elwazri, R Gauvin, S Yue, McGill University 710CD
Study of Self-assembled InAs Quantum Dots on InP Nano-templates by Low Voltage Scanning Electron Microscopy Cathodoluminescence - N Lahkak, D Drouin, J Beerens, P Magny, Universite de Sherbrooke; J Lefebvre, Conseil national de recherches Canada 712CD
New Tools for Micro-Characterization at Low Beam Voltages (The Right Tools for the Right Job) - D Redfern, J Nicolosi, EDAX Inc 714CD
Fish-eye Optics for the Scanning Electron Microscope - T Agemura, D C Joy, University of Tennessee 716CD
Energy Filtered Imaging in a FEG-SEM for Enhanced Dopant Contrast - C Schönjahn, C Humphreys, University of Cambridge, UK; M Glick, Marconi Labs, UK 718CD
A Study of the Effectiveness of the Removal of Hydrocarbon Contamination by Oxidative Cleaning Inside the SEM - N Sullivan, T Mai, S Bowdoin, Schlumberger Technologies; R Vane, XEI Scientific 720CD
Low Temperature Scanning Electron Microscopy of Irregular Snow Crystals - W P Wergin, U.S. Department of Agriculture; A Rango, New Mexico State University; J Foster, NASA; E F Erbe, C Pooley, U.S. Department of Agriculture 722CD

Electron Backscatter Diffraction in the SEM: Orientation Mapping and Phase Identification for Materials Science

Size Selective Synthesis of Colloidal Platinum Nanoparticles for Use as High Resolution EM Labels - D Meyer, R Albrecht, University of Wisconsin 124
Methodology Advancements in Electron Microscopy Immunolabeling of Hydrated Brain Tissue Using Subnanometer Colloidal Gold Conjugates - H Yi, Emory university; R Peijper, J Leunissen, Aurion Immunogold Reagents, The Netherlands 126
Bioengineering of Reporter Transgenes for Integrated Imaging with Magnetic Resonance, Fluorescence, and Electron Spectroscopy - M Malecki, SABA University School of Medicine 128
Comparing Confocal Immunofluorescence Microscopy (CIM) vs. Freeze Fracture Replica Immunogold Labeling (FRIL) for Cell-specific Localization of Plasma Membrane Proteins in the Mammalian CNS - T Yasumura, K G V Davidson, C S Furman, Colorado State University; J I Nagy, University of Manitoba, Canada; J E Rash, Colorado State University 130
Colloidal Gold Conjugates of Cholera Toxin B-Subunit of Alexa Fluor(r) Fluorescent Dyes for Use in Correlative Studies - E Rosa-Molinar, J L Serrano-Vélez, University of Puerto Rico-Río; P Oshel, R M Albrecht, University of Madison 132
Electron Backscatter Diffraction in the SEM: A Tutorial - J R Michael, Sandia National Laboratories 724CD

New Developments in Immunolabelling

Carbon Nanotube Membrane Probes:Immuno-labeling by LM, AFM, TEM, & FESEM - B Panessa-Warren, Brookhaven National Laboratory; S Wong, B Ghebrehiwet, G Tortora, State University of New York; J Warren, Brookhaven National Laboratory 726CD

PML Nuclear Body Identification and Ultrastructure in Rodent Tissues and Cultured Cells by Post-Embedding Immunogold Labeling. - S Hearn, The Hospital for Sick Children, Toronto Canada; E Heard, Curie Institute, Paris, France; E Querido, S Lowe, D Spector, Cold Spring Harbor Laboratory 728CD
Immunofluorescent Detection and Quantification of Hepatitis A Virus Using Scanning Confocal Microscopy - I Kukavica-Ibrulj, A Darveau, I Fliss, Laval University 730CD
Immunocytochemical Strategies for Signal Enhancement in Post-Embedding Immunogold Procedure - G Grondin, Université de Sherbrooke, Canada 732CD
Rapid Processing of Cultured Cells for LR White Embedding - G Ning, P Traktman, Medical College of Wisconsin 734CD
Microwave-Assisted MAP-2 Immunoreactivity in Formalin-Fixed, Paraffin-Embedded Guinea Pig Brain - R Kan, C M Pleva, T A Hamilton, J P Petrali, U.S. Army Medical Research Institute of Chemical Defense ... 736CD

Microscopy, Microanalysis and Image Analysis in the Pharmaceutical Industry

Quantitation in Image Analysis: Practical Considerations for Drug Discovery - M Esterman, J Hanson, Lilly Research Labs 134
A Novel Application of Solids Characterization by Environmental Scanning Electron Microscopy (ESEM) Utilizing a Peltier Stage - R J Maxwell, J A Hanko, Pharmacia Corporation 136
Immunolocalization of Phosphodiesterase Isoenzymes in Rat Tissues using Confocal Microscopy - B E Maleeff, R C Mirabile, T K Hart, H C Thomas, L W Schwartz, S J Newsholme, GlaxoSmithKline 138
A Field Emission Scanning Electron Microscopy Method to Assess Recombinant Adenovirus Stability - L J Obenauer-Kutner, P M Ihnat, T Yang, B J Dovey-Hartman, M J Grace, Schering-Plough Research Institute 140
Relevant Applications of Scanning Electron Microscopy in a Pharmaceutical Development Laboratory - R L Mueller, GlaxoSmithKline 142
Design of a Digital Microscopy Imaging Platform for Pathology Studies in Pharmaceutical Research and Development - X Ying, T Monticello, Aventis Pharmaceuticals 738CD
A Quantitative Image-Based in vitro Assay for Induction of Phospholipidosis in Hepatocytes - G D Gagne, R J Gum, R A Jolly, M A Heindel, J A Fagerland, Abbott Laboratories 740CD

Corporate Members Session

Applications of Atomic Force Microscopy in the Pharmaceutical Sciences - J T Thornton, Digital Instruments, Veeco Metrology Group 742CD
Laser Scanning Cytometry for the Pharmaceutical Industry - E Luther, CompuCyte Corporation 744CD

Advances in Microwave Technology – Creating a Revolution in Biological Specimen Processing for Light and Electron Microscopy

Microwave-assisted Embedding of Tissue Culture Cell Monolayers - K L McDonald, University of California, Berkeley 144
Microwave Tissue Processing in a Teaching Laboratory - R S Demaree, California State University 146
Microwave Assisted Decalcification with Recirculation of Temperature Controlled Solutions - S P Tinling, R Kular, University of California, Davis; R T Giberson, Ted Pella, Inc. 148
Microwave Assisted Rapid Tissue Processing for Disease Diagnosis in a Veterinary Diagnostic Laboratory - R Nordhausen, B Barr, R Hedrick, University of California-Davis 150
Microwave Processing in Diagnostic Electron Micreoscopy - R G Gerrity, G W Forbes, Medical College of Georgia 152

The Use of Microwave Technology in a Clinical E.M. Laboratory - R Austin, Louisiana State University Medical Center 154
Immunocytochemistry: A New Microwave Application - J Day, R Demaree, California State University; R Giberson, Ted Pella, Inc.; T E Munoz, California State University 156
Recent Advances in Microwave Assisted Specimen Processing: Keeping it Cool. - M Sanders, University of Minnesota 158
Microwave Processing and Pre-Embedding Nanogold Immunolabeling for Electron Microscopy - J Buchanan, K Micheva, S J Smith, Stanford University School of Medicine 160
The Use of New Microwave Techniques to Facilitate the Immunostaining of Paraffin Sections on Glass Slides - R Giberson, Ted Pella 162

Biominerals

A Basic Strategy for Biomineralization: Taking Advantage of Disorder - S Weiner, Y Levi-Kalisman, S Raz, I M Weiss, L Addadi, Weizmann Institute, Israel 164
Hydroxyapatite Formation and Its Interaction with Osteoblastic Cells - H Vali, P Ghiabi, McGill University; J Henderson, Jewish General Hospital; M D McKee, E Chevet, S K Sears, McGill University 166
Magnetic and Structural Characterization of Biogenic Magnetite - M R McCartney, Arizona State University; R E Dunin-Borkowski, Cambridge University, UK 168
Scanning Electron Microscopy and Atomic Force Microscopy of the Ring Structures in Human Calcium Oxalate Urinary Stones - P Rez, Arizona State University; H Fong, M Sarikaya, University of Washington 746CD
A Novel Titanium Nitride (TiN) Coating Enhances Early Osteointegration of Titanium Alloy Pins in Rat Femora - G Sovak, Technion-Israel Institute of Technology 748CD
TEM Approach in Investigations of Microbially Assisted Uranium Reduction - A C Dohnalkova, D W Kennedy, J W Fredrickson, Pacific Northwest National Laboratory 750CD
Botanical Iron Biominerals: Electron Diffraction and Microscopy Identification - M Gajdardziska-Josifovska, M Schofield, D Robertson, R McClean, W Kean, C Sommer, University of Wisconsin-Milwaukee 752CD
Searching for Phytoferritin in Unstained Plant Sections - M Gajdardziska-Josifovska, H Owen, University of Wisconsin Milwaukee 754CD
Chemical and Morphological Changes in an Arsenopyrite Crystal Induced by Thiobacillus ferrooxidans - M A Makita, Instituto Technologico de Chihuahua II; A Duarte, Centro de Investigacion en Materiales Avenzados; S Arevalo, Unversidad Autonoma de Chihuahua; E Orrantia, Centro de Investigacion en Materiales Avenzados 756CD
Characterization of Biodegradable/Bioresorbable Polymer Scaffolds for Osteoblast Cell Growth by SEM and TEM - S Iadarola, A Crugnola, R Joshi, J Tessier, C Sung, University of Massachusetts 758CD
In Situ ESEM Investigation of the Drying of Demineralized Dentin - V M Dusevich, A G Glaros, J D Eick, University of Missouri - Kansas City 760CD

Scanning Probe Microscopy: Technical Advances and Applications

Global and Local DNA Structure and Dynamics. Single Molecule Studies with AFM - Y Lyubchenko, L Shlyakhtenko, Arizona State University; V Potaman, R Sinden, Texas A&M University 170
Microelastic Mapping of Living Cells: Changes in Relative Elasticity Between Nuclear and Cytoplasmic Regions of Mitotic MDCK Cells - E A-Hassan, J Hoh, Johns Hopkins School of Medicine 172
Touching In Biological Systems: A 3D Force Microscope - R Superfine, G Bishop, J Cummings, J Fisher, K Keller, G Matthews, D Still, R M Taylor, L Vicci, C Weigle, B Wilde, University of North Carolina-Chapel Hill 174
Atomic Force Microscopy Studies of Initiation Steps of DNA Mismatch Repair - H Wang, Y Yang, D A Erie, University of North Carolina 762CD
Atomic Force Microscopy as a Tool for the Investigation of Cellular Cytoplasmic Membrane Dynamics - G Fried, S Rubakin, J Sweedler, University of Illinois at Urbana Champaign 764CD

Synthesis and STM Imaging of Substituted Phenylalkyl Ethers: Towards Functional Group Discrimination - A J M Lubag Jr, K Kangasniemi, I H Musselman, University of Texas at Dallas 766CD
Frictional Properties of Hydrogenated and Deuterated Alkanethiols - J R Garcia, University of Texas at El Paso; D C Christensen, R W Carpick, University of Wisconsin-Madison 768CD
Investigation of Self-Assembled Nanofibers Using Atomic Force Microscopy - M E Salmon, P E Russell, North Carolina State University; E B Troughton Jr., Lord Corporation 770CD
Scanning Probe Microscopy of Silicone Treated Cellulose Fibers: A Comparison of In-Situ, LFM, and Phase Imaging for Characterization of Adsorbed vs. Covalently Bonded Silicone Coatings - J Teetsov, M Butts, S Stoessel, K Shaffer, M Burrell, General Electric 772CD
Thin Film Thickness and Grain Structure Determination of Ferroelectric SrBi2Ta2O9 with Cross-sectional Atomic Force Microscopy - D Pechkis, C Caragianis-Broadbridge, Southern Connecticut State University; A Hein Lehman, K Klein, Trinity College; J P Han, T Ma, Yale University 774CD
Construction with Collagen - Insight through Atomic Force Microscopy - J K Rainey, University of Toronto; M F Paige, Stanford University; C K Wen, A C Lin, M C Goh, University of Toronto 776CD
Atomic Force Microscopy Studies of Human Recombinant Bone Morphogenetic Protein OP-1 - L Siperko, Northeastern Ohio Universities College of Medicine; S Chubinskaya, Rush Medical College; D Rueger, Stryker Biotech; W Landis, Northeastern Ohio Universities College of Medicine 778CD

Scanned Probe Microscopy Joint Physical and Biological Sciences Tutorials

Scanned Probe Microscopy: A Brief Tutorial - P Russell, North Carolina State University 780CD

Industrial Applications of Microscopy – Techniques for the Real World

Microstructural Characterization of Automotive Materials - W Donlon, A Chen, L Gonzalez, J Hangas, E Lee, M Peck, Ford Motor Company 176
Applied Microscopy for the Paper Industry - D R Rothbard, Institute of Paper Science and Technology 178
Transmission Electron Microscopy Applications in the Semiconductor Industry - Challenges and Solutions for Specimen Preparation - Y Xu, C Schwappach, Intel Corporation 180
The Unique Diversity of Electron and Confocal Imaging Applications in a Natural History Museum Setting - A V Klaus, American Museum of Natural History 182
3D Imaging of Polymer-based Materials by Laser Scanning Confocal Microscopy - L Liang, DuPont Co. 184
Application of the UHREM Technique in Atomic Modelling of Growth Defects in CVD-Diamond Films - D Dorignac, Centre National de la Recherche Scientifique, France 186
Contributions of Microscopy to Advanced Industrial Materials and Processing - S Dionne, G Carpenter, G Botton, T Malis, CANMET-Materials Technology Laboratory; M Phaneuf, Fibics Incorporated 188
AFM / SEM Backscattered Imaging of Slip Bands in Titanium - S Okerstrom, Medtronic Inc 782CD
Application of X-ray Micro-Tomography to Study the Morphology and Porosity of Pharmaceutical Granules - L Farber, Merck &Co, Inc; G Tardos, The City College of the City University of New York; J N Michaels, Merck & Co, Inc 784CD
Low Temperature SEM as a Tool for Understanding Dynamic Events in Consumer Products Research Manufacture and Use - D Jacobs, Y Boissy, S Lindberg, Procter & Gamble Co 786CD
Novel Rapid Nondestructive Technique for Locating Tiny Voids in Metallization Line - Z H Gan, C M Tan, Nanyang Technological University, Singapore 788CD
Characterization of Si in a W Matrix Using Diffraction Contrast in the TEM - B Prenitzer, Agere Systems; B Kempshall, University of Central Florida; J McKinley, W Stinebaugh, I Wylie, Agere Systems 790CD
A New Automated Method for Fast and Reliable Tilt-Series Acquisition in Electron Tomography - C Kuebel, W Voorhout, A van Balen, D Hubert, M Otten, FEI Company 792CD
Measurement and Visualisation of Micron - Scale Strain Distributions in Aluminum Alloys - D C Steele, D J Lloyd, Alcan International 794CD
Developing a Mechanistic Understanding of CO2 Mineral Sequestration Process for Power Plants - R Sharma, M J McKelvy, H Bearat, A V G Chizmeshya, R W Carpenter, Arizona State University 796CD

Polarization Micorscopy Study of Jet Fuel Crystalization - M Vangsness, The University of Dayton Research Institute 798CD
Microscopy and Microanalysis of Hematite Precipitates from the Zinc Industry - T T Chen, J E Dutrizac, Natural Resources Canada 800CD
One Button Wear Debris Analysis - R J Lee, H P Lentz, D G Kritikos, RJ Lee Group; A M Toms, JOAP-TSC 802CD
Microscale Morphology and Micro-Fluorescence of Oil Sands Extraction Froth from Poorly Processing Ores - R Mikula, V Munoz, Natural Resources Canada 804CD
SEM Studies in Textile Finishing Processes - W Goynes, D V Parikh, V Edwards, T Vigo, US Department of Agriculture 806CD
Measurement of Multiple Lumens in Glass Rods - W Mershon, CamScan USA; J Edwards, Becton Dickinson Accu-Glass 808CD
Examination of Water-Formed Deposits in Steam Boilers By Scanning Electron Microsopy - J N Williard, M J Esmacher, BetzDearborn Inc 810CD
Coating for High-Magnification with Your Au/Pd Sputter Coater - K Pham, Medtronic Inc 812CD
Size Dependence to The Metal Support Interaction: Pd/g-Al2O3 Catalysts - K Sun, University of Illinois at Chicago; L Liu, Monsanto Company; N Nag, Engelhard Corporation; N D Browning, University of Illinois at Chicago 814CD

Quality Systems for Microscopy and Microanalysis: ISO 9000 and More

Quality Systems for Microscopy & Microanalysis: ISO 9000 and More - E Steel, National Institute of Standards and Technology 816CD

Special Staining Techniques for Biological/Materials Samples

The Staining of Polymers - R Smith, Lake Havasu City, AZ 190
Staining and Other Microscopic Techniques for Textiles - E Boylston, United States Department of Agriculture 192
All That Glitters is Not Gold: Approaches to Labeling for EM - R Albrecht, University of Wisconsin 194
Stains for the Determination of Paper Components and Paper Defects - J H Woodward, Buckman Laboratories, Inc. 196
Fluorescent Specimen Preparation Techniques for Confocal Microscopy - J Drazba, The Cleveland Clinic Foundation 198

Technologists' Forum Special Topics: Immunology 101: Back to Basics

Immunology 101: The Basics of Immunoglobulins and Immunostaining - W G Jerome, Vanderbilt University School of Medicine 818CD

3-D Electron Microscopy of Macromolecules

Transmembrane Signalling of the Insulin Receptor: 3D Reconstruction from STEM Imaging, Crystallography and NMR Spectroscopy - P Ottensmeyer, A Oh, R Luo, University of Toronto; A Fernandes, D Beniac, Ontario Cancer Institute; C Yip, University of Toronto 200
Structure of the Eukaruotic Transcription Machinery: Insights into the Mechanism of Transcription Initiation and Regulation - F J Asturias, The Scripps Research Institute 202
Functional Architecture of a Protein-Degradation Machine - A C Steven, T Ishikawa, M R Maurizi, NationalIn-stitutes of Health 204

The Ribosome-Ligand Interactions and Dynamics as Inferred by Cryo-EM - J Frank, J Sengupta, M Valle, R Agrawal, State University of New York at Albany 206
Sindbis Virus Reconstruction at 11Å Resolution Reviews Details of Glycoprotein and Nucleocapsid Organization - W Zhang, M Mukhopadhyay, S V Pletnev, R J Kuhn, M G Rossmann, T S Baker, Purdue University 208
The Rigor Structure of Acto-Myosin and Its Implications for Motor Function - R R Schroeder, I Angert, W Jahn, K C Holmes, Max-Planck-Institute for Medical Research 820CD
Pseudo-atomic Structure of Coxsackievirus A21 Complexed with Its Cellular Receptor, ICAM-1 - C Xiao, C M Bator, P Chipman, TS Baker, R J Kuhn, Purdue University; E Wimmer, State University of New York; A Craig, John Radcliffe Hospital; M G Rossmann, Purdue University 822CD
Three-Dimensional Architecture of Latent and Active Meprin B - M T Norcum, K B Labat, University of Mississippi Medical Center; G P Bertenshaw, J S Bond, The Pennsylvania State University College of Medicine 824CD
X-Ray Cryo-Tomography of Whole Yeast at 60 nm Resolution - C Larabell, University of California at San Francisco; M A LeGros, Lawrence Berkeley National Laboratory 826CD
Capturing Transient Molecular Structures on the Millisecond Time Scale for EM Imaging - F Q Zhao, R Craig, University of Massachusetts Medical School 828CD
TEM Analysis of B-amyloid Fibrillogenesis: New Strategy-Old Problem - R P Apkarian, J Dong, D Lynn, Emory University 830CD
NICKEL-NTA-NANOGOLD BINDS HIS-TAGGED PROTEINS - J Hainfeld, Brookhaven National Lab; W Liu, V Joshi, R D Powell, Nanoprobes, Inc. 832CD
Strategies to Optimize Order Within Planar Arrays of Myelin Basic Protein for Electron Crystallography - C M Hill, I R Bates, C E Antler, G F White, F R Hallett, G Harauz, University of Guelph 834CD
Electron Crystallography of the E. Coli Outer Membrane Protein WzaK30 - C M Hill, J Nesper, C Whitfield, G Harauz, University of Guelph 836CD
Three-dimensional Imaging of Toxoplasma gondii-Host Cell Membrane Interactions Reveals Numerous Bridges and Fission Pore with High Resolution Low Voltage Field Emission Scanning Electron Microscopy on De-embedded Thick Sections - H Schatten, University of Missouri-Columbia; H Ris, University of Wisconsin 838CD
GroEL: A Proteinaceous "Surfactant"? - J Deaton, C Savva, J Sun, Texas A&M University; S Sharma, Texas A&M System HSC Houston ; A Holzenburg, J Sacchettini, R Young, Texas A&M University 840CD
Automated Electron Tomography Software for High-Precision, 3-D Reconstructions - T Oikawa, H Nishioka, JEOL Ltd Japan; H Furukawa, M Shimizu, Y Suzuki, JEOL System Technology Ltd Japan; B L Armbruster, JEOL USA 842CD

Electron Cryo-microscopy of Biological Macromolecules

Cryo-EM and X-ray Crystallographic Studies on the Monomeric Kinesin Motor KIF1A - M Kikkawa, University of Texas; E Sablin, University of California, San Francisco; Y Okada, H Yajima, University of Tokyo; R J Fletterick, University of California, San Francisco; N Hirokawa, University of Tokyo 210
Electron Cryo-Microscopy and Image Reconstruction of Adeno-Associated Virus Type 2 Empty Capsids - B Buttcher, EMBL-Heidelberg; S Kronenberg, J R Kleinschmidt, DKFZ-Heidelberg 212
Molecular Structure of an Icosahedral Pyruvate Dehydrogenase Complex - S Subramaniam, D Shi, N Perham, V Cambridge, J L S Milne, National Cancer Institute 214
Merging Focal Pairs for Improved Particle Selection and Orientation Determination - S Ludtke, W Chiu, Baylor College of Medicine 216
Challenges in the Automation of Cryo-microscopy of Macromolecular Structure - C Potter, D Fellmann, R Milligan, J Pulokas, C Suloway, Y Zhu, B Carragher, The Scripps Research Institute 218
Herpes Simplex Virus Capsid Maturation Visualized by Time-Lapsed Cryo-EM - J Heymann, B Trus, N Cheng, National Institutes of Health; W Newcomb, J Brown, Univ. of Virginia; A Steven, National Institutes of Health 844CD
Structure of the Human Reovirus Virion at 9.6Å Resolution - X Zhang, S B Walker, Purdue University; M L Nibert, Harvard Medical School; T S Baker, Purdue University 846CD

Improvements in single particle cryo-EM: 11A Structure of 290 kDa XDH Refined from Multiple 3D Reconstructions - D Beniac, Ontario Cancer Institute; T Iwasaki, Nippon Medical School; B Eger, University of Toronto; E Pai, P Ottensmeyer, Ontario Cancer Institute 848CD
Conformational Changes of Dynamin-Lipid Tubes upon GTP Addition: A Time-Resolved Study Using Digital-Imaging Cryo-TEM - D Danino, J Hinshaw, National Institutes of Health 850CD
Cucumber Mosaic Virus-Fab Complex Examined by Electron Cryo-Microscopy and Three-Dimensional Image Reconstruction - V D Bowman, P R Chipman, T J Smith, E S Chase, AW Franz, K L Perry, T S Baker, Purdue University 852CD
HHMI Tecnai F30 Helium Microscope: Initial Results and Observations - R Grassucci, Z Liu, T Wagenknecht, J Frank, Wadsworth Center 854CD
Multivariate Statistical Analysis and Tomographic Processing of Helical Objects - R R Schroeder, I Angert, Max-Planck-Institute for Medical Research; J Frank, Howard Hughes Medical Institute; K C Holmes, Max-Planck-Institute for Medical Research 856CD
Dynamical Scattering and Protein Reconstruction by Electron Crystallography of Multi-Layered Crystals - O Vossen, R R Schroeder, Max-Planck-Institute for Medical Research 858CD
Electron diffraction and microscopy studies of an amyloid forming peptide - R Diaz-Avalos, C Long, E Fontano, D L Caspar, Florida State University 860CD
Low-Dose Electron Diffraction of Catalase Crystals Dried Within a Matrix of the Disaccharide, Trehalose: Is a "Dry Protein Crystallography" Feasible? - W H Massover, UMDNJ - New Jersey Medical School 862CD
Enhancing Contrast of Weak Phase Objects Using a Zernike-Type Phase Plate in Phase Contrast TEM - E Majorovits, Max-Planck-Institute for Medical Research; K Nagayama, National Institute for Physiological Sciences; R R Schroeder, Max-Planck-Institute for Medical Research 864CD
Monitoring the Temperature of a Cryogenic Stage for Cryo-EM - D Fellmann, J Puloka, C Conway, C S Potter, B Carragher, The Scripps Research Institute 866CD
Microstructural Characterization of Molecular Assemblies by Digial-Imaging Cryo-TEM - D Danino, National Institutes of Health 868CD
A Modularized Program For A Fully Automated Image Acquisition - S Meyer, D Typke, J M Plitzko, W Baumeister, Max-Planck-Institute for Biochemistry 870CD
Remote Scripting for Microscope Control Applications on the Tecnai TEM - C Suloway, J Pulokas, The Scripps Research Institute; A van Balen, FEI Electron Optics B.V.; B Carragher, C S Potter, The Scripps Research Institute 872CD
Controlled Production of Unsupported Lipid Bilayers As Supports For Electron Microscopy of Single Membrane Particles. - M Strauss, D McAlduff, P Ottensmeyer, University of Toronto 874CD

The Nuts and Bolts of Electron Tomography

The Nuts and Bolts of Electron Tomography: Imaging of Big and Messy Biological Structures - G Sosinsky, University of California, San Diego 876CD

Biological Ultrastructure (Cells, Tissues, Organ Systems)

The Microvasculature of The Brain of The Sterlet, Aciprnser ruthenus L. A Scanning Electron Microscope (SEM) Study of Vascular Corrosion Casts - M Klein, B Stöttinger, B Minnich, W D Krautgartner, A Lametschwandtner, Universty of Salzburg, Austria 220
Development of Subretinal Venous Sphincters in Spontaneously Hypertensive Rats (SHR) - S Aharinejad, U Firbas, University of Vienna 222
Microvasculature of the Urinary Bladder of the Dog Studied with Light Microscopy, Electron Microscopy and Vascular Corrosion Casts - C W Ridner, R L Kao, F E Hossler, East Tennessee State University 224
Blood Flow Regulating Structures in the Avian Kidney - H Ditrich, University of Vienna 226
Dynamics of Rat Endometrium as Studied with SEM and Vascular Corrosion Casts - U M Spornitz, I Bartuskova, University of Basel, Switzerland 228

Electron Tomography of Frozen Hydrated Sections - C E Hsieh, M Marko, J Frank, C Mannella, Wadsworth Center 878CD
High-Voltage Electron Tomography of the Centrosome in Caenorhabditis elegans - E O'Toole, University of Colorado; K McDonald, University of California; A Hyman, T Mueller-Reichert, Max Planck Institute of Molecular Cell Biology and Genetics 880CD
Limits of Specimen Thickness for Energy-Filtered Tomography of Stained Plastic Sections at 120 kV Beam Voltage - E Kocsis, A Weisberg, B Moss, X Chen, T S Reese, R D Leapman, National Institutes of Health . . 882CD
New Embedding Formulations Using Quetol 651 - E A Ellis, Texas A & M University 884CD
Hormonal Regulation of the Glycogen/Glucose Balance in Xenopus Laevis - D R Atar-Zwillenberg, M Atar, U M Spornitz, University of Basel, Switzerland 886CD
Changes in Steroidogenic Ultrastructural Features of Corpus Luteum in the Turtle Chelydra Serpentina Relative to Hormonal Levels Under Natural Conditions - T Ba-Omar, A Al-Kindi, I Mahmoud, Sultan Qaboos University 888CD

Advances and Applications in Vascular Corrosion Casting in Microvascular Research

The Circulatory System Of Decapod Crustaceans and Its Functional Role In Cardiovascular Dynamics - I McGaw, University of Nevada, Las Vegas 230
Blood Supply of the Symphysis Pubis - R C G da Rocha, R P Chopard, University of Sao Paulo, Brazil 232
Corrosion Casting of the Microvasculature in Normal Limbs and Limbs with Venous Ulceration - M N Phillips, A M van Rij, M Zhang, G T Jones, University of Otago, New Zealand 234
Quantitative Analysis of Cerebral Corrosion Casts from Endoglin +/+ and +/− Mice - R J Mount, J Satomi, A D Paterson, The Hospital for Sick Children; K G terBrugge, M Wallace, Toronto Western Hospital; R Harrison, M Letarte, The Hospital for Sick Children 890CD

Cellular Dynamics Using Atomic Force Microscopy

Cellular Dynamics Observed at Sub-Nanometer Resolution Using Atomic Force Microscopy - D Müller, Max-Planck-Institute of Molecular Cell Biology and Genetics 892CD

Pathology

The Distribution of NPC-1 Protein in Macrophages is Altered after Oxidized LDL Lysosomal Accumulation - W G Jerome, B Cox, J B Vaughan, Vanderbilt University Medical Center 894CD
Compositional Analysis of Atherosclerotic Lesions in a Mouse Model: Validation of a new Method using Brightfield, Fluorescence and Polarized Light Microscopy in Conjunction with Computer-Assisted Image Analysis - M Wadsworth, D Schneider, B Sobel, D Taatjes, University of Vermont 896CD
Characterizing Visible Lipids Imaged by the Real Time Microscope in Fibroblasts from Patients with Lysosomal Storage Diseases - N A Pham, B Chue, T Richardson, Richardson Technologies Inc.; J W Callahan, The Hospital for Sick Children 898CD
Apoptotic Cell Death: A Factor in Mustard Gas-Induced Dermal Pathology - J P Petrali, R K Kan, T Hamilton, C Pleva, U.S. Army Medical Research Institute of Chemical Defense 900CD
Confocal Arthroscope Development For in vivo Knee Joint Diagnosis - D Smolinski, T B Kirk, University of Western Australia; P Delaney, OptiScan Pty Ltd; J P Wu, K Miller, M Zheng, University of Western Australia 902CD
Role of EM in the Diagnosis of Smallpox and Other Causes of Rash Illness - C S Goldsmith, I K Damon, Y Ichihashi, S Schmid, S R Zaki, Centers for Disease Control and Prevention (CDC) 904CD
Immunohistochemical Identification of a Malignant Tumor of the Heart. - S Siew, Professor, Michigan State University; N C Caliman, Ingham Regional Medical Center 906CD
Microscope or MACROscope - Which System Provides a Better Scope on Image Analysis?? - J Woo, T Nicklee, D Hedley, P Constantinou, Ontario Cancer Institute 908CD

Morphometric and Ultrastructural Assessment of Bronchial Mucous Glands in Sheep Following Smoke Inhalation and Burn Injury - R A Cox, Shriners Hospital for Children; PC Moller, University of Texas Medical Branch; A S Burke, Shriners Hospital for Children; A Chandra, K Shimoda, L D Traber, F C Schmalstieg, D L Traber, H K Hawkins, University of Texas Medical Branch 910CD
The STEM and Retrovirus Structure - M N Simon, B Y Lin, JS Wall, Brookhaven National Laboratory 912CD
Processing Cell Cutures, Cytospins, Smears and Epoxy, Parraffin or Frozen Sections on Glass or Polystyrene Supporting Subsrates for TEM:A Review - K Chein, M L heathershaw, R C Heusser, H Shirolski, R Gonzalez, C C Nast, A H Cohen, Cedars-Sinai Medical Center 914CD
Enzymatic Metallography: A Simple New Staining Method - J F Hainfeld, Nanoprobes; R N Eisen, Greenwich Hospital; R R Tubbs, Cleveland Clinic Foundation; R D Powell, Nanoprobes 916CD
Detection of Beryllium in Human Lung Using Enery Dispersive X-ray Analysis - P Ingram, K J Butnor, T A Sporn, Y L Roggli, Duke University 918CD
Degranulation of Mast Cells in Dengue Patients - L S Asher, A S Norton, Walter Reed Army Institute of Research; S Krivda, Walter Reed Army Medical Center; H Wong, M Mammen, A Lyons, Walter Reed Army Institute of Research; S Thomas, Walter Reed Army Medical Center; W Sun, K Eckels, D Vaughn, Walter Reed Army Institute of Research 920CD
Image Analysis of Egyptian Mummy Hair - W M Hess, T B Ball, W Griggs, M Kuchar, R Phillips, Brigham Young Univesity 922CD
Unexpected Disturbance of Ultrastructure of Porcine Oocytes by Very Low Concentrations of the Solvent DMSO - C Campagna, J L Bailey, M-A Sirard, Université Laval, Canada; P Hyttel, Royal Veterinary and Agricultural University, Denmark 924CD
Adrenal Cortex Ultrastructural Alterations Caused by Zootoxins - M Pulido-Mendez, A Rodriguez-Acosta, H J Finol, Universidad Central de Venezuela 926CD
Glycogen Storage Disease in a Mexican Baby girl:Ultrastructural and Light Microscopy Findings. - M A Ponce-Camacho, Hospital y Clínica OCA; J P Flores-Gutierrez, E Ramírez-Bon, J Ancer-Rodríguez, Hospital Universitario Dr. Jose Eleuterio Gonzalez 928CD
Light and Electron Microscopy in a Case of Myophosphorylase Deficiency (Glycogenosis V or McArdle Disease) - M G Hadfield, J Perez-Berenguer, Virginia Commonwealth University Health Sciences/MCV; E Westbrook, Virginia State University 930CD
Ultrastructural Defects in Developing Enamel of Diabetic Mice - M Atar, University of Basel; P Verry, Hoffmann-La Roche Ltd.; D R Atar-Zwillenberg, U M Spornitz, University of Basel 932CD
Angio-Tumoral Laminin Expression in Human Gastrointestinal Adenocarcinomas - P Tonino, H J Finol, Central University of Venezuela; C Hidalgo, Venezuelan Institute for Scientific Research; L Sosa, Central University of Venezuela 934CD
Ultrastructure of Granular Cell Change in a Wilm's Tumor - V Edwards, C Senger, J Hwang, H Rosenberg, C Smith, The Hospital for Sick Children Toronto 936CD
Ganglioglioma: Unusual Ultrastructural Features of Neuronal Cells - K L Ho, Henry Ford Hospital 938CD
Ultrastructural Characterization of the Neuromuscular Junction in Diaphragm of the Acetylcholinesterase Knockout Mouse - T Hamilton, U.S. Army Medical Research Institute of Chemical Defense; R E Sheridan, S S Deshparde, O Lockbridge, University of Nebraska; M Adler, U.S. Army Medical Research Institute of Chemical Defense 940CD
The Expression of Caveolin by HDL in Cholesterol-Loaded Aortic Endothelial Cells - W T Chao, V C Yang, Tunghai University 942CD
Atherosclerosis: The Apolipoprotein E-Deficient Mouse Model Revisited - R Coleman, T Hayek, S Keider, M Aviram, Technion-Israel Institute of Technology 944CD
Identification of Actin-Based Stress Fibers with a Morphometric Shape Factor - C A Heckman, J M Urban, Y Li, M L Cayer, J A Barnes, Bowling Green State University 946CD
Evidence of an Intraerythrocytic Secretory Pathway in Erythrocytes Infected With Plasmodium Falciparum - T Schneider, M E O'Donnell, T F Taraschi, Jefferson Medical College 948CD
Fluorescence Imaging of Mitochondrial Responses to Glucose Challenge and Probes in Mitochondrial DNA-Deficient Osteosarcoma Cells - E Kohen, J Hirschberg, C Ornek, M Monti, J Berry, University of Miami 950CD
Zwitterionic Detergents Promote the Formation of Atypical Aß40 Fibrils - R W McLaughlin, R J Chalifour, X Kong, L Lavoie, P Sarazin, D Stéa, Neurochem Inc.; H Vali, McGill University; X Wu, P Tremblay, F Gervais, Neurochem Inc. 952CD

P53 and Statistical Analysis of the Seminiferous Tubules Epithelium in Experimental Unilateral Cryptorchid Testes in Rats - F Al-Bagdadi, R Downer, R Stout, T Gillis, D Hillmann, P Crawford, B Eilts, Louisiana State University 954CD
Scanning Electron Microscopy Morphology of the Uterine Epithelium of Mares Infused with Gentamicin - F Al-Bagdadi, B Eilts, Louisiana State University; G Richardson, University of Prince Edward Island; D Hillmann, Louisiana State University 956CD

The Microstructural Approach to Food Processing and Engineering

Experimental Microscopy of Food Systems - E Kolodziejczyk, M Michel, Nestlé Research Center, Switzerland 236
Microstructural Analyses to Study Ingredient Functionality, Interactions and Quality in Frozen Foods - H D Goff, University of Guelph 238
Effect of Acidity on Optical Properties of Isolated Skeletal Muscle Fibers - H J Swatland, University of Guelph 240
Fat Crystal Networks - A Marangoni, University of Guelph 242
The Effects of a Non-polar Surfactant on the Crystallization Behaviour and Physical Properties of the High-Melting Triglyceride Fraction of Milkfat - J W Litwinenko, A G Marangoni, University of Guelph ... 244
Light Microscopy and TEM to Study the Effect of Biopolymers on Ice Recrystallization in Ice Cream - A Regand, D Goff, University of Guelph 246
Methodologies for the Study of Food systems by Environmental Scanning Electron Microscopy (ESEM) - D J Stokes, B L Thiel, A M Donald, University of Cambridge, UK 958CD
Application of Environmental SEM to the Study of Food Systems - B Thiel, D J Stokes, A M Donald, University of Cambridge, UK 960CD

Plant-microbes Interactions at the Cellular and Molecular Levels

Cytological Features of Programmed Cell Death in Nicotiana tabacum Cells in Relation to the Expression and Localization of Death Regulators - L Brisson, N Bolduc, M Ouellet, F Pitre, I Fortin, University Laval 248
In Situ Localization of AOS in Host-Pathogen Interactions - K B Tenberge, M Beckedorf, B Hoppe, Westfälische Wilhelms-Universitét Münster, Germany; A Schouten, Wageningen University, Netherlands; M Solf, M von den Driesch, Westfälische Wilhelms-Universitét Münster, Germany 250
Use of High Pressure Freezing and Freeze Substitution to Study Host-Pathogen Interactions in Fungal Diseases of Plants - C W Mims, E A Richardson, University of Georgia 252
Cytochemical Localization of Fungal Wall Components in Host-Pathogen Interactions: Particular Labeling wilth Gold-complexed Probes - G B Ouellette, Canadian Forest Service; R P Baayen, Plant Protection Institute; H Chamberland, Universite Laval; M Simard, Canadian Forest Service; P M Charest, Universite Laval 254
Ultrastructural Investigation of the Mycoparasitic Interaction Between Stachybotrys Elegans and its Host Rhizoctonia Solani - P M Charest, Universite Laval; G Taylor, S H Jabaji-Hare, McGill University 256
H2O2 In Interspecies Signaling: A New Role in Host Detection - W J Keyes, D G Lynn, W K Erbil, J V Taylor, R P Apkarian, Emory University 962CD
Biological Cycle of Helminthosporium solani: An Overview using Microscopy - C Martinez, R Tweddell, Université Laval, Canada 964CD
TMV Infection Cycle in Plant Cells Revealed by 3-D Microscopy - R H Berg, S Asurmendi, R N Beachy, Danforth Plant Science Center 966CD
Ultrastructural Study of Interactions Between Phytophthora Fragariae and a Biological Contral Agent Streptomyces hygroscopicus var Geldanus - M Paquet, Universite Laval; S Agbessi, C Beaulieu, Universite de Sherbrooke; P M Charest, Universite Laval 968CD
The Fusion Protein BnBI-1GFP is Localized to the Endoplasmic Reticulum (ER) and Allows Visualization of ER Reorganization Following Treatment with Salicylic Acid - N Bolduc, L Brisson, Laval University 970CD
Inhibition of Turnip Vein Clearing Virus Movement in Seeds of Infected Arabidopsis thaliana Plant; A Microscopic Study - L Gallegos, K Fambrough, New Mexico State University; R Lartey, US Department of Agriculture; S Ghoshroy, New Mexico State University 972CD

Microbiology

Diffusion of Bacteriophage Capsids on a Glass Surface During Single-particle Fluorescence Microscopy - P Serwer, I Wu, S Huang, G A Griess, The University of Texas 974CD
Escherichia Coli and Metal: How Much Will Be Sequestered and Where?; a STEM-EDX Study - J J Goldberg, T E Jensen, Lehman College 976CD
Uptake and Release of Phosphorus from Polyphosphate Bodies by Synechococcus leopoliensis. - J Hagan-Brown, T E Jensen, Lehman College 978CD
Observations of Nom-Human Primate Diarrhea Viruses by Negative Stain Electron Microscopy - C Humphrey, B Jiang, United States Centers for Disease Control and Prevention; H McClure, Emory University 980CD
Dynamic Atomic Force Microscopy of RecA/DNA System - B Sattin, C Goh, University of Toronto 982CD
Ultrastructural Characterization of ATP Synthase Subunit b Overexpression in E coli - T Gales, M Mazzulla, H Kallender, GlaxoSmithKline; F Hill, Avidis SA; B Maleeff, GlaxoSmithKline 984CD
Origin of Cytoplasmic Nucleocapsids in MDBK-cells Infected With Bovine Herpesvirus 1 (bHV-1) - P Wild, E M Schraner, E Loepfe, University of Zürich; P Walther, M Müller, University of Ulm, Germany; M Engels, University of Zürich, Switzerland 986CD
Observation and Investigation of Plasimid DNA Using an Atomic Force Microscope - Y R Ma, H C Chiang, National Dong Hwa University, Taiwan; Y D Yao, Academia Sinica, Taiwan 988CD
Imaging Soil Bacteria in an Environmental Scanning Electron Microscope - J L Saleta, P Holden, University of California, Santa Barbara 990CD

Biomedical Applications

Detection and Localization of Gd-DTPA Following Dynamic Contrast Enhanced Magnetic Resonance Imaging (dMRI) Using Cryo Field Emission Scanning Electron Microscopy (FESEM) and Energy Dispersive X-ray Spectrometry - C A Ackerley, M D Noseworthy, A Tilups, G A Wright, L E Becker, Hospital for Sick Children, Toronto Canada 992CD
The Expression of Ets-1 and c-Jun in Morphogenesis of Aortic Endothelial Cells in vitro - Y C Hsu, V C Yang, Tunghai University 994CD
Entrapment of Streptomycete Spores in a Chitosan-Polyphosphate Matrix - G Jobin, G Grondin, C Beaulieu, Université de Sherbrooke, Canada 996CD
Ultrastructural damage of CHO AA8 cells in the presence of uranyl acetate (UA): The cytotoxicity of hexavalent uranium (U(VI)). - M Salanga, D Stearns, Northern Arizona University 998CD
Imaging Physical Properties and Dynamics of Dendritic Spines by Atomic Force Microscopy - B A Smith, P H Grütter, McGill University; Y De Koninck, Laval University 1000CD
mtCLIC/CLIC4, a Chloride Channel Protein, Participates in Apoptosis and is Localized to the Inner Membrane of Mitochondria - V V Speransky, K S Suh, E Fernández-Salas, S H Yuspa, A C Steven, National Institutes of Health 1002CD
Morphological Evidence for Adhesive Role of Thin-Symmetric Junctions in Undulating Membranes of Lens Nuclear Fiber Cells - M J Costello, K O Gilliland, C D Freel, University of North Carolina at Chapel Hill . 1004CD
Mitochondrial Internal Structure Correlated with Respiratory Activity in Cultured Human Cells - J M L Selker, R W Gilkerson, K Snyder, R Rossignol, R A Capaldi, University of Oregon 1006CD
Acid Phosphatase Activity in the Chick's Inner Ear - G Cohen, Troy State University; T Kido, Yamaguchi School of Medicine, Japan 1008CD
Distribution of Keratin Intermediate Filaments in Cultured Thymic Epithelial Cells (TEC) is Dependent Upon Growth Medium Calcium Content - S S Sands, W D Meek, Oklahoma State University; J Hayashi, University of Maryland; R J Ketchum, Oklahoma State University 1010CD
Epidermal Growth Factor (EGF) Induces a Phenotypic Switch of a Human Glioblastoma Cell Line to Neurospheres - T Bargar, H El-Refaey, I Ahmad, University of Nebraska; M Ebadi, University of North Dakota; J Rodriguez-Sierra, University of Nebraska 1012CD
Ultrastructure of Glassy-winged Sharpshooter Mouthparts and Salivary Sheaths - T Freeman, North Dakota State University; R Leopold, J Buckner, D Nelson, US Department of Agriculture 1014CD

Cry-electron Microscopy of Large Complexes

Cry-electron Microscopy of Large Complexes: A Cold Look at Fusion - S Fuller, University of Oxford 1016CD

Advances in Linking Structure to Function in Biomaterials

Directing Protein Assembly At Interfaces: Balancing Electrostatic And Hydrophobic Forces and the Role of Epitaxy. - C Yip, University of Toronto 258
Spreading of Fibrinogen at Model Surfaces Studied by AFM - C Siedlecki, A Agnihotri, Pennsylvania State University 260
Linking Atomic Force Microscope Images of Proteins to Their Genetic Sequence - S Eppell, B A Todd, Case Western Reserve University 262
The Effect of Pepsin Digestion on Type II Collagen Monomers - J Rammohan, S J Eppell, Case Western Reserve University 1018CD
Surface Organization and Nanopatterning of Collagen by Dip-Pen Nanolithography - D L Wilson, R Martin, M Cronin-Golomb, Tufts University; C A Mirkin, Northwestern University; D L Kaplan, Tufts Universtiy 1020CD
Electron-Beam Micropatterning of Bioactive Surfaces - P Krsko, M Libera, Stevens Institute of Technology; R Clancy, New York University; J Kohn, Rutgers University 1022CD
Vascularity of a Tissue-Engineered Model of Human Phalanges - A Yanke, Miami University; J Hillyer, J Killius, N Isogai, S Asamura, R Jacquet, W J Landis, Northeastern Ohio Universities College of Medicine 1024CD

Correlative Microscopy

Parallel Transmission Electron and Atomic Force Microscopy: Direct and Repetitive Correlation of TEM and AFM images by a Novel Sample Holder - A C Lin, M C Goh, University of Toronto 1026CD
Ultrasound Biomicroscopy as a Probe of Cellular Ultrastructure - Y M Heng, The Hospital for Sick Children Toronto; M Butler, University of Toronto; M Kolios, Ontario Cancer Institute; G Czarnota, Princess Margaret Hospital Toronto, Canada 1028CD
Combined ALEXA-488 and Nanogold Antibody Probes - W Liu, J F Hainfeld, R D Powell, Nanoprobes, Incorporated 1030CD
Correlative Microscopy of Cereballar Bergmann Glial Cells - O J Castejon, Universidad del Zulia Maracaibo; M Dailey, Iowa University; R P Apkarian, Emory University; H V Castejon, Universidad del Zulia Maracaibo 1032CD

Basic Confocal Microscopy

Basic Confocal Microscopy: A Tutorial - J Jerome, Vanderbilt University Medical Center; R Price, University of South Carolina School of Medicine 1034CD

Confocal and Deconvolution for Biologists

Laser Scanning Microscopy: Seeing More by Imaging Less - N S White, Oxford University, UK 264
Characterization and Use of Wide-Field Fluorescence Microscopy and Image Restoration in Quantitative Live Cell Imaging - M Platani, A Lamond, J Swedlow, University of Dundee 266
Biological Photonic Crystals - Revealed by Multi-photon Nonlinear Microscopy - P C Cheng, University of Buffalo; C K Sun, National Taiwan University; B L Lin, Development Ctr for Biotechnology, Taiwan; S W Chu, I S Chen, T M Liu, National Taiwan University; S P Lee, Development Ctr for Biotechnology, Taiwan; H L Liu, M X Kuo, D J Lin, Natl. Taiwan Normal University 268

Fluorescent Probes for Ultrasensitive Cytochemical and Histochemical Imaging - I Johnson, Molecular Probes 270
Confocal Microscopy System Performance: Foundations for Measurements, Quantitation and Deconvolution. - R Zucker, T Stoker, US Environmental Protection Agency 272
Spatial and temporal assays to determine the dynamics of protein localisation and organelle movement in single living cells - R J Errington, P J Smith, S C Chappell, W H Evans, A Fajardo-Bermudez, P E M Martin, University of Wales 274
Medical Diagnosis Using Miniaturised Confocal Microscopes - A Hibbs, BIOCON 276
Optimization of Illumination Pulse Duration Increases Flexibility and Performance of Multiphoton Microscopes for Multidisciplinary Research - K Garsha, G Fried, University of Illinois at Urbana-Champaign 1036CD
FFT 'light' - Image Deconvolution based on Real to Real-Space Frequency Transformations - F Margarine, University of Sydney; P C Cheng, University at Buffalo 1038CD
Automatic 3D detection and quantification of co-localization - S Costes, E Cho, M Catalfam, T Karpova, J McNally, P Henkart, S Lockett, National Institutes of Health 1040CD
Multi-View Three-Dimensional Image Montaging & Signal Attenuation Correction for Maximizing the Imaging Depth and Lateral Extent of Confocal Microscopes - O Al-Kofahi, A Can, Rensselaer Polytechnic Institute; S Lasek, D Szarowski, J Turner, The Wadsworth Center; B Roysam, Rensselaer Polytechnic Institute 1042CD
Mapping Organism Expression Levels at Cellular Resolution in Developing Drosophila - D W Knowles, S Keranen, M D Biggin, D Sudar, Lawrence Berkeley National Laboratory 1044CD
VEGF-Induced Cytoskeletal Alterations Relating to Vascular Permeability in Endothelial Cells - K Spencer, D Cheresh, The Scripps Research Institute 1046CD
An Estimate of the Contribution of Spherical Aberration and Self-shadowing in Confocal and Multi-photon Fluorescent Microscopy - P C Cheng, University at Buffalo; A R Hibbs, Biocon; H Yu, P C Lin, National University of Singapore; W Y Cheng, Williamsville East High School, NY 1048CD
Structures Inside Living Cells Limit Optical Sectioning Precision - J Pawley, University of Wisconsin 1050CD
Yeast Meiotic Chromosome Structure Revealed by Deconvolution Microscopy - Z Zhang, University of Wyoming; M N Conrad, M E Dresser, Oklahoma Medical Research Foundation 1052CD
Genetic Analysis and Age Determination of Chinook Salmon in the California Central Valley - J J Youngblom, J Mullins, J H Youngblom, California State University Stanislaus; T Heyne, California Department of Fish and Game 1054CD
Mitochondria Abnormalities in Prostate Cancer Cells and Tissue - H Schatten, University of Missouri-Columbia; A Chakrabart, Cleveland Clinical Foundation 1056CD
Zebrafish Neuromast Hair Cell Nuclei are Labeled In Vivo by Uptake of Monomeric Cyanine Dyes - G MacDonald, D Raible, E Rubel, University of Washington 1058CD

Advances in Ultrastructural and Non-invasive Imaging of Skin

New Aspects of the Skin Barrier Organisation Assessed by Diffraction and Electron Microscopic Techniques - J Bouwstra, Y Grams, Leiden/Amsterdam Center for Drug Research; G Pilgram, Utrech University, The Netherlands; H Koerten, Leiden University Medical Center, The Netherlands 278
Non-Invasive diagnosis of skin sructure and biochemistry based on non-linear optical microscopy & spectroscopy - P T C So, K H Kim, L H Laiho, K Bahlmann, MIT; C Buehler, Paul Scherrer Institute, Switzerland; C Y Dong, National Taiwan University, Taiwan 280
Detectability of Reflectance and Fluorescent Contrast Agents for Real-Time in Vivo Confocal Microscopy - M Rajadhyaksha, S Gonzalez, Massachusetts General Hospital 282
Mapping Inter-Cellular Water in Skin - A Aitouchen, Stevens Institute; S Shi, Unilever Research; M Libera, Stevens Institute; M Misra, Unilever Research 284
Monitoring Skin Hydration and Product Induced Changes by Near-Infrared Spectroscopic Imaging - S L Zhang, T M Hancewicz, D J Palatini, P Kaplan, M Misra, Unilever; E M Attas, NRC Canada 286
In Vivo Confocal Raman Spectroscopy of the Skin - P J Caspers, Erasmus University, Rotterdam; G W Lucassen, Philips Research, The Netherlands; H A Bruining, G J Puppels, Erasmus University, Rotterdam 288

Imaging of Cellular Trafficking in Skin Using Multiphoton and Handheld Confocal Microscopy Techniques - S C Watkins, G C Papworth, University of Pittsburgh 290
Electron Tomography Study on Junctions in Skin - W He, Skirball Institute of Biomolecular Medicine; P Cowin, New York University Medical Center; D L Stokes, Skirball Institute of Biomolecular Medicine 1060CD
In vivo confocal Fluorescence imaging of skin surface cellular morphology - D T Leeson, Unilever Research 1062CD
Two-Photon 3-D Mapping of Tissue Endogenous Flourescence Specias Based in Flourescence Excitation and Emission Data - L H Laiho, Massachusetts Institute of Technology; T M Hancewicz, P D Kaplan, Unilever; P T So, Massachusetts Institute of Technology 1064CD
Skin Hydration Disrupts the Stratum Corneum Barrier Lipids with Formation of Large Internal Water Pockets - R R Warner, K J Stone, Y L Boissy, Procter & Gamble Co 1066CD

Botany

Irradiation-Induced Development of Nanoscale Features in Steel: Complementary 3D-APFIM and FEG-STEM Characterization - M G Burke, Bettis Atomic Power Laboratory; M Watanabe, D B Williams, Lehigh University; J M Hyde, AEA TEchnology 292
On the Control of Atomic Clustering, Segregation and Partitioning: Nanoscale Materials Technology - S P Ringer, University of Sydney, Australia 294
Nanoscale Materials for Information Storage - A K Petford-Long, University of Oxford, UK; P Shang, IBM Storage Division; Y G Wang, N Owen, University of Oxford, UK 296
Future Hard X-ray Nanoprobe at the Advanced Photon Source - D C Mancini, J Maser, G B Stephenson, Argonne National Laboratory 298
Scanning Probe Microscopy in TEM : an In-Situ Approach for Nano-scale Property Measurements - Z L Wang, Georgia Institute of Technology 300
TEM Characterization of Thin, Epitactic Ni2MnGa films on GaAs - S McKernan, J W Dong, C J Palmstrøm, University of Minnesota 302
Mechanics of Nanowires - R S Ruoff, X Chen, D Dikin, W Ding, Northwestern University; MF Yu, Advanced Technologies Groups; G J Wagner, Northwestern University 304
The Study of Maize Epidermal Replica by Oblique Illumination Microscopy - V Cheng, Transit Middle School; W Y Cheng, Williamsville East HS; D B Walden, University of Western Ontario; P C Cheng, University at Buffalo 1068CD
The Characteristics of Early Helladic II Period Wood Recovered From an Underwater Shipwreck Site near Dokos Greece - M W Pendleton, T C Stephens, A Ellis, Texas A&M University; G Fox, California State University 1070CD
3-Dimensional Visualization of na2/na2 Stem in Maize - W Y Cheng, Williamsville East High School; D B Walden, University of Western Ontario; P C Cheng, University at Buffalo 1072CD
Subcellular Localization of a Novel Transcription Factor in Watered and Drought Stressed Phaseolus acutifolius - L Rodriguez-Uribe, S Ghoshroy, M O'Connell, New Mexico State University 1074CD
Application of Stain FM4-64 and Confocal Microscopy to Investigate the Nature of The Lemon, Citrus limon (L.) Burm. F. Oil Glands - D A Margosan, L H Aung, US Department of Agriculture 1076CD
Changes in Thallus and Algal Cell Components of Two Lichen Species in Response to Low-Level Air Pollution at Pacific Northwest Forests - H S Ra, R F E Crang, University of Illinois at Urbana-Champaign 1078CD
The Study of Airflow Pattern Around a Maize Plant by Schlieren Optics - W Y Cheng, Williamsville East High School NY; P C Cheng, University at Buffalo; D B Walden, University of Western Ontario 1080CD
Elemental Distribution in Leaves of Sporobolus virginicus using Nuclear Microprobe - Y Naidoo, G Naidoo, University Durban 1082CD
Immunolocalization of a Recombinant Cellulase in Transgenic Tobacco Plants - H J Bae, H L N Chamberland, Université Laval; S Laberge, Centre de Recherches et de Agriculture et Agroalimentaire Canada; YS Kim, Chonnam National University, Korea 1084CD
Annual Sea Slug Population's Life Cycle is the Result of Apoptosis - W Mondy, S K Pierce, University of South Florida 1086CD
SEM Examination of Conductive Tissues of Pinus Koraiensis Needles - Z H Ning, Southern University; X He, Chinese Academy of Sciences; K Abdollahi, Southern University 1088CD

Structure and Function of Hornwort Stomata - J R Lucas, K S Renzaglia, Southern Illinois University Carbondale 1090CD

Advances in Nanoscale Technology

Surface Imaging by Self-propelled Nanoscale Probes - H Hess, J Clemmens, University of Washington; J Howard, Max Planck Institute, Dresden, Germany; V Vogel, University of Washington 1092CD
First Data from a Commercial Local Electrode Atom Probe - T T Gribb, J D Olson, R L Martens, J D Shepard, S A Wiener, T C Kunicki, R M Ulfig, D R Lenz, E M Strennen, E X Oltman, J H Bunton, D R Strait, T F Kelly, Imago Scientific Instruments 1094CD
Subnanoscale characterization of lamellar interfaces in a complex TiAl alloy - S S A Gerstl, Northwestern University; Y W Kim, Wright-Patterson Air Force Research Laboratory; D N Seidman, Northwestern University 1096CD
Influence of W on the Temporal Evolution of the Microstructure of a Ni-Al-Cr Superalloy on a Nanoscale - C K Sudbrack, D Isheim, Northwestern University; R D Noebe, NASA Glenn Research Center; D N Seidman, Northwestern University 1098CD
A Subnanoscale Study of Mg Segregation at Al/Al3Sc Interfaces - E A Marquis, D N Seidman, Northwestern University 1100CD
Composition Modulation in GaAs/GaSb Short Period Superlattices - C J Wauchope, University of Michigan Electron Microbeam Analysis Lab; C Dorin, J M Millunchick, University of Michigan Materials Science and Engineering 1102CD
3-D Electron Microscopy for Nano-Technology and the IC Industry - C Kuebel, D Hubert, W F Voorhout, M T Otten, FEI Company, The Netherlands 1104CD
A New High-Resolution Electron Microscope with Easy Operation System for Nano Analysis - M Matsushita, M Ohsaki, Y Kondo, M Naruse, T Honda, JEOL Ltd Japan; M Kersker, JEOL USA Inc. 1106CD
Atomic Resolution Images Using Forbidden Reflections on (111) Face Centered Cubic Nanoparticles - SC Y Tsen, P A Crozier, Arizona State University; M Gajdardziska-Josifovska, University of Wisconsin Milwaukee 1108CD
HRTEM Studies of Morphological and Interfacial Changes of Nanodiamond in Field Emission Experiments - T Tyler, A V Kvit, V V Zhirnov, J J Hren, North Carolina State University 1110CD
TEM Study of Metal/Support Interaction in Pd/CexZr1-xO2 Model Auto Catalyst - H P Sun, University of Michigan; G W Graham, Ford Research Laboratory; C H F Peden, S Thevuthasan, Pacific Northwest National Laboratory; X Q Pan, University of Michigan 1112CD
Atomic Scale Structural Analysis of Sn-Si Quantum Dots - Y Lei, P Möck, T Topuria, N D Browning, University of Illinois at Chicago; R Ragan, H A Atwater, California Institute of Technology 1114CD
Virus Nanoblocks for Molecular Electronics - M J Kim, University of North Texas; A S Blum, Naval Research Laboratory; B Gnade, University of North Texas; B R Ratna, Naval Research Laboratory 1116CD
On-Chip NanoFabricated Collagen Membranes Observed by High-Voltage Electron Microscopy - J N Turner, D H Szarowski, W Shain, K Buttle, W F Tivol, New York State Department of Health; H Bagle, A J Spence, S Retterer, L Lapek, T Richards, M Isaacson, M Spencer, Cornell University 1118CD
Enhanced Quality Nanocrystalline Metal Films Produced By Dendrimer Mediated Thin Film Growth - F Xu, L Li, J C Yang, University of Pittsburgh; S C Street, University of Alabama; J A Barnard, University of Pittsburgh 1120CD
Ag Nanostructure Evolution on H-terminated Si(111) Surfaces - B Q Li, Y F Shi, H Chen, J M Zuo, University of Illinois 1122CD
Spontaneous Self-Organisation Of Gold Nanoparticles Into Ordered Two-Dimensional Arrays - C J Kiely, Lehigh University; C S Cheung, M Brust, University of Liverpool 1124CD
Atom Probe Tomography: A Technique for Nanoscale Characterization - M K Miller, E A Kenik, Oak Ridge National Laboratory 1126CD
TEM Characterization of WS2 Nanotubes - R Rosentsveig, A Margolin, Y Feldmann, R Popovitz-Biro, R Tenne, The Weizmann Institute of Science, Israel 1128CD
TEM Nanostructure Manipulation Device for in Situ Study of Elastic Behavior of Nanoparticle Chain Aggregates - Y J Suh, S V Prikhodko, S K Friedlander, University of California, Los Angeles 1130CD

Low Voltage FESEM Evaluation of Nano and Microfabricated Materials - E J Basgall, The Pennsylvania State University Nanofabrication Facility, University Park, PA 16802 .... 1132CD
TEM Image Analysis of Self-Organized Large Gold Nanoparticle Arrays - S L Tripp, B Kim, A Wei, Purdue University 1134CD
Nanoscale Structural Manipulation of Ion Irradiated Pyrochlore - J Lian, L M Wang, R C Ewing, University of Michigan 1136CD
AFM Analysis of Gas Cluster Ion Impact Craters and Smoothing - C Santeufemio, Epion Corporation 1138CD
Field Effects on Particle Assembly in Composite Needle Systems - D L Jaeger, T Tyler, A V Kvit, R S Sanwald, V V Zhirnov, J J Hren, North Carolina State University 1140CD
Focused Ion Beam Assisted Nanofabrication - Patterned Growth of Carbon Nanotubes - J Jiao, L Dong, S Foxley, C L Mosher, D W Tuggle, Portland State University 1142CD
Characterization of Nanoparticle Films and Structures Using Focused Ion Beam Milling and Transmission Electron Microscopy - C R Perrey, C B Carter, University of Minnesota; P G Kotula, J R Michael, Sandia National Laboratories 1144CD
Nanocrystalline Diamond in Ru-doped DLC Films - G D Lian, E C Dickey, Pennsylvania State University; M Ueno, M K Sunkara, University of Louisville 1146CD
Deposition of Au Nanocrystals on TiO2 Crystallites - D A Blom, T G Schaaff, Oak Ridge National Laboratory . 1148CD
Composition-Size Diagrams fo Supported Pt-Sn Catalyts - L Bednarova, Norwegian University of Science and Technology; C E Lyman, Lehigh University; E Rytter, A Holmen, Norwegian University of Science and Technology 1150CD
High Resolution TEM Observation and EELS Analysis of Carbon Nanotubes at Elevated Temperatures - T Yaguchi, T Sato, T Kamino, Hitachi Science Systems, Japan; T Hashimoto, Hitachi High-technologies cooperation, Japan; K Motomiya, K Tohji, A Kasuya, Tohoku University, Japan 1152CD
Characteristics of Palladium Particles on Tin Dioxide Thin Films Studied by Transmission Electron Microscopy - J E Dominguez, University of Michigan; G W Graham, Ford Motor Co; X Q Pan, University of Michigan 1154CD

Interfaces

Microstructural Studies of Copper Sulfide Film Growth: Influence of Humidity - M J Campin, New Mexico State University; J C Barbour, J Braithwaite, Sandia National Laboratories; J Zhu, New Mexico State University 1156CD
Electron Microscopy Study of Fe3O4(111)/MgO(111) Polar Oxide Interface - V Lazarov, M Gajdardziska-Josifovska, University of Wisconsin-Milwaukee 1158CD
HRTEM Characterization of Interface between Iso-structural Thin Solid Film and Substrate - C Wang, S Thevuthasan, F Gao, V Shutthanandan, D E McCready, S A Chambers, C H F Peden, Pacific Northwest National Laboratory 1160CD
Strain Relaxation by Misfit Dislocations in Nanoscale Epitaxial Ferroelectric BaTiO3 Films Grown on SrTiO3 Substrate - H P Sun, W Tian, University of Michigan; J H Haeni, D G Schlom, Pennsylvania State University; X Q Pan, University of Michigan 1162CD
Layer and Defect Structures of BaF2/CaF2 Multilayers - N Y Jin-Phillipp, N Sata, J Maier, C Scheu, M Rühle, Max-Planck-Institut fuer Metallforschung 1164CD
Nanoscale Grain Boundary Dissociation: Role of Shockley Partial Dislocations - D L Medlin, D Cohen, G Lucadamo, S Foiles, Sandia National Laboratories 1166CD
Aliovalent Dopant Distribution in Nanocrystalline Tin Dioxide Studied by X-Ray Energy Dispersive Spectroscopy - J E Dominguez, H P Sun, X Q Pan, University of Michigan 1168CD
Grain Boundary Segregation in Titanium Dioxide - Q Wang, G Lian, E Dickey, Pennsylvania State University 1170CD

Semiconductors

Charging of a Structured Material During Electron Beam Exposure - M Kotera, Y Ishida, Osaka Institute of Technology, Japan 1172CD

Inelastic Electron Scattering Observation using Energy Filtered Transmission Electron Microscopy for Silicon-Germanium Nanostructures Imaging. - R Pantel, S Jullian, D Dutartre, ST Microelectronics 1174CD
Observation of Device Cross-Sectional Thin Films Prepared by FIB Using JEM-2500SE, an Electron Microscope for Nano-Analysis - N Endo, T Suzuki, E Okunishi, Y Kondo, JEOL ltd. Japan 1176CD
Nanoscale Compositional Characterization of Silicon Oxide-Nitride-Oxide Stacks - I Levin, National Institute of Standards and Technology 1178CD
TEM Observation on Single Defect in SiC - J Q Liu, M Skowronski, Carnegie Mellon University; P G Neudeck, J A Powell, NASA Glenn Research Center 1180CD
Microanalysis of Nano-Crystalline Diamonds - M A Stevens-Kalceff, University of New South Wales, Australia.; S Prawer, J O Orwa, J Peng, J McCallum, D Jamieson, L Bursill, University of Melbourne, Australia; W Kalceff, University of Technology, Australia. 1182CD
Assessment of Integrated Sub-micron Polysilicon Fuses for Low Voltage CMOS Applications - J Schaper, T Hopson, A VanVianen, Motorola 1184CD
Ion Channeling Contrast Imaging of Aluminum Wire Bonds - K D Dye, R A Youngman, Medtronic Inc. 1186CD
Quantum Wire Arrays in Compositionally Modulated InAs/AlAs Superlattices - D M Follstaedt, J L Reno, S R Lee, Sandia National Labs 1188CD
Z-Contrast Imaging of InAs Quantum Wires in GaAs/AlAs Quantum Wells - G Lian, E C Dickey, Pennsylvania State University; J Wu, Chinese Academy of Sciences 1190CD
Energy Dispersive Spectrometry Calibration For The HD-2000 STEM - C B Vartuli, Agere Systems; F A Stevie, North Carolina State University; B Rossie, S Anderson, M Jamison, M Decker, J McKinely, C Darling, Agere Systems; R Irwin, Texas Instruments 1192CD
EFTEM Mapping of Copper - SiLK Structures - A Myers, S Panglre, Advanced Micro Devices 1194CD
TEM Study of the Microstructure of Si Thin Films Deposited by Hot Wire CVD - K M Jones, M H Al-Jassim, D H Levi, B P Nelson, National Renewable Energy Laboratory 1196CD
Screw Dislocations in GaN - Z Liliental-Weber, J Jasinski, J Washburn, M O'Keefe, Lawrence Berkeley National Laboratory 1198CD
Evolution of GaSb/GaAs Quantum Dot Strain Relaxation - V Fink, O J Pitts, S Watkins, K L Kavanagh, Simon Fraser University 1200CD
Cracking of GaN Based III-Nitride Heterostructures Grown by MOVPE on (0001)-6H-SiC - A Hasenkopf, Max-Planck-Institut fur Metallforschung; F Scholz, Universitat Stuttgart; F Phillipp, Max-Planck-Institut fur Metallforschung 1202CD
Characterization and Optimization of Semiconductor Specimen Preparation for QHREM - N Jin-Phillipp, M Kelsch, F Phillipp, M Rühle, Max-Planck-Institut für Metallforschung 1204CD
Electrical and Structural Characterization of GaN p-n Heterostuctures by Scanning Probe Microscopy - M da Silva, J Gonzalez, P Russell, North Carolina State University 1206CD
Measurements of GaN-based Heterostructures with Electron Beam Induced Current - K L Bunker, J C Gonzalez, A D Batchelor, P Russell, North Carolina State University 1208CD
A Novel Method for Direct TEM Studies of the Microstructure of Polysilicon Films Crystallized With and Without Underlying Oxide - X Z Bo, Princeton Unversity; N Yao, Princeton University; J Sturm, Princeton Unversity 1210CD
Applications of a Novel FIB-SIMS Instrument in SIMS Image Depth Profiling - G McMahon, Fibics Incorporated; J Nxumalo, Semiconductor Insights Inc.; M W Phaneuf, Fibics Incorporated 1212CD
CuPt-Type Ordering in MOCVD In0.49Al0.51P - T H Kosel, D C Hall, University of Notre Dame; R D Dupuis, R D Heller, University of Texas Austin; R E Cook, Argonne National Laboratory 1214CD

Modulated Structures and Quasicrystals

Electron Microscopy and its Application to the Study of Incommensurately Modulated Compositionally and/or Displacively Flexible Phases - R L Withers, L Norén, Y Liu, F Brink, Australian National University 306
Electron Crystallographic Study of Incommensurate Modulated Structures - H F Fan, Y Li, Z H Wan, Z Q Fu, Y D Mo, T Z Cheng, F H Li, Chinese Academy of Sciences, Beijing 308
Electron Microscopy Study of Misfit Layer Structures in the Sb-Nb-S and Bi-Nb-S Systems - L C Otero-Díaz, Univerisdad Complutense de Madrid, Spain 310

Scanning Tunneling Microscopy of Modulated Surface Structures - A Prodan, H J P van Midden, N Jug, Institute Jozef Stefan, Slovenia; F W Boswell, University of Waterloo; J C Bennett, Acadia University, Canada; H Böhm, Johannes Gutenberg University, Germany 312
Quasi-periodic Materials - Crystal Redefined - D Shechtman, Technion, Haifa, Israel 314
Electron Diffraction Evidence For An Elastic XY Model Phase In Niobia-Zirconia Ceramic Alloys - J R Sellar, Monash University, Australia 1216CD
Local Symmetry and Phason Fluctuations of an Ideal Al-Ni-Co Quasicrystal Studied by Atomic-resolution HAADF-STEM - E Abe, S J Pennycook, Oak Ridge National Laboratory; A P Tsai, National Institute for Materials Science, Japan 1218CD
Direct Observation of Icosahedral Clusters in Quasicrystals and Crystals - E Abe, S J Pennycook, Oak Ridge National Laboratory; A P Tsai, National Institute for Materials Science, Japan 1220CD
In Segregation and Phase Separation in Multilayer Structures - C Dorin, C Wauchope, J Mirecki Millunchick, University of Michigan; C A Pearson, University of Michigan Flint; Y Chen, B G Orr, University of Michigan 1222CD
Atom Clusters in Decagonal Quasicrystals: Local Structures and Long-Range Arrangements Revealed by High-Resolution Electron Microscopy - W Sun, K Hiraga, Tohoku University, Japan 1224CD

Microscopy and Microanalysis of Self-Organized Soft Condensed Matter

Imaging of self-assembly and self-assembled materials - P V Braun, University of Illinois at Urbana-Champaign 316
Self Assembled Phenylene Vinylene Materials - M U Pralle, G N Tew, Ion Optics, Inc.; M Sayar, L Li, S I Stupp, Northwestern University 318
Self-ordered colloidal arrays as photonic crystal hydrogels for trainable metal ion sensor and as superparamagnetic matierals - A C Sharma, X Xu, M S Ward, L Li, J C Yang, S A Asher, University of Pittsburgh 320
Near-Field Optical Imaging of Microphase Separated and Semi-Crystalline Polymer Systems - M J Fasolka, L S Goldner, NIST - Gaithersburg; A M Urbas, MIT; J Hwang, K Beers, NIST - Gaithersburg; P DeRege, E L Thomas, MIT 322
In Situ Imaging of Langmuir Films Using Environmental Scanning Electron Microscopy - A F Miller, University of Cambridge; S J Cooper, University of Durham, UK 1226CD
Imaging of Biologically Derived Anisotropic Fluids using Environmental Scanning Electron Microscopy - A F Miller, A M Donald, University of Cambridge 1228CD

Polymer Characterization: It's Not Just For Microscopes Anymore

Ultramicrotomy of Polymers Using an Oscillating Diamond Knife; Improving Polymer Morphology - J S Vastenhout, Dow Benelux B.V.; H Gnägi, Diatome Ltd 324
Total Microscopy of a Tire - R Smith, RWS Consulting 326
Microstructures of Electrospun Polycarbonate Fibers with Solvent Mixture THF and DMF by SEM/TEM - J Shawon, C Sung, University of Massachusetts 1230CD
A TEM Investigation of the Network Structure of Electron Beam Cured Epoxy Polymers - R Schalek, B Defoort, L Drzal, Michigan State University 1232CD
EELS Characterization of Highly Irradiated PVC - A Duarte Moller, Centro de Investigación en Materiales Avanzados, S. C.; G A Hirata, M Avalos Borja, L Cota Araiza, E Adem, Instituto de Física - UNAM, México 1234CD
The Affects of Lightning Strikes and High Current on Polymer Composites - L M Gammon, Boeing Materials Technology; B S Hayes, University of Washington 1236CD

Practical Methods for Transmission Electron Microscopy of Polymers

Practical Methods for Transmision Electron Microscopy of Polymers - J S Vastenhout, Dow Benelux B.V., The Netherlands 1238CD

Electron Microscopy of Macro-, Micro- and Meso-Porous Materials

Self-Assembled Nanostructures: From Nanocrystals to Mesopores and to Nanobelts - Z L Wang, Georgia Institute of Technology 328
Three Dimensionally Ordered Macroporous Bioactive Glasses - D C Bell, K Zhang, H Yan, L F Francis, A Stein, University of Minnesota 330
Energy Loss Spectroscopy and Electron Microscopy of Photoluminescent p-type Porous Silicon Treated with NaOH and NH3 Solutions - M FuruyaSong, Y Yu, Y Fukuda, K Furuya, National Institute for Materials Science, Japan 332
Applications of SEM and Con-focal Laser Microscopy in Developments of Macro-Porous Materials Produced by Sintering - K Ishizaki, M Ohyagi, K Jodan, K Matsumaru, M Nanko, Nagaoka University of Technology, Japan 334
Pore Hierarchies in High-Temperature Composite Refractories - W E Lee, S Zhang, S Hashimoto, University of Sheffield 336
Mesoporous, microporous and nanowired: electron microscopy of aerogel composites - R M Stroud, J W Long, J J Pietron, E M Lucas, Naval Research Laboratory; M L Anderson, Naval Surface Warfare Center; K E Swider Lyons, C I Merzbacher, D R Rolison, Naval Research Laboratory 1240CD
The Impact of Porosity Upon the Cathodoluminescence from III-V Compounds - M A Stevens-Kalceff, University of New South Wales, Australia.; I M Tiginyanu, Technical University of Moldova, Moldova; S Langa, H Föll, Christian-Albrechts University, Germany 1242CD
TEM Characterization of Textured Alumina Produced by Templated Grain Growth (TGG) - B J Hockey, National Institute of Standards and Technology; G L Messing, The Pennsylvania State University 1244CD
HAADF-STEM and HRTEM of Porous Alumina - M E Brito, Synergy Materials Research Center, AIST 1246CD
Microstructural Observations on the Pore Structure Development in Carbon-Carbon Composites During Processing - J H Steele, Pine CO 1248CD
Grain Boundary Issues in Porous Silicon Nitride Ceramics - M E Brito, Synergy Materials Research Center, AIST 1250CD

Microstructural Examination and Imagery of Engineering Materials

Understanding Complex Microstructures with High-Resolution Microanalysis in the Transmission Electron Microscope - G A Botton, McMaster University; J A Gianetto, National Resources Canada; C V Hyatt, Defense Research and Development, Halifax; M W Phaneuf, Fibics, Inc., Ottawa 338
Characterization of Dislocation Structures in Hexagonal Close-Packed Metals by X-ray Line-Broadening Analysis - M Griffiths, D Sage, Atomic Energy of Canada Ltd.; D Galindo, McGill University 340
Neutron Diffraction as a Probe of Microstructure: Surveying the Forest Before Examining the Trees - J H Root, National Research Council of Canada 342
TEM Microstructure Examination of Weld HAZ in Microalloyed Steels - K Poorhaydari-A, B M Patchett, D G Ivey, University of Alberta 1252CD
Cross-Sectional Examination of Crystallinity of Carbon Fibers by Transmission Electron Microscopy - Z P Luo, J Sue, O Ochoa, A Holzenburg, Texas A&M University 1254CD
Preparation to Target: Hit Target +/− 10 mm and a line of Target with in +/− 25 mm with Common Metallographic Equipment. - B Rasmussen, Struers A/S, Denmark 1256CD
NANOCHARACTERIZATION OF GALLIUM ANTIMONIDE SUBSTRATE SURFACE BY TEM/AFM - X Li, University of Massachusetts; L Allen, Epion Corporation; W Goodhue, C Sung, University of Massachusetts 1258CD
Observing the Microstructure of Rapidly Solidified Powders with the SEM - J H Steele, Consultant, Pine 1260CD
Examining the Transformation Products of Austenite in Steels Using Backscatter Electron Imaging in the SEM - J H Steele, Consultant, Pine 1262CD
Characterization of Directionally Recrystallized Cold-rolled Nickel Using EBSP - B Iliescu, J Li, I Baker, Dartmouth College 1264CD
Site-specific TEM Specimen Preparation of Grain Boundary Corrosion in Nickel-Based Alloys Using the FIB "Plan-View Lift-Out" Technique - M W Phaneuf, R J Patterson, Fibics Incorporated 1266CD

The Microstructure of MgO Refractory Brick via Backscatter Electron Imaging - J H Steele, Consultant, Pine 1268CD
Characterization of the Effects of Particle Size on the Microstructure of MoSi2/TiB2 Composites Produced by Elemental in-situ Reactions Using Scanning Electron Microscopy (SEM) and Electron Probe Microanalysis (EPMA) - L A Dempere, M J Kaufman, University of Florida 1270CD
Application of Cathodoluminescence Technique in Light Microscopy to Crystallisation Study of Mold Fluxes - E Paransky, E Divry, M Rigaud, Ecole Polytechnique Montreal 1272CD
MEMS Stage and Piezoelectric Motor-controlled TEM Holder for Quantitative in-situ Testing of Thin Film Specimens - S J Robinson, G Fried, Beckman Institute; A Haque, University of Illinois 1274CD
TEM Analysis of a Thermal Sprayed Steel - J Hangas, A D Roche, Ford Motor Company 1276CD
Electron Microscopy Characterization of Aluminum Alloy - Fly Ash Composites - D P Robertson, M Gajdardziska-Josifovska, J K Kim, R Q Guo, P K Rohatgi, University of Wisconsin-Milwaukee 1278CD
Grain Boundary Precipitation in Aged Ni-23Cr-16Mo Alloy Resolved by TEM - E S M Nicoletti, PUC-Rio, Rio de Janeiro; P D Portella, BAM, Berlin; F A Darwish, I G Solórzano, PUC-Rio, Rio de Janeiro 1280CD

Practical Applications of Metallography

Failure Analyses of Three 6061-T6 Aluminum Alloy Turbo-Expander Wheels Exposed to Natural Gas Environments, - B Bavarian, California State University 344
Failure Analysis of Small Gap Brazing of a Stainless Steel Heat Exchanger - M Neff, Michael Neff Associates 346
Grain Size Measurements Variables to Consider - J Klansky, Buehler 348
Automatic Phase Segmentation of Spectrum Images - J Friel, R Batcheler, Princeton Gamma-Tech 350
Critical Pitting Potential and Stress Corrosion Cracking of Aluminum Alloy in Chloride Media - M Elboujdaini, M T Shehata, CANMET; E Ghali, Laval University 1282CD
Initiation of Stress Corrosion Cracking on X-65 Linepipe Steels in Near-Neutral pH Environment - M Elboujdaini, M Shehata, W Revie, CANMET 1284CD
Black Pad Metallography in Electronics - P Snugovsky, Celestica 1286CD
Metallography Observations of Delayed Hydride Cracking in Zr-2.5Nb - Z Pan, A Lockley, Atomic Energy of Canada Ltd. 1288CD
Microstructure vs. Impact Toughness Relationship in Handfield's Austenitic Manganese Steel - R Zavadil, S Kuyucak, Natural Resources Canada 1290CD
Characterization of a Mo Diffusion Barrier for Au/Sn Solder Bonding of Micro/Optoelectronic Devices to Carriers - A He, D G Ivey, University of Alberta 1292CD
The Microstructure of an Aged TiAlW Alloy - H Zhang, L L He, H Q Ye, Chinese Academy of Sciences 1294CD
A Comparative Look at Microstructures of Iron Meteorites - F Hogue, Hogue Metallography; S Sheybany, Pacific Metallurgical Company 1296CD
Standard Reference Material (SRM)482: A Metallographic Challenge - E S Windsor, National Institute of Standards and Technology; R A Carlton, Elan Pharmaceuticals; S A Wight, G Gillen, National Institute of Standards and Technology 1298CD
Microinhomogeneity of Liquid Alloys: Microscopy characterization and new production methods. - L Shepelev, V Manov, Advanced Metal Technologies Ltd., Israel 1300CD
Metallographic Preparation Imaging and Analysis of High Purity Refractory Metals - G Lucas, Buehler; J Spanos, Williams Advances Materials/PureTech 1302CD
Comparison of Partly Revealed Anisotropic Microstructures Using Grid Intersepts as Applied to Zirconium Tubes - M Lagacé, L Rodrigue, M Trudeau, Hydro-Québec Research Center 1304CD
Extraction of Quantitative Data from Lithium Polymer Battery Micrographs - M Lagacé, P Hovington, Hydro-Québec Research Center; C Baril, LTEE Research Center; E Dupuis, P Noel, Hydro-Québec Research Center 1306CD
Measuring a-Zr grain Size in extruded Zr-2.5Nb Pressure Tubes by using linear intercept lengths from SEM images - A Lockley, R Mayville, Atomic Energy of Canada 1308CD
Incorporation of Actinide Elements into Iron-Zirconium Intermetallic Phases in Metallic Waste Forms for High-Level Nuclear Waste - D E Janney, Argonne National Laboratory-West 1310CD
Vapor Hydration Testing of Nuclear Waste Glasses Using D2O and H2O - A C Buechele, C T Mooers, I L Pegg, The Catholic University of America 1312CD

Metallography and Microstructural Evaluation of Contemporary Materials

A Transmission Electron Microscopy Study of Dual Phase High Strength Steels - I Yakubtsov, D Boyd, Queen's University; D Emadi, CANMET 352
Fractographic Evaluation of Medium Carbon Steels with Low Hot Ductility - O Dremailova, D Emadi, E Essadiqi, J Brown, CANMET 354
Recovery and Recrystallization of Ferrite in Warm Forging of a Medium Carbon Steel - P Zhao, D Boyd, Queen's University, Canada 356
Characterization of the Inhibition Layer on Galvanized Interstitial Free Steels - S Dionne, G Botton, M Charest, CANMET; F Goodwin, International Lead Zinc Research Organization 358
Metallographic Methods for Troubleshooting of Roll Problems in the Finishing Train of a Ferrous Hot Strip Mill - R Webber, M Lalik, Dofasco Inc. 360
Macro-etching of Continuous Cast Steel - J Casey, Dofasco Inc. 362
Mechanical Properties of an Austempered High Carbon, High Silicon and High Manganese Steel - S K Putatunda, Wayne State University 1314CD
Microstructural Characterization of Metal Injection Molded (MIM) AISI 316L - S R Collins, Swagelok Company 1316CD
Oxidation Behavior of FeCrAlY Felt at Elevated Temperature - B Johnson, Y H Chin, Pacific Northwest National Laboratory 1318CD
Microstructure and Characterization of CuAlNi Shape Memory Thin Films - A C Kneissl, K Kutschej, X Wu, University of Leoben 1320CD
Transient Liquid Phase Bonding of Titanium Aluminide Alloys - A Microstructural Investigation - W F Gale, D A Butts, T Zhou, Auburn University 1322CD
Microstructure Evolution during Semi-Solid Processing of Magnesium AZ91D - M T Shehata, V Kao, E Essadiqi, CANMET; C A Loong, C Q Zheng, National Research Council Canada 1324CD
Evaluation of Microstructural Factors Affecting the Mechanical Properties of Thin-Wall Ductile Iron Castings - A Javaid, K Davis, CANMET 1326CD
Metallographic Procedure to Microstructural Characterization of a Multiphase Steel Applied to Pipelines Industry by Optical Microscopy - M D S Pereira, P E L Garcia, T M Hashimoto, Sao Paulo State University, Brazil 1328CD
Identification of Retained Austenite by Optical Microscopy and Its Correlation with Mechanical Properties in API-5L-X80 Steel Applied in Pipelines Industry - M D S Pereira, E M Orue, T M Hashimoto, Sao Paulo State University, Brazil 1330CD
In-Situ Metallography and Replication of Microstructures for Condition Assessment and Remaining Life Analysis - Y Ranaware, G Shejale, Thermax Babcock & Wilcox Limited, India 1332CD

Magnetic Materials and Super-conducting Materials

Metallic Magnetic Nanocrystals - Shapes, Self-Assembly and Phase Transformation - Z L Wang, Z Dai, Georgia Institute of Technology; S Sun, IBM 364
The Magnetism-Nanostructure Interface in Advanced Magnetic Materials - D Sellmyer, University of Nebraska-Lincoln 366
TEM Microstructure Studies of Thin Film Magnetic Recording Media - R Sinclair, U Kwon, J Risner, Stanford University 368
Determination of Disordered Magnetic Structures in High-Coercivity Nd-Fe-Based Glassy Alloys - N Lupu, H Chiriac, National Institute of R&D for Technical Physics 370
EELS Analysis of Magnetic Materials - C G Trevor, P J Thomas, R Harmon, R Alani, H A Brink, Gatan 372
Characterization of Magnetic Materials by Means of Neutron Scattering and Future Possibilities at a Next Generation Spallation Neutron Source - F Klose, Oak Ridge National Laboratory; G Ehlers, Institute Laue-Langevin 374
Quantitative Measurements Of Magnetic Vortices Using Position Resolved Diffraction In Lorentz STEM - N J Zaluzec, Argonne National Lab 376

Advanced Magnetic Force Microscopy Tips for Domain Images of Soft Magnetic Materials under Magnetic Field - S H Liou, University of Nebraska 378
Dynamic Observation of Vortices in High-Tc Superconductors by Lorentz Microscopy - A Tonomura, Hitachi, Ltd. 380
HREM Characterization of Magnetic Thin Films and Multilayers - D J Smith, Arizona State University 382
STEM Investigations of Defects and Interfaces In Complex Oxides - S J Pennycook, M Varela, Oak Ridge National Laboratory; J Santamaria, Universidad de Madrid; D Kumar, North Carolina A & T State University; G Duscher, Oak Ridge National Laboratory, North Carolina State University 384
A Challenging to Characterization of Superconductors: Accurate Measurements of Charge Distribution and Interfacial Displacement - Y Zhu, L Wu, Brookhaven National Laboratory; J Tafto, University of Oslo 386
High-Resolution and Low Temperature TEM Study of Superconducting Cuprates and CMR-Manganites - Y Matsui, National Institute for Materials Science 388
In-Situ Lift-Out FIB Specimen Preparation for TEM of Magnetic Materials - B Kempshall, L Giannuzzi, University of Central Florida 390
Crystal Structure Determination of Superconductors and Related Compounds by Combining High-Resolution Electron Microscopy and Electron Diffraction - F H Li, H B Wang, H Jiang, Chinese Academy of Science 392
Exploring the Surface of Pb-doped Manganese Perovskite - C Borca, University of Colorado; P Dowben, University of Nebraska 394
Mossbauer Studies of Fe-doped La-Ca-Mn-O Colossal Magnetoresistive Perovskites - Z H C Cheng, Z H Wang, N L Di, R W Li, Chinese Academy of Sciences; R A Dunlap, Dalhousie University; B G Shen, Chinese Academy of Sciences 396
Domain Structure of Magnetic Nanocrystalline Materials Studied by Electron Holography - D Shindo, Tohoku University, Japan 398
Quantum Well Interference and Exchange Coupling in Double Quantum Well Thin Films - Z D Zhang, Shenyang National Lab, China 400
Solving the Structural Mysteries of Magnetic Materials in TEM - Y Liu, University of Michigan at Ann Arbor . 402
Interfacial Segregation in Modified Fe-Nd-B Permanent Magnets by Analytical Electron Microscopy - J Bentley, Oak Ridge National Laboratory; J E Shield, University of Nebraska-Lincoln 1334CD
Influence of Microstructure on the Magnetic Properties of Co Alloy Thin Films for 100 Gbit/in2 Longitudinal Recording - J E Wittig, Vanderbilt University; J Bentley, Oak Ridge National Laboratory; J Ma, J Al-Sharab, Vanderbilt University; N D Evans, Oak Ridge National Laboratory 1336CD
Low Temperature Magnetic Force Microscopy Studies of a Superconducting Nb Film - M Roseman, P Grutter, McGill University 1338CD
Scanning Thermal Ferromagnetic Resonance Microscopy of Fe-GaAs Heterostructures and Ni-nanodots - R Meckenstock, D Spoddig, J Pelzl, Ruhr-Unversitaet, Germany 1340CD
Characterization of Microstructure in Ag-doped La2/3Sr1/3MnO3 Films - Q Zhan, L L He, D X Li, Chinese Academy of Sciences 1342CD
Magnetic Induction Mapping in TEM of Micro- and Nano-Patterned Co/Ni Arrays - VV Volkov, Y Zhu, M Malac, J Lau, M Schofield, Brookhaven National Laboratory 1344CD
Preparation of Cross-Sectional TEM Specimens of Obliquely Deposited Magnetic Thin Films on a Flexible Tape; Electron Transparency Beyond 6 Micron - E Keim, L Nguyen, C Lodder, University of Twente, Netherlands 1346CD
Study of Nano-Granular Co-Zr-O Thin Films by Holography and HRTEM - Z Liu, D Shindo, Tohoku University; S Ohnuma, H Fujimori, The Research Institute for Electric and Magnetic Materials 1348CD
Surface Nano-Oxidation of Ferromagnetic Thin Films Using Atomic Force Microscope - Y Takemura, Yokohama National University; J Shirakashi, Akita Prefectural University, Japan 1350CD
On the Half Unit Cell Intergrowth of Bi2Sr2Ca3Cu4Ox with Other Superconducting Phases in Two-step Annealed LFZ Fibers - L Yang, F M Costa, A B Lopes, R F Silva, J M Viera, University of Aveiro, Portugal 1352CD
Selected Reflection Imaging: A Useful Tool for Imaging Nanocomposite Magnetic Materials - Y Liu, University of Michigan; Y Qiang, M J Yu, J P Liu, D J Sellmyer, University of Nebraska 1354CD
Elemental Mapping of Co-Pr Nanostructured Powders by EELS Image Filtering - Y Liu, University of Michigan; C Nelson, Lawrence Berkeley National Laboratory; H Tang, D J Sellmyer, University of Nebraska 1356CD
TEM of Nanostructure of Cu and Ti Doped Sm-Co Magnetic Materials - Y Liu, University of Michigan; J Zhou, R Skomski, D J Sellmyer, University of Nebraska 1358CD

Application of Energy-Filtered Imaging and HREM in the Study of Terbium Nanoparticles - Y Zhang, University of Delaware; C Nelson, Lawrence Berkeley National Laboratory; Z Yan, V Skumryev, G Hadjipanayis, University of Delaware 1360CD
Preparation and Characterization of Core-Shell Cobalt Silver Nanoparticles - M Giersig, Hahn-Meitner-Institut Berlin, Germany 1362CD
Transmission Electron Microscopy Studies of Epitaxial Superconducting MgB2 Thin Film - W Tian, University of Michigan; C B Eom, University of Wisconsin; X Q Pan, University of Michigan 1364CD
Mossbauer and XPS analysis of Fe-SiO2 and Fe-SiO2/SiO2 granular films - S Honda, T Shimizu, Shimane University, Japan; I Sakamoto, AIST Tsukuba, Japan; T Une, K Kawabata, Hiroshima Inst. Tech., Japan 1366CD
TEM Characterization of Self-Assembled Magnetic Nanowires - X Zhao, University of Michigan; Y Liu, M Zhen, Z Hao, S Bandyopadhay, D Sellmyer, University of Nebraska 1368CD
Microscopy and Magnetoresistance studies in zigzag and semi-circle-in-series Permalloy wires - C Yu, S F Lee, Y D Yao, Academia Sinica, Taiwan; Y R Ma, Dong Hwa University, Taiwan; J L Tsai, C R Chang, Academia Sinica, Taiwan 1370CD
Low Temperature Dependence of HF-Magnetic Properties of Soft Nanostructured Films - I A Ryzhikov, L A Alekseeva, A L Djachkov, S A Maklakov, M V Sedova, T A Furmanova, Russian Academy of Science; N S Perov, Moscow State University, Russia 1372CD
TEM Study of Epitaxial Growth of La0.65Pb0.35MnO3 on LaAlO3 and Its Relation to Electronic Structure and Spin Polarization - L Yuan, University of Nebraska; Y Liu, University of Michigan; P A Dowben, S H Liou, University of Nebraska 1374CD
Structure Analysis of CoPt Nanoparticles - Y Huang, Y Zhang, University of Delaware; C E Nelson, Lawrence Berkeley National Laboratory; G C Hadjipanayis, University of Delaware; D Weller, Seagate Technology 1376CD
Crystallographic Texture Study in Melt-Spun Pr-Fe-B 2:14:1 Based Nanocomposite Magnet - Y Zhang, Z Lin, H Wang, G Hadjipanayis, University of Delaware 1378CD
[010] Oxygen Ordering at Grain Boundaries in MgB2 - J C Idrobo, R F Klie, N D Browning, University of Illinois at Chicago; A C Serquis, F M Mueller, Los Alamos National Lab 1380CD

In Situ Electron Microscopy Techniques and Applications / Reactions

Progress Towards More Realistic In-Situ Microscopy Observations - A Howie, University of Cambridge 404
In-situ Observation of Alloy Phase Formation in Isolated Nanometer-sized Particles - H Mori, J G Lee, Osaka University, Japan; H Yasuda, Kobe University, Japan 406
High Resolution In-Situ SEM of Competitive Particle Sintering and Other Surface Processes - E D Boyes, DuPont Company 408
Real Time UHV-HRTEM Observation of Si(111)root3xroot3-Pd Surface and Dynamic Motion of Pd Clusters - M Takeguchi, K Mitsuishi, M Tanaka, K Furuya, National Institute for Materials Science, Japan 410
In Situ Molecular Imaging of Heterogeneous Catalytic Processes in Liquid Environments - P Gai, DuPont 412
In-situ UHV-Electron Microscopy with Scanning Tunneling Microscope - K Takayangi, Y Ohshima, K Mohri, Y Naitoh, H Hirayama, Y Tanishiro, Tokyo Institute of Technology; Y Kondo, JEOL Ltd., Japan 414
In Situ HREM of Crystallization Reactions - R Sinclair, K H Min, Stanford University 416
Aberration Correction for Analytical In Situ TEM - the NTEAM Concept - B Kabius, C W Allen, D J Miller, Argonne National Laboratory 418
In situ Transmission Electron Microscopy of Copper Electrodeposition - F M Ross, IBM; M J Williamson, University of Virginia; R M Tromp, IBM; R Hull, University of Virginia; P M Vereecken, IBM 420
Local Measurement of Reaction Kinetics Using in situ Transmission Electron Microscopy - R Sharma, P Crozier, Arizona State University 422
The Use of Stereomicroscopy in Conjunction with In Situ Straining TEM for Studying Dislocation Behavior - R J McCabe, A Misra, T E Mitchell, Los Alamos National Laboratory 1382CD
Time Resolved Electron Energy Loss Spectroscopy as a Tool for Controlling and Monitoring the Early Stages of Electron Beam Induced Transformations - S Trasobares, Rensselaer Polytechnic Institute; O Stephan, C Colliex, Universite Paris Sud, France 1384CD

Combined Confocal Raman Microscope With Scanning Electron Microscope; a Parallel Analysis of Inorganic and Organic Materials. - Y Aksenov, A A van Apeldoorn, University of Twente; J D de Bruijn, C A van Blitterswijk, Isotis NV; J Greve, C Otto, University of Twente, The Netherlands 1386CD
In Situ SEM Analysis of Lithium Metal Polymer Battery - P Hovington, Hydro-Quebec Research Institut; M Simoneau, Avestor Corp.; M Lagacé, P Noel, E Dupuis, Hydro-Quebec Research Institut, Canada 1388CD
Microstructural Studies of the Chromia Stabilized Iron Oxide Water Gas Shift Catalyst - C J Kiely, Lehigh University; M Edwards, D Whittle, Liverpool University, UK; C Rhodes, G Hutchings, University of Wales, UK .. 1390CD
Electron Microscopy of Cr/Silica Catalyst for Ethylene Polymerization - R J Liu, P Crozier, Arizona State University 1392CD
Simulations of TEM Images of Nano-particles Embedded in Amorphous Ice - J O Malm, N Pettersson, L R Wallenberg, J O Bovin, Lund University, Sweden 1394CD
Atomic Scale Characterization of Oxygen Vacancy Dynamics by In-Situ Reduction and Analytical Atomic Resolution STEM - R Klie, N Browning, University of Illinois; Y Zhu, Brookhaven National Laboratory ... 1396CD
SEM/EDS Studies of Impurities in Natural Ice - D Cullen, D Iliescu, I Baker, Dartmouth College 1398CD
Morphological Evolution and Junction Dynamics at Faceted Grain Boundaries - D Medlin, Sandia National Laboratories 1400CD
In situ Transmission Electron Microscopy Study of Dislocation Emission at Junctions Between Sigma=3 Grain Boundaries in Gold Thin Films - G Lucadamo, D L Medlin, Sandia National Laboratories 1402CD
Influence of Germanium Interdiffusion on the Morphological Evolution of Sigma3 Grain Boundaries in Gold Thin Films - T Radetic, U Dahmen, Lawrence Berkeley National Laboratory 1404CD
Initial Oxidation Kinetics of Copper Films Investigated by In-Situ UHV-TEM - G Zhou, J C Yang, University of Pittsburgh 1406CD
Ex-Situ TEM Study of Au Islands - C E Kliewer, J L Robbins, A Malek, ExxonMobil Research and Engineering 1408CD
In Situ TEM Observations on Morphological Instability of Ultrathin Pb Film - L H Zhang, M L Sui, L Zhang, D X Li, Chinese Academy of Sciences, China 1410CD
Structural Variations in Nanocrystalline Nickel Films - P Gai, DuPont Central Research; R Mitra, J Weertman, Northwestern University 1412CD
A New Form of MgTa2O6 Obtained by the Molten Salt Method - S Nangia, M Thirumal, A K Ganguli, Indian Institute of Technology, India; P Gai, DuPont Corp. 1414CD
A New Formulation of the Diffraction Contrast Theory of Dislocations and its Application to the Weak Beam Images - H S Kim, Kyungsung University, South Korea 1416CD
Study on the Irradiation Effects on Iron Nitride by High Energy Electrons and Ions at the Atomic Level - H Hashimoto, Z Liu, Okayama University of Science; T Sakata, H Mori, Osaka University; M Song, H Yasuda, K Furuya, National Institute for Materials Science, Japan 1418CD
Ion Irradiation Damage in Titanate Ceramics as a Function of Dose - M Blackford, G R Lumpkin, K L Smith, H Li, M Colella, Australian Nuclear Science & Technology Organisation 1420CD
Comparison of In- and Ex-Situ Analysis of Post-Irradiation Annealing - J Busby, G Was, University of Michigan; E Kenik, Oak Ridge National Laboratory 1422CD
In Situ TEM Study of Order-Disorder Transition in Murataite Ceramics - J Lian, L M Wang, R C Ewing, University of Michigan 1424CD
Quantitative Analysis of the Brownian Motion of Small Liquid Lead Inclusions in Solid Aluminum - U Dahmen, T Radetic, J Turner, Lawrence Berkeley National Laboratory; S Prokofjef, Russian Academy of Sciences; M T Levinsen, E Johnson, University of Copenhagen, Denmark 1426CD

Phase Transformation in Metals, Alloys and Ceramics

Microstructure Progress in Pressureless Sintered AlN Polytypes - Y D Yu, I L Tangen, R Høier, T Grande, Norwegian University of Science and Technology 1428CD
Dislocation Dissociation and Short-Range Ordering in ZrN - P Li, J M Howe, University of Virgina 1430CD
Size Distribution of Gamma' Precipitates in Ni-Cr-Co-Al-Ti Alloys - S K Menon, UES Inc.; J P Simmons, D M Dimiduk, Air Force Research Laboratory 1432CD
Size-Dependent Equilibrium Shape of Co-Cr Particles in Cu - T Fujii, T Tamura, M Kato, S Onaka, Tokyo Institute of Technology 1434CD

TEM studies of TbNiAl in the Disproportionation Stage of the HDDR Process - C M Andrei, Norwegian University of Science and Technology; J Walmsley, SINTEF Materials Technology Applied Physics; Y Yu, Norwegian University of Science and Technology; H Brinks, Institute for Energy Technology; R Holmestad, Norwegian University of Science and Technology; B Hauback, Institute for Energy Technology 1436CD
TEM Studies of Microstructural Transformations in Thin Iron Films Induced by Vacuum Annealing - E Keim, University of Twente, The Netherlands; W Lisowski, Polish Academy of Sciences 1438CD
Microstructure and Wear Properties of Fe Surface Alloyed Al Alloy 319 - J W Carroll, Y Liu, J Mazumder, University of Michigan 1440CD
Selectivity of Grain Boundary Precipitation in Al Alloys - D D Perovic, A Perovic, University of Toronto, Canada 1442CD
The GP-Zone to Beta" Transformation in the Al-Mg-Si System - C D Marioara, S J Andersen, SINTEF Materials Technology, Norway; J Jansen, H W Zandbergen, Delft University of Technology, The Netherlands 1444CD
An Analytical Electron Microscopy Study of the Initial Stage of Formation Processes of Aged Omega Phase Crystals in a Ti-15Mo Alloy due to Aging at 323 K. - M Hajime, E Sukedai, H Hashimoto, Okayama University of Science, Japan 1446CD
Beta to Omega Phase Transformation in a Ti-Mo Alloy Deformed in an Impact Compression Mode Due to Aging. - E Sukedai, D Yoshimitsu, Okayama University of Science, Japan; M Kiritani, Hiroshima Institute of Technology, Japan; H Hashimoto, Okayama University of Science, Japan 1448CD
TEM Study of Structure and Deformation Mechanism of Superplastic Titanium Based Alloy - A Kumao, K Nishio, Kyoto Institute of Technology, Japan 1450CD
TEM Study of AA 6111 Weld by Pulsed YAG Lasers - S Liu, Y Liu, J Mazumder, University of Michigan ... 1452CD
Nb-Silicide Phase Stabilization in Cast and HIP in-situ Composites - B P Bewlay, M Larsen, P R Subramanian, M R Jackson, General Electric 1454CD
Microstructure Characterization of Sn-Ag-Cu Lead-Free Solder Solidified at Different Cooling Speeds - A Zbrzezny, University of Toronto 1456CD
Observation of Substructure in Steels and Ni200 Using Electron-Channeling Contrast Imaging - J Steele, Pine 1458CD

Quantitative X-Ray Microanalysis in the Microprobe and in the SEM: Theory and Practice

X-Ray Microanalysis of Light Elements - G F Bastin, H J M Heijligers, Eindhoven University of Technology 424
Charging at the Steady State in EPMA; SEM and ESEM - J Cazaux, DTI Faculty of Sciences Reims France .. 426
Capability and Uncertainty in Multilayer Quantitative Procedure with Electron Probe Microanalysis - C Merlet, University Montpellier 2 428
On the Simulation of True EDS X-Ray Spectra - R Gauvin, McGill University; E Lifshin, Albany Institute for Materials 430
The Influence of X-ray Counting Statistics on Trace Analysis and Spatial Resolution - E Lifshin, University at Albany; R Gauvin, McGill University 432
Low-Overvoltage Microanalysis an Alternative High Resolution Strategy to Low-Voltage Microanalysis - D E Newbury, National Institute of Standards and Technology 434
X-ray emission induced by low energy electrons - C Bonelle, P Jonnard, Université Pierre et Marie Curie 436
A Simple Method for Determining Optimum Corrections for High-Accuracy EPMA in Difficult Chemical Systems - J T Armstrong, R B Marinenko, J M Davis, National Institute of Standards and Technology 438
Spectral Imaging: Towards Quantitative X-ray Microanalysis - P G Kotula, M R Keenan, Sandia National Laboratories 440
Artifacts From the Electric Field Build Up in the Microbeam Analysis of Insulating Materials - O Jbara, M Belhaj, S Fakhfakh, DTI Faculty of Sciences Reims France 1460CD
X-Ray Microanalysis of a Coated Non-Conductive Specimen: Monte Carlo Simulation - H Demers, R Gauvin, McGill University 1462CD
Standardless Quantitative Electron Beam X-ray Microanalysis The Situation Remains caveat emptor - D E Newbury, National Institute of Standards and Technology 1464CD
Quantitative Microanalysis at Low kV : Precautions and Validation - P J Statham, Oxford Instruments Analytical Ltd, UK 1466CD

Standardless Eds Quantitative Analysis at High Tilt Angles - A O Sandborg, R A Anderhalt, EDAX, Inc.; J M Dijkstra, EDAX Europe, The Netherlands; R B Shen, EDAX, Inc. 1468CD
Evaluation of Current Standardless Quantitative Analysis Programs using Dispersive Spectrometry in the SEM - H Campbell, R Gauvin, McGill University 1470CD
X-RAY ANALYSIS OF ROUGH SURFACES AT LOW ENERGY - P Hovington, M Lagacé, L Rodrigue, Hydro-Quebec Research Institut 1472CD
Quantitative X-Ray Microanalysis with a Low Voltage Scanning Electron Microscope - P Horny, R Gauvin, McGill University; E Lifshin, D Wu, Albany Institute of Materials University at Albany 1474CD
Pu-Ga Standards for Microanalysis and Matrix Correction Development - C C Davis, R E Lakis, Los Alamos National Laboratory 1476CD
X-ray Microanalysis of Insulators in a Variable Pressure Environment - M Toth, J P Craven, University of Cambridge; M R Phillips, University of Technology, Sydney; B L Thiel, A M Donald, University of Cambridge 1478CD
Testing EDS Performance in ESEM - V M Dusevich, J D Eick, University of Missouri - Kansas City 1480CD
SEM-EDS Quantitative Analysis of Aerosols >= 80nm: Impacts on Atmospheric Aerosol Characterization Campaigns - M A Carpenter, E Lifshin, University at Albany; R Gauvin, McGill University 1482CD
Quantitative Analysis of Yttrium Barium Copper Oxide Films on Strontium Titanate - E Lifshin, M S Hatzistergos, University at Albany; J L Reeves, IGC Superpower, Schnecdaty; R Gauvin, McGill University 1484CD
Measurements of Absolute X-ray Generation Efficiency - M S Prasad, D C Joy, University of Tennessee 1486CD
Chemical Characterization of Optical Data Storage Materials by EPMA - S Richter, M Bückins, S Kyrsta, R Cremer, RWTH Aachen 1488CD
Chemical Characterization of Silicon-Germanium Single Crystals - Initial Evaluation of the Extent of Heterogeneity - R B Marinenko, J T Armstrong, S Turner, E B Steel, National Institute of Standards and Technology; F A Stevie, North Carolina State University 1490CD
EPMA Characterization of Residual Al Content in Oxidized High-Temperature Alloys - B A Pint, L R Walker, I G Wright, Oak Ridge National Laboratory 1492CD
Low Background Glass Substrates for Microanalysis - E S Windsor, C J Zeissler, S A Wight, E B Steel, D H Blackburn, National Institute of Standards and Technology 1494CD
Absorption Correction of Fe La,b Emission from Iron Oxides - G Rémond, University of Technology, Australia; G M Fialin, Université Pierre et marie Curie, France; C E Nockholds, M R Phillips, University of Technology, Australia; C Roques-Carmes, Ecole Nationale de Mécanique et des Microtechniques, France 1496CD
WinX-Ray: A New Monte Carlo Program for the Simulation of X-Ray and Charging Materials - H Demers, P Horny, R Gauvin, McGill University; E Lifshin, University at Albany 1498CD
Tantalum Planchettes as a Substrate to Collect and Analyze Ambient Air Particles - R Stearns, P Ruiz, J Lawrence, P Koutrakis, J J Godleski, Harvard School of Public Health 1500CD

State of the Art Infrared and Raman Microanalysis

Infared Microscopic analysis of tissues: a comparison of Methodologies - M Jackson, J Dubois, National Research Council Canada; R Baydak, T Booth, Canadian Science Center for Human and Animal Health . . . 442
Discriminating vital Tumor from necrotic tissue in human glioblastoma samples by Raman microspectroscopy - S Koljenovic, LP Choo-Smith, T C Baker Schut, Erasmus University, The Netherlands; J M Kros, H J van den Berge, University Hospital Rotterdam; G J Puppels, Erasmus University, The Netherlands 444
Spatially Resolved Improved FT-IR Microspectroscopy of Deuterated Species in Tissue - D L Wetzel, Kansas State University; S M LeVine, University of Kansas Medical Center 1502CD
Cancer Diagnosis and Detection via Infrared Microspectroscopy of Cells and Thin Tissue Sections. What Have we Learned? - R A Shaw, S L Ying, K C McCrae, G Steiner, R Salzer, F B Guijon, H H Mantsch, National Research Council of Canada 1504CD
Looking at Prion Diseases in situ with Infared Microscopy - J Dubois, National Research Council Canada; R Baydack, University of Manitoba, Canada; M Jackson, National Research Council Canada 1506CD
Biomedical Applications of Flurorescence-Assisted Synchrotron Infrared Micro-Spectroscopy - L M Miller, Brookhaven National Laboratory 1508CD
Explanatory Analysis Strategies for High Deimensional Mid-Infared Microspectroscopy Data from Tissue Sections - M G Sowa, L Leonardi, M D Hewko, B Schattka, J Dubois, M Jackson, K Z Liu, H H Mantsch, National Research Council Canada 1510CD

Analysis of Bone Utilizing Infrared and Raman Chemical Imaging - T J Tague, C P Schultz, Bruker Optics, Inc.; L Miller, National Synchrotron Light Source 1512CD
Application of Fourier Transform Infrared Spectroscopic Imaging to Biomedical Analyses - R Bhargava, D C Fernandez, S W Huffman, M D Schaerble, I W Levin, National Institutes of Health 1514CD
Near Infrared Spectroscopic Imaging: A Paradigm Shift in Quantitative Analysis - E N Lewis, L H Kidder, E Lee, Spectral Dimensions, Inc. 1516CD
Raman selection rules in the presence of an electric field gradient - C L Jahncke, St. Lawrence University; E J Ayars, Walla Walla College; H D Hallen, North Carolina State University 1518CD
Raman Microspectroscopy of Some High Explosives – Revisited - E S Etz, S V Roberson, G Gillen, National Institute of Standards and Technology 1520CD
Micro Raman Spectroscopy: An Appropriate Method for the Characterization of Orientation and Crystallinity of Polypropylene - A Gupper, P Wilhelm, G Kothleitner, Graz University, Austria; P Zipper, Karl Franzens Universität, Austria; D Gregor-Svetec, University of Lubljana, Slovenia; M Gahleitner, Borealis GmbH, Linz, Austria 1522CD
Increasing Spatial Resolution and Extending Spectral Range of Synchrotron Infrared Microscopy - G D Smith, G L Carr, National Synchrotron Light Source 1524CD
A New Approach to Infrared Microspetroscopy: Adding FT-IR to a Light Microscope - J A Reffner, D K Wilks, K C Schreiber, R V Burch, SensIR Technologies 1526CD
Applying Correlative Brightfield, Fluorescence, and Raman Microscopy to Determine and Track Component Locations in Formulations - D J Palatini, S L Zhang, B Aral, H Crookham, Unilever 1528CD
Spectral Imaging with Near-Field Infrared Spectroscopy and Microscopy - C A Michaels, National Institute of Standards and Technology; D B Chase, Dupont; S J Stranick, National Institute of Standards and Technology 1530CD
Raman Microscopy and Remote Laser Raman Spectroscopy in Art History and Conservation Science: Analysis of Three Gutenberg Bibles - G D Smith, T D Chaplin, R J H Clark, University College London, U.K.; K Jensen, D Jacobs, The British Library U.K. 1532CD

Image Contrast Mechanisms in the Variable Pressure SEM: the New Imaging Dimension

Time Resolved Analysis of the Positive Ion Dynamics in the Variable Pressure Scanning Electron Microscope - M R Phillips, S W Morgan, University of Technology, Australia 446
Considerations for Secondary Electron Imaging of Dielectric Materials in Low-Vacuum and Environmental SEM - B L Thiel, M Toth, University of Cambridge, UK 448
Charge Contrast Imaging in Variable Pressure SEM: Correlation, Optimisation and Application - B J Griffin, A S Suvorova, University of Western Australia, Australia 450
Modelling contrast in Variable Pressure Scanning Electron Microscopes - R Gauvin, H Demers, K Robertson, J Finch, McGill University, Canada 452
Residual Surface Potentials in the Variable Pressure/ Environmental Scanning Electron Microscope - M A Stevens-Kalceff, University of New South Wales, Australia 1534CD
Charging Contrast Imaging of Gibbsite in the Variable Pressure SEM - K Roberston, R Gauvin, J Finch, Mcgill University, Canada 1536CD
ESEM Beam Current Measuring Devive based on Planar Schottky Diode - A S Aubin, D Drouin, M R Phillips, Universite de Sherbrooke, Canada 1538CD
The Effect of Gas Atmospheres on Charging - X Tang, University of Tennessee; D C Joy, Oak Ridge National Laboratory 1540CD
Measurement of Elastic Cross-Sections for Gases - J He, University of Tennessee; D C Joy, Oak Ridge National Laboratory 1542CD

Geology/Mineralogy

Structure, Chemistry and Properties of Grain Boundaries in H2SO4-Doped Ice - D Iliescu, D Cullen, C Muscat, I Baker, Dartmouth College 1544CD

Scanning Electron Microscopy of Vostok Accretion Ice - D Cullen, I Baker, Dartmouth College 1546CD
Rietveld Refinement and HRTEM Simulation of Calcium-Lead Apatites - Z Dong, T White, Environmental Technology Institute, Singapore 1548CD
Electron Microscopy of In Situ Presolar Silicon Carbide - R M Stroud, Naval Research Laboratory; M O'Grady, Vanderbilt Univeristy; L R Nittler, C M O Alexander, Carnegie Institution of Washington 1550CD
SEM Syudy and X-RAY Microanalysis of Lateritics from "Los Pijiguaos" Bauxite ore, Estado Bolívar, Venezuela. - D Espinoza, W Meléndez, D Iapicca, C Urbina, Universidad Central de Venezuela 1552CD
Automated X-ray Spectral Image Analysis of a Large Area of a Geologic Material - P Kotula, P Hlava, M Keenan, Sandia National Laboratories 1554CD
Application of Spatial and Feature Analysis to Electron Microscope Petrography - S Lowther, A Wisher, University of Puget Sound 1556CD
Electron Microscopy Study of Mineral Colloids in the Ground Water Near Nevada Test Site - L P You, L M Wang, S Utsunomiya, R C Ewing, University of Michigan; A B Kerstiing, P Zhao, Lawrence Livermore National Laboratory 1558CD
A Practical Application of Scanning Electron Microscopy to Characterize a Fine-Grained Sulfide Gold Ore - L R P De Andrade Lima, McGill University; D Hodouin, Laval University 1560CD
Electronprobe Microanalysis of Volcanic Glass at Cryogenic Temperatures - S Kearns, N Steen, E Erlund, University of Bristol 1562CD
The WebSEM in K-12 Classrooms: Lessons Learned - L S Chumbley, D J Eisenmann, A E Chumbley, T Frizell, T Andre, C P Hargrave, Iowa State University 454
Microscopic Digital Imaging in Introductory Biology - J Ekstrom, Phillips Exeter Academy 456
Developing and Implementing a Unified Imaging and Lab Information Management WorkFlow System - Lessons Learned and Insights Gained - C Yip, University of Toronto, Canada 458

Teaching and Learning, Creating Effective, Innovative Solutions in Microscopy, Imaging and Analysis (MIA)

Introduction to Sophisticated Instrumentation in Education Through the Use of a Scanning Electron Microscope Simulator - G Casuccio, H Lentz, S Kennedy, RJ Lee Group; C Staun, J Lang, West Greene School District; S Chumbley, C Hargrave, Iowa State University; T Nolan, Oak Ridge National Laboratory 1564CD
VSEM: From Technology Demonstrator Towards Integrated Educational Tool - N H M Caldwell, B C Breton, D M Holburn, R P Robertson, Cambridge University, UK 1566CD
Tele-Tutoring - From Learning to Earning Part IV: The School-to-Work Program; The Student's Perspective - L Bosco, J Johnson, West Greene School District; G Casuccio, T Lersch, S Kennedy, S Schlaegle, R J Lee Group 1568CD
Electron Microscopy Images As An Interactive Tool For Cell Biology Modeling And Education - T C Araujo-Jorge, T S Cardona, C L S Mendes, A Henriques-Pons, Institute Oswaldo Cruz-Fiocruz, Brazil; L E Aguiar, CEFETEQ, Brazil; C M L M Coutinho, R Santa-Rita, C N Spiegel, R M S Meirelles, M N L Meirelles, S L De Castro, H L Barbosa, Institute Oswaldo Cruz-Fiocruz, Brazil 1570CD
ctfExplorer: Interactive Software for 1d and 2d Calculation and Visualization Of TEM Phase Contrast Transfer Function - M V Sidorov, Advanced Micro Devices 1572CD
Particle-Like Fractal Images for Testing Algorithms that Measure Boundary Fractal Dimension - D S Bright, National Institute of Standards and Technology 1574CD
Development of a Web Based Data Storage and Project Management System for Biological Electron Cryo-microsopy - L Nason, W Chiu, S J Ludtke, Baylor College of Medicine 1576CD
Do-It-Yourself Multimedia Teaching and Training Tools For the Laboratory and Classroom - S Barlow, San Diego State University 1578CD
XMRBS: A Web Based Facilities Sheduler - D Fellmann, B Carragher, C S Potter, The Scripps Research Institute 1580CD
"Ugly Bug" Contest - J Ekstrom, Phillips Exeter Academy 1582CD
Using Wavelets to Adjust Focusing of a Scanning Electron Microscope - C Morgan, California State University Hayward; S Vikas, University Planet Inc.; L Sun, California State University Hayward 1584CD

Elmar Zeitler Symposium: Analytical Electron Microscopy - Past and Future

Valence Excitations in Electron Microscopy: Pursuing Zeitlerian Initiatives - A Howie, University of Cambridge 460

A Comparison of Microcompositional Methods - M Isaacson, Cornell University 462

The Future of EELS - R F Egerton, University of Alberta 464

Locating Atoms in Small Crystals by Combining Convergent Beam Electron Diffraction and Electron Channeling - J Tafto, S Foss, A Olsen, University of Oslo, Norway; C Simensen, SINTEF, Norway 1586CD

Comparison of Detection Limits for Elemental Mapping by EF-TEM and STEM-XEDS - M Watanabe, Dept. Mater. Sci. & Eng. Lehigh University; D B Williams, Lehigh University; Y Tomokiyo, Kyushu University, Japan 1588CD

The relevance of imaging and analytical electron microscopy in the understanding of heterogeneous selective oxidation catalysis - R Schlögl, Fritz-Haber-Institut der MPG, Germany 1590CD

Comparative ELNES Measurements on Selected Transition Metal Oxides on a New High Energy-Resolution Spectrometer / Monochromator TEM - G Kothleitner, F Hofer, Graz University of Technology, Austria; D S Su, R Schlögl, Fritz-Haber Institut, Germany; B H Freitag, P C Tiemeijer, FEI Electron Optics, The Netherlands 1592CD

Density Functional Theory as a Tool for the Electron Microscopist - C Hébert, P Schattschneider, University of Technology, Wien, Austria 1594CD

Nanostructures and Defects in Several Materials Under Electropulsing - W Zhang, M L Sui, D X Li, Shenyang National Laboratory for Materials Science, Chinese Academy og Sciences 1596CD

Sb Grain Boundary Segregation in Rapidly Solidified Cu-Sb Alloy - C Li, M Watanabe, Lehigh University; J Li, IBM Microelectronics; D W Ackland, D B Williams, Lehigh University 1598CD

Reduction of PtO2 Powders (Adam's Catalyst) Under Electron Beam Irradiation - J Liu, Monsanto Company 1600CD

Phonon Scattering in Quantitative High-Resolution Electron Microscopy - Effects, Problems and Approaches - Z L Wang, Georgia Institute of Technology 1602CD

Z-Contrast Imaging of Dislocation Cores at the Si/GaAs Interface - S Lopatin, J Narayan, North Carolina State University; G Duscher, North Carolina State University, Oak Ridge National Laboratory 1604CD

The Study of Intergranular Segregation and Elemental Partitioning in Partially Molten Olivine-bearing Geological Composites by STEM-EDX - I M Anderson, Oak Ridge National Laboratory; T Hiraga, D L Kohlstedt, University of Minnesota 1606CD

Nanobelt Thickness and Mean-free Path Determination by CBED and PEELS - Y Berta, C Ma, Z L Wang, Georgia Institute of Technology 1608CD

EFTEM and its Application in Cryo Electron Microscopy - J M Plitzko, W Baumeister, Max-Planck-Institute for Biochemistry, Germany 1610CD

The Optimization of EDX Performance in Tecnai TEMs - H S von Harrach, B Freitag, W Gerits, E van Cappellen, FEI Electron Optics, The Netherlands; A Sandborg, Edax Inc. 1612CD

Imaging Single Dopant Atoms and Nanoclusters in Highly n-type Bulk Si - P M Voyles, D A Muller, J L Grazul, P H Citrin, Bell Labs, Lucent Technologies; HJ L Gossmann, Agere Systems 1614CD

Multi-Electrode Samples for TEM Studies of Corrosion - D S Elswick, J J Hren, North Carolina State University; P G Kotula, F D Wall, Sandia National Labs 1616CD

Trends in AEM Over the Years in Four Materials Science Journals - T Malis, Natural Resources Canada 1618CD

Structure Characterization of ZnSe/GaMnAs Quantum Well On GaAs Substrate - G D Lian, E C Dickey, S H Chun, N Samarth, Pennsylvania State University 1620CD

Stanley L. Erlandsen, President
Microscopy Society of America

Greg Meeker, President
Microbeam Analysis Society

Raynald Gauvin, President
Microscopy Society of Canada /
Société de Microscopie du Canada

Richard K. Ryan, President
International Metallographic Society

FOREWORD

It is our pleasure to welcome you to the Microscopy and Microanalysis 2002 Meeting on behalf of the Microscopy Society of America, the Microbeam Analysis Society, the Microscopy Society of Canada, and the International Metallurgical Society.

This year marks several milestones for MSA and the Proceedings. Bill Bailey, the MSA Proceedings Editor, will be stepping down after 35 of service. Our sincere congratulations to Bill and his successor, Stuart McKernan. The Proceedings this year has a new look with printed invited papers and a fully searchable CD-ROM of all papers created from electronic submission. Many thanks to Edgar Voelkl and others who worked so hard to bring this major change about. We also would like to welcome Ed Barnas and our new journal publisher, Cambridge University Press, who have facilitated our transition into electronic abstract submission.

This year the Program Chair, Edgar Voelkl, and Co-Chairs David Piston (MSA), Raynald Gauvin (MAS/MSC), and Alan Lockley (IMS) have compiled an outstanding technical program. Highlights include symposia held on 3-D electron microscopy and cryo-electron microscopy of macromolecules, the quantitative aspects of X-ray microscopy, confocal microscopy, biomaterials, biological and materials specimen preparation. Special sessions will be held on holography, phase imaging, on deep tissue imaging, on (S)TEM instrumentation, developments in focused-ion beam instruments and imaging, metallographic specimen preparation from start to finish, and on the changing role of atom probe microscopes in the nanotechnology era. A special analytical EM session honoring the work of Elmar Zeitler is scheduled this year.

A special pre-meeting workshop "*Future of Materials Characterization of Charging Materials using Microbeam Analysis*" organized by Dr. Raynald Gauvin will be held at McGill University in Montreal on August 2–3. During the course of the meeting a series of special symposia, tutorials and presentations sponsored by the Technologists' Forum, MSA Education Committee, and also various exhibitors will be held during the meeting. Ceremonies for presentation of MSA awards will occur at the Presidential Happenings. Award winners honored at the meeting include Marc Adrian and Ryuichi Shimizu selected for MSA Distinguished Scientist Awards in Biological and Physical Sciences, respectively. Nigel Browning is the Burton Medallist, while the MSA Outstanding Technologist Award has been given to José Mascorro. The MAS Presidential Science Award goes to Klaus Keil. The K. F. J. Heinrich Award will be presented to David Wollman, and the recipient of the MAS Presidential Service Award is John F. Mansfield. Congratulations are also due to our other award winners: MSA Professional Technical Staff awardees, MSA Presidential Scholars, and MAS Distinguished Scholars.

Our Local Arrangements Committee, headed by Pierre Charest, has scheduled the scientific sessions and the commercial exhibits in the Convention Center in Quebec City. In addition to this excellent venue, a number of attractive social events were scheduled. These included the opening Sunday reception at the Quebec Museum of Civilization, Whale Watching, a dinner at the Chapel of the Amérique Francaise Museum, a dinner at the Manoir Montmorecy, and the Grand Finale of the Quebec City International Fireworks Festival on the St. Lawrence River. Quebec City, renown for its heritage, culture, enchanting cafes and fine food, spectacular scenery in the old town, has provided a wonderful venue for this year's meeting. Working closely with the Local Arrangements and Program Committee, the meeting and exhibits have been ably managed by The Rebedeau Group directed by Mary Beth Rebedeau.

We would like to extend our warm thanks for "joie de vivre"—the warm hospitality of the people of Quebec. We also want to thank all the organizers and participants who contributed to making Microscopy and Microanalysis-2002 the renown meeting for microscopy in the world. Now we look forward to San Antonio, Texas, the exciting site of Microscopy and Microanalysis 2003.

Stanley L. Erlandsen, President, Microscopy Society of America
Greg Meeker, President, Microbeam Analysis Society
Raynald Gauvin, President, Microscopy Society of Canada
Richard Ryan, President, International Metallurgical Society

Letter from the Retiring Editor

FAREWELL

In 1966, John Watson asked Claude Arceneaux and me to develop and arrange a Proceedings Publication for the 1967 Meeting of the Electron Microscopy Society of America, to commemorate the 25th Annual Meeting of the Society. Thus, the Proceedings was born. We thought it would be a one year deal—now we are publishing our 2002 Proceedings. The EMSA/MSA Proceedings has become an internationally recognized publication of the Society and has been enhanced by the other participating Societies such as the Microbeam Analysis Society, the Microscopical Society of Canada, the Histochemical Society, and the International Metallographic Society. As participation has grown, the Proceedings has been able to adapt to the changes. I believe that we have a first class Proceedings that serves all participating Societies to the fullest! I am proud of what the Proceedings has done for MSA, for the science of microscopy, and for all of the participating Societies.

Beginning in 1967, I have been a volunteer worker. Although I have never received nor expected any monetary payment for being Proceedings Editor, I have received the most valued reward that could ever be expected: serving a great society, MSA, and experiencing the fellowship, friendship, and above all, the respect of my many friends in MSA and in the participating Societies. How could I be so fortunate? It has been wonderful and in my now Southern tradition, thanks to "you all".

The M&M Proceedings will take on a new look for 2002. We will have a combination hard copy Proceedings for invited papers and a CD for all papers—invited papers and contributed papers. We are fortunate in having a new Proceedings Editor, Stuart McKernan, and a new Publisher, Cambridge University Press, represented by Ed Barnas. They will both do an excellent job.

At the risk of not acknowledging someone special, I want to say thanks to some people who have made my "career" as Proceedings Editor something special: Ron Anderson—we have worked together for 30 plus years; Charlie Lyman—I trained him as a Program Chair and now he is my Editor-in-Chief; Dale Johnson, Bob Fisher, and Barbara Reine who got me through the 1990 International Meeting; Charles Susskind, who was our San Francisco Press Publisher for many years; Herb Niemerow, our most recent Publisher's representative who gave us a great five years; to the Rebedeau Group: to the Bostrom organization; to Mort and Larry Maser; and to my wife, Pat, who did all of my typing, offered support, and helped me to always have the material to the publisher on schedule. Above all, I thank MSA for many great years and hope that MSA, the Proceedings, and the Baileys have many great years ahead.

G. W. Bailey, Proceedings Editor

MICROSCOPY SOCIETY OF AMERICA

Established 1942

OFFICERS 2002

Executive Council

President	Stanley L. Erlandsen
President Elect	Alwyn Eades
Past President	Ron Anderson
Secretary	Janet H. Woodward
Treasurer	Kathleen Alexander
Directors, Physical Sciences	Thomas F. Kelly
	John Mansfield
	Nestor Zaluzec
Directors, Biological Sciences	Jay Jerome
	Sara E. Miller
	Steven Samuelsson
Director, Local Affiliated Societies	Ev Osten

Appointed Officers

Archivist	Michael Marko
Awards Committee Chair	James Bentley
Book Reviews and Calendar	JoAn Hudson
Certification Board Chair	John P. Chandler
Education Committee Chair	Mary Grace Burke
International Committee Chair	C. Barry Carter
Journal Editor	Charles E. Lyman
Local Arrangements	Pierre Charest
	David Olmos
Marketing and Communications Chair	Paul Fischione
Placement Officer	Pamela Lloyd
Proceedings Editor	Stuart McKernan
Program Committee Chair	Edgar Voelkl
Program Committee Vice-Chair	David Piston
Program Committee Vice-Chair	John Bruley
Public Policy Committee Chair and Legal Liaison	Peter Ingram
Sustaining Members	Dennis Masaki
	Michael Bode
Technologists' Forum	Jeannette Killius
Business Office	Bostrom Corporation
Administrative Manager	Judy Janes
Managing Director	Philip Lesser
Meeting Manager	Mary Beth Rebedeau

MSA PAST PRESIDENTS

1942. G.L. Clark[1]
1943. R. Bowling Barnes[2]
1944. R. Bowling Barnes
1945. James Hillier
1946. David Harker
1947. William G. Kinsinger
1948. Perry C. Smith
1949. F.O. Schmitt
1950. Ralph W.G. Wyckoff
1951. Robley C. Williams
1952. R.D. Heidenreich
1953. Cecil E. Hall
1954. Robert G. Picard
1955. Thomas F. Anderson
1956. William L. Grube
1957. John H.L. Watson
1958. Max Swerdlow
1959. John H. Reisner
1960. D. Gordon Sharp
1961. D. Maxwell Teague
1962. Keith R. Porter
1963. Charles Schwartz
1964. Sidney S. Breese
1965. Virgil G. Peck
1966. Walter Frajola
1967. Joseph J. Comer
1968. John H. Luft
1969. W.C. Bigelow
1970. Russell Steere
1971. Robert M. Fisher
1972. Daniel C. Pease
1973. Benjamin Siegel
1974. Russell J. Barnett
1975. Gareth Thomas
1976. Etienne de Harven
1977. T.E. Everhart
1978. Myron Ledbetter
1979. John Silcox
1980. Michael Beer
1981. John Hren
1982. Lee Peachey
1983. David Wittry
1984. J. David Robertson
1985. Dale Johnson
1986. Robert Glaeser
1987. Linn W. Hobbs
1988. John-Paul Revel
1989. Ray Carpenter
1990. Keith R. Porter
1991. Charles Lyman
1992. Patricia Calarco
1993. Michael S. Isaacson
1994. Robert R. Cardell
1995. Terence E. Mitchell
1996. Margaret Ann Goldstein
1997. C. Barry Carter
1998. Ralph M. Albrecht
1999. David Joy
2000. Kenneth Downing
2001. Stanley Erlandsen

[1] Chair of committee to arrange first meeting

[2] Temporary (pre-constitution)

MSA CORPORATE SPONSORS

4pi Analysis, Inc
Advanced Micro Devices
Advanced MicroBeam, Inc.
Advanced Microscopy Techniques
Allied High Tech Products, Inc.
Anatech Ltd.
ASPEX Instruments
Cameca Instruments
Carl Zeiss, Inc.
Carnegie Mellon University
Chroma Technology Corp.
Coherent Laser Division
Columbian Chemicals Co.
Denton Vacuum, LLC
Diatome U.S.
Digital Instruments, Inc.
E. A. Fischione Instruments, Inc.
Eastman Chemical Co.
Eastman Kodak Co.
EDAX Inc.
Electro Image, Inc.
Electron Microscopy Sciences
Emispec Systems, Inc.
EMSL Analytical, Inc.
Energy Beam Sciences, Inc.
Ernest F. Fullam, Inc.
ETP-USA/Electron Detectors
Evex Analytical
ExxonMobil Research & Engineer Co.
FEI Beam Technology Division
FEI Company
Gatan Inc.
GW Electronics, Inc.
Halcyonics GmbH
HKL Technology, Inc.
IBM Microelectronics Div.
JEOL USA, Inc.
Kluwer Academic/Plenum Publishing
K-Tek International, Inc.
Ladd Research Industries
Laurin Publishing
LEO Electron Microscopy Inc.
M. E. Taylor Engineering, Inc.
Mager Scientific, Inc.
Mastology Centers, Inc.
Materials Analytical Services
Materials Technology Lab
McCrone Research Institute
Media Cybernetics LP
Micro Photonics Inc.
Micro Star Technologies, Inc.
Microcosm, Inc.
Micron, Inc.
Microscopy Today
Nanomics Imaging Ltd.
Nikon Inc.
NSA/Hitachi Scientific Instruments
Optronics Engineering
Oxford Instruments, Inc.
Polaroid Corp.
Princeton Gamma-Tech, Inc.
RMC-Boeckeler Instruments Inc.
RONTEC USA, Inc.
Scientific Inst Services, Inc.
SEMICAPS Inc.
Soft Imaging System Corp.
Sonoco Products Co.
South Bay Technology, Inc.
SPI Supplies
Technotrade International
Ted Pella, Inc.
Thermo NORAN
TissueInformatics, Inc.
TM Microscopes, VEECO
Tousimis Laboratories
UCF Materials Characterization Facility
Universal Imaging Corp.
Western Michigan University
XEI Scientific

DISTINGUISHED SCIENTIST AWARDS

Biological Sciences
Marc Adrian

Physical Sciences
Ryiuchi Shimizu

Marc Adrian was born in May 1945 in Paris, France. He was educated in Paris and studied biochemistry at the Ecole Supérieur de Chimie Appliquée. His first contact with an electron microscope was as an assistant at the Institute of Cancer and Immunogenetics. in 1963. In 1966 he entered National Service in Germany in the "Service de micropscopie electronique" at the F. Picaud Hospital in Buhl Baden, returning to civilian life two years later at the Cancer Institute in the Electron Microscopy group.

He completed his thesis in 1976 after a stay at the Pasteur Instiute in Paris at the "Station Centrale de Microscopie Electronique".

After a three-year hiatus from science he started at the European Molecular Biology Laboratory (EMBL) in Heidelberg, Germany in 1982, where his first contact with cryogen and vitrification took place. His work on cryoelectron microscopy of vitrified specimens has continued to evolve, and he is now Laboratoire d'Analyse Ultrastructurale (LAU) at the University of Lausanne, Switzerland.

Ryuichi Shimizu was born on 2 January 1937. He took Dr. Eng, at Osaka University in 1964 under the supervision of Professor Gunji Shinoda who had been an honorary member of the Microbeam Analysis Society. Since 1965 Ryuichi Shimizu had been a faculty member of the Department of Applied Physics, Osaka University until he retired in 2000, mainly involved in studies on the interaction of charged particles with solids.

He is a professor in Information Science at the Osaka Institute of Technology, an emeritus professor of Osaka University and an honorary member of the Microbeam Analysis Society.

Burton Medal
Nigel Browning

Nigel D. Browning is currently an Associate Professor of Physics at the University of Illinois at Chicago. He received his Ph.D from the University of Cambridge in 1992 and was a post-doctoral fellow in the Solid State Division at Oak Ridge National Laboratory from 1992–1995. He has authored over 80 publications and has co-edited one book entitled "Characterization of High-Tc materials and devices by Electron Microscopy".

Prof. Browning's research interests center on the development of atomic resolution imaging and spectroscopic techniques to analyze the structure-property relationships at materials interfaces and defects. His current interests lie in the atomic scale properties of semiconducting quantum dots, dislocations in GaN, grain boundaries in oxide materials and heterogeneous catalysts. He has organized several symposia at the Microscopy and Microanalysis meetings related to both technique development and interface properties. He has previously received an NSF CAREER award and a University of Illinois Scholar award.

Morton D. Maser Distinguished Service Award
Beverly E. Maleeff

Beverly E. Maleeff received her B.S. in Microbiology from the Pennsylvania State University, certification as an Electron Microscopy Technologist from (E)MSA, and a Certificate in Computer Graphic Design from the Philadelphia College of Textiles and Science. She joined the Philadelphia Electron Microscope Society in 1980 and served as President, Secretary-Treasurer and Newsletter Editor. Bev joined EMSA in 1982. At her first Annual Meeting she was introduced to the Technologists' Forum. Since then, she has created the TF logo and the original TF booth backdrop. Bev served as TF Vice-Chair, Exhibit Booth Coordinator and ultimately, as TF Chair from 1997–1999. Bev has been a member of the Nominating and Awards Committees and still serves on the Program Committee. In 1999, Bev volunteered to chair an *ad hoc* committee to design and commission the MSA booth that has been the centerpiece of the Exhibit Hall at Microscopy & Microanalysis ever since. Bev was Secretary-Treasurer of the Local Arrangements Committee for Microscopy & Microanalysis 2000, held in Philadelphia. In addition, she designed the official M&M2000 logo. Bev is currently an Investigator in preclinical Safety Assessment at GlaxoSmithKline in King of Prussia, PA, where she manages the Cellular Imaging group.

MSA DISTINGUISHED SCIENTIST AWARDS

Biological Sciences

1977. Keith Porter
1978. L.L. Marton
1979. Robley Williams
1980. Thomas Anderson
1981. Daniel Pease
1982. George Palade
1983. Sanford Palay
1984. Richard Eakin
1985. Hans Ris
1986. Cecil Hall
1987. Gaston Dupouy
1988. F.O. Schmitt
1989. Marilyn Farquhar
1990. Morris Karnovsky
1991. Don W. Fawcett
1992. Audrey M. Glauert
1993. Hugh E. Huxley
1994. Fritiof Sjöstrand
1995. Jean-Paul Revel
1996. Andrew Somlyo
1997. Shinya Inoué
1998. Myron C. Ledbetter
1999. S.J. Singer
2000. Avril V. Somlyo
2001. Sir Aaron Klug
2002. Kiyoteru Tokuyasu
2003. Patrick Echlin

Physical Sciences

1977. Robert Heidenreich
1978. Albert Crewe
1979. James Hillier
1980. V.E. Cosslet
1981. John Cowley
1982. Gareth Thomas
1983. Vladimir Zworykin
1984. Benjamin M. Siegel
1985. Otto Scherzer
1986. Sir Charles Oatley
1987. Ernst Ruska
1988. Peter Hirsch
1989. Jan LePoole
1990. Hatsujiro Hashimoto
1991. Elmar Zeitler
1992. Gertrude F. Rempfer
1993. Archie Howie
1994. Oliver Wells
1995. Ken Smith
1996. Dennis McMullan
1997. David B. Wittry
1998. John Silcox
1999. Peter Swann
2000. Michael J. Whelan
2001. Takeo Ichinokawa
2002. Severin S. Amelinckx
2003. Thomas Mulvey

Burton Medalists

1975. James Lake
1976. Michael Isaacson
1977. Robert Sinclair
1978. David Joy
1979. Norton B. Gilula
1980. John Spence
1981. Barbara Panessa-Warren
1982. Nestor Zaluzec
1983. Ronald Gronsky
1984. David B. Williams
1985. Richard Leapman
1986. J. Murray Gibson
1987. Ronald Milligan
1988. A.D. Romig, Jr.
1989. Laurence D. Marks
1990. W. Mason Skiff
1991. Joseph R. Michael
1992. Kannan Krishnan
1993. Joseph A.N. Zasadzinski
1994. Jan M. Chabala
1995. Joanna L. Batstone
1996. Vinayak P. Dravid
1997. P.M. Ajayan
1998. Ian M. Anderson
1999. Zhong Lin Wang
2000. Eva Nogales
2001. J.M. Zuo

THE MORTON D. MASER DISTINGUISHED SERVICE AWARD

1992. Ronald Anderson
G.W. "Bill" Bailey
Frances Ball
Blair Bowers
Deborah Clayton
Joseph Harb

1992. Kenneth Lawless
Morton Maser
Caroline Schooley
John H.L. Watson
1993. E. Laurence Thurston
1994. Richard F.E. Crang

1995. Raymond K. Hart
1996. José A. Mascorro
1997. William T. Gunning, III
1998. Nestor J. Zaluzec
1999. Charles E. Lyman
2000. Barbara A. Reine

OUTSTANDING TECHNOLOGIST AWARD

1977. Ben O. Spurlock
1978. Bernard J. Kestel
1979. Kai Chien
1980. David W. Ackland

1981. John P. Benedict
Stanley J. Klepeis
1982. Hilton H. Molenhauer
Charles J. Echer

1983. John M. Basgen
John C. Wheatley
1984. Nancy C. Smith
1985. Conrad Bremer

OPTICAL IMAGING ASSOCIATION AWARD FOR ACHIEVEMENT IN OPTICAL MICROSCOPY

2000 Gregg G. Gundersen

2001 Clare Waterman-Storer

MSA PRESIDENTIAL SCHOLARS

Omar Al-Kofahi — Rensselaer Polytechnic Institute
"Multi-View Three-Dimensional Image Montaging & Signal Attenuation Correction for Maximizing the Imaging Depth and Lateral Extent of Confocal Microscopes"

J. E. Dominguez — University of Michigan
"Aliovalent Dopant Distribution in Nanocrystalline Tin Dioxide Studied by X-Ray Energy Dispersive Spectroscopy"

Lilia Gallegos — New Mexico State University
"Inhibition of Turnip Vein Clearing Virus Movement in Seeds of Infected Arabidopsis Thaliana Plant a Microscopic Study"

John Keyes — Emory University
"H_2O_2 in Interspecies Signaling: a New Role in Host Detection"

Peter Krsko — Stevens Institute of Techology
"Electron-Beam Micropatterning of Bioactive Surfaces"

Vlado Lazarov — University of Wisconsin-Milwaukee
"Electron Microscopy Study of Fe_3O_4 (111) / MgO (111) Polar Oxide Interface"

Peng Li — University of Virginia
"Dislocation Dissociation and Short-range Ordering in ZrN"

Hyung-Shim Ra — University of Illinois at Urbana Champaign
"Changes in Thallus and Algal Cell Components of Two Lichen Species in Response to Low-level Air Pollution at Pacific Northwest Forests"

Chantal Sudbrack — Northwest University
"Influence of W on the Temporal Evolution of the Microstructure of a Ni-Al-Cr Superalloy on a Nanoscale"

Xiaohu Tang — University of Tennessee
"The Effect of Gas Atmospheres on Charging"

Xiaodong Tao — Lehigh University
"Another Way to Implement Diffraction Contrast in SEM"

Donna Wilson — Tufts University
"Surface Organization and Nanopatterning of Collagen by Dip-Pen Nanolithography"

Chun Xiao — Purdue University
"Pseudo-atomic Structure of Coxsackievirus A21 Complexed with its Cellular Receptor, ICAM-1"

2002 PTSA AWARD WINNERS

Biological Professional Technical Staff Awards Winners

Richard Mount — Hospital for Sick Children, Toronto, Canada
"Quantitative Analysis of Cerebral Corrosion Casts from Endoglin $^{+/+}$ and $^{+/-}$ Mice"

Marilyn Wadsworth — University of Vermont
"Compositional Analysis of Atherosclerotic Lesions in a Mouse Model: Validation of a New Method using Brightfield, Fluorescence and Polarized Light"

Physical Professional Technical Staff Awards Winners

Donald Robertson — University of Wisconsin
"Electron Microscopy Characterization of Aluminum Alloy—Fly Ash Composites"

Don Steele — Alcan International, Canada
"Measurement and Visualisation of Micron–Scale Strain Distributions in Aluminum Alloys"

MICROBEAM ANALYSIS SOCIETY

Established 1966

OFFICERS 2002

Executive Council

President	Greg Meeker
President Elect	Edgar S. Etz
Past President	Richard W. Linton
Secretary	Inga Holl Mussellman
Treasurer	Harvey A. Freeman
Directors	Richard D. Leapman
	Valerie Woodward
	Ernie Hall
	Stacie Kirsch
	Paul Kotula
	Charles C. H. Nielson

Appointed Officers

Archivist, Finance Committee Chair .... Gordon Cleaver
Computer Activities Committee Chair .... John F. Mansfield
Corporate Liaison Committee Chair .... Thomas G. Huber
International Liaison .... Raynald Gauvin
Long Range Planning committee .... John A. Small
Membership Services .... Lou M. Ross
MicroNews Editor .... Ryna B. Marinenko
Sustaining Members Committee Chair .... Cathy Johnson

MAS Business Office .... William S. Thompson

MAS PAST PRESIDENTS

1968. L. S. Birks
1969. K. F. J. Heinrich
1970. R. E. Ogilvie
1971. A. A. Chodos
1972. K. Keil
1973. D. R. Beaman
1974. P. Lublin
1975. J. W. Colby
1976. E. Lifshin
1977. J. I. Goldstein
1978. J. D. Brown
1979. D. F. Kyser
1980. O. C. Wells
1981. J. R. Coleman
1982. R. Myklebust
1983. R. Bolon
1984. D. C. Joy
1985. D. E. Newbury
1986. C. G. Cleaver
1987. C. Fiori
1988. W. F. Chambers
1989. D. B. Wittry
1990. A. D. Romig, Jr.
1991. J. T. Armstrong
1992. D. B. Williams
1993. T. G. Huber
1994. J. Small
1995. J. McCarthy
1996. D. E. Johnson
1997. Joseph R. Michael
1998. Ryna B. Marineko
1999. John J. Friel
2000. Charles E. Lyman
2001. Richard Linton

MAS CORPORATE SPONSORS

4pi Analysis, Inc
Advanced MicroBeam, Inc.
ASPEX Instruments
Cameca Instruments
Denton Vacuum, LLC
EDAX Inc.
Electron Microscopy Sciences / DIATOME US
Emispec Systems, Inc.
ETP-USA/Electron Detectors
FEI Company
Gatan Inc.
Geller MicroAnalytical Laboratory
GW Electronics, Inc.
Halcyonics GmbH
JEOL USA, Inc.
Lehigh University
LEO Electron Microscopy Inc.
Materials Analytical Services
Micron, Inc.
Microscopy Today
NSA/Hitachi Scientific Instruments
Oxford Instruments, Inc.
Princeton Gamma-Tech, Inc.
RONTEC USA, Inc.
SEM / TEC Laboratories, Inc.
ThermoMicroscopes
ThermoNORAN

Presidential Science Award
Klaus Keil

Klaus Keil received his Doctorate at Johannes-Gutenberg University in Mainz, Germany. His early work on meteorites was as a Research Associate at Max-Plank Institute for Chemistry, University of California San Diego, La Jolla, and at NASA Ames Research Center, Moffett Field, California. While at La Jolla Dr. Keil, Ray Fitzgerald and Kurt Heinrich developed the first energy dispersive x-ray detector for use on the electron microprobe. In 1966 he was a member of the eight-man panel that established the Electron Probe Analysis Society of America, now called The Microbeam Analysis Society, and in 1972 he became President of the Society. In 1968 Dr. Keil became Director of the Institute of Meteoritics and Professor of Geology at the University of New Mexico. While at UNM he became a Presidential Professor and served as Chair of the Department of Geology. In 1990 he accepted a position at the University of Hawaii as Head of the Planetary Geosciences Division and Professor of Geology. He is currently the Director of the Hawaii Institute of Geophysics and Planetology, University of Hawaii, Manoa. Over the years Professor Keil has served on numerous scientific review panels and advisory committees for NASA, the Lunar and Planetary Institute and the Lunar Science Institute. He has been the recipient of many honors and awards including the first George P. Merrill Award given by the National Academy of Sciences, and the Leonard Metal of the Meteoritical Society. In 1993 Asteroid 5054 was renamed Asteroid Keil by the International Astronomical Union. This year the International Mineralogical Association approved the name keilite for a newly discovered extraterrestrial mineral.

K. F. J. Heinrich Award
David Wollman

David A. Wollman received his undergraduate degree in Physics from Michigan State University in 1990, and his Ph.D. in Physics in early 1996 from the University of Illinois at Urbana-Champaign. He received a National Research Council Postdotoral Research Associateship at the National Institute of Standards and Technology (NIST) in Boulder, Colorado, and subsequently joined the permanent staff of NIST's Electronics and Electrical Engineering Laboratory. At NIST, David has worked with colleagues to develop high-energy-resolution cryogenic x-ray microcalorimeters for energy-dispersive (EDS) x-ray microanalysis. He has demonstrated the usefulness of microcalorimeter EDS to address important microanalytical problems including the need for improved low-voltage particle analysis and thin film analysis in the semiconductor industry. He shares several microcalorimeter-related patents (issued and filed) with co-workers, part of a suite of patents that NIST has licensed to support commercialization of this technology. David has been a MAS tour speaker and has received many awards including the Department of Commerce's Gold Medal, NIST's Applied Research Award and MAS's Macres Award. David is a member of Beta Kappa and Sigma Xi and serves on the Executive Committee of the American Physical Society's Forum on Industrial and Applied Physics. For the past year, David has been on detail in the NIST Director's Office in Gaithersburg, MD.

MAS Service Award
John F. Mansfield

John Mansfield is the manager of the University of Michigan North Campus Electron Microbeam Analysis Laboratory (EMAL). Mansfield specializes in structural and chemical analysis of materials at the nanometer level with the aid of analytical electron microscopy, including convergent beam electron diffraction, X-ray energy dispersive spectrometry, electron energy loss spectroscopy and environmental scanning electron microscopy. Following his study at the University of Bristol, where he received a PhD in 1983, he worked as a post-doctoral researcher at Argonne National Laboratory, in Argonne, Illinois, from 1984 to 1986. He was a Visiting Scientist at the Microelectronics Center of North Carolina from 1986 to 1987 and he joined the University of Michigan in 1987. Mansfield is an active member of both the Microbeam Analysis Society (MAS) and the Microscopy Society of America (MSA). He is a past director of MAS and a current director of MSA. He also maintains the MAS Website and Microanalysis Listserver.

MAS Distinguished Scholars

Catalina Dorin	University of Michigan *"In Segregation and Phase Separation in Multilayer Structures"*
Jing He	University of Tennessee *"Measurement of Elastic Cross-sections for Gases"*
Paula Horny	McGill University *"Quantitative X-Ray Microanalysis with a Low Voltage Scanning Electron Microscope"*
Yuanyuan Lei	University of Illinois at Chicago *"Atomic Scale Structural Analysis of Sn-Si Quantum Dots"*
Jie Lian	University of Michigan *"In Situ TEM Study of Order-Disorder Transition in Murataite Ceramics"*
Angelo Lubag	University of Texas at Dallas *"Synthesis and STM Imaging of Substituted Phenylalkyl Ethers: Towards Functional Group Discrimination"*
Emmanuelle A. Marquis	Northwestern University *"A Subnanoscale Study of Mg Segregation at Al/Al_3Sc Interfaces"*
M. Satya Prasad	University of Tennessee *"Measurements of Absolute X-ray Generation Efficiency"*
Kevin Robertson	University of Michigan *"Charging Contrast Imaging of Gibbsite in the Variable Pressure SEM"*
Haiping Sun	University of Michigan *"Strain Relaxation by Misfit Dislocations in Nanoscale Epitaxial Ferroelectric $BaTiO_3$ Films Grown on $SrTiO_3$ Substrate"*

PREVIOUS AWARDS Microbeam Analysis Society

Science

1977. R. Castaing
1978. K. F. J. Heinrich
1979. P. Duncumb
1980. D. B. Wittry
1981. S. J. Reed
1982. R. Shimizu
1983. J. Philibert
1984. L. S. Birks
1985. E. Lifshin
1986. R. Mykleburst
1987. O. C. Wells
1988. J. D. Brown
1989. J. Hillier
1990. T. E. Everhart
1991. J. I. Goldstein
1992. G. Lorimer
G. Cliff
1993. D. E. Newbury
1994. D.C. Joy
1995. G. Bastin
1996. A. V. Somlyo
A. P. Somlyo
1997. D. B. Williams
1998. F. H. Schamber
1999. R. A. Sareen
2000. R. F. Egerton
2001. P. E. Batson

Service

1977. P. Lublin
1978. D. R. Beaman
1979. M. A. Giles
1980. A.A. Chodos
1981. R. Myklebust
1982. J. Doyle
1983. D. Newbury
1984. J. I. Goldstein
1985. M. C. Finn
1986. V. Shull
1987. D. C. Joy
1988. G. Cleaver
1989. W. F. Chambers
1990. C.E. Fiori
1991. T. G. Huber
1992. E. Etz
1993. H. A. Freeman
1994. J. L. Worrall
1995. R. W. Linton
1996. P. F. Hlava
1997. J. A. Small
1998. J. J. McCarthy
1999. T. G. Huber
2000. R. B. Marinenko
2001. C. E. Lyman

K.F. J. Heinrich

1986. P. Statham
1987. J. T. Armstrong
1988. D. B. Williams
1989. R. Leapman
1990. R. W. Linton
1991. A. D. Romig, Jr.
1992. S. Pennycook
1993. P. E. Russell
1994. J. R. Michael
1995. N. Lewis
1997. R. Gauvin
1998. V. P. Dravid
1999. J. Bruley
2000. H. Ade
2001. C. Jacobsen

MICROSCOPICAL SOCIETY OF CANADA
SOCIÉTÉ DE MICROSCOPIE DU CANADA

Established 1972

OFFICERS 2002

Executive Council

President	Raynald Gauvin
1st Vice-President	Elaine Humphrey
2nd Vice-President	Gianluigi Botton
Secretary	Pierre-Mathieu Charest
Treasurer	Nancy Clark
Past President	George Harauz
Councilors-at-Large	Susan Belfry
	Craig Bennett
	Rakesh Bhatnagar
	Odette DesBiens
	Glynis de Silveira
	Douglas Holmyard
	Doug Ivey
	Karen Rethoret
	Hélène Roberge
	Alexandra Smith

Ex Officio

Executive Secretary .... Frances Leggett
Bulletin Editor .... Michael Robertson

Section Chairmen

Alberta .... Ray Egerton
Atlantic .... Ian Meinertzhagen
Ontario .... Peter Ottensmeyer
Pacific .... James Drummond
Québec .... Raynald Gauvin

MSC PAST PRESIDENTS

1972–1975 A. F. Howatson
1975–1977 E. J. Chatfield
1977–1979 G.T. Simon
1979–1981 G. H. Haggis
1981–1983 F. P. Ottensmeyer
1983–1984 D. O. Northwood
1984–1985 J. M. Sturgess
1985–1987 D. A. Craig
1987–1989 R. F. Egerton
1989–1991 P. J. Lea
1991–1993 G. L'Esperance
1993–1995 L. Arsenault
1995–1997 R. Sherburne
1997–1999 J. Corbett
1999–2001 G. Haurauz
2001–2003 R. Gauvin

MSC CORPORATE SPONSORS

BOC Canada Limited
Canberra Packard Canada
Canemco Inc.
EDAX Inc.
Electron Microscopy Sciences
FEI Systems Canada Inc.
Gatan Inc.
JEOL USA Inc.
Leica Microsystems Inc.
Marivac Limited
Mc Crone Research Institute
Meridian Scientific Services
Nikon Canada Instruments Inc.
Nissei Sangyo Canada
Osram Sylvania Ltd.
Oxford Instruments
Pelco International
Soquelec Ltd
Spectra Research Corporation
SPI Supplies Canada
Systems for Research Corp.

MSC—Gerard T. Simon Award

Materials Science

Nadia Lahkak — Université de Sherbrooke
"*Study of Self-assembled InAs Quantum Dots on InP Nano-templates by Low Voltage Scanning Electron Microscopy Cathodoluminescence*"

Biology

Natalie Bolduc — Laval University
"The *Fusion Protein BnBI-1GFP is Localized to the Endoplasmic Reticulum (ER) and Allows Visualization of ER Reorganization Following Treatment with Salicylic Acid*"

INTERNATIONAL METALLOGRAPHIC SOCIETY

Established 1968

OFFICERS (2001–2003)

President	Mr. Richard K. Ryan
Vice President	Mr. Allan J. Lockley
Financial Officer	Mr. David J. Fitzgerald
Secretary	Dr. Dennis W. Hetzner
Past President	Dr. Elliot Clark
Executive Director	Dr. William W. Scott, Jr.
Directors	Richard A. Blackwell (2001–2003)
	Luther Gammon (2001–2003)
	Stephen Glancy (2001–2005)
	Frauke Hogue (2001–2003)
	Janice Klansky (2001–2005)
	Nat Saenz (2001–2005)

Appointed Officers

Editor, *Materials Characterization* .... Chris Bagnall
2000 Convention General Chair .... Chris Bagnall
IMS Website .... Jeff Stewart
International Metallographic Contest .... Jeff Stewart
Editor, SlipLines .... Carla Sly
Coordinator, IMS Services .... Suzanne Campbell

PAST PRESIDENTS

1968–1971 John H. Bender Jr.
1971–1973 Arthur E. Calabra
1973–1975 E. Daniel Albrecht
1975–1977 James H. Richardson
1977–1979 Robert J. Gray
1979–1981 P.M. French
1981–1983 George Vander Voort
1983–1985 James E. Bennett
1985–1987 William E. White
1987–1989 M.R. Louthan, Jr.
1989–1991 Donald W. Stevens
1991–1993 Ian LeMay
1993–1995 Japnell D. Brown
1995–1997 E. Daniel Albrecht
1997–1999 Mahmoud T. Shehata
1999–2001 Elliot A. Clark
2001–2003 Richard K. Ryan

IMS CORPORATE SPONSORS

IMS Benefactor .... Struers Inc.
Buehler LTD
IMS Patron .... LECO Corporation
IMS Associate .... Precision Surfaces International, Inc
Unitron, Inc.
IMS Sponsor .... Carl Zeiss MicroImaging, Inc.

IMS AWARDS

International Metallographic Society
President's Award (Service to IMS)

1977 Carus K. H. DuBose
1978 Richard D. Buchheit
1979 Arthur E. Calabra
1980 James L. McCall
1981 E. Daniel Albrecht
1982 James H. Richardson
1983 Robert J. Gray
1984 Japnell D. Braun
1986 P. Michael French
1987 George F. Vander Voort
1988 Robert S. Crouse
1989 Ian Le May
1990 William E. White
1991 Chris Bagnall
1992 Gary W. Johnson
1993 Donald W. Stevens
1994 MacIntyre R. Louthan, Jr.
1995 Gunter Petzow
1996 James Nelson
1997 John Wylie
1998 John W. Simmons
1999 William Forgeng, Jr.
2000 Natolio Saenz
2001 William W. Scott, Jr.

Sorby Awards

1976. Georg L. Kehl
1977. Cyril Stanley Smith
1978. Adolph Buehler
1979. Frederick N. Rhines
1980. Len E. Samuels
1981. Robert J. Gray
1982. Gunter Petzow
1983. William D. Forgeng
1984. Ervin E. Underwood
1985. Alan Price
1986. Robert W. K. Honeycombe
1987. Gareth Thomas
1988. Franz Jeglitsch
1989. Tanjore R. Anantharaman
1990. E. Daniel Albrecht
1991. W.C. Leslie
1992. Charles S. Barrett
1993. Raimond B. Castaing
1994. F. Brian Pickering
1995. Erhard Hornbogen
1996. Peter Duncumb
1997. Robert T. DeHoff
1998. Kay Geels
1999. Joseph Goldstein
2000. Hans Eckhart Exner
2001. Brian Ralph

PROGRAM COMMITTEE

Edgar Voelkl
Program Chair

Raynald Gauvin
MAS/MSC/SMC Program Co-Chair

Alan Lockley
IMS Program Co-Chair

David Piston
Program Vice-Chair

Executive Members

Program Chair: E Voelkl, *nLine Corp*

MAS/MSC/SMC Program Co-Chair: R Gauvin, *McGill University*

IMS Program Co-Chair: A Lockley, *Atomic Energy of Canada*

Program Vice-Chair: D Piston, *Vanderbilt University*

Committee Members and Session Chairs

S Aharinejad, University of Vienna
R Albrecht, University of Wisconsin
L. Allard, Oak Ridge National Lab
I Anderson, Oak Ridge National Lab
K Baker, Baker and Associates (Acton) Ltd
GW Bailey, Baton Rouge, LA
S Barlow, San Diego State University
C Bennett, Acadia University
D Blom, Oak Ridge National Lab
G. Botton, McMaster Univ
M Brito, AIST, Japan
J Bruley, IBM
V. Bryg; Ferro Corp
B Carnagher, Scripps Research Institute
P Charest, Universite Laval
W Chiu, Baylor College of Medicine
J Corbett, University of Waterloo
E Dickey, Penn State University
V Dravid, Northwestern University
S Eppell, Case Western University
D. Erie, University of North Carolina
S Erlandsen, University of Minnesota
E Etz, NIST
J Frank, State University of New York, Albany
B Frost, University of Tennessee
P Gai, DuPont
M Gajdardziska-Josifovska, University of Wisconsin Milwaukee
A. Geary, Metallographic Consulting Ltd
L Giannuzzi, University of Central Florida
R Giberson, Ted Pella, Inc
B Griffin, University of Western Australia
W Gunning, Medical College of Ohio
M Haider, CEOS, Germany
B Hartman, Schering-Plough Research Institute
C. Hearne; University of Wyoming
T. Hirayama, Japan Fine Ceramics Center
F Hossler, East Tennessee State University
E Humphrey, University of British Columbia
S Jabaji-Hare, McGill University
M Jackson, NRC Canada
J Jerome, Vanderbilt University
C Johnson, Gates Rubber Co
D Joy, Oak Ridge National Laboratory
L Kerr, Woodshole Marine Biological laboratory
M Kersker, JEOL, USA
J Killius, NEOU College of Medicine
P Kotula, Sandia National Lab
O Krivanek, NION, Inc.
T Kelly, Imago Scientific Instruments
A Lametschwandtner, Universityof Salzburg
D Larson, Seagate Technologies
D Li, Shenyang National Lab
Z Li, DuPont
S Liou, University of Nebraska
Y Liu, University of Michigan

E Lifshin, State University of New York, Albany
S Ludtke, Baylor College of Medicine
C Lyman, Lehigh University
T. MacPherson, Dofasco Inc
B Maleeff, GlaxoSmithKline
J Mansfield, University of Michigan
A Marangoni, University of Guelph
J Mascorro, Tulane University
W Massover, UMDNJ–NJ Medical School
M McCartney, University of Arizona
K McDonald, University of California, Berkley
S McKernan, University of Minnesota
G Meeker, US Geological Survey
D.A. Meyer, University of Wisconson
J Michael, Sandia National Laboratory
D Miller, Argonne National Laboratory
M Misra, Unilever Research
K Moore, John Hopkins University
H. Mori, Res. Ctr. for Ultra High Voltage EM
I. Musselman, University of Texas, Dallas
D Newbury, NIST
P Ottensmeyer, University of Toronto
X Pan, University of Michigan
M Phaneuf, Fibics Inc
R Price, University of South Carolina
C. Potter, Scripps Research Institute
W Rau, LEO Electron Micrscopy
D. Rocheleau, Transportaion Safety Board of Canada
H. Schatten, University of Missouri–Columbia
C Schooley, Caspar, CA
J Scott, NIST
S. Shaybany, Pacific Metallurgical
D. Sherman, Purdue University
D Shindo, Tohoku University
C. Siedlecki, Hershey Medical Center
R Sinclair, Stanford University
W Sinkler, UOP LLC
J Small, NIST
D Smith, Arizona State University
G. Sosinsky, University of California
U Spornitz, University of Basel
E Stach, National Center for Electron Microscopy
D Su, FHI Berlin
K Takayanagi, Tokyo Institute of Technology
M Thompson, FEI Company
G. Vander Voort, Buehler Ltd
Z Wang, Georgia Tech
S Watkins, University of Pittsburgh
O Woo, Atomic Energy of Canada Ltd
J Woodward, Buckman Laboratories, Inc
V Woodward, Noveon, Inc
J Yang, University of Pittsburgh
N Yao, Princeton University
S Yue, McGill University
N Zaluzec, Argonne National Lab
Y Zhu, Brookhaven National Laboratory
J Zuo, University of Illinois

Local Arrangements Committee

Dr Pierre M. Charest, Chair, Université Laval, Québec
Dr Jim Corbett, Treasurer, Naramata, BC
Mr Ken Baker, Microscopy, Imaging, Analysis, Acton, ON
Dr Louise Brisson, Université Laval, Québec
Dr Hélène Chamberland, Université Laval, Québec
Ms Odette Desbiens, Université Laval, Québec
Dr Raynald Gauvin, McGill University, Montréal
Dr Anja Geitmann, Université de Montréal, Montréal
Mr Yves Giroux, Nissei Sangyo Canada Inc., Terrebone
Mr Gilles Grondin, Université de Sherbrooke, Sherbrooke
Dr Pierre Hovington, IREQ, Hydro–Québec, Varennes
Ms Line Mongeon, Noranda, Pointe–Claire
Ms Diane Montpetit, CRDA, St-Hyacinthe
Mr Jean-François Pageau, System for Research, Longueil
Mr Aristide Pusterla, Université Laval, Québec
Ms Hélène Roberge, National Research Council, Boucherville
Ms Marie Simard, Canadian Forestry Service, Québec

Society Logos

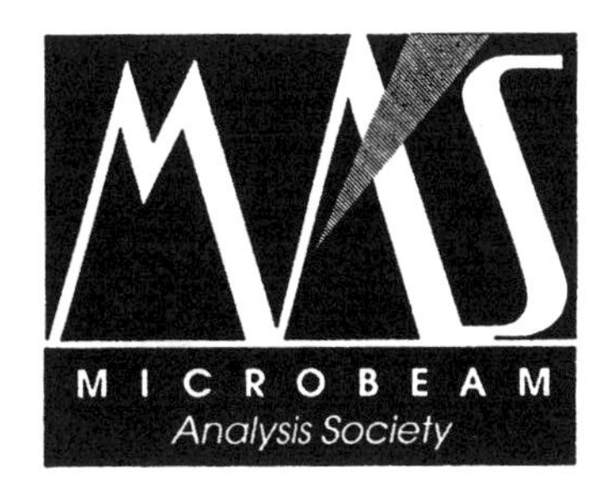

International
Metallographic
Society, Inc.
An Affiliate Society of ASM International

Microsc. Microanal. 8 (Suppl. 2), 2002
DOI 10.1017.S1431927602101449

The Emergence of Aberration Correctors for Electron Lenses.

John Silcox

School of Applied and Engineering Physics, Cornell University, Ithaca NY 14853

In 1936, Otto Scherzer published his famous theoretical paper[1] on the correction of aberrations in round magnetic electron lenses. For fundamental reasons, the aberration coefficients are positive and to introduce negative corrections, the electron optical elements have to produce either charges or poles on the optic axis, or contain a mirror or be a multipole system. In 1997, three papers appeared with actual demonstrations of the correction of aberrations in imaging lenses,[2] probe forming lenses[3] and electron mirrors.[4] It took almost two thirds of a century for this seemingly straightforward development to occur and for the promise of wavelength limited electron imaging to emerge as a practical proposition. In this paper I will put this achievement into perspective, i.e., try to explain why this was so difficult, give some indication of the implications for our field and provide a sense of the contributions of two scientists, Albert Crewe and Harald Rose, whose work has immeasurably influenced the work we do now and, in my view, has even more influenced the future of this field.

Before aberration correction can actually have an impact, the electron optical systems should be operating at or near the theoretical limits. The first instrument that I encountered (in 1961), the Siemens Elmiskop lA, had a spatial resolution of 15 Å. Fortunately, this was more than adequate for the identification and study of dislocations and other important microstructural features in materials. Environmental and manufacturing capabilities almost certainly determined the final performance levels of this instrument. Given the 100 keV electron wavelength of 0.037 Å, the 15 Å value was so far away from the performance limit that serious improvements in design, manufacturing and environment of the basic instrument had to happen before aberration correction could be considered essential to further progress in electron microscopy. In other words there was plenty to do with the existing instrument.

These considerations did not prevent efforts at understanding how to do this and in looking for neat ways to circumvent the aberration limitations. For example, Denis Gabor invented optical holography (and won the 1971 Nobel Prize for physics) as a means of correcting electron micrographs for aberration distortions. Only very recently have serious efforts been undertaken to seriously exploit electron holography as a method of improving information at the atomic scale (See, for example the symposium at this meeting in honor of Hannes Lichte and Akira Tonomura). The prospect of atomic scale imaging resolution, albeit a distant goal, still served to stimulate efforts to circumvent Electron optics was undertaken as a study in its own right and substantial commitments of time, funds and ingenuity were expended in making progress towards this eventual goal. Thus a serious background to the subject was developed even though working electron microscopes did not operate at the level that was limited by the electron optical performance of the lenses.

I date the modern era from the arrival of Albert Crewe on the electron microscopy scene in the late to early seventies. Professor Crewe made his reputation in accelerator physics and used insights from his experience in that field in his venture into electron optics. On becoming Director of Argonne National Laboratory, the story goes that on wandering through the laboratory he was struck by how many of the staff used electron microscopy and wanted to help. In the intervening period both at

Argonne and later at the University of Chicago, he and his group introduced the field emission gun that is now common, added a high quality (i.e., small spherical aberration coefficient) lens to form a high resolution scanning transmission electron microscope (STEM), used an Annular Dark Field detector for "Z-contrast" imaging and added an electron spectrometer for Electron Energy Loss Spectrometry. In this period atoms were imaged for the first time and in short he and his team changed the field of electron microscopy as we know it today. For a period, that team included Elmer Zeitler and Harald Rose as well as a number of bright graduate students who have gone on to make their own independent marks.

Once the new instrument had been established, serious work began on the very difficult problem of correcting aberrations to reduce the probe even more. Work was under way in several laboratories around the world in the UK, Germany and US throughout this period. Space forbids an exhaustive survey in this abstract so only the work of Jiye and Crewe analyzing the electron optics of multipole systems and sextupole/roundlenses/sextupole[5] systems will be noted here.

In the meantime, Harald Rose went back to Germany and took up the chair in Darmstadt following in the long history and tradition established by Otto Scherzer. There he has trained a number of students in electron optics (such as Max Haider[2]) and has been particularly concerned with the problems of practical construction of the many devices invented by himself and his students. His 1990 paper[6] laid out in detail the design of a spherically corrected semi-aplanatic TEM that was the harbinger of the instrument later reported.

In the final analysis, this has been a long path to realization and almost certainly, had to await the arrival of the computational capability as applied in simulation of the optical paths through the lenses, control of the machining capability and control of the lens excitations and overall operations. Has it been worth it? I am an unabashed advocate of the new era and look forward eagerly to these instruments.

References

[1] O. Scherzer, *Z.Physik* **101,** 593-603 (1936).
[2] M. Haider, S. Uhlemann, E. Schwan, H.Rose, B. Kabius, K. Urban, *Nature*, **392,** 768-769 (1998).
[3] O.L. Krivanek, N. Dellby, A.J.H. Spence, R.A. Camps and L.M. Brown, *EMAG '97 I.o.P. Conference series (ed. J.M.Rodenburg)* **153,** 35-40 (1997).
[4] G.F. Rempfer, D.M. Desloge, W.P. Skoczylas, and O.H. Griffith, *Microscopy and Microanalysis* **3**, 14-27 (1997).
[5] X. Jiye and A.V. Crewe, Optik 70 37-42 (1985)
[6] H. Rose, Optik 85 19-24 (1990).

Microsc. Microanal. 8 (Suppl. 2), 2002
DOI 10.1017.S1431927602101450

Some Chicago Aberrations

A. V. Crewe

Enrico Fermi Institute and Department of Physics, University of Chicago, 5630 Ellis Avenue, Chicago, IL 60637

The account that follows is based on my own memory although some of it can even be documented.

The story begins in 1966. At that time I was a beginner in electron microscopy and we (Isaacson, Wall and Johnson) were attempting to get a field emission tip to function as the source of electrons in the first STEM. There was not much encouragement to think that it would work, nothing except the skill of those students. Nevertheless one could hope. I recall reading Zworykin's book with its account of the ideas of Scherzer on correction of aberrations. There was also a review on the subject by Septier. It looked like a good idea and the STEM was clearly the instrument of choice for a corrector. I asked David Cohen to investigate the concept and he produced a set of guidelines for a device with four quadrupoles and three octupoles. When I applied this to a 30 kV STEM, the whole thing looked feasible - except for the mechanical tolerances. A few months later I left Argonne and found Walter Mankawich at the UofC, a man who could machine to micron tolerances. We began a long journey together and he used every spare moment to build jigs and fixtures.

Vernon Beck took over the project when he joined our group. He changed the design somewhat and added new features [1] and we built a new STEM to test the device. I recall that both Michael Thomson and Harold Rose spent a year or so in our laboratory in those years, but although both of them had experience in the correction business they had other interests and did not get involved in the work.

We attempted to operate it in 1976, but we were not able to make it work [2, 3]. In spite of the 40 trim coils we could not find a suitable setting. The effective centers of the quadrupoles and octupoles could not be brought together. My own conclusion at the time - right or wrong - was that the iron of the pole pieces was not sufficiently uniform. In any case, no other student would take a chance on it and so the work on aberration correction stopped.

The next event came when I gave a series of lectures to my group on electron optics. After discussing guns and round lenses I also talked about quadrupoles and octupoles. After finishing the course one of the students (I think it was Isaacson) pointed out that I had not mentioned sextupoles. This prompted me to calculate their properties and to my astonishment the third order aberrations were cylindrically symmetric about the axis and of the opposite sign to those of round lenses! (This result had also been noted by Hawkes but I was unaware of it).

My calculations are notoriously full of errors so I asked David Kopf to check them. He found that I was correct.

The difficulty in using sextupoles as a corrector is that of finding an electron optical system that eliminates the second order focusing effect while retaining the third order aberrations. I am embarrassed by the length of time it took me to devise such a system. It was only when I recalled the use of field lenses in light optics that the solution appeared. It was to place a non-rotating round lens (actually two) between two sextupoles, imaging one on to the other. With a suitable physical rotation between the two, the desired result is obtained. I submitted a paper on this idea in July 1981 [4]. Shao later added to the concept by considering the fifth order aberrations as well and devising a way to control them [5].

One of the advantages of using sextupoles is that the required excitation is small enough that one can use air coils - no iron is needed. With funding from the NSF we began construction of a 200 kV STEM with a sextupole corrector. Unfortunately, when the system was more than half finished, their enthusiasm ran out and we had to abandon the project. I believe that it would have worked.

The third - and maybe final - attempt at correction occurred as a result of an academic exercise in which I am trying to develop electron optics "ab initio", starting with uniform fields. Surprisingly this is not readily available in the literature.

It was of academic interest to consider a uniform magnetic field superimposed on a uniform electrostatic field with an arbitrary angle between them. The focusing properties are remarkably simple and the most interesting case was the magnetically focused electrostatic mirror [6]. It turned out to have third order aberrations with opposite sign to those of a round lens. It was then possible to design a system with zero aberrations. Frank Tsai was the one who operated the device and showed that he could eliminate both chromatic and spherical aberration [7, 8]. This concept remains a curiosity because it is difficult to design a useful optical system. The only one appears to be a STEM configuration.

We obtained DOE support for such a system and once again the funding was terminated when the project was incomplete.

The sum total of our efforts is not very encouraging, one technical failure, two funding failures and one success. But perhaps we did stimulate ideas and encouraged competition.

References

[1] A.V. Crewe and V. Beck, Proc 32nd EMSA meeting (1974, p. 426
[2] V. Beck and A.V. Crewe, Proc 34th EMSA meeting (1976), p. 578.
[3] V. Beck, Proc. 35th EMSA meeting (1977), p. 90.
[4] A.V. Crewe, Optik, 60 (1982) 271.
[5] Z. Shao, Rev. Sci. Instr. 59 (1988) 2429.
[6] A.V. Crewe, Ultramicroscopy, 41 (1992) 279.
[7] A. V. Crewe et al., Journal of Microscopy, 197 (2000) 110.
[8] F. C. Tsai, Journal of Microscopy, 197 (2000) 118.

Microsc. Microanal. 8 (Suppl. 2), 2002
DOI 10.1017.S1431927602101462

Correction of Aberrations—Past, Present and Future

H. Rose

Materials Science Division, Argonne National Laboratory, Argonne IL 60439, USA

The performance of static rotationally symmetric electron lenses is limited by unavoidable chromatic and spherical aberrations. In 1936, Scherzer demonstrated that the integrands of the integral expressions for the coefficients of these aberrations can be written as a sum of positive quadratic terms [1]. Hence these coefficients can never change sign. This important result is called the Scherzer theorem, the only theorem existing in electron optics. Employing variational methods, Tretner determined the field of magnetic and electrostatic round lenses, which yields the smallest spherical aberration coefficient for particular constraints [2]. Unfortunately, these coefficients are still too large for realistic boundaries to enable sub-Ångsrtöm resolution at medium voltages of about 200 to 300 kV. Therefore, the only possibility to directly reach this limit is the correction of the troublesome aberrations. It was again Scherzer who showed different procedures for cancelling these aberrations [3]. The most promising is the incorporation of a corrector consisting of multipole elements or of a tetrode mirror in the case of low voltages. Although the mirror is rotationally symmetric, a non-rotationally symmetric beam splitter is needed to separate the incident beam from the reflected beam.

The first corrector proposed by Scherzer consists of electrostatic elements, and was built and tested by Seeliger [4]. He first showed that spherical aberration could be eliminated by properly adjusting the octopoles. Since the resolution of his microscope was limited by instabilities, the correction did not improve the actual resolution. Nevertheless, his experiments clearly revealed that spherical aberration can be eliminated by means of a corrector. In order to demonstrate the effect of this correction, Moellenstedt increased the illumination angle to 0.02 rad, thereby enlarging the spherical aberration to such an extend that it blurred the image and limited the resolution. After adjusting the octopoles of the Seeliger corrector appropriately, the resolution was improved considerably accompanied by a striking increase in contrast [5]. The latter results from the reduction of delocalisation, because the correction of spherical aberration eliminates the extended base of the point-spread function. The importance of this behavior for artefact-free imaging of non-periodic objects, such as interfaces, has been recognized only recently [6].

The Seeliger corrector and the quadrupole-octopole corrector investigated later by Deltrap [7] are only applicable for probe-forming systems because they introduce large off-axial aberrations. The same holds true for the two Chicago correctors developed by Beck and Crewe, which were intended to reduce the probe-size in their STEM [8,9]. For imaging extended objects with high resolution in a TEM, an electron-optical aplanat has been designed, built and tested at Darmstadt over a period of almost 10 years starting in 1972. By increasing the number of multipole elements and employing symmetry conditions for their fields and the paths of the paraxial rays, it was possible to correct axial chromatic aberration, spherical aberration, chromatic distortion and coma [10]. It was shown by Koops [11] and Hely [12] in the course of the project that the chromatic and spherical correction worked perfectly. Nevertheless, an actual improvement in resolution could not be achieved because it was not possible to align the system with the required accuracy or to achieve the necessary electrical and mechanical stability at the time. Owing to a lack of adequate technology, all attempts have failed over a period of almost 45 years to improve the performance of electron microscopes by correcting the aberrations. As a result it was widely believed that a posteriori methods such as holography and digital image processing would provide more realistic and easier solutions.

This pessimistic view was proven wrong by Zach and Haider [13] who were the first to achieve an actual improvement in resolution with a multipole corrector in a dedicated low voltage SEM. Due to the rapid advancement in electronic technology and in computer-assisted alignment of systems consisting of many elements, the correction of aberrations can now routinely be performed in the TEM [14] and in the STEM [15]. Aplanatic systems have the further advantage to allow large tilt angles for the incident beam because tilting its axis introduces neither an appreciable image shift, nor defocus, astigmatism or axial coma. Aberration correction is also important to improve the performance of imaging energy filters. It has been demonstrated recently that a corrected 90-degree imaging energy filter is capable of filtering and transferring large-angle diffraction patterns without any appreciable distortion up to scattering angles of about 150 mrad.

At present the deleterious effect of chromatic aberration is best avoided by employing a monochromator, because it requires a significantly smaller expenditure than the correction of the chromatic lens defect by means of electric/magnetic quadrupoles. In this case it suffices to correct only for spherical aberration and off-axial coma by means of a hexapole corrector, because hexapoles do not affect the paraxial path of rays. However, the need for chromatic correction has recently been revived in the context of in situ high-resolution energy-filtering TEM and high-throughput electron projection lithography. The latter requires a system which is corrected for chromatic aberration and all third-order geometrical aberrations. A feasible system is proposed which consists of a demagnifying telescopic round lens doublet and a corrector composed of two identical symmetric quadupole-octopole septuplets. The realization and the alignment of this multi-element system is significantly facilitated by symmetry conditions imposed on the corrector as a whole and the two parts of it.

References:

[1] O. Scherzer, Z. Phys. 101 (1936) 593.
[2] W. Tretner, Optik 16 (1959) 155.
[3] O. Scherzer, Optik 2 (1947) 114.
[4] R. Seeliger, Optik 5 (1949) 490.
[5] G. Moellenstedt, Proc. Int. Conf. on EM, London (1954) 694.
[6] M. Haider et. al., J. Electr. Micr. 47(1998) 395.
[7] J.H.M. Deltrap. PhD Thesis, Univ. Cambridge,1964
[8] V. Beck and A. Crewe, EMSA 34 (1976) 578.
[9] V. Beck, Optik 53 (1979) 241.
[10] H. Rose, Optik 34 (1971) 285.
[11] H. Koops, G. Kuck, O. Scherzer, Optik 48 (1977) 225.
[12] H. Hely, Optik 60 (1982) 353.
[13] J. Zach and M. Haider, Nucl. Instr. and Meth. Phys. Res. A 363 (1995) 316.
[14] M. Haider, H. Rose, S. Uhlemann, E. Schwan, B. Kabius, K. Urban, Ultramicroscopy 75 (1998) 53.
[15] O.L. Krivanek, N. Dellby, A.R. Lupini, Ultramicroscopy 78 (1999) 1.
[16] Research supported by US Department of Energy under Contract W-31-109-Eng-38.

DOI 10.1017.S1431927602101474

Application of Aberration-Corrected Transmission Electron Microscopy to Materials Science

K. Urban and M. Lentzen

Institute for Solid State Research, Research Center Jülich GmbH, D-52425 Jülich, Germany

Recently an electromagnetic hexapole system for the correction of the spherical aberration of the objective lens of a 200 kV transmission electron microscope was constructed by Haider and coworkers [1]. Integrating this system in a Philips CM200 FEG ST an increase in point resolution from 0.24 to better than 0.13 nm could be demonstrated [2]. However, transmission electron microscopy involving an aberration corrector offers additional features of interest for materials science application.

By appropriately exciting the hexapole elements it is possible to adjust specific values of the spherical aberration coefficient C_S ranging from the value of the original objective lens to zero. Thus C_S enters as an additional variable, besides the defocus value Z, into the wave aberration function

$$\chi(\vec{g}) = \tfrac{1}{2} Z \lambda g^2 + \tfrac{1}{4} C_S \lambda^3 g^4,$$

where λ is the electron wave length and $\vec{g}$ the spatial frequency.

Adjusting the corrector for $C_S=0$ and applying $Z=0$ the phase-contrast term $-i\sin 2\pi\chi(g)$ in the transfer function vanishes. On the other hand, the amplitude-contrast term $\cos 2\pi\chi(g)$ adopts its maximum value 1. This means that in the fully aberration-corrected mode the microscope images atomic structures by amplitude rather than by the usual phase-contrast [3]. In the electron exit-plane wave function local amplitude variations are produced by electron diffraction channeling. The projected atom columns are imaged by bright contrast on a dark background. As demonstrated by the simulation for Ge [110] in Fig. 1 maximum contrast occurs at thicknesses of half the extinction distance defined by the two most excited Bloch waves and odd multiples of it. This mode has the special advantage that contrast delocalization defined by the radius of the point-spread function ($g_{\max}$ - information limit) [4]

$$R = \max\left|\frac{\partial\chi}{\partial\vec{g}}\right|, \; [0, g_{\max}],$$

is zero and optimum imaging of defects, boundaries and interfaces can be achieved.

Besides this pure amplitude-contrast mode other imaging modes are available in which phase-contrast contributes to the image in addition. In order to determine the optimum values for Z and C_S for this kind of imaging an optimization problem has to be solved which yields an optimum defocus and spherical-aberration setting. This pair of values represents the equivalent to Scherzer's defocus in idealized weak-phase object imaging. The optimization problem arises from the fact that finite values for Z and/or C_S induce contrast delocalization. A detailed treatment of this problem [3] yields for the optimum values yielding maximum phase contrast at a minimum of delocalization

$$C_{S,opt} = \tfrac{64}{27} \lambda^{-3} g_{\max}^{-4}; \; Z_{opt} = -\tfrac{16}{9} \lambda g_{\max}^{-2}.$$

For the CM200 FEG ST one obtains for λ=2.5 pm and an information limit of 7.3 nm^{-1} the values $C_{S,opt}$=53 µm and Z_{opt}= -13.3 nm. Under these conditions the contrast delocalization amounts to R_{opt}=0.081 nm. In such a calculation, in the same way as in conventional transmission electron microscopy and the derivation of the idealized Scherzer formula, only phase shifts introduced by the electron optics are taken into account. Phase shifts already contained in the electron exit-plane wave function are not considered. In practice it has proven to be advantageous to take a through-focus series of images and to accompany the practical high-resolution work by image simulation or exit-wave function reconstruction. This not only allows to define the conditions for Z=0, it also permits to select and understand the images taken under optimum conditions.

The following examples of images are taken at $Z=0/C_S=0$. Figure 2 (courtesy K. Tillmann) shows an application for the determination of Al-concentration profiles in low-temperature GaAs/AlAs heterostructures. Due to the absence of contrast delocalization the profile can be directly derived from the (002) component of the Fourier transform of the image intensity distribution. Figure 3 (courtesy J.L. Jia) shows a tilt boundary in $(BaSr)TiO_3$ along the [100] zone axis demonstrating the high quality of delocalization-free interface imaging under amplitude contrast conditions. Figure 4 (courtesy J.L. Jia) demonstrates the high resolution of 0.14 nm in $(BaSr)TiO_3$ along the [110] zone axis.

References

[1] M. Haider, H. Rose, S. Uhlemann, E. Schwan, B. Kabius, K. Urban, *Nature* 392 (1998) 768.
[2] M. Haider, H. Rose, S. Uhlemann, E. Schwan, B. Kabius, K. Urban, *Ultramicroscopy* 75 (1998) 53.
[3] M. Lentzen, B. Jahnen, C.L. Jia, A. Thust, K. Tillmann, K. Urban, *Ultramicroscopy*, in press.
[4] H. Lichte, *Ultramicroscopy* 38 (1991) 13.

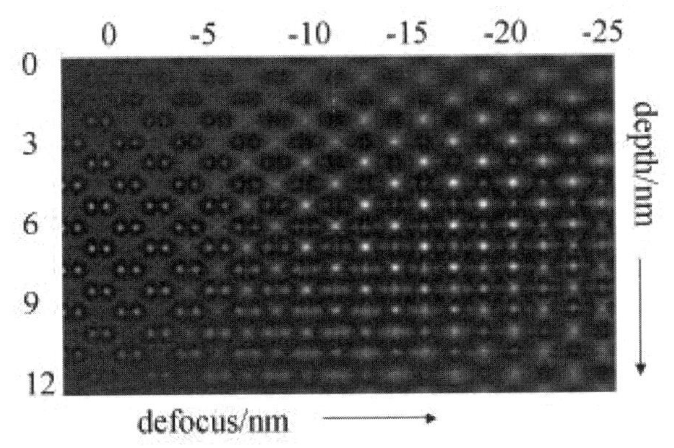

Fig.1 Calculated through-focus and through-thickness image series of Ge [110] for $C_S=0$.

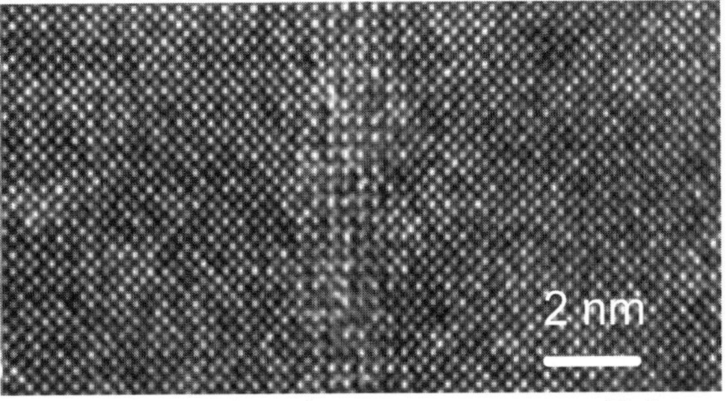

Fig. 2 GaAs/AlAs heterostructure at $Z=0$ and $C_S=0$.

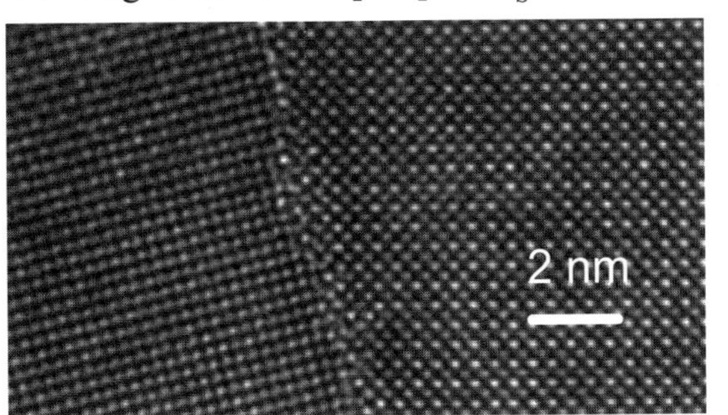

Fig. 3 Tilt boundary in $(BaSr)TiO_3$ at $Z=0$ and $C_S=0$. Viewing direction [100].

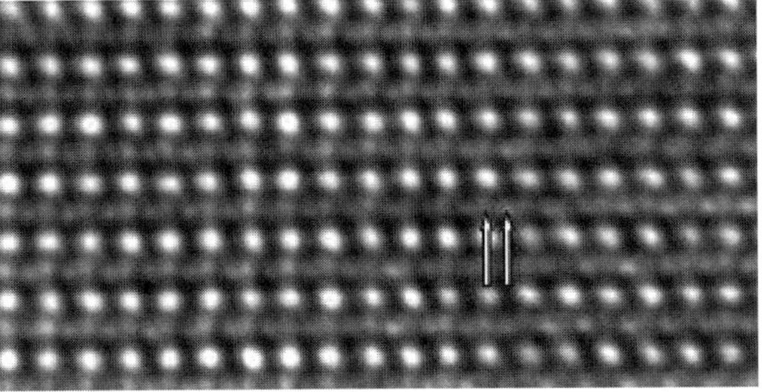

Fig.4 $(BaSr)TiO_3$ [110] at $Z=0/C_S=0$. The separation of TiO/O columns (arrows) is 0.14 nm.

Microsc. Microanal. 8 (Suppl. 2), 2002
DOI 10.1017.S1431927602101486

Applications of a C_s Corrected HRTEM in Materials Science

J.L. Hutchison,* J.M. Titchmarsh,* D.J.H. Cockayne,* G. Möbus,* C.J. Hetherington,* R.C. Doole,* F. Hosokawa,** P. Hartel**& and M. Haider***

*Department of Materials, Oxford University, Parks Road, Oxford, OX1 3PH, UK
**JEOL Ltd., 1-2 Musashino 3-chome, Akishima, Tokyo 196, Japan
***CEOS GmbH, Engelstr. 28, D-69126 Heidelberg, Germany

The development of aberration-corrected lenses [1,2] for electron microscopy offers significant improvements for higher spatial resolution in both imaging and nanoanalysis. This will have a considerable impact in materials science applications, particularly those involving nanostructures.. A new TEM with aberration correction of both condenser and objective lenses is being developed for installation at Oxford.[3] This instrument, based on the JEOL 2010 FEF with a Schottky field emission source. With optimised beam energy spread this unique instrument aims to provide HREM information limit and HAADF resolution close to 0.1nm. Additional analytical features include EELS and energy-filtered imaging, EDS and holography. In order to minimise external interference and to maximise thermal and mechanical stability, the microscope column, power supplies and operating console will all be installed in separate, adjacent rooms within a new laboratory, now under construction. Components of the new instrument are now being developed and tested prior to final assembly.

The objective lens C_s corrector has been succesfully developed and tested on a separate JEOL 2010F instrument and we will illustrate here some of the initial results. Tests have confirmed the stability of the corrector over long periods. This corrector system [1] uses extended "Zemlin tableaux" of diffractograms [4] to measure the main aberrations present, i.e. defocus, 2-fold astigmatism, coma, 3-fold astigmatism, and spherical aberration, in addition to 4-fold and star aberration. Fig.1 illustrates tilt tableaux from an amorphous Ge film, without aberration correction and with all main aberrations (i.e. 2-fold and 3-fold astigmatism, coma and spherical aberration) correcteded. These results demonstrate an information limit in the corrected state close to 0.13 nm, consistent with the energy spread in the beam (0.8eV) and chromatic aberration coefficient of 1.2mm at 200keV. The montages of diffractograms are recorded with a range of incident beam tilt semi-angle up to 18mrad, demonstrating the capability of the corrector and the absence of significant aberrations.

There are several benefits arising from the use of a C_s corrected objective lens for image formation. These include an improved "point resolution" limit, coinciding with the instrument's information limit [1] With an optimised electron beam energy spread, achievable by using a monochromator in the electron gun, this value approaches 0.1 nm. Contrast delocalisation at interfaces and grain boundaries is also greatly reduced [1], as shown in Fig. 2.

Another significant benefit of C_s corrected imaging is illustrated in Fig.3 which shows a small CdSe quantum dot on an amorphous carbon support film. In an uncorrected image at Scherzer defocus the presence of strong Fresnel fringes and "ghost" images would obscure the fine scale structural features, which are further upset by strong phase contrast arising from the carbon film. However, following C_s correction, these effects are almost completely removed while the strong amplitude contrast of the heavier CdSe is enhanced by using a C_s of virtually zero and Gaussian focus. The atomic structure of the dot is thus clarified [5].

Acknowledgement

This project is supported by a major JIF grant from HEFCE and EPSRC.

References

1. M. Haider et al., Ultramicroscopy 75 (1998) 53.
2. O.L. Krivanek, N. Delby and A.R. Lupini, Ultramicroscopy 78 (1999) 1.
3. G. Möbus et al. (2001) Electron Microscopy and Analysis, M.Aindow and C.J. Kiely, eds., IoP Publishing, Bristol and Philadelphia, 27
4. F. Zemlin et al., Ultramicroscopy 3 (1977) 49.
5. N.A. Allsop, J.L. Hutchison and P.J. Dobson, to be published (2002).

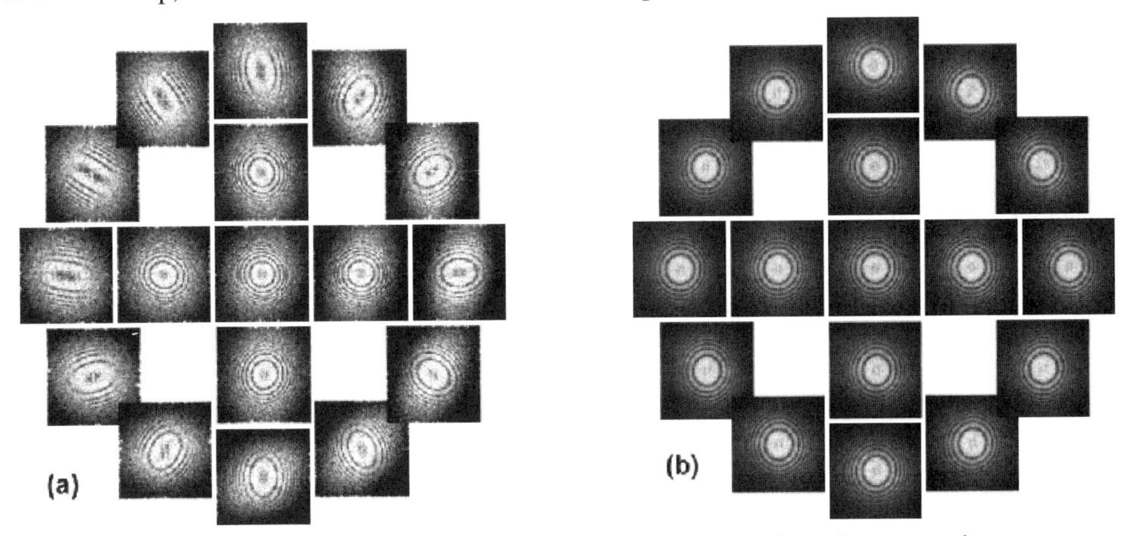

Fig. 1.Tilt tableaux of diffractograms (a) without, and (b) with aberration correction.

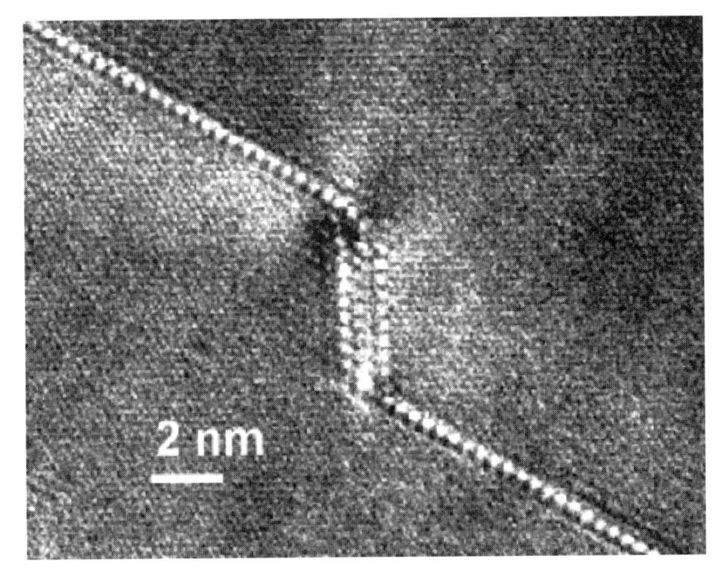

Fig.2. Grain boundary in <111> Au foil showing localised contrast at grain boundary

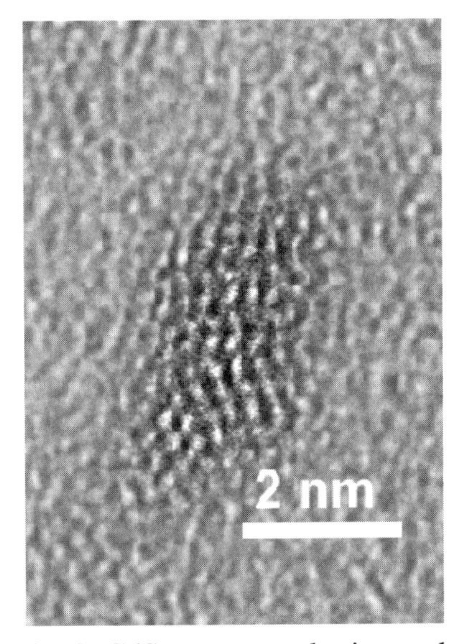

Fig. 3. CdSe quantum dot imaged at Gaussian focus and $C_s \sim$ zero.

Microsc. Microanal. 8 (Suppl. 2), 2002
DOI 10.1017.S1431927602101498

Benefits and Possibilities of C_C–Correction for TEM / STEM

Heiko Müller, Stephan Uhlemann, and Maximilian Haider

CEOS GmbH, Engler Straße 28, 69126 Heidelberg, Germany

During the last decade aberration correction became possible for several types of electron optical instruments. Correctors for the spherical aberration of TEM / STEM objective lenses are now commercially available [1,2]. In case of the SEM [3] and the LEEM / PEEM [4] the correction of both the spherical and the chromatic aberration has been demonstrated in experiments. In spite of these great advancements aberration correction for TEM / STEM is still restricted to the spherical aberration. An improvement of the information limit by C_c–correction for the TEM / STEM has not yet been demonstrated.

For a C_s–corrected 200kV FEG–TEM equipped with a hexapole corrector an information limit of $d = 0.12$ nm has been measured. For such an instrument the information limit is determined both by the chromatic aberration of the objective lens and that of the hexapole corrector. In order to reach this limit a very accurate alignment and a sufficient mechanical and electrical stability of the microscope are necessary. For a thin phase object the chromatic information limit can be quantified by a visibility criterion based on the chromatic envelope of the contrast transfer function

$$H_c(\theta) = \exp\left(-\frac{\pi^2 \sigma_E^2 C_c^2 \theta^4}{2\lambda^2 E_0^2}\right) \overset{!}{>} 0.3 \ ,$$

where $\Delta E_{1/2} = \sqrt{8 \ln(2)}\, \sigma_E$ denotes the FWHM of the energy width. The influence of the lateral incoherence on the information limit is strongly suppressed for a C_s–corrected microscope. A further improvement of the information limit is only possible either by reducing the energy width or by chromatic correction. Different types of gun monochromators have been proposed to reduce $\Delta E_{1/2}$ below 0.3 eV [5]. A TEM equipped with a gun monochromator and a high–resolution pole piece optimized for a minimum C_c should be able to approach an information limit of about 0.08 nm. Unfortunately, a high–resolution pole piece with a very narrow magnetic gap is inadequate for many applications, especially for in-situ electron microscopy. In order to improve the resolution of an electron microscope equipped with a pole piece with a sufficiently large gap C_c–correction is the only possibility, if we refuse to accept a dramatic loss in beam current by monochromatization down to $\Delta E_{1/2} < 0.1$ eV. Several methods for correcting the chromatic aberration have been proposed. While a mirror corrector [4] or a purely electro-static corrector [6] are not feasible for a high–voltage TEM / STEM an electric–magnetic quadrupole C_c/C_s–corrector is a possible choice. Unlike a hexapole C_s–corrector a quadrupole corrector strongly alters the path of the paraxial rays. If we use the linear combinations $u_\alpha = \frac{1}{2}(x_\alpha + y_\beta)$ and $u_{\overline{\alpha}} = \frac{1}{2}(x_\alpha - y_\beta)$ for the axial fundamental rays x_α and y_β the contribution of the quadrupoles to the chromatic aberration C_c and to the chromatic two–fold astigmatism $\overline{C}_c$ adopts the simple form

$$C_c = \int_{z_O}^{z_I} 2\,(\Psi_{2s} - \Phi_{2c})\, u_\alpha u_{\overline{\alpha}}\, dz \ , \quad \overline{C}_c = \int_{z_O}^{z_I} (\Psi_{2s} - \Phi_{2c})\left(u_\alpha^2 + u_{\overline{\alpha}}^2\right) dz \ .$$

The normalized multipole strengths are chosen imaginary $\Psi_2 = i\Psi_{2s}$ for the magnetic and real $\Phi_2 = \Phi_{2c}$ for the electrostatic quadrupoles in order to separate the paraxial ray equations. We find

that the path of rays inside the corrector must be astigmatic ($u_{\overline{\alpha}} \not\equiv 0$). The "minimum" quadrupole C_c–corrector requires two magnetic and two combined electric–magnetic quadrupoles arranged anti–symmetrically with respect to the mid-plane of the system in order to avoid a two–fold chromatic astigmatism $\overline{C}_c$. The correction of the spherical aberration can be achieved by additional octupole elements which act symmetrically on the path of rays. The path of the fundamental rays for each section is determined by the sum of the focussing and defocussing strength of the magnetic and electric quadrupoles. Hence, for each element a negative chromatic aberration is introduced in one section. The dependence of the refraction power on the beam energy is different for electric and magnetic quadrupoles, since only magnetic interaction depends on the velocity of the electron. For small energy deviations the change of the refraction power of a magnetic quadrupole is only half as large as in the case of an electric one.
Unfortunately, the minimum quadrupole corrector described above introduces strong off-axial aberrations. Therefore, it can be applied only in a low–voltage SEM [3]. For the TEM more advanced designs with reduced off–axial aberration have been proposed. All these correctors are symmetrically arranged with respect to their mid-plane. Recently, we have seized a suggestion by Rose [7] and investigated a new class of correctors based on a double–symmetric arrangement of electric–magnetic quadrupoles. For these highly symmetric systems the geometrical aberrations of third order can be controlled completely.
A second, even more critical property of all C_c/C_s–correctors based on the superposition of electric–magnetic quadrupole fields is the stability requirement for the power supplies of the combined electric–magnetic elements. In order to produce a sufficiently large negative chromatic aberration they must be rather strong. Their electric and magnetic refraction powers compensate each other to a large extent. However, this is not true for small uncorrelated instabilities of voltage and current. For a perfectly designed and aligned corrector instabilities result in a variation of defocus σ_{C_1} and two–fold astigmatism σ_{A_1} avoiding image displacement. These incoherent perturbations decrease the information limit, which can be approximated by an additional damping envelope of the contrast transfer function. Just like in the case of the chromatic aberration the tolerable variation of defocus and two–fold astigmatism can be estimated by a visibility criterion

$$H_D(\theta) = \exp\left(-\frac{\pi^2|\sigma_D|^2\theta^4}{2\lambda^2 E_0^2}\right) \overset{!}{>} 0.3\ , \quad D \in \{A_1, C_1\}\ .$$

This shows that any C_c–corrector is useful for a maximum aperure θ_A only if the stability condition $|\sigma_D| < 0.5\,\frac{\lambda}{\theta^2}$, with $D \in \{A_1, C_1\}$ can be fulfilled.

References

[1] N. Dellby, et al., J. Electron Microsc. 50(3) (2001) 177.
[2] M. Haider et al., Nature 392 (1998) 769.
[3] J. Zach and M. Haider, Nucl. Instr. Meth. A363 (1995) 316.
[4] P. Hartel et al., Proc. EUREM Brno/Cz, Vol.III (2000) 1459.
[5] F. Kahl, Proc. Microsc. Microanal. 7, Suppl. 2 (2001) 922.
[6] C. Weißbäcker and H. Rose, J. Elecrton Microsc. 50(5) (2001) 383.
[7] H. Rose, private communication.

Microsc. Microanal. 8 (Suppl. 2), 2002
DOI 10.1017.S1431927602101504

Sub-Angstrom Probe Size in HADF-STEM at 120KV

P.E. Batson, N. Dellby*, and O.L. Krivanek*

IBM T.J. Watson Research Center, Yorktown Heights, New York 10598
* Nion, Inc., Kirkland, Washington

We have recently installed a spherical aberration corrector of the quadrupole-octupole type in the VG Microscopes HB501 STEM in an attempt to improve the spatial resolution of the instrument. [1] In evaluating the achieved probe size, it has become apparent that use of the presence of spatial periodicities in the images, determined by Fourier analysis, may not be a reliable method for estimating the spatial resolution for this very small probe size. [2]

Therefore, we have attempted to characterize the probe by examination of the image of single atoms. As Figs. 1 and 2. show, a standard gold resolution test sample turns out to have not only 10-50nm Au islands, but also a sub-monolayer of gold atoms loosely arranged in "rafts," isolated clusters and single atoms. The image size of the single atoms is comparable to, or smaller than, 1Å as illustrated in Fig. 2.

In Figure 3, we have isolated one atom, comparing it with probes having widths of 2Å (for Cs = 1.2mm at 120kV in the uncorrected STEM) and 0.78Å (calculated for a set of experimental aberration coefficients, measured to 4rd order with the corrector operating). The only adjustable parameter is the total probe intensity, scaled to match the image. The first observation we can make is that single atom detection in the presence of the carbon background is vastly improved using the small probe. In fact, single Au atoms are not visible at all in this sample using the 2Å probe. Secondly, the apparent size of the calculated probe is a good match to the Au atom image. In the second part of Fig. 3, we have compared line profiles from the image with an expected image profile including contributions from: a)the probe aperture diffraction limit of 0.68Å for the 25mR half angle aperture, 2) the measured optical aberrations, yielding a limit of about 0.72Å, c) a 0.3Å source size, consistent with that expected from a probe current measurement of 40 picoamps, yielding 0.78Å, and d) the projected potential for the Au atom, obtained using the programs of Kirkland, [3], which is convoluted with the probe shape to produced a 0.92Å image width, in good agreement with the measurement.

We thus believe that this instrument is now limited by electron optics to a probe size of about 0.78Å at 120kV. In Fig. 4, we show an analysis of an image of the SiGe $< 110 >$ projection, wherein the Si "dumbbell" is clearly resolved. Interestingly, the apparent resolution of the image now has a significant component from the size of the atoms themselves.

[1] N. Dellby, O.L. Krivanek, P.D. Nellist, P.E. Batson, and A.R. Lupini, *J. Electron Microscopy*, 50 (2001)177.
[2] Z. Yu, P.E. Batson, and J. Silcox, in *Microscopy and Microanalysis*, edited by G.W. Bailey, R.L. Price, E. Voelkl, and I.H. Musselman (Springer-Verlag, New York, 2002), in press.
[3] E.J. Kirkland, *Advanced computing in electron microscopy* (Plenum Press, New York, 1998).

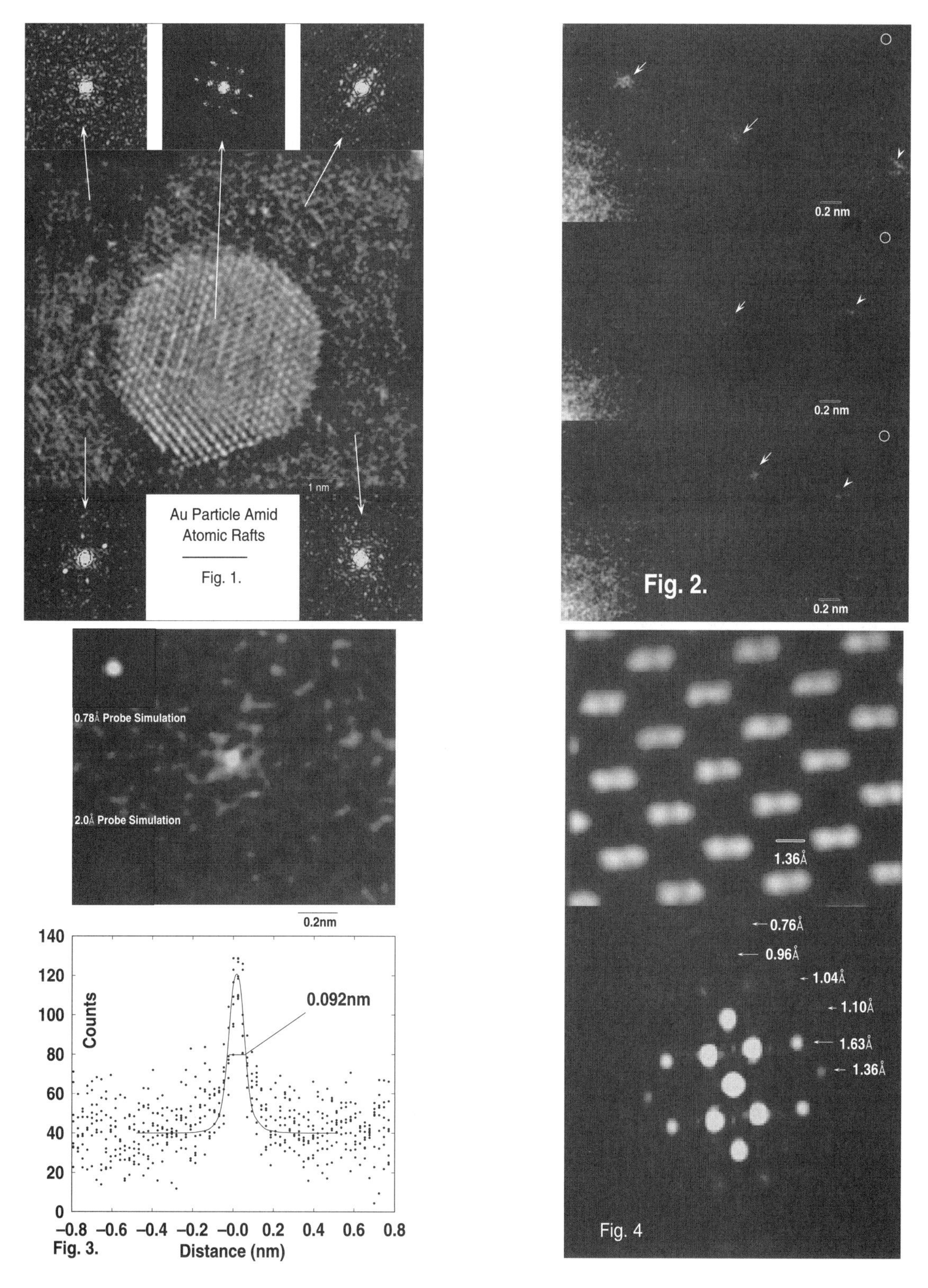

Fig. 1. Au island on carbon film with rafts of Au atom arrays. Fig. 2. Several single Au atoms compared with 1Å circle. Fig. 3. Analysis of a single Au atom image width. 0.92Å width is consistent with a 0.78Å diameter probe. Fig. 4. Analysis of Si< 110 > projection, showing periodicities to 0.76Å.

Microsc. Microanal. 8 (Suppl. 2), 2002
DOI 10.1017.S1431927602101516

The Ultimate Resolution in Aberration-Corrected STEM

S. J. Pennycook[1], A. R. Lupini[1], and P. D. Nellist[2]

[1] Solid State Division, Oak Ridge National Laboratory, Oak Ridge, TN 37831-6030 USA.
[2] Nion Co., 1102 8th St., Kirkland, WA 98033, USA.

Aberration correction on the STEM offers the potential to reach the fundamental quantum-mechanical limit for resolution in zone-axis crystals. In free space, plane waves are good quantum mechanical stationary states to describe the propagation of an electron, but not in a zone axis crystal. Here the electron must take on the periodicity of the crystal and it propagates as Bloch states. The fundamental resolution limit is therefore the smallest Bloch state. Figure 1 shows the first few Bloch states for Si ⟨110⟩, where the most localized state is the 1s state. These states are the most deeply bound in the potential well of the columns, and are typically 0.5 – 0.8 Å in diameter dependent on the strength of the potential well. This is comparable to the probe sizes predicted after correction of aberrations.

For axial illumination the antisymmetric p-type Bloch states are not excited in a perfect crystal. In phase contrast microscopy, imaging with 1s Bloch states can be achieved by selecting a sample thickness in which the contribution of the 1s state to the exit face wave function is maximized. This occurs at a thickness of $\xi/4$, where ξ is the extinction distance. But this thickness is different for columns of different composition; at thicknesses greater than $\xi/4$, the column reverses contrast. In Z-contrast imaging it is the detector that provides the filtering. The inner detector angle in increased until only the most localized states contribute to the intensity. The large angular integration ensures transverse incoherence, and phonon scattering ensures longitudinal incoherence. It was recognized long ago that the 1s states were the dominant contribution to the image, and their non-dispersive nature was necessary for the incoherent nature of the image.[1] However, by assuming the image intensity to be proportional to the intensity at the atom sites, the detector geometry was not included. Recently it has been shown that the detector provides more perfect Bloch state filtering than originally thought.[2] The 1s states are responsible for the image contrast even when the 2s states are more highly excited. As the probe size is reduced in size, eventually, the Z-contrast image will become a direct image of the 1s Bloch states.

For EELS, there has been much discussion on delocalization, that inner shell excitation could be achieved from a point charge passing at a distance. Classical expressions for the impact parameter are velocity-dependent, but quantum mechanical predictions are not. Furthermore, use of the dipole approximation is invalid in the present context. For atomic resolution EELS, large acceptance angles are necessary and we are interested in the response at large distance. Thus we cannot expand $e^{i\mathbf{q}\cdot\mathbf{r}}$ by $1 + i\mathbf{q}.\mathbf{r}$. Doing so suggests significant delocalization (Fig 2) [3] but this is not seen with the full calculation.[4] In this case the ultimate resolution for a single atom is very close to the geometric size of the inner shell. Delocalization is negligible. For zone axis crystals the ultimate resolution is again the 1s Bloch state.

References:

[1] S. J. Pennycook and D. E. Jesson, Phys Rev Lett, **64** (1990) 938, Ultramicroscopy, **37** (1991) 14.
[2] P. D. Nellist and S. J. Pennycook, Ultramicroscopy, **78** (1999) 111.
[3] D. A. Muller and J. Silcox, Ultramicroscopy, **59** (1995) 195.
[4] B. Rafferty and S. J. Pennycook, Ultramicroscopy, **78** (1999) 141.
[5] V. W. Maslen and C. J. Rossouw, Philos Mag A, **49** (1984) 735.
[6] This work was supported by the USDOE under contract DE-AC05-00OR22725 managed by UT-Battelle, LLC.

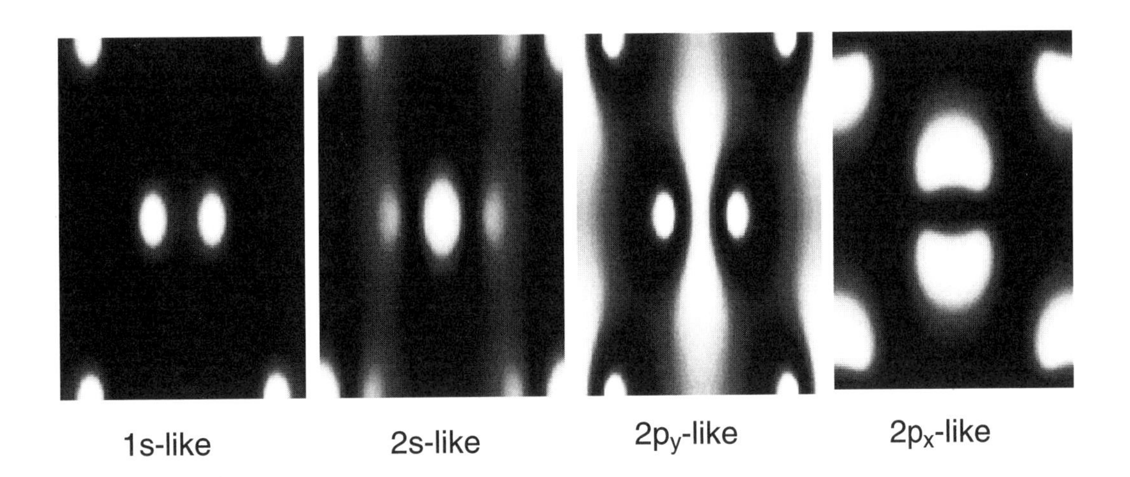

Fig. 1. Bloch state intensities for Si⟨110⟩.

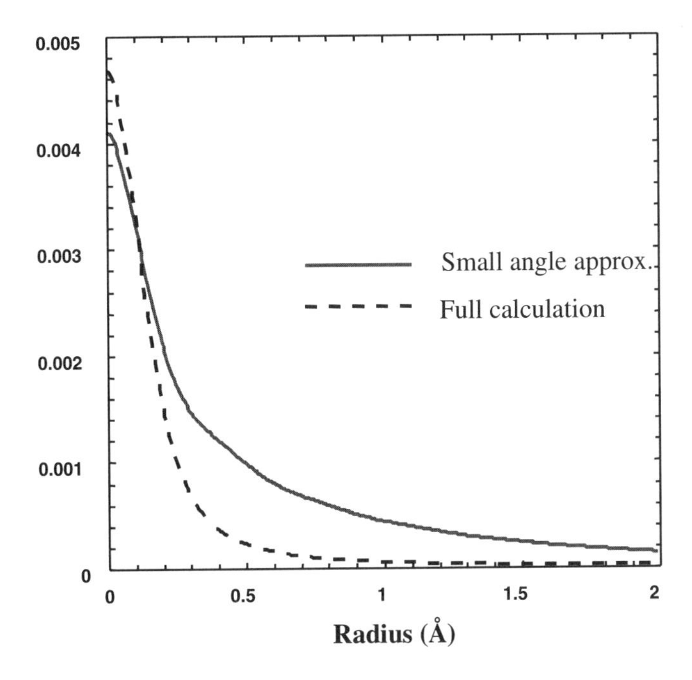

Fig. 2. Spatial distribution of EELS intensity for an infinitesimal probe comparing the dipole expansion to the full calculation.

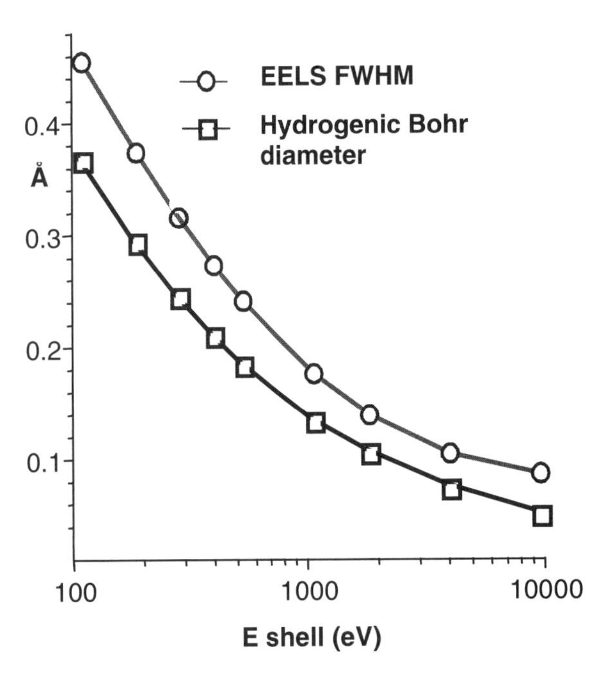

Fig. 3. Full width half maximum of the EELS object function compared to the diameter of the inner shell. Calculations use the hydrogenic model. [5]

Microsc. Microanal. 8 (Suppl. 2), 2002
DOI 10.1017.S1431927602101528

Enhancing the Resolution and Sensitivity of STEM by Aberration Correction

N. D. Browning[1], K. Sun[1], R. F. Klie[1], J. Liu[2], M. M. Disko[3], P. D. Nellist[4], N. Dellby[4], O. L. Krivanek[4]

[1]University of Illinois, Department of Physics, 845 W. Taylor St, Chicago, IL 60607, USA.
[2]Monsanto Company, 800 N. Lindbergh Avenue, U1E, St Louis, MO 63167, USA.
[3]ExxonMobil Res. & Eng. Corp, Corporate Strategic Research, Annandale, NJ 08801. USA
[4]Nion Company, 1102 8th Street, Kirkland, WA 98033, USA

The ability to quantify the atomic structure, composition and local bonding at interfaces and defects is key to developing a fundamental understanding of the properties of many nanoscale systems. The combination of Z-contrast imaging and electron energy loss spectroscopy (EELS) in the scanning transmission electron microscope (STEM) [1] provides the ability to generate precisely this type of characterization. For these techniques, the spatial resolution is limited primarily by the probe size and the sensitivity is limited by the probe current. Enhancing the performance of the microscope and our ability to characterize the key properties relevant to the development of nanoscale materials, devices and processes therefore requires only a modification of the electron optics for probe formation.

Improvements in the probe size and probe current have recently been afforded by the advent of aberration correction [2]. Cs-correctors can now be purchased to upgrade the performance of 100kV dedicated VG STEM, pushing the probe size close to 0.12nm. Provided the room and intrinsic microscope instabilities can be removed, this probe size translates to an image resolution at 100kV that is comparable with the best STEM resolution at 200kV [3]. This improvement in resolution is coupled with an increase in probe current, a smaller energy spread of the beam (achieved by using a cold field-emission source rather than a Schottky source), and a lower accelerating potential; all features that will improve the analysis of "real" nanoscale systems.

One application where the decrease in probe size and increase in probe current can significantly increase our understanding of nanoscale phenomena is in the area of supported heterogeneous catalysts [4,5]. Here, understanding the origin of the activity and selectivity of a given catalytic system requires an intimate understanding of this metal-support interaction as a function of cluster size and reduction/oxidation treatments. For example, figure 1 shows an image of a larger (~200nm) copper particle in a Cu/Al_2O_3 catalyst calcined at 1073K. Image contrast and the energy loss spectra shown in figure 1b, indicate that a covering oxide layer forms on this particle despite being heated in flowing hydrogen, i.e. a highly reducing atmosphere. Although the oxide layer can be observed in these larger particles, the sensitivity of the current microscope (JEOL 2010F [3]) does not allow the fine-structure of the edge to be analyzed in the particles ~few nanometers in size to determine whether a crystalline oxide is formed or whether the signal originates simply from adsorbed oxygen on the surface.

An aberration corrected 100kV dedicated STEM will significantly improve the characterization of these systems. Although in its infancy, the VG HB601 C_s corrected STEM at UIC will be used primarily for the study of these systems. Initial results from the factory tests of the corrector are shown in figure 2. Here a resolution of ~0.118nm is achieved with an objective aperture of 22 mrad and a probe current ~60pA. Both the probe size and the current are improved over what is currently available in the JEOL 2010F at UIC (0.13nm probe size and ~40pA probe current). In this presentation, the latest results from the corrected STEM will be discussed with particular attention being paid to the practical advantages/disadvantages of such a system over a conventional TEM/STEM instrument [6].

References

[1] N. D. Browning, M. F. Chisholm, and S. J. Pennycook, *Nature* **366** (1993) 143.
[2] N. Dellby et al, *J. Electron Microscopy* **50** (2001) 177.
[3] E. M. James and N. D. Browning, *Ultramicroscopy* **78** (1999) 125.
[4] R. F. Klie et al, *J. Catalysis* **205** (2002), 1
[5] K. Sun et al, in press *Applied Catalysis B*
[6] This work was sponsored in part by the National Science Foundation under grant numbers DMR 9601792 and DMR 007364, by Monsanto Company, and by the Petroleum Research Fund.

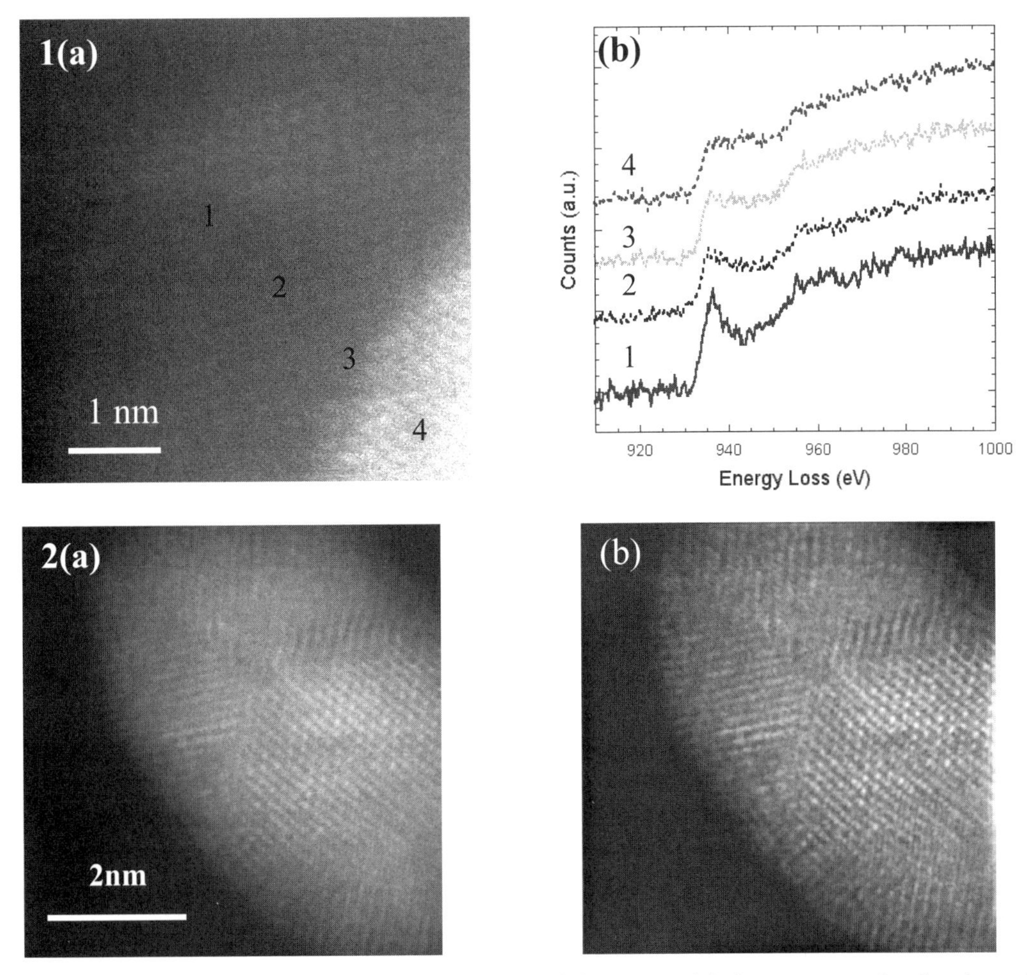

Figure 1: (a) Z-contrast image of a 200nm Cu particle on an Al_2O_3 support obtained on the JEOL 2010F. The surface shows a lower intensity, indicative of the formation of an oxide, which is confirmed by the presence of a white line in the copper L-edge spectrum (b).

Figure 2: Figure 2: Z-contrast image of an Au particle, (a) raw data, (b) low-pass Fourier filtered to improve fringe visibility. The largest observable spacing corresponds to the 0.23nm (111) fringes. Fringes with much lower spacings, such as the 0.118nm (222) fringes, are also visible.

Microsc. Microanal. 8 (Suppl. 2), 2002
DOI 10.1017.S143192760210153X

STEM Aberration Correction: Where Next?

O.L. Krivanek, N. Dellby, M. Murfitt, P.D. Nellist and Z. Szilagyi

Nion Co., 1102 8th St., Kirkland, WA 98033, USA

Present-day aberration-corrected scanning transmission electron microscopes (STEMs) improve the attainable resolution at a given operating voltage by 2-3x compared to their uncorrected cousins [1]. There are now four such instruments in laboratories in North America. Sub-Ångström probe size has become available even at 100-120 keV operating energies [2], and the current available in the ultra-small probe has risen considerably compared to uncorrected instruments. In this paper we summarize some of these recent achievements and look ahead at expected future developments.

Depending on the complexity and the precision of the corrector, the probe shape produced by the STEM depends on 25 or more different aberration coefficients. Further influential factors are the chromatic aberration and energy spread, the geometrical size of the projected electron source, the size, position and cleanliness of the beam-defining aperture and the mechanical and electrical stabilities of the whole instrument. Successfully managing this complexity was one of the key factors that made practical aberration correction possible. Recent modifications of our Ronchigram-based autotuning software [3] have enabled it to determine all the aberration coefficients up to 5th order (25 coefficients total) in about 20 seconds, and improved its precision.

As an example of performance now routinely possible, Fig. 1 shows unprocessed high angle annular dark field (HAADF) images of gold particles and single gold atoms on a carbon film recorded with our second generation quadrupole-octupole corrector [1] at 100 kV. Even though the supporting carbon film is over 10 nm thick, single Au atoms can be readily distinguished once the contrast is suitably expanded (Fig. 1 b and c). In such images, atomic rafts one monolayer thick which connect the particles become readily apparent. Unlike the particles which are fairly stationary, the rafts and the single atoms are highly mobile, facilitating continuous transport of matter from particle to particle. Profiles taken through images of the single atoms and the rafts show that the image levels are quantized, and that the single atom images are tail-free and typically 1.0 to 1.5 Å wide.

The most important remaining limits on the performance of the corrected STEMs arise from a) higher order (fifth and above) aberrations, b) chromatic aberration, and c) the overall stability of the whole electron microscope. We are addressing them as follows:

a) The resolution limits due to fifth and seventh order aberrations are given by:

$$d_5 = 0.4\, C_5^{1/6}\, \lambda^{5/6} \qquad \text{and} \qquad d_7 = 0.4\, C_7^{1/8}\, \lambda^{7/8}.$$

Our present (2nd generation) corrector nulls aberrations up to 4th order but produces C_5 of about 10 cm, which gives a resolution limit of 0.8 Å at 100 kV. At 300 kV, however, d_5 is lowered to less than 0.5 Å for the same corrector. We are now building a second generation C_s corrector for the 300 kV VG HB603 [4]. More radically, our 3rd generation corrector [5] will null all geometric aberrations up to 6th order, and will give C_7 of about 20 cm. The resolution limit due to geometric aberrations with this system will be just 0.3 Å even at 100 keV.

b) The principal effect of chromatic aberration on STEM HAADF images is to transfer intensity from the probe center into the probe tail, without greatly affecting the shape of the central maximum [1]. This is best described by the fraction f_s of the total electron flux that is shifted away from the central maximum:

$$f_s = (1 - w)^2, \quad \text{where} \quad w = 2\, d_g^2\, E_o / (?E\, C_c\, \lambda) \ \text{or} \ w = 1, \text{ whichever is smaller.}$$

d_g is the resolution in the absence of chromatic aberration. At a resolution d_g = 0.8 Å, energy spread ?E = 0.5 eV, coefficient of chromatic aberration C_c = 1.5 mm and primary energy E_0 = 100 keV, the above gives f_s = 30% as the fraction of the electron flux shifted out of the probe maximum into the probe tail. This shows that with the low energy spread of a cold field emission gun, the present-day 100 kV performance is not strongly limited by chromatic aberration. At higher primary energies the influence of chromatic aberration on HAADF imaging willbe even less.

c) Overall stability of the whole instrument becomes increasingly important as the electron-optical resolution is improved to 1 Å and beyond. Present-day electron microscopes typically use columns that were designed several decades ago when 2-3 Å was the best resolution expected. In order to reach sub-Å resolution, they tend to use incremental fixes that fall short of the required overall performance. We are addressing this problem by developing a whole new STEM column, specifically designed for sub-Å stability even in mildly problematic operating environments.

In summary, aberration-corrected STEM has progressed from a curiosity to an accepted scientific instrument. The performance of the corrected STEMs is now limited by obstacles other than spherical aberration. Overcoming the obstacles will likely be accomplished in the next few years. Beyond them lies the ultimate resolution limit – the size of an atom – and an era of electron microscopy in which limitations due to electron optics will no longer be the major concern.

References

[1] N. Dellby et al., J. Electron Microscopy 50 (2001) 177.
[2] P.E. Batson et al., these proceedings.
[3] O.L. Krivanek et al., US patent application (2000).
[4] S.J. Pennycook et al., these proceedings.
[5] O.L. Krivanek et al., to be published.

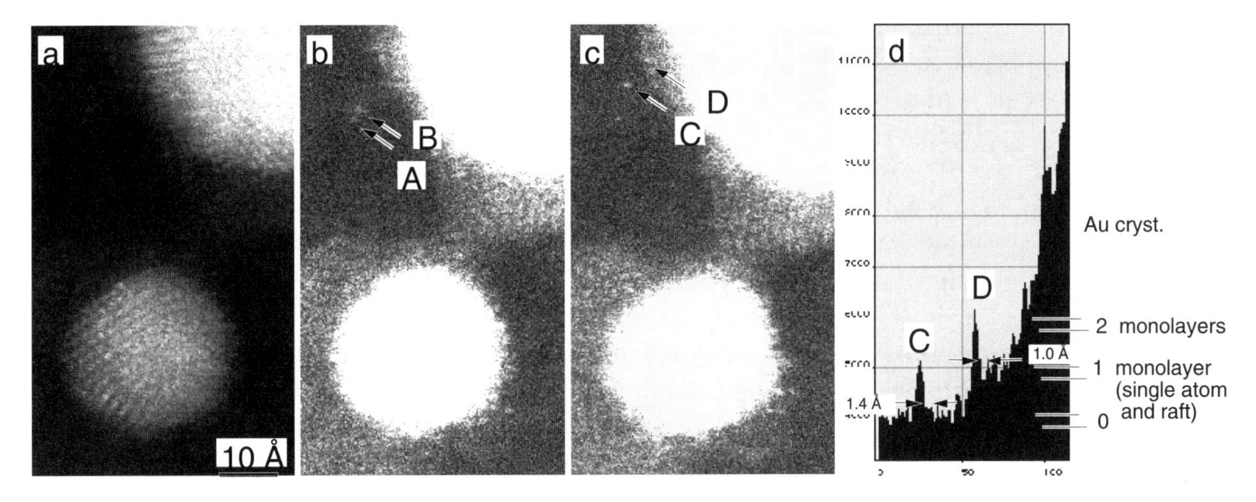

Fig. 1. Unprocessed HAADF images of gold particles, monolayer rafts and single Au atoms. 32 msec per pixel, pixels 0.18 Å wide. a) HAADF image, b) same as (a) but with boosted contrast, c) HAADF image taken under the same conditions as (b) but 100 seconds later, d) line profile through atoms C and D plus adjacent raft and crystal in image (c). Note that atoms A and B have moved elsewhere during the 100 seconds and that the image of the single atom D sitting at the edge of a monolayer raft is just 1.0 Å wide.

Microsc. Microanal. 8 (Suppl. 2), 2002
DOI 10.1017.S1431927602101334

Electron holography – where we are and where to go

Hannes Lichte

Institute of Applied Physics, Dresden University, D 01062 Dresden, Germany, www.triebenberg.de

Visions:

The ultimate goal of the different TEM-methods is to answer the most essential questions in solid-state physics and materials science, such as *Which atom is where? Which magnetic or electric fields are around? Which is the binding structure?* However, the restricted transfer properties of conventional TEM imaging prevent one from reaching this goal: shortcomings are immanent, such as nonlinearity, phase loss in the image intensity, irretrievable cross talk between amplitude and phase of the object exit wave in the image intensity, delocalisation of seemingly resolved structures, and mixing of elastic and inelastic information. Furthermore, the relation between real space and reciprocal space is not sufficiently unique, wave optical tools for analysis of the results are poor, and the Stobbs-factor problem hampers a reliable interpretation by means of simulation calculations.

Off-axis holography offers solutions:

Holography is strictly linear and pure zero-loss (10^{-15}eV)-imaging; the successful recording of inelastic holograms may shed new light on this interesting point. Furthermore, it provides amplitude and phase separately; one can correct all the coherent aberrations, also the ones to come up with improving resolution; there is an unique relation between real and reciprocal space, since everything is reconstructed from one hologram; the reconstructed wave can be exploited by means of all thinkable wave optical tools like nanodiffraction and analysis of diffracted waves; all data are quantitative, and, according to our latest findings, probably due to the inherent zero-loss-imaging, the Stobbs-factor is of minor importance. Holography seems the ideal method for reaching the goals, however, it is applicable to real-world problems only after developing the general performance.

Optimum Performance:

Holographic imaging requires specific parameters for optimum performance as described by resolution, field of view and noise properties. The biprism has to be installed at an optimum position in the TEM. Furthermore, there is an Optimum Focus optimising the gradient of the wave aberration in the TEM to give both maximum information limit and smallest possible point spread function, as well as minimal artefacts; also at the Optimum Focus, one achieves optimum sampling of the hologram by the CCD-camera both in real space and in Fourier space. Correction of aberrations needs a high accuracy of at least 10 aberration parameters. The noise level in the reconstructed wave determines the detection limit for weak objects; it is given by the contrast of the hologram fringes transferred to the computer and by the collected dose; besides a high brightness gun, one needs careful sampling by the CCD-camera and high stability of the whole microscope system including the environment; an optimum specimen thickness of twice the inelastic mean free path - whenever affordable e.g. at medium resolution - yields optimum signal/ noise ratio.

In summary, today's limits of holographic performance are given by brightness and stability, point spread function, pixel number, and accuracy of aberrations.

Problems of Applications

After many years of development, electron holography has reached such a performance that, besides showing the "in-principle"-facilities, it has surmounted the state of basic development of the method and is applied to the solution of increasingly sophisticated real-world problems in materials science.

Medium Resolution:

For structure components larger than about 5 times the Scherzer resolution, aberrations can be neglected; for finer details one has to be very careful about delocalisation effects. With this in mind, the reconstructed phase images can be interpreted as the object phase structure, which is not at all discernible by conventional TEM methods. In a wide variety, magnetic microstructures are analysed in the range down to about 50nm. Trying to analyse finer magnetic structures, one finds that the reachable lateral resolution is limited by the strength of the phase shift: since the phase shift is given by the enclosed magnetic flux, it shrinks down with the size of the object details. For example, the magnetic phase shift due to a single Bohr Magneton in a unit cell can be estimated as $2\pi\,10^{-5}$, which, alas, is far beyond the phase detection limit of about $2\pi/30$ found at atomic resolution. The situation is much more advantageous with the electric phase shift integrating the electric potential locally along the electron trajectory. This is true for the electric fields measured e.g. in semiconductors, for dopant profiling, and for the analysis of polarisation in ferroelectrics. Optimising the object thickness, the phase shift can be evaluated well above noise also in the nanometer range, even down to atomic resolution.

High Resolution:

At high resolution, aberrations have to be corrected prior to any meaningful evaluation of the reconstructed wave. Correction of aberrations re-sorts the cross talk between amplitude and phase and heals delocalisation; it opens the effective objective aperture, consequently, it not only achieves improvement of resolution, additionally it enhances the measurable phase shift e.g. due to single atoms. Since the phase shift correlates with the atomic number, the phase image allows distinguishing between different atom species. As the Stobbs-factor seems only a minor problem with holography, the evaluation of the phase data by comparison with simulated ones opens a way for holographic materials analysis at an atomic scale.

Outlook:

After the first experiments by Moellenstedt and Wahl showing the principle of off-axis electron holography in 1968, the different holography groups around the world have developed electron holography as a meanwhile indispensable tool in the field of nanoanalysis. Much has been reached; still much is left to be done, for example:

- The fields in the specimen leak out producing 3D-fields, which influence both object wave and reference wave. In any case, the phase represents the whole 3D field, not only the inner object.
- The phase shift stems from many different contributions, e.g. magnetics, mean inner potential, semiconductor potentials, inner charges at interfaces, surface charges probably induced by the electron beam, also from thickness variations, and from dynamic interaction. For unique interpretation, these have to be sorted out, possibly by in-situ experiments.
- With the reconstruction of the object exit wave, we have not yet reached the final goal: the step back from the wave to the underlying object structure is a difficult Inverse Problem.

Microsc. Microanal. 8 (Suppl. 2), 2002
DOI 10.1017.S1431927602101346

Effects of Fresnel Corrections for Phase-Shifting Electron Holography

K. Yamamoto, T. Hirayama*, T. Tanji** and M. Hibino***

Japan Science and Technology, Domestic Research Fellow, c/o Japan Fine Ceramics Center, 2-4-1, Mutsuno, Atsuta, Nagoya, Japan, 456-8587

* Japan Fine Ceramics Center, 2-4-1, Mutsuno, Atsuta, Nagoya, Japan, 456-8587

** CIRSE, Nagoya University, Furo-cho, Chikusa, Nagoya, Japan, 464-8603

*** Aichi Institute of Technology, 1247, Yagusa-cho, Toyota, Japan 470-03

Phase-shifting electron holography [1] is a useful technique for detecting a small phase change of an electron wave. In this method, however, Fresnel diffraction at an electron biprism causes two kinds of problems to the reconstructed phase. One is the influence on the amplitude of the electron wave, i.e. Fresnel fringes, resulting in non-uniform contrast of interference fringes. The non-uniform fringes bring numerical phase errors in reconstructed images. In order to remove the phase errors, we calculated the envelopes of the fringes, and normalized the non-uniform contrast. [2] The other problem is that Fresnel diffraction directly distorts the phase of the electron waves. We prepare the reference holograms that were taken in a vacuum without specimens, and then correct the distortion by subtracting the reconstructed phase of the reference holograms from that of holograms. [2] In this report, we evaluate the effects of the above correction methods (Fresnel corrections) using a computer simulation and show the possibility of observing a very small amount of electric charges.

In this simulation, we assumed a charged latex sphere placed on a uniform carbon film. We also assumed that the size of the sphere was 60 nm in diameter and that the amount of electric charges was +5 e (e = 1.6×10^{-19} C). Figure 1(a) shows an interference micrograph directly calculated from the phase changes due to the electric charges. Figure 1(b) shows the simulated hologram of this sphere. In this Fig. 1(b), we considered the influence of Fresnel diffraction, quantum noise and electron-wave coherency. [3] Figure 1(c) shows the interference micrograph reconstructed without Fresnel corrections from 100 holograms and 100 reference holograms. Figure 1(d) shows the micrograph with Fresnel corrections. These interference micrographs are phase-amplified by a factor of 300.

In Fig. 1(c), the phase errors due to Fresnel diffraction obscure the electric field. Using Fresnel corrections, however, the phase errors decreased to 5% of those in Fig. 1(c) and the phase-measurement precision reached 2π / 410 rad. Consequently, the electric field can be observed as shown in Fig. 1(d).

References

[1] Q. Ru, G. Lai, K. Aoyama, J. Endo and A. Tonomura, Ultramicroscopy **55**, (1994) 209-220.

[2] K. Yamamoto, I. Kawajiri, T. Tanji, M. Hibino and T. Hirayama J. Electron Microsc. **49**, (2000) 31-49.

[3] K. Yamamoto, T. Tanji and M. Hibino, Ultramicroscopy **85**, (2000) 35-39.

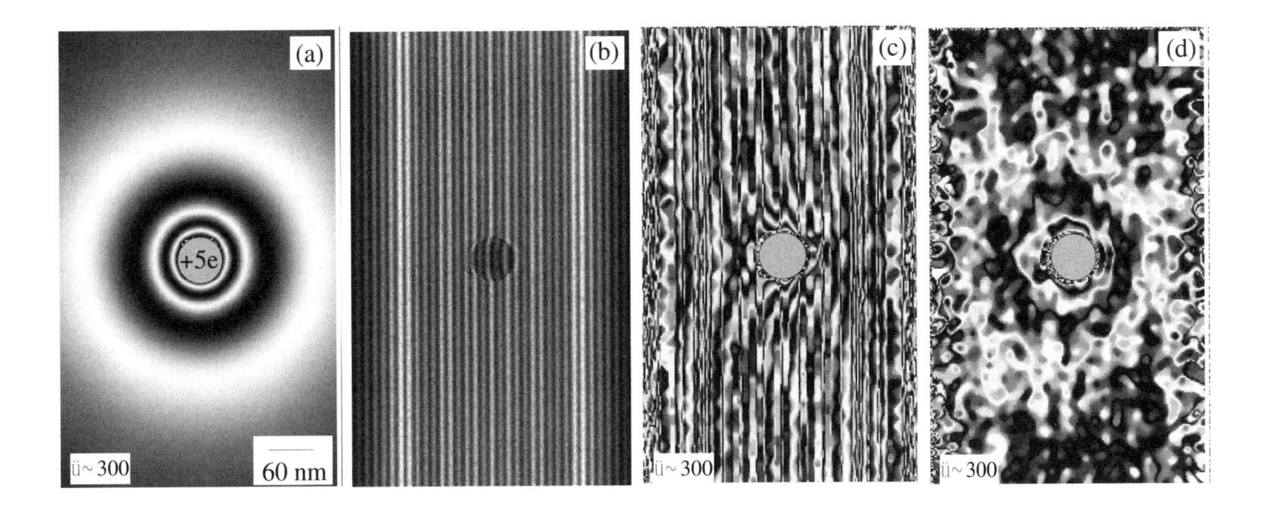

Fig. 1(a) Interference micrograph of a charged latex sphere. It is directly calculated from the phase change due to positive charges of 5 electrons. Phase-amplified factor is 300. (b) Simulated hologram of the sphere. The influence of Fresnel diffraction, quantum noise and electron-wave coherency are considered. (c) Interference micrograph reconstructed from 100 holograms and 100 reference holograms without Fresnel corrections. (d) Interference micrograph with Fresnel corrections.

Microsc. Microanal. 8 (Suppl. 2), 2002
DOI 10.1017.S1431927602101358

Bright Electron Beams and Their Applications to Electron Phase Microscopy

Akira TONOMURA

Advanced Research Laboratory, Hitachi, Ltd. Hatoyama, Saitama, Japan
SORST, Japan Science and Technology Corporation (JST), Tokyo, Japan
Frontier Research System, The Institute of Chemical and Physical Research(RIKEN), Saitama, Japan

Bright beam sources, like lasers and synchrotron radiation, have brought about new ways of investigating microscopic structures. The same is true with electron sources, especially when their phase information is used for microscopy.

We have continued to develop brighter electron beams in order to raise the level of performance of electron microscopy and to enable new applications to it since 1968 [1], when we became convinced through our experience with interference experiments that a bright electron beam was indispensable for obtaining high-quality images through electron holography. As it turned out, every time beam brightness was increased, new applications opened up.

In 1978, using a 80-kV field-emission electron beam, we improved beam brightness by two orders of magnitude compared to that of thermal electron beams from pointed filaments, enabling us to directly observe the magnetic lines of force in *h/e* flux units as an interference micrograph [2]. With our 250-kV microscope, we conclusively confirmed the AB effect by a series of experiments from 1982 [3] to 1986 [4]. In 1992, we dynamically observed [5] magnetic quantized vortices in superconductors with our 350-kV microscope. In Spring 2000, we completed a 1-MV microscope [6] (Fig. 1) that has a beam brightness higher than that of electron beams from pointed filaments by four orders of magnitude and a lattice resolution of less than 0.5Å.

We have just begun obtaining various new results with this microscope. For example, we have distinguished two kinds of vortices *inside* Bi-2212 film; one trapped along tilted columnar defects and the other penetrating perpendicularly to the film plane in the form of two different Lorentz images (Fig. 2). We have also obtained information about the experimental conditions where vortices are trapped along the defects and strongly pinned [7]. Furthermore, a special arrangement of vortices, the chain-lattice state in a Bi-2212 thin film, reflecting the layered structure of the material was observed by Lorentz microscopy, and the disappearance of the chain-vortex images was found to occur at temperatures well below the melting temperature [8] (Fig. 3).

With this microscope we will be able to clarify other various unconventional behaviors of vortices in high-T_c superconductors.

References

[1] A. Tonomura et. al., Jpn. J. Appl. Phys., 7 (1968) 295.
[2] A. Tonomura et. al., Phys. Rev. Lett. 44 (1980) 1430.
[3] A. Tonomura et al., Phys. Rev. Lett. 48 (1982) 1443.
[4] A. Tonomura et al., Phys. Rev. Lett. 56 (1986) 792.
[5] K. Harada et al., Nature 360 (1992) 51.
[6] T. Kawasaki et al., Appl. Phys. Lett. 76 (2000) 1342.
[7] A. Tonomura et al., Nature 412 (2001) 620.
[8] T. Matsuda et al., Science 294 (2001) 2136.

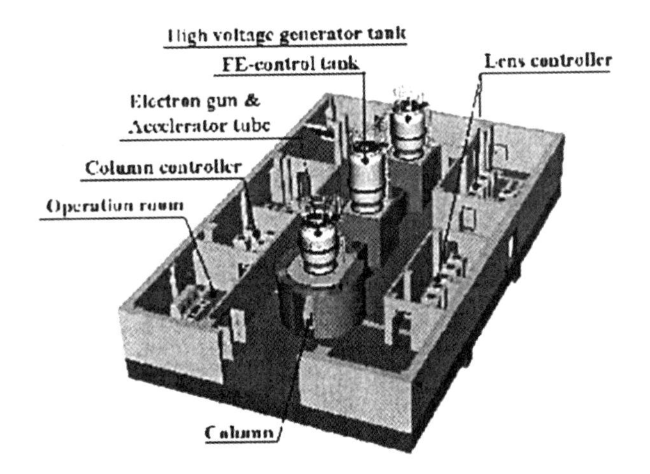

FIG. 1. Schematic of 1-MV field-emission electron microscope [6].

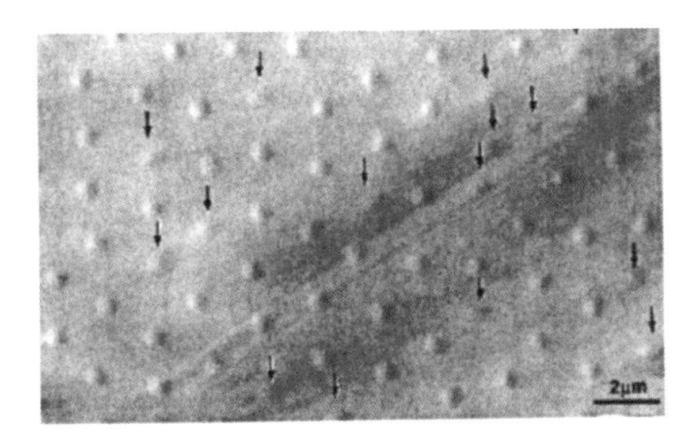

FIG. 2. Lorentz micrographs of vortices in Bi-2212 film with tilted columnar defects [7]. One can see two kinds of vortex images, circular images and elongated ones (indicated by the arrows). The circular images correspond to vortex lines penetrating perpendicularly to the film, and the elongated images correspond to vortex lines trapped along tilted columnar defects.

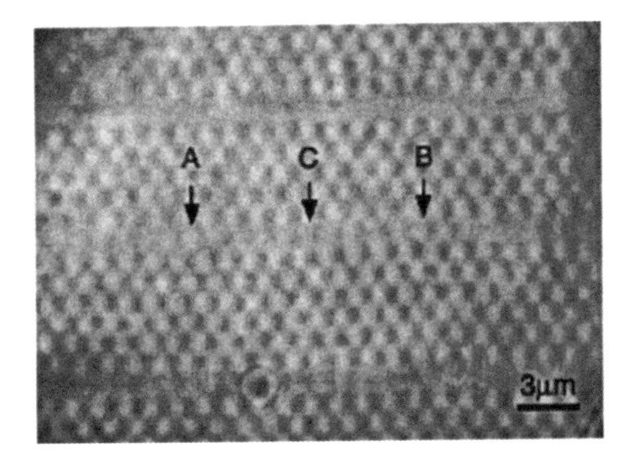

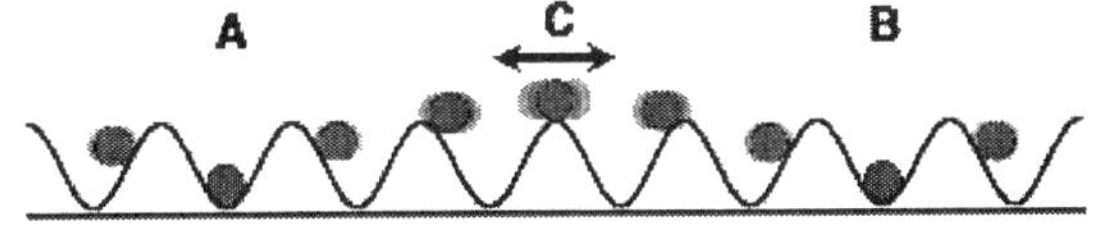

FIG. 3. Disappearance of chain-vortex images in Bi-2212 at 70 K [8]. When a magnetic field was applied at a grazing angle to the layer plane, two domains of chain vortices and lattice vortices were alternately produced. When the temperature was increased, only the images of chain vortices begin to disappear. This is attributed to the oscillation of the vortices along the chains. Note that unstable chain vortex C begins to disappear while stable vortices A and B can still be seen.

Microsc. Microanal. 8 (Suppl. 2), 2002
DOI 10.1017.S143192760210136X

Low Voltage Electron Holography - High Voltage Electron Holography

B.G.Frost, A.Thesen and D.C.Joy

University of Tennessee, M407 Walters Life Sci, Knoxville, TN 37996-0840

The typical set-up for low voltage holography introduced by Morton and Ramberg [1] in 1939 consists of only an electron emitter, a sample and an electron detector placed in a vacuum chamber (Fig.1). They described the most important features of the microscope, which they called point projector microscope (ppm), as follows: "As this type of microscope involves no electron- optical lens elements, the images obtained are free from the ordinary aberrations. The limit of resolution depends solely on the distribution of initial velocities of the field electrons and on Fresnel diffraction by the object, making it possible to proceed beyond the resolution of the light microscope by some orders of magnitude." At that time, however, the performance of the available electron sources was very limited. After the fabrication of a highly coherent nanotip field-emitter with great brightness [2] which operates at a very low extraction voltage (100V or less) the ppm was subject to new research [3,4]. An out of focus image of a holey carbon film acquired at 350 Volts is displayed in Fig.2. We estimate from that image the lateral coherence of our field emitter and the diameter of the tip apex using the contrast of both Fresnel fringes and Youngs fringes. In order to apply the ppm to bulk materials we design a special reflection mode of operation. Our ray tracing simulations show that we have to replace the emitter by an electron gun.

We perform high voltage electron holography using a Hitachi HF-2000 FEG TEM operated at 200kV. The electron biprism can be placed either between the objective lens and the first intermediate lens or between the first and second intermediate lenses. For both positions a medium magnification range (10kX - 100kV) can be achieved by only weakly exciting the first intermediate lens such that it acts as a magnifier. A ray diagram with the biprism between the objective and first intermediate lens is shown in Fig.3. Here the sample is imaged by the objective lens to the first image plane. In this step the objective lens is normally excited which preserves its good imaging qualities. Then the image is further magnified by the first intermediate lens which, in combination with the negatively biased biprism, forms the hologram. We routinely obtain 100 fringes or more with a fringe contrast of 10% or better and a field of view of about 350nm. These numbers are obtained from the hologram of a CMOS device (Fig.4).

References

[1] G.A. Morton and E.G.Ramberg, Phys.Rev. 56 (1939) 705.
[2] H.W. Fink, IBM J. Res. Develop. 30 (1986) 460.
[3] J.C.Spence et al. in Electron Holography, A.Tonomura et al. (Eds.), Elsevier, p267-276.
[4] This research was supported by SRC under contract number LJ413.003 and was carried out as part of the user program sponsored by the U.S. DOE under contract DE-AC05-00OR22725 with the ORNL, managed by UT-Battelle, LLC.

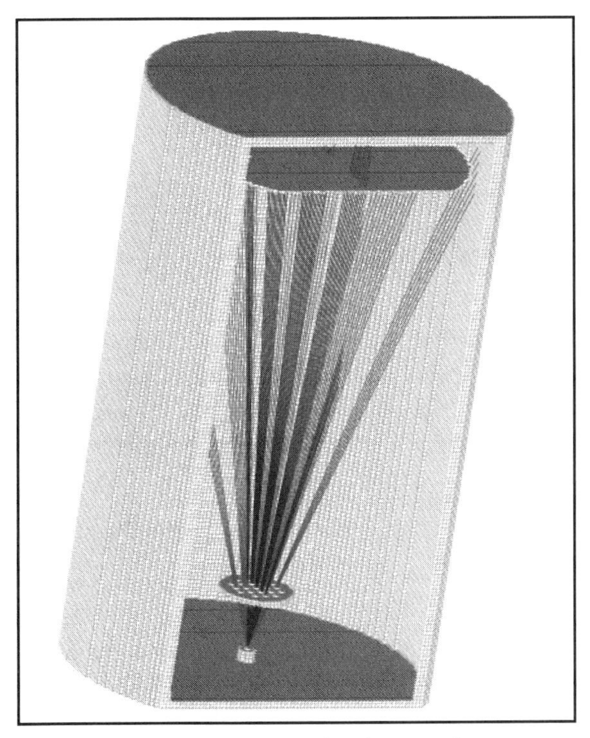

FIG.1. Ray diagram of a low voltage point projection microscope.

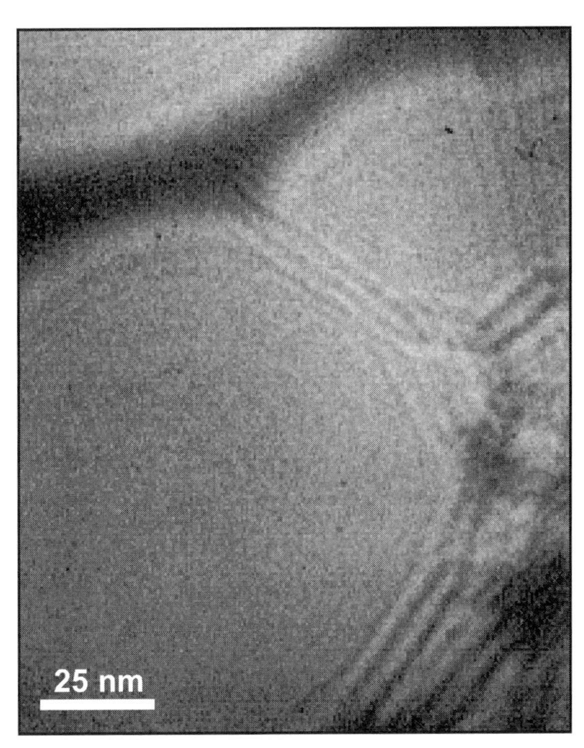

FIG.2. In-line hologram with Fresnel - and Youngs fringes.

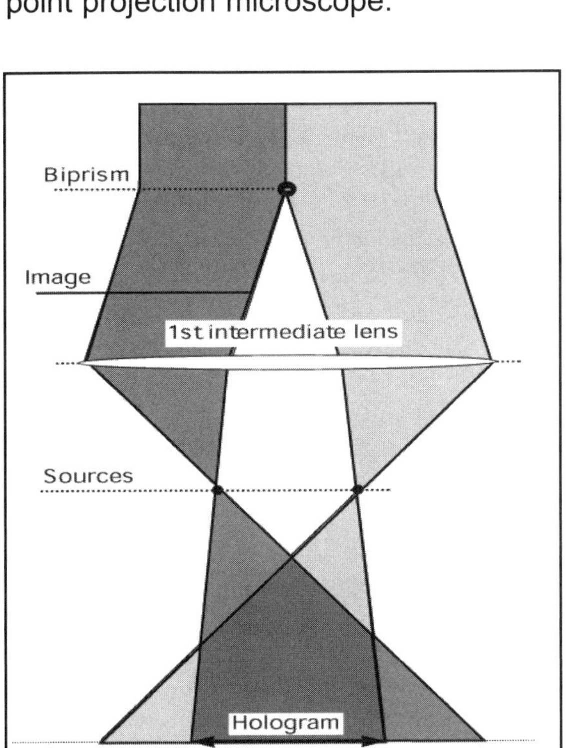

FIG.3. Ray diagram of high voltage TEM for medium magnification holography.

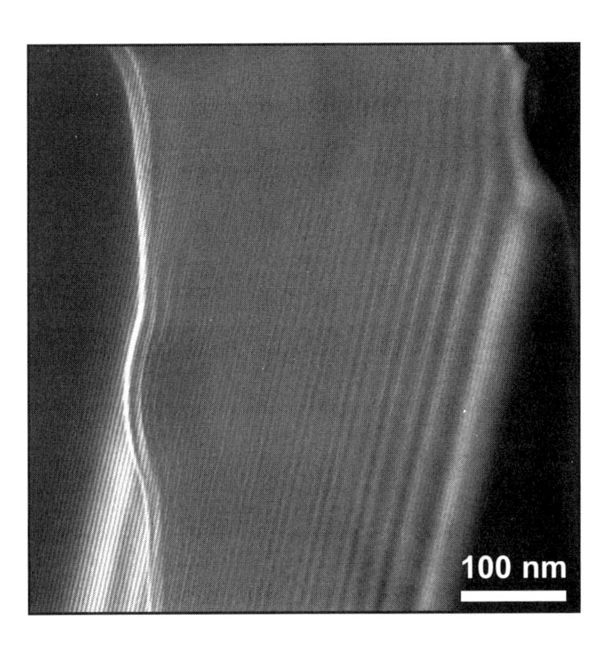

FIG.4. Hologram of transistor acquired adapting diagram of Fig.3.

Microsc. Microanal. 8 (Suppl. 2), 2002
DOI 10.1017.S1431927602101371

Holographic Observation of Magnetic Fine-Structures in New Magnetic Materials

T. Tanji*, S. Hasebe**, and T. Suzuki**

*CIRSE, Nagoya University, Chikusa, Nagoya 463-8603, Japan
**School of Engineering, Nagoya University, Chikusa, Nagoya 463-8603, Japan

New magnetic materials, metal multilayers (Co/Cu) and granular films (Fe-Zr), have been observed by electron holography. Giant magnetoresistance (GMR) is attributed to the spin-dependent scattering of conduction electrons, which arises from the magnetic coupling between each adjacent ferromagnetic layers through a nonmagnetic spacer layer (multilayers), or between neighboring ferromagnetic particles in nonmagnetic matrix (granular films). Direct observation of such a coupling, however, has hardly been reported so far.

Co/Cu multilayers were deposited onto Si(111) wafers in a rf magnetron sputtering system. The thickness of a Cu layer was varied 1.5-3.5nm, while that of a Co layer was fixed to 4.0nm. The sample of Co(4.0nm)/Cu(2.0nm) shows the magnetic curve most like to indicate an antiferromagnetic ordering of Co layers (Fig.1). Therefore cross sections of $Si[Co(4.0nm)/Cu(2.0nm)]_n$, n=4,5,6 were observed by electron holography. Specimens were prepared by ordinary ion-milling. Fe-Zr granular films were prepared by annealing amorphous films of $Fe_{80}Zr_{20}$ which were deposited by dc magnetron sputtering at –170°C. An as-sputtered Fe-Zr film shows scarcely clear ferromagnetism, and after annealing it shows a typical ferromagnetic curve (Fig.2). Particles precipitated were 10-30nm in diameter. They were ascertained to consist mainly of Fe by a high spatial-resolution EDX.
Electron holograms were recorded by Hitachi HF-2000 equipped with a low magnetic field objective lens and reconstructed on a PC using Gatan DigitalMicrograph scripts.

Figure 3 shows TEM image (a), reconstructed phase distribution (b), their profiles (c, d) and the magnetization direction (e) of $[Co(4.0nm)/Cu(2.0nm)]_4$. Those of $[Co(4.0nm)/Cu(2.0nm)]_5$ are shown in Fig.4. It is found that the magnetization of Co layers in $[Co(4.0nm)/Cu(2.0nm)]_4$ have ferromagnetic ordering despite the antiferromagnetic form of the curve in Fig.1. While the magnetization in $[Co(4.0nm)/Cu(2.0nm)]_5$ is separated to antiparallel two parts, the first two layers and the next three layers in which inside layers order ferromagnetically as shown in Fig.4. Such antiferromagnetic ordering has been found always in $[Co(4.0nm)/Cu(2.0nm)]_4$ and sometimes in $[Co(4.0nm)/Cu(2.0nm)]_5$ so far. The reason of this difference between the holographic observation and the magnetization curve is still under investigation.

Figure 5a shows phase distribution inside and outside of a Fe-Zr granular film, and the magnetic connection among the particles are drawn schematically in Fig.5b. A large particle at the center of Fig.5b seems to have a magnetic direction with a high angle against the film, so lines of magnetic force come into and out almost perpendicularly.

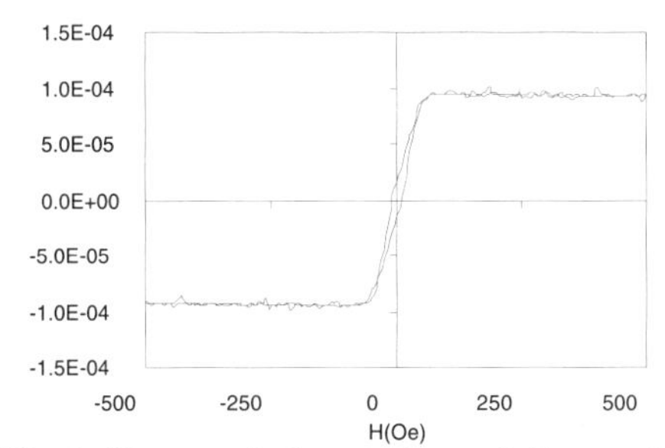

Fig.1: Hysteresis loop of a multilayered film $Si[Co(4.0nm)/Cu(2.0nm)]_4$.

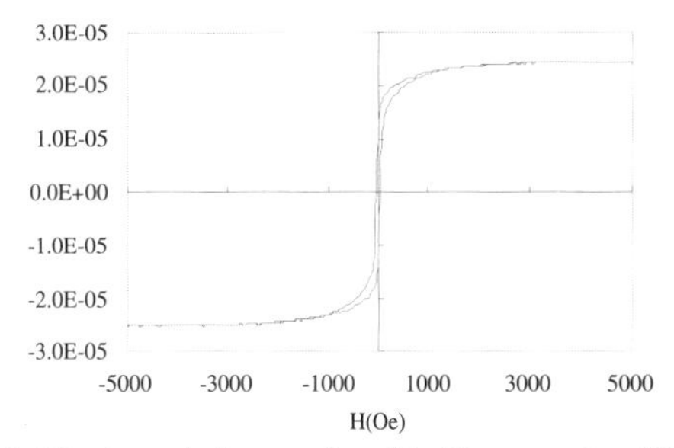

Fig.2: Hysteresis loop of an Fe-Zr granular film.

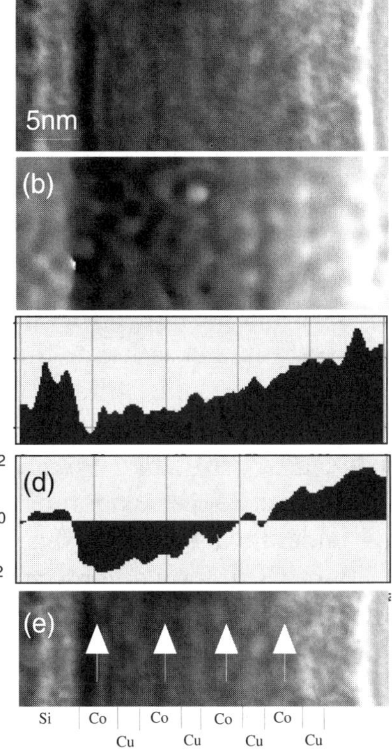

Fig.3: TEM image (a), reconstructed phase distribution (b), their profiles (c, d) and the indication of the magnetization direction (e) of $[Co(4.0nm)/Cu(2.0nm)]_4$.

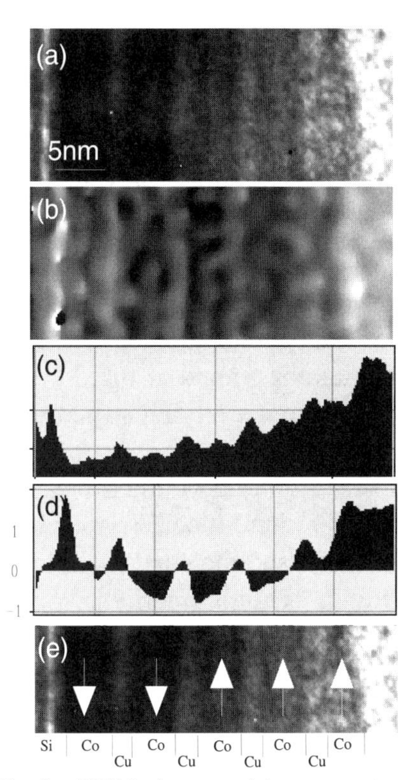

Fig.4: TEM image (a), reconstructed phase distribution (b), their profiles (c, d) and the indication of the magnetization direction (e) of $[Co(4.0nm)/Cu(2.0nm)]_5$.

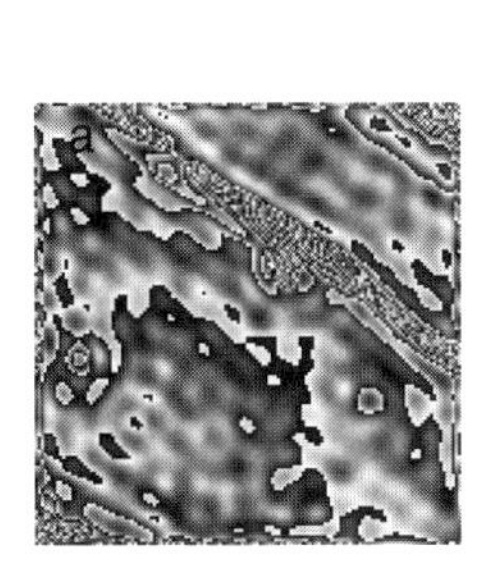

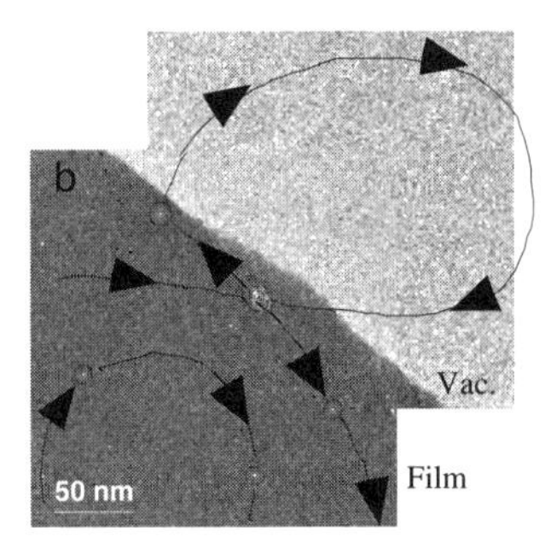

Fig.5: Phase distribution inside and outside of a Fe-Zr granular film (a), and the magnetic connection among the particles (b).

Microsc. Microanal. 8 (Suppl. 2), 2002
DOI 10.1017.S1431927602101383

Fourier Methods for Field and Phase-shift Calculations of Long-range Electromagnetic Fields

G. Pozzi

Dept. of Physics and INFM, University of Bologna, viale B.Pichat 6/2, 40127 Bologna, Italy.

Recently, by analyzing the problem of the observation of superconducting fluxons by transmission electron microscopy, it has been found that the calculation of the electron optical phase shift can been carried out successfully by a new approach[1,2]. First the vector potential is decomposed into its Fourier components and then the phase shift is calculated for each component separately. In this way, once the problem of finding the vector potential in analytical form has been solved, the Fourier transform of the phase shift is immediately obtained, and can be inverted either analytically or numerically. The main advantages of this approach are that the case of a periodic array of fluxon can be easily analyzed [1], a troublesome problem in the former real space approach owing to the long-range behaviour of the fluxon magnetic field, and that new superconducting structures, like pancake vortices present in high-T_C materials [2,3], which were beyond the scope of the flux tube model and its implementations, can be successfully investigated. In this way, it is possible to interpret recently-obtained experimental results by means of the new 1-MV holographic microscope relative to fluxons pinned at tilted columnar defects [4,5] and to correlate the image features to the anisotropy of the underlying structure, as also shown in other contributions [6].

In this work it will be outlined how this approach can be profitably extended also to other cases. Mansuripur [7] introduced Fourier methods for the numerical calculation of the magnetic field, its vector potential and the corresponding phase shift, under the assumption that the distribution of the magnetization is doubly periodic so that the Fast Fourier Tranform algorithm can be safely applied. These requirements are not met by the case of antiparallel magnetic stripe domains, lying in a semi-infinite specimen, where the question is how the presence of the edge and of the associated fringing field influences the field and corresponding phase shift. It can be shown that by means of our Fourier approach this problem can be solved analytically, obtaining the solution in closed form when the domain wall width is negligible [8].

In the case of electrostatic fields, Fourier methods were employed by Vanzi [9] in his investigation of electric fields to prove important relations in the real space. Let us focus our attention on the Fourier space. The general solution of the Laplace equation, e.g. in the vacuum region above the specimen, $z > 0$, can be written as

$$V(x,y,z>0) = \frac{1}{4\pi^2}\iint \tilde{V}_+(k_x,k_y)\, e^{ik_x x + ik_y y - k_\perp z}\, dk_x\, dk_y$$

where $k_\perp = \sqrt{k_x^2 + k_y^2}$ and the Fourier transform $\tilde{V}_+(k_x,k_y)$ refers to the potential distribution at the upper specimen surface. From this expression it is easy to ascertain that the corresponding contribution to the phase shift is given by

$$\varphi_+(x,y) = \frac{\pi}{\lambda E}\int_0^\infty V(x,y,z)\,dz = \frac{1}{4\pi\lambda E}\iint \frac{\tilde{V}_+(k_x,k_y)}{k_\perp}\, e^{ik_x x + ik_y y}\, dk_x\, dk_y$$

The former equation allows us to extract a simple and significative relation in the Fourier space between the Fourier transform of the phase shift and potential:

$$\tilde{\varphi}_+(k_x,k_y) = \frac{1}{4\pi\lambda E}\frac{\tilde{V}_+(k_x,k_y)}{k_\perp}$$

These considerations can be extended also to the vacuum region below the specimen, emphasizing that the calculation of the external phase shift in the Fourier space, at least formally, is a very simple matter, once the potential distribution on the two surfaces of the specimen is known.

These results have been applied to the analytical model for the electric field associated to a periodic array of alternating p- and n-doped stripes lying in a half-plane, tilted with respect to the specimen edges [10]. The solution of this problem has been found in the real space, by exploiting the striking similarity with the well-known optical problem of the diffraction of an inclined plane wave by a perfectly conducting half-plane [11]. As it is possible to calculate the Fourier transform of the potential at the specimen surface, it turns out that the the time-consuming integration along z of the potential (whose expression in the whole space, although analytical, is much more complicated with respect to its value on the specimen plane) is replaced by the division by $k_\perp$. The inverse Fourier transform can be subsequently carried out by a mixed analytical-numerical method which allows a substantial reduction of the computation time for the phase shift.

My coworkers, M. Beleggia and P.F. Fazzini, and I are presently exploiting the capabilities of Fourier methods when applied to the investigation of long range electromagnetic fields and our results [12] seem to indicate that this is a powerful approach to find the solution of otherwise unmanageable problems and even when the solution of the problem is known by real space methods, it can offer a useful different perspective or at least lead to computational benefits.

The stimulus to use Fourier methods originated with the interpretation problems related to the experimental observations of superconducting fluxons, a research carried out within a collaboration scheme with Dr. A. Tonomura and his group at the Hitachi Advanced Research Laboratory, Japan. Useful discussions with Dr. A. Tonomura, the members of his group, and with Professors H. Lichte and M. Vanzi are gratefully acknowledged.

References

[1] M. Beleggia and G.Pozzi, Ultramicroscopy 84 (2000) 171 .
[2] M. Beleggia and G.Pozzi, Phys. Rev. B 63 (2001) 054507.
[3] M. Beleggia and G. Pozzi, J. Electron Microscopy 51 (Suppl. in press) (2002).
[4] A. Tonomura et al., Nature 412 (2001) 620.
[5] A. Tonomura et al., These Proceedings (2002).
[6] M. Beleggia et al., These Proceedings (2002).
[7] M. Mansuripur, The Physical Principles of Magneto-optical recording. Cambridge University Press, Cambridge 1995.
[8] M. Beleggia et al., Proceedings ICEM 15 (2002).
[9] M. Vanzi, Optik (1984) 319.
[10] M. Beleggia et al., Phil. Mag. B 80 (2000) 1071.
[11] F. Gori, Optics Comm. 48 (1983) 67.
[12] M. Beleggia et al., in preparation.

Microsc. Microanal. 8 (Suppl. 2), 2002
DOI 10.1017.S1431927602101395

Contributions of Elastically and Inelastically Scattered Electrons to High-Resolution Off-Axis Electron Holograms – a Quantitative Analysis

Michael Lehmann and Hannes Lichte

University of Dresden, Institute of Applied Physics (IAPD), D-01062 Dresden, Germany

Off-axis electron holograms incorporate not only the information of conventional image intensity but also the perfect zero-loss information of the object wave encoded in the interference fringes. This is due to the setup of off-axis electron holography: Since the reference wave passes the object plane unscattered, only electrons which are elastically scattered in the object are still coherent hence can contribute to the interference pattern. For coherent detection, the energy loss in the object must be less than $4 \cdot 10^{-15}$ eV, which is an energy width far beyond the possibilities of current energy filters. After Fourier transformation of the hologram, the information of the conventional HRTEM image including inelastics is found in the centerband whereas the perfect zero-loss information of the complex object wave is well-separated in the sidebands; therefore, off-axis electron holography allows a quantitative comparison at exactly the same specimen area. The conventional HRTEM image is gained by isolation of the centerband and inverse Fourier transformation. Centering and isolating one sideband, inversely Fourier transforming, and calculating the intensity of the image wave yields the perfect zero-loss image intensity. The normalization of both image intensities is normally performed at an image area without object. If this area is not available within the field of view, the reference hologram can be used due to the high stability of both microscope and Triebenberg lab hence high constancy of fringe contrast over nearly all exposures.

Fig. 1 (top) shows the image intensities of a 90-degree GaAs-wedge in [100]-orientation reconstructed from centerband and sideband, respectively. The image intensity reconstructed from the sideband shows a higher contrast than the other one. However, due to different noise contributions to centerband and sideband, contrast determination of lattice fringes is much more reliable when calculated from nano-diffractograms (fig. 1 bottom) by division of intensity of reflections by the zero-beam intensity. The ratio of corresponding lattice fringe contrasts in sideband and centerband reconstructed image intensity, respectively, is plotted over the sample thickness (fig. 2). With increasing sample thickness beyond the first extinction thickness, the plot shows an up to 4.3 times higher lattice fringe contrast in the zero-loss image. The drop of contrast at subimages 5 and 6 is probably due to increasing noise in the sideband. Recently, a series of off-axis electron holograms taken at different defocus has been recorded and analyzed showing nearly the same lattice fringe contrasts in amplitude and phase when reconstructed from sideband but less lattice fringe contrast when the focal-series is reconstructed from centerband [1].

Making use of both centerband and sideband information suggest that the contrast mismatch between simulated and experimental HRTEM image intensities, often referred to as "Stobbs-Factor", is mainly due to contributions from inelastically scattered electrons. Since the lattice fringe contrast reconstructed from sideband is substantially larger compared with contrasts of HRTEM images, we assume that off-axis electron holography is only less affected by the Stobbs-Factor.

[1] M. Lehmann, D. Geiger, I. Büscher, H.W. Zandbergen, D. Van Dyck, H. Lichte, submitted to ICEM-15 (2002), Durban, South Africa

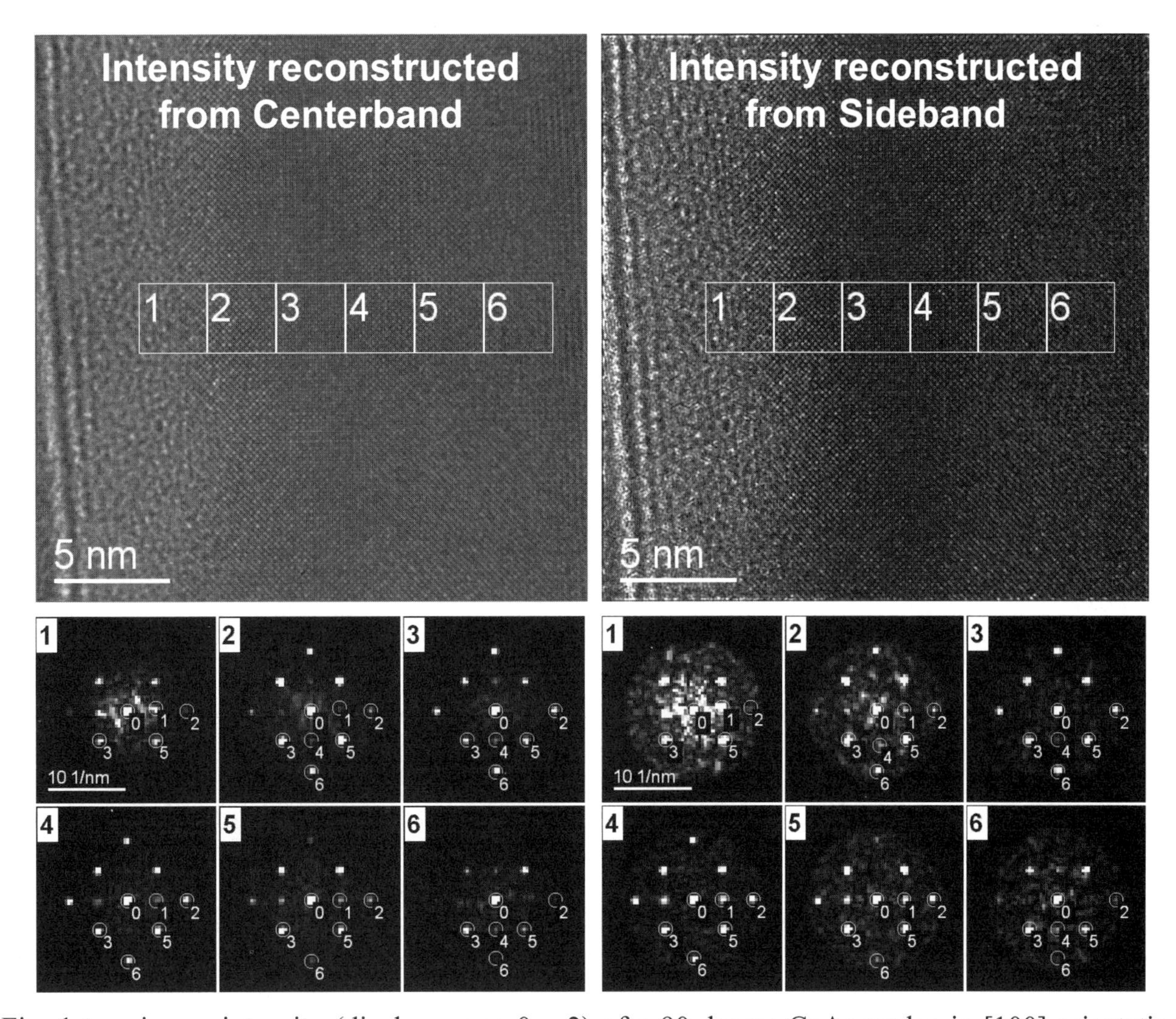

Fig. 1 top: image intensity (display range 0 .. 2) of a 90-degree GaAs-wedge in [100]-orientation reconstructed from centerband (left) and sideband (right), respectively. Bottom: nano-diffractograms determined from subimages of the image intensities shown above. The increasing subimage number corresponds to increasing sample thickness.

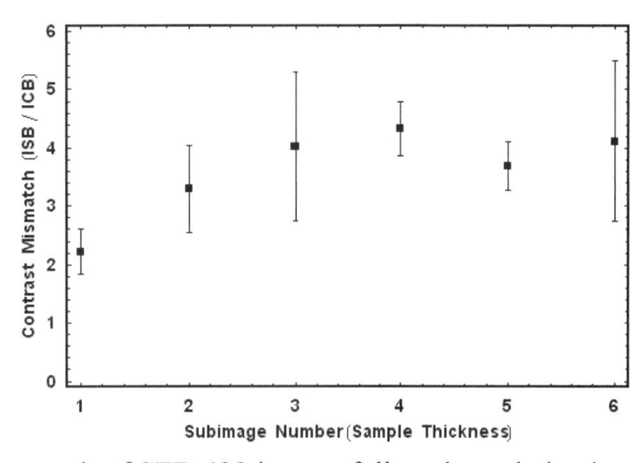

Fig. 2: Contrast mismatch of lattice fringes reconstructed from sideband and centerband, ISB and ICB, respectively, plotted over the subimage number i.e. the sample thickness. The sample thickness increases from left to right. Subimages 3 and 6 correspond to the first and second extinction thickness.

Thanks are due to Ides Büscher and Henny W. Zandbergen not only for providing the GaAs-wedge sample but also, together with Dirk Van Dyck, for the close cooperation. The financial support by the DFG within the framework of SFB 422 is gratefully acknowledged.

Microsc. Microanal. 8 (Suppl. 2), 2002
DOI 10.1017.S1431927602101401

Mapping of Process Induced Dopant Redistributions by Electron Holography

Wolf-Dieter Rau, Alexander Orchowski
LEO Electron Microscopy, Carl-Zeiss-Str 56, D-73447 Oberkochen, Germany[3]

We present dopant mapping examples in semiconductors by electron holography and outline their potential applications for experimental investigation of 2D dopant diffusion. Moreover, we address the technical challenges of the technique when applied to device structures with respect to quantification of the results in terms of the 2D pn junction potential.
Semiconductor devices consist of precisely placed foreign atoms to produce doping and composition variations, and hence a microscopically tailored electrostatic potential. Externally applied signals can then be used to modify this electrostatic "landscape", and hence control charge transport through the device. From electron holographic phase images one can directly map this electrostatic potential distribution after various stages of device fabrication, which is highly desirable in order to understand and control the fundamental solid state processes that govern device fabrication [1,2]. Of main interest here are the lateral redistributions of locally implanted doping profiles during typical anneal conditions.
An investigation of an anneal sequence in ultra shallow junction formation is shown in fig. 1. An n-doped Si substrate material has be locally implanted with B through a patterned mask. A series of samples have been prepared at different anneal stages. The lateral and vertical evolution of the highly p doped regions can be seen in the phase images of unmasked regions shown in fig1 (a-c), allowing for a qualitative evaluation of the respective junction formation process. Unfortunately, further quantification of these images in terms of local electrostatic potential distribution is not possible, due to charging of the nitride mask still residing on top of the Si substrate material, as can be seen from the non-flat phase distribution in the vacuum region. Removing any sources of charge build-up is therefore absolutely essential for quantitative interpretation.
An example for successful potential mapping by holography is shown in figs. 2/3, where mask layers have been removed before sample preparation. Here the goal was the investigation of lateral outdiffusion from a 5nm B doping peak located within a 20 nm thick $Si_{0.8}Ge_{0.2}$ layer after implanting an additional dose of B through a patterned surface mask. Fig. 2a shows a TEM image of an area around an opening in the mask in an implanted and annealed sample. Residual defects can be seen in the implanted region. The electron holographic phase image of the masked region on the right is shown in fig. 2b. The SiGe layer is visible as a bright stripe due to the higher mean inner potential of Ge. The dark stripes on either side of the layer are due to the electrostatic potential change in the depletion zones of the n-p and p-n junction on top and bottom of the SiGe layer, caused by outdiffusion of B from the SiGe region. The broadening does not cease abruptly at the edge of the mask opening, extending significantly beneath the mask. Fig. 3a shows the quantitative map of the potential distribution. A comparison with process simulation shown in fig 3b can now be used to refine according parameters of dopant diffusion models [2].

References

[1] W.-D. Rau et al., *Phys. Rev. Lett.* 82, 19 (1999) 2614.
[2] A. Orchowski et al., *accepted by Appl. Phys. Lett.* (2001).
[3] This work has been performed at the IHP for Microelectronics, Frankfurt (Oder) and the University of Dresden, Germany. We gratefully acknowledge the support, valuable contributions and input of H. Rücker, P. Schwander, A. Ourmazd, H. Lichte and M. Lehmann

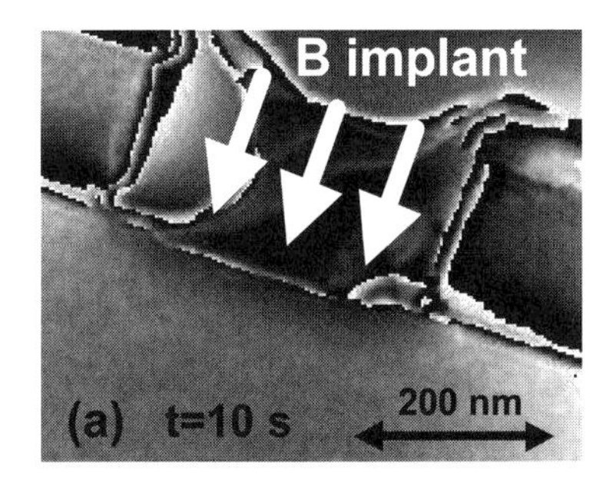

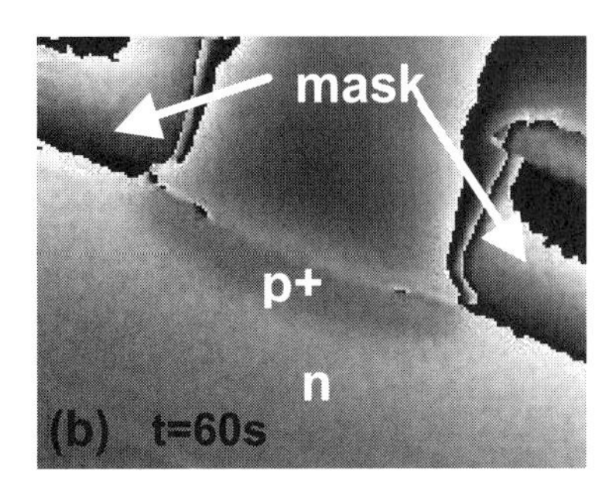

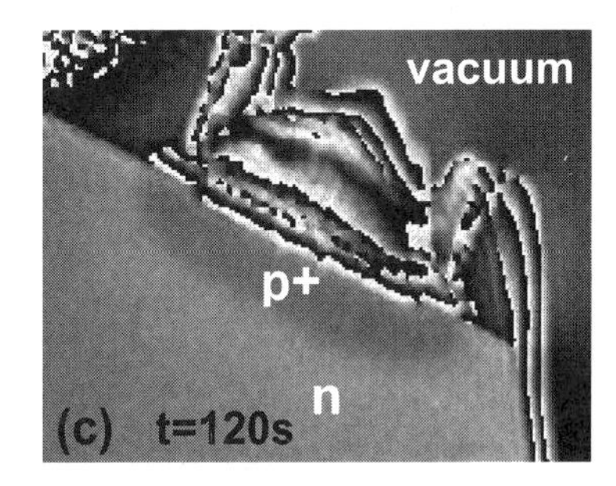

FIG. 1. Anneal sequence for Ultra Shallow Junction formation. The phase images reveal the vertical and lateral boron redistribution under the mask opening after a) 10 sec b) 60 sec c) 120 sec anneal.

FIG. 2. Mapping of lateral boron distribution at mask edge after boron contact implant in SiGe test structure. (a) Conventional TEM micrograph of implanted region. The mask was removed before sample preparation. (b) Phase image of the region boxed in (a). The SiGe layer appears as a bright stripe. The reduced phase in the dark regions reveals the electrostatic potential change due to B doping. B outdiffusion from the SiGe layer extends beneath he masked region.

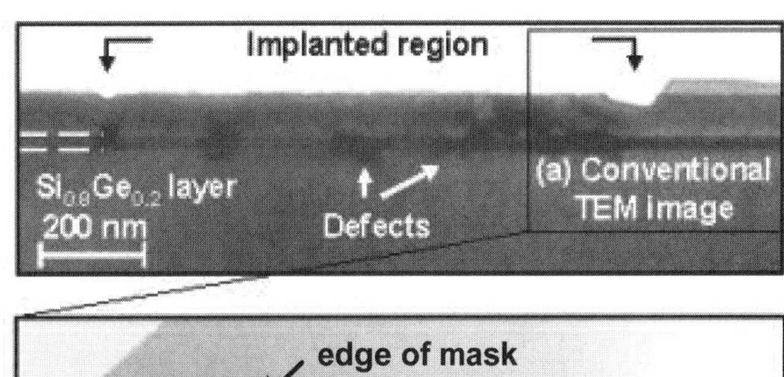

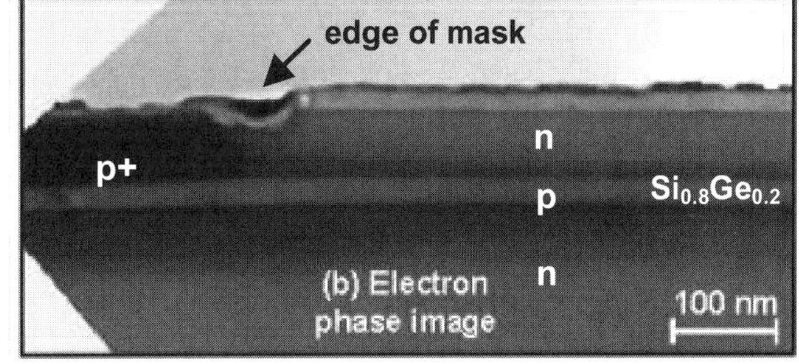

FIG. 3. (a) Experimental potential map derived from the phase image in fig. (b) Best matching process simulation of electrostatic potential distribution. The according dopant distribution can be used to extract diffusion model parameters for lateral B diffusion [2]

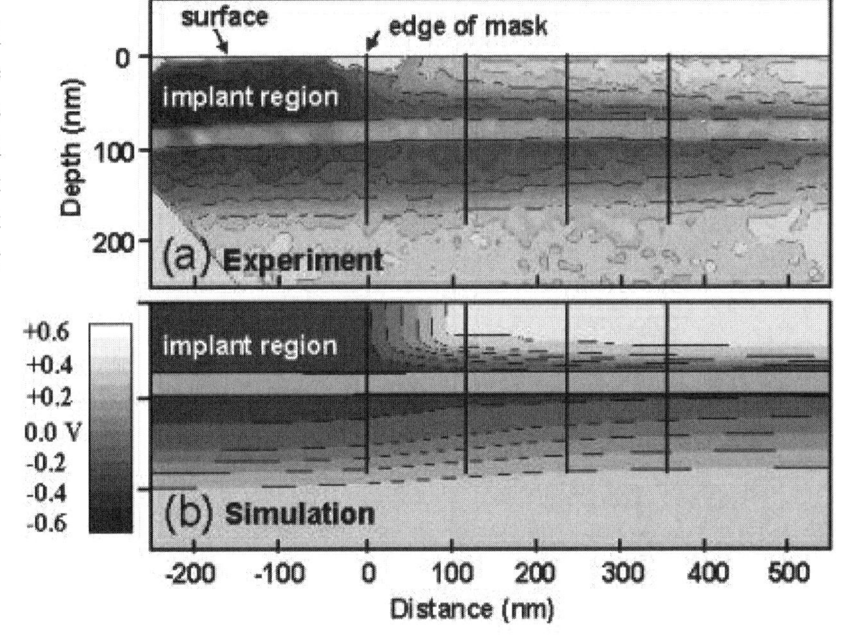

Microsc. Microanal. 8 (Suppl. 2), 2002
DOI 10.1017.S1431927602101413

New Developments and Applications of Electron Holography

Tsukasa Hirayama*, Zhouguang Wang*, Kazuo Yamamoto**, Takeharu Kato*, Naoko Kato***, Katsuhiro Sasaki****, and Hiroyasu Saka****

* Japan Fine Ceramics Center (JFCC), Nagoya, 456-8587, Japan
**Japan Science and Technology Corporation, c/o JFCC, Nagoya, 456-8587, Japan
*** ITES, IBM Japan, Yasu, Shiga-ken, Japan
***** Department of Quantum Engineering, Nagoya University, Nagoya 464-8603, Japan

Even though electron holography was invented to correct spherical aberration of electron lenses, it can also be used for observing electromagnetic micro-fields. [1] Since electromagnetic properties of materials or devices are strongly related to their electromagnetic micro-fields, observation of such micro-fields is an important experiment for materials science or device engineering. In this paper, we present an application of conventional electron holography, two-dimensional visualization of electrostatic potential distribution in semiconductors, and a new interferometry based on amplitude-division three-wave interference to directly display electromagnetic micro-fields.

Figures 1(a) and 1(b) show an electron micrograph and a phase map of a cross-sectioned silicon-metal oxide semiconductor field-effect transistor (MOSFET), respectively. This MOSFET was fabricated from a silicon wafer with a boron concentration of 10^{15} cm^{-3}, and the TEM specimen was carefully prepared by the focused ion beam (FIB) method. The two dimensional electric potential distribution is clearly discernible in this figure. [2]

Figure 2 shows an amplitude-division three-wave interference pattern of a latex particle. In this method, a thin crystal of silicon prepared by ion-milling is installed at the standard specimen position, and a thin carbon film to which latex particles are adhered is placed at the selected aperture position. By decreasing the electric current of the objective lens, a lattice image of Si is formed below the selected area aperture position. This lattice image can be observed at the final imaging plane by overexciting the first intermediate lens. With this method, we obtained three defocused images of a latex particle as shown in Fig. 2. The electric field around the particle is represented by the intensity modulation of Si lattice fringes.

We would like to thank Mrs. T. Yanaka, K. Shirota, K. Moriyama and Ms. T. Morino of Topcon for their technical assistance in performing the amplitude-division three-wave interference experiment. We also thank Prof. N. Tanaka and Prof. T. Tanji of Nagoya University and Dr. Craig Fisher of JFCC for their valuable suggestions. Sections of this work were performed as part of the Active Nano-Characterization and Technology Project, using Special Coordination Funds from the Ministry of Education, Culture, Sports, Science and Technology of the Japanese government.

References

[1] A. Tonomura, Rev. Mod. Phys. **59** (1987) 639.
[2] Z. Wang et al., Appl. Phys. Lett. **80** (2002) 246.

FIG. 1. Cross -sectional view of a MOSFET. (a) Electron micrograph. (b) Phase image obtained by electron holography.

FIG. 2. Amplitude-division three-wave interference pattern of a latex particle.

Microsc. Microanal. 8 (Suppl. 2), 2002
DOI 10.1017.S1431927602101425

ELECTRON HOLOGRAPHIC CHARACTERIZATION OF NANOSCALE MAGNETIC AND ELECTROSTATIC FIELDS

Martha R. McCartney

Center for Solid State Science, Arizona State University, Tempe, AZ USA 85287

In addition to improving the resolution limit of electron microscopes[1], off-axis electron holography at moderate resolution can be used to characterize electrostatic and magnetic fields[2] down to the nanometer scale. We have investigated samples with one- and two-dimensional dopant profiles and obtained maps of electrostatic potential with a sensitivity of 0.1V and spatial resolution approaching 5nm. The magnetization behavior of patterned magnetic nanostructures at high sensitivity and spatial resolution has also been studied.

In off-axis electron holography the hologram is formed after overlapping part of the electron wave that has traveled only through vacuum with the sample wave by means of a positively charged electrostatic biprism. The instrumentation required for this research includes a transmission electron microscope (TEM) equipped with a field emission gun in addition to the biprism. Digital acquisition of the hologram allows for rapid and quantitative reconstruction. The Philips CM200-FEG instrument at ASU is also equipped with an extra coil below the lowerpole piece of the objective lens that allows for magnetic samples to be examined in a field-free environment at magnifications of up to 70KX with the objective lens turned off. This electron-optical arrangement is also advantageous in terms of a convenient field-of-view for imaging dopant depletion regions in current semiconductor devices.

The semiconductor community is intensely interested in obtaining 2-D dopant profiles at the nanometer scale. For example, the ITRS Metrology 2000 Update, table 84a – " 2000 Front End Processes Technology Requirements – Near Term", lists this requirement as one with "No known solution". Electron holography has been shown to be capable, both in principle 3 and in practice 4,5, of providing high resolution (~3nm), high sensitivity (0.1V) 2-D maps of electrostatic potential in semiconductor devices. Over the past few years, we have performed holographic experiments on a variety of two-dimensional device structures as well as 1-D doped samples. Two-dimensional potential maps of transistor structures can provide essential information for accurate modeling parameters for device engineers. Figure 2 shows the reconstructed holographic amplitude and phase images from a 0.13micron transistor structure.6 Analysis of the electrostatic potential derived from the phase image can be used to determine both the junction depth and the width under the gate after process annealing. Quantitative analysis of the junction potential allowed details of the lateral diffusion of the implanted dopant to be used for comparison with process simulations.

As an example of holographic imaging of magnetic fields, Fig. 2 shows spin-valve structures consisting of two ferromagnetic layers separated by a thin non-magnetic spacer layer[7]. The shaped elements consist of Co(10nm)/Au (5nm)/Ni (10nm). The figure shows the magnetic contributions to the reconstructed phase over part of an experimental hysteresis cycle along with micromagnetic simulation. The images reveal a reduction in the number of phase contours on approaching zero applied field. The simulations show that the magnetization direction of the Ni in each element reverses well before the external field is reduced to zero as a result of the presence of the demagnetizing field of the magnetically more massive Co layer. Interestingly, the simulations were unable to replicate the vortices observed experimentally in the diamond- and elliptical-shaped elements, stressing the need for direct imaging method for magnetic element at reduced size[8].

References

1 H. Lichte, Ultramicroscopy, 20 (1986) 293.
2 A. Tonomura, Electron Holography (Springer, Heidelberg 1986)
3. M. R. McCartney, et al., Appl. Phys. Lett., **65** (1994) 2603.
4. W.D. Rau, et al., Phys. Rev. Lett., **82** (1999) 2614.
5. M.R. McCartney, et al., Appl. Phys. Lett., in press.
6. M. A. Gribelyuk, et al., Phys. Rev. Lett., in press.
7 R. E. Dunin-Borkowski, et al., J. of Microscopy, 200 (2000) 187.
[8] MRM gratefully acknowleges contributions to this work and on-going collaborations with D.J. Smith, Jing Li, R. E. Dunin-Borkowski, M.R. Scheinfein and M. Gribelyuk.

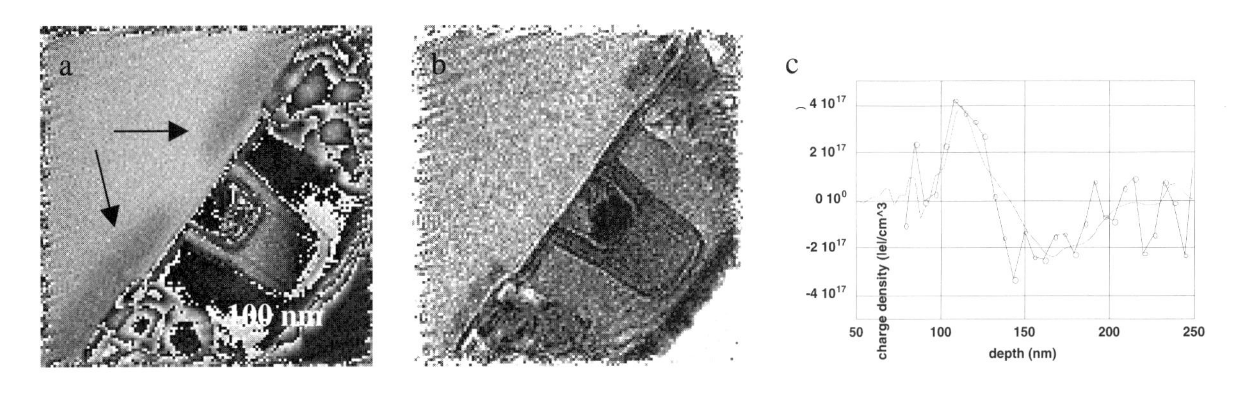

FIG. 1. Holographic reconstructed phase (a) and amplitude (b) of a 130nm p-MOS transistor structure. Contrast due to the depletion regions associated with p-n junctions is visible as darker regions (arrowed) under the Si surface in the phase image. (c) Comparison of dopant distributioı and junction position derived from second derivative of hologram (open circles) with that calculated from SIMS profile (solid line).

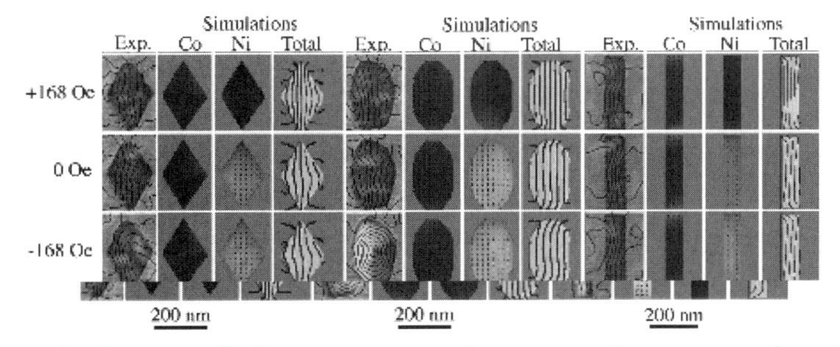

FIG. 2. Composite image of phase contours for magnetic nano-scale spin-valve structures along with simulations for the individual layers and total phase shift.

Microsc. Microanal. 8 (Suppl. 2), 2002
DOI 10.1017.S1431927602101437

The Determination and Interpretation of Electrically Active Charge Density Profiles at Reverse Biased p-n Junctions from Electron Holograms

R.E. Dunin-Borkowski*, A.C. Twitchett* and P.A. Midgley*

* Dept of Materials Science and Metallurgy, University of Cambridge, Cambridge CB2 3QZ, UK

It is important to understand the contrast seen in off-axis electron holograms of doped semiconductors. Here, we show that although electrostatic potential profiles measured from holograms of p-n junctions appear to agree with predictions, detailed charge density profiles across the junctions may be different from those expected for bulk samples. We have examined p-n junctions in Si, prepared for TEM examination by focused ion beam (FIB) milling, as a function of both reverse bias and sample thickness. Contacts were applied to the front and back surfaces of cleaved wedges, on which uniform thickness membranes had been prepared using FIB milling (Fig. 1a). Figure 1b shows a phase image from such a sample, in which the dark and bright contrast correspond to p and n-type regions in the sample, respectively. Qualitatively, phase profiles across the junction (Figs. 1c, d) are consistent with predictions as a function of bias and thickness, and several parameters can be inferred directly from the data. For example, if the phase change across the junction is plotted as a function of bias (Fig. 1e), the gradient of this graph can be used to infer that the 'electrically active' sample thickness is 340 nm. (The crystalline thickness was measured to be 390 nm). Similarly, the intercept of this graph provides a value for the built-in voltage of 0.9 V. Although the phase change plotted as a function of thickness is found to deviate from a straight line, this behavior can also be understood if it is assumed that the thickness of the 'electrically dead' layer on each sample surface increases slightly at the lowest sample thicknesses.

Unexpected results are obtained when simulations are used to obtain electric field and charge density profiles across the junction. Figures 2a and b show best fits to the data as a function of reverse bias for a symmetrical model that assumes a diffuse junction profile. The results are surprising for two reasons. Firstly, the charge density in an unbiased sample ($\sim 3 \times 10^{17} cm^{-3}$) is lower than the nominal value ($\sim 4 \times 10^{18} cm^{-3}$). Secondly, the charge density in the depletion region *increases* with bias voltage rather than remaining constant. As a result, the depletion width increases more slowly with applied bias than expected for a bulk sample (Fig. 3a). This behavior, which is not understood at present, may result from the effect of either the high-energy electron beam or sample preparation on the charge density in the sample. The insensitivity of the data to such charge density information, and the consequent need for simulations, is highlighted by the fact that a less physically realistic model (Fig. 2c) also provides a reasonable match to the data. Additional information can be obtained by comparing simulations with phase profiles as a function of sample thickness; the depletion width in an unbiased sample is found to increase slightly at the lowest thicknesses (Fig. 3b). A comparison of sample thickness measurements from holographic amplitude images with convergent beam electron diffraction reveals the presence of a 30 nm thick amorphous layer on each surface of the sample (Fig. 3c). The experimental data and simulations show that beneath these amorphous layers are ~25 nm thick crystalline but electrically dead layers and within these is the active junction, across which the phase change is broadly correct but whose charge density profile appears to be affected by the electron beam or sample preparation [1].

References

[1] We are grateful to the Royal Society, the EPSRC and the Worshipful Company of Armourers and Brasiers for financial support, and Philips Research Laboratories (Eindhoven) for samples.

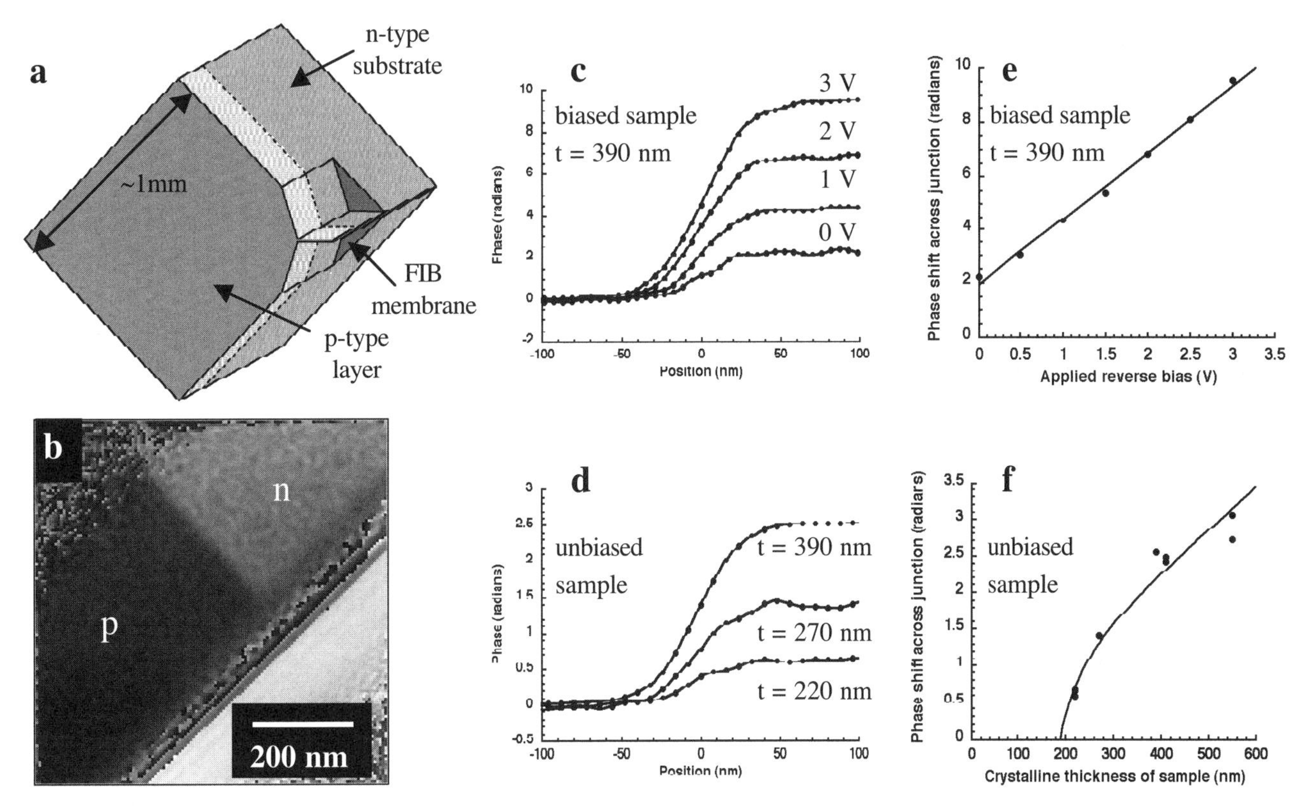

FIG. 1. (a) Sample geometry after cleaving and FIB. (b) Phase image of Si p-n junction. Phase profiles across junction (c) vs. reverse bias for a crystalline sample thickness of 390 nm and (d) vs. sample thickness for an unbiased sample. (e, f) Magnitude of step in phase across junction in (c) and (d).

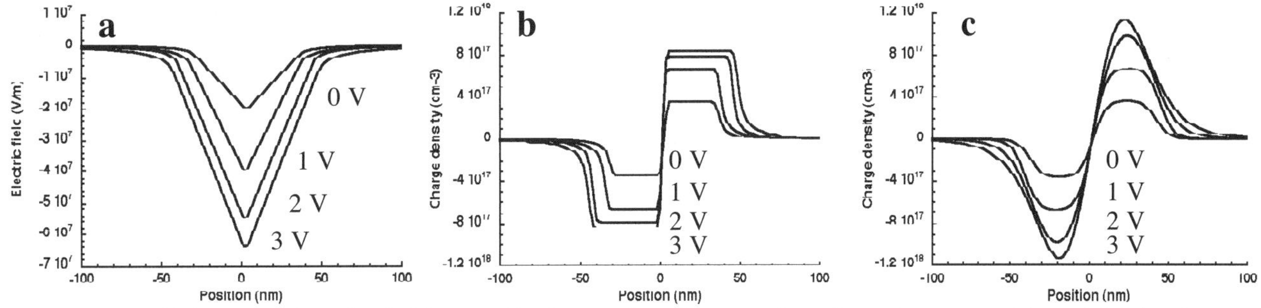

FIG. 2. Best-fitting (a) electric field and (b) charge density as a function of reverse bias; (c) Fits for an alternative charge density model that also provides a reasonable fit to the experimental phase images.

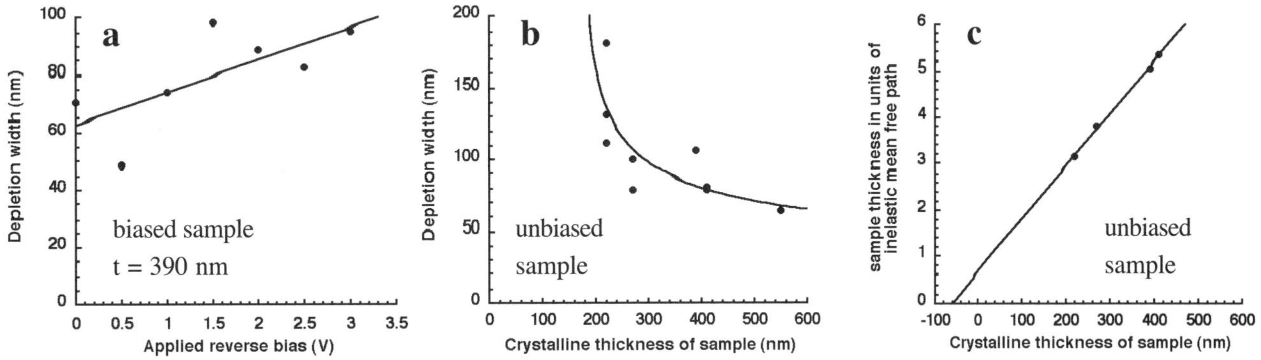

FIG. 3. Fitted depletion width as a function of (a) reverse bias and (b) sample thickness. (c) Sample thickness measured from holographic amplitude image and using convergent beam diffraction.

Microsc. Microanal. 8 (Suppl. 2), 2002
DOI 10.1017.S1431927602101541

COMPARISON OF FIB TEM SPECIMEN PREPARATION METHODS

Ron Anderson

Microscopy Today*

The increasingly wide-spread adoption of the focused ion beam (FIB) method for making TEM specimens in the semiconductor industry and elsewhere has spawned several competing methodologies. This paper compares the pros and cons of the various methods used today in a table. Important points are italicized.

In the first column we see the original method proposed to make TEM specimens, [1] which some researchers call the "H-bar" method because a SEM view of the finished sample looks like the capitol letter "H." Briefly, the sample is located, rough cut from its matrix, and polished to a thickness of about 50 to 100 microns. The specimen is mounted on a half-grid (a large, single-hole, aperture grid with one side cut away to allow the FIB beam to strike the edge of the specimen). In the FIB, the location of the finished TEM specimen is coated with a W or Pt line and large trenches are cut on either side of the desired location using large apertures. Final cuts are made to thin and clean the resulting specimen. Because of the TEM tilt limitations and the probability of the specimen suffering severe FIB artifact contamination, this method has little to recommend it in the face of newer protocols.

The combined tripod and "H" pattern method is actually a trivial modification of the conventional "H" pattern method [2]. The difference between the two protocols is that the specimen is initially polished to a thickness less than 10 microns instead of 50 to 100 microns. The end result of making the specimen thinner before FIBing, besides shorter FIB times, is that the resulting specimen geometry allows the specimen to be ion milled after thinning from the substrate side. The benefits of ion milling the FIBed specimen are substantial. First, the resulting specimen can be tilted to the full range of the TEM instrument's tilt capability. Second, the specimen is made thinner, frequently making HRTEM imaging of atom columns in the specimen possible. Third, the thick (20 nm or so) amorphous layer created by the 30 to 50 keV FIB beam can be replaced by the negligible, order-of-magnitude thinner amorphous layer created by argon ion milling at 2 or 3 keV at near grazing incidence. And lastly, any FIB back sputtered artifactual material on the specimen can be removed. Both "H" pattern methods have the disadvantage of requiring a mechanical polishing step but the times to make specimens compare well with the lift-out methods.

The lift-out methods [3] are well covered in considerable detail in the recent literature. Ex-situ removal is when the specimen is plucked from the substrate outside the FIB tool via electrostatic pick-up on a glass filament and then placed on a carbon-film coated TEM grid. In-situ mounting is accomplished in the FIB tool via fastening the specimen to a transfer fixture, plucking it from the substrate, and then fastening it to a TEM grid all inside the FIB. The lift-out, ex-situ grid mounting method is very fast, produces specimens that do not limit the TEM tilt capability, and automation of the initial preparation steps is possible. The main disadvantage is that no further thinning or artifact

	Conventional "H" Pattern	Combined Tripod and "H" Pattern	Lift-Out Method External Grid Mounting	Lift-Out Method Internal Grid Mounting
❖ **Pros**	❖ **Can put back into FIB for additional thinning** ❖ **Multiple specimen sites per initial prep**	❖ **Can put back into FIB for additional thinning** ❖ ***Can be ion milled to make thinner and to remove FIB artifacts*** ❖ ***Full TEM tilt capability*** ❖ **Multiple specimen sites per initial prep**	❖ ***Full TEM tilt capability*** ❖ **Serial sectioning, 3D reconstructio n** ❖ ***Automated initial prep on multiple specimen sites unattended*** ❖ **Bulk specimen may be returned for further processing**	❖ ***As for Lift-Out external mounting*** ❖ ***Ion milling possible for additional thinning and artifact reduction***
❖ **Cons**	❖ **Destructive to bulk specimen** ❖ **Limited tilting capability** ❖ **Preliminary prep needed** ❖ **FIB artifacts**	❖ **Destructive to bulk specimen** ❖ **Preliminary prep needed**	❖ **No additional FIB or ion milling possible** ❖ **FIB Artifacts** ❖ **Inapplicable for fragile specimens**	❖ **Additional ion milling may be one-sided only** ❖ **Possible ion mill artifacts**
❖ **Speed**	❖ **1 to 4 Hours depending on target size**	❖ **1 to 4 Hours depending on target size**	❖ ***1 to 2 Hours depending on target size***	❖ ***1 to 4 Hours depending on target size***

removal is possible on the specimen sliver resting on the thin carbon film. Whatever you get is all there is. If you aren't happy with the TEM specimen you must start over again. The in-situ method is newly developed and very promising. It combines all of the advantages of the ex-situ method but offers the possibility of subsequently ion milling the resulting specimen to further thin the specimen and to remove artifacts. The times cited are for experienced operators and include all pre- and post-FIB-tool specimen site selection, cutting or polishing as appropriate, and mounting the resulting specimen. All the methods are capable of high target specificity, it just takes longer to prepare very small pre-specified locations. Likewise, very hard materials, like jet engine turbine blades, take far longer to process in every phase compared to silicon specimens.

References

[1] Kirk EC et al. (1989) Inst. Phys. Conf. Series, 100, p 501
[2] Klepeis SJ (1998), Proc. of the 14th Int. Cong. on Electron Microscopy, Cancun, Mexico.
[3] Overwijk MHF et al. (1993) J. Vac. Sci. Technol. B 11(6), pp. 2021-2024.

*Formerly IBM East Fishkill (retired) presently Editor-in-Chief of Microscopy Today magazine

Microsc. Microanal. 8 (Suppl. 2), 2002
DOI 10.1017.S1431927602101553

Focused Ion Beam Based Sample Preparation Techniques

R. M. Langford, A. K. Petford-Long and *P. Gnauck

Department of Materials, University of Oxford, United Kingdom, OX1 3PH
* LEO Elektronenmikroskopie GmbH, Oberkochen, Germany

Focused ion beam (FIB) systems are routinely used for the preparation of site specific transmission electron microscopy (TEM) specimens of a diverse range of materials such as semiconductors, ceramics and metals. In this article we report on some of the FIB sample preparation methods that we have developed or are developing to improve upon or extend the current FIB sample preparation techniques.

The 'trench' and the 'lift-out' techniques can be used for the preparation of plan-view specimens. A specimen is cleaved/or polished and mounted in the FIB system with the cleaved edge vertical. The plan-view specimens are then made at the cleaved surface using either one of the two techniques. A limitation with this approach is that because it is not easy to cleave or polish a sample close to a feature of interest, it is difficult to use it to make site specific plan-view specimens. The approach we use to make such specimens is to FIB mill a wedge shaped piece of material which is free from the substrate. (Three cuts are made in the shape of a 'U' and a fourth is made at 45 ° such that it cuts through the bases of the other three). A micromanipulator and glass needle are used to lift the wedge out from the substrate and to orientate it with the surface of interest vertical, as shown in Figure 1. The plan-view specimen is then made from the wedge using the lift-out technique.

For some samples, such as crack tips, the region of interest can be more than 20 - 30 μm beneath the surface and therefore the required FIB milling time is prohibitively long. To reduce the required FIB milling time we have used broad ion beam (BIB) milling on the top surface of the sample to decrease the distance to the feature of interest [(2)]. A slice containing the crack tip is mounted onto a copper disc and then BIB milled as shown in Figure 2. It is repeatedly examined using a light microscope to determine the distance of the feature of interest to the surface of the sample. Once the distance has been reduced to around 5 μm the specimen is mounted in the FIB system with the support grid vertical and the TEM sample is made using the trench technique. Another advantage of this approach is that the BIB milling also reduces the width of the slice, and therefore the lengths of the trenches, which reduces the degree of geometric shadowing when imaging in the TEM.

FIB milling using a 30 keV ion beam can introduce a considerable amount of damage at the sidewalls of the specimen which can affect high resolution imaging. Low energy (5 keV) milling with the sample tilted by 10° reduces the amount of damage but may result in preferential milling of the different materials or layers. The approach we use to reduce both the thickness of the specimen and depth of the damage [(3)] is to prepare the specimen using the lift-out technique and to place it onto a 0.5 μm thick silicon support membrane into which an array of windows has been patterned. The top side of the specimen is sputtered in a BIB system using a low incident energy (3 keV) and shallow angle (2 - 3°). The specimen is then turned over using a glass needle and micromanipulator so that the other side can be sputtered.

Automatic preparation of a specimen considerably increases the utilization of the FIB system, enabling it to be run over night. We are currently developing a procedure for the automatic thinning of specimens which warp during the final thinning stages due to intrinsic stress. Relaxation of the specimen in the ion beam can result in non-uniform thinning and also in the milling of the features of interest. During the line milling/cleaning stage, the intensity of the secondary electron signal is monitored and if it dramatically increases the thinning is stopped. An image of the specimen is taken and the positions of its edges, relative to FIB milled alignment marks, are determined. The FIB is then scanned along the edges of the specimen to further thin it uniformly. This method is shown diagrammatically in Figure 4. It is realised that this approach will not suitable for regions of the specimen which warp into the plane perpendicular to the ion beam but is suitable for thin films at the top of the specimens.

References

[1] R. M. Langford et al., *J. Vac. Sci. Tech.* A 19 (2001) 982

[2] S. Lozano-Perez et al, *Inst. Phys. Conf. Ser. No. 168 (2001)* 191

[3] R. M. Langford et al. *J. Vac. Sci. Tech.*, B 19 (2001) 755

[4] The authors gratefully acknowledge the support of EPSRC

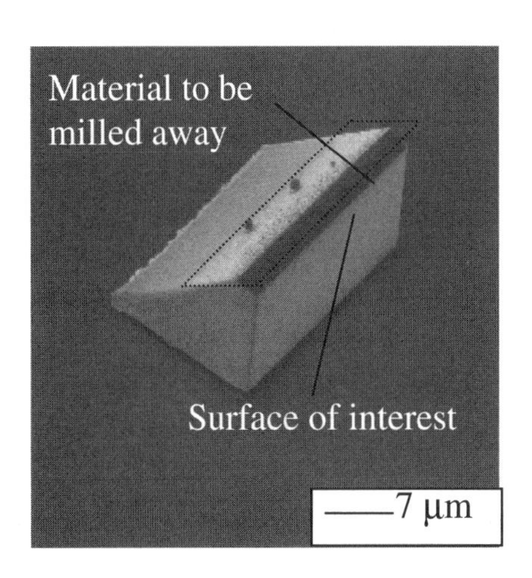

Fig. 1. FIB milled wedge orientated on the substrate with the surface vertical.

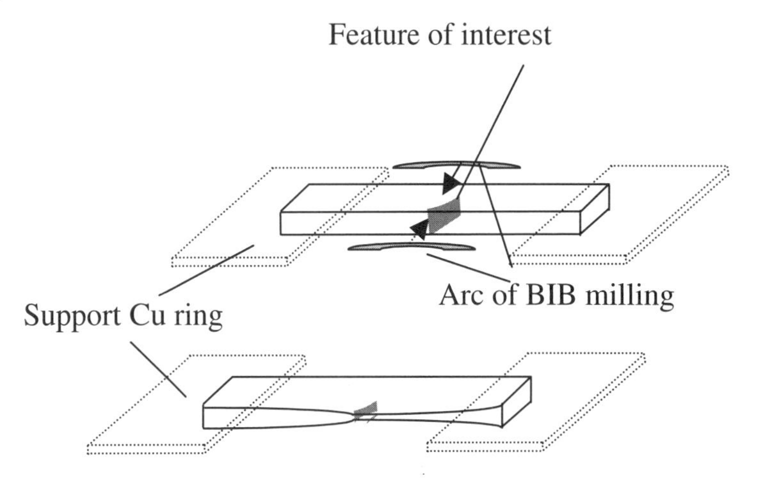

Fig. 2. BIB milling of the slice

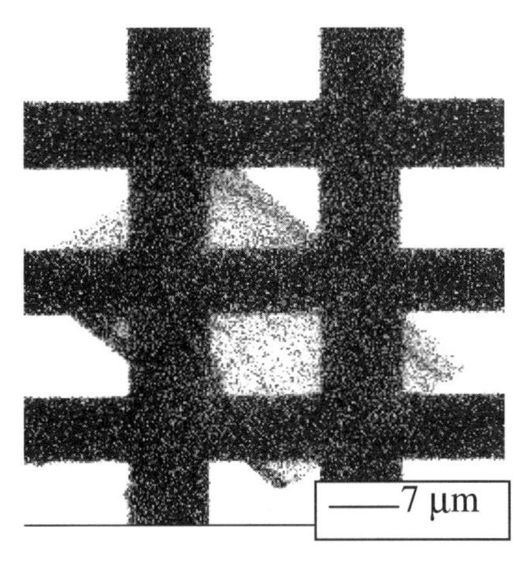

Fig. 3. TEM specimen on Si support membrane for BIB milling and imaging.

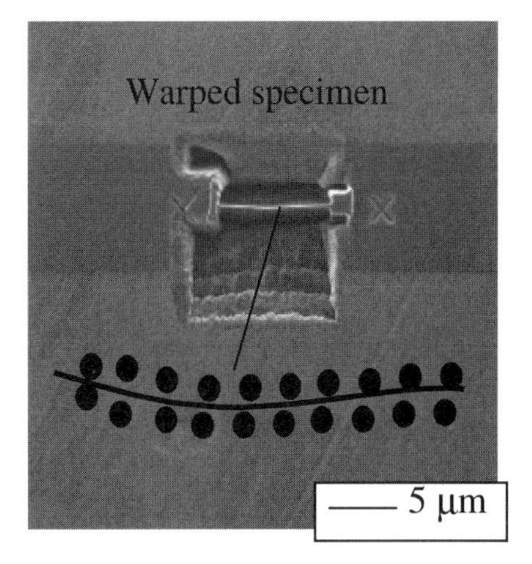

Fig. 4. Software determines the edges of the warped specimen so that the ion beam can be scanned along its edges for further thinning

Microsc. Microanal. 8 (Suppl. 2), 2002
DOI 10.1017.S1431927602101565

A NEWLY DEVELOPED FIB SYSTEM FOR TEM SPECIMEN PREPARATION

T.Kamino*, T.Yaguchi*, Y.Kuroda *, T.Hashimoto **, T.Ohnishi** , T.Ishitani**, K.Umemura***, K.Asayama****

*Hitachi Science Systems, Ltd., 882 Ichige, Hitachinaka, Ibaraki, 312-8504 Japan
** Hitachi High-Technologies Corp., 882 Ichige, Hitachinaka, Ibaraki, 312-8504 Japan
*** Central Research Laboratory, Hitachi, Ltd., Kokubunji, Tokyo 185-8601 Japan
****Semiconductor and Integrated Circuits Division, Hitachi, Ltd., Oume, Tokyo, 198-8512 Japan

In the materials characterization using transmission electron microscope(TEM), FIB technique is demanding more and more as the method to prepare electron transparent specimen[1)]. We have developed a dedicated FIB system FB-2000A employing FIB-TEM(STEM) compatible specimen stage[2)] and an FIB micro-sampling tehnique[3-5)]. The FIB-TEM(STEM) compatible specimen stage allowed site specific TEM specimen preparation with a positional accuracy of 0.1 m or better[2)]. The FIB micro-sampling allowed extraction of TEM specimen directly from bulk sample without any pre-FIB preparation.

Recently, an FIB system FB-2100 with a newly designed 40kV ion optics has been developed to perform high-speed TEM specimen preparation. Accelerating voltage is variable from 10kV to 40kV at a minimum step of 5kV. The maximum ion beam current and maximum ion beam current density of the new ion optics are 30nA and 25A/cm^2 , respectively. The ion milling speed in TEM specimen preparation is approximately twice as high as that of previous model(FB-2000A). No noticeable increase of beam damage for the TEM specimen preparation at 40kV has been confirmed. Figure 1 shows a high resolution TEM image of Si single crystal specimen prepared at 40kV. Crystal lattice and the dumb-bell structure with the spacing of 0.136nm are clearly observed. The requirements for a high-speed TEM specimen preparation are rapidly increasing. One of solutions is employment of automatic fabrication for rough milling of a TEM specimen. We have developed a new automatic fabrication system employing a new marker detection unit based on a phase only correlation(POC) method. The marker detection unit worked for various brightness and sharpness of the images. Figure 2 shows a series of test marker images observed at various images brightness. Images 2-8 were allowable in the marker detection. Figure3 shows a series of test images observed at various Z-position. All markers in these images were identified as the memorized marker. The newly designed ion optics allowed reduction of ion beam diameter together with increase of ion beam current. Minimum probe size and the probe current for the milling are 12nm and 0.02nA, respectively. Additionally, some optimization have been made in settings of voltages, lens for various sizes and accelerating voltages. Those improvements were effectively worked out to improve scanning ion microscopy(SIM) image quality. The SIM image resolution is 6nm or better at 40kV.

References

1. T.Ishitani, et al., J. Electron Microsc.**43**, 5 (1994)322-326
2. T.Kamino, et al., Proc. Microsc. and Microanal. **6**, Supp.l2 (2000) 510-511
3. K.Umemura, et al., Electron Microscopy, 34, Supp. 1 (1999) 510-511
4. T.Ohnishi,et al., Proc.25[th]Int.Symp. for Testing and Failure Analysis (1999) 449-503
5. T.Yaguchi, et al., Proc. Microsc. and Microanal.7, Supp. 2 (2001) 938-939

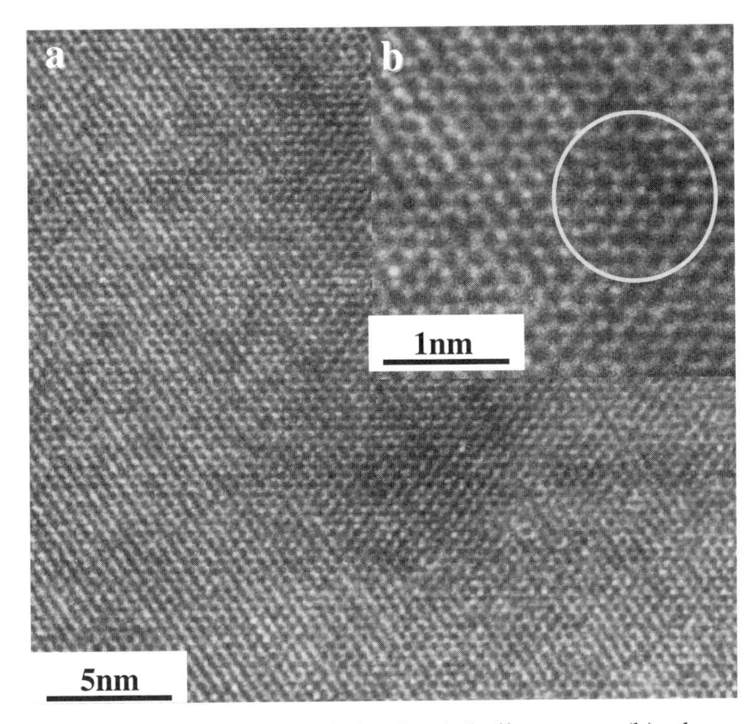

Figure 1. Crystal lattices image (a) and the dumb-bell structure(b) observed on a Si single crystal specimen prepared by FIB milling at 40kV.

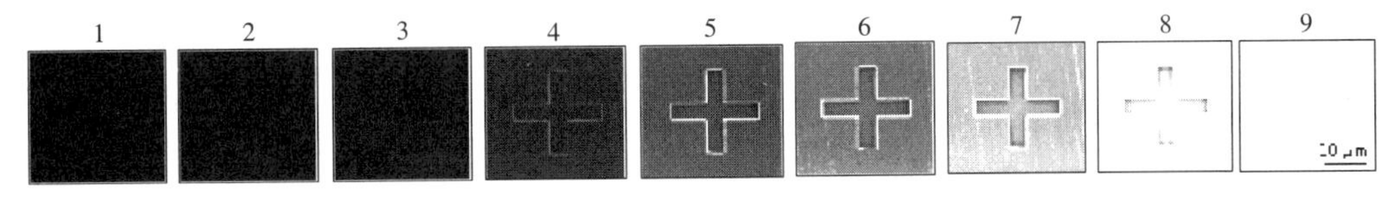

Figure 2. A series of SIM images of a cross marker observed at various brightness. Images 2-8 were allowable in the marker detection.

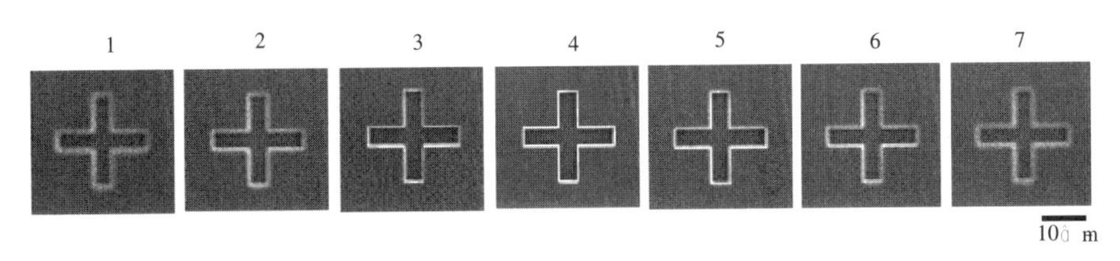

Figure 3. A series of SIM images of a marker at various Z-position. All markers in these images were identified as the memorized marker.

Microsc. Microanal. 8 (Suppl. 2), 2002
DOI 10.1017.S1431927602101577

FIB Damage in Silicon: Amorphization or Redeposition?

S. Rajsiri, B. W. Kempshall, S.M. Schwarz, and L. A. Giannuzzi

Department of Mechanical, Materials, and Aerospace Engineering, University of Central Florida, 4000 Central Florida Blvd., Orlando, FL, 32816-2450

The Focused Ion Beam (FIB) instrument has been utilized for site-specific specimen preparation for a wide range of analytical techniques due to its ability to achieve high spatial resolution imaging, milling, and deposition [1,2]. The understanding of FIB damage is important to ensure that the region being analyzed is indeed representative of the material, and is not due to a specimen preparation artifact. The interaction between the incident ions (e.g., Ga^+) and the target material during FIB operation (e.g., imaging or milling mode) may lead to surface damage and consequently limit the ability to achieve high quality high-resolution TEM images. Amorphization of a FIB milled crystalline surface may occur due to sufficient atom displacement within the collision cascade resulting in the loss of long-range order when the density of point defects reaches a critical value [3]. Redeposition of sputtered atoms has also been reported as a result of FIB milling [4]. The propensity for redeposition increases when FIB milling is performed in a confined and/or a high aspect ratio trench, or when FIB conditions are used that contribute to factors that increase the sputtering rate (e.g., using a higher beam current) [5]. Observations have shown that FIB milling with Ga at an energy of 30 keV will produce amorphization damage along a Si side-wall that is ~ 28 nm thick and up to 20 wt% Ga may be present within the damage region [6]. Previous work in our lab has shown that the side-wall damage thickness in Si varies with beam current [7]. In addition, while significant amounts of Ga were observed in the side-wall damage [8], Ga was not detectable in side-wall damage when Si was FIB milled using gas assisted etching (GAE) [9]. The following study was performed in an attempt to better understand FIB damage. In this study, three square trenches (2x2 μm^2, 4x4 μm^2, and 6x6 μm^2) were FIB milled to 1 μm in depth in a (100) Si wafer using an FEI 200TEM FIB workstation equipped with a Ga liquid metal ion source and an Omniprobe in-situ W probe. An accelerating voltage of 30 keV and a beam current of 1000 pA was used to mill the trenches. The specimen was removed from the FIB and sputter-coated with Cr to preserve the FIB milled damage layers. The specimen was put back into the FIB and the trenches were filled with CVD Pt deposition using a beam current of 100 pA. A cross-section TEM specimen was prepared across the trenches using the in-situ FIB lift-out method [10]. The specimen was observed using a Philips EM430 operating at 300 keV. A bright field (BF) TEM image of the trenches is shown as FIG. 1 (a). A BF image from the top of the wafer is shown in FIG. 1(b). A BF image from the side-wall of the middle trench is shown in FIG. 1(c). Note that the side-wall damage clearly consists of two regions with different contrast. This layer clearly indicates that the side-wall damage consists of two regions: (i) an amorphization layer and (ii) a redeposition layer. X-ray energy dispersive spectrometry (XEDS) results showed that significant amounts of Ga (up to ~ 30 wt%) was observed in the redeposition layer, while the presence of Ga is barely detectable within the limits of the XEDS technique in the amorphization region. The darker region in the redeposition region is consistent with atomic number contrast due to the presence of Ga and Si in the redeposition region. The redeposition region increases toward the mouth of the trench. This is consistent with the cosine distribution of sputtered material from within the trench. The thickness of the amorphization layer on the top of the wafer (~ 70 nm) is thicker than just the amorphization damage on the side-wall (~ 14 nm) which is consistent with ion beam/interactions theory.

Therefore, the amorphization layer may be considered as an intrinsic behavior of the ions interacting with the target. However, any significant amount of Ga (e.g., > 1 wt %) that is observed in the FIB damage region is a redeposition artifact. This is consistent with previous observations whereby Ga was not observed in FIB trenches prepared in conjunction with GAE, since GAE aids in removing redeposition artifacts [9]. In addition, it is noted that no Ga was observed in the TEM specimen using the in-situ lift-out method since any sputtered material may readily be removed by the FIB vacuum system. Since ex-situ lift-out specimens are prepared to electron transparency within a trench, they are more likely to suffer from redeposition artifacts. Thus, we conclude that any Ga observed in TEM specimens may be do to redeposition artifacts [11].

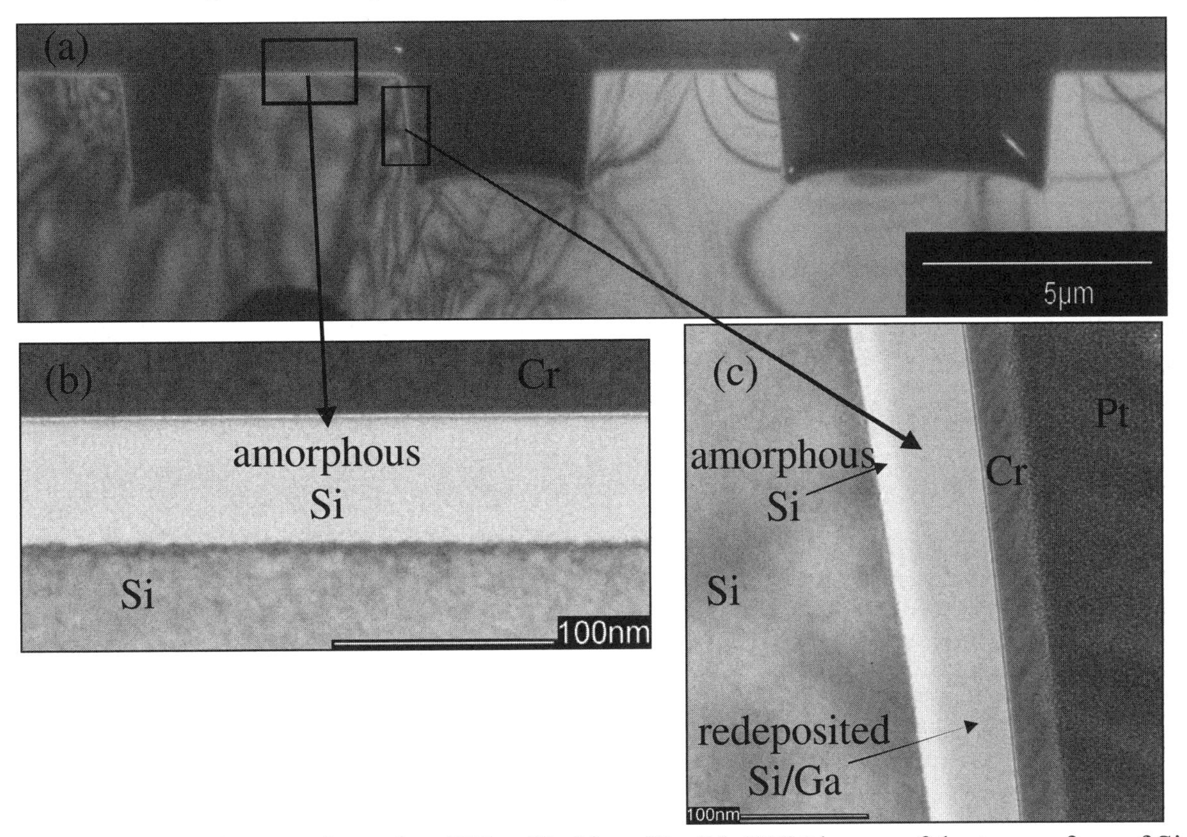

FIG. 1. (a) TEM image of trenches FIB milled into Si. (b) TEM image of the top surface of Si. (c) TEM image of a trench side-wall.

References

[1] L.A. Giannuzzi, et al., Mat. Res. Soc. Symp. Proc., 480 (1997) 19.
[2] F.A. Stevie et al., Surf. Interface Anal. 31 (2001) 345.
[3] H. Cerva et al., Inst. Phys. Conf. Ser 134 (1993) 133.
[4] Abramo et al., Int. Symp. For Testing and Failure Analysis 20 (1994) 439.
[5] B.I. Prenitzer et al., Microsc. Microanal. 6 (Suppl. 2) (2000) 502.
[6] D.W. Susnitzky et al, Microsc. Microanal. 4 (Suppl. 2) (1998) 656.
[7] C.Urbanik Shannon et al., Microsc. Microanal. 5 (Suppl. 2) (1999) 740.
[8] C. Urbanik Shannon , et al., Proceedings of the Second Conference of the IUMAS (2000) 177.
[9] C. Uranik Shanno, M.S. Thesis, University of Central Florida
[10] T. Kamino et al., Microsc. Microanal. 6 (Suppl. 2) (2000) 510.
[11]The support of NSF DMR #9703281 and Omniprobe is greatly appreciated.

Microsc. Microanal. 8 (Suppl. 2), 2002
DOI 10.1017.S1431927602101589

Gallium Phase Formation in Cu and Other FCC Metals During Near-Normal Incidence Ga-FIB Milling and Techniques to Avoid this Phenomenon

M.W. Phaneuf*, J. Li* and J.D. Casey Jr.**

* Fibics Incorporated, 556 Booth St. Suite 200, Ottawa, Ontario, Canada K1A 0G1 (info@fibics.com)
** FEI Company, One Corporation Way, Centennial Park Peabody, Massachusetts, USA 01960-7990

Focused ion beam (FIB) circuit modification of devices with Cu-based interconnect is well known to be problematic [1, 2] due to the vast difference in sputter rate as a function of Cu grain orientation. Measurements of FIB sputter rates on single crystal Cu specimens [3, and Table I] show sputter rate variation of 3.6 times between fast milling orientations such as (111) and slow milling orientations such as (110). This difference in sputter rate poses severe difficulties in uniformly cutting Cu interconnect, but at first glance produces only an annoyance in terms of a requirement of extra milling time to produce TEM specimens by FIB. However, the reasons for this differential sputter rate conceal more alarming issues that need to be taken into account by those producing TEM specimens, particularly as these effects do not seem to be limited only to Cu, but have been observed at Fibics to appear to occur in certain Au and Ni based systems.

It is tempting to attribute this sputter rate difference to ion channeling effects, nevertheless it appears that the slow sputtering of the (110) orientation is not solely due to differences in channeling. In fact this effect may result, to a significant extent, from the formation of an anomalous metal–gallium (M_xGa_y) phase during FIB milling under conditions in which the incident FIB beam hits the specimen at angles far from glancing and closer to normal incidence.

During FIB milling of Cu, "stubborn" grains frequently take on a "dark" appearance when viewed in FIB secondary electron mode; these "dark" grains grow and spread as the ion dose increases. In the case of the single crystal (110) Cu, these "dark" regions first appear at a low dose, and persist and grow throughout the milling operation (Figure 1).

TEM cross-sections of FIB "craters" have been prepared through these dark grains in (110) Cu. These specimens revealed a layer of Cu material rich in Ga, approximately 85 nm in thickness, at the floor of the sputter crater (Figure 2). The electron beam in the TEM was focused to a fine spot and convergent beam electron diffraction (CBED) patterns were obtained from both the substrate Cu and the Ga-rich layer at the crater floor. Analysis of the patterns identified the Ga-rich layer to be Cu_3Ga. (matching JCPDS powder diffraction file 44-1117). This suggests that not only was the (110) crystal slow to mill because of channeling, but also enough Ga was implanting into the sample to transform the bottom floor of the crater into Cu_3Ga, which significantly resisted sputtering.

When examining the edge of the sputter crater in the TEM, it was noticed that the Cu_3Ga phase was not evident until the point at which the crater floor was perpendicular to the incident ion beam (Figure 3). This suggests that the Cu_3Ga phase would only be produced when the incident ion beam was effectively parallel to the (110) surface normal. These are the same conditions where substantial channeling would be expected into pure Cu. These conditions are quite common during FIB circuit edit of Cu interconnect but are rare during conventional FIB-TEM specimen preparation ***unless*** the specimen's protective surface layer (usually FIB-deposited W or Pt) is lost, or the specimen is tilted towards the beam for coarse machining which can occur when performing the "Lift-Out" preparation technique.

A patented and patent pending set of techniques known collectively by FEI as "CoppeRx" [2, 3, 4] can successfully prevent this anomalous phase formation, and care to avoid implantation conditions during TEM specimen preparation will also reduce this problem, but microscopists should be aware of the potential existence of this artifact. Due to the extremely high concentration of Ga present in this phase, the artifact is readily detected by both EDX and diffraction in the TEM. Before panic ensues, it should be noted that Fibics has produced over two hundred TEM specimens of Cu, Ni or Au, rarely observing anomalous phase formation, nor are there references in the FIB literature, so it is clear that standard FIB-TEM specimen preparation is not in jeopardy. This presentation will demonstrate techniques to identify and avoid anomalous gallium phase formation at the early stage of specimen preparation while observing the sample in the FIB system.

References
1. S. Herschbein, L.S. Fisher, T.L. Kane, M.P. Tenney, A.D. Shore, Proc. ISTFA, 127, (1998).
2. J.D. Casey, K.E. Noll, R. Shuman, C. Chandler, M. Megordan, M. Phaneuf and J. Li, LSI Testing Symposium 2000, Osaka, Japan (2000).
3. M.W. Phaneuf, J. Li, R.F. Shuman, K. Noll, J.D. Casey Jr., U.S. Patent Application 20010053605 (March 2000).
4. R.F. Shuman, K. Noll, J.D. Casey Jr., U.S. Patent 6,322,672 B1 (Issued November 2001).
5. The authors wish to thank Dr. Shigeo Saimoto of Queen's University at Kingston, Ontario, for providing the single crystal copper specimens (Dr. Saimoto is supported by Materials and Manufacturing Ontario) and also Dr. Graham Carpenter, Emeritus TEM Scientist, Materials Technology Laboratory (MTL), Natural Resources Canada and Dr. Gianluigi Botton formerly of MTL and now at McMaster University, for many useful discussions and TEM analyses over the years.

	Depth of FIB Crater (μm)		
Trial Number	(1 1 1)	(1 1 0)	(1 0 0)
1	2.46	0.66	1.60
2	2.46	0.69	1.44
3	2.52	0.66	1.60
4	2.41	0.72	1.60
5	2.41	0.66	1.60
Average Depth (μm):	**2.45**	**0.68**	**1.57**
Std. Deviation (μm):	**0.04**	**0.02**	**0.06**

TABLE I

Measured sputter crater depths into electropolished single crystal copper of known orientation for a dose of 5 nC / μm^2 per crater at constant beam current and pixel spacing.

Depth ratios (Proportional to Sputter Rates)

(1 1 1) / (1 1 0) = 3.62 (1 0 0) / (1 1 0) = 2.31

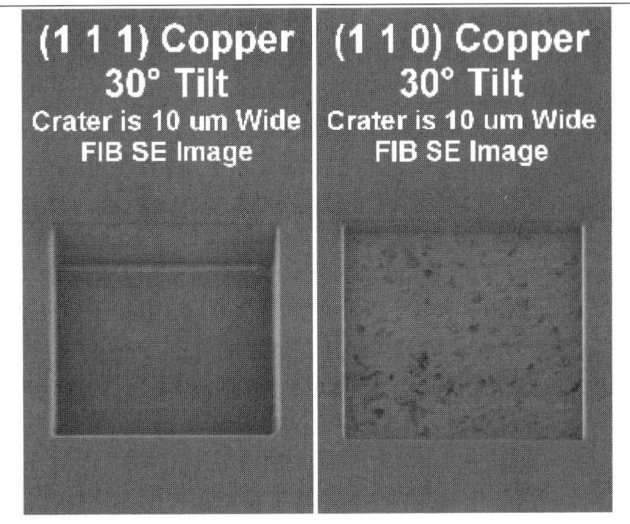

FIG 1. Comparison of Cu single crystals after equal milling. Note the mottled and dark grains in the (1 1 0) crater floor.

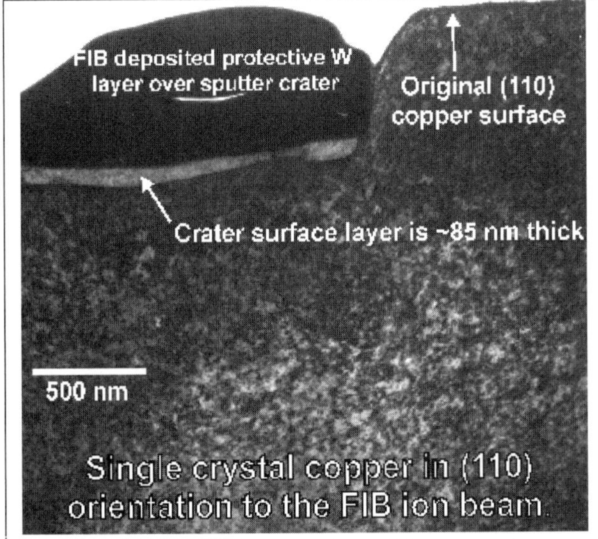

FIG 2. TEM BF image of a FIB-prepared cross-section from the edge of a (110) sputter crater prepared as in FIG 1. A thin layer of gold was sputter deposited on the surface of the crater prior to FIB deposition of tungsten. The ~85 nm thick "surface layer" proved (by CBED) to be Cu_3Ga.

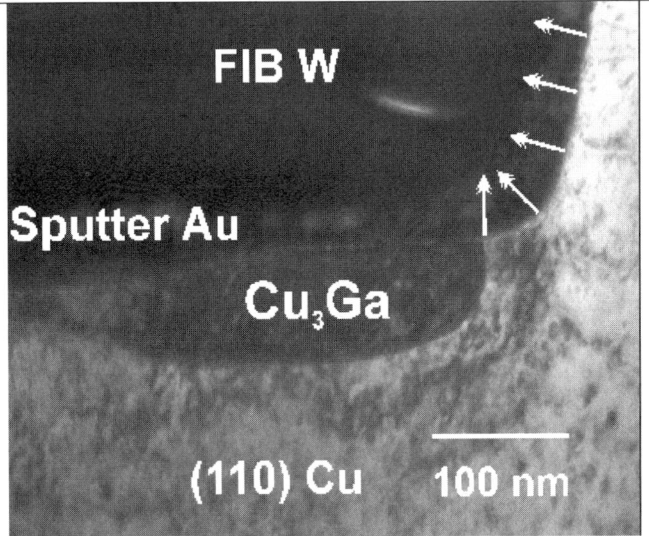

FIG 3. TEM BF image of the crater edge. The protective Au and W coatings preserved the original surface of the crater. Note that the Cu_3Ga phase only begins to form when the surface normal of the crater wall (white, double headed arrows) becomes essentially parallel to the incident FIB Ga beam during sputtering.

Microsc. Microanal. 8 (Suppl. 2), 2002
DOI 10.1017.S1431927602101590

Focused Ion Beam: Much More Than a Sample Preparation Tool

P.E. Russell[1], T.J. Stark[2], J.P. Viterelli[2], A.R. Guichard[1], J. Wang[1], K. L. Bunker[1], J.C. Gonzalez[1], and D.P. Griffis[1]

[1]Analytical Instrumentation Facility, North Carolina State University, Raleigh, NC 27502-7531, [2]Materials Analytical Services, Raleigh, NC 27606

The Ga^+ Focused Ion Beam (FIB) is a very versatile tool which, although established as the preeminent tool for microcircuit edit, has recently evolved into the preferred TEM sample preparation tool for site specific applications. However, the applications of FIB extend far beyond the tasks for which FIB is most commonly employed. FIB can be utilized for micro and nano structural creation and/or modification. This can be accomplished by taking advantage of FIB's capability to very precisely remove material (via physical sputtering or using beam induced chemistry for enhancing material removal rate), to deposit material (using beam induced chemistry), to provide localized ion implantation, and to locally induce sample structural damage. In addition to the above capabilities, FIB provides the ability to image the sample via secondary electrons or ions before, after and during micromachining. In many cases, the ability to image while removing or depositing material provides invaluable feedback for process evaluation and control.

In this presentation, an overview of FIB will be presented with emphasis on non traditional applications. Examples presented will include a wide range of FIBbased nano and micro fabrication techniques including stamp/mold fabrication (Fig. 1), shaping of a diamond indenter tool (Fig. 2), AFM tip characterization structure fabrication (Fig. 3), sharpening of an STM tip (Figure 4), MEMS device cross section (Fig. 5), etc.

In some instances where micro or nano material removal or deposition is required, it is possible to perform these tasks using an electron rather than an ion beam. This can be especially advantageous when the Ga "stain" left behind during allFIB processes is deleterious to the desired result e.g. during EUV mask repair. The electron beam when use in conjunction with appropriate chemical precursors, can initiate chemical reactions resulting in either material removal or material deposition. Examples will be presented showing the ability to precisely remove and deposit materials using an electron beam.

In summary, FIB is a versatile tool that combines both nanofabrication and microscopy capabilities.

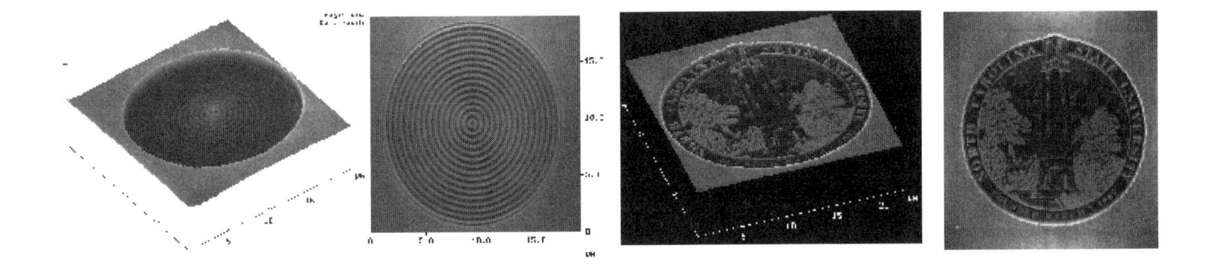

Fig. 1. SEM and AFM images of a micro fabricated Fresnel lens and the NCSU seal. These micro fabricated structures could be used for either stamps or molds.

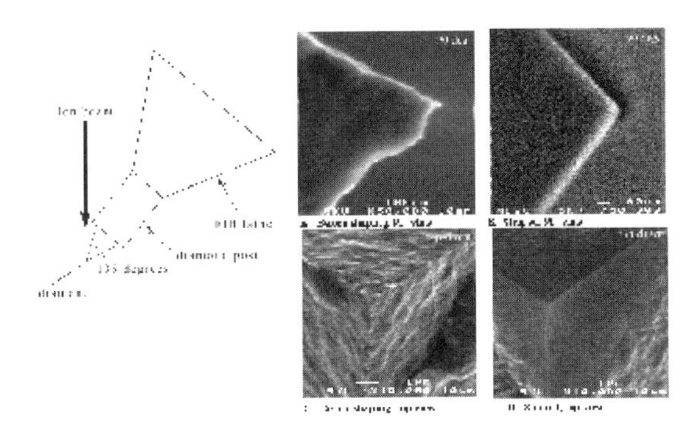

Fig 2. SEM micrographs of a diamond indenter tip before and after shaping using FIB

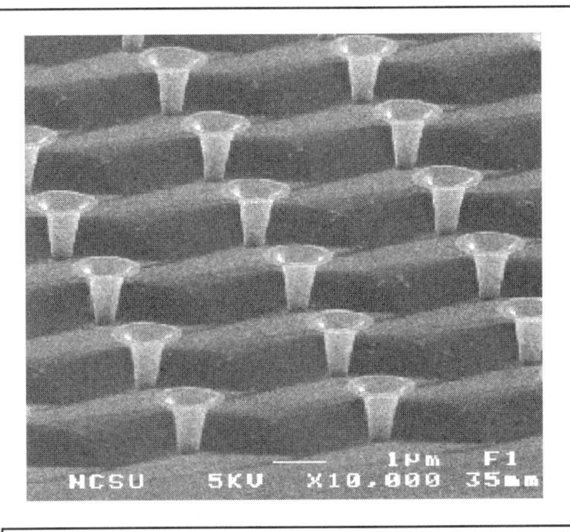

Fig 3. SEM micrographs of a structure fabricated using FIB implantation in Si followed by etching.

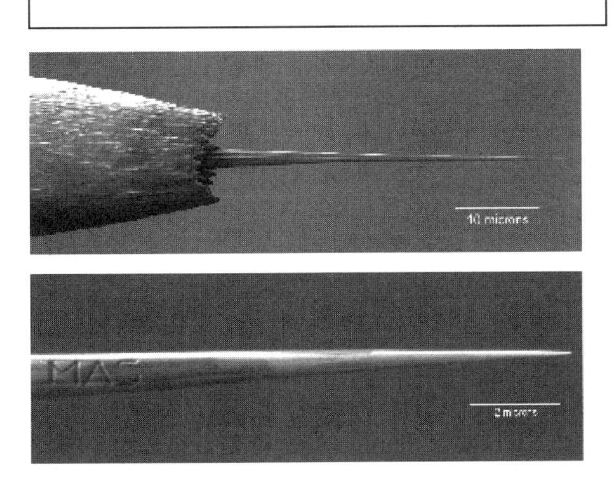

Fig 4. SEM micrographs of an STM tip sharpened using FIG. Note inscription on tip in lower image.

Fig 5. SEM micrographs of a MEMS motor after FIB cross sectioning.

Microsc. Microanal. 8 (Suppl. 2), 2002
DOI 10.1017.S1431927602101607

Dual-beam Focused Ion Beam: A Multifunctional Tool for Nanotechnology

N. Yao, E. Kung, S. Allameh, W. O. Soboyejo

Princeton University, Princeton Materials Institute, Princeton, NJ 08540

New materials are now being fabricated with greater geometric complexity and smaller feature sizes. Such complexity is inherent to most biomaterials and their synthetic analogues, as well as to nano technology. Beyond complexity, phenomena occurring on even smaller length scales often adversely affect performance and reliability. Foremost among these phenomena are interface properties within the composite materials, resulting in chemical and structural gradients. Understanding and improving the performance and durability of these systems requires high resolution structural, chemical and geometric analysis of cross sections through the layers. The ability to perform such analysis on precisely located cross sections is one of the key attributes of the dual beam focused ion beam (FIB). This FIB system has been developedto combine the imaging capabilities of a typical scanning electron microscope (SEM) with the imaging and fabrication capabilities of a highly focused ion beam. As focused ion beam technology has advanced, focused ion beam systems have been gaining more attention as an alternative method for micro- and nano-scale machining and fabrication. Presently, the preparation of sections by mechanical polishing, dimpling and ion milling is either excessively time consuming or possible only with highly trained metrology experts or, in some cases, totally impossible. In other cases, the compliance of the constituents prevents the formation of undamaged/undistorted cross sections: certain shell-like biological structures exemplify this problem. The FIB system obviates these problems and also provides unique high precision etching and deposition for micromanipulation as well as imaging and analysis. This instrument provides unique capabilities for advanced materials research connected to nanoscale, biological, photonic and multifunctional materials.

The dual beam FIB offers unprecedented capability when new understanding requires that imaging be performed on cross sections made at precise geometric locations. This is essential at sites where failures are known to originate. Additionally, it may be used to cross-section through the exact center of an impression, or along planes parallel to a set of microstructural features. Standard methods are incapable of preparing cross sections with the requisite spatial precision. A number of novel imaging capabilities, recently developed for the SEM, may be used on cross sections created by the FIB to locate and identify the defects and microstructural features of interest. These imaging modes include Electron Beam Induced Current (EBIC), Orientation Imaging Microscopy (OIM) and Cathodoluminescence (CL). Each of these capabilities can be interfaced with the imaging column of the dual-beam FIB system.

The use of FIB technology has seen much application in research but its practical use in industry is yet to be determined. Challenges arise when the FIB is applied on the large scale because of the high doses that would be required and the relatively long fabrication times compared to equivalent lithography techniques. Efforts have been made to further develop FIB technology beyond current limitations to increase the usable resolution of the instrument so that work may be done around the 10 nm range reproducibly [1]. Investigations into its inspection, metrology, and failure analysis functions demonstrate its capability to self-diagnosis [1,2]. The direct write capability of the FIB has been applied using several different methods and compounds in order to cause deposition onto the surface [3,4]. Deposition using the FIB occurs through the introduction—by a gas feed system—of a cloud of atoms directly above the surface of the sample in the region of ion bombardment, a process know as chemical vapor deposition (CVD). Ion interaction volumes in the sample surface are similar to those of electrons and are important to understanding how the surface morphology can be modified by impacting ions and the energy thus imparted. While the energy of the incoming ion beam could cause the sputtering of atoms from the surface, it could also be used to "push" atoms, near the sample surface, onto the surface, creating a deposited feature. Direct-write capabilities and more precise and accurate FIBs offer the possibility of creating usable nanoscale structures controllably and reproducibly. This paper presents some of the potential of the dual beam FIB as well as of the FIB as a general, multifunctional tool.

References

[1] J. Gierak et al., *Microel. Eng.* 57-58 (2001) 865-875.
[2] S. Reyntjens, R Puers, *J. Micromech. Microeng.* 11 (2001) 287-300.
[3] J. Melngailis, *J. Vac. Sci. Technol. B* 5 (1987) 469-495.
[4] T. Chikyow, *Surf. Sci.* 386 1-3 (1997) 254-258.

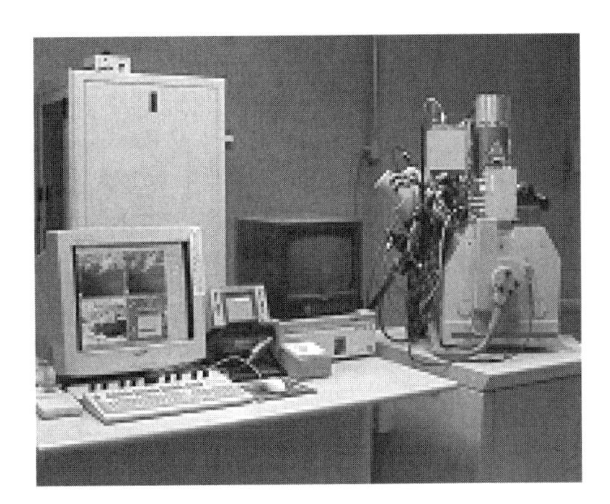

FIG. 1. The FEI Strata™ DB 235 at the Princeton Materials Institute incorporates both a focused gallium ion beam as well as a scanning electron microscope. The SEM column is above the chamber while the ion beam column is attached towards the side at 52° angle. The precision provided by the exactness of the ion beam machining and the high resolution imaging of the SEM provide the ability to prepare various samples and to perform *in situ* fabrication with observation.

Microsc. Microanal. 8 (Suppl. 2), 2002
DOI 10.1017.S1431927602101619

The DualBeam FIB in a Materials Science Laboratory

Hendrik O. Colijn

Campus Electron Optics Facility, The Ohio State University, 116 W. 19th Ave. Columbus, OH 43210

Focused ion beam instruments have been widely used in the semiconductor industry for sample preparation [1] and are also beginning to show their usefulness in materials science laboratories. The DualBeam FIB/SEM instrument provides many new opportunities for not only sample preparation, but also materials analysis in the materials science environment.

The DualBeam FIB, with the built-in electron column, provides much more flexible preparation possibilities than the single-beam FIB. One useful feature is the ability to directly measure the endpoint of sample thinning by using the penetration of the electron excited volume through the backside of the thin foil. When this occurs, the secondary electron image becomes much brighter and loses contrast (Fig. 1). If we use a 5kV primary electron beam, the penetration depth is quite small and the foil is sufficiently thin to view in the TEM. This method works regardless of the atomic number and density of the material and is particularly helpful for materials with higher atomic number than Si.

The electron imaging function of the Dual-Beam system also makes possible the "Slice and View" method of visualizing the 3-D structure of a sample. With a single beam instrument, serial sectioning, while possible, would be extremely tedious. By taking electron images of the face of a trench as it is being milled into the sample (Fig. 2.), we can serially reconstruct the 3-D structure (Fig. 3.) [2]. We can correlate the surface and depth features by imaging both the surface and milled faces. Because the ion and electron columns are 52° apart, tilting the sample surface to 38° allows us to use the same tilt correction for both the sample and cut surface.

Metals show more defects due to ion damage than do most semiconductor materials. If the ion beam creates a large number of defects, it becomes difficult to separate the defects of interest from those created by the ion beam in both lift-out and H-bar samples (Fig. 4 & 5). TRIM calculations suggest that the ion damage layer due to both forward and lateral scattering is much less than the TEM sample thickness [3]. At this time it seems that the best remediation is to prepare samples using the H-bar technique followed by a brief low-angle low energy ion polish.

The addition of a STEM detector to the Dual-Beam can reduce the number of samples that go into the TEM. Although the resolution is not as good as a standard materials science TEM, for analytical work it may be sufficient to solve many of the problems which cannot be handled in a regular SEM. Also, by judicious use of milled trenches, one can improve the EDX resolution without having to lift the membrane out of the surface (Fig. 5).

References

[1] L.A. Giannuzzi, F.A. Stevie, *Micron* 30 (1999) 197.
[2] reconstruction courtesy of Mike Uchic, Wright-Patterson AFB.
[3] B. Prenitzer, Ph. D. dissertation, University of Central Florida, 1999, p. 169f.

FIG. 1. Electron beam penetration of the thin membrane.
FIG. 2. Section of an alpha/beta Ti alloy .
FIG. 3. Serial reconstruction of gamma prime precipitates in a Ni superalloy.
FIG. 4. Ion damage in alpha Ti using the lift-out technique.
FIG 5. Ion damage in alpha Ti using the H-bar (traditional) technique.
FIG. 6. In-situ membrane technique for improved x-ray resolution.

Microsc. Microanal. 8 (Suppl. 2), 2002
DOI 10.1017.S1431927602101620

Advances in Dual Beam TEM Sample Preparation

M.V. Moore

FEI Company, 7451 NW Evergreen Parkway, Hillsboro, OR 97124

Today's samples present many challenges for transmission electron microscope (TEM) sample preparation. Focused ion beam (FIB) technology techniques continue to advance to provide solutions to meet the challenges due to new materials, structures of shrinking dimensions, and new TEM applications. While the FIB provides the advantages of site-specificity and wide material applicability to TEM sample preparation, a dual beam instrument with its added electron column gives expanded capabilities (Fig.1). The electron column adds higher-resolution endpointing along with non-destructive imaging and surface protection of the sample (via electron-beam deposition); having coincidence of the ion and electron beams gives the operator immediate feedback about the sample being prepared.

The dual beam also enables the use of electron-beam-deposited Pt (from an organo-metallic precursor) for surface-sensitive samples. This type of deposition grows conformally over high aspect ratio structures and provides good contrast with many materials because of its high carbon content. This coating is useful, for example, in examining Cu seed layers for process development. Sub-100 nm-thick TEM membranes of Cu seed structures have been made automatically with recently enhanced automation software. Automated TEM sample preparation provides the advantage of reproducibility as well as increase in throughput to process development.[1]

Semiconductor design rules are approaching 100 nm (with gate lengths half that size); this increases the challenge in endpointing on these structures, centering ever-thinner TEM membranes on interconnects, and making thinner samples of increasing quality for high-resolution imaging. To improve the quality of TEM membranes for high-resolution work, low kV (Ga ion) beam cleaning can be used to reduce the amorphous layer of Si to enhance lattice imaging, as shown in Fig. 2.[2] The shrinking dimensions of structures to be cross-sectioned has stimulated further development of this technique. For sub-50 nm TEM cross-sections, low kV-assisted milling is necessary to avoid amorphizing the Si substrate during the milling process.

The relatively new application of TEM holography to dopant profiling of semiconductors presents demanding sample requirements. The sample must be very uniform in thickness (in the 200-300 nm range), have an edge or hole in the same field of view as the area of interest (vacuum area for the reference beam), and be free from charging. Fig. 3 shows such a sample of a semiconductor transistor, that was successfully made in a dual beam using the high resolution of the ion beam for precision milling and the electron beam for endpointing and surface-quality monitoring.[3]

[1] R.J. Young, *Microsc. Microanal.* 6 (Suppl. 2) (2000) 512.
[2] R.J. Young et al., *Proc. ISTFA* (1998) 329.
[3] The author gratefully acknowledges assistance of FEI colleagues Sean Da, Jay Jordan, YC Wang and Richard Young, and collaboration with C. Kisielowski of LBL and S. Rozeveld and E.Beach of DOW Chemical. *SiLK is a trademark of the DOW Chemical Company.

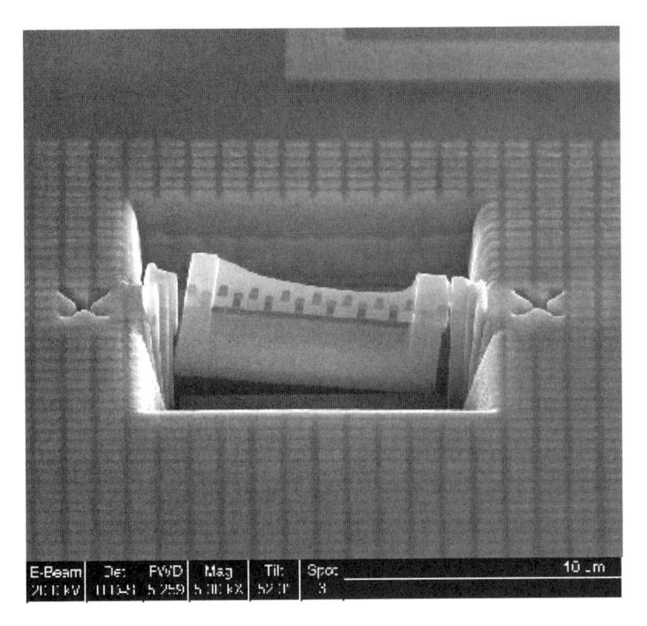

FIG. 1. Dual beam-prepared TEM sample of Cu damascene structure with SiLK*, a low-k dielectric. (Sample, courtesy IMEC. Image, S. Rozeveld and E. Beach)

FIG. 2. Low kV Ga-ion cleaned Si lattice; image size is 20 nm (M.V.Moore, X. Xu, C. Kisielowski)

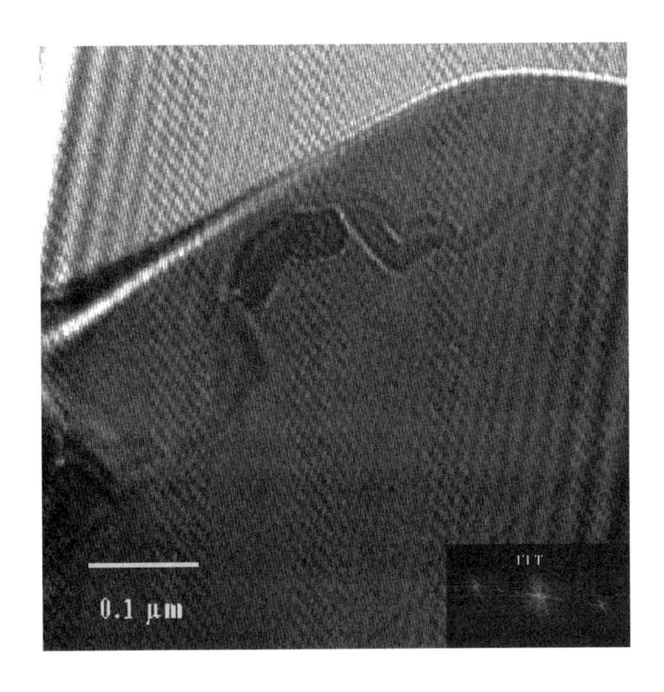

FIG. 3a. Hologram of semiconductor transistor.

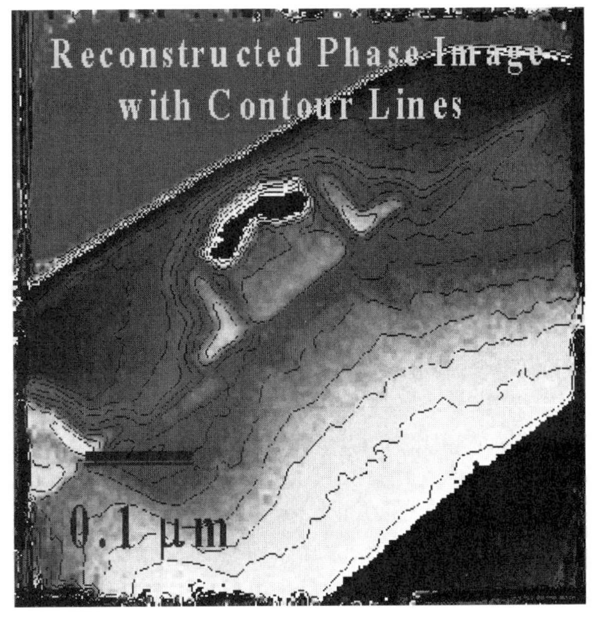

FIG. 3b. Reconstructed phase image of transistor region, useful for dopant profiling.

Microsc. Microanal. 8 (Suppl. 2), 2002
DOI 10.1017.S1431927602101802

Probing the Electronic Structure of Transition Metal Oxides using Electron Energy-Loss Spectroscopy

D.W. McComb

Department of Chemistry, University of Glasgow, Glasgow, G12 8QQ, Scotland, UK

Electron energy-loss spectroscopy (EELS) carried out in the nano-analytical electron microscope (nano-AEM) is a powerful probe of chemistry, bonding and electronic structure in a wide range of materials. Analysis of energy-loss near-edge structure (ELNES) present on ionization edges in EELS is now recognized as the only technique that can be used to obtain such information with close to atomic-scale spatial resolution. The information on coordination number, local symmetry and oxidation state derived from ELNES data can provide key insights into chemical and structural inhomogeneities in advanced materials. However the electron-specimen interactions that result in ELNES are complex. In order to develop and advance the use of the technique it is important to combine fundamental investigations of the ELNES in known structures with detailed theoretical modeling of the electron-specimen interactions.

This report will focus on the analysis and interpretation of the oxygen K-ELNES in a range of technologically relevant materials. Experimental ELNES data from complex ternary structures such as yttria-stabilized zirconia and ternary oxides with the spinel structure clearly demonstrate that the anion edges can be used to obtain information about BOTH the anion and cation sites. The oxygen K-edges in a range of ternary spinels as the cation the octahedral and tetrahedral site is varied systematically are shown in Figure 1 [1]. Similar results have been obtained in the study of the oxygen K-edge in zirconia powders doped with varying amounts of aliovalent dopants such as Y_2O_3 (Fig. 2) [2]. While it is clear that the oxygen K-ELNES is influenced by altering the electronic structure of neighboring cations the strength of the interaction is not easily predictable – it is a sensitive function of the ion type, the crystallography of the site and local coordination environment.

Previously such data has been interpreted in terms of a fingerprint for the local coordination environment. To go beyond such an interpretation it is necessary to perform electronic structure calculations to model the observed ELNES. Such calculations are an essential step towards a complete interpretation of the experimental data. However, in performing such calculations it is crucial to use a realistic model for the structure under investigation. In the case of Y_2O_3-doped ZrO_2 it is not only crucial to include vacancies and dopant atoms in the model but also to identify the correct atomic positions by permitting lattice relaxations to occur (Fig. 2) [3, 4]. In the case of the ternary spinels the influence of the degree of inversion and the effect of electron correlation on the band structure calculations must be considered (Fig. 3) [5].

References:

[1] F. T. Docherty et al., *Ultramicroscopy* **86** (2001) 273-288.
[2] D. Vlachos et al., *J. Phys Cond. Matter* **13** (2001) 10799.
[3] S. Ostanin et al., *Phys. Rev. B* **62** (2000) 14728-14735
[4] S. Ostanin et al., *Phys. Rev. B* (2002 – In Press)
[5] L. Chioncel (Private communication)
[6] This research was supported by the EPSRC, Johnson Matthey and MEL Chemicals.

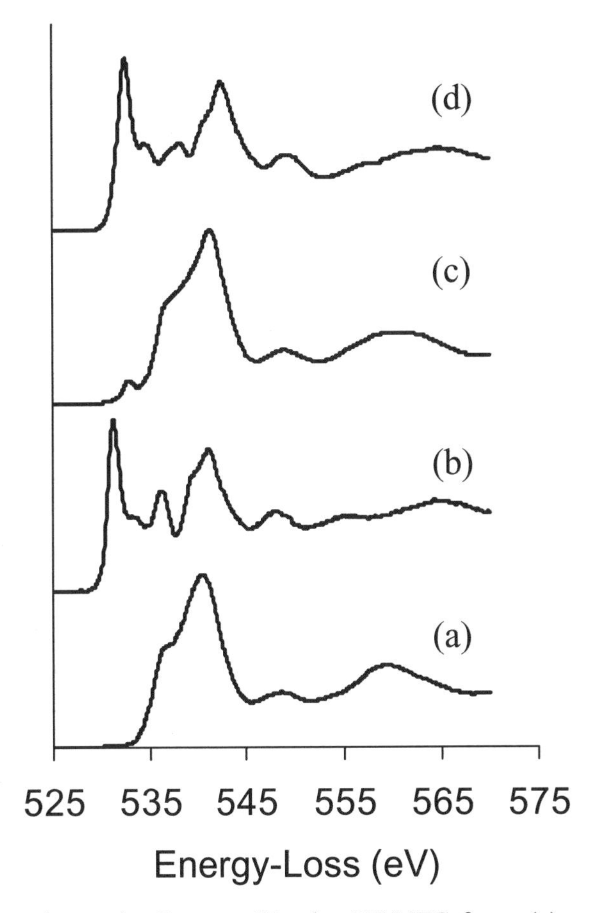

Figure 1. Oxygen K-edge ELNES from (a) $MgAl_2O_4$, (b) $MgCr_2O_4$, (c) $NiAl_2O_4$ and (d) $NiCr_2O_4$ to show the effect of systematically changing the tetrahedral (A-) and octahedral (B-) cations in the AB_2O_4 structure.

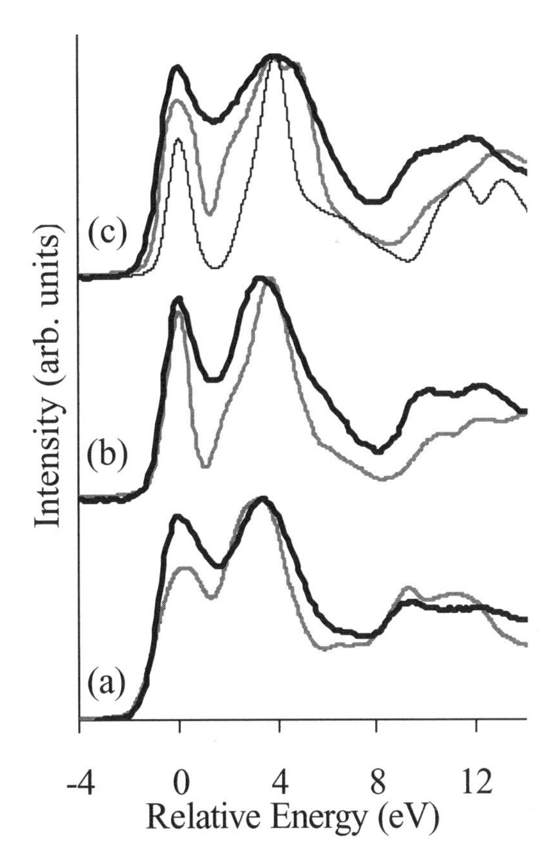

Figure 2. Oxygen K-edge ELNES from (a) pure ZrO_2, (b) 3mol% Y_2O_3-ZrO_2, (c) 10mol% Y_2O_3-ZrO_2. The calculated ELNES obtained after atomic relaxation is shown in each case (grey). In (c) the result of electronic structure calculations without any lattice relaxation is shown (thin black line).

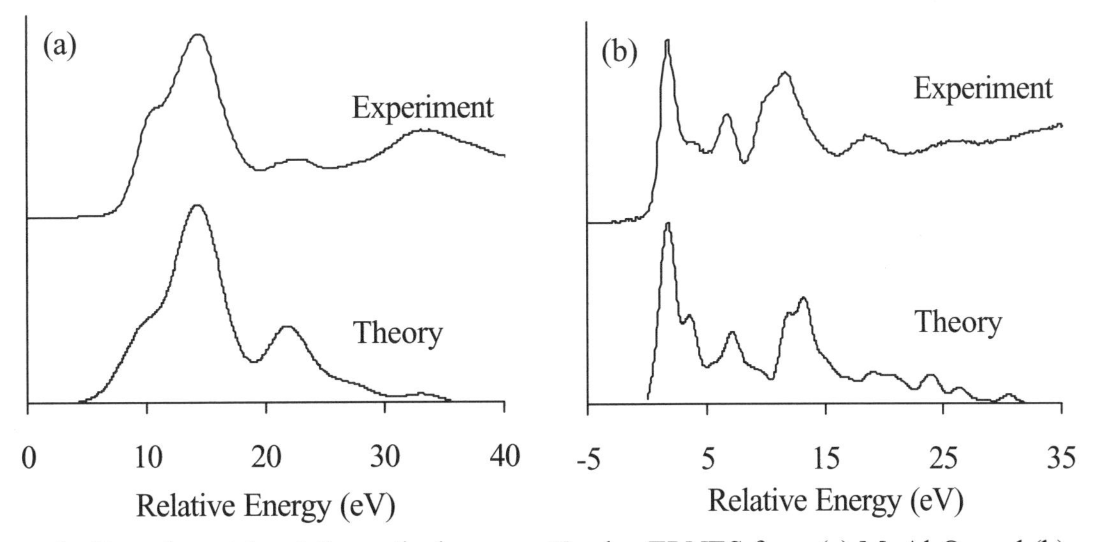

Figure 3. Experimental and theoretical oxygen K-edge ELNES from (a) $MgAl_2O_4$ and (b) $MgCr_2O_4$. Both calculations were performed within the local density approximation but in the case of (b) spin polarisation was required in order to obtain reasonable agreement with experiment.

Microsc. Microanal. 8 (Suppl. 2), 2002
DOI 10.1017.S1431927602101814

Structures and Energetics of Interfaces in Materials – Ab-initio Local-Density-Functional Theory –

Christian Elsässer

Max-Planck-Institut für Metallforschung, Seestrasse 92, D-70174 Stuttgart, Germany.

Ab-initio density-functional theory of computational condensed-matter physics provides microscopic insights into structural and functional properties of real materials with high accuracy and predictive power. Topic of this contribution is the ab-inito mixed-basis pseudopotential approach [1] for fundamental theoretical investigations of energetic stabilities and microscopic structures of internal boundaries in crystalline materials. The method is based on the density functional theory of inhomogeneous electron gases, with exchange and correlation treated in the local-density approximation (LDFT) [2]. Interfaces and other structural defects of crystals are modelled by means of supercells with periodic boundary conditions. The interactions of atomic nuclei and closed-shell core electrons with electrons in valence and conduction bands are described by norm-conserving pseudopotentials. The valence-band and conduction-band one-electron states are represented by a mixed basis of plane waves and localised functions.

With this theoretical approach, microscopic quantities are accessible for comparison, interpretation and augmentation of experimental observations by transmission electron microscopy (TEM). The formalism yields total energies and atomic forces for structure optimisations, for a quantitative comparison with experimental high-resolution images by TEM, as well as energies and wavefunctions of one-electron states, for an interpretation of experimental electron excitation spectra in TEM. Concerning the latter, calculated site- and angular-momentum-projected local densities of conduction-band states were found useful for a site-specific analysis of experimental energy-loss near-edge structures (ELNES) in core-electron ionisation-edge [3,4]. The capabilities of the approach will be illustrated by selected results from two representative case studies of grain boundaries in ceramics and of metal/ceramic contacts, respectively.

For α-Al_2O_3 as a paradigm ceramic material, four symmetrical tilt grain boundaries were studied: basal $\Sigma3$ (0001) [5] and prismatic $\Sigma3$ $(10\bar{1}0)$ [6] twins, rhombohedral $\Sigma7(\bar{1}012)$ [7] and pyramidal $\Sigma13(10\bar{1}4)$ [8] twins. The investigated properties are geometric translation states and local atom arrangements, adhesion energetics and electronic structures. The theoretical results are related to experimental TEM observations for the same twin interfaces in bicrystals of α-Al_2O_3 produced by diffusion bonding in ultra-high vacuum [9]. For the rhombohedral $\Sigma7(\bar{1}012)$ twin, the preferred microscopic interface structure was obtained consistently from LDFT, HRTEM and ELNES [7,9], validating the synergistic approach of ab-initio theory and experimental TEM imaging and spectroscopy. In the latter, the achieved spatial and energetical resolutions were approximately 2 nm and 1 eV, respectively. For the prismatic $\Sigma3$ $(10\bar{1}0)$ twin, the LDFT calculations yielded two competing interface-stucture variants [6]. For the discrimintation of these by experimental ELNES, it is predicted [6] that a spatial resolution of about 0.2 nm will be required, recommending this twin interface as a promising benchmark case for atom-column-resolution spectroscopy in TEM.

The adhesion energetics, atomistic and electronic structures of thin Pd and Mo films on $SrTiO_3$ (001) substrates were studied [11-13]. Both Pd and Mo are preferably bound on top of O on the substrate surface terminated by a TiO_2 layer. Non-reactive Pd films tend to form atomically abrupt interfaces [11,12]. For Mo, because of its high O affinity, theoretical evidence was found for an energetic instability of the pure film towards the formation of an interfacial oxide reaction layer [13]. The microscopic origins for the different behaviours were analysed in terms of real-space bonding electron densities, local densities of states, orbital bond orders and bond strengths. Experimental TEM results of interfacial O-K ELNES [14] for the same metal/ceramic contacts produced by molecular beam epitaxy [15] are used for comparison and discussion of the theoretical findings.

References

[1] C. Elsässer et al., J. Phys.: Condens. Matter 2 (1990) 4371; C. Elsässer, doctoral thesis, Universität Stuttgart (1990); K. M. Ho et al., J. Phys.: Condens. Matter 4 (1992) 5189; B. Meyer, doctoral thesis, Universität Stuttgart (1998); B. Meyer et al., Fortran90 Program for Mixed-Basis Pseudopotential Calculations for Crystals, Max-Planck-Institut für Metallforschung Stuttgart (unpublished).
[2] P. Hohenberg and W. Kohn, Phys. Rev. 136 (1964) B864; W. Kohn and L. J. Sham, Phys. Rev. 140 (1965) A1133.
[3] S. Köstlmeier and C. Elsässer, Phys. Rev. B 60 (1999) 14025.
[4] C. Elsässer and S. Köstlmeier, Ultramicroscopy 86 (2001) 325.
[5] A. G. Marinopoulos et al., Phys. Rev. B 63 (2001) 165112.
[6] S. Fabris et al., submitted to Phys. Rev. B (2001).
[7] A. G. Marinopoulos and C. Elsässer, Acta mater. 48 (2000) 4375.
[8] S. Fabris and C. Elsässer, Phys. Rev. B 64 (2001) 265117.
[9] S. Nufer, doctoral thesis, Universität Stuttgart (2001).
[10] S. Nufer et al., Phys. Rev. Lett. 86 (2001) 5066.
[11] T. Ochs, doctoral thesis, Universität Stuttgart (2000).
[12] T. Ochs et al., Integrated Ferroelectrics 32 (2001) 959.
[13] T. Classen, diploma thesis, Universität Stuttgart (2001).
[14] K. van Benthem, doctoral thesis, Universität Stuttgart (2002).
[15] T. Wagner et al., J. Appl. Phys. 89 (2001) 2606.
[16] Gratefully I acknowledge the substantial contributions of my present and former coworkers at the MPI-MF Stuttgart to these theoretical investigations: S. Fabris, A. Marinopoulos and S. Nufer for the twin boundaries in α-Al_2O_3, T. Classen, T. Ochs and K. van Benthem for the Pd/$SrTiO_3$ contacts, and S. Köstlmeier for the LDFT approach to ELNES. I thank Professor M. Rühle for his continuous interest and encouragement in the theoretical studies. The projects were supported by the Deutsche Forschungsgemeinschaft (Grants No. 155/4-1 and No. EL155/7-1, the latter within the Priority Program Wetting and Structure Formation at Interfaces).

Microsc. Microanal. 8 (Suppl. 2), 2002
DOI 10.1017.S1431927602101826

Electron Energy-Loss Spectroscopy of Alternative Gate Dielectric Stacks

S. Stemmer*, D. Klenov*, Z. Chen*, J.-P. Maria**, A. I. Kingon**, D. Niu***, G. N. Parsons***

* Department of Mechanical Engineering and Materials Science, Rice University, Houston, TX 77005-1892
** Department of Materials Science and Engineering, North Carolina State University, Raleigh, NC 27695-7919
*** Department of Chemical Engineering, North Carolina State University, Raleigh, NC 27695-7905

Continued scaling of silicon technology (Moore s Law) requires a paradigm shift in the materials used as gate dielectric in complementary metal-oxide-semiconductor (CMOS) devices. In the near future, the SiO_2 thickness is projected to be thinner than 1 nm. In this thickness range tunneling currents through the SiO_2 become unacceptably high. Currently, alternative dielectrics, such as ZrO_2, HfO_2, Y_2O_3 and their alloys with SiO_2 or Al_2O_3, a re being investigated to replace SiO_2. These oxides have greater dielectric constants (*k*) than SiO_2, and are potentially stable in contact with silicon. High-resolution analytical capabilities afforded by scanning transmission electron microscopy techniques are essential in analyzing the interface and bulk stability of these ultrathin (< 5 nm) layers.

Here, we apply high-resolution electron energy-loss spectroscopy in combination with atomic resolution Z-contrast imaging to investigate the stability of alternative gate dielectric layers at high temperatures, and under reducing and oxidizing conditions. Z-contrast images are used to image the chemical homogeneity of the layers and to position the probe for electron energy-loss spectroscopy (EELS). EELS is used to measure composition and bonding across the gate dielectric with sub-nanometer spatial resolution. We use oxygen K-edges and Si L-edges to investigate interfacial SiO_2 and silicate formation. In addition, the near-edge fine-structure of these edges and comparison with bulk reference spectra are used to fi ngerprint phase formation and nonstoichiometry. Conventional high-resolution transmission electron microscopy is used to investigate crystallization.

Using the combination of these methods we are able to determine the stability of these ultrathin layers with *a-priori* unknown structures. For example, we show that ZrO_2/Si layers annealed under moderately oxidizing conditions (oxygen partial pressure > ~ 10^{-4} torr), form a low-*k* interfacial SiO_2 layer through oxygen diffusion through the ZrO_2 and silicon consumption at the interface. Layers annealed under moderately reducing conditions (oxygen partial pressure ~ 10^{-5} torr) do not show extensive SiO_2, formation, whereas layers annealed under even lower oxygen partial pressures (~ 10^{-7} torr) form an interfacial silicide, consistent with predictions from thermodynamic estimates. In contrast to ZrO_2 layers, CVD grown Y_2O_3 films show extensive silicate formation upon annealing, through Si diffusion into the dielectric. We show that thin films transform to an amorphous yttrium silicate upon annealing, whereas thicker films form an interfacial silicate and crystalline Y_2O_3 on the surface. We will discuss possible mechanisms, in particular the role of crystallization and Si diffusion to explain the observed results. We also show that pre-nitridation of the Si surface impedes the Si diffusion.

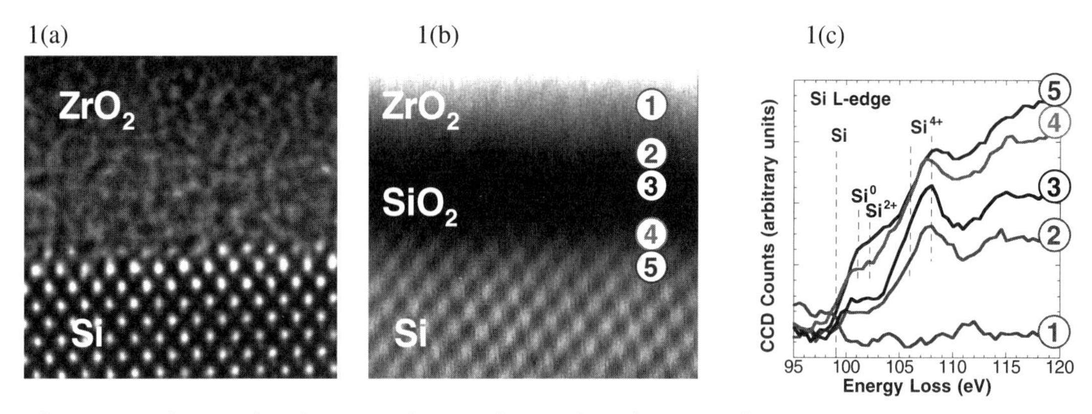

Figure 1: (a) Conventional HRTEM image of an as-deposited ZrO_2 film, (b) Z-contrast image and (c) EELS spectra through the thickness of the gate stack. No Si can be detected in the ZrO_2 layer. Note the atomic number sensitivity of the Z-contrast image that clearly shows the interfacial SiO_2.

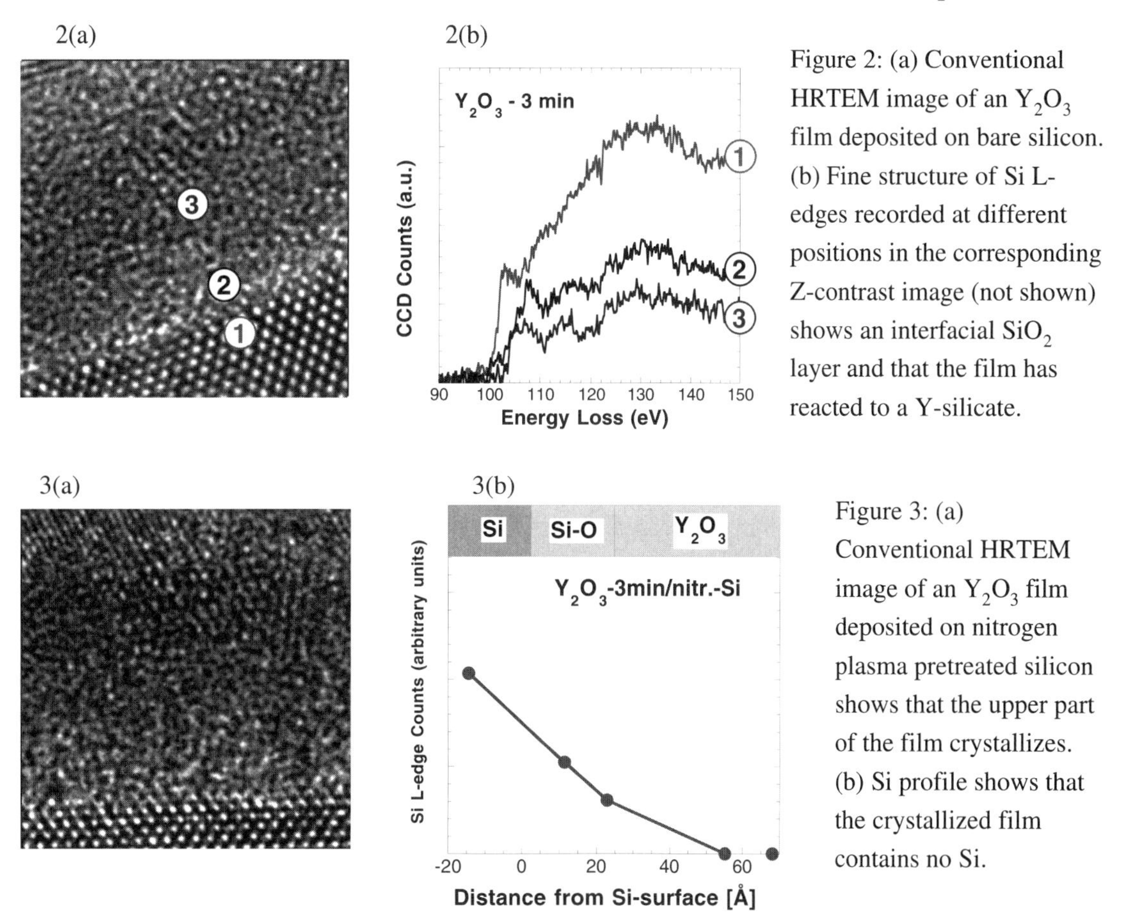

Figure 2: (a) Conventional HRTEM image of an Y_2O_3 film deposited on bare silicon. (b) Fine structure of Si L-edges recorded at different positions in the corresponding Z-contrast image (not shown) shows an interfacial SiO_2 layer and that the film has reacted to a Y-silicate.

Figure 3: (a) Conventional HRTEM image of an Y_2O_3 film deposited on nitrogen plasma pretreated silicon shows that the upper part of the film crystallizes. (b) Si profile shows that the crystallized film contains no Si.

This research was supported by the SRC/Sematech Front End Process Center.

Microsc. Microanal. 8 (Suppl. 2), 2002
DOI 10.1017.S1431927602101838

Development of An 0.2eV Energy Resolution Analytical Electron Microscope

M. Tanaka, M. Terauchi, K. Tsuda, K. Saitoh, M. Mukai[*], T. Kaneyama[*], T. Tomita[*], K. Tsuno[*], M. Kersker[**], M. Naruse[*], and T. Honda[*]

Institute of Multidisciplinary Research for Advanced materials, Tohoku University, Sendai, Japan
[*] JEOL LTD., 1-2 Musashino-3 Akishima, Tokyo, Japan
[**]JEOL USA, INC., 11 Dearborn Road, Peabody, MA 01960

We developed a high energy-resolution electron energy-loss spectroscopy (EELS) microscope (JEM-HREA80) to investigate detailed electronic structures of materials [1,2]. The microscope is equipped with an octapole-type Wien-filter monochromator and analyzer. The energy and spatial resolutions were 50meV-0.2eV and 30-100nm in diameter, respectively. On the other hand, we developed an Ω-filter electron microscope (JEM-2010FEF), which enables us to perform precise crystal structure refinement in a sub-nanometer scale using the CBED method [3,4]. Since it does not have a monochromator, its energy resolution of EELS spectra remains at about 1eV, which is not sufficient for the detailed study of electronic structures of materials.

We have started to manufacture a new 200kV electron microscope under a project "MIRAI-21", which enables us to investigate both crystal- and electronic-structures of advanced materials in nanometer scale areas [5]. MIRAI means "Future" in Japanese and the abbreviation of Microscope for Innovative Research and Advanced Investigation. The microscope has a point resolution of 0.19nm and an energy resolution of 0.2eV at a less than 2nm diameter probe. In this paper, the basic design of the microscope and the test result of the monochromator are reported.

Figure 1 shows the appearance of the MIRAI-21 microscope. This microscope is constructed based on the JEM-2010FEF, and equipped with a newly developed Wien-filter monochromator and an improved Ω–filter analyzer. The monochromator is located between the extraction anode for the ZrO/W emitter and the acceleration tube. The monochromator consists of two octapole-type Wien-filters (Fig.2) and an energy-selection slit, which exists between the two filters. . Astigmatic focus is used to reduce the Boersch effect, as shown in Fig. 3. The filter 1 (WF1) disperses the incident electron beam and forms a line-focused image on the energy selection slit. The filter 2 (WF2) cancels out the energy dispersion of WF1 and forms a stigmatic electron beam at the exit of the monochromator.

The basic performance of the monochromator was tested using a remodeled 120kV-type electron microscope with a LaB_6 filament. Figure 4(a) shows the astigmatic beam shape, which elongated in the direction, perpendicular to the energy-dispersion direction at the energy-dispersion plane (the slit position). The electron beam energy ingoing to the monochromator was 800eV. The electron beam was shifted about 70μm by a 4eV change of the electron-beam energy. Thus, the energy dispersion was about 17μm/eV. Figure 4(b) shows the electron-beam shape at the exit of the monochromator under the same incident beam condition as Fig.4 (a). It should be noted that the beam has an almost round shape.

The present study is supported by a project "Technology Transfer D98-04" of Japan Science and Technology Corporation. (JST).

References

[1] M. Terauchi, M. Tanaka, K. Tsuno and M. Ishida: J. Microscopy, 194 (1999) 203.
[2] M. Terauchi and M. Tanaka: J. Surface Analysis, 3 (1997) 240.
[3] M. Tanaka et al.: J. Microscopy, 194 (1999) 219.
[4] K. Tsuda and M. Tanaka: Acta Cryst., A55 (1999) 939.
[5] M. Tanaka et al.: Inst. Phys. Conf. Ser., 165 (2000) 217.

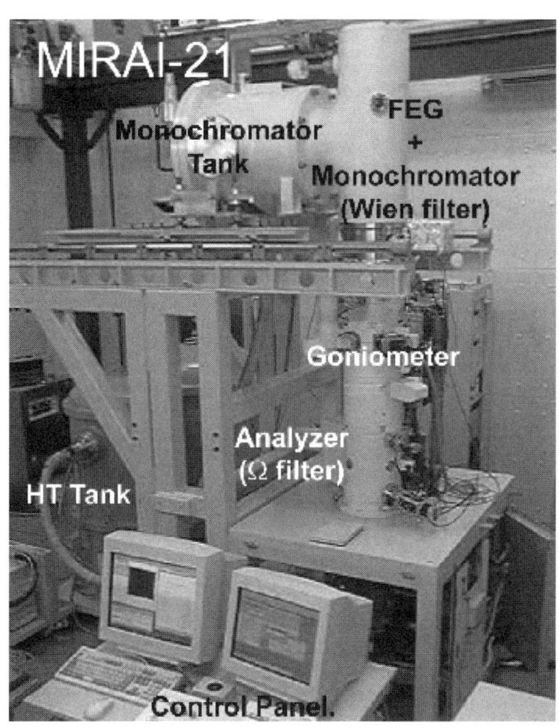

FIG. 1. Appearance of MIRAI-21

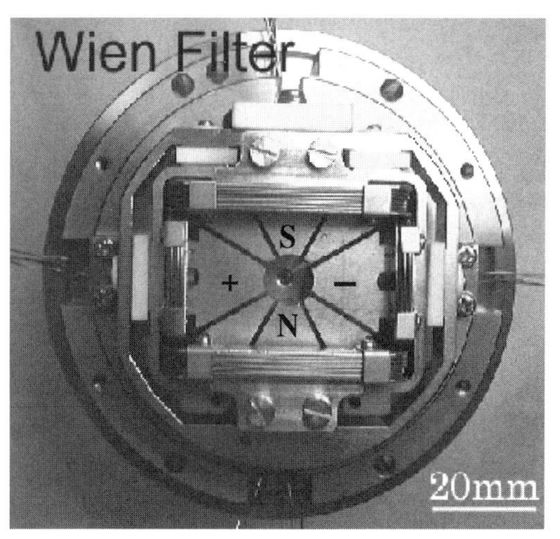

FIG. 2. Octapole-type Wien filter

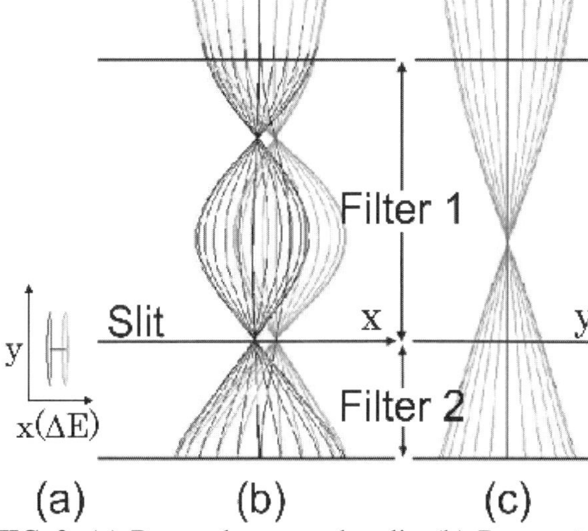

FIG. 3. (a) Beam shape on the slit. (b) Ray-path in dispersion direction (X-Z). (c) Ray-path in non-dispersion direction (Y-Z).

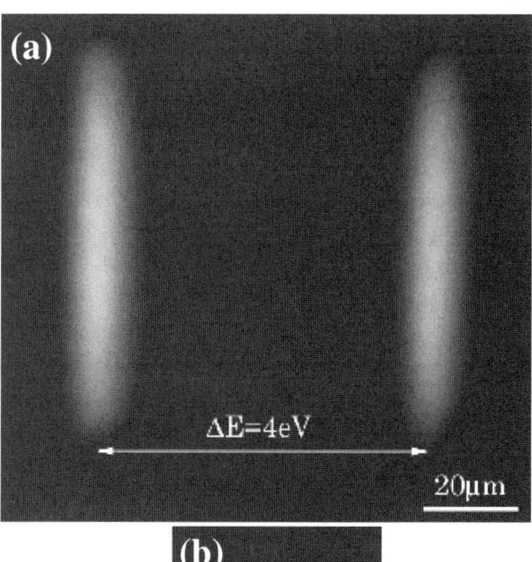

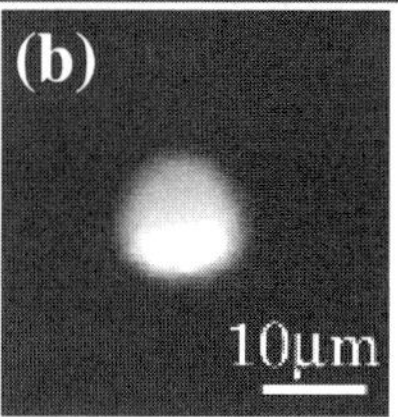

FIG. 4. (a) Astigmatic beam shape on the slit with an energy dispersion of about 17 μm/eV. (b) Stigmatic beam shape at the exit of the monochromator.

Microsc. Microanal. 8 (Suppl. 2), 2002
DOI 10.1017.S143192760210184X

Monochromized 200kV (S)TEM

P.C. Tiemeijer, J.H.A. van Lin, B.H. Freitag, and A.F. de Jong

FEI Electron Optics, P O Box 80066, 5600 KA Eindhoven, The Netherlands

FEI recently completed the development of a monochromator for a Tecnai F20ST by successfully installing a monochromized 200kV (S)TEM at the National Centre for High Resolution Electron Microscopy at the University of Delft. This instrument features a special 200kV supply, a dedicated High Resolution Gatan Imaging Filter (HR-GIF) [1], and a monochromator [2]. This instrument is primarily intended for high resolution Electron Energy Loss Spectroscopy (EELS), in order to enable analysis of bonding states on a nanometer scale, and chemical analysis on a sub-nanometer scale. When the monochromator is on, the system can do EELS with an energy resolution down to 0.10 eV at a spatial resolution $\lesssim$2 nm. When the monochromator is not used, the microscope has 0.50 eV resolution at a spatial resolution of 0.2 nm.

When the monochromator is off, the optics and performance of the microscope is very similar to that of a standard Tecnai F20ST. When the monochromator is switched on, the optics are as sketched in Figure 1. The Wien-filter type monochromator is located directly behind the field emission gun (FEG) and creates a small dispersed image of the source at its exit plane. The accelerator magnifies this dispersed image to the energy selection slit, which is integrated in the C1 aperture holder in front of the first condenser lens. The first condenser lens and objective lens can be used to demagnify the filtered image at the selection slit on to the specimen.

From the outside, this microscope looks just like a standard 300kV TEM: the HT supply has the size of a 300kV supply (because of additional RC filters which reduce the 25 kHz ripple of the high tension generator and its power converters down to 10 mV at 200 kV), the accelerator has the size of a 300 kV accelerator (in order to accomodate the monochromator), the monochromator and FEG supplies (redesigned in small SMD technology) all fit in the small vessel floating at HT close to the FEG at the back side of the column, and the monochromator controls are completely integrated in the Tecnai software.

The user aligns the monochromator as follows: first he sets the condenser system such that the dispersive plane is in focus. Next he sets the desired dispersion and monochromator potential. Beam shift and astigmatism in the FEG or monochromator or accelerator can be corrected with an electrostatic multipole in the monochromator. Astigmatism is judged simply by looking at the dispersed image. The shift and astigmatism can be stored such that they are automatically recalled when the same monochromator dispersion and potential are chosen again. With these, one can switch between monochromator off and on within fractions of minutes. After that, the energy selection slit of the monochromator is inserted and centered.

An important parameter for the performance of the microscope is the quality of its environment, especially with respect to stray magnetic fields and temperature variations. This is nicely illustrated in Figure 3 which shows the FWHM of the zero loss peak as measured as a function of exposure time. At the shortest exposure time (see Figure 2), the resolution is limited only by the HF ripple and the aberrations and point spread function of the monochromator and GIF. When the exposure time is increased, contributions of stray magnetic fields (especially 50 Hz and 150 Hz), noise and drift start to worsen the resolution.

Figures 4 and 5 show a typical example of how the information in an energy loss spectrum can be increased by the monochromator. Figure 4 shows the Ti-L_{23} peaks in rutile (TiO_2) as measured on a standard F20ST with 0.70 eV resolution, whereas Figure 5 shows the same peaks measured on the monochromized F20ST tuned to 0.25 eV resolution. As discussed in Ref. [3], the peaks visible in Figure 5 are caused by spin-orbit interactions, the crystal field effect, bonding symmetry breaking, and by transitions to a triplet state. In this way, with the improved EELS resolution a better understanding of local bonding states can be obtained.

References

1. M. Barfels et al., *elsewhere in these proceedings.*
2. P.C. Tiemeijer, *Inst. Phys. Conf. Ser.* 161(1999)191.
3. C. Mitterbauer et al., *contribution to ICEM15* (2002).

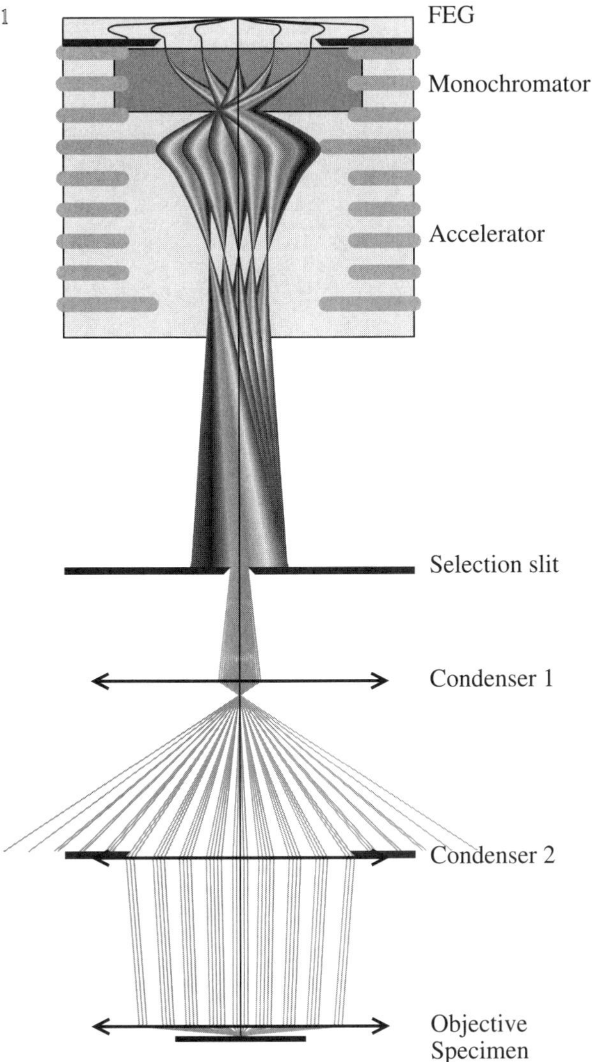

FIG. 1. Typical electron trajectories from source to specimen in the monochromized TEM. Different energies are indicated by different colors.

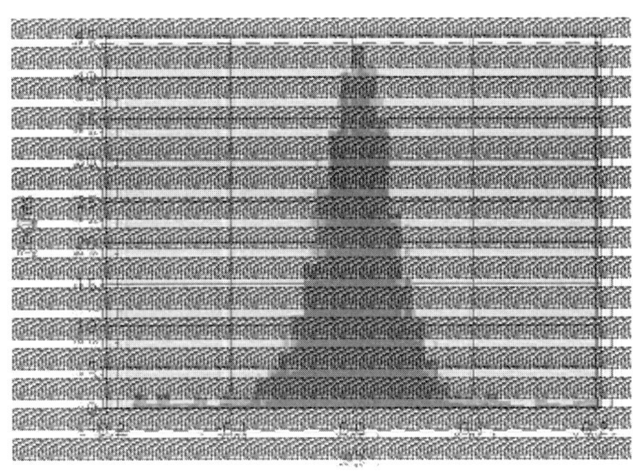

FIG. 2. Zero-loss peak measured at maximum dispersion of the monochromator, smallest GIF entrance aperture, and 1 ms exposure time.

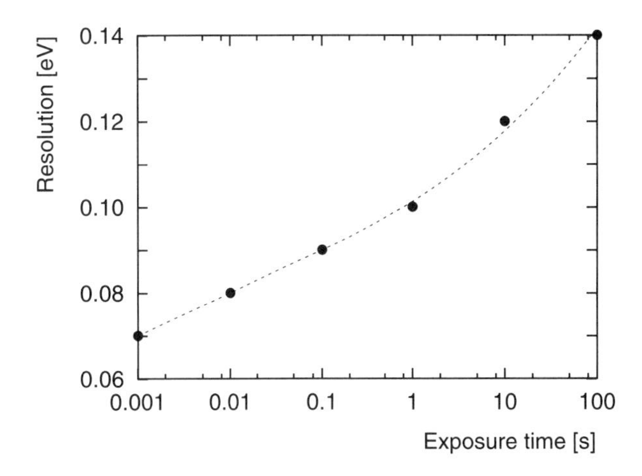

FIG. 3. Full Width at Half Maximum of the zero-loss peak as a function of exposure time, at maximum dispersion of the monochromator.

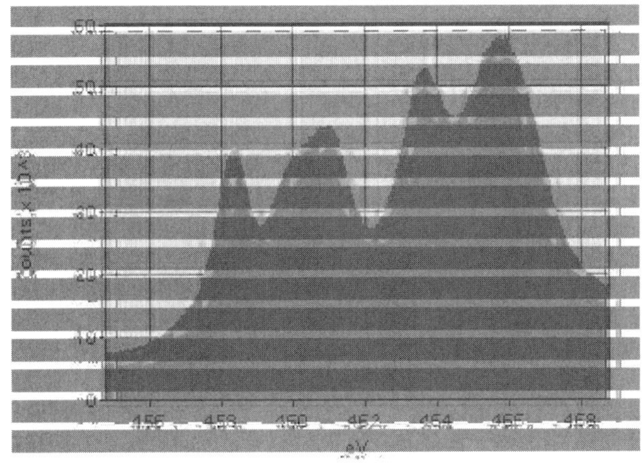

FIG. 4. EELS spectrum of the Ti-L_{23} peaks in TiO_2 measured on a standard Tecnai F20ST with 0.70 eV resolution.

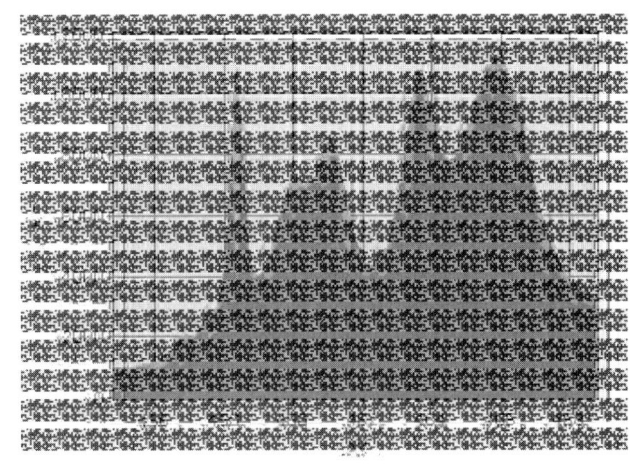

FIG. 5. EELS spectrum of the Ti-L_{23} peaks in TiO_2 measured on the monochromized Tecnai at 0.25 eV resolution (sample courtesy of F. Hofer, Felmi Graz).

Microsc. Microanal. 8 (Suppl. 2), 2002
DOI 10.1017.S1431927602101851

EFTEM at High Magnification: Principles and Practical Applications

Werner Grogger*, Kannan M. Krishnan**, Ferdinand Hofer*

* Research Institute for Electron Microscopy, Graz University of Technology, Steyrergasse 17, A-8010 Graz, Austria
** NCEM, Materials Sciences Div., Lawrence Berkeley National Lab., Berkeley, CA 94720, USA; present address: Dept. of Materials Sciences, University of Washington, Seattle, WA 98195, USA

Energy-filtering TEM (EFTEM) has proven to be an efficient technique for determining elemental distributions on a nanometer length scale. Especially, post-column energy-filters in combination with high resolution TEMs can provide the capability to detect and separate small features down to subnanometer regions [e.g. 1, 2, 3; fig. 1]. Far more than in high-resolution TEM, the attainable spatial resolution is influenced by the experimental setup: The energy and the shape of the ionization edge, as well as experimental parameters like the collection angle, influence the obtainable resolution together with instrumental parameters (aberration coefficients of the TEM objective lens) [4, 5]. Additionally to these theoretical limitations, however, there are practical aspects, that also limit the resolution in EFTEM: maximum electron dose, exposure time, specimen drift and instrumental instabilities. When considering, that at high magnifications only a small number of atoms (typically 10 – 1000) contribute to the signal in a pixel of the final image, it is obvious that accurate instrumental alignments and a careful experimental setup are paramount for good quality results. It should also be noted, that a certain amount of elastic contrast is always transferred to energy-filtered images [6], which can lead to artifacts in elemental distribution images.

In this study we present results obtainable with high resolution EFTEM. Fig. 1 shows an elemental map (three window technique) of a Mn/PdMn multilayer using the Mn-L_{23} ionization edge acquired on a 200 kV FEG-TEM with a post-column energy filter. The Mn layer thicknesses are between 2.6 and 0.47 nm, the thickness of the PdMn spacer layer is constant (3.7 nm). The integrated profile across the image shows peaks for every single layer. As the layers get thinner, the widths of the peaks remain constant for layer thicknesses below about 1.5 nm. For thinner layers the area under the peak decreases, until it may reach a level, where it cannot be longer distinguished from noise (estimated to be around 0.16 nm for the acquisition parameters used in fig. 1).

For many applications it is important to characterize thin oxide layers, e.g. in semiconductor devices, which nowadays reach thicknesses of just a few monolayers. The thickness of such layers in semiconductor devices plays an essential role in the functionality of the device. Using the contrast difference between the Si substrate and the thin silicon oxide layer at certain energy losses, the oxide layer can be imaged at nm resolution (fig. 2). At the low energy-losses used (30 – 90 eV) the signal to noise ratio is very high, which allows short acquisition times. However, at these low energy-losses delocalization is decreasing the resolution to values of a couple of nm. Our work will be presented how to account for this broadening, and to remove it digitally in order to improve the resolution.

The dimensions in many materials science devices become smaller and smaller, that's why the reliability of compositional profiles is an important issue. EFTEM provides a quick analytical tool for such investigations, but in some cases its applicability may be limited due to the limited spatial resolution. In this paper we highlight these limitations as well as the prospects of the technique.

[1] B. Freitag and W. Mader, J. Microsc. 194 (1999) 42
[2] K.T. Moore et al., Phil. Mag. B 82 (2002) 13
[3] M. Varela et al., Phys. Rev. Lett. 86 (2001) 5156
[4] O.L. Krivanek et al., J. Microsc. 180 (1995) 277
[5] R.F. Egerton, J. Electron Microsc. 48 (1999) 711
[6] T. Navidi-Kasmai and H. Kohl, Ultramicroscopy 81 (2000) 223
[7] W.G. acknowledges financial support from the Max Kade foundation for his stay at the NCEM, Berkeley, where part of this work was done. We thank Austriamicrosystems AG, Schloss Premstätten, A-8141 Unterpremstätten, Austria for supplying the semiconductor sample.

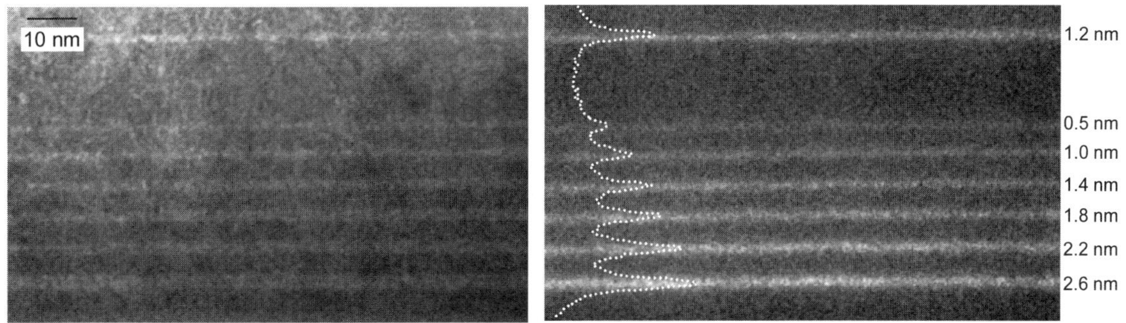

Fig. 1: Bright field image (left) and Mn-L_{23} elemental map (right) of a Mn/PdMn multilayer. The thinnest Mn layer is 0.5 nm thick. The elemental map was calculated from energy-filtered images at 607, 627 and 650 eV with a 20 eV slit and an exposure time of 20 s.

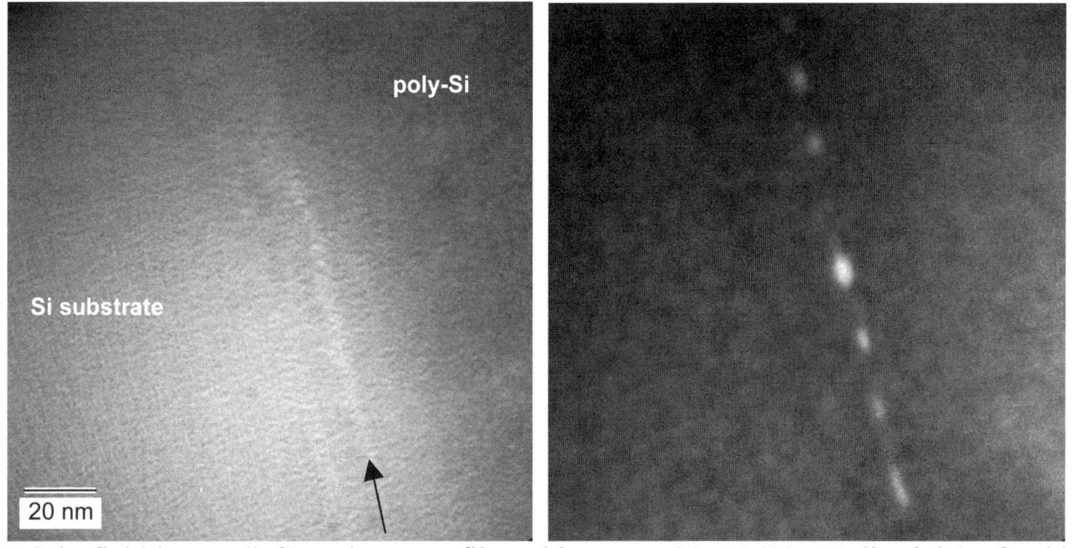

Fig. 2: Bright field image (left) and energy-filtered image at 30 eV (10 eV slit, right) of a thin oxide layer (arrow) on a Si substrate underneath an originally amorphous silicon layer, which was subsequently recrystallized by annealing at 1000 °C. Using EFTEM the nominally 1 nm thin oxide layer could be detected and an insular growth confirmed.

Microsc. Microanal. 8 (Suppl. 2), 2002
DOI 10.1017.S1431927602101863

Understanding DNA Organization in the Nucleus By Fluorescence Microscopy and Energy Filtered Transmission Electron Microscopy

D.P. Bazett-Jones

Programme in Cell Biology, The Hospital For Sick Children, 555 University Avenue, Toronto, ON M5G 1X8, Canada

Molecular imaging techniques provide structural information, often down to the sub-nanometer level of spatial resolution, complementing x-ray crystallography and magnetic resonance spectroscopy. Molecular imaging offers some advantages; for example, complexes that have not been crystallized can be examined, and very small amounts of purified material are often sufficient. Energy filtered transmission electron microscopy is a very sensitive analytical electron microscopic technique that can generate high resolution distribution maps and quantification of particular chemical elements. The technique has allowed us to map DNA or RNA in nucleoprotein complexes and to determine stoichiometric relationships of protein and nucleic acid in these complexes. One of these complexes is condensin, a five-subunit protein complex essential for mitotic chromosome condensation from yeast to humans, introduces positive supercoils into DNA in an ATP-dependent manner in vitro [1]. We report here the direct visualization of this supercoiling reaction by electron spectroscopic imaging. In the presence of ATP, a single condensin complex is capable of introducing $\geq$2 compensatory supercoils into the protein-free region of a closed circular DNA. Within the condensin-bound region, ~190 bp of DNA is organized into a compact structure with two distinct domains, indicative of the formation of two oriented gyres. The current results suggest that the action of condensin is more dynamic and more efficient than that postulated before, providing fundamental insight into the energy-dependent mechanism of higher-order chromatin folding.

Regulated transcription requires accurate temporal assembly of complex machines involving scores of components onto chromatin templates of specific genes at precisely the right time. Microscopical evidence indicates that the nucleus is organized into compartments and sub-domains, which could provide local environments that, at least fine tune transcription from a particular gene promoter, if not dictate whether transcription from that promoter will proceed at all. More fundamental questions, however, are: How do proteins move throughout the nucleus? What is the physical basis for nuclear sub-domains? How are these sub-domains assembled and disassembled, and maintained at the proper size, volume and location relative to chromosome territories within the nucleus?

Energy filtered transmission electron microscopy, in combination with fluorescence and live cell microscopy, is well suited to such *in situ* studies of nuclear domains because elemental mapping can be used to delineate biochemical features. Two subnuclear domains are shown in the figures, identified as local accumulations of a protein called promyelocytic leukemia protein (PML) and glucocorticoid receptor (GR). By comparisons of nitrogen and phosphorus maps, it is apparent that the PML has a protein-based core (nitrogen-rich, phosphorus-poor), whereas the core of the GR body contains both a fibrous protein-based structure and nucleic acid accumulations. A combination of biochemical and microscopical techniques should lead to an understanding of the structure of these domains.

References:

[1] K.Kimura et al., Cell 98 (1999) 239.
[2] This research was supported by operating grants from the Cancer Research Society, Inc., the Canadian Institutes of Health Research and the Natural Sciences and Engineering Research Council. The author holds a Canada Research Chair in Molecular and Cellular Imaging.

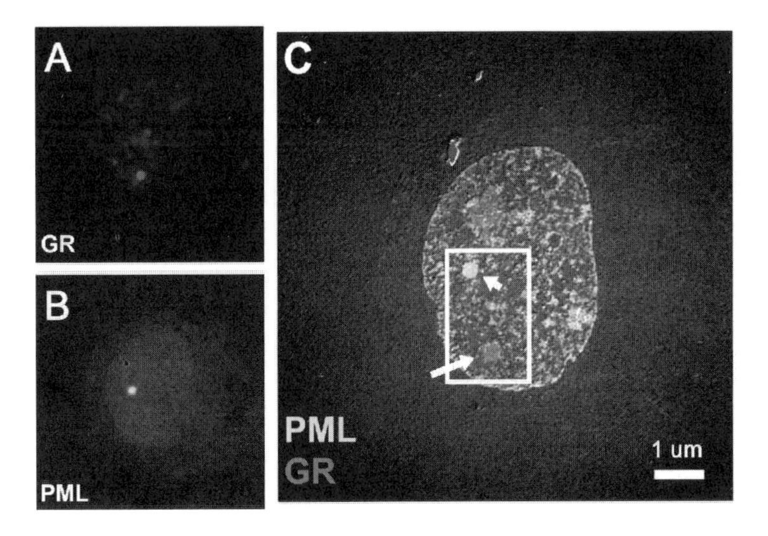

FIG. 1 Fluorescence image of thin section of cell labelled with labelled antibodies against PML protein and GR protein (A, B). The fluorescence image was superimposed on an energy filtered (C) (150 eV energy loss) image of the same section to identify the PML (short arrow) and GR (long arrow) accumulations in the nucleus.

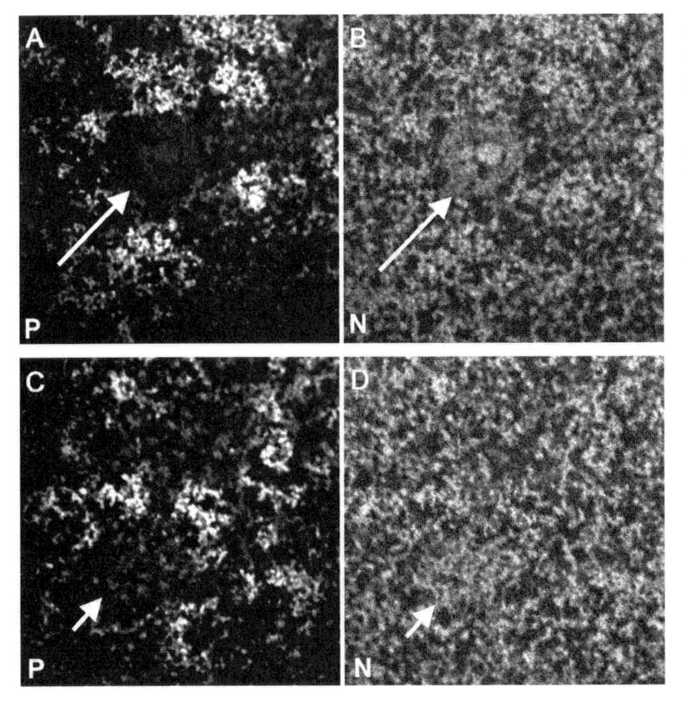

FIG. 2 Energy filtered images at higher magnification of the PML (A, B) and the GR (C, D) accumulations imaged in FIG. 1. Net phosphorus maps (A, C) and net N maps (B, D) can be compared to identify protein-based and nucleic acid-based structures. Both nuclear bodies are composed of a protein-based core (arrows) and are surrounded by blocks of chromatin.

Microsc. Microanal. 8 (Suppl. 2), 2002
DOI 10.1017.S1431927602101759

X-ray Spectrometry In The Fast Lane: An Introduction To High Speed Digital Processing Techniques And Their Application To Emerging EDS Technologies

P.M. Grudberg*, W.K. Warburton* and J.T. Harris*

* X-ray Instrumentation Associates, 8450 Central Ave., Newark, CA 94560-3430

Silicon drift detectors (SDDs), a new class of energy dispersive detector that operates at near-room temperature, have just started to become commercially available from companies such as Photon Imaging (PI) [1] and Röntec [2]. Using a design that employs applied electric fields to laterally drift charge deposited in the detector from absorbed x-rays to a tiny, small capacitance anode, SDDs achieve energy resolutions comparable to the best Si(Li) detectors but at much shorter peaking times. Recent studies by X-ray Instrumentation Associates' (XIA) personnel have shown that rather remarkable results can be obtained using these detectors: 215 eV FWHM at Mn Kα using a peaking time of 300 ns, and output counting rates (OCRs) exceeding 500 kcps using a 150 ns peaking time, using the Röntec XFlash SDD. In addition, with the PI SDD, the detection of C Kα x-rays (at 277 eV) has been demonstrated using peaking times as short as 2 μs [3].

The exceptional performance of these detectors at short peaking times allows running at count rates that are unheard of in conventional EDS systems based upon Si(Li) detectors. The sub-microsecond peaking times can handle input count rates (ICRs) approaching 1 Mcps with reasonable deadtimes, with the ability to resolve elements as light as Silicon (1.74 keV) or possibly even lighter. For light element work, a peaking time of 2 μs allows OCRs in excess of 75 kcps with the ability to work with Oxygen or even Carbon; for clean Carbon spectrometry, a peaking time of 4 μs achieves OCRs in excess of 35 kcps, well beyond the capabilities of current Si(Li) systems. These count rates present a significant challenge to the processing electronics in terms of raw processing speed. A greater challenge is to maintain spectral quality under those conditions.

XIA has already demonstrated preliminary SDD measurements achieving output counting rates exceeding 500,000 cps using one of our DXPs (digital x-ray processors), so the raw processing power to handle the SDD rates exists. The DXP achieves the exceptional throughput using a patented approach where the digital filtering (including triggering and pileup inspection) is done in real time in fast programmable logic, and the event by event processing is handled by a digital signal processor, which adds no deadtime beyond the pileup inspection interval. With this approach, the DXP achieves optimal throughput for a given peaking time. Furthermore, since the DXP has the processing power to run at very short peaking times (100 ns minimum), the throughput can be very high. However, the quality of the spectrum does degrade with rate, especially using the ultra-short peaking times, primarily due to pileup effects and baseline tracking problems. Figure 1 shows results from early work with the PI SDD at the Stanford Synchrotron Radiation Laboratory, demonstrating the spectral distortions at high rates [4].

Pileup effects are the most noticeable problem with extreme rate spectrometry. The typical pulse-pair resolution for the fast trigger signal in a spectrometer is 200 ns; at an ICR of 1 Mcps, nearly 20% of all pulses occur within 200 ns of another pulse. To the spectrometer, these pulses are indistinguishable, and they form pileup. The pileup issues are slightly different when using longer peaking times (2 –4 μs) for light element work, at ICR's ranging from 50 kcps to 200 kcps. Under these conditions, the pileup rate is not large as long as the fast filter can detect the energy pulse; however, the noise associated with the short trigger filter is too large for light element detection, and pileup inspection must be based upon longer filters.

Baseline tracking becomes a problem when running at high rates, due to the fact that the energy filter rarely gets down to baseline. In a digital spectrometer, discrete samples are averaged to determine the signal baseline level; under high deadtime conditions, the number of samples available to track the baseline is small, and the baseline average is not well correlated with the current baseline value, leading to peak broadening effects. Furthermore, high rates increase the chances of erroneous baseline sampling, where some energy pulses go undetected and are included in the baseline average. This effect biases the baseline average, leading to peak shifts in the spectrum.

Work is progressing towards improving the quality of EDS spectra taken at extreme count rates. After a brief introduction to the general principles of high speed spectrometry, several new approaches will be presented, along with the most recent results of work with SDDs. [5]

References

[1] J.S. Iwanczyk et al., *IEEE Trans. Nucl. Sci.* 46 (1999) 284-288.

[2] L. Strüder et al., *Microsc. Microanal.* 4 (1998) 622-631.

[3] Based on testing at SEAL Laboratories in Los Angeles, CA and at NIST in Gaithersburg, MD.

[4] SSRL beamline 9-1, 1999. The aid of the SSRL beamline support staff is appreciated.

[5] An SBIR grant has been submitted to NIST (solicitation 7.17.02) to support new development work.

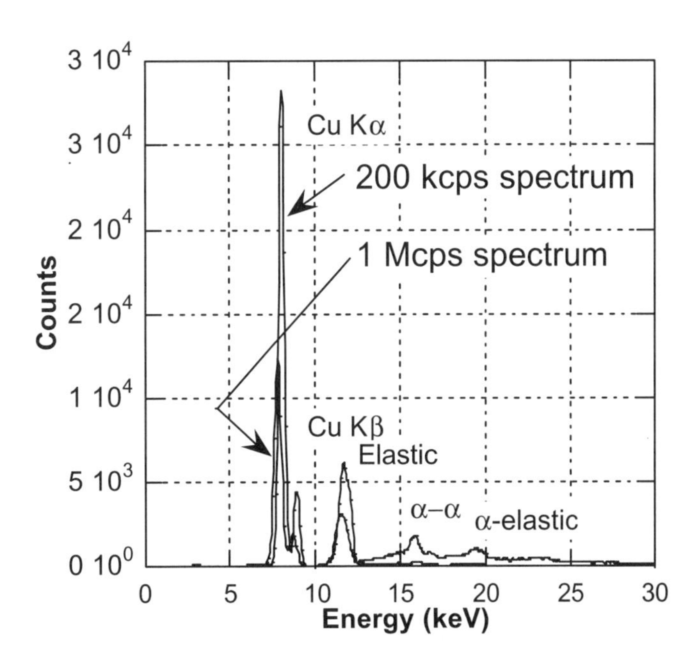

FIG. 1. Normalized spectra taken using PI SDD detector and XIA DXP-X10P using a 20 MHz ADC, 40 MHz DSP. The ICRs are 200 and 1,100 kcps. OCRs are consistent with a 900 ns deadtime. Note the spectral distortions caused by pileup, which increase at the higher rates.

Microsc. Microanal. 8 (Suppl. 2), 2002
DOI 10.1017.S1431927602101760

The Development of Microcalorimeter EDS Arrays

K.D. Irwin,* J. A. Beall,* S. Deiker,* G. C. Hilton,* L. King, ** S. W. Nam,* D. E. Newbury,** C. D. Reintsema,* J. A. Small,** and L. R. Vale*

* Electronics and Electrical Engineering Laboratory, NIST, Boulder, CO 80305
** Chemical Science and Technology Laboratory, NIST, Gaithersburg, MD 20899

High-energy-resolution cryogenic microcalorimeters are a powerful new tool for x-ray microanalysis.[1] With demonstrated energy resolution ~20 times better than withconventional semiconductor EDS, microcalorimeters are useful in applications such as nanoscale particle analysis. Unfortunately, single x-ray microcalorimeters are limited by low count rate (~500 s^{-1}) and small area (~ 0.16 mm^2). Both the count rate and the area can be improved by the implementation of arrays of microcalorimeters. In principle, this improvement in count rate and area comes without degraded energy resolution. The implementation of small microcalorimeter arrays will lead to improvements in the minimum detectable size of nanoscale particles and in the trace element concentration detection limit for a fixed analysis time. The development of kilopixel arrays capable of acquiring hundreds of thousands of counts per second, and with collecting areas of order 50 mm^2, will make it possible to collect high-statistics spectra in small fractions of a second. This will make new applications possible, including real-time process-stream monitoring and the study of the evolution of film properties during deposition with x-ray fluorescence.

Before fabricating an array, it is useful to demonstrate the operation of a single-pixel microcalorimeter system in different environments. We have now transferred a single-pixel microcalorimeter system from Boulder, Colorado to the Chemical Science and Technology Laboratory in Gaithersburg, Maryland. In the process of transferring the system, we discovered that our shadowmask-fabricated microcalorimeters based on aluminum-silver bilayers were not stable in the humid environment of Maryland. We now fabricate microcalorimeters photolithographically, using the immiscible molybdenum-copper material system, which we have found to be stable and robust. An example spectrum taken with the system in Gaithersburg, Maryland, is shownin Figure 1. The microcalorimeter is a 0.16 mm^2 molybdenum-copper bilayer with a 3-μm-thick bismuth absorber. The energy resolution with real-time analog signal processing is ~7 eV. This result compares well with the ~4 eV achieved with aluminum-silver bilayers, since the saturation energy of the new detectors is significantly higher. We will use molybdenum-copper bilayers in arrays.

In order to fabricate large arrays, it is necessary to develop an array structure that thermally isolates the adjacent pixels, while still allowing the leads to be extracted from the array to instrument the microcalorimeters. We fabricate single pixels on a silicon-nitride-coated silicon wafer. To create the needed thermal isolation, we remove the silicon beneath the microcalorimeter using a wet KOH etch, leaving the microcalorimeter on a suspended silicon-nitride membrane. Unfortunately, fabricating large arrays with thermally isolated pixels using this approach is difficult. To make it possible to fabricate large arrays, we are developing surface micromachining techniques to make arrays of silicon nitride platforms suspended over a silicon wafer using a sacrificial layer approach. A small array of such structures is shown in Figure 2. We are now developing the process to fabricate microcalorimeters on these platforms.

Another difficulty in the development of kilopixel arrays is the implementation of the cryogenic wiring and electronics to instrument a large number of channels. The heat load and complexity involved in running wires from room temperature to every pixel is prohibitive. To overcome this difficulty, we are developing superconducting electronics to multiplex the readout of microcalorimeters at cryogenic temperatures.[2] A 32-channel Superconducting Quantum Interference Device (SQUID) multiplexer chip is shown in Figure 3. While considerable work remains in the development of the cryogenic circuits and room-temperature electronics, in the future kilopixel arrays of microcalorimeters will be instrumented using only 32 SQUID multiplexer output channels.

References:

1. D. A. Wollman et al., *J. Microscopy* 188 (1997) 196.
2. J.A. Chervenak et al., *Appl. Phys. Letters* 74 (1999) 4043.
3. This work was supported in part by the NIST Office of Microelectronic Products (OMP).

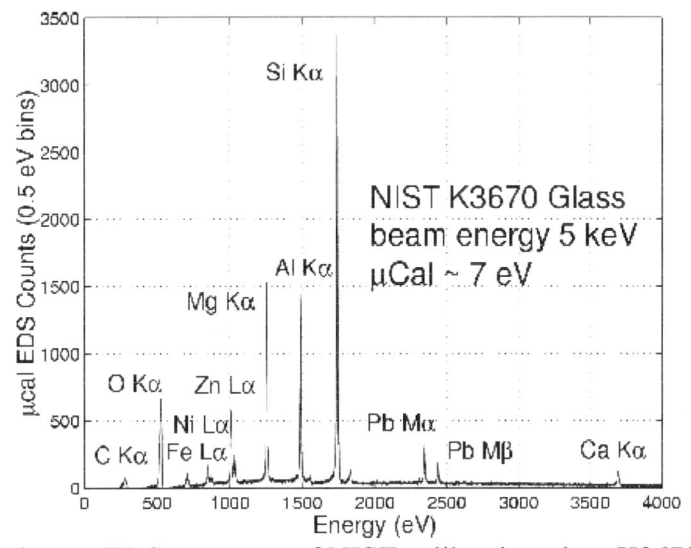

Figure 1. Microcalorimeter EDS spectrum of NIST calibration glass K3670 taken by the microcalorimeter system transferred to Gaithersburg, Maryland.

Figure 2. A 6x6 array of suspended silicon nitride platforms to thermally isolate separate microcalorimeter pixels

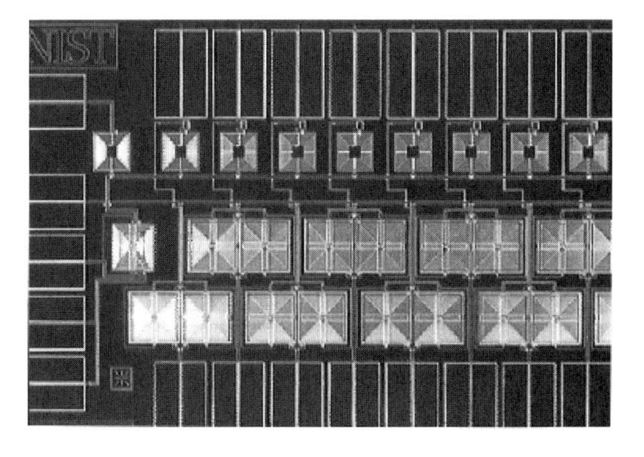

Figure 3. Photograph of part of a 32-channel SQUID multiplexer to instrument microcalorimeter arrays

DOI 10.1017.S1431927602101772

The Latest Experiences Using A Cryogen Free Microcalorimeter Energy Dispersive X-Ray Spectrometer

Del Redfern[*], Joe Nicolosi[*], Jens Höhne[**], Rainer Weiland[***], Birgit Simmnacher[***] and Christian Hollerith[****]

[*] EDAX Inc. Mahwah, NJ, USA
[**] CSP Cryogenic Spectrometers GmbH, Munich, Germany
[***] Infineon Technologies AG, Munich, Germany
[****] Technische Universität München

The Polaris Microcalorimeter Energy Dispersive X-Ray Spectrometer has been in use at Infineon Technologies AG, Munich since April 2001. The microcalorimeter is presently in use as an every day tool, producing resolutions <15eV at 1.5keV.

This paper will discuss the first year of operation, review the goals of the project and how they have been met and review potential opportunities of this exciting new technology. The acceptance criteria that will be discussed include energy resolution, count rate and temperature hold time, all being required for suitable analysis of microelectronic devices.

The microcalorimeter is being used for routine analysis, including Ti/TiN, WSi2, Si/Ta/W and boron, phosphorus, silicon glass (BPSG) samples. Data will be presented showing the capabilities of the system to perform simple qualitative analysis on these materials. Small volume analysis is also regularly performed and data from thin layers, such as SrBi2Ta2O9/Pt/ and small particles will also be presented.

There are two main areas of concern to develop the microcalorimeter energy dispersive x-ray spectrometer into a commercial product; the count rates (acquisition time) and vibration. At present the acquisition time for a spectrum to ensure the scientist is fully confident of the results is between 3 to 5 minutes. This paper will describe the actions proposed and undertaken to bring the acquisition time down to the more acceptable levels of around 60 seconds. Images at 100,000x magnification have shown little effect of vibration with the microcalorimeter running under normal operation. The target for vibration free operation is 200,000x; the work to reduce the level of vibration will be discussed.

Another development area to improve the suitability of the Polaris for industrial applications is the hold time (working time) of the adiabatic demagnetization refrigerator (ADR). The present beta unit has a hold time of approximately 8 hours before the ADR has to be recharged. A new ADR using the extended hold time unit is being evaluated, this would increase this hold time to over 24 hours and the outcome of the evaluation will be described in the paper.

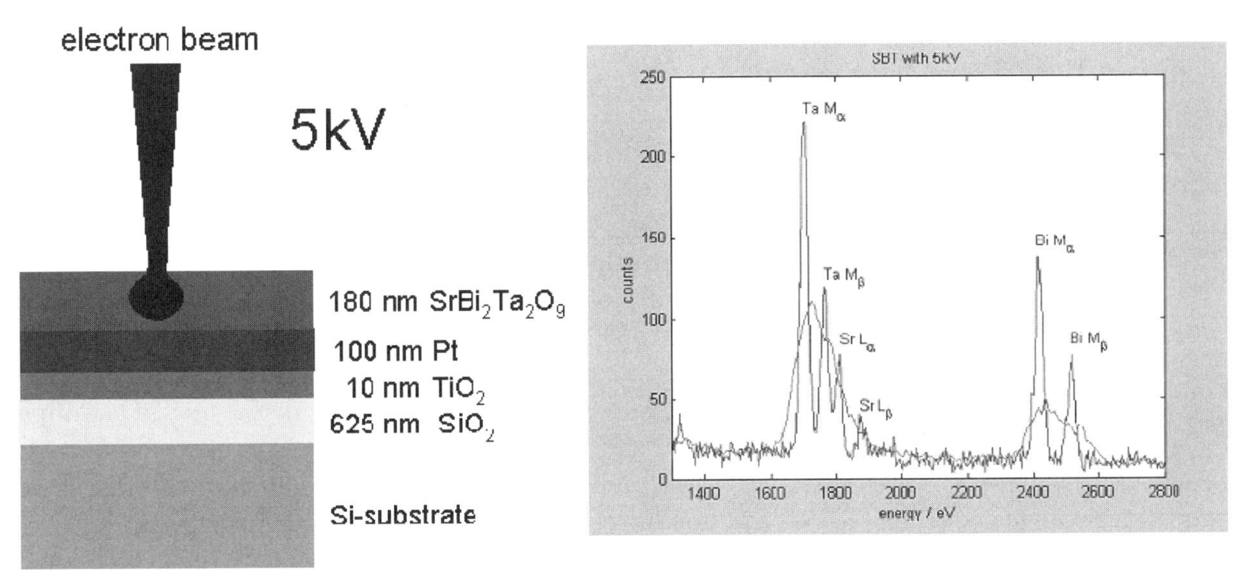

Spectra from the Microcalorimeter Energy Dispersive X-Ray Spectrometer of a layered sample using 5kV excitation voltage.

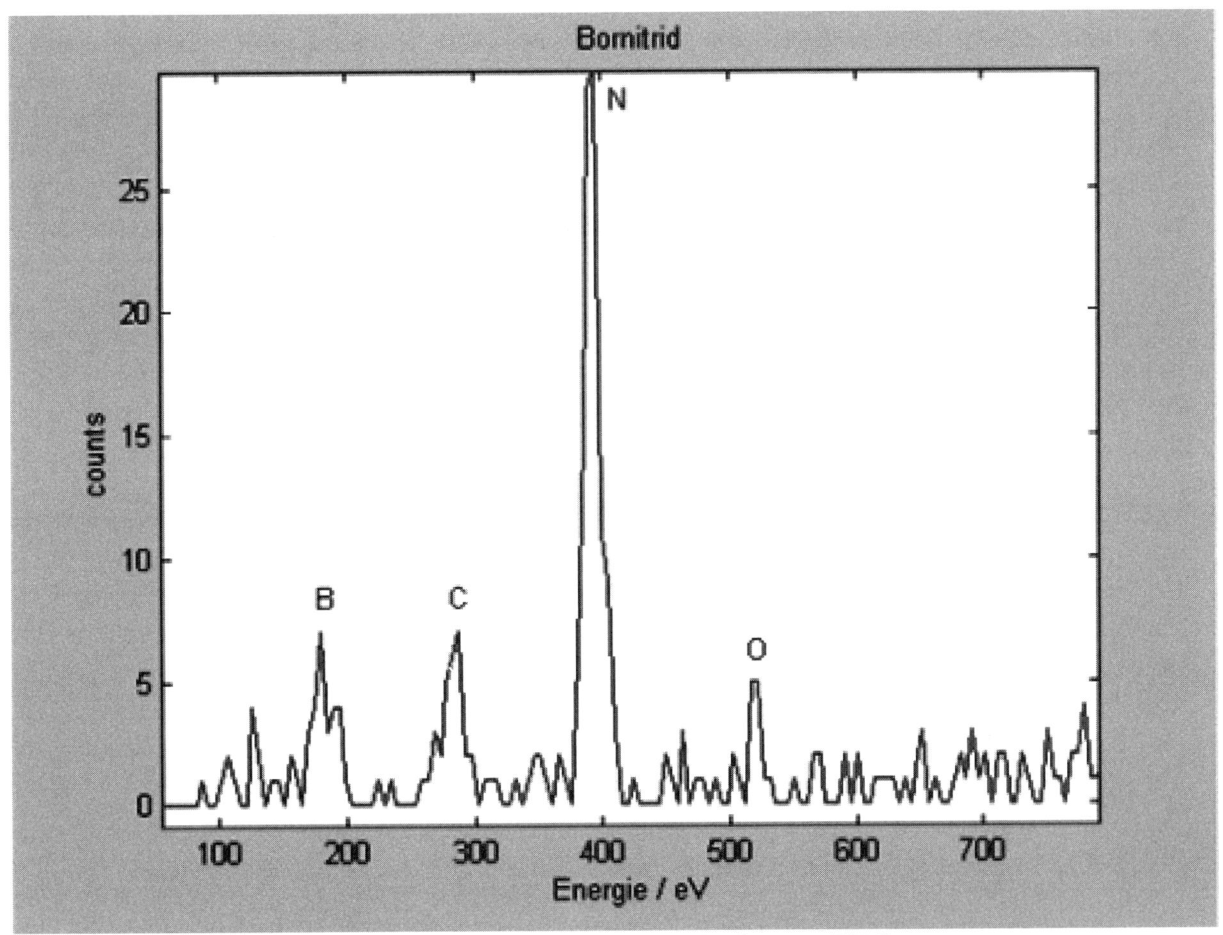

Spectra of a BN sample acquired using the Microcalorimeter EDS system

Microsc. Microanal. 8 (Suppl. 2), 2002
DOI 10.1017.S1431927602101784

A Well Dressed Microscope: Practical Experience with Microcalorimeter and Silicon Drift Detector Systems

John A. Small*, Dale E. Newbury*, John Henry J. Scott*, Lance King*, Sae Woo Nam**, Kent Irwin**, Steve Deiker**, Shaul Barkan,*** and Jan Iwanczyk***

*Surface & Microanalysis Science Division, NIST, Gaithersburg, MD 20899
** NIST, EEEL, Boulder, CO 230940
***Photon Imaging Inc., Northridge, CA 91324

NIST, Gaithersburg has recently installed a first generation silicon drift detector (SDD) from Photon Imaging and the NIST Boulder microcalorimeter energy dispersive x-ray spectrometer (μcal-EDS) on a JEOL 840 SEM[1], as shown in Fig. 1.[1,2] The instrument is also equipped with a conventional Si-Li x-ray detector (LINK ISIS 3 position turret) and a JEOL wavelength dispersive x-ray spectrometer. NIST, Gaithersburg staff have had the opportunity to work with these new detector technologies as "users" and to compare preliminary results with conventional systems.

<u>μcal-EDS</u>: The NIST μcal-EDS has been outfitted with a second-generation Mo-Cu sensor with a Bi absorber replacing the first-generation Al-Ag sensor and Bi absorber. We obtained 7 eV resolution on Al as shown in Fig. 2. The maintenance of the cryogenic systems includes the filling of the LN_2 and LHe reservoirs and the magnetization cycle of the adiabatic demagnetization refrigerator (ADR). Although fairly complex, these tasks became routine after only a few operations. A minor drawback of the existing cryostat is the need to fill the LN_2 reservoir every day (this is in the process of being automated) and the LHe reservoir (which does not currently lend itself to automation) every other day. At the NIST Gaithersburg labs this requires two people (for safety) to be present during the LHe transfer. After the initial cool down and operation of the detector system, a vacuum leak caused the ADR to warm up to room temperature. The standard procedure for obtaining good performance is to initially cool the detector system in a low magnetic field and attachthe system to a probe after the system is cooled to 4K (LHe). However, we attempted to cool the system down while it was attached to the microprobe. Our attempt to obtain at least 10 eV resolution was not successful, and the resolution degraded to about 50 eV. There are several possible reasons that are being explored such as inadequate magnetic shielding and ground loops.

<u>SDD Detector</u>: Operation of the SDD was straightforward with similar requirements to operating a conventional Si-Li detector. The Peltier cooler requires about 10 minutes to reach the operational temperature and is then stable indefinitely. Spectra collected with the SDD (resolution of 175 eV at MnKα for a 4 μs time constant) are in general comparable to a conventional Si-Li detector. One interesting exception is the low-energy response of the SDD, which appears to be significantly better than that of a conventional Si-Li detector with an ultrathin window, as shown in Fig. 3 for Mn. The spectrum from the LINK detector was collected immediately after conditioning to eliminate any ice. The limiting count rate for the SDD appears (from oscilloscope traces) to be on the order of 1.5 MHz to 2.5 MHz. However, the peak-processing electronics limit the practical output count rate to about

[1] Certain commercial equipment, instruments, or materials are identified in this abstract to specify adequately the experimental procedure. Such identification does not imply recommendation or endorsement by the National Institute of Standards and Technology, nor does it imply that the materials or equipment identified are necessarily the best available for the purpose.

100 kHz. To take full advantage of the detector's counting speed will require a redesign of these electronics.

References

[1] D.A. Wollman, et al., Nucl. Instrum. Meth. A 444:(1-2) (2000) 145-150.
[2] B.E. Patt, J.S.Iwanczyk, C.R. Tull et al., Microsc. Microanal. 6 (Suppl. 2) (2001) 728.

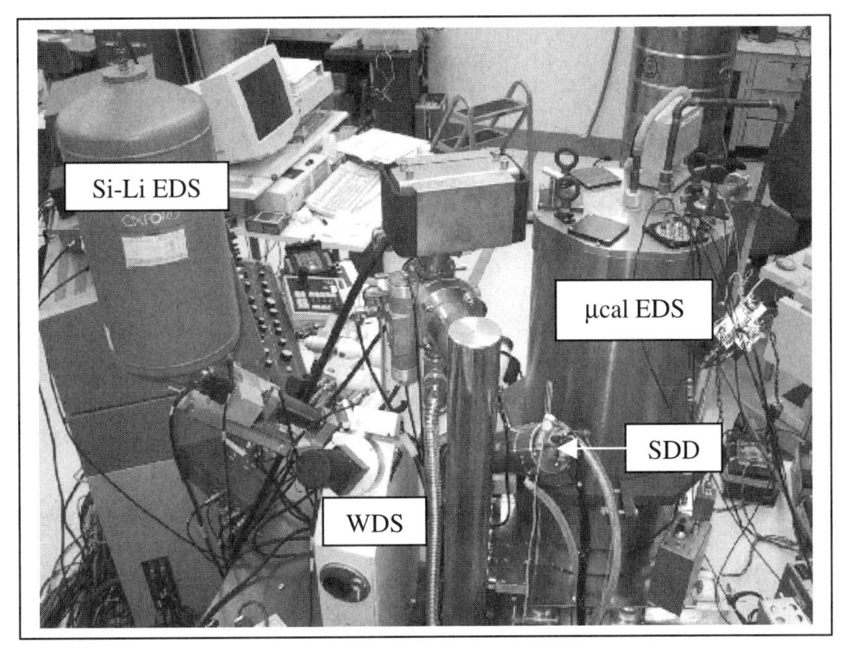

FIG 1. Image of JEOL 840 SEM showing various x-ray detectors.

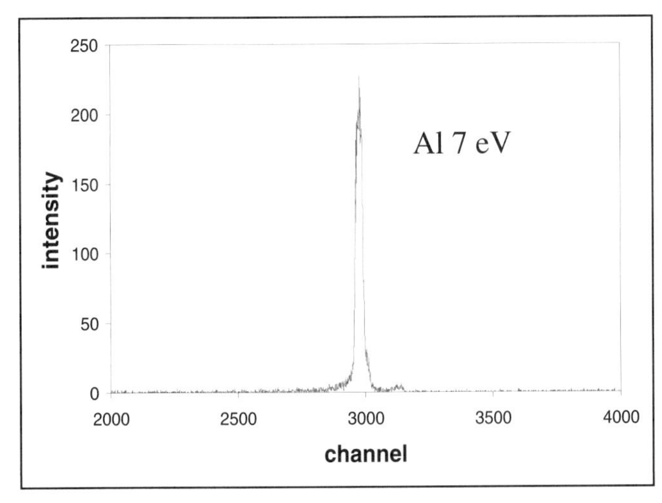

FIG. 2. Al Spectrum from NIST μcal EDS.

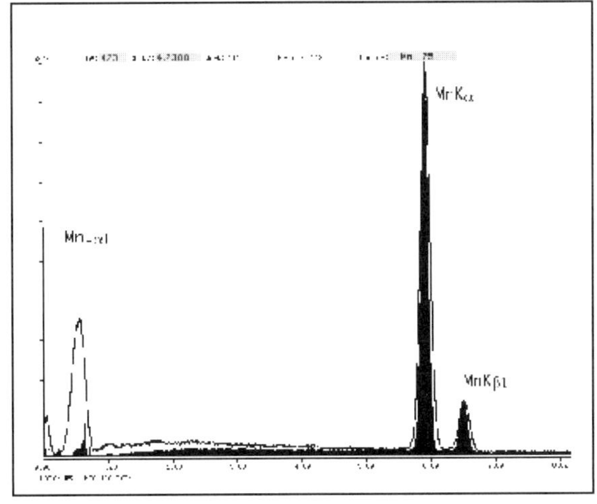

FIG. 3. Mn spectra from the SDD and an ultra thin window Si-Li detector.

Microsc. Microanal. 8 (Suppl. 2), 2002
DOI 10.1017.S1431927602101796

Real Time Color Scans with Scanning Electron Microscopes – A New Application of the XFlash® X-ray Detector Technology

Gabriele Mäurer, Thomas Schülein, Gert Kommichau

RÖNTEC GmbH, Schwarzschildstr. 12, D-12489 Berlin, Germany

In recent years, x-ray detector technologies based on silicon drift diodes (SDD) have been employed for microanalysis (EDX) on scanning electron microscopes in increasing numbers. Compared to conventional Si(Li) or HP Ge crystal detectors, these new detectors are superior in terms of performance, reliability and speed. Furthermore, an integrated thermoelectric cooling element eliminates the need for liquid nitrogen or other consumables for operation, resulting in a very compact, lightweight detector [1].

The second generation of RÖNTEC XFlash® detectors marks a breakthrough in the development of SDD detector technologies. Its energy resolution of 127 eV now exceeds that of the widely used Si(Li) detector. Moreover, the low energy detection capability has been extended to include light elements. With its high X-ray pulse throughput capabilities, XFlash® opens the door for enhanced EDX applications and at the same time is very well suited for other analysis methods (e.g., XRF, PIXE).

One of the most outstanding XFlash® applications in the field of electron microscopy is the RÖNTEC ColorSEM system which was specifically designed to provide a new imaging capability at the SEM. Within seconds it generates precise composite color images based on the characteristic X-rays of the specimen. These brilliant, high resolution images are built from up to 8 high-speed, single-color X-ray maps with each color representing an individual element. The acquisition time of a composite image (up to 2048 x 1536 pixels) is similar to that required for a high quality SE/BSE image on the SEM. In contrast, conventional EDX maps at 256 x 256 pixels can often take up to a half hour to acquire[2].

The rapid acquisition is due to the high processing capabilities of the RÖNTEC XFlash® detector and its electronics unit. Capable of handling up to 400,000 cps, the system is 13 times faster than conventional mapping systems. Individual X-ray “raw data” maps (each with 16 bit pixel resolution) are blended with the SE/BSE image electronically in order to produce the spatial impressions. ColorSEM is a stand-alone imaging system that can be easily fitted to SEMs of any type and model.

ColorSEM images are exceptional in speed, contrast and resolution and have become a powerful tool for the microanalyst examining compositional features as small as one micrometer. The display of element gradients provides detailed information on grain boundaries, phase details or micro inclusions as well as the identification of particles and fibers.

The ability to produce “color scans” at the scanning electron microscope has great potential to revolutionize electron microscopy. The fast and easy visualization of element distributions in a specimen and the identification of microstructural features and compositional variations of different materials can be more easily interpreted, particularly by persons unfamiliar with electron microscopy. In addition to secondary and backscattered electron images “color scans” introduce rapid color with maximum resolution to the SEM.

References:
[1.] J. Gannon et al., A new methodology for element imaging in the scanning electron microscope, Microsc. Microanal. 7 (Suppl 2: Proceedings) 884
[2.] M. Procop: Fast Elemental Mapping in Materials, Microscopy and Analysis 1 (2002) 17

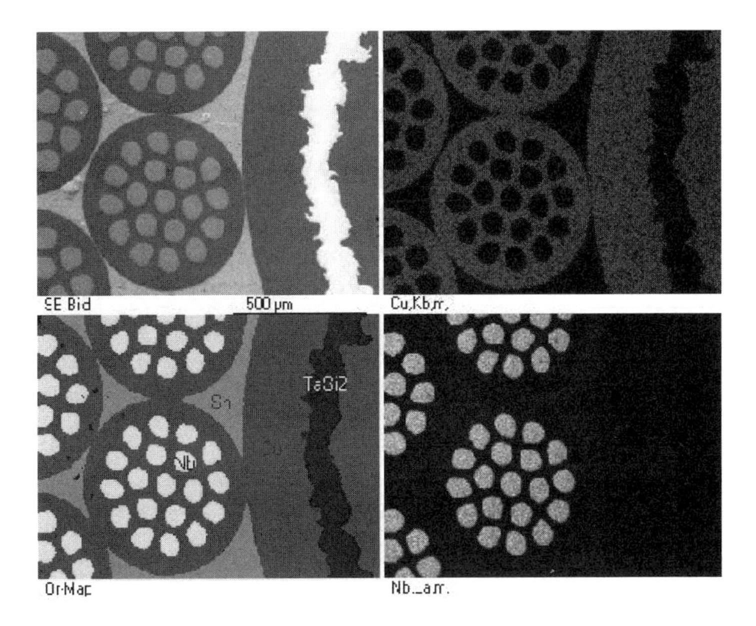

Figure1 (above). Superconductor (R&D Application). Cross-section of a superconducting cable. The superconducting filaments (wires) are embedded in a Cu matrix. This 256x192 pixel image was acquired in 20 seconds (one scanning cycle). Courtesy of Dr. M. Procop and K. Meyer (BAM).

All colors (here presented in grayscale) become more apparent in the original color image.

Figure 2 (below). Composite element image of a MEMS micromotor. Left: two "raw data" maps (Si, Ni) with 16 bit data depth each. Center: Automatically generated color scan image.

Microsc. Microanal. 8 (Suppl. 2), 2002
DOI 10.1017.S1431927602101942

Exploring the Valence Electron Distribution in High Temperature Superconductors with a Focused Electron Probe

J. Tafto[*,**], Lijun Wu[*] and Yimei Zhu[*]

[*] Materials Science Department, Brookhaven National Laboratory, Upton, NY11973, USA
[**] Department of Physics, University of Oslo, P.O. Box 1048 Blindern, 0316 Oslo, Norway

Since the discovery of the high temperature superconductors there has been a great deal of experimental and theoretical effort to relate their superconducting properties and electronic structure. We have used a focused 300 keV electron probe to study important features associated with the electronic structure of these oxides, in particular electron transfer away from the CuO_2 planes that creates the electron holes that are believed to be responsible for the superconductivity of these oxides. Our experimental procedure differs from conventional convergent electron diffraction, CBED, in that we focus the electron probe some 100 micrometers above the specimen, rather than at the specimen level. Thus, when the beam divergence is small enough to avoid overlap between the diffraction disks, each diffraction disk becomes a dark-field shadow image of the specimen. We use wedge-shaped thin specimens, enabling us to record the intensity variation with thickness for many reflections at the same time. We refer to this method as PArallel Recording Of Dark-field Images, PARODI [1,2].

Before starting with the cuprates, we first take a look at the recently discovered superconductor MgB_2. Fig. 1 shows conventional CBED patterns and fig. 2 shows PARODI patterns of the $00l$ row where the planar spacing is 0.352nm. For this material conventional CBED and PARODI give virtually the same value and accuracy for the structure factor of the 001 reflection. Given the small unit cell and the low Z elements of MgB_2, this material is also well suited for x-ray diffraction using a synchrotron beam-line [3]. From powder x-ray diffraction experiments, many structure factors, and thus sufficient data for a electron density map is available from a single experiment. Nonetheless quantitative electron diffraction is an important supplement in that, for selected structure factors, very high accuracy can be achieved using CBED and PARODI. Measuring just a few of these provides a useful test of electronic structure calculations.

Electron diffraction is particular powerful for reflections with short g-vectors that occur for crystals with large unit cells. The reason is that the electrons interact with the electrostatic potential, and it follows from the Mott formula that for short g-vectors the Fourier components of the electrostatic potential are extremely sensitive to the valence electron distribution. This sensitivity is clearly illustrated by considering $Bi_2Sr_2CaCu_2O_{8+\delta}$. The innermost, i.e. 002, reflection in this material corresponds to a planar spacing of 1.54 nm. Calculated values of the x-ray structure factor with the origin chosen at the Ca atom is –150.7e based on the ionic model and -157.9e from electronic structure calculations [4]. Converting these structure factors to electron structure factors gives –147.9 Å and 15.8 Å, respectively i.e. they differ by a factor 10 in absolute values for the two models, and their signs are opposite. Fortunately dynamical electron diffraction patterns also provide information about the phase of the structure factors.

Conventional CBED is not easily applied to dense reciprocal lattice rows because the angle spanned by the convergent beam disks has to be very small to avoid overlap between them. On the other hand PARODI works well because we then study the intensity variation with thickness, and the thickness range in the shadow images varies with the distance from the focused probe to the specimen. Fig 3 shows a PARODI pattern for the (00l) row of $Bi_2Sr_2CaCu_2O_{8+\delta}$. Based on such patterns we arrived, for the 002 electron structure factor, at the value 15 ± 4 Å [2]. After conversion to x-ray structure factor this becomes -157.9 ± 0.2 e which is the same as from electronic structure calculations.

References

[1] J. Tafto, Y. Zhu and L. Wu, Acta Cryst. A **54**, 532 (1998)
[2] L. Wu, Y. Zhu and J. Tafto, Phys. Rev. B **59**, 6035 (1999)
[3] E. Nishibori et. al. J. Phys. Soc. Japan. **70**, 2252 (2001)
[4] R. P. Gupta and M. Gupta, Phys. Rev. B **49**, 13154 (1994)
[5] Work supported by US DOE under contract No DE-AC02-98CH10886 and the Norwegian Research Council under the FIN project.

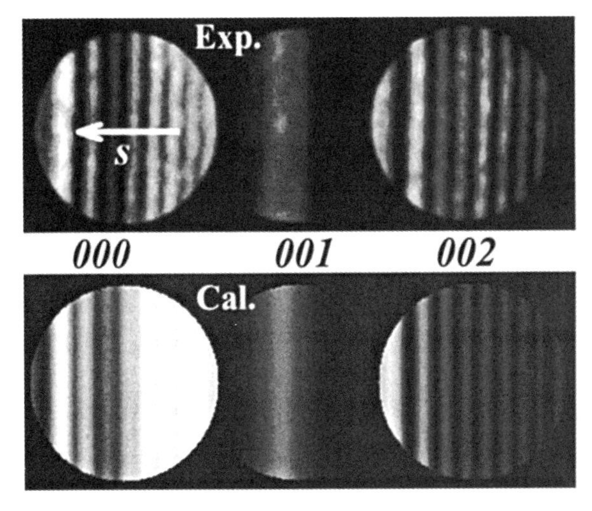

FIG.1. Experimental and calculated CBED pattern of MgB_2

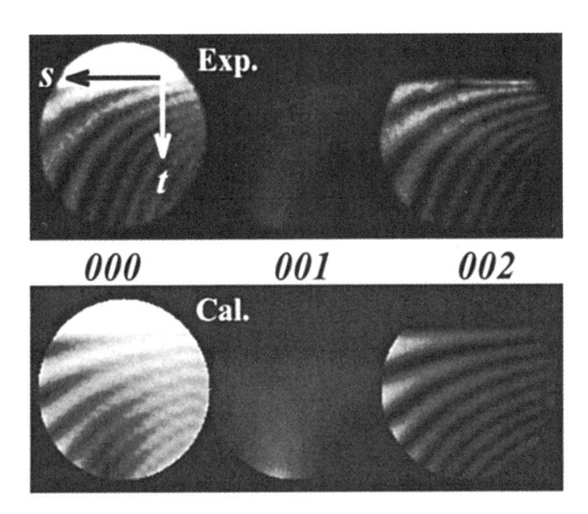

FIG.2. Experimental and calculated PARODI pattern of MgB_2.

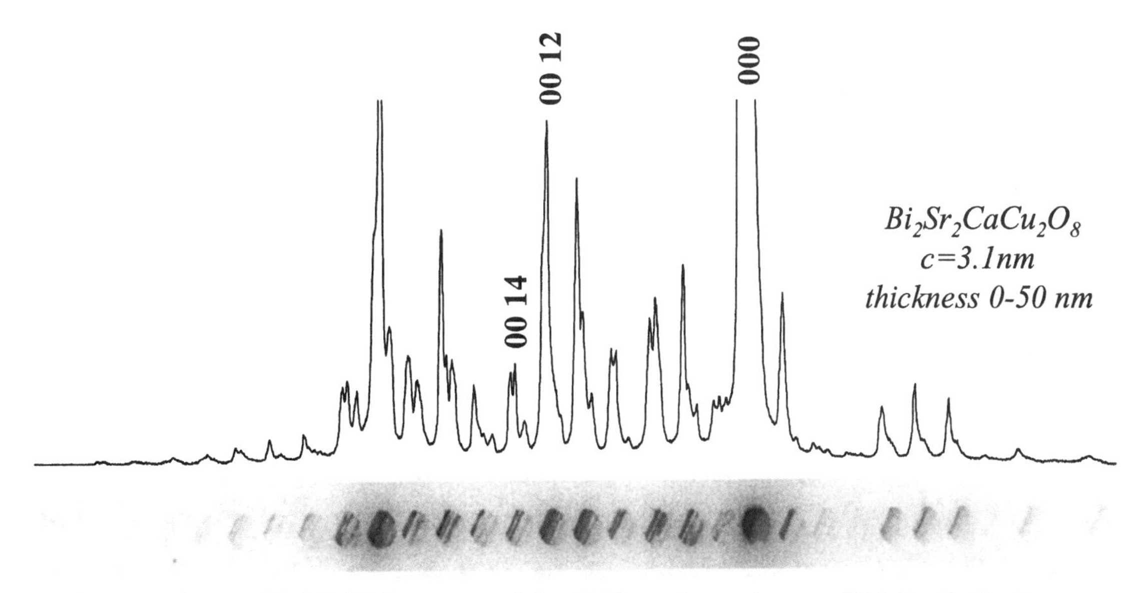

FIG.3. Experimental PARODI pattern of the (00l) reciprocal row of $Bi_2Sr_2CaCu_2O_{8+\delta}$.

Microsc. Microanal. 8 (Suppl. 2), 2002
DOI 10.1017.S1431927602101954

Refinement of Crystal Structural Parameters and Charge Density using Convergent-Beam Electron Diffraction

K. Tsuda, Y. Ogata and M. Tanaka

Institute of Multidisciplinary Research for Advanced Materials, Tohoku University, Sendai 980-8577, JAPAN

We have developed a new method to refine crystal structural parameters using convergent-beam electron diffraction [1]. The method is based on the fitting between theoretical calculations and experimental intensities of energy-filtered *two-dimensional* CBED patterns containing both of zeroth-order Laue-zone (ZOLZ) reflections and higher-order Laue-zone (HOLZ) reflections. The use of HOLZ reflections is essential for the present method because small displacements of atoms can be sensitively detected using HOLZ reflections with large reciprocal vectors. For this purpose, we developed an Omega-filter transmission microscope *JEM-2010FEF* [1], [2], which can take energy-filtered CBED patterns up to a high angle to include HOLZ reflections with a small distortion, and an analysis program *MBFIT* to refine structural parameters, which is based on many-beam Bloch-wave calculations and nonlinear least-squares fitting. Using the method, we refined the atom positions and anisotropyic Debye-Waller factors of the rhombohedral phase of $LaCrO_3$, which is a perovskite-type material for interconnector of solid oxide fuel cells. Clear anisotropy of the thermal vibration of the oxygen atoms was successfully detected for the first time [3].

The present method can be applied to the determination of charge density distribution because the low-order Fourier coefficients of the electrostatic potential (low-order crystal structure factors for electron diffraction), which are sensitive to valence electrons, can be refined together with the atom positions and Debye-Waller factors. Through Poisson's equation, the structure factors for electron diffraction are related to those for X-ray diffraction, or the Fourier coefficients of the charge density. According to the nature of Poisson's equation, a small change in the low-order structure factors for X-rays causes a large change in those for electrons. It should be noted that the accurate determination of Debye-Waller factors, which can be performed successfully by the present method, is crucial to obtaining high-precision low-order structure factors for X-rays from the structure factors for electrons. Figure 2(a) shows the charge density of the CrO_2 plane determined by the present method. Figure 2(b) shows the deformation charge density of the CrO_2 plane, or the difference between the charge density of Fig. (a) and the charge density calculated with the neutral atoms. The charge transfer from the Cr atoms to the O atoms is clearly seen.

The anisotropic charge density of the orbital-ordering phase of $LaMnO_3$ determined by the present method will be also presented in the conference.

References

[1] K. Tsuda and M. Tanaka, Acta Cryst., **A55** (1999) 939.
[2] M. Tanaka et al., J. Microsc., **194** (1999) 219.
[3] K. Tsuda and M. Tanaka, Microsc. Microanal., 6 (Suppl. 2) (2000) 152.

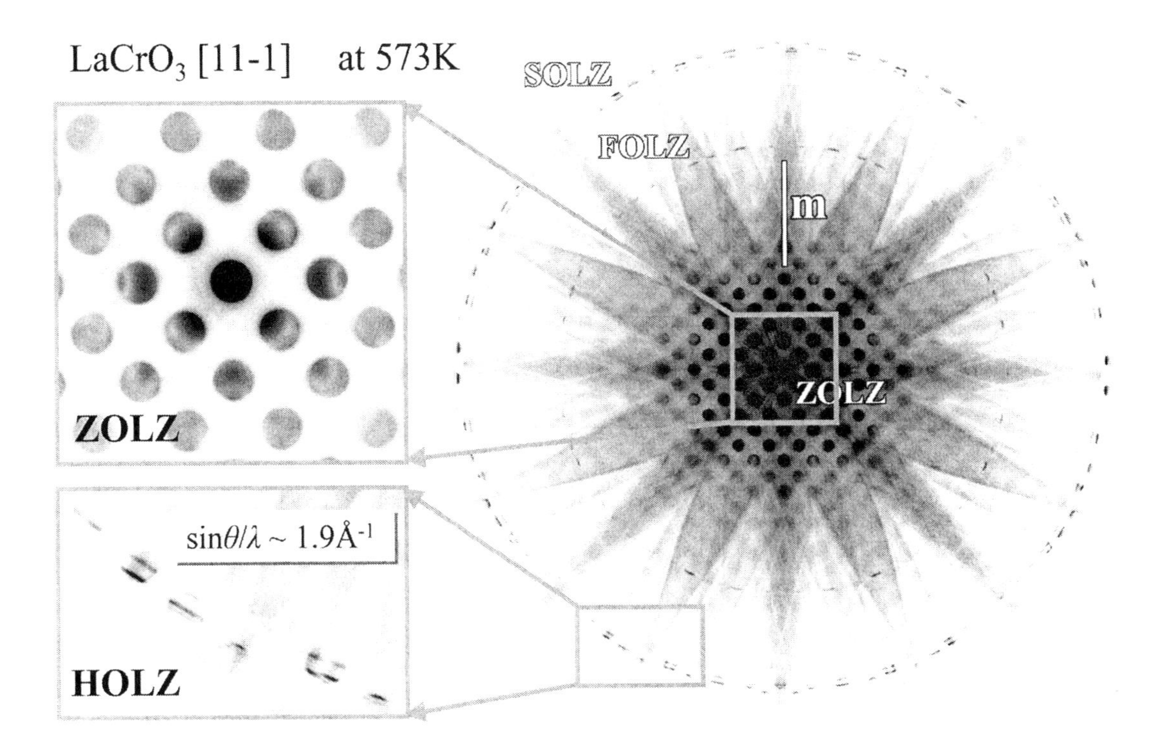

Fig. 1: An energy-filtered CBED pattern of the rhombohedral phase of $LaCrO_3$. The HOLZ reflections with large reciprocal vectors ($\sin\theta/\lambda$~1.9Å^{-1}) are seen.

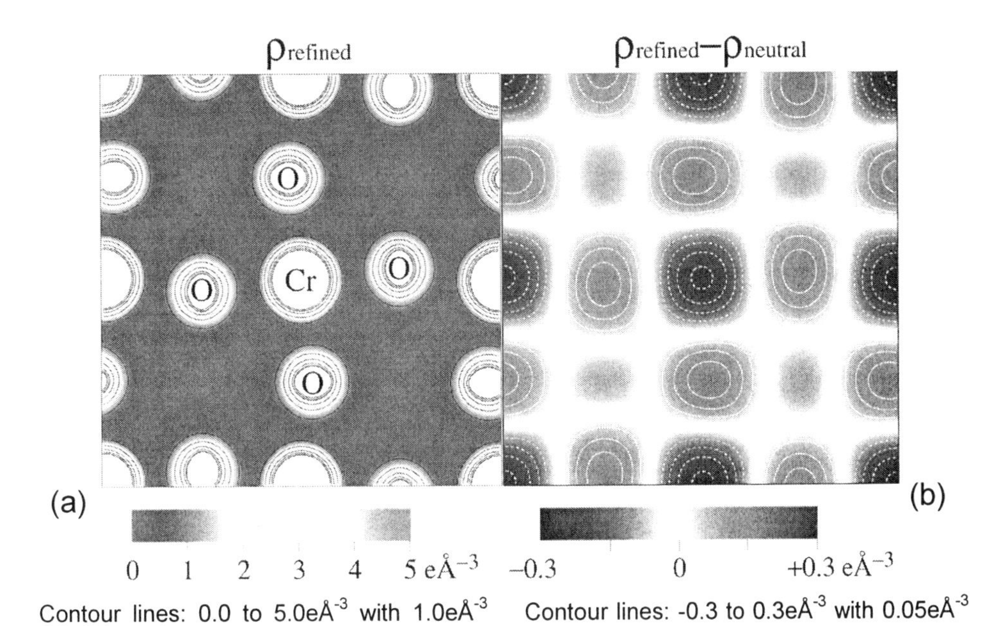

Fig. 2: Charge density (a) and deformation charge density (b) at the CrO_2 plane determined by the present method. The solid lines and dotted lines of the contour maps indicate positive and negative values, respectively.

Microsc. Microanal. 8 (Suppl. 2), 2002
DOI 10.1017.S1431927602101966

Quantitative Convergent Beam Electron Diffraction

R. Holmestad* and J. Friis*

*Dept. of Physics, Norwegian University of Technology and Science (NTNU), N-7491 Trondheim, Norway

Over the last few years, energy filtering, new digital recording systems, developments in theory and tremendous advances in computer power have opened up a new field of quantitative analysis of Convergent Beam Electron Diffraction (CBED) patterns [1-5]. Quantitative CBED techniques are still in an early stage of development and more research is required to fully understand their potential. This paper intends to give a review of the quantitative CBED methods used to determine low order structure factors from inorganic crystals and discuss some of the future prospects. Different examples of QCBED used in materials science will be shown.

The major advantage of the CBED method is the nanoscale electron probe size, which can give information from minuscule regions beyond the reach of other diffraction methods. Using such a small probe, one can almost always find a perfect crystal region where the theory of dynamical diffraction in a perfect crystal area is valid. Atomic resolution transmission electron microscopy (TEM) of the region the data is collected from can ensure that no defects are present. Another advantage is the fast electrons' strong interaction with matter, which gives rise to dynamical diffraction effects, making electron diffraction very sensitive to the crystal potential and the related charge distribution. This makes electron diffraction a powerful method for accurate measurements of low order structure factors (Fourier components of the crystal potential) and the study of crystal bonding.

Quantitative analysis consists of appropriate processing of experimental data and extraction of quantitative information from the processed data using a refinement method. This is done by pixel-by-pixel comparisons of experimental and theoretical intensities, obtaining the best fit by adjusting the parameters in the theoretical model using a goodness of fit criterion [2]. The intensity in one point in each disk in the experimental CBED pattern corresponds to exactly one diffraction condition (one incident beam direction), which makes the intensity distribution well suited for comparisons with theory. The theoretical intensities are usually based on the Bloch wave dynamical theory of high energy transmission electron diffraction [1]. A limited number of beams are included in the calculations, and additional ones are accounted for using the Bethe perturbation. Parameters refined are structure factors, absorption potentials, sample thickness, beam direction and scaling. Several diffraction geometries have been proposed [3-5]. Figure 1 shows an example of a systematic row CBED pattern with extracted line scans and the corresponding fit.

In quantitative work, energy filtering is crucial. Electron diffraction theory refers only to the elastically scattered electrons. Absorption is included in the calculations, but only as a removal of flux from the elastically scattered wave. Another important point, which has to be handled with great care, is the removal of the point spread function (PSF) in the digital recording systems (slow scan CCD cameras or imaging plates) [6,7]. The PSF comes from the averaged response of a parallel detector to a point signal.

In the refinement of low order structure factors, one needs prior information about the crystal structure. Lattice parameters, Debye-Waller factors, atom positions and operating voltage of the microscope are all fixed parameters in the refinements, and have to be known in advance. They remain fixed and errors will transfer to the refined values as systematic errors and limit the accuracy. Many of these parameters can be found using other CBED techniques [1].

The charge density associated with bonding is a very small fraction of the total charge in a solid. The bonding charge densities in a representative covalent crystal are typically of the order of 0.01% of the total charge around the core regions of the atoms. This illustrates the big challenge that the QCBED methods face and how accurate they need to be. The aim is refinement of the structure to such an extent that the deformation charge density can be found and predictions can be made about the nature of bonding, whether covalent, ionic or metallic. QCBED can give a small number of low order structure factors to very high accuracy (approaching 0.1 %). Bragg X-ray and QCBED methods complement each other [8]. Using the advantage of each method, QCBED can be used to measure accurately the absolute values of a few low-order structure factors and then be combined with X-ray results for weak and high order reflections. This has e.g. been done successfully in the case of CuO_2 [9]. [10]

References

[1] J.C.H. Spence and J.M. Zuo, *Electron Microdiffraction,* Plenum, New York, 1992.
[2] J.M. Zuo, *Mat. Trans. JIM*, 39 (1998) 938.
[3] R. Holmestad and C.R. Birkeland, *Phil.Mag* A, 77 (1998) 1231.
[4] M. Saunders et al., *Acta Cryst.,* A55 (1999) 471.
[5] K. Tsuda and M. Tanaka, *Acta Cryst.,* A55 (1999) 939.
[6] A.L.Weickenmeier et. al., *Optik*, 99 (1995) 147.
[7] J.M. Zuo, *Ultramicr.* 66 (1996) 21.
[8] B. Jiang et al., *Acta Cryst.,* A58 (2001) 4.
[9] J.M. Zuo et al., *Nature,* 401 (1999) 49.
[10] This work is supported by the Norwegian Research Council. The authors want to thank Prof. R. Høier (Norwegian University of Science and Technology), Prof. JM. Zuo (University of Illinois), Prof. John Spence and Dr. B. Jiang (Arizona State University) for fruitful discussions.

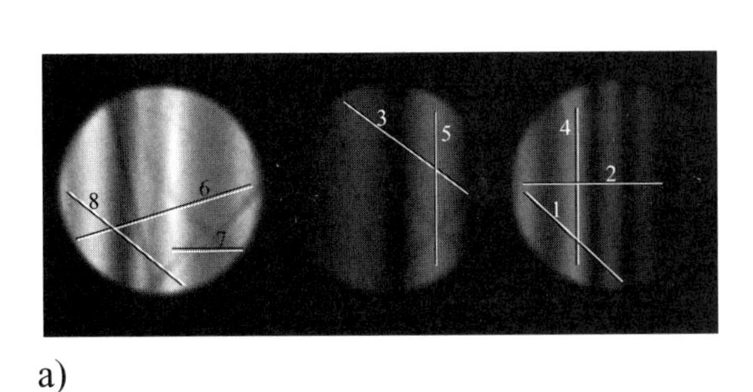

a)

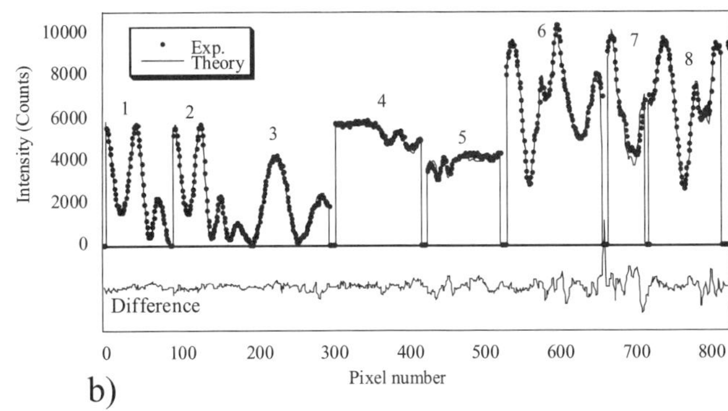

b)

Figure 1. a) Experimental CBED pattern from TiAl showing the (101) systematic row. Extracted line scans are indicated. b) Experimental line scans and best theoretical fit. The difference between theory and experiment is shown.

Microsc. Microanal. 8 (Suppl. 2), 2002
DOI 10.1017.S1431927602101978

Bloch Wave Degeneracies and Critical Voltage Effects in CBED patterns.

H. Matsuhata,* and J. Gjϕnnes

* National Institute of Advanced Industrial Science and Technology (AIST), Central 2, 1-1-1, Umezono, Tsukuba, Ibaraki 305-8568, Japan

Centre for Materials Research, University of Oslo, Gaustadalleen 0371 Oslo, Norway

In electron diffraction, the scattering factor is described as below. $f^{el}(s)$ and $f^{x}(s)$ are the scattering factors for electrons and for x-rays, respectively.

$$f^{el}(s) = (me^2/2h^2)[Z - f^{x}(s)]/s^2.$$

Where $s=\sin\theta/\lambda$, θ is the Bragg angle, λ is the wavelength of an electron , Z is the atomic number. This indicates scattering factors for low angle reflections obtained by the electron diffraction have advantage in accuracy than the ones obtained from x-ray diffraction.
In the critical voltage effect, one measures the accelerating voltage at which a particular contrast feature in Kikuchi, or CBED pattern appears, along a line or at a point. The effect depends strongly on the structure factors for low angle reflections, and is thus sensitive to the rearrangement of outer electron. The critical voltage effect is explained as an accidental degeneracy of Bloch waves. In diffraction patterns, we can observe various degeneracies of Bloch waves. Not all of degeneracies are the accidental degeneracies which can be used for structure factor estimations. In this report we describe the classifications of various degeneracies of Bloch wave observed in the CBED pattern in SnO_2, ZnS and $SrTiO_3$, [1-3] and shows some of obtained structure factors of low angle reflections.

Degeneracies observed in CBED patterns are classified as follow.
a) The degeneracies observed at the 3m, 4mm, 6mm special points. When the projected point of centre of Ewald sphere on to the zeroth order Laue zone is at 3m, 4mm, 6mm point of the zone, we can observe the degeneracies of Bloch waves. However, this is not the accidental degeneracies.
b) When a crystal has a glide-plane or screw axis, we can observe the GM-lines in the CBED discs of forbidden reflections. In the solid-state physics, it is known that the bands degenerate at the Brillouin zone boundary in crystals which belong to the nonsymorphic space group.[4] The B -line of the GM-line corresponds to this degeneracy. This is not the accidental degeneracy.
c) The degeneracies observed on mirror lines. [5] When the projected point of centre of Ewald sphere is on a mirror line of the zeroth order Laue zone, often the degeneracies between mirror and anti-mirror type Bloch waves are observed. These degeneracies are defined as accidental degeneracies in solid state physics textbook. The position of this degeneracy depends on the structure factors as well as the accelerating voltage. Some of the degeneracies move along the mirror-lines by the change of accelerating voltage, and a pair of these degeneracies disappears at the 2mm point of the Laue zone. This accelerating voltage is defined to be non-systematic critical voltage.
d) Degeneracy observed on general point caused by the three beam dynamical interaction. [6] This type of degeneracy is considered to be a general case of the accidental degeneracy on mirror-line type c. The degeneracy points move among the general points by the change of the accelerating voltage, and sometimes a pair of the degeneracies points disappear at the two-fold-rotation symmetry

point on the zeroth order Laue zone.
e) The critical voltage effect of systematic reflections [7], and zone-axis critical voltage effect at 4mm and 6mm zone axis [9]. This accidental degeneracy is observed at a fixed diffraction condition and at a certain accelerating voltage.
f) Similar phenomenon to critical voltage effect which is observed in non-centrosymmetric crystals. [8] In non-centrosymmetric crystal, often decrease in intensity by the three-beam interaction, similar to the case of critical voltage effect, is observed. In this case the dispersion surfaces branches come closer by the change of accelerating voltage. However, two branches do not touch each other. This type of phenomenon has been used for phase angle estimation of structure factors.[8] The similar phenomenon is known in molecular physics. In an asymmetric molecule energy levels of bonding outer electrons do not degenerate. [10]

By the analysis of systematic critical voltage (e), non -systematic critical voltages (c) and (d), we have estimated the rearrangement of outer -electron in ZnS, SnO2 and SrTiO3. The example of results for $SrTiO_3$ are summarized in Table 1. The results clearly show the deviation from the neutral state due to the rearrangement of the outer electrons.

Table 1

SrTiO3	U_{110}	0.0428(3) ü= $^{-2}$	, experiment 321 reflection at 133 zone axis
	U_{110}	0.0432(3) ü= $^{-2}$	, experiment 111 zone axis critical voltage
	U_{110}	0.0402 ü= $^{-2}$	, theoretical neutral state Doyle & Turner[11]
	U_{110}	0.0455 ü= $^{-2}$	, theoretical ionized state Rez *et al.*[12]
	U_{200}	0.0521 ü= $^{-2}$	, experiment 100 systematic reflections
	U_{200}	0.0516 ü= $^{-2}$	, theoretical neutral state Doyle & Turner

Reference

[1] H.Matsuhata and J.Gjϕnnes, Acta Cryst. **A50** (1994) 115.
[2] H.Matsuhata and J.Gjϕnnes, Acta Cryst. **A52** (1996) 686.
[3] H.Matsuhata and J.Gjϕnnes, Proceedings of the Int. Centennial Sympo. on Electron (1997) p470.
[4] G.Burns "Introduction to group theory with applications" New York, Academic press 1977.
[5] J.Taftϕ and J.Gjϕnnes, Ultramicroscopy **17** (1985) 445.
[6] J.Gjϕnnes and R.Hϕier, Acta Cryst. **A27** (1971) 313.
[7] D.Watanabe, R.Uyeda and M.Kogiso Acta Cryst. **A24** (1968) 249.
[8] K.Mathinsen and R.Hϕier, Acta Cryst. **A44** (1988) 558.
[9] M.D.Shannon and J.W.Steeds Philos Mag. **36** (1977) 279.
[10] W.A.Harrison "Electronic structure and properties of solids: the physics of chemical bonds"San Francisco, W.H.Freeman and Company, 1980.
[11] P.A. Doyle and P.S.Turner Acta Cryst **A24** (1968) 390.
[12] D.Rez, P.Rez and I.Grant, Acta Cryst **A50** (1994) 481.

Microsc. Microanal. 8 (Suppl. 2), 2002
DOI 10.1017.S143192760210198X

How to Select the Items for the Shopping List of Future High Resolution Electron Microscopists?

D. Van Dyck,* S. Van Aert, ** A.J. den Dekker, ** and A. van den Bos**

* University of Antwerp, Department of Physics, Antwerp, 2020, Belgium
**Delft University of Technology, Department of Applied Physics, Delft, 2628 CJ, The Netherlands

Instrumental developments continue to push the resolution of electron microscopes beyond 1 Å. Apart from the continuous improvement in resolution of the classical high resolution electron microscopes (HREM), new possibilities emerge, such as, correction of the spherical aberration both in transmission and scanning transmission electron microscopy (TEM and STEM, respectively), combination of TEM with high angle annular dark field (HAADF) STEM, monochromators, and off-axis holography. Furthermore, in our view, the new electron microscope will show a large versatility in experimental settings under computer control, such as, TEM or STEM, imaging or diffraction techniques (such as convergent beam electron diffraction, precession, and ptychography), focus, voltage, spherical aberration, beam tilt, and crystal tilt. The main limiting factors in the experiment will be the incident electron dose, that is, the amount of electrons that interact with the object during the experiment, and the recording time, because of the radiation sensitivity of the object and the specimen drift, respectively. The question then arises which instrument and which experimental settings are optimal given the incident electron dose or the recording time available. To answer this question, an optimality criterion that meets the purpose of future HREM is needed.

The last decades are characterized by an evolution from macro- to micro- and more recently to nanotechnology. In the future, it will even become possible to compose nanostructures atom by atom. Most of the interesting properties of materials are mainly related to their nanostructure. In parallel, there is an evolution in solid state theory where materials properties are increasingly better understood from first principles theoretical calculations. The merging of these fields will enable materials science to evolve into materials design, that is, from describing and understanding towards predicting of materials properties. If this evolution is to be continued, it is imperative that characterization techniques keep pace. In order to correlate real properties with simulations, atom positions in aperiodic structures should be determined with a precision of the order of 0.01 Å.

In principle, quantitative HREM is the most appropriate technique to provide the required precision. The reason for this is that from all possible imaging particles, electrons interact most strongly with matter, hereby providing most information for a given amount of radiation damage [1]. Furthermore, electrons make observations in real space possible. This is an asset for the study of aperiodic structures. Extraction of structure parameters from the images can be formulated as a parameter estimation problem. A parametric model describing the expectations of the intensity observations is needed. The model includes the electron object interaction, the transfer in the microscope, and the image detection. The unknown structure parameters are estimated by fitting the model to the experimentally obtained images using a criterion of goodness of fit. Then, the precision of the estimates of the parameters is limited by the presence of noise.

Following the lines of thought mentioned above, the obvious optimality criterion to be used is the precision of the estimates of the atom positions. Statistical parameter estimation theory allows the derivation of an expression for the highest precision of the parameter estimates [2]. Thus, it is possible to compute the lower bound on the standard deviation of the position of an atom. This lower bound, which is called the *Cramér Rao Lower Bound* (CRLB), is a function of both object and microscope parameters. The availability of the CRLB allows quantitative evaluation, comparison, and optimization of different experimental settings. This process, called statistical experimental design, can be illustrated as follows. Suppose that the microscope is able to visualize an isolated atom and that σ is the width of the image of the atom, that is, the Rayleigh resolution. The thus defined resolution will depend on object and microscope parameters only. The CRLB on the standard deviation of the position of the atom, on the other hand, may be shown to be of the order of $\sigma/\sqrt{N}$, with N the total number of detected electrons. It is now clear that it is not only the Rayleigh resolution that matters but the electron dose as well.

Recently, progress has been made in the derivation of the optimal probe and detector configuration in ADF STEM [3]. From the evaluations of the CRLB, it follows that the optimal probe is not the narrowest probe possible. Furthermore, it has been found that the radius of the hole in the detector should be as large as the optimal aperture radius. Moreover, by the same token, a monochromator usually doesn't pay off in terms of precision [4]. For example, Fig. 1 shows the CRLB on the standard deviation of the position of a gold [100] atom column as a function of the electron energy spread. Two particular cases have been distinguished. Either the recording time or the incident electron dose has been kept constant, presuming that specimen drift or radiation damage puts a practical limit on the experiment, respectively. Currently, STEM and HRTEM are compared along the same line in order to find out which method is to be preferred.

References
[1] R. Henderson, *Quarterly Reviews of Biophysics* 28 (1995) 171.
[2] A. van den Bos and A.J. den Dekker, *Advances in Imaging and Electron Physics* 117 (2001) 241.
[3] S. Van Aert et al., *Ultramicroscopy* 90 (2002) 273.
[4] A.J. den Dekker et al., *Ultramicroscopy* 89 (2001) 275.

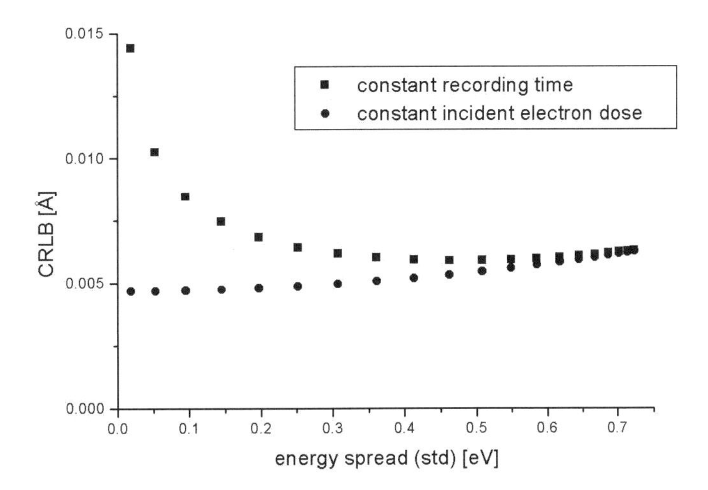

FIG. 1. Lower bound on the standard deviation of the position of a gold [100] atom column as a function of the electron energy spread associated with a constant recording time and a constant incident electron dose, respectively.

Microsc. Microanal. 8 (Suppl. 2), 2002
DOI 10.1017.S1431927602101991

The 3D structure of a complex quasicrystal approximant determined by electron crystallography

Xiaodong Zou*, Zhimin Mo*, Sven Hovmöller*, Xingzhong Li** and Kehsin Kuo***

*Structural Chemistry, Stockholm University, SE-106 91 Stockholm, Sweden
**Center for Materials Research & Analysis, Univ. of Nebraska-Lincoln, Lincoln, NE 68588-0656
***Beijing Laboratory of Electron Microscopy, P.O. Box 2724, 100080 Beijing, China

Electron crystallography can be applied for solving structures of crystals too small for X-ray diffraction. The advantage of electron crystallography over X-ray crystallography is that the phases of the crystallographic structure factors, which are lost in X-ray diffraction, are present in HREM images. Electron crystallography has been demonstrated on many structures with one short axis [1, 2]. However, for crystals lacking of short axes, atoms may overlap in any directions so several images from different directions must be combined in order to solve the 3D structure. One of such examples was shown on a mineral staurolite [3]. Here the 3D reconstruction is applied to a very complex quasicrystal approximant - ν-AlCrFe with the space group $P6_3/m$ and $a = 40.687$ and $c = 12.546$ Å. The structure was too complicated to be solved directly by single crystal X-ray diffraction and part of the structure was first deduced from the related phases by comparing the HREM images. A complete structure model was then obtained from difference Fourier maps and refined by single crystal X-ray diffraction [4].

HREM images and electron diffraction patterns of the ν-AlCrFe phase were collected from different zone axes and digitised by a CCD camera (Fig. 1). For each image, Fourier transform was calculated from the thinnest part of the crystal using the program CRISP (Fig. 2). The defocus value and astigmatism under which the image was taken were estimated from the position(s) of the dark ring(s) in the Fourier transform and their effects were compensated for. Amplitude and phase were extracted from each reflection and the symmetry restrictions were then applied. Finally a potential map was calculated by inverse Fourier transformation using the corrected amplitudes and phases (Fig. 2c). Intensities of electron diffraction spots were extracted using the program ELD.

Amplitudes and phases from 11 zone axes were merged into a 3D data set. Amplitudes from the HREM images were replaced by those from the ED patterns. A 3D potential map was calculated from the 3D data by XtalView. Atoms are clearly resolved from the map (Fig. 3). The structure is layered and contains one unique flat layer (F) and one unique puckered (P) layer. In each unit cell, six such layers are stacked along the c-axis, with the stacking sequence $PFP^m(PFP^m)'$, where P and P^m are related by a mirror, PFP^m and $(PFP^m)'$ are related by a 2_1 axis. 136 unique atomic positions could be located from the two unique layers. The structure obtained from 3D reconstruction is very similar to that obtained by X-ray diffraction [4].

References

1. Weirich, T.E., Ramlau, R., Simon, A., Hovmöller, S. & Zou, X.D. (1996) Nature, 382, 144-146.
2. Zou, X.D., Sundberg, M., Larine, M. & Hovmöller, S. (1996) Ultramicroscopy, 62, 103-121.
3. Wenk, H.-R., Downing, K.H., Meisheng, H. & O'Keefe, M.A. (1992) Acta Cryst. A48, 700-716.
4. Mo, Z.M., Zhou, H.Y. & Kuo, K.H. (2000) Acta Cryst. B56, 392-401.

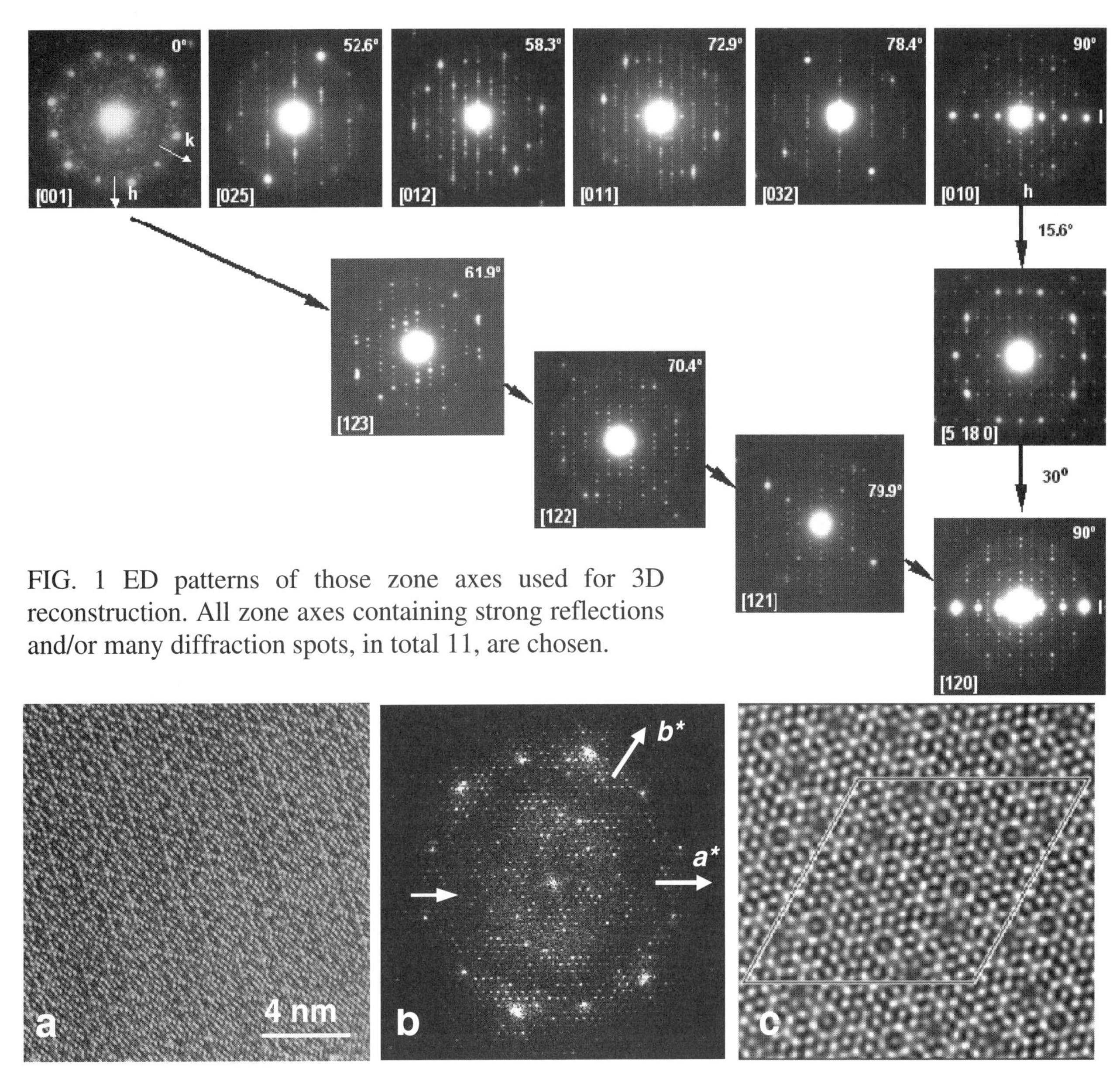

FIG. 1 ED patterns of those zone axes used for 3D reconstruction. All zone axes containing strong reflections and/or many diffraction spots, in total 11, are chosen.

FIG. 2 a) A [001] HREM image of the ν-AlFeCr phase and b) its Fourier transform. The defocus was determined from the dark ring (marked by an arrow) that corresponds to the zero cross-over. c) Projected potential maps after imposing the symmetry and compensating for the CTF.

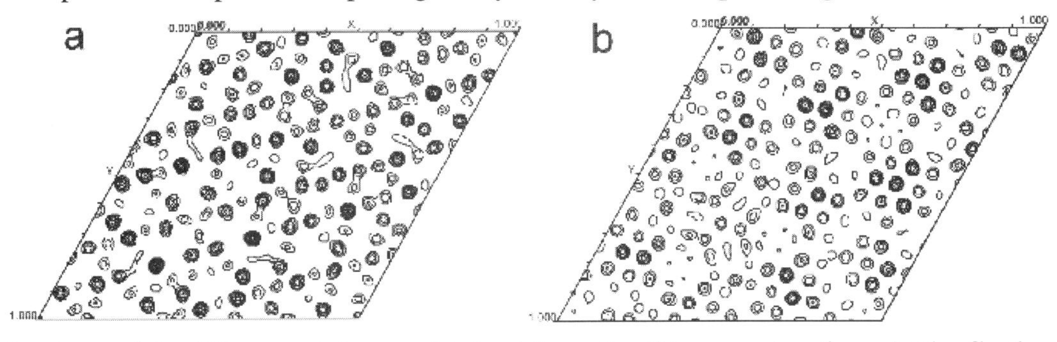

FIG. 3 Sections of the 3D potential map obtained from the 3D reconstruction. a) The flat layer (F) at z = 0.25 and b) the puckered layer (P) with z =0.35-0.45.

Microsc. Microanal. 8 (Suppl. 2), 2002
DOI 10.1017.S1431927602102005

HRTEM Resolution Extension for Interface by Gerchberg-Saxton Algorithm with Supported Constraint

F.R. Chen* , U. Dahmen** and J. J. Kai*

* Dept. of Engineering and System Science, National Tsing Hua University, HsinChu, Taiwan, ROC
** National Center for electron Microscopy-LBL, Berkeley, CA 94720, USA

When the electron beam passes through the thin specimen and objective lens, it forms diffraction pattern in back focal plane (reciprocal space) and high resolution image in the image plane (real space). Nevertheless, only partial information is preserved in either space. As in the case of x-ray diffraction, the modulus of the diffraction beams is recorded and the information of the phases lost in the back focal plane, although, it contains diffraction amplitude up to 2 $Å^{-1}$. On the contrary, high resolution TEM (HRTEM) retains distorted phases in the low spatial frequency region (less than information limit) and the information in higher spatial frequency region were cut off by the lens contrast transfer function (CTF). For a medium high voltage TEM, the phase information contains in a HRTEM image may only up to about 0.5 $Å^{-1}$, so that direct correlation of the HRTEM image of crystal structure is not trivial. Gerchberg-Saxton algorithm has been utilized to recover the information from partial data in real and reciprocal spaces [1-3]. The Gerchberg-Saxton algorithm restores spatial resolution by operating real space and reciprocal space projections cyclically. In our methodology, a generalized maximum entropy method (Kullback-Leibler cross entropy) dealing with weak object case is used as a real space (P1) projection. After P1 projection, not only the phases within the input spatial frequencies are improved, but also the phases in the higher frequencies are extrapolated. The optimum solutions from P1 projection can be further improved by a P2 projection that square root of diffraction intensities from a diffraction pattern are then substituted to complete a cycle operation of Gerchberg-Saxton algorithm. Sometimes, diffraction data may be incomplete or not available from irregular local defect such as dislocation or reconstructed interface which may make the performance of Gerchberg-Saxton algorithm harder, if not impossible. The same difficult exists in the case of surface direct method [4]. Marks has employed a prior knowledge called "support constraint" which was suggested in the image reconstruction literatures [5] to enhance the strength of the direct method for solving surface structure [4]. The concept of support constraint can be applied to the case of irregular local defect at the interface. A prior knowledge of known crystal structure in the perfect crystal region can be used as "supported constraint" by separating it from the local defect. New method has been developed to define quantitatively the support constraint region. The final process of Gerchberg-Saxton algorithm has to finish in the real space operator P1 in order to reveal the local defect at the interface. The detail procedures and mathematics will be discussed in detail in the talk. A HRTEM image of $\Sigma=5$ {210} boundary at TiO_2 is shown in FIG. 1(a). FIG. 1(b) is the final Gerchberg-Saxton solution of the square marked region in FIG. 1(a). The red and blue dots represent Ti and O, respectively. Example of a 2x1 interfacial reconstruction in the NiSi2/Si interface is shown in the FIG. 2(a). Super-periodicity of 2x1 necking contrast in FIG. 2(a) is exhibited. Without supplying the diffraction amplitude for the 2x1 interfacial reconstruction, the 2x1 super-periodicity can still be revealed. Gerchberg-Saxton solution of FIG. 2(a) is shown in FIG. 2(b). The necking contrast corresponds to deficient of Si column in the interface.

References

[1] R. W. Gerchberg and W. O. Saxton, Optik **35** (1972) 709

[2] F.-R. Chen et al., *J. of Electron Microscopy* **48** (1999) 827-836

[3] F.-R. Chen et al., *J. of Electron Microscopy* (2002) in press

[4] L. D. Marks, *Phys. Rev. B*, **60** (1999) 2771

[5] J. C. Dainty and J. R. Fienup, (1987) Chap. 7, Restoration from phase and magnitude by generalized projections, In: *Image Recovery: Theory and Application,* ed. Stark H, pp231-275 (Academic Press, London)

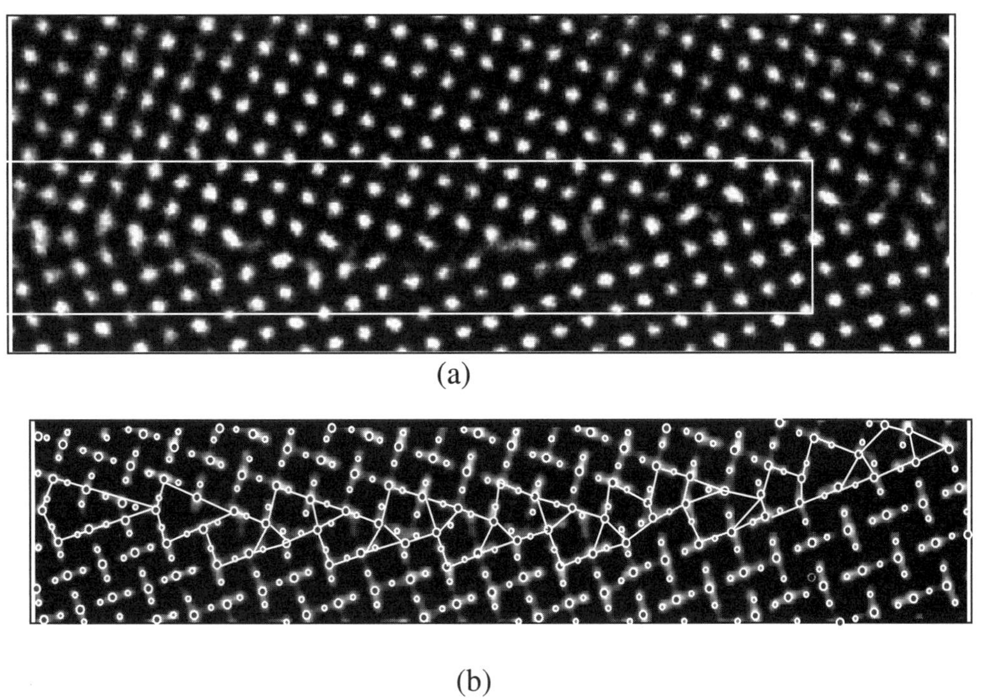

(a)

(b)

Fig. 1 a HRTEM image of Σ=5 {210} boundary in TiO2 (b) The Gerchberg-Saxton solution of (a). The red and blue dots represent Ti and O, respectively.

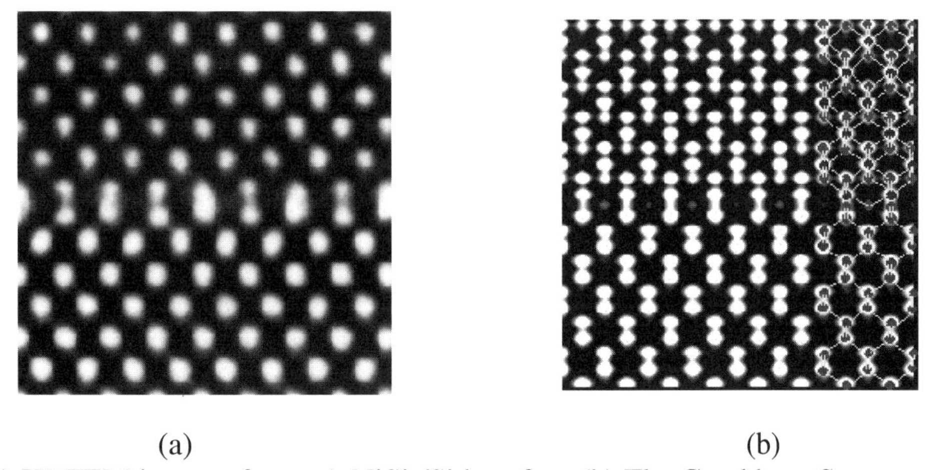

(a) (b)

FIG. 2 (a) HRTEM image of type A $NiSi_2$/Si interface (b) The Gerchberg-Saxton solution for (a). The blue and red dots represent Ni and Si, respectively.

Microsc. Microanal. 8 (Suppl. 2), 2002
DOI 10.1017.S1431927602102017

The collection of electron diffraction intensity data and their use in structure determination

J. Gjonnes*, V. Hansen** and X.Z. Li***

*Center for Materials Science, University of Oslo, Gaustadalleen 21 N-0349 Oslo, Norway
**Stavanger University College, Departement of Technology and Natural Sciences,
P.O. Box 2557 Ullandhaug N-4091 Stavanger, Norway
*** Walter Scott Engineering Center, University of Nebraska, Lincoln, NE 68588-0656. USA

Electron diffraction crystallography has re-emerged in recent years, as a realistic option for structure solution, by itself or in combination with high-resolution imaging (HRTEM). This is connected with several developments, of improved detection (imaging plates, SSCCD, energy filter), of new measuring concepts, and better crystallographic procedures. Further progress in the field will depend on the extent and precision of intensity data that can be collected. We shall discuss three principles for collecting electron diffraction intensities:

1) the usual way - the SAED or microdiffraction spot pattern, collected with a stationary beam along a zone axis works best for thin crystals in dense zones. Structures have been solved mainly in projection, often in combination with HRTEM [1,2,3]; satisfactory refinement may be obtained when dynamical scattering is included.

2) the precise way - the CBED-profile, or high-resolution electron diffraction (in analogy with current X-ray terminology) has been developed for precise refinement of small-unit cell structure factors. The technique can also be used *ab initio* for unknown structures, including accurate structure factor phases in the non-centrosymmetrical case [4,5].

3) *the X-ray way* - integrating intensities through the Bragg angle. In electron diffraction this can be achieved with the aid of scanning systems, or by integrating across features in CBED patterns. The precession technique [6,7] emulates the precession photograph in X-ray diffraction: the incident beam is tilted off a zone axis and precessed around this, combined with de-scanning of the pattern below the specimen. A Laue circle of reflections is swept through the zero Laue zone, Fig 1; a spot pattern similar to SAED is obtained. There are several advantages: sensitivity to variations in local thickness and orientation is reduced. Effects of dynamical scattering are reduced, especially those involving non-systematic interactions. This is important when several patterns are merged by normalization to common rows of reflections into a three-dimensional data set. The data can extend to high scattering angle, crystals need not be very thin.

The three-dimensional data can be treated as kinematical in the early stages of a structure determination, for calculations of Patterson projections and sections, or in direct method programs. All atoms may not be located, the structure can be completed by chemical considerations and/or by Fourier refinement, and confirmed by least squares refinement. The fit may be poor, but the general experience is that atomic arrangement tend to be correct. Improvement of the fit must be based on dynamical calculations. A confirmation/refinement strategy can utilize approximate, two-beam like expressions for integrated intensities in off-axis configurations:

$$\int I_g^{two-beam}(s_g,t)ds_g = U_g \int_0^{U_g t} J_0(x)dx$$

where s_g is the excitation error, t thickness, U_g the Fourier potential and J_0 the zero order Bessel function. This can be extended to cover multiple beam cases by introduction of dynamical or effective potentials - assuming two-beam like profile, and suitable approximations for the U_g^{eff}, *e.g.* the Bethe potentials: $U_g^{eff} = U_g - \Sigma_h U_h U_{g-h}/2ks_h$, where the excitation errors s_h refer to the position at the precession circle, *i.e.* s_g=0. By combining these two expressions in a statistical procedure, applicable to the centrosymmetric case, we can retrieve U_g, (with signs) in a projection from experimentally determine integrated intensities, as described in [8]. The result can be used for calculation of difference Fourier maps, or a least squares refinement. Application to aluminum alloy system will be referred, including combiation with high-resolution diffraction (CBED).

Coherent precipitates are common in alloy system. When these are large enough, the incident beam can be focussed on a single precipitate extending through the thickness. In other cases, as in the Al-Zn-Mg alloys, the coherent metastable are smaller and embedded in the Al matrix. SAED diffraction patterns are superpositions of four orientations, with extensive overlap of reflections. Measurement is a compromise between low-order matrix zones, with strong multiple diffraction, and projections, with fewer reflections and contributions from upper layers. Structure model has been proposed from HRTEM images.

References

[1] H.W. Zandbergen, *et al.* Micron 30 (1999) 395
[2] I.G. Voigt-Martin, *et al.* Acta Cryst A56 (2000) 436
[3] D.L. Dorset, Structural Electron Crystallography. Plenum, New York, 1995
[4] Y.F.Cheng, *et al.* Acta Cryst. A52 (1996) 923
[5] R. Høier, *et al.* Acta Cryst. A55(1999) 188
[6] R.Vincent and P Midgley, Ultramicroscopy 53 (1994) 271
[7] E. Sørbrøden *et al*, Proc ICEM 14 Cancun, Mexico 1998. Vol I, 425
[8] K. Gjønnes, *et al.* Acta cryst A54 (1998)102

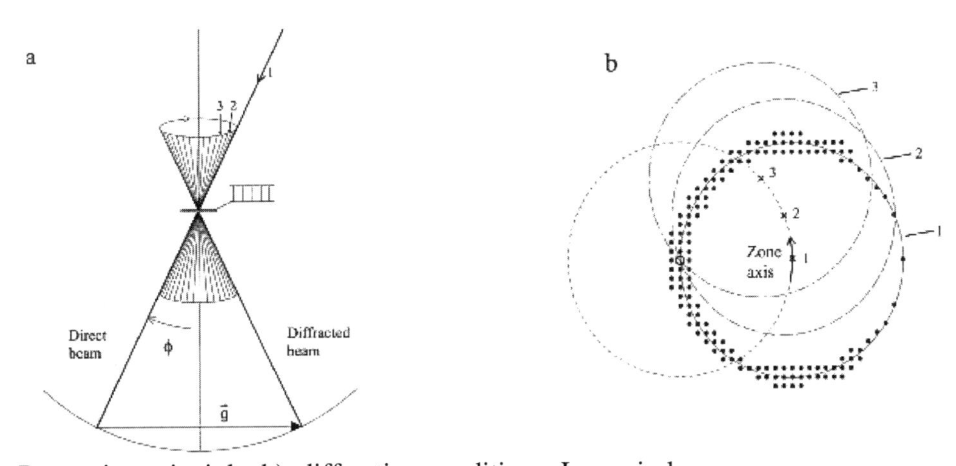

Fig. 1a Precession principle, b) diffraction conditions: Laue circles.

Microsc. Microanal. 8 (Suppl. 2), 2002
DOI 10.1017.S1431927602102029

The Accuracy of Crystal Structure Refinement from Electron Diffraction Data using Parallel Beam Illumination

J.Jansen

Nationaal Centrum voor HREM, Technische Universiteit Delft, Rotterdamseweg 137, 2628AL Delft, Nederland.

Crystal structure refine ments using electron diffraction data, obtained by using a parallel electron beam, gives accurate results if a full dynamical calculation is performed to calculate the elastic diffracted intensities from the atomic model[1]. Apart from the dynamical diffra ction it is of utmost importance to take even a slight tilt from the zone axis and the crystal thickness into account during calculation. Our refinement program MSLS, based on the Multi -Slice algorithm, can cope with all those geometrical constraints. Sinc e only a tiny crystal is used (typically 100x100x100Å), the method is extremely powerful in the area of small precipitates, thin layers and multi-phased samples. MSLS has been applied successfully to any type of material whether it being inorganic (i.e. superconductors), metallic or organic compounds. Comparison between the calculated and observed intensities is usually measured in terms of R -values, whose values appear to be of the same order of magnitude as when single crystal X-ray diffraction would have been used.

The definition of the R-values is not unique. MSLS uses R-values of the form:

$$R = \Sigma\, w_i(A_{obs,i} - A_{calc,i})^2 / \Sigma w_i A_{obs,i}^2$$

The symbol A represents here either the (observed and calculated) intensities -in this case we call the R-value R_1- or the square root of the intensities. In the latter case we refer to the R -value as R_2. w_i represents the weighting scheme. Up to now only unit weights and $1/\sigma(A)^2$ are used to refine structures with MSLS. In the example below R_1 and R_2 and the two weighting schemes are compared, as a first try to set the R -value criterion in such that the best accuracy o f the determined parameters is ensured.

Besides the choice of R -value the calculation of the intensities introduces some uncertainty. It is known that the calculation method (i.e. Multi -slice as in MSLS, Bloch waves) does not have a significant effect on the resulting intensities. However, all these methods start from the same atomic scattering factors. Currently, for simulations of electron diffraction patterns and HREM images, the Doyle and Turner table[2] is most frequently used. Several other, more re cent, tables are available in literature [3-5]. In addition some authors modified the atomic scattering factors in order to include an approximation for absorption of the electrons in the crystal[6,7]. Changing from one to the other table gives some change s to the resulting crystal structure obtained by MSLS as shown in the example below.

To illustrate the effect of the different R -values, weighting schemes and atomic scattering factor tables we use diffraction patterns of the [001] -zone of $Ce_5Cu_{19}P_{12}$ from nine different areas of which the crystal thickness varied from 80 to 200 Å. The crystallographic data are listed in table 1. For each of the different R -values and scattering factor tables the structural parameters, the crystal thicknesses and the orientation of the zones in respect to the electron beam were refined starting with the structure parameters from the original publication of the structure[1]. An example of the results the x-coordinate of one of the Ce -atoms is plotted in figure 1 as function o f the R -value/scattering factor table. It is clear that inclusion of absorption gives different results compared to the others and that the Doyle and Turner table is even different from the others which do not include absorption. The difference of the coor dinate is of the order of 0.06 Å. In the presentation of this paper it will be argued that Doyle and Turner may be the worst choice. It is obvious that the errors estimated from the diagonal elements of the refinement matrix underestimate the errors in the parameters especially in case of $1/\sigma^2$ weights.

[1] J. Jansen et al., *Acta Cryst.* A54 (1998) 91.
[2] P.A. Doyle et al., *Acta Cryst.* A24 (1986) 390.
[3] D.T. Cromer et al., *Acta Cryst.* A24 (1986) 321.
[4] D. Rez et al., *Acta Cryst.* A50 (1994) 481.
[5] E.J. Kirkland, *Advanced Computing in Electron Microscopy,* Plenum, New York, 1998.
[6] D.M. Bird et al., *Acta Cryst.* A46 (1990) 202.
[7] A. Weickenmeier et al., *Acta Cryst.* A47 (1991) 590.

atom	x	y	z		atom	x	y	z
Ce1	2/3	1/3	0		Cu4	0.6354(8)	0.1165(8)	0.5
Ce2	0.8112(9)	0	0.5		Cu5	0.4467(6)	0	0
Cu1	0	0	0		P1	0.1734(8)	0	0
Cu2	0.2868(7)	0	0.5		P2	0.6457(6)	0	0
Cu3	0.3789(6)	0.1776(7)	0		P3	0.3171(4)	0.8329(5)	0.5

Table 1 *Crystallographic data of $Ce_5Cu_{19}P_{12}$. Spacegroup : P-62m, a=12.4Å, c=4.0 Å*

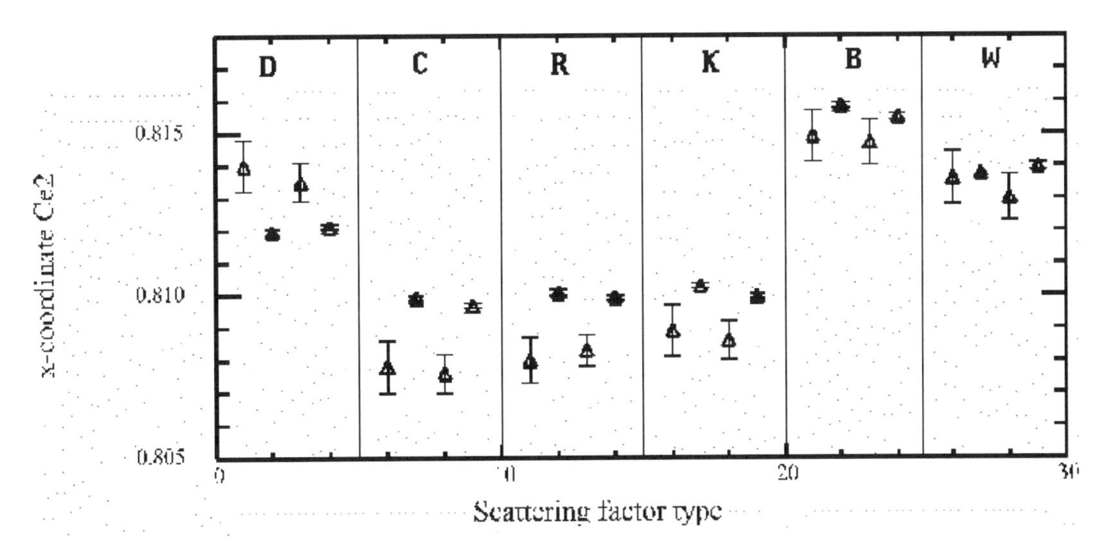

Figure 1 *Refinement results for the X -coordinate of Ce2. Each group of 4 points represen ts one of the scattering factor tables: from left to right Doyle, Cromer, Rez, Kirkland, Bird, Weickenmeier. Within each group of 4 points the first 2 are obtained using R_1 and the last 2 using R_2. Point 1 and 3 in each group are obtained using unit weights, the other 2 using $1/\sigma^2$ weights.*

Microsc. Microanal. 8 (Suppl. 2), 2002
DOI 10.1017.S1431927602102030

Single Crystal Electron Crystallography on Organic Molecules

Ute Kolb*, Galina Matveeva**

* Institut für Physikalische Chemie, Johannes Gutenberg Universität, Welderweg 11, 55099 Mainz, Germany.
** Institute of Macromolecular Compounds of the Russian Academy of Sciences, Bolshoi pr. 31, 199004 St. Petersburg, Russia.

Many compounds crystallize not only in small crystals non-suitable for single crystal x-ray structure analysis, they also exist in different modifications. The structures of these materials can be accessed by single crystal electron diffraction and x-ray powder measurements.

X-ray powder diffraction delivers two-dimensional data, which suffers often from overlapping reflections, inadequate crystal quality, unknown impurities and preferred orientation. Electron diffraction on the other hand provides us with three-dimensional data. Unfortunately, it suffers from an incomplete reciprocal space (missing cone problem), elongated reflections (spike function) and is effected by multiple scattering and dynamical scattering for thick samples. Because dynamical scattering is less serious for organic materials, structure determination can be carried out based on kinematical procedures in the first step and the resulting model can be refined dynamically in a second step. The combination of both approaches provides us with the tool for structure analysis of organic molecules [1].

The lack of reliable intensities, which x-ray powder diffraction and single crystal electron diffraction have in common, leads often to a failure of approaches, which use only intensity information for an "ab initio" structure analysis such as "direct methods" or Patterson approach. Examples for successful use of the tangent formula or Sayre equation for structure determination from electron diffraction data are worked out intensely by D. Dorset [2]. Other direct methods like maximum entropy can provide us with an envelope of the molecules in the cell, which gives an idea of its orientation [3-5].

To optimize their physical properties which are strongly influenced by the structure we investigated a number of non-linear optical active compounds for laser applications. All crystallize in non-centrosymmetric space groups: 1-(2-furyl),3-(4-benzamidophenyl)-2-propene-1-one (FAPPOBE) has a non-linear optical activity of approx. 100 x urea whereas (E)-1(4 aminophenyl)-3-(4 bromophenyl)-2-propen-1-one (U800) shows a non-linear optical activity up to 800 times urea. The third compound 1-(2-furyl),3-(4-aminophenyl)-2-propene-1-one (FAPPO) crystallized in three modifications which could not be purified. The structure analysis for two of the three identified modifications was performed only based on electron diffraction data.

Approx. 100-200 Å thick crystals were grown from hot ethanol or toluene. In diffraction mode three dimensional data sets were measured on a Philips EM300 at 100keV and a FEI TECNAI F30ST by tilting the sample with a rotation-tilt holder around suitable axes (max. tilt angle ±60°). Determination of the orthorhombic unit cells ($Pna2_1$, $Pca2_1$ and Fdd2) were performed from approx.15 zones originating from a tilt around main axes. Intensities were collected using an Nikon AF-4500 scanner with 2400 dpi resolution and 12 bit grey level or a 1024x1024 slow-scan CCD.

Refinement of cell parameters we used x-ray powder diffraction pattern measured in transmission and fitted with a Pawley Fit with DASH or Cerius2 [6, 7].

The conformation of the molecules in the gas phase as well as dipole μ and hyperpolarisability β were calculated by semiempirical, quantummechanical methods (MOPAC6.0 ; PM3)[8]. Alternative approaches are the search of the Cambridge Database [9] or "ab initio" calculations [10].

For simulation methods cell dimensions, space group, the initial model of the molecule and structure factor amplitudes were used. Packing energy minimisation [11] and simulation of diffraction patterns [7] were performed alternatingly gaining a good R-factor for each zone. Based on the obtained atomic coordinates, hyperpolarizability tensors were calculated using quantum mechanical methods and related to the crystal by appropriate co-ordinate transformation [12]. Separately, a 3D-data set was buildt up and statistical ab initio methods such as „Maximum Entropy"[13, 14] were used.

[1]I.G. Voigt-Martin, U. Kolb, International School of Crystallography, 26th course: Electroncrystallography, Erice (1997).
[2] D.L. Dorset, Structural Electron Crystallography, New York: Plenum Press 1995.
[3] G. Bricogne, Acta Cryst. A Found Cryst. A46:830, 1984.
[4] G. Bricogne and C. Gilmore, Acta Cryst. A Found Cryst. A46:284, 1990.
[5] C.J. Gilmore, G. Bricogne and G. Bannister, Acta Cryst. A Found Cryst. A47:830, 1991.
[6] DASH – CCDC Software Ltd., Cambridge, GB.
[7] Cerius2 version 4.2MS, Molecular modeling environment from Accelrys Inc., 9685 Scranton Road, San Diego, CA 92121-3752, USA.
[8] J.J.P. Stewart, MOPAC6.0 *A General Purpose Molecular Orbital Package*, QCPE.
[9] CSD – Cambridge structural database, CCDC Software Ltd., Cambridge, GB.
[10] TURBOMOL v235, Biosym Technologies, San Diego, 1993.
[11] DREIDING2.21, S.L. Mayo, B.D. Olafson, W.A. Goddard III J. Phys. Chem. 94:8897-8909, 1990.
[12] A.V. Yakimanski et al., Acta Cryst. A53, (1997) 603-614.
[13] I.G. Voigt-Martin, et al. Ultramicroscopy 68, (1997) 43-59.
[14] I.G. Voigt-Martin, et al. Phys. Rev. B59, 10, (1999) 6722-.

Microsc. Microanal. 8 (Suppl. 2), 2002
DOI 10.1017.S1431927602102042

Discussion of Ways to Energy-Filter the Electron Backscattering Pattern (EBSP) in the Scanning Electron Microscope (SEM).

Oliver C. Wells, Research Staff Member Emeritus, IBM Research Division, PO Box 218, Yorktown Heights, NY 10598

An electron backscattering pattern (EBSP) is formed by the electrons that are scattered with the highest energies from a crystalline target that is illuminated with an electron beam (EB) of kilo-electron-volt (keV) energy. In appearance it closely resembles the electron channeling pattern (ECP) that is formed in the scanning electron microscope (SEM) when the incident EB is rocked about a point. According to one possible explanation, it is assumed that in either case the pattern is formed by the electrons that leave the specimen with a minimum loss of energy in the reflected electron peak following a single wide-angle Rutherford-type scattering event (typically through about a right angle) close to the surface. Now if ECP and EBSP are reciprocal, they can be related by reversing the direction of the (zero-loss) electrons in the specimen. It can then be argued that these patterns must be formed because the probability of such a Rutherford-type wide-angle scattering event is modulated by both the incoming (ECP) and outgoing (EBSP) channeling conditions [1-6]. The slower scattered electrons undergo scattering, channeling and diffraction events that are in addition to this basic process. The related reflection high-energy electron diffraction (RHEED) pattern is formed with grazing incident and exit directions with electrons that are diffracted by Bragg planes parallel to the surface.

Generally the EBSP is formed over a limited solid angle by placing a detecting screen in line-of-sight from the specimen or (in principle) with a retarding-field energy filter [6]. An alternative approach that is suggested here is to magnetically filter the scattered electrons by mounting the sample between the polepieces of a magnetic immersion lens, with the limitations: (a) The sample must be totally non-magnetic. (b) It will be irradiated more heavily than with the standard EBSP.

Fig. 1 shows a solid sample mounted between the polepieces of a magnetic immersion lens. The magnetic field is adjusted so as to focus the incident EB onto the specimen. This same magnetic field also deflects the scattered electrons to follow spiral paths that periodically return to the lens axis. The fastest scattered electrons (reflected electron peak) can reach a limiting surface at the end of the first half-turn that cannot be reached by the slower scattered electrons. The slower scattered electrons can, however, go beyond this limiting surface on subsequent turns of the spiral path if not prevented from doing so [7]. It has been demonstrated that a detector that is placed at a short distance inside this limiting surface will collect only the fastest scattered electrons and can be used to give a magnetically filtered low-loss electron (LLE) image in the SEM [8-10].

The next step is to make use of this situation to give an energy-filtered EBSP, based on the idea that there is a one-to-one correspondence between the direction in which the fastest scattered electrons leave the specimen and the point at which these electrons reach (just touch) the limiting surface:

(a) Thus, if a curved detecting screen can be placed at exactly this boundary surface over a sufficiently large area, then it will show an energy-filtered EBSP by detecting the fastest scattered electrons as they just touch this surface. However, it mightnot be easy to align an image screen that must be curved in two directions with the limiting surface that will move as the incident EB is moved on the specimen.

(b) Another method would be to place a flat image-forming screen (similar to that in a digital camera) so that it intersects the limiting surface at an angle. The image is then analysed by a computer to locate the single line of pixels that lies along the edge of the illuminated area at the boundary. The sample is then either rotated about some axis or the screen is moved in some way to give a series of lines of pixels that can be placed side by side to form an EBSP. The incident EB can be scanned and/or deflected under computer-control so as to always land at the same point on the specimen as it rotates (for example) about the axis of the lens. (The idea of rotating the sample and merging the relevant information as the image changes can in principle also be applied to the conventional method for recording EBSP.) It is believed that these techniques should be investigated for the purpose of recording an energy-filtered EBSP [11].

References:

[1] L. Reimer, Scanning Electron Microscopy, Physics of Image Formation and Microanalysis, Springer-Verlag, 1985.
[2] L. Reimer, Private communication: "...scattering is more likely to occur into a favorable direction than into an unfavorable one..." 1994.
[3] D.C. Joy, Proc. 52nd. Ann. Mtg. MSA, (1994) 592.
[4] O.C. Wells, Scanning 21 (1999) 368.
[5] ECP with a retarding-field energy filter: E.D. Wolf, P.J. Coane and T.E. Everhart, Cong. Int. Micro. Electron., Grenoble, 2, (1970) 595.
[6] RHEED pattern with a retarding-field energy filter: staib-us@staib-instruments.com
[7] O.C. Wells and E. Munro, Ultramicroscopy 47 (1992) 101.
[8] O.C. Wells, F.K. LeGoues and R.T. Hodgson, Appl. Phys. Lett. 56 (1990) 2351.
[9] O.C. Wells, E. Munro and I.M. Fisher, Inst. Phys. Conf. Series No. 119, Section 5 (EMAG91), (1991) 181.
[10] O.C. Wells, Proc. ALC'01 (Nara, Japan, Nov. 2001) in press, 2002.
[11] I would like to thank P.E. Batson, J. Del Vecchio, D.J. Dingley, L. Dylla, R. von Gutfeld, M. Kammler, D.E. Newbury, J.R. Michael, M.T. Postek and R.M. Tromp for helpful discussions.

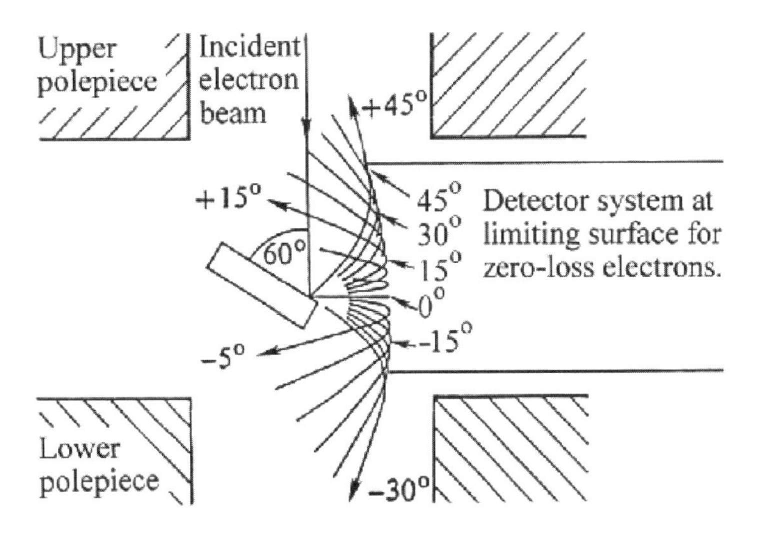

FIG. 1. Sample mounted between the polepieces of a magnetic immersion lens, showing the limiting surface just reached by the fastest electrons that are scattered from the sample at the angles shown above or below the horizontal plane (r-z plots of trajectories; modified from ref. 10).

Microsc. Microanal. 8 (Suppl. 2), 2002
DOI 10.1017.S1431927602102054

EBSD Analysis Optimised for Twin-related Boundaries

Valerie Randle

Dept. Materials Engineering, University of Wales Swansea, Swansea SA2 8PP, UK.

The nickel-based superalloy Nimonic PE16 is precipitation strengthened by the presence of gamma-prime. Overaging the alloy results in an increase in the pinning force on the boundary of at least a factor of two, and the grain size is very stable when the alloy is annealed below the gamma-prime solvus, namely 880°C [1,2]. Another microstructural feature of PE16 is that annealing twins are readily formed because of its low stacking-fault energy. Experiments have been conducted to explore the evolution of grain boundary parameters in PE16. Specimens of commercially heat treated PE16 were annealed in air at 850°C, i.e. 30°C below the gamma-prime solvus, for times of 1h, 10h and 100h. Misorientation data across interfaces were obtained by use of an electron back-scatter diffraction (EBSD) system from HKL Technology in a Philips XL30 SEM [3]. Several orientation maps with a grid step size of 0.5μm were collected and analysed from each specimen.

Figure 1 shows a typical example of an orientation map from overaged PE16. In the coincidence site lattice (CSL) system boundaries are categorised by Σ, the reciprocal density of coinciding sites. In figure 1 Σ3 boundaries are represented as thick white lines, Σ9 and Σ27 boundaries are thin white lines and other boundaries are black. The statistics for the $\Sigma 3^n$ family are shown in figure 2. Apart from Σ3, plus some Σ9 and Σ27, there were almost no other Σ boundaries recorded. There was a small increase in Σ3 proportion from 43% after 1h annealing to 50% after 100h. More than 90% of the total length of Σ3 in the map is within 2° of the exact reference misorientation, which indicates that these Σ3s are predominantly annealing twins. The remaining 10% of Σ3 length deviates by more than 2° from the reference structure. It is evident from the morphology of these higher deviation Σ3s that they are more akin to grain boundaries than to twins.

Higher $\Sigma 3^n$ boundaries in the microstructure are usually generated by impingement of two appropriate $\Sigma 3^n$ boundaries, e.g. $\Sigma 3+\Sigma 3 \rightarrow 9$, $\Sigma 3+\Sigma 9 \rightarrow 27$. The proportions of Σ9 and Σ27 in the present data are particularly low, which is an indication that there has been little movement and interaction of Σ3s, hence disallowing their impingement. This also explains why most Σ3s are twins: there has been very little impingement to produce Σ3s via 'back' reactions of the type $\Sigma 9+\Sigma 3 \rightarrow \Sigma 3$, which is expounded in the 'Σ3 regeneration model' [4]. This near-stagnation in the grain boundary population is a result of the strong grain boundary pinning by gamma-prime precipitates which inhibited grain boundary migration.

References

[1] W. Betteridge and J. Heslop, *The Nimonic Alloys,* Edward Arnold, London, 1974.
[2] V. Randle and B. Ralph, *Acta Metall.* 34 (1986) 891.
[3] V. Randle and O. Engler , *'Introduction to Texture Analysis: Macrotexture, Microtexture and Orientation Mapping'*, Gordon and Breach, London, 2000.
[4] V. Randle, *Acta Mater.* 47 (1999) 4187.

Figure 1. Example of an orientation map from PE16 after annealing for 10h. Σ3 boundaries are represented as thick white lines, Σ9 and Σ27 boundaries are thin white lines and other boundaries are black.

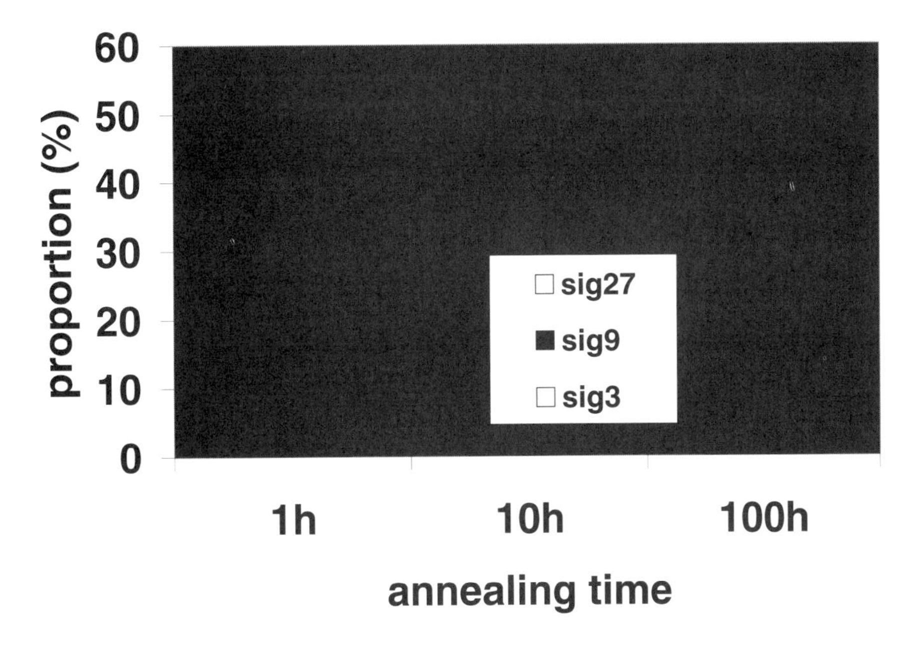

Figure 2. Proportions of $\Sigma3^n$ boundaries after 1h, 10h and 100h annealing below the gamma-prime solvus. The scale bar is 100μm.

Microsc. Microanal. 8 (Suppl. 2), 2002
DOI 10.1017.S1431927602101875

An Assessment of the Pros and Cons of Low Voltage X-ray Analysis in the SEM

E D Boyes

DuPont Company, CR&D, PO Box 80356-383, Wilmington, DE 19880-0356, USA.

Low voltage analysis is now widely used in many industrial applications. It has two main characteristics. The primary advantages (pros) derive from the strong sensitivity to voltage of the electron beam penetration range into the sample. The depth (Fig. 1) and spatial resolution (Fig. 2, 3) of analysis broadly follow the Bethe relationship where for a beam voltage E_o the range $R = F(E_o)^{5/3}$. For the same x-ray line energy the improvement from 20kV to 5kV is about one order of magnitude (10x). The escape range of the x-rays through the sample is reduced by a similar amount; leading to a large reduction and in many, but not all, cases effectively elimination of the matrix absorption (A) and fluorescence (F) corrections. As a practical consequence light elements are generally more faithfully represented, and sometimes even overly so, in low voltage analyses (Fig.3). The downside is a similarly enhanced sensitivity of the analysis to surface coatings, oxides or contamination; although in practice this problem can be controlled quite well. For micron and sub-micron sized features the geometrical factors may dominate the corrections required for accurate analysis. Restricting the analysis volume to smaller features which are more likely to be single phase meets one of the prime criteria for accurate analysis of heterogenous materials and simplifies interpretation.

The primary disadvantages (cons) of low voltage analysis lie in the characteristics of the x-ray emissions. Firstly, the beam voltage must exceed that of the target x-ray line by at least a factor of 1.3x. This may require the selection of lower energy x-ray lines with different, less favorable and often less well known parameters. Secondly, even with the same x-ray line energy (E_x) the peak to background and kcps/nA sensitivity are reduced at lower beam voltages (Fig. 4). This is a fundamental limitation of the method. For a given series of x-ray lines, e.g. K, the problem can be partially generalized on the basis of overvoltage ($U = E_o/E_x$). The characteristics of the different types of x-ray detectors also play a role. A typical EDS system has an energy resolution and a P/B ratio inferior to a WDS system by a factor of ~10x. Although the practical EDS energy resolution improves at low x-ray line energies (e.g. ~60eV at 300eV vs ~130eV at 6kV) the opportunity for improvement with new types of detector combining improvements in both sensitivity and energy resolution is unquestionably substantial.

Complications arise when the two primary characteristics interact. An example is the sensitivity for analysis of small (sub-micron and even nm) surface particles as is typical in root cause analyses of process defects. At lower voltages the beam is more concentrated in the particle, by a mass factor of ~1000x in the previous example, but the P/B is at the same time reduced by a factor of ~10x; for a nett gain in sensitivity of ~100x. However, if light elements are involved (as the SiO_2 example in Fig. 3) the relative sensitivity for oxygen analysis with respect to Si gains back ~5x due to greatly reduced differential mass absorption of the O(K) x-ray line along the shorter x-ray escape path and the disproportionate attenuation of the Si signal with beam energy. The relative mass sensitivity for O is therefore ~500x improved. The detailed numbers will depend on the particular experiment but these data are representative of the big gains which can typically be achieved in low voltage analysis; even if they are not quite as large as may at first be promised.

The balance of corrections is transformed at low beam voltages. The conventional ZAF corrections are replaced with something more like the Cliff-Lorimer model for thin sections in the analytical transmission electron microscope, but with much modified Z factors. Ideally, and in most but not all practical cases, the A and F corrections are reduced to insignificance which greatly simplifies quantization, especially where samples are heterogeneous on a fine scale. However the Z term combining x-ray fluorescence with detector sensitivity becomes much more complex and a strong function of both electron beam and x-ray line energies. Fortunately these issues can be addressed for a particular system with a limited number of standards. A simplification is to fix the beam voltage for quantitative analysis at one value, say 5kV, as we have mostly done. For analysis at 5kV non-conducting samples may require coatings not necessary for imaging at ~1kV. The acute surface sensitivity of 1-10nm and >100x range at a beam energy of 1.5kV requires great care in the preparation, preservation and transfer of sample surfaces. At 5kV continuous coatings <5nm should stabilize the imaging and analysis conditions without adding appreciably to the spectra. At this time there is also a serious lack of data to support quantitative analysis at low beam voltages and we are still working to determine exactly which parameters are important. In principle low voltage analysis should be more sensitive and more precise than the conventional approach; but this is so far probably only realized in selected examples where light elements are analyzed directly.

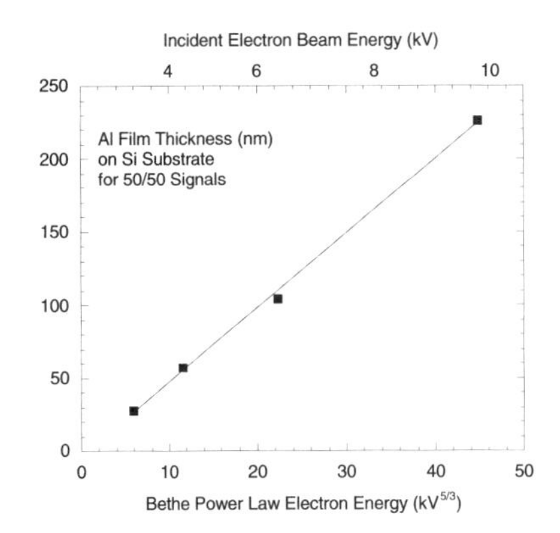

Fig. 1 : Depth resolution Al thin film on Si wafer substrate.

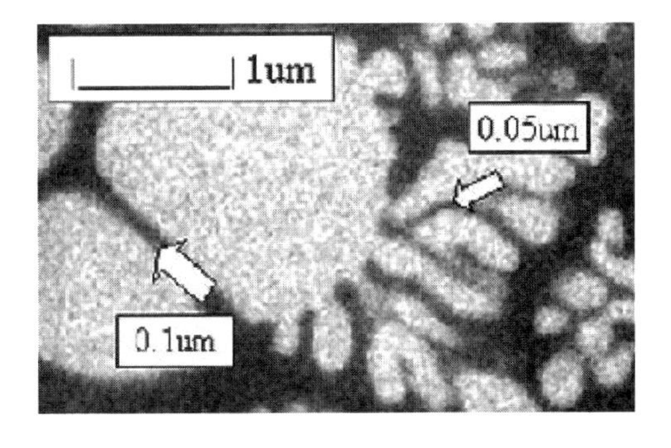

Fig. 2 : High lateral resolution ZrO_2 in SiO_2 matrix.

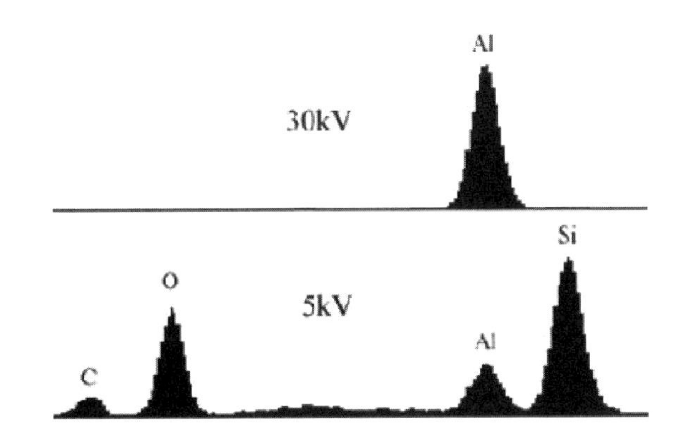

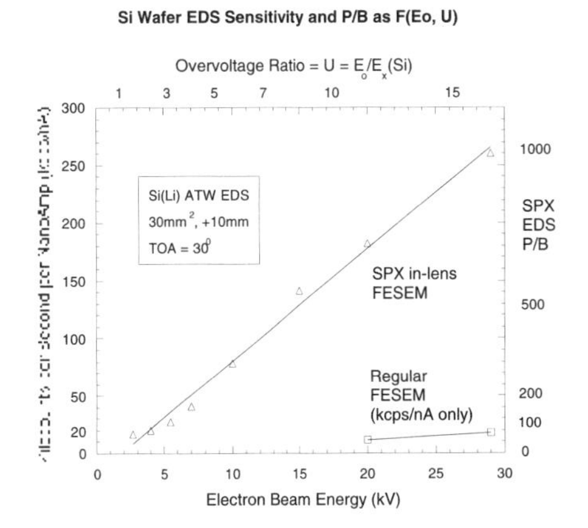

Fig. 3 : SiO_2 on Al substrate at 3 and 30kV.

Fig.4 : kcps/nA sensitivity and P/B for Si

Reference : G Cliff and G Lorimer, *J Microscopy*, 103 (1975) 203

Microsc. Microanal. 8 (Suppl. 2), 2002
DOI 10.1017.S1431927602101887

About the Topographic Contrast in LVSEM

J. Cazaux

DTI, Faculty of Sciences, BP 1039, 51687, Reims Cedex 2, France

In SEM, at primary beam energies E_0 above 5 keV, the observed topographic contrast may easily be explained from the change of the SEE yield, δ, with the angle of incidence, i, when standard Everhart-Thornley (E-T) detectors are used [1,2]. As shown in fig 1, a&b (diamond sample) more complicated is to establish the same correlation in LVSEM (E_0 <1 keV) with an in-lens detector because the difficulty lies also in the architecture of the detection system.

The constant loss theory accounts fairly well for the yield from materials[1-3]. By a simple change of the range of incident electrons R into R/cos i; it expresses the evolution of δ with i (fig 2a &[4]): $\delta(E_0)=B\ [E_0/E(e.h)]\ [s/R\cos i]\ [1-\exp.-(R\cos i/s)]$ (1), [B: the escape probability; s: empirical attenuation length of the secondary electrons; E(e.h): energy required for generating an electron-hole pair]. For Rcos i>>s (or E_0>5 keV), eq. 1 may be written in the form $\delta(i)= \delta(0)/\cos i$ (2) which accounts for the topographic contrast obtained with a E-T detector. When E_0 is decreased down to R<<s, $\delta(i)$ takes the linear form, independent from i,: $\delta(i)=B\ [E_0/E(e.h)]$ which may explain the decrease in contrast previously observed with this type of detector [5]. In addition, the role of the emission angle α has to be taken into account when in-lens detectors are used (fig. 2b). This role may explains the main contrast obtained at 0.2 and at 0.5 keV but the remaining difference between the two images (indicated by arrows in fig 1; 0.2keV in) require a deeper analysis.

To go deeper into the subject, various instrumental and theoretical causes may be considered. In modern instruments (ex.: fig 3a), an efficient in-lens detection is associated to a cross-over position being a function of the primary beam energy and of the working distance[6]. A non-Lambert-type emission of the SE's (fig 2c) may be due to a residual electric field on the specimen surface, to charging effects (insert fig 3a) but also to the physics of the SE emission induced by sub-KeV beams. In particular, the transport of the SEs in the specimen is less simple than suggested by eq.1; it involves isotropic random walks that may lead to escape probabilities, B, larger than 50% (fig.3b & [7]). Sophisticated electron optical calculations for the SE trajectories in and out of the specimen are always possible but the first step is to control the stability of the microscope. For this goal, the use of the scatter diagram technique seems very promising (fig 1d & [4]) and it may also be applied to two images acquired after an azimuth rotation (to control the axis-symmetry of the equipment).

References

[1]. D.C. Joy, J. Microscopy, 147 (1987) 51
[2] H. Seiler, .J. Appl. Phys. 53 (1983) R1
[3] J. Cazaux, J. Appl. Phys. 89(2001) 8265
[4] J.M. Patat et al, Scanning, in press.
[5] D.C Joy & C.S Joy, Micron. 27(1996)247
[6] E. Plies. in Proc. 12th EUREM, Brno, Czech Rep., P. Tomanek & V. Kolarik ed. vol I,(2000)423
[7] J. Cazaux, Nucl. Instr. & Meth in Phys. B ,in press.

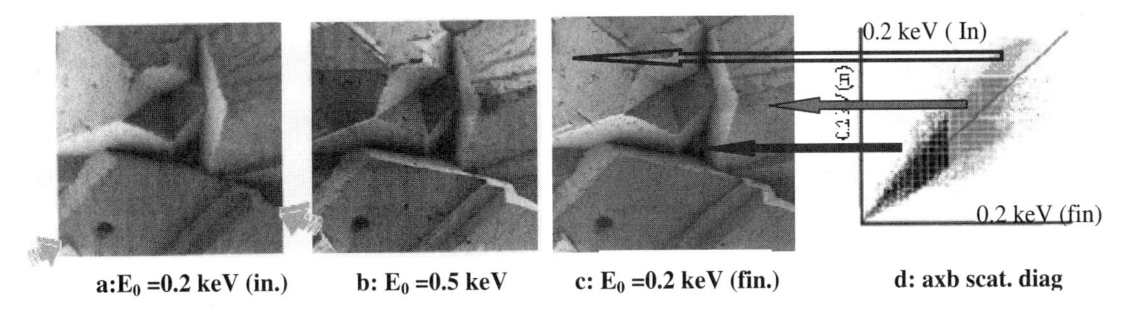

FIG 1a&b: Obtained at E_0 = 0.2 and 0.5 keV (left) with a SEM (LEO, Gemini DSM 982:in-lens detection; magnification x1000; w.d.: 2.5 mm), images of a micro-structured diamond specimen. After a series up to E_0 = 15 keV, a final image, c, has also been taken at 0.2 kV(fin). d: use of the scatter diagram technique for correlating the initial and the final images acquired at 0.2 kV [4].

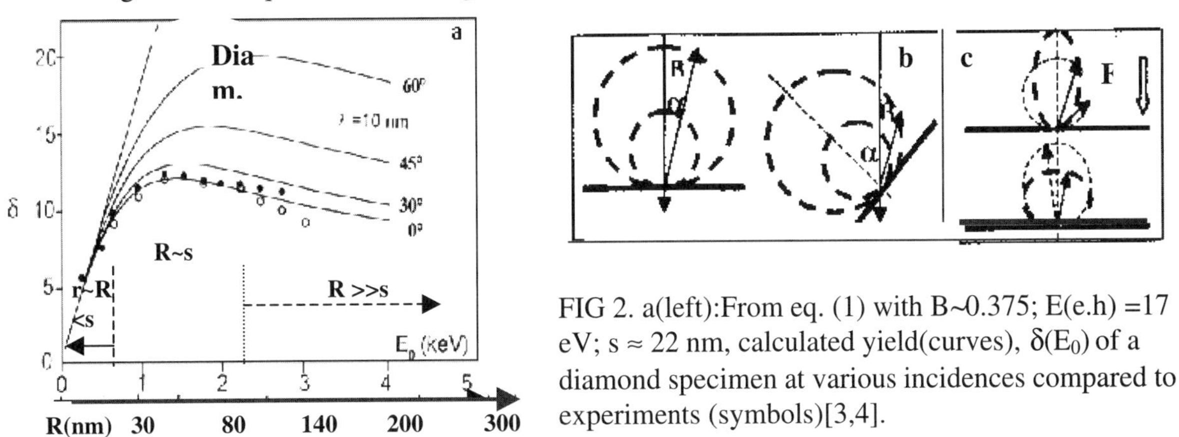

FIG 2. a(left):From eq. (1) with B~0.375; E(e.h) =17 eV; s ≈ 22 nm, calculated yield(curves), $\delta(E_0)$ of a diamond specimen at various incidences compared to experiments (symbols)[3,4].

FIG 2b(medium): Influence of the detection angles α & β in the case of a Lambert emission. c(right) :possible causes of deviation of a cos α emission.

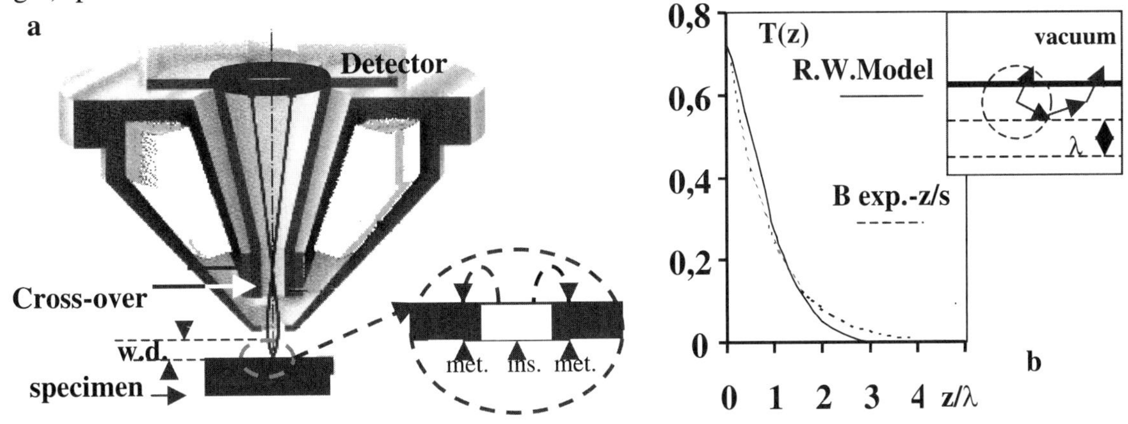

FIG. 3.a(left):Influence of the architecture of the instrument on the SE detection efficienc y; insert: local field effect in differential charging. b(right): Comparison between SE transmission functions, T(z), based on an exponential decay (dashed line) and on a random walk model, R.W.M, involving a transport mean free path λ. Note that T(0) (or B) is > 50%[7].

Microsc. Microanal. 8 (Suppl. 2), 2002
DOI 10.1017.S1431927602101899

HIGH RESOLUTION EXAMINATION OF BIOLOGICAL SAMPLES USING FIELD EMISSION SCANNING ELECTRON MICROSCOPY

S. Erlandsen*, J. Detry**, C. Ottenwaelter*, and C. Frethem*

*Department of Genetics, Cell Biology, and Development, University of Minnesota Medical School, Minneapolis, MN, 55455
**Honeywell, Plymouth, MN 55447

The development of field emission SEM (FESEM) in the last two decades has revolutionized the examination of biological samples. Advances in the design of field emission guns (cold field and Schottkey guns), the improvements attained with immersion lens design for both in-lens and below-the-lens instruments, and the preparation of thin metal coatings (~1 nm) all contributed to rapid advances in achieving high resolution imaging of features on biological surfaces.

Resolving power of a SEM is related to the size of the electron probe rastered over the sample, and probe size is usually inversely proportional to keV. High voltage provides smaller probe size (improves resolution) but also produces more SE2 which reduce contrast. Low voltage FESEM, defined as less than 5 keV, has been proposed (1) as desirable for examination of biological samples due to decreased penetration of the sample surface by the primary beam, resulting in higher contrast of surface features due to a reduction in SE2. Use of low voltage (<5 keV) FESEM seems to result in less specimen contamination and radiation damage when compared to use of higher keV (1).

Biological samples for FESEM can be divided into two main types: a) infinitely thin, and b) bulk samples. Infinitely thin samples, e.g. viruses, macromolecules, can be examined at higher keV due to the thinness of the samples (<50nm) and are usually coated with chromium for SE imaging. Bulk samples, e.g. cells, tissues, can be examined by SE imaging but often show signs of charging artifact. Advances in BSE detectors permit examination of bulk biological samples with simultaneous imaging of coating metals together with discrimination of high atomic number tracers, such as colloidal gold (Figs. 1-4).

Development of the double coating method of Walther et al (2,3) together with cryo-immobilization of specimens have facilitated examination of molecular topography (Fig. 3). Retention of the shell of hydration on surface molecules prevents molecular collapse seen with complete freeze-drying. The double coating method permits use of higher keV for improved BSE imaging with smaller probe diameters, and problems frequently encountered in specimen damage in cryo-SEM, e.g. signs of cracking and contamination, are seen to a much lesser degree with this double coating method.

BSE imaging can be accomplished by different types of BSE detectors: 1) solid state silicon or proprietary phosphorus detectors are the least costly, 2) YAG crystals doped with cerium are expensive, but offer the highest resolution, or 3) newly introduced ExB filters function well, particularly at low magnification, at very low voltages (~1 keV) where silicon and YAG are insensitive due to threshold limitations (Figs. 5-6). Due to low BSE yield, chromium is not a desirable coating for high resolution BSE imaging and Pt or W are preferred. Detection of colloidal gold (~10 nm diameter) can be accomplished routinely in FESEM at magnifications <50,000X with most BSE detectors using 3-5 keV. High resolution BSE imaging at high magnification (>100,000X) is optimal with the YAG detector (Fig. 4).

Detection of colloidal gold probes with BSE has become routine in low voltage FESEM, and in conjunction with cryo-immobilization methods, FESEM can now provide direct information about molecular topography of cell surface molecules.

(1) J. Pawley et al *The Science of Biological Specimen Preparation for Microscopy and Microanalysis*, Scan. Micros. Intl., Chicago, 1989.
(2) P. Walther et al, Scanning 19 (1997) 343.
(3) S. Erlandsen, BioTechniques 31 (2001) 300.

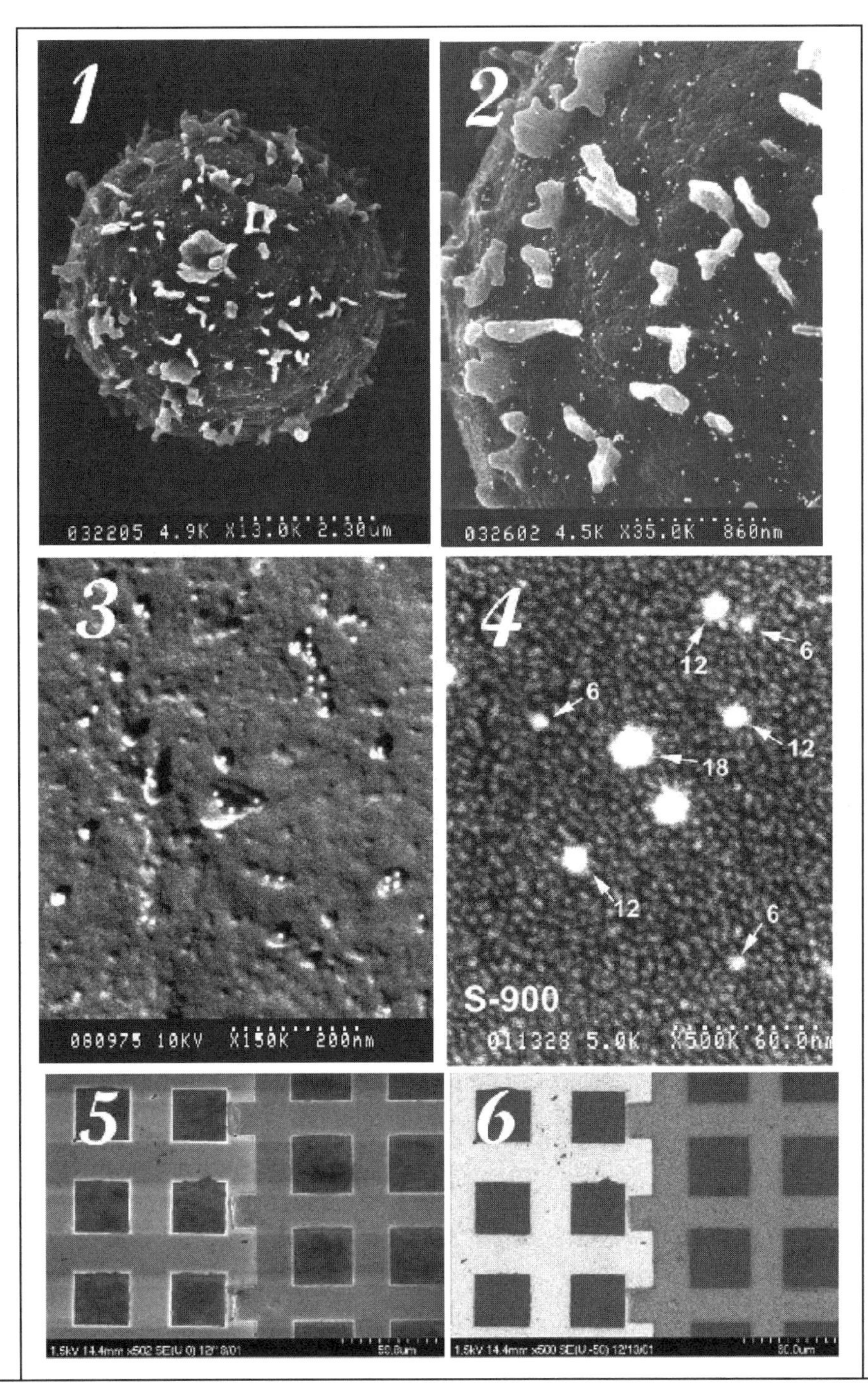

Figures 1 and 2. Low voltage FESEM immunogold localization (10nm) of CD43 on human neutrophil by backscatter electron imaging (YAG detector). Figure 3. Double layer coating of cryo-immobilized human platelet showing immunogold localization (10nm) of P-selectin molecules on membrane surface (YAG detector). Figure 4. Colloidal gold standard coated with ~1 nm of Pt and imaged by BSE with YAG detector in an in-lens FESEM. Figures 5 and 6. Test grid of gold and copper imaged at 1.5 keV by SE in figure 5, or in figure 6 with ExB filter (-50eV) in a S-4700 FESEM. The gold grid is on the left in each figure.

Microsc. Microanal. 8 (Suppl. 2), 2002
DOI 10.1017.S1431927602101905

Physics of Low Voltage Scanning Electron Microscopy

Raynald Gauvin

Department of Mining, Metals and Materials Engineering, McGill University, Montréal, Québec, Canada, H3A 2B2

With the advent of the Field Emission Scanning Electron Microscopes (FE-SEM), images can be acquired routinely with incident electron energies as low as 500 eV. Recently, images acquired in a FE-SEM with incident electron energy of 50 eV have been reported[1]. In this context, it is critical to model electron scattering in solids at such low electron energy. Despite the success of Monte Carlo simulations based on the Continuous Slowing Down Approximation[2], it is clear that a different model should be used to compute energy loss for electron energy below 1 keV.

To model electron scattering below 1 keV, the probability of inelastic collision must be computed at each collision and if an inelastic collision occurs, energy loss must be computed using an appropriate distribution obtained from an appropriate physical model. The dielectric theory allows computing such an energy loss distribution function if the complex dielectric constant of the simulated material is known. This complex dielectric constant can be determined for a specific material from an Electron Energy Loss Spectroscopy (EELS) experiment[3].

In order to validate a Monte Carlo program that computes discrete energy loss at each inelastic collision, comparison with measured signals from this materials must be performed. Figure [1] shows all the measured backscattering coefficients reported so far for gold[4]. Below 2 keV, the scattering of the experimental measurements is about 50 %. It is clear that the validation of Monte Carlo programs for energy smaller than 2 keV is impossible with such a lack of good experimental data. As seen in figure [2], the situation is even worse for the measured secondary electron yields of silicon[4]. The scatter of experimental data below 2 keV is about 200 %. Clearly, very accurate measurements of fundamental parameters like backscattering coefficients and secondary electron yields are needed in order to validate Monte Carlo models of electron scattering at low electron energy.

References

1. D. C. Joy and C. S. Joy (1996), Proceedings of Microscopy & Microanalysis, pp. 144 – 145.
2. P. Hovington, D. Drouin and R. Gauvin (1997), Scanning, Vol.19, pp. 1-14.
3. D.C. Joy, S. Luo, R. Gauvin, P. Hovington and N. Evans (1996), Scanning Microscopy, Vol. 16, pp. 209-220.
4. D.C. Joy (2001), Microscopy & Microanalysis, Vol. 7, pp. 159– 167.

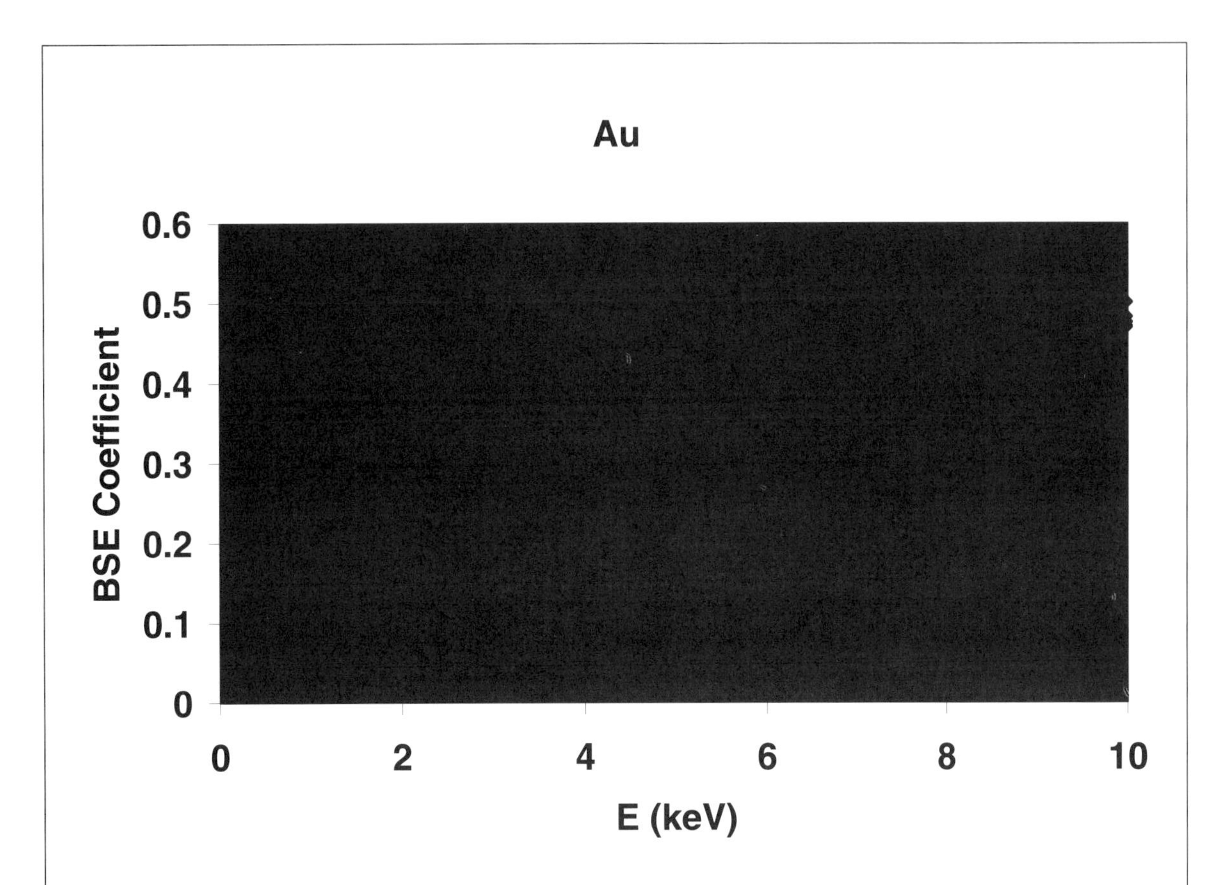

Figure 1 Measured Backscattered Electron coefficient of Gold as a function of E_0.

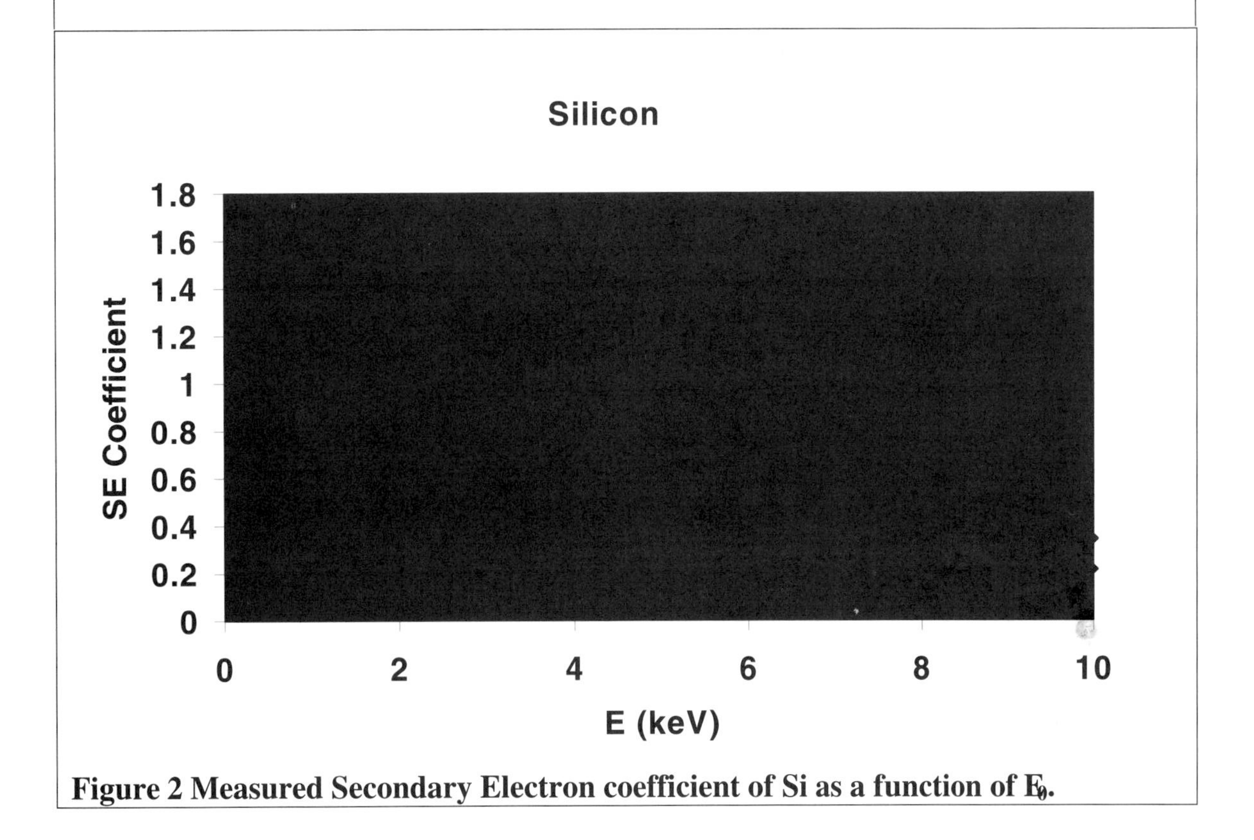

Figure 2 Measured Secondary Electron coefficient of Si as a function of E_0.

Microsc. Microanal. 8 (Suppl. 2), 2002
DOI 10.1017.S1431927602101917

Advanced Instrumentation for Low Voltage Scanning Microscopy

David C Joy

EM Facility, University of Tennessee, Knoxville, TN 37996, USA and

Oak Ridge National Laboratory, Oak Ridge, TN 37831, USA

The basic electron optical design of the scanning electron microscope (SEM) derives from its distant ancestor the high voltage oscillograph and has remained essentially unchanged for 70 years. Similarly most of the components of the SEM retain a close resemblance to those found in the first commercial instruments 40 years ago. However, the continuous pressure to further enhance the performance of the low voltage SEM in the face of fundamental physical limitations is forcing a reappraisal of both individual components and of the basic electron-optical design.

Since the brightness of all electron sources varies linearly with their energy low voltage guns compare poorly with those used at conventional energies. Consequently there is much interest in new ultra-bright sources, such as nanotip field emitters, which can provide as much, or more, brightness at 1keV or below as older sources operated at 20keV. The energy spread of the source is also a major issue because chromatic aberration results in enlarged spot sizes and poorly shaped probe profiles at the lowest energies. Sources, such a cold field emitters which have an energy spread of ~0.3eV and negative affinity sources which can have an energy width below 0.1eV, will become increasingly important. In order to further minimize the effects of chromatic aberration the convergence angle of the beam must also be restricted. At the lower beam energies, however, this means that the optical performance becomes diffraction limited because of the large wavelength of the electrons. To achieve the highest levels of performance the chromatic and spherical aberrations of the probe lens must also be significantly reduced. This can now be achieved by the use of an aberration corrector[1]. Such a device results in both a significantly smaller probe diameter and higher beam current from the same electron source. However, this is achieved at the expense of a drastically reduced depth of field, which may make conventional operation of the SEM very difficult. Alternatively a cathode lens configuration, in which the incident electron is rapidly decelerated to the desired landing energy just prior to striking the sample, can be employed[2] This arrangement is simple and efficient, however the high electric fields normal to the specimen surface can lead to significant problems with charging. In either case it now becomes possible to maintain the instrumental resolution essentially constant down to beam energies as low as 20eV or so (figure 1) provided that adequate screening against external disturbances can be provided. Advances are also possible and desirable being in the design of electron detectors. The familiar Everhart-Thornley detector is of relatively poor efficiency (DQE~0.1) and collects a signal which a mixture of secondary and backscattered electrons, leading to poor image contrast. Through-the-lens (TTL) detectors, which use the post-field of the lens to collect the secondary signal, have not only a much higher efficiency (DQE~0.8) but are also more selective in the energy spectrum of the electrons that they accept. In the most advanced design the TTL detector can , in effect, be tuned so as to collect either a pure SE signal, or a backscattered signal, or some mixture of the two[3]. This provides great flexibility in imaging, avoids the necessity of a separate BSE detector, and permits backscattered operation at very short working distances. The ultimate goal remains a detector which is both efficient and can be used to select any arbitrary energy window in the emitted spectrum for imaging.

Radically new designs which seek to avoid some of these problems are also now being pursued by several groups. Low Energy Electron Microscope (LEEM) based systems use flood beams of low energy electrons (or light or X-rays) to excite secondary electrons which are then accelerated and imaged through a lens systems onto a CCD or similar recording device[4]. This approach removes the problem of generating a small probe at low energies, and can provide very high sensitivity to surface topography, potentials, and chemistry, although the broad energy spread of the secondary emission means that aberration corrected imaging lenses are still required for high resolution. An additional benefit is that image formation is a parallel rather than a sequential operation which provides a significant speed advantage for some applications. An alternative approach is the point projection microscope (PPM) which consists only of an electron source, the sample, and a detector[5]. Because there are no lenses, the resolution is limited only by the size of the electron source and by the electron wavelength. Since the source must be physically very small, to eliminate penumbral blurring, it is also highly coherent and so the image becomes an in-line Fresnel hologram (figure 2).

References

1. J Zach and M Haider, (1995), Nucl.Instrum. and Methods **A363**, 316
2. I,Mullerova and L. Frank, (1993), Scanning **15,** 193-201
3. S. Joens, (2001), Micros. and Microanalysis **7**, Suppl. 2, 875
4. M. Mundschau, (1991), Ultramicroscopy **36**, 29
5. B G Frost, A Thesen, and D C Joy, (2002), Proc. SPIE Conference on Microlithography, D Herr, ed: San Jose, CA (in press)

6. This work has been partially supported by SRC Contract LJ-413.004

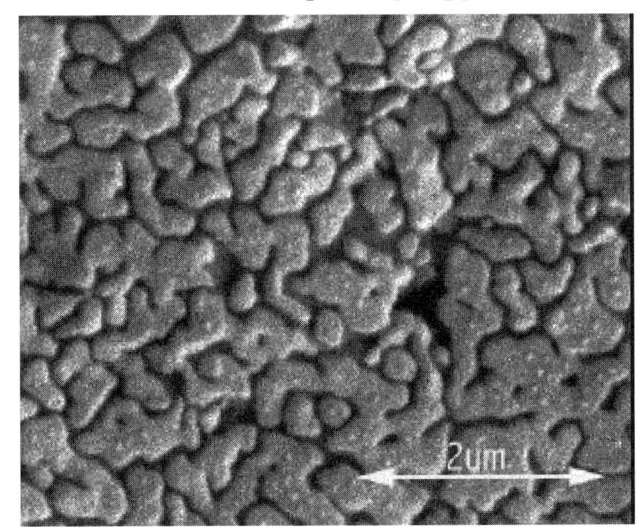

Figure 1. SE image of a gold film at a landing energy of 30eV in a field emission SEM operated with a retarding cathode lens

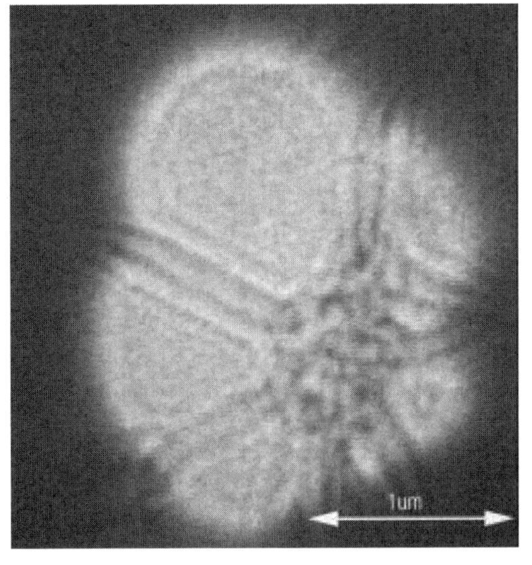

Figure 2. In-line Fresnel hologram of a holey carbon film. 350eV in a point projection microscope

Microsc. Microanal. 8 (Suppl. 2), 2002
DOI 10.1017.S1431927602101929

Low Voltage Energy Dispersive Quantitative X-Ray Microanalysis of Inorganic Light Elements in Bulk Frozen Hydrated Biological Specimens

Patrick Echlin

Cambridge Analytical Microscopy, 65 Milton Road, Cambridge CB4 1XA, United Kingdom.

Although there are many advantages to working at low voltages and low temperatures [1] [2], the topic discussed here remains one of the most challenging areas of quantitative x-ray microanalysis. The specimens, which have naturally low concentrations of elements, are beset with problems associated with sample preparation, peak overlaps, uncertain analytical algorithms and radiation damage. It is not feasible to use ultra-thin (50-100nm) frozen hydrated sections as they are quickly destroyed by the electron beam and there are problems maintaining the fully hydrated state in thicker (500-1000nm) frozen hydrated sections. This paper will discuss some of the critical issues that arise during low voltage quantitative x-ray microanalysis of frozen hydrated bulk biological material and suggest ways we may hope to improve the analytical procedures.

An accelerating voltage of 5keV is suitable for most studies but there are advantages to working down to 3keV which is now accepted as the lowest practical voltage for x-ray microanalysis [3]. However at 3keV there is not enough energy to produce sufficient x-rays from elements much above Z=15 [4]. With frozen hydrated samples, there is a 17 fold decrease in the size of the interactive volume between 10 and 5 keV and a further 15 fold decrease between 5 and 3keV. The sample volume from which x-rays may be detected is smaller than the volume penetrated by the primary electrons It is important to carefully monitor the landing energy of the primary electron beam and avoid any charging. The non-invasive charge balance procedures [5] are only practical for SEM imaging of organic material up to E2=2keV and do not appear to be applicable to x-ray studies carried out at 3keV and above. It is unwise to rely just on a visual examination of the sample as a check that there is no charging. It is far better to check the Duane-Hunt limit, first on the calibration standard and then on the specimen throughout spectral acquisition [6]. The most practical way to avoid charging is to coat the sample with a thin continuous layer of a conducting metal bearing in mind the caveats of peak overlaps, electron and x-ray photon energy attenuation and specimen damage. Because the volume from which the detected x-rays is so small, the coating layer has to be very thin (2-3nm), must not form an oxide layer and must not be contaminated. The metals Be, Al and Cr together with C, despite various advantages and disadvantages, may be used as coating materials [7]. Radiation damage is a real problem. There is nothing that can be done to ameliorate the primary events and secondary physical and chemical reaction associated with radiation damage as these occur anywhere between 10^{9} to 10^{-18} sec. The tertiary events that centre on the production and movement of free radicals, occur between 10^{-2} to 10^{-4} sec at room temperature, and their diffusion rate is temperature dependent. A study made of radiation damage in the environmental SEM, confirmed that liquid and gaseous water acts as a rich source of small, highly mobile free radicals, of which the hydroxyl (.OH) radical is the most dominant and reactive species [8]. There is an increase in the quantities of reactive species with beam energy. The increased viscosity of ice slows the movement of free radicals, and their rates of reaction decrease with temperature resulting in a 3-8 fold decrease in radiation damage. [1]. The consequences of beam damage in ice, even at 100K, are mass loss, loss of crystallinity, structural damage and the production of volatile materials that could influence the landing energy of the primary electron beam. The two analytical algorithms, ZAF and

P/B ratio method, form the basis of quantitation. At the low voltages we are discussing here, the count rate decreases by x100, the minimum detectable limit is only between 1-10%, and there are severe peak overlap problems with the Ca and K L lines and the C and N K lines [3] [6].

The specimen preparation and analytical procedures currently being used are as follows. Small samples (1-3mm^3) are quench cooled at between 2-10^3 K s^{-1}, cryofractured at 110K at 5 x 10^{-6} torr (667µPa), plasma magnetron sputter coated with 2nm Cr at 120K using high purity argon at 5 x 10^{-3} torr (667mPa) and transferred to the cold stage of the microscope held at 120K and 1 x 10^{-6} torr (133µPa). The sample is kept under high vacuum throughout the whole process and the collected x-rays analysed by the P/B ratio algorithm using a ATW germanium ED detector. At the present time, using a 200s live time and a beam current of between 250-400pA, quantitative analysis to within +/-5%, has been successfully carried out on plant material, with a total inorganic elemental content of between 2-4% FW, at 5keV for Na, Mg, Al and Si; at 6keV for P and S; at 7keV for Cl; at 8keV for K and at 9keV for Ca, all at a spatial resolution of 50-100nm [9].

The elements C and N make up a substantial part of the 10-12% FW of the sample organic material Their K lines overlap the much smaller L lines of K and Ca and it not possible to use these two lines for analysis and thus reduce the accelerating voltage closer to 5keV for the analysis of all elements Z=11-20. The two L lines could be resolved using a WD detector or the newer bolometric detector, although it remains to be seen whether the two very small peaks could be used to obtain quantitative data. Water, which is 88% oxygen, makes up 80-90%FW of the samples and for quantitation, the O K line is used as a measure of the water (ice) content. The O K line is overlapped by the Cr L line and although considerable progress has been made in measuring the contribution of Cr to the O peak [10] further studies are needed on how to maintain an anoxic vacuum environment in the microscope. We need to calculate the oxidation rates of the other metals that have been used to coat frozen hydrated samples. Although preliminary studies suggest that Be, Mg and Al oxidize at a faster rate than Cr, we need to find out whether Mg might be a better coating material than Cr. We need to look for more accurate ways of continuously measuring the landing energy of the low voltage beam on the specimen. We need more information on the fundamental constants used for quantitative analysis in order to improve the matrix corrections that are used in quantitative analysis [11]. We need to explore the effects on quantitation of reducing the overvoltage from the usual 2 to closer to unity.

References

1. Echlin, P. (1992) Low Temperature Microscopy and Microanalysis, Plenum, New York.
2. Joy, D.C. & Joy, C.S. (1996) Micron 27 247.
3. Joy, D.C. (1999) Microscopy & Microanalysis 5 (Supp 2) 302.
4. Echlin, P. (2001) Microscopy and Microanalysis 7 211.
5. Joy, D.C. & Joy, C.S. (1995) Microscopy and Microanalysis 1 109.
6. Newbury, D.E. (2001) Microscopy and Microanalysis 7 (Supp 2) 702.
7. Goldstein, J. I. et al (1992) SEM and X-ray Microanalysis. Plenum, New York.
8. Royal, C.P. et al (2001) Journal of Microscopy 204 185.
9. Echlin,P. (2000) Microscopy and Microanalysis 6 (Supp 2) 758.
10. Echlin, P. (2000) Proc. Micros. Soc. Southern Africa 30 1.
11. Joy, D.C. (2001) Microscopy and Microanalysis 7 (Supp 2) 664.

Microsc. Microanal. 8 (Suppl. 2), 2002
DOI 10.1017.S1431927602101930

New FESEM Design for 1nm at 1kV Imaging, EDS and BSE Nanoanalysis, and a Discussion of Diffraction Limits, Depth of Field and the Future

E D Boyes

DuPont Company, CR&D, PO Box 80356-383, Wilmington, DE 19880-0356, USA

A novel approach to FESEM design integrates new levels (>2x) of low voltage image resolution (~1nm at 1kV and <0.5nm at 20-30kV) with greatly improved (15x/0.3sr/30mm^2) sensitivity for EDS elemental nanoanalysis and chemical imaging, with spatial resolutions down to <100nm and in favorable cases with sensitivity limits of 1-10nm, while retaining the many advantages of robust and representative wide area (mms) bulk specimens. High resolution and sensitivity backscattered (Z) imaging and BSE nanoanalysis are being added. The goal of this project is to extend the practical capabilities of the FESEM as far as possible with component technologies available at the end of the 20th century, and with our current understanding of the laws of physics as they apply to SEM.

The heart of the novel DuPont/Hitachi S5000SPX or X-scope instrument combines a cold field emission gun (FEG) aggressively processed for high emission stability (<1% per hour variation in probe current can be achieved) and a new design of final lens with record breaking (in any electron optics ?) low levels of aberrations down to Cs~0.1mm and Cc<0.2mm (Fig.1). These low values are achieved by natural means; i.e. without correctors. The position of the specimen in the lens is changed to optimize selected experimental parameters. Figs.1 and 2 illustrate the parameter space. The highest tilt range (currently +/-25^0) is achieved at the gap center with Cs=0.8mm and a resolution of ~1.5nm at 1kV. The corresponding collection angle with the special 30mm^2 Noran ATW EDS detector is 0.3sr; and about 15x that of a typical conventional SEM. To optimize low voltage (<5kV) imaging with limited tilt and restricted EDS the sample is positioned <0.5mm from the face of the upper polepiece with Cs~0.1mm and a secondary electron image resolution ~1nm. The very low aberrations are achieved by focusing the probe with only a small fraction of the highly excited lens field.

In principle much better natural aberrations (Cs, Cc) can be achieved at low voltages in SEM than at high ones in TEM. Absolute excitations (AT=4200) and the accompanying thermal and magnetic saturation issues remain reasonable with relative excitations ($NI/V_R^{1/2}$ >100) which would be astronomical for a 200kV STEM. In the third and fourth zone modes, which may be practical only at low voltages, natural aberrations can be achieved which are superior to anything possible without correctors at 100-200kV. The final limit in LVSEM is access for the sample, stage and detectors. The SEM application is greatly simplified in that the electron beam forming the probe enters on the same side of the lens field as the electron imaging signals exit towards the (in-lens) detectors in the upper column. With a bulk specimen only about 10% of the total lens field is used to focus the probe and what would have happened if the electron beam had continued further into or through the specimen is of no concern. In comparison to a 1Δ STEM at 200kV, the 1nm LVSEM with a cold field emitter is in principle even more sensitive to AC magnetic fields. Other engineering and environmental factors are also important. Another significant limitation on very short focal length and high resolution operations is the need to maximize the aperture size to avoid diffraction limits intruding, especially at low and very low voltages. The serious trade off is in depth of field. This remains workable at ~20nm at 1kV (and ~1nm resolution at 300,000x, Fig.3) but it obviously is no longer as extensive as in classical SEM operations at more modest resolution. This would seem to set a practical limit on what we could hope to achieve even with an aberration corrected system; although the greater flexibility in specimen access and WD this would make possible, with some other complications, at the current resolution would have value, especially for in-situ experiments.

The TEM-style jeweled rod specimen holders for the X-scope are innovative, defy conventional wisdom at no more than 1mm thick, are mostly home built, and with some operational restrictions are reverse compatible with older S5000 systems. Samples are preferably <0.5mm thick and 2mm is the maximum. The end sections of the holders are electrically isolated with provision for probe current measurements and continuous touch alarms and stage motor interrupts ensure safe tilting to optimum angles. The regular holders for 3mm disks or grids allow +/-25^0 of tilt as do the ones for 4.5 x 8.5mm plates, and a stage limited 60^0 of tilt is available with 2.3mm wide plates and disks. For operation at the highest z positions with very low Cs (~0.1mm) and very short WD (<0.5mm) special holders offset the specimen position above the eucentric tilt axis of the stage.

The design of (Fig.2) of the optics, vacuum system, specimen chamber, stage and specimen holders, and the SEI, BSE and ATW EDS detectors, have been fully integrated in the ab initio design of the new system to great benefit and minimum compromises; except explicitly in specimen size which so far is not proving to be very restrictive in applications with particles, pigments, fibers, catalysts.

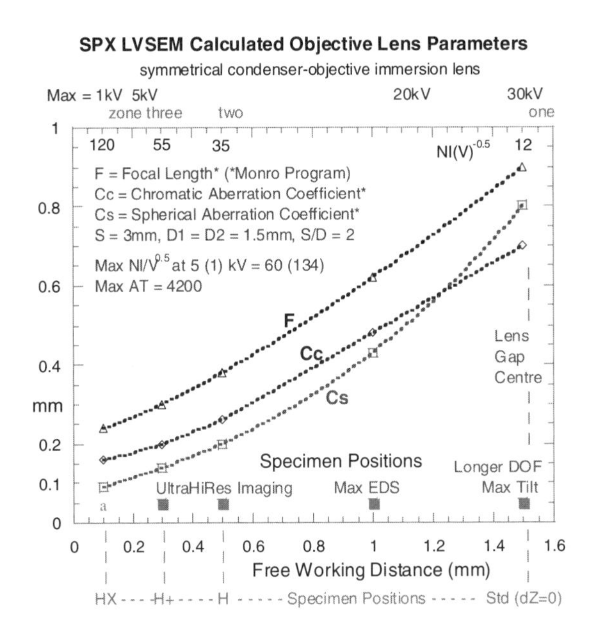

Fig.1 : Cs, Cc, F calculated for new X-scope

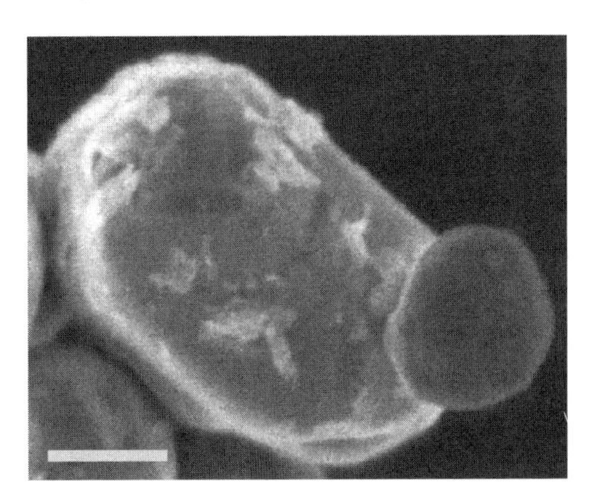

Fig.3 : Partially coated ceramic particle at 1kV
Scale is 60nm, original mag 150,000x

Fig.2 : X-scope specimen area schematic ➔

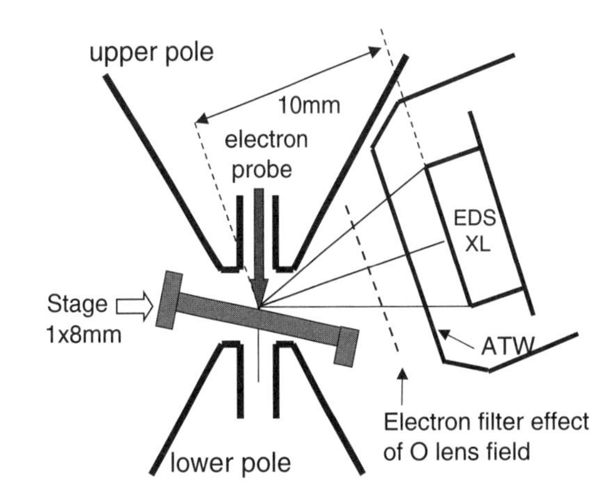

Microsc. Microanal. 8 (Suppl. 2), 2002
DOI 10.1017.S1431927602101632

Size Selective Synthesis of Colloidal Platinum Nanoparticles for Use as High Resolution EM Labels

D.A. Meyer and R.M. Albrecht

Department of Animal Sciences, University of Wisconsin, Madison, WI 53706

In the 30 years since Faulk and Taylor introduced the immunogold technique [1], labeling with colloidal gold (cAu) particles has become the predominant method for electron microscopy (EM). Frens subsequently developed a procedure to prepare monodisperse Au suspensions over a broad range of particle diameters [2], which enabled multiple labeling with cAu particles of different sizes. However, the number of different cAu labels that can be used simultaneously for multiple labeling is limited by a number of factors. The minimum particle diameter is determined by the instrument resolution, and each label must correspond to a set of particles with average diameters sufficiently distinct from one another to prevent size overlap and confusion between different labels. As the diameter increases, larger particles decrease the spatial resolution of the labeling and sterically hinder other particles from gaining access to their attachment sites. Both of these factors decrease the labeling efficiency. In order to address these concerns, recent efforts in this lab have led to the development of two novel approaches to EM multiple labeling, employing colloidal particles of similar sizes but of different shapes and elemental compositions [3].

The drawbacks inherent with using colloidal metal particles of different sizes for multiple labeling do not negate the utility of having available particles in a wide range of sizes in order to maximize the versatility of the labeling technology. Frens demonstrated with cAu that exquisite control over the average particle diameter could be achieved by means of a standard condensation technique in which gold chloride is reduced by varying concentrations of trisodium citrate in aqueous solution. The citrate reduces Au atoms to zero valency, which then nucleate. Subsequent growth of the colloidal particles occurs as the remainder of the Au atoms in solution crystallizes around the nuclei. Higher citrate concentrations maximize the rate of nucleation relative to crystallization, leading to the formation of a large number of nuclei and, hence, particles with smaller diameters. On the other hand, lower concentrations of the reducing agent induce the formation of smaller numbers of gold nuclei and particles with larger diameters. However, Frens' citrate method has thus far proven inadequate for the synthesis of colloidal particles composed of metals other than Au over a broad range of particle diameters. Like the citrate method, other techniques that control the independent processes of nucleation and crystallization, such as manipulation of the reaction temperature, work only for specific metals, in this case palladium.

One promising, simple, and practical avenue for the size-controlled synthesis of colloidal particles, composed of a potentially wide variety of different metals, was developed by Zsigmondy during the early years of the past century [4]. His method involves the reduction of gold chloride in the presence of previously synthesized cAu particles one to three nanometers in size that act as nuclei around which additional Au crystallizes. Particle size is controlled by varying the concentration of cAu nuclei relative to the Au salt concentration, much as Frens controlled particle size by adjusting the amount of reducing agent added. Voigt and Heumann applied Zsigmondy's method to the formation of uniform cAg particles [5]. We have recently utilized this approach to prepare cPt particles of varying diameters by reducing aqueous solutions of chloroplatinic acid with tannic acid in the presence of cPt nuclear particles with an average diameter of 2.5 nm (table 1 and figure 1). Future work will apply this method to other noble metals, including Pd, Rh, and Ru.

Of course, Zsigmondy's nuclear method is not limited solely to the synthesis of colloidal particles composed of single metals. In order to develop nanoparticles with enhanced catalytic properties, researchers in this field have prepared bimetallic clusters using similar techniques as well as co-reduction processes [6,7]. Such bimetallic particles might also have application in the field of EM labeling if they prove to be distinguishable from their homogenous counterparts by analytic means, such as electron energy loss spectroscopy, as we have shown previously for particles composed of single metals [3].

References

[1] W.P. Faulk and G.M. Taylor, *Immunochemistry* 8 (1971) 1081.
[2] G. Frens, *Nature Physical Science* 241 (1973) 20.
[3] D.A. Meyer and R.M. Albrecht, *Microscopy and Microanalysis* 7 (Suppl. 2) (2001) 1032.
[4] R.A. Zsigmondy, *Zeitschrift für Anorganische Chemie* 99 (1917) 105.
[5] J. Voigt and J. Heumann, *Zeitschrift für Anorganische Chemie* 169 (1928) 140.
[6] A. Henglein, *Journal of Physical Chemistry B* 104 (2000) 6683.
[7] J.H. Hodak et al., *Journal of Chemical Physics* 114 (2001) 2760.

TABLE 1. Average diameters of cPt particles prepared with varying amounts of cPt nuclear sol. For sols 1-5, 100ml cPt was prepared by adding nuclear sol to solutions of 500μM H_2PtCl_6 stabilized with 0.04% trisodium citrate and reduced with 0.05% tannic acid at 100°C.

Sol	Volume Nuclear Sol Added (ml)	Average Particle Diameter (nm)
Nuc. Sol	—	2.5 ± 0.03
1	10.0	5.5 ± 0.06
2	5.0	7.1 ± 0.06
3	2.5	9.0 ± 0.09
4	1.0	11.6 ± 0.12
5	0.5	13.4 ± 0.16

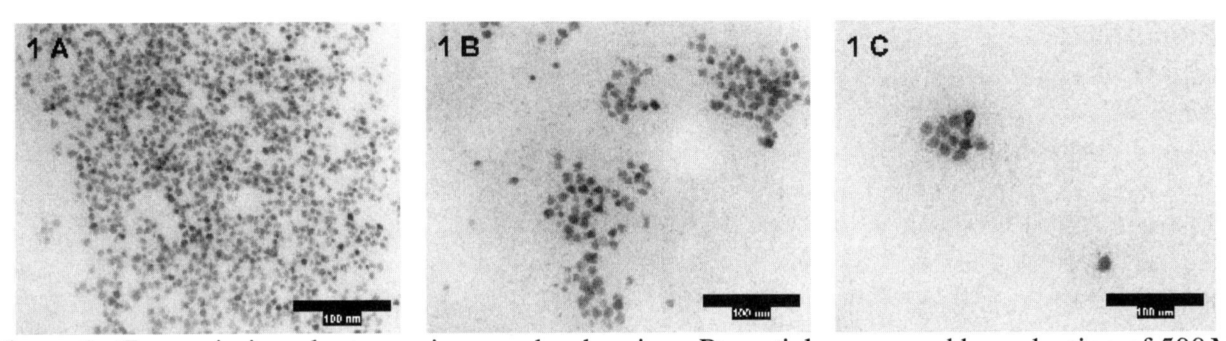

Figure 1. Transmission electron micrographs showing cPt particles prepared by reduction of 500μM H_2PtCl_6 at 100°C with 0.05% tannic acid and stabilized with 0.04% trisodium citrate. Various volumes of nuclear sol were added prior to reduction, and the total reaction volume in each case was 100ml. Magnification in each image is 230,000 times at an accelerating voltage of 80KeV. Bars are each 100nm. 1A. cPt particles prepared in the presence of 10ml nuclear sol. Average particle diameter is 5.5nm. 1B. cPt particles prepared in the presence of 2.5ml nuclear sol. Average particle diameter is 9.0nm. 1C. cPt particles prepared in the presence of 0.5ml nuclear sol. Average particle diameter is 13.4nm.

Microsc. Microanal. 8 (Suppl. 2), 2002
DOI 10.1017.S1431927602101644

Methodology Advancements in Electron Microscopy Immunolabeling of Hydrated Brain Tissue Using Subnanometer Colloidal Gold Conjugates

Hong Yi, Rahmat Peijper **, and Jan LM Leunissen **
***Emory Neurology Microscopy Core Facility, Emory University, Atlanta, GA, USA,**
****Aurion Immunogold Reagents, Accessories, and Custom Labeling, Wageningen, the Netherlands**

The ultrastructure of the central nervous system is characterized by the abundance of membrane enclosed cellular and sub-cellular elements, including neuronal and glial cell bodies and processes. Maintaining membrane ultrastructural integrity is essential for identification of these elements, and is a challenge in brain ultrastructural localization experiments. The availability of ultrasmall gold conjugates and silver enhancement solutions allows the penetration of the immunoreagents into hydrated brain tissue without compromising ultrastructure quality, making pre-embedding immunogold labeling a valuable method for high resolution localization of intracellular antigens in brain tissue. Further refinement of the reagent and technique has also made it possible to co-localize multiple antigens using exclusively ultrasmall gold conjugates and sequential silver enhancements.

We present here a few examples of our immunogold localization studies in brain tissue (Figure 1 and 2) through which issues such as localization resolution, reagent penetration, antigen accessibility, and particle size control will be addressed. We will also discuss the development of new conjugates, and demonstrate preliminary localization results using these new conjugates.

References

1) H. Yi, et al., J. Histochem. Cytochem. 49, 279 (2001)

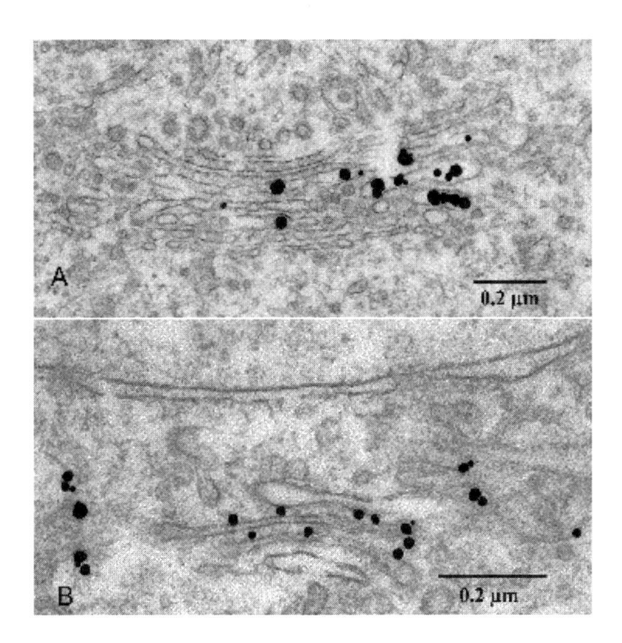

Figure 1. Pre-embedding immunogold labeling of MGP-160 (A) and Huntingtin Interaction Protein Interacter (B) in brain tissue. Note the localization of the gold particles in the inner (A) and outer (B) surface of Golgi membrane

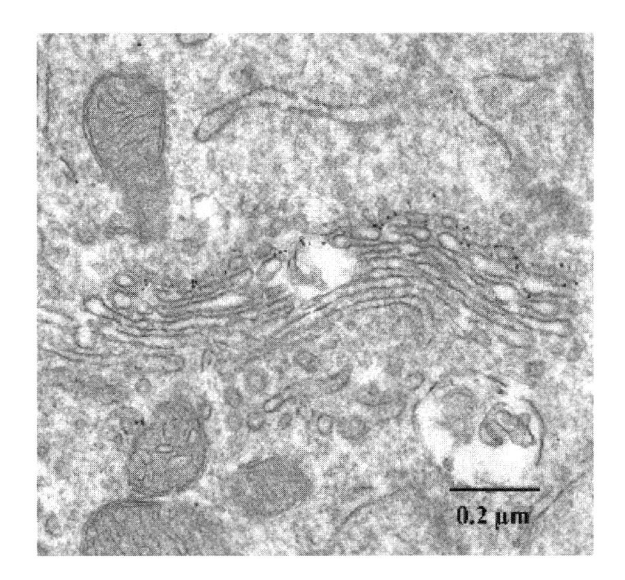

Figure 2. Pre-embedding immunogold labeling of MG-130, a *cis*-Golgi protein in mouse brain tissue.

Microsc. Microanal. 8 (Suppl. 2), 2002
DOI 10.1017.S1431927602101656

Bioengineering of Reporter Transgenes for Integrated Imaging with Magnetic Resonance, Fluorescence, and Electron Spectroscopy.

Marek Malecki.

SABA University School of Medicine.

For efficient monitoring of the new transgene constructs expression, it is necessary not only to monitor their life-time distribution within a whole body, but also their spatial distribution within tissues and cells as well as location within bio-macromolecular assemblies. This is accomplished through either labeling with antibodies tagged with reporter molecules or expressing fusion proteins containing reporting domains. Unfortunately, the imager allowing us to reveal the transgene products over such a wide spectrum of approaches does not exist. Hence, there is a strong need for correlations between various imaging modes [1]. Recently, we bioengineered the recombinant single-chain variable fragment (scFv) antibodies designed for simultaneous detection with magnetic resonance imaging (MRI) and positron emission tomography (PET) followed by molecular imaging with electron energy loss spectroscopy (EELS) [2]. In our previous work, we bioengineered the plasmid constructs driving organelle targeted expression of the transgene for a fusion protein with the integrated reporter molecules [3]. Here, we report the results of bioengineering of the constructs for transgenes, which products are detectable with MRI, fluorescence, and EELS.

The plasmid constructs were bioengineered into the previously described frame [2] in which the scFv site was replaced with the occludin coding sequence. The constructs were delivered into cultured human umbilical vein endothelial cells via receptor mediated gene transfer [3]. The cells were grown either as monolayers or within polystyrene sponges or as clusters injected into a nude mice. For preliminary evaluation of the transfection efficiency with fluorescence imaging, the cultured cell whole-mounts were suitable. For magnetic resonance imaging, the cell clusters and polystyrene sponges were appropriate. For electron energy loss spectroscopy, the cells from all strategies were rapidly frozen, freeze-substituted, embedded in LRW, and ultra-thin sectioned.

These newly bioengineered constructs open new avenues not only for studying dynamics of angiogenesis, but also endothelial cell response to oxidative stress or administration of pharmaceutics. These studies can now be pursued through monitoring of transgene expression via integrated imaging with magnetic resonance, fluorescence, and electron spectroscopy [4].

References.

[1] R. Albrecht et. al., Meth. Enzymol. 215, (1992) 456.
[2] M. Malecki et. al., Proc. Natl. Acad. Sci. U S A., 99 (2002) 213.
[3] M. Malecki et. al., Microsc. Microanal. 4 (Suppl. 2) (1998) 994.
[4] This research was supported by the NSF grants 9420056, 9522771, 9902020, and 0094016.

Figure legends.

The transgene constructs for fusion protein (occludin and integrated reporter molecule) were expressed in cultured human umbilical vein endothelial cells (HUVEC). The images were acquired with ultra-low-light, wide-field, computer deconvolution fluorescence microscopy. Horizontal field width = 50 microns.

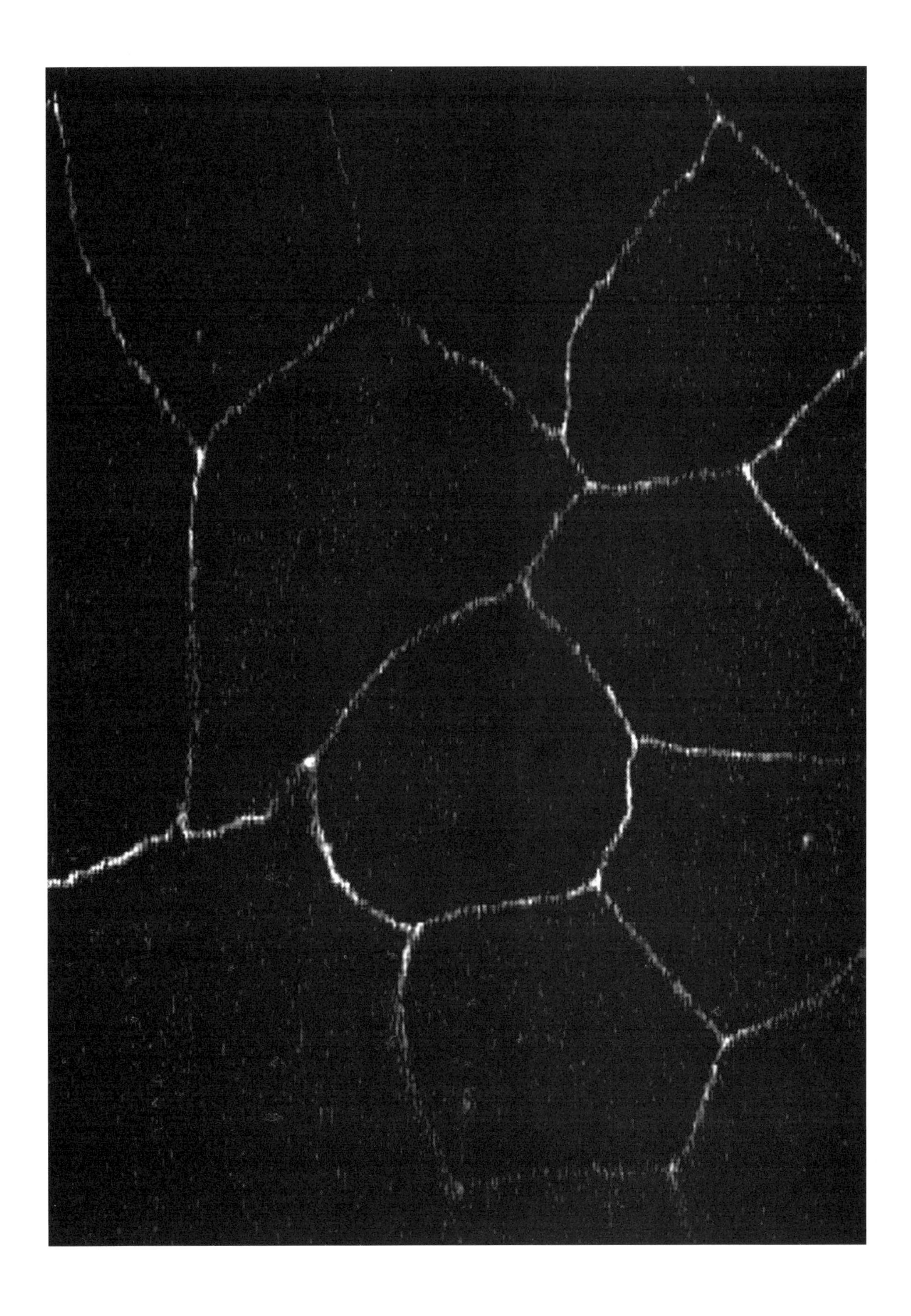

Microsc. Microanal. 8 (Suppl. 2), 2002
DOI 10.1017.S1431927602101668

Comparing confocal immunofluorescence microscopy (CIM) *vs.* freeze fracture replica immunogold labeling (FRIL) for cell-specific localization of plasma membrane proteins in the mammalian CNS.

T. Yasumura*, K.G.V. Davidson*, C.S. Furman*, J.I. Nagy** and J.E. Rash*

*Department Biomedical Sciences, Colorado State University, Fort Collins, CO 80523 and
**Department of Physiology, University of Manitoba, Winnipeg, Manitoba Canada, R3E 3J7

In the past decade, light microscopy has undergone a resurgence based on the use of confocal microscopy and immunofluorescence labeling, particularly the localization and three-dimensional mapping of membrane proteins. For example, connexins have been mapped to gap junctions in liver hepatocytes, where cells are essentially cuboidal, and identification of the cells forming a particular gap junction is not at issue. However, in the mammalian CNS, where most neuronal and glial processes are smaller than the limit of resolution of light microscopy, and where there are many cell types and possible combinations of coupling partners, assignment of connexin proteins to individual cell types is almost never possible by confocal microscopy. Localization of a specific membrane protein to a specific cell or cell pair by CIM faces three major difficulties: 1) With a maximum of three or four fluorphores available for use at one time, it is not possible to use fluorescent markers simultaneously to identify the four major cell types, and in addition, to label connexins/gap junctions. As a consequence, most investigators have used only one of the available fluorescent markers to visualize a single cell type (e.g., neurons) and one or two fluorophores to visualize target connexins. With other cells and cell processes not labeled or visualized, the images often have been erroneously interpreted as demonstrating the fluorescent connexin within the plasma membrane of the fluorescently-visualized cells. 2) The limit of resolution of confocal microscopy (0.2μm in blue wavelengths and 0.4μm in red wavelengths, and double those values in the Z-axis) is greater than the diameter of most cell processes in the CNS. This means that three or more cell processes and six or more plasma membranes are often present in the smallest volume resolvable by CIM (Fig. 1).

In contrast, FRIL has several distinct advantages over CIM: A) In most instances, cells are positively identified based on ultrastructural features, reserving immunogold labels for identification of membrane proteins. B) With the development of immunogold labels of at least five discrete size classes (e.g., 5nm, 10nm, 15nm, 22nm and 30nm), as many as five or more membrane proteins may be localized to ultrastructurally-identified cells. C) In addition, FRIL has the distinct advantage of direct visualization and unambiguous identification of ultrastructurally-defined gap junctions in ultrastructurally-identified cell processes. D) The ability to visualize and quantify connexins and to quantify immunogold labels by FRIL allows us to answer questions not addressable by CIM, including determination of the relative labeling efficiency of immunogold labels vs. fluorescent labels for each class of connexins present in individual gap junctions (Fig. 2). E) Because each immunogold bead represents a separate labeling event, and because the number of labeled gap junctions and immunogold beads for each connexin may number in the hundreds and tens of thousands, respectively, one may use FRIL to ascertain the relative amount of a particular connexin in gap junctions of a particular cell class. For example, we have labeled more than 5000 astrocyte gap junctions with more than 100,000 gold beads for Cx43, with no Cx43 detected in gap junctions of neurons or oligodendrocytes. Thus, we conclude that <u>if</u> Cx43 is present in neuronal gap junctions, it is present at a density at least five orders of magnitude lower than in astrocytic gap junctions. Similarly, if Cx32 is present in neuronal gap junctions, it is present at a density at least three orders of magnitude lower than in oligodendrocyte gap junctions. We conclude that of the six CNS connexins tested to date, Cx36 is present in most neuronal gap junctions, and that Cx26, Cx29, Cx30, Cx32, and Cx43 are not present in neuronal gap junctions. Thus, we have used FRIL to identify the connexins present in gap junctions of each cell type, and then redesigned our CIM experiments to confirm that each connexin is obligately associated with the appropriate cell type.

References:

1. Fujimoto, K., J. Cell Sci. (1995) 3443.
2. T. Yasumura et al., Microsc. Microanal. (Suppl. 2) (1997) 345.
3. J.E. Rash et al., Cell Tissue Res. (1999) 307.
4. J.E. Rash et al., J. Neurosci. (2001) 1983.

5. This research was supported by NIH NS38121 and NS39040.

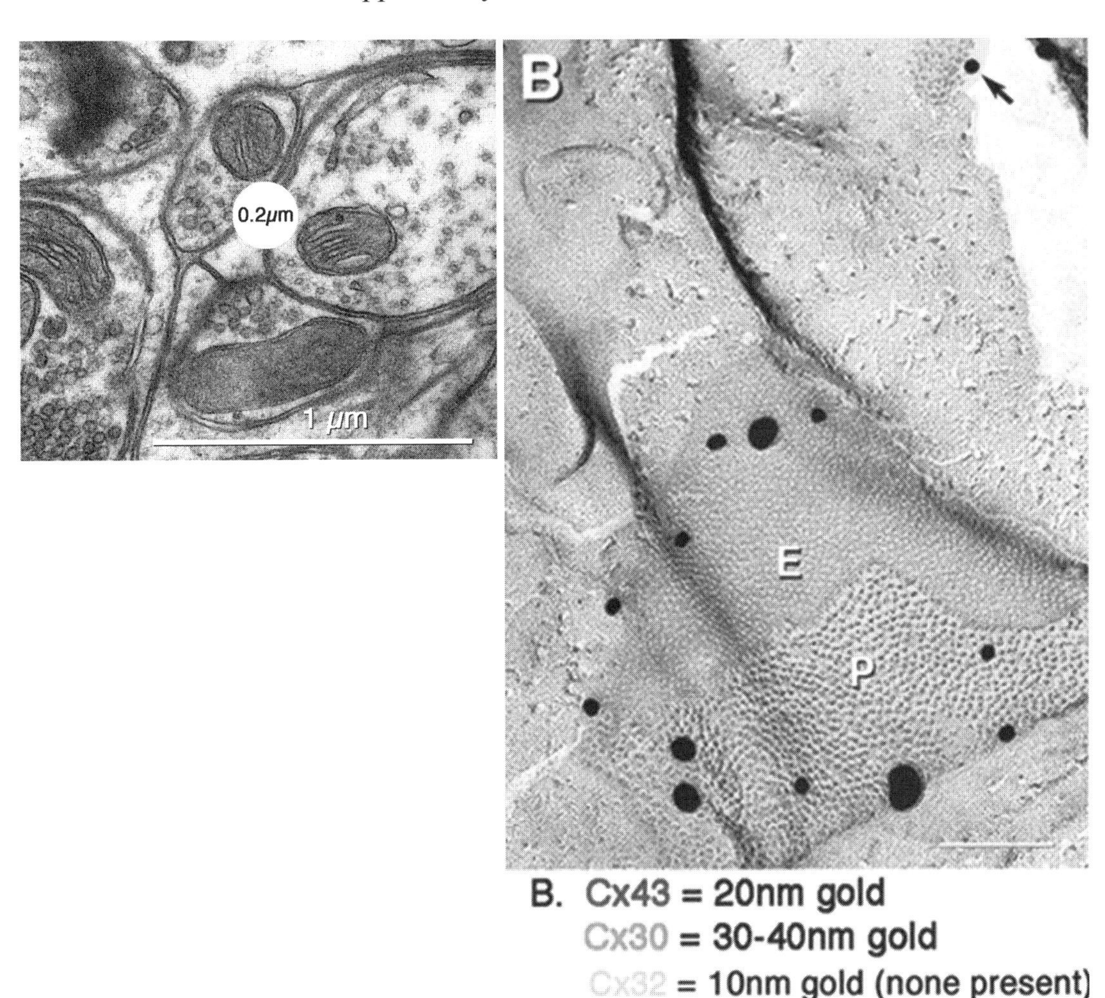

Figure 1. Limit of resolution of confocal microscopy in blue wavelength (0.2μm) superimposed on thin section image from adult rat spinal cord.
Fig 2. (B) Triple-immunogold labeling of astrocyte gap junction. Cx30 (30-40nm gold) and Cx43 (20nm gold) are present, but Cx32 (10nm gold) is not present.

Microsc. Microanal. 8 (Suppl. 2), 2002
DOI 10.1017.S143192760210167X

Colloidal Gold Conjugates of Cholera Toxin B-Subunit of Alexa Fluor® Fluorescent Dyes for Use in Correlative Studies

Eduardo Rosa-Molinar*, José L. Serrano-Vélez*, Philip Oshel**, and Ralph M. Albrecht**

*Julio García Díaz Center for Investigations in Biology, University of Puerto Rico-Río Piedras, Rio Piedras, PR 00931-3360
**Department of Animal Health and Biomedical Sciences, University of Wisconsin, Madison WI 53706

Conjugations of colloidal gold (18.0 nm) to the cholera toxin B-subunit (CTB) of green fluorescent Alexa Fluor® 488 (ex 495 nm/em 519 nm) and to the CTB of the red-fluorescent Alexa Fluor® 594 (ex 590 nm/em 617 nm) show promise for correlative light (*i.e.* differential interference contrast [DIC]), fluorescence, and electron microscopy. Commercially available (Molecular Probes, Inc., Eugene, OR) Alexa Fluor 488 conjugated to CTB and Alexa Fluor 594 conjugated to CTB were separately dialyzed in distilled water for 1hr and separately added to 18.0 nm colloidal gold (pH 8.18) to a concentration of 200 µgm per ml of gold, spun down, and re-suspended in 0.1M phosphate buffer at pH 7.4. The final concentrations of Alexa Fluor 488 and Alexa Fluor 594 conjugates of colloidal gold CTB were approximately 5 x 10 twelfth particles/ml.

Filter paper saturated with Alexa Fluor 488 or Alexa Fluor 594 colloidal gold conjugates of CTB was immediately applied to both ends of peripheral nerve stumps of the anal fin appendicular support of newborn, immature and adult Western Mosquitofish, *Gambusia affinis affinis*. The cut peripheral nerves were exposed for 1 minute, the time found to be sufficient to retrogradely label even the most finely axonal fibers and dendritic branches (Figs. 2-3). *G. a. affinis* were revived, allowed to survive for 6-24 hours, then euthanized, perfusion fixed with 4% paraformaldehyde, washed, bleached, cleared, and visualized using DIC and fluorescence microscopy.

Alexa Fluor 488 and Alexa Fluor 594 colloidal gold conjugates of CTB labeled Mauthner cells and their axons within the hindbrain (Figs. 1-2) and secondary spinal motor neuron cell bodies and their extensive dendritic arbors (Figs. 2-3). The tyramide signal amplification procedure for Alexa Fluor 594 (Molecular Probes, Inc., Eugene OR) combined with a previously described whole-mount clear and stain procedure [1] labeled peripheral nerve fibers of the ano-urogenital plexus at all stages and made them easy to visualize (Fig. 4).

Our results demonstrate that the conjugations of colloidal gold to the CTB of Alexa Fluor 488 and to the CTB of Alexa Fluor 594 have properties that permit their identification in light and fluorescence microscopy and show promise for electron microscopy and correlative studies.

References

[1] Brain Res. Brain Res. Protoc. 1999 (4): 115-123
[2] This research was supported by the National Science Foundation (IBN-0091120 [ER-M]) and the National Institute of General Medical Sciences (GM-063001 [RMA]).

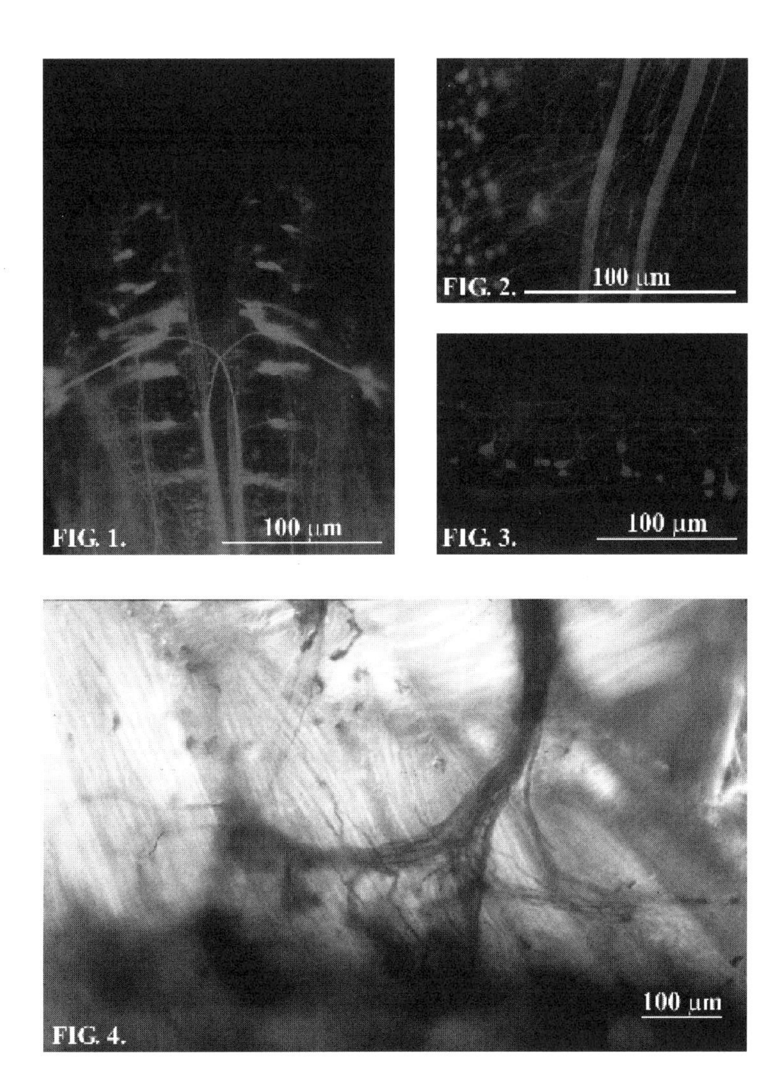

FIG. 1. Mauthner cells and axons in a whole-mount preparation of paraformaldehyde-fixed male Western Mosquitofish, *Gambusia affinis affinis* hindbrain visualized using green-fluorescent Alexa Fluor 488 colloidal gold conjugate of Cholera Toxin B-subunit.

FIG. 2. Mauthner axons (green) and secondary motor neurons (red) in a whole-mount preparation of paraformaldehyde-fixed male Western Mosquitofish, *Gambusia affinis affinis* spinal cord visualized using green-fluorescent Alexa Fluor 488 and red-fluorescent Alexa Fluor 594 colloidal gold conjugates of Cholera Toxin B-subunit.

FIG. 3. Secondary motor neurons in a whole-mount preparation of paraformaldehyde-fixed Western Mosquitofish, *Gambusia affinis affinis* spinal cord visualized using red-fluorescent Alexa Fluor 594 colloidal gold conjugate of Cholera Toxin B-subunit. Note the extensive dendritic arborization.

FIG. 4. Peripheral nerve fibers of the ano-urogenital plexus in a whole-mount preparation of paraformaldehyde-fixed male Western Mosquitofish, *Gambusia affinis affinis* anal fin appendicular support visualized using red-fluorescent Alexa Fluor 594 colloidal gold conjugate of Cholera Toxin B-subunit and detected using Tyramide Signal Amplification Kit # 5 with horseradish peroxidase goat anti-mouse IgG (Molecular Probes, Inc., Eugene, OR).

DOI 10.1017.S1431927602102066

Quantitation in Image Analysis: Practical Considerations for Drug Discovery

Michail A. Esterman* and Jeffrey C. Hanson

* Discovery Information Technology, Lilly Research Labs, Lilly Corporate Center Drop Code 0545, Indianapolis, IN 46285

The traditional disciplines of science are grounded in the observation and measurement of object properties, static and dynamic. The development of imaging tools such as the microscope extended the range of natural vision and enabled scientist to observe, record and measure otherwise inaccessible object properties by means of their images.[1] Over the last decade, the development of computer-based imaging systems resulted from technological advances in computer science and light detectors. These technologies have increased the number of object properties that can be observed as well as the accuracy and reproducibility with which they can be measured.

The biological sciences have entered the post-genomic era where the focus has shifted from cataloging the genetic complement to understanding how individual proteins participate in the biochemical pathways that govern cell and tissue functions. Imaging technology can be used for analysis of complex cellular properties, allowing phenotype, rather than the activity of individual targets, to drive lead generation. This approach aims to identify molecules that alter specific cellular properties, allowing identification of drugs that may act at a variety of points in functional target dependent and independent cellular pathways.[2]

Imaging deals with the principles, strategies and methods of image formation and the measurement of object properties which can be characterized by their numerical values. Imaging requires consideration of physics, biophysics, mathematics, statistics, computer science, and engineering. These disciplines must be applied to the physical system being imaged.

Digital imaging systems provide: Objective measure of spatial and temporal values, measurement of multiple variables in real time; programmatically and explicitly defined analysis; and sub population analysis, which can be crucial for defining the biological activity of a candidate compound. The ultimate goal of digital imaging in drug discovery is to convert digital representations, images, with multiple dimensions (size, shape, brightness, color,) to one-dimensional data values, which can easily be tabulated, compared and understood.

In any scientific detection / measurement system, certain principles must be followed. For digital imaging, close attention must be paid to the principles involved in sample preparation, detection and acquisition, image enhancement, segmentation, and measurement of the experimental entity. These principles are illustrated by examples like object counting, surface area measurement, coefficient of co-localization, and protein compartmentalization.

1. Beck, R., 1993. Proc Natl Academy Sci USA. Vol 90: 9746-9750.
2. Trask, O.J. and Large, T. 2001. Current Drug Discovery,

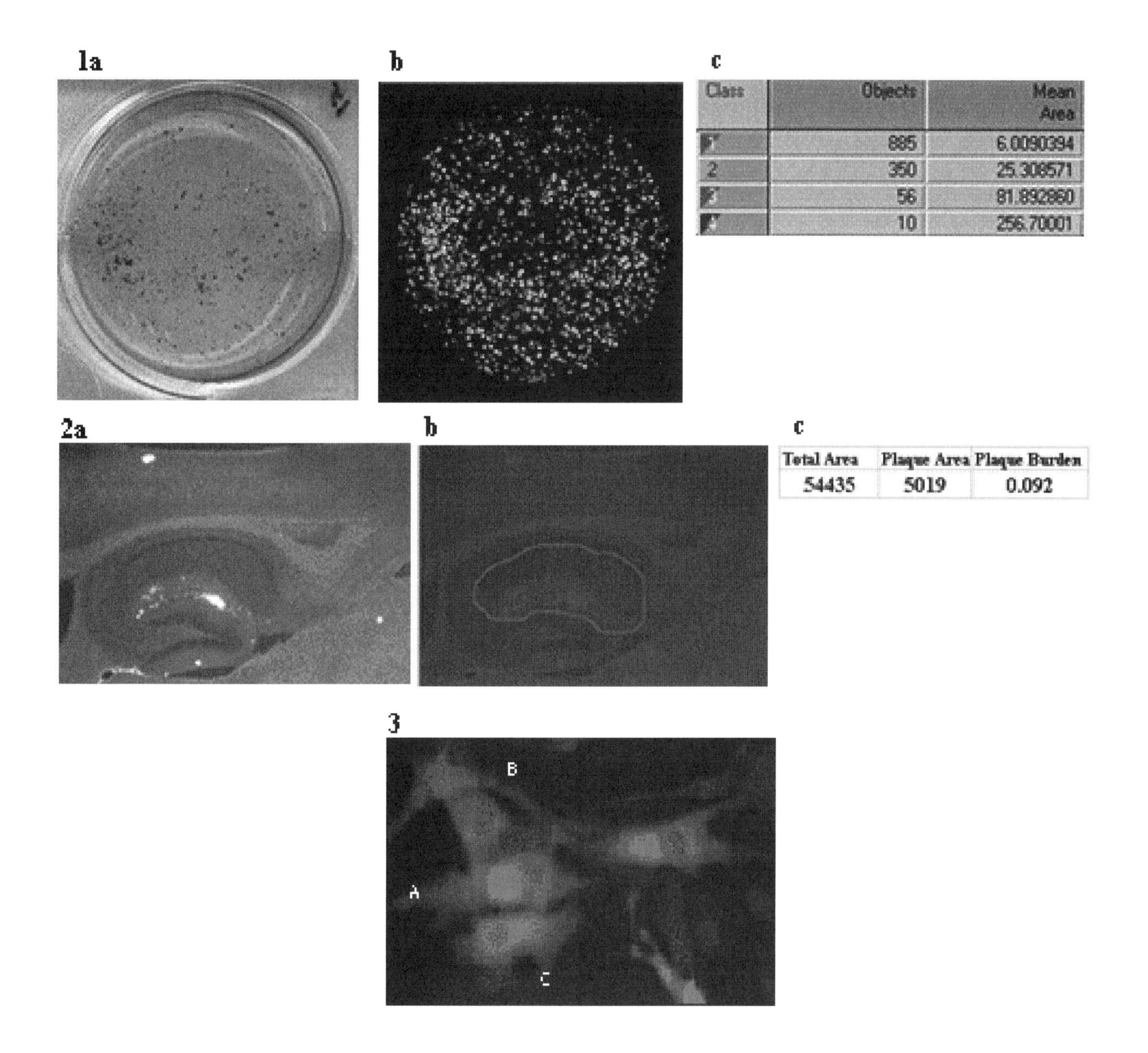

Figure legends

Figure 1. Image of cell colonies (a), image with interfering objects removed, analyzed and classified by colony size (b) and data table of count by size classification (c)

Figure 2. Image of mouse brain containing Beta Amyloid plaque (a), Region of interest and segmented plaque (b), and analysis results of surface area measurement.

Figure 3. Staining for receptor (green) and nuclei (blue) image demonstrates three sub populations of cells A strongly staining receptor, B lightly staining receptor and C cells with no receptor evident.

Microsc. Microanal. 8 (Suppl. 2), 2002
DOI 10.1017.S1431927602102078

A Novel Application of Solids Characterization by Environmental Scanning Electron Microscopy (ESEM) Utilizing a Peltier Stage

R.J. Maxwell* and J.A. Hanko**

*Trace Analysis, Pharmacia Corporation, 7000 Portage Road, Kalamazoo, MI 49001

**Global Pharmaceutical Sciences, Pharmacia Corporation, 4901 Searle Parkway Skokie, Il 60077

The advent of the Environmental Scanning Electron Microscope (ESEM) has brought about new opportunities to utilize this technique to characterize pharmaceutically active compounds. The ability to examine materials under a partial pressure of water vapor allows for imaging an uncoated specimen. This flexibility allows for the routine nondestructive examination of a variety of materials that could not be achieved with a coated specimen (e.g., hydrated and/or solvated crystal lattices. By utilizing an uncoated specimen, dynamic imaging of the surface during dehydration and or desolvation can now be routinely observed. With the addition of a Peltier stage, the specimen can be exposed to a wide range of relative humidity (RH) conditions by varying the stage temperature and/or pressure of the gas within the environmental chamber.

Using this technique for *in situ* monitoring a pharmaceutical sample environment, specifically a solid dosage form, was of considerable interest to understand the affects of humidity on sample integrity. Given the typical tablet size, the Peltier stage sample capacity, and the concern about a suitable cross-section to represent the bulk sample, a fundamental question arose: "Is the effective RH at the analysis site on a tablet surface the same as at the tablet/Peltier contact site for a given temperature/pressure combination?" Scheme 1 summarizes this question. The high degree of temperature control provided by the Peltier stage coupled with ESEM pressure controls, allows the operator to span the entire RH scale and maintain conditions for prolonged periods of time. However, the instrumental data output for temperature/pressure/RH conditions are only accurate within the Peltier stage environment. For situations where samples extend beyond the Peltier stage, a method for determining RH at the site of analysis is useful.

Thus a quick and sensitive technique for estimating the actual temperature/RH at the analysis site is critical for a better understanding of observed specimen effects. This was achieved by placing a few salt grains (see Figure 1A for established deliquescence conditions) on the elevated surface of a compressed tablet and recording the deliquesce point. By recording the output conditions at which deliquesce was achieved and factoring in the known deliquescence conditions for the salt, the temperature/RH at the analysis site (see Figure 1B) can be calculated.

Scheme 1:

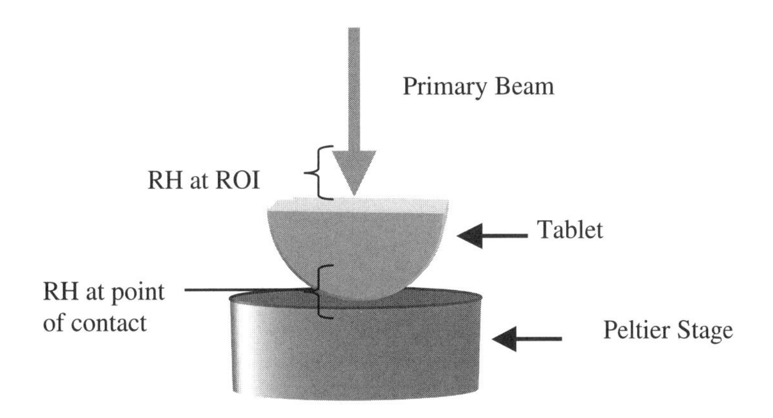

Figure 1: (A) NaCl crystals on the Peltier Stage starting to deliquesce. (9.9Torr/15°C/~77%RH). (B) View of NaCl crystals on Surface of a Tablet at 10.2Torr/15°C corresponding to 80%RH at the sample/Peltier contact point

Microsc. Microanal. 8 (Suppl. 2), 2002
DOI 10.1017.S143192760210208X

IMMUNOLOCALIZATION OF PHOSPHODIESTERASE ISOENZYMES IN RAT TISSUES USING CONFOCAL MICROSCOPY

Beverly E. Maleeff, Rosanna C. Mirabile, Timothy K. Hart, Heath C. Thomas, Lester W. Schwartz and Stephen J. Newsholme

Department of Safety Assessment, GlaxoSmithKline, King of Prussia, PA

Cyclic nucleotide phosphodiesterases (PDE) are a family of structurally related enzymes responsible for degrading cAMP and cGMP in eukaryotic cells[1]. Many tissues express members of multiple PDE families[2]. PDE type 3 (PDE3) has two known isoforms and has been purified from a number of tissues[3]. One of its known functions is relaxation of vascular smooth muscle. PDE type 4 (PDE4) has four known isoforms, has been purified from tissues including germ cells and vascular smooth muscle, and is believed to play a role in blocking contraction of airway smooth muscle[4]. PDE4 inhibitors are among several drug classes that induce segmental medial necrosis of mesenteric arteries after administration to rats. The nature and distribution of the lesions suggest a consequence of extreme localized disturbance of smooth muscle tone. We hypothesized that regional distribution of PDE3 and PDE4 isoenzymes predisposes the rat mesenteric artery to damage; therefore, we explored the distribution of these isoenzymes in a panel of tissues that may be implicated in these drug interactions.

Mesenteric, femoral and carotid arteries, aorta, liver and lung were collected from untreated male Sprague-Dawley rats. Frozen transverse sections were prepared and fixed in ethanol:acetone, blocked with BSA, incubated with primary antibody (Ab) (rabbit anti-rat PDE3B; rabbit anti-rat PDE4A, B, or C; mouse anti-rat PDE4D), followed by secondary antibodies conjugated to the fluorochrome Cy5, and counterstained with DAPI (fluorescent nuclear stain). Whole mounts of mesenteric arteries, briefly fixed with formaldehyde, were stained as above and enrobed in agar. All tissues were examined with a Zeiss LSM-510 confocal laser scanning microscope using 488 nm and 568 nm excitation lines of an Ar/Kr laser, the 633 nm excitation line of a HeNe laser and appropriate emission filters, or in differential interference contrast (DIC) mode.

In vascular sections all PDE3 and PDE4 isoenzymes were localized, to varying degrees, throughout the medial smooth muscle layer and within the endothelium. In whole vessel mounts, PDE3B isoenzymes were found only on the outer surface associated with adventitial adipocytes, whereas all 4 PDE4 isoenzymes were localized diffusely throughout the medial smooth muscle; in all cases, there was no evidence of segmental distribution along the vascular wall. In sections of liver and lung, PDE4 isoenzymes were localized to varying degrees in hepatic arteries and alveolar capillaries.

The principal PDE4 isoenzymes (A, B, C and D) are expressed in medial smooth muscle and endothelium of rat mesenteric arteries and therefore could be involved in mediation of altered tone, either dependent upon or independent of endothelial activity. Diffuse expression of PDE4 isoenzymes along the vascular wall suggests that the segmental lesion induced by PDE4 inhibitors is unlikely related to local differences in PDE4 enzyme distribution. Similar findings with PDE3B expression support the rejection of our hypothesis. Therefore, there is still no clearcut mechanism for the segmental nature of damage occurring in rat mesenteric vasculature following exposure to PDE4 inhibitors.

References
1. Müller T, Engels P and Fozard JR. Trends in Pharmacol. Sci. 17(8): 294-298, 1996.
2. Liu H, and Maurice DH. British J. Pharmacol. 125(7): 1501-1510, 1998.
3. Degerman E, Belfrage P and Manganiello, VC. J. Biol. Chem. 272(11): 6823-6826, 1997.
4. Saldou N, et al. Cellular Signaling 10(6): 427-440, 1998.

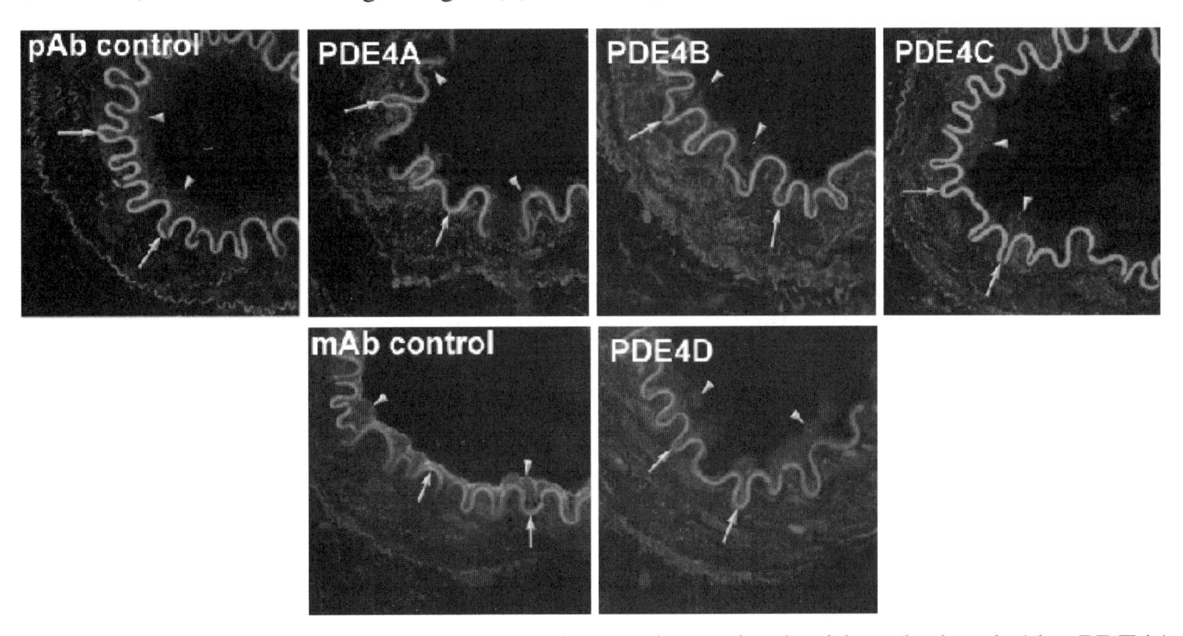

Figure 1. Transverse sections of mesenteric arteries stained with polyclonal Abs PDE4A-C or monoclonal Ab PDE4D, with negative controls. Elastin in the internal elastic lamina (arrows) is autofluorescent. Endothelial cells line the lumen of each vessel (arrowheads).
Original magnification 100x.

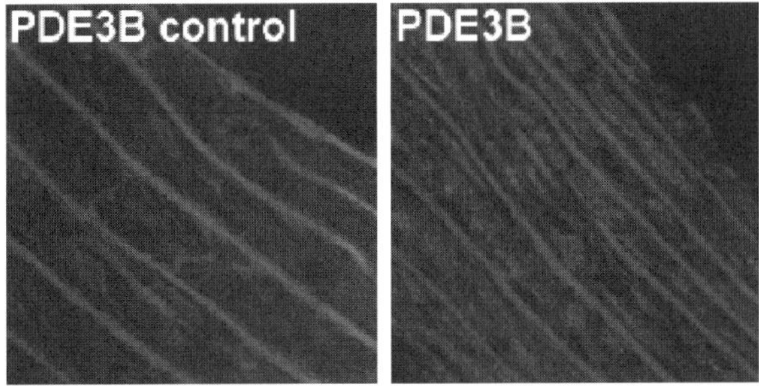

Figure 2. Transverse sections of aorta stained with anti-PDE3B Ab.
Original magnification 100x.

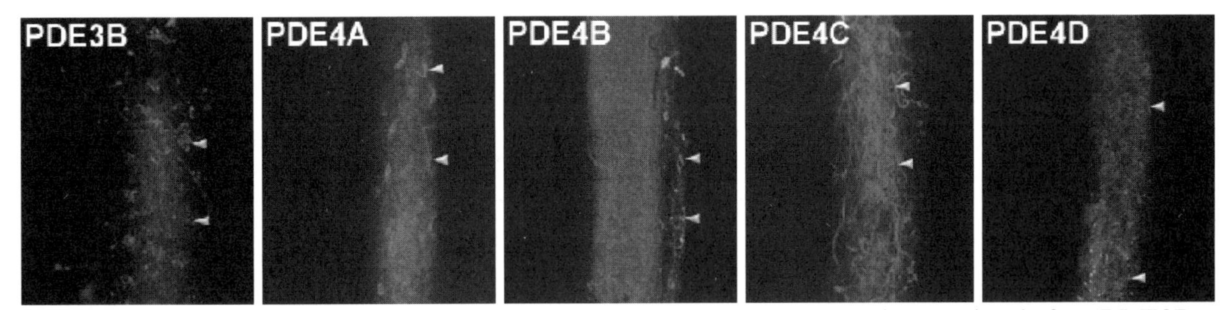

Figure 3. Adventitial surface of whole-mount mesenteric arteries, stained for PDE3B and PDE4A-D. Note the stained surface structures (arrowheads). Original magnification 40x.

Microsc. Microanal. 8 (Suppl. 2), 2002
DOI 10.1017.S1431927602102091

A Field Emission Scanning Electron Microscopy Method to Assess Recombinant Adenovirus Stability.

L. J. Obenauer-Kutner,* P. M. Ihnat,** T-Y. Yang,* B. J. Dovey-Hartman,*** and M. J. Grace.*

*Schering-Plough Research Institute (SPRI), Biotechnology Development, Union, NJ 07083
**SPRI, Pharmaceutical Development, Kenilworth, NJ 07033
***SPRI, Drug Safety and Metabolism, Lafayette, NJ 07848

A field emission scanning electron microscopy (FESEM) method was developed to assess recombinant adenovirus (rAd/p53) stability. This method was designed to simultaneously sort, count, and size the total number of rAd/p53 objects observed in an image. To test the method we treated a rAd/p53 preparation [1,5] with thermal incubation at 37°C for 0, 0.5, 1.0, 1.5, 2.0, 3.0, and 4.0 hours, and then monitored the effect on stability by assessing the anatomy of the virions using FESEM, with automated image-analysis (Image-Pro Plus v.4.1, Media Cybernetics, Silver Spring, MD), and transmission electron microscopy (TEM). In addition to electron microscopy (EM), the infectious activity of the thermally stressed rAd/p53 samples was quantitated using an established flow cytometry method. [2,3]

Unconjugated gold particles were mixed with each sample post-incubation to facilitate focusing and to provide an internal control for sizing of the virus particles. Viral specimens were subsequently fixed and stained on carbon coated copper grids, and then prepared for EM image-analysis. The T=0 time point showed that virions were evenly distributed over the grid, and the icosahedral geometry of the virus was evident. [4] As shown in Figure 1, FESEM image-analysis revealed a decrease in the total number of detectable single rAd/p53 particles and an increase in apparent micro-aggregates composed of multiple viral particles (multiplets) as early as 2 hours. In addition there was an observed decrease in the size of the single rAd/p53 particles and an increase in multiplet size with time at 37°C. The described changes predominated after 4 hours of incubation. The changes noted in virus morphology were concomitant with the observed loss in viral infectivity (Figure 2). The FESEM results were reproduced with TEM.

In conclusion, the results reported in this study suggest a novel method of assessing the stability of recombinant adenovirus products using FESEM image-analysis. The decrease in single rAd/p53 particles and the increase in higher order multiplets suggest that the method may be useful for monitoring or assessing stability.

References

[1] B. Huyghe, et al., Human Gene Therapy. 6 (1995) 1403.
[2] M. L. Musco, et al., Cytometry. 33, (1998) 290.
[3] C. Nyberg-Hoffman, et al., Nature Medicine. 3 (1997) 808.
[4] L. Philipson, Curr. Top. Microbiol. Immunol. 109 (1983) 1.
[5] P. W. Shabram, et al., Human Gene Therapy. 8 (1997) 453.

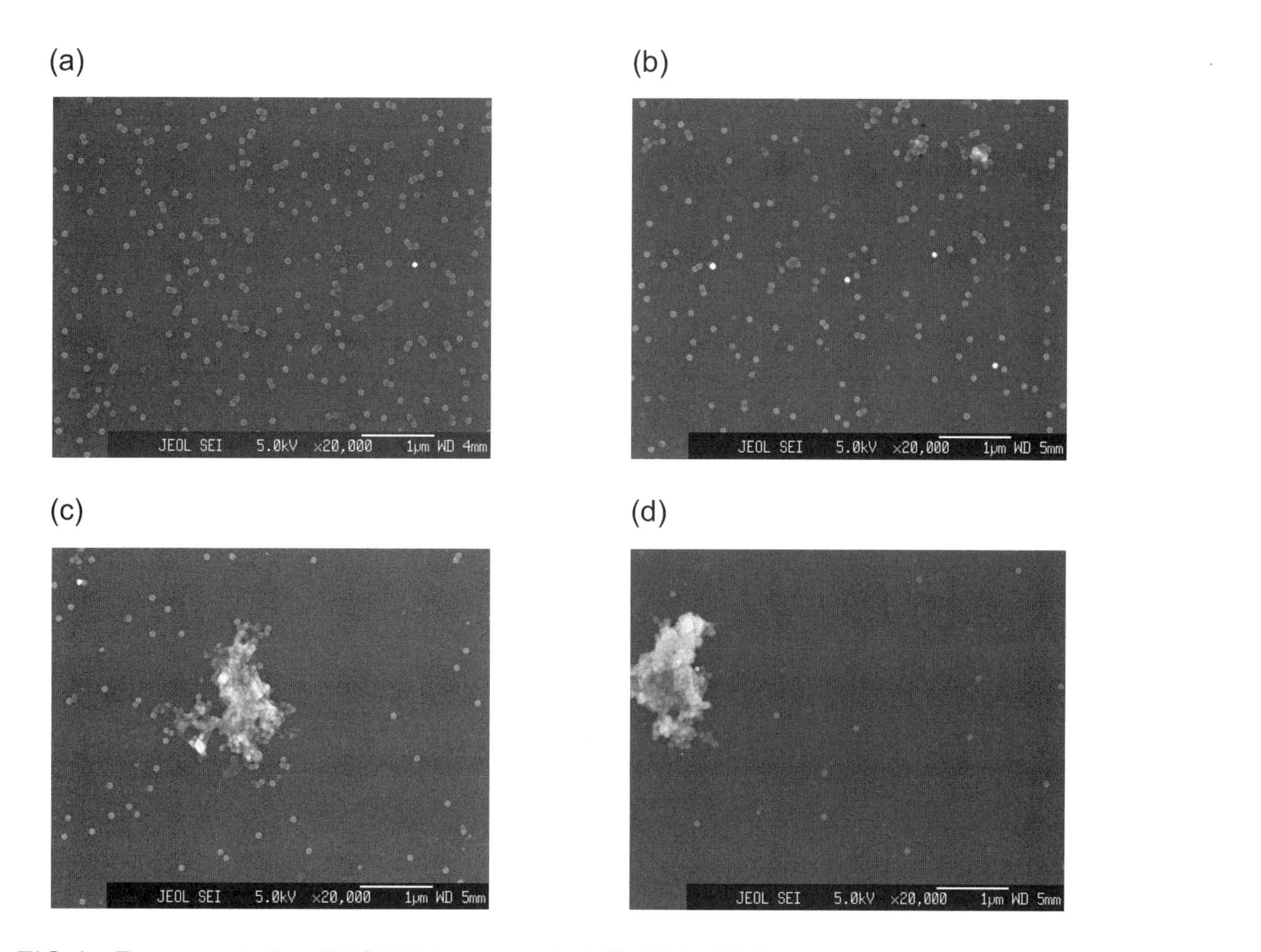

FIG.1. Representative FESEM images of rAd/p53 in PBS incubated at 37°C at the various times indicated. (a) T=0 (b) T=0.5 hours (c) T=2.0 hours (d) T=4.0 hours

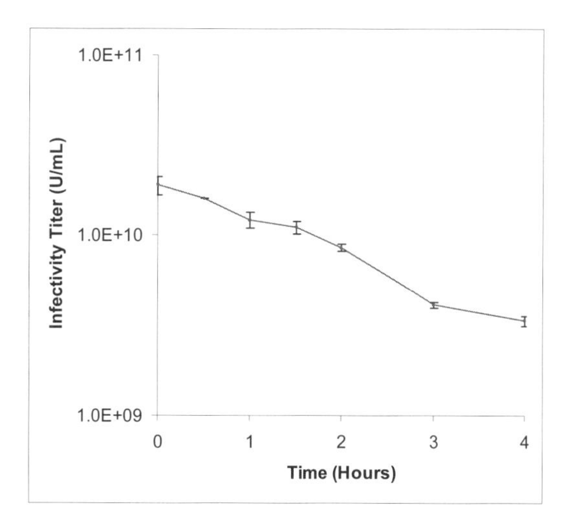

FIG. 2. Representative plot showing kinetics of rAd/p53 infectivity during thermal stress at 37°C.

Microsc. Microanal. 8 (Suppl. 2), 2002
DOI 10.1017.S1431927602102108

Relevant Applications of Scanning Electron Microscopy in a Pharmaceutical Development Laboratory

Ronald L. Mueller*

*GlaxoSmithKline, Worldwide Physical Properties-Pennsylvania, King of Prussia, PA 19406

This presentation provides examples where conventional scanning electron microscopy (SEM) contributed significantly in the drug development process. It also presents a general overview of how the SEM, in conjunction with other various analytical techniques, functions in a physical characterization group within the Pharmaceutical Industry. When SEM is used in a multi-disciplinary approach, it becomes a powerful characterization tool for the development of pharmaceutical materials.

Three case studies are presented. Firstly, an efficient mixing process for a drug substance with various excipients was required.[1] However, the drug substance was an oncological agent and red iron oxide was used as a substitute since it could be easily handled during the processing investigations, had a similar particle size distribution and tended to agglomerate during the mixing process. Under the same processing conditions in a high shear mixer, calcium phosphate dibasic excipients were evaluated for their mixing efficiency with 1% w/w red iron oxide. The mixing efficiency was followed by colorimetry and was collaborated by using energy dispersive x-ray (EDX) spectroscopy. Figure 1 displays EDX overlay maps for phosphorous and iron at two different time points during the processing investigation of A-Tab. Two of the three calcium phosphate dibasic excipients investigated mixed well.

Secondly, a laser light diffraction method was modified so that it could discriminate between "as is" and micronized materials. The laser light diffraction data was complimented by the image analysis results obtained by optical microscopy. Figure 2 displays the "as is" and micronized materials used in this study.

Thirdly, by using EDX spectroscopy, it was demonstrated why some core tablets manufactured from newer drug substance batches failed content uniformity. Whereas core tablets produced from older drug substance batches or reworked newer drug substance batches passed content uniformity. Figure 3 displays EDX overlay maps for carbon and sulfur for core tablets manufactured from a newer drug substance batch which failed and then passed content uniformity after reworking the core tablets.

References

[1] R. Abramowitz et al., *"Effect of Carrier on the Mixing Efficiency of a Cohesive Powder: Red Iron Oxide,"* presented at the 1999 AAPS Annual Meeting and Exposition.

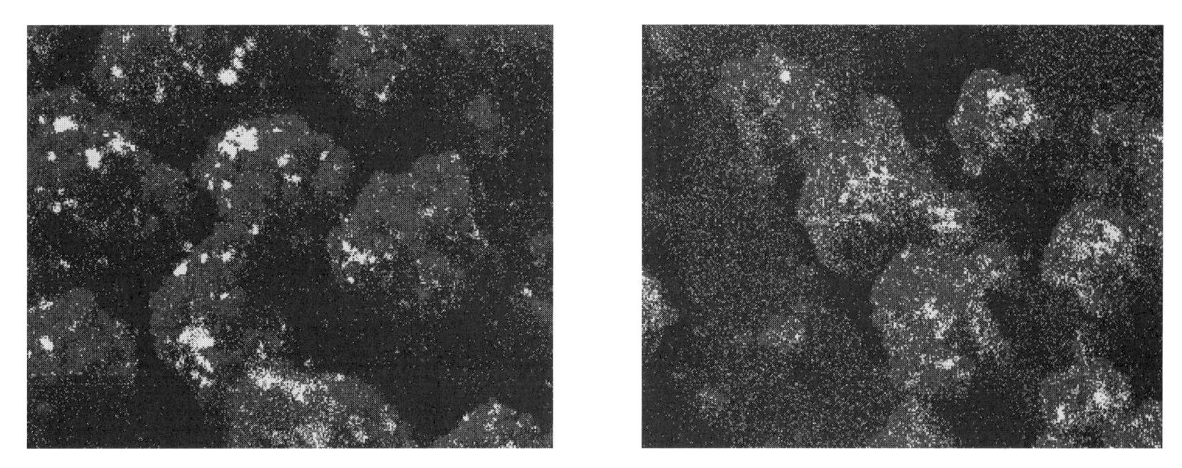

Figure 1. EDX overlay maps at 200X for phosphorous (red) and iron (yellow) at 15 seconds (left image) and 5 minutes (right image) during the processing investigation of "A-Tab."

Figure 2. SEM images of "as is" (left image) and micronized (right image) materials.

Figure 3. EDX overlay maps for carbon (red) and sulfur (yellow) for core tablets manufactured from a newer drug substance batch which failed (left image) and then passed (right image) content uniformity after reworking the core tablets.

Microsc. Microanal. 8 (Suppl. 2), 2002
DOI 10.1017.S143192760210211X

Microwave-assisted Embedding of Tissue Culture Cell Monolayers

Kent L. McDonald

Electron Microscope Lab, University of California, Berkeley, CA 94720-3330

When cells are grown as a monolayer on a substrate, a useful method of embedding is to retain the monolayer orientation in the polymerized resin. One can then use a good compound light microscope to screen the samples and select certain cells for sectioning..There are a number of ways to do this for conventional embedding methods with epoxy resins [1], but it is more of a challenge when using the microwave for polymerization because resin polymerization is best done under water. However, if cells are grown on plastic, they can be embedded in TAAB polyethylene flat-bottom capsules (Ted Pella, Inc., Cat. No. 133) and when the resin is polymerized, the plastic is peeled off and the cells remain in the resin.

Microwave set-up. The microwave used was a Pelco Model 3440, to which had been added a Pelco Model 3430 Power Controller for variable wattage, a Pelco Model 3435 Microwave Vacuum Chamber, Pelco 3420 Microwave Load Cooler, and a Pelco ColdSpot™ for continuous water circulation in the microwave cavity. All equipment was from Ted Pella, Inc., Redding, CA. The load cooler temperature was set to 25°C, the vacuum chamber put directly on the Coldspot, and the variable wattage controller set to 1, about 100W.

Cell preparation and handling. Cells are grown on 6-7 mm disks of Aclar plastic film (Ted Pella, Inc., Cat. No. 10501). In our experience, Thermanox or other plastics do not work as well as Aclar. The disks were made by using a standard single hole paper punch, then they were sterilized by UV irradiation overnight. When cells were at about 60% confluency, they were fixed and processed according to the procedure outlined in Table 1. This is a modification for microwave of a previously published procedure[2]. To facilitate handling, the samples were put into a 6-well PREP-EZE specimen holder (Ted Pella, Inc. Cat. No.36157-1) which fits into a 60 X 15 mm polypropylene Petri dish (Ted Pella Inc., Cat. No. 36135). Ten ml of solution is used to cover the cells. It is important to use the polypropylene dishes because they are resistant to the solvents used in electron microscopy specimen preparation.

Polymerization. After the last change of pure resin during the infiltration steps, transfer the cells to a flat-bottom capsule that has been partially filled with resin. Make sure that the cells are facing up, i.e., away from the bottom of the capsule. Fill the capsule with resin and seal with Parafilm and a snap cap from a BEEM capsule. Load into a Teflon holder and place in water for polymerization as shown in Giberson [3].

Aclar removal. Once the resin is hardened, remove it from the capsule then use a razor blade to remove excess resin from around the edges of the Aclar™. Insert the razor blade between the Aclar and resin and it should lift off easily with the cells remaining in the resin.

Other applications. This basic strategy can be used to embed cells in LR White for immunolabelling. The fixatives used would be different and the resin could be cured in about half the time. Another strategy for LR White microwave thin-layer embedding can be found in Lonsdale et al. [4].

[1] Kingsley RE, Cole NL. J of Electron Microsc. Tech., 10, 77-85 (1988).
[2] McDonald, K. J. ultrastruct. Res. 86, 107-118 (1984).
[3] Giberson, R.T. In, Microwave Techniques and Protocols, R. Giberson and R. Demaree (Eds.), 13-23, (2001).
[4] J.E. Lonsdale et al., In, Microwave Techniques and Protocols, R. Giberson and R. Demaree (Eds.), 139-153, (2001).

TABLE 1. Microwave Processing Steps for Tissue Culture Cells on Aclar

Processing Step	Power	Time	Vacuum	Comments
Fixation in 2% glutaraldehyde in 50 mM Cacodylate buffer, pH 7.2	100W	I min on 5 min off [a]	yes	
Rinse in buffer	none	5 X 1 min	no	Not in microwave
Fixation in 0.05% OsO_4 + 0.08% $K_3Fe(CN)_6$ in buffer	100W	1 min on 5 min off [a]	yes	
Rinse in buffer	none	5 X 1 min	no	Not in microwave
Fixation in 0.15% tannic acid in buffer	none	2 min	no	Not in microwave
Rinse in buffer	none	2 X 1 min	no	Not in microwave
Rinse in dH_2O	none	3 X 1 min	no	Not in microwave
Dehydration in 35, 50, 75, 95, 100,100 100% acetone	100W	1 X 40 sec each step	no	
Infiltration with 1:1 Eponate 12[b] resin: Acetone	250W	1 X 5 min	yes	
Pure Eponate 12 resin	350W	2 X 5 min	yes	
Polymerization in flat-bottom capsules	650W	100 min	no	Underwater [c]

[a] For this fixation step and for the osmium step, it is important to let the cells rest at least 5 minutes before going to the rinse steps.
[b] 23.5 g Eponate 12, 12.5 g DDSA, 14 g NMA, 0.37 ml DMP-30.
[c] No need to take out the vacuum chamber, just put the container on top.

DOI 10.1017.S1431927602102121

MICROWAVE TISSUE PROCESING IN A TEACHING LABORATORY

R. S. Demaree, Jr.

Department of Biological Sciences, California State University, Chico, CA 95929

Many investigators have used microwave ovens for various aspects of biological sample processing for light microscopy (LM), transmission electron microscopy (TEM.) and scanning electron microscopy (SEM.)[1]

Microwave-assisted processing also can play a significant role in the teaching laboratory. Students routinely prepare samples for SEM in a single 3 hour laboratory period. They fix, dehydrate and dry with hexamethyldisilazane in less than 1 ½ hours. The only part of the process not utilizing microwave assist is the final 15 minute drying step in a conventional oven (Fig. 1.).

TEM samples may now be processed in a 3 hour laboratory period (Fig. 2.) We routinely fix, dehydrate and infiltrate samples during one laboratory period. Embedding molds are placed in a conventional drying oven until the next laboratory period. Blocks are trimmed, sectioned and viewed during the next laboratory session.

Lengthy immunological staining protocols requiring one or more days to perform may now be completed in a single laboratory session with microwave assist. A bench protocol for LM requiring more than 12 hours is now routinely completed in 2 hours (Fig. 3.) The final example is a TEM immunological label using microwave assist (Fig. 4.) The initial fixation through infiltration was done in a single lab period; blocks were then placed in a conventional oven overnight. The next day, ultrathin sections were immunolabelled in less than 2 hours.

In our hands, microwave-assisted biological sample processing is our primary method of sample preparation. All micrographs presented here are student projects.[2]

References

1. *R. T. Giberson and R. S. Demaree, Jr. Microwave Techniques and Protocols, Totowa, NJ, Humana (2001).*
2. *Appreciation is expressed to Norm Fox, Charles Heise, Teresa Munoz and Julie Janes for use of their micrographs.*

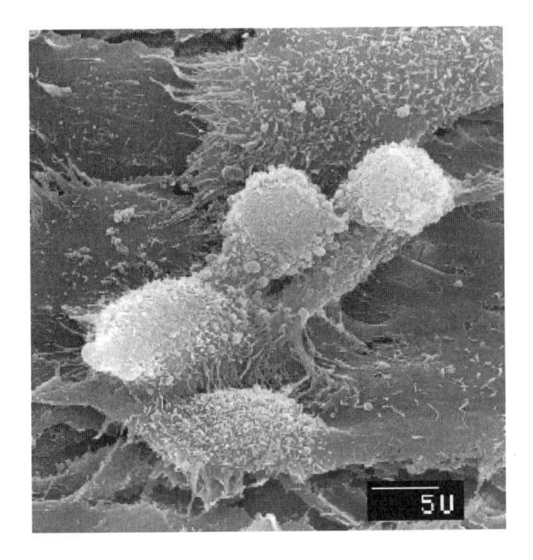

Microwave SEM prep of MODE-K cells prepared is less than 1 1/2 hours. Bar equals 5 micrometers.

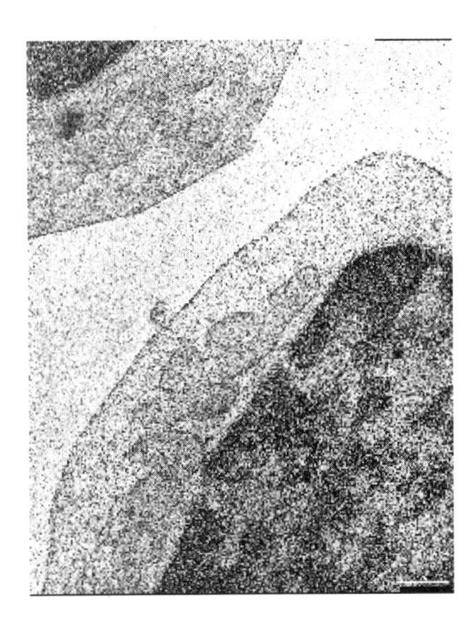

Microwave TEM prep from fix to plastic in a single 3 hour lab session. Bar equals 0.5 micrometers.

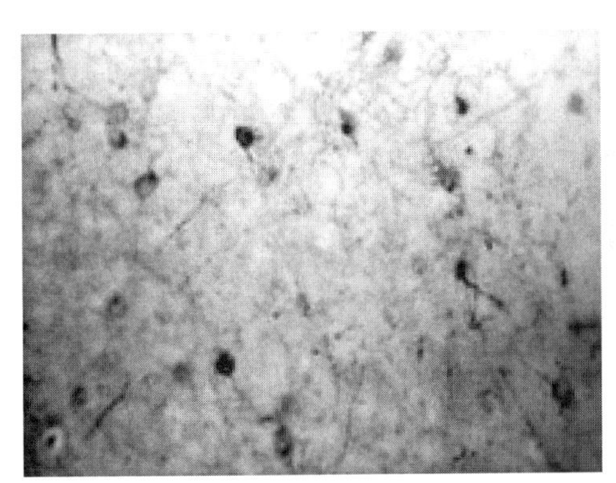

Microwave LM immuno label (GluR1) of frozen rat brain section in 2 hours. Normal staining time 12 hours. X 25.

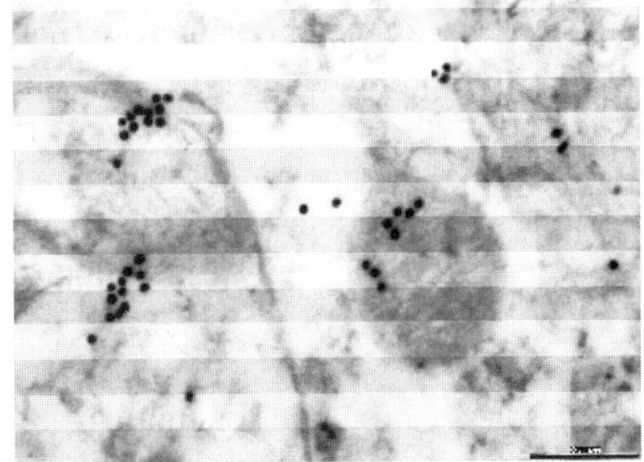

Microwave TEM immuno label (GluR1) completed in two 3 hour lab periods. Bar equals 0.4 micrometers.

Microsc. Microanal. 8 (Suppl. 2), 2002
DOI 10.1017.S1431927602102133

Microwave Assisted Decalcification with Recirculation of Temperature Controlled Solutions.

S. P. Tinling*
R. Kular*
R. T. Giberson**

*Otolaryngology Research Laboratory, University of California, Davis. 95616
**Research and Development, Ted Pella, Inc., Redding, CA 96003

Recent advances in microwave processing of tissue for histological analysis have shown its efficacy[1]. Improvements have been demonstrated in histologic detail, ease of processing and sample throughput time for both soft and hard (undecalcified) tissue [1,2]. Soft tissue has the advantage of being easy to sample and limited in size. This results in protocols of short duration with minimal sample heating at intermediate power settings (450W) and almost no heating(1 to 10EC) depending on solution type and volume. However, calcified tissue presents unique challenges to microwave assisted processing. Calcified samples are often large and cannot be trimmed without detrimental fractures through internal structuresof interest (for example: the Organ of Corti of the cochlea embedded within the temporal bone). These large samples also present a significant diffusion gradient against the removal of hydroxyapatite. In non-microwave assisted protocols, samples require daily changes of large volumes of solution over several weeks to months to achieve decalcification at room temperature.

Several studies have been published for microwave assisted processing of calcified tissue[3,4,5]. Variables include decalcifying solution type and concentration, time, temperature and type of microwave processing. Microwave factors have varied with type of oven (household to laboratory) wattage settings, temperature control and magnetron cycle times. These studies have led to a disagreement over the effect of solution temperature during microwave irradiation. The goal of this study was to evaluate the effect of microwaves on the rate of decalcification independent of temperature.

We compared the decalcification rate for standardized samples of gerbil calvarium by the following methods: 1) Constant rotation of 10 ml vials (16 rpm at 30E of inclination) at 20 EC with daily changes of 5 ml of decalcifying solution (standard non-microwave processing). 2) Constant recirculation of 9 L of decalcification solution at 20 EC. 3) Constant recirculation of 9 L of decalcification solution at 20 EC with simultaneous exposure to microwave irradiation using a Pelco Model 3470 laboratory microwave oven. Random samples from each group were removed at various intervals, dehydrated and embedded in epon/araldite plastic. Sections were cut at 1 um and evaluated for the degree of decalcification. Microwave exposure significantly increased the rate of decalcification, independent from temperature, when compared with the other two methods.

References:
1. R.T. Giberson and R. Demaree JR., Microsc. Res. Tech. 32(3) (1995) 246.
2. V.J. Madden and M.M. Henson., Hearing Research 111 (1997) 76.
3. I Louw et al., Histochem. J. 26 (1994) 487.
4. M. Kaneko et al., Biotech Histochm 74(1) (1999) 49.
5.C. D. Cunningham III et al., Laryngoscope. 111 (2001) 278.

Fig. 1 Fig. 2

Fig. 1. Cross section of gerbil calvarium after 21.5 hours of constant recirculation of decalcifying solution at 20 EC. Note remaining unstained undecalcified bone

Fig. 2. Cross section of gerbil calvarium after 20 hours of constant recirculation of decalcifying solution at 20 EC and constant microwave irradiation. Note complete decalcification.

Microsc. Microanal. 8 (Suppl. 2), 2002
DOI 10.1017.S1431927602102145

Microwave-Assisted Rapid Tissue Processing For Disease Diagnosis In A Veterinary Diagnostic Laboratory

R.W. Nordhausen,* B.C. Barr,* and R.P. Hedrick**

*University of California, Davis, California Animal Health and Food Safety Lab, School of Veterinary Medicine, Davis, CA 95616
**University of California, Davis, Dept. of Medicine and Epidemiology, School of Veterinary Medicine, Davis, CA 95616

Recent advances in microwave (MW) technology have elucidated a unified procedure for rapid microwave-assisted thin section electron microscopy tissue preparation incorporating all aspects of specimen preparation from aldehyde fixation through resin polymerization. This procedure drastically reduces tissue preparation time without compromising specimen quality (1-3).

The California Animal Health and Food Safety Laboratory is a large state veterinary diagnostic laboratory system whose primary goal is rapid accurate diagnosis of livestock and poultry diseases within the state of California. As such the laboratory encounters a wide range of diseases in domestic and wildlife species. Diseases are often encountered for which there are no available commercial diagnostic test reagents. These may include both known or established livestock diseases, as well as previously unrecognized diseases. Within this diagnostic environment, thin section electron microscopy (TEM) has been an extremely useful tool, particularly in the diagnosis of infectious diseases. However, staff commitment and turnaround time previously limited the practical usefulness of TEM on a regular basis. With the adoption of rapid MW processing, we have been able to expand sample volume without the addition of FTE. Also, the rapid MW processing technique has allowed us the flexibility to achieve 4-5 hour turnaround on high profile cases. This is extremely important not only in regards to highly contagious endemic disease diagnosis, but also allows us to offer TEM as a further adjunct in foreign or exotic animal disease surveillance, a topic that is very timely in today's global environment. The following is offered as a recent example of the usefulness of this technique in the rapid diagnosis of a contagious viral disease hatchery raised fish.

A sample of formalin fixed and fresh unfixed sturgeon skin was received from a large fish hatchery raising endangered Pallid sturgeon fingerlings. Skin lesions noted on the affected fish were sent with a request to rule out Lymphocystis, a contagious disease caused by a member of the iridovirus family. While viral isolation was possible a rapid diagnosis was needed to institute immediate control measures. Rapid (MW) processing allowed for examination of the sample within 5 working hours with a specific diagnosis of iridovirus infection. Based on this diagnosis, immediate control measures were undertaken by the hatchery to eliminate further spread of the disease. Iridoviruses are known to cause serious disease in many vertebrate mammalian, reptilian and amphibian animal species [4]. Pigs are susceptible to the highly contagious foreign animal disease, African swine fever and many varieties of fish are susceptible to a couple of different iridoviruses, one of which causes lymphocystosis, resulting in unsightly skin lesions. Other iridoviruses cause viral erythrocytic necrosis in oceanic and fresh water species that can be problematical, especially in fish culture operations [5]. Figure 1 illustrates this case material showing large 240 nm hexagonally shaped iridovirus particles from skin of the hatchery reared Pallid sturgeon that is infected with iridovirus and in the early stages of development of lymphocystosis. The virus spreads through the water as the epithelial cells are sloughed. Lymphocystis is considered a serious and unacceptable disease in fish aquaculture and as a result of our electron microscopy the hatchery was depopulated and disinfected.

References

[1] R.T. Giberson et al., Microwave Techniques and Protocols, Humana Press, New Jersey, 2001.
[2] R.T. Giberson et al., Microsc. Res. Tech. 32 (1995) 246.
[3] R.T. Giberson et al., J. Vet. Diagn. Invest. 9 (1997) 61.
[4] F.W. Doane et al., Electron Microscopy in Diagnostic Virology, Cambridge Univ. Press, New York, 1987.
[5] F.A. Murphy et al., Veterinary Virology (3rd Ed.), Academic Press, Boston, 1999.

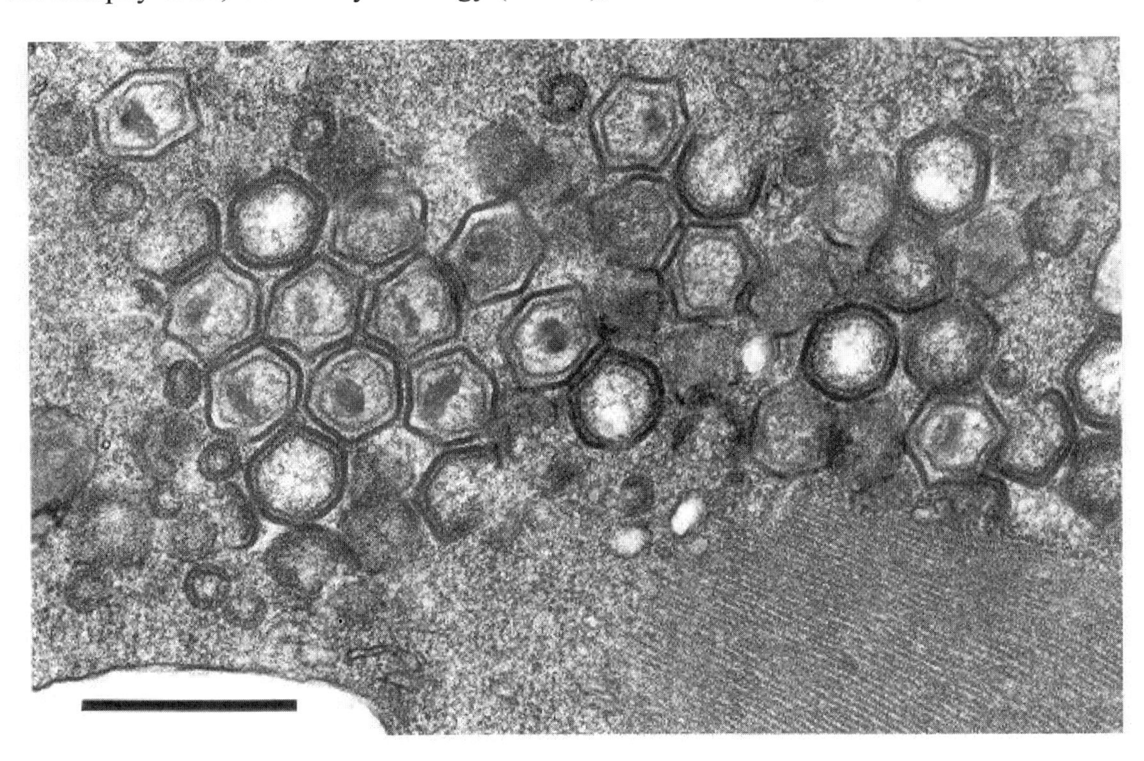

FIG.1 High magnification image of the Pallid sturgeon skin infected with the hexagonally shaped iridovirus. Bar marker is 500nm.

Microsc. Microanal. 8 (Suppl. 2), 2002
DOI 10.1017.S1431927602102157

Microwave Processing in Diagnostic Electron Microscopy

Ross G. Gerrity and George W. Forbes
Department of Pathology, BF 152, Medical College of Georgia, Augusta, Georgia 30912

Transmission electron microscopy (TEM) continues to play an important role in diagnostic surgical pathology [1,2], particularly in such areas as kidney pathology and tumor diagnosis, among others. Diagnostic TEM is subject to unique time constraints and other problems not seen in other TEM applications. The diagnostic TEM laboratory must produce high-quality electron microscopy on small samples which frequently are suboptimal in fixation and tissue quality due to the pathology involved and time factors associated with biopsy and surgery. Despite these problems, the diagnosis must be done as rapidly as possible, and rapid "turnaround" times of samples are a high priority, even in conditions of high caseload. Thus technology which reduces the long processing procedures for TEM samples could be of significant benefit in reducing turnaround time in the diagnostic TEM laboratory. We therefore compared turnaround times of pathology cases processed with traditional routine methods with those processed using a microwave oven for all tissue processing stages (fixation, dehydration, embedding, polymerization).

As shown in Table I, a total of 605 cases each were processed by microwave and routine methods, and the respective turnaround times noted. The cases in each category were processed within the same time span, and the figures have been corrected for non-working days (holidays and weekends). The average turnaround times calculated from these data were 2.5±1.1 days for microwave samples vs. 3.5±1.4 days for routinely-processed samples. Although the difference in means was not statistically significant, 83% of microwave cases were completed in 3 days, 59% in 2 days and 16% in 1 day, compared to 58%, 25% and 0%, respectively, for routinely-processed cases. Additionally, less than 1% of microwave cases took >5 days to complete compared to10% for routine processing.

Thus, from the practical standpoint, the microwave technique significantly reduced the maximum turnaround time and increased the number of cases completed in 2 days or less by 3-fold. It also allowed cases to be completed in less than 1 day on a routine basis, which is not possible using routine methods. In our hands, microwaved blocks were indistinguishable from routinely-processed blocks in sectioning, staining and ultrastructural appearance. Of particular importance, ultrastructural features necessary for diagnoses were identical to those seen in routinely-processed samples (Fig. 1a-d). Blood cells and granules (Fig. 1a), plasma membranes, microvilli and cilia (Fig. 1b,c), microtubules and cytoplasmic filaments (Fig. 1d) as well as all other cellular organelles showed excellent preservation. Moreover, in cases difficult to embed, such as tumors with high lipid and collagen content or skin punch biopsies, microwave embedment was superior to that obtained by routine processing. The major drawback of microwave processing in the diagnostic TEM laboratory is that it is so rapid that cases are prepared more rapidly than they can be sectioned and examined, and a technician must be committed full-time to the microwave during processing and is thus not available for other duties, including sectioning. Thus to optimize the benefits of microwave processing, it must be scheduled according to the technical priorities of the day.

[1] J. Lloreta-Trull et al., Ultrastruct. Pathol. 24 (2000) 105-108.
[2] R. Kandel et al., Ultrastruct. Pathol. 22 (1998) 141-146.

153

TABLE 1: Turnaround Time: Microwave vs. Routine Processing

	Number of Days							Total Cases
	1	2	3	4	5	6	>6	
No. Microwave Cases	94	262	148	58	38	5	0	605
No. Regular Cases	0	150	200	133	59	36	27	605

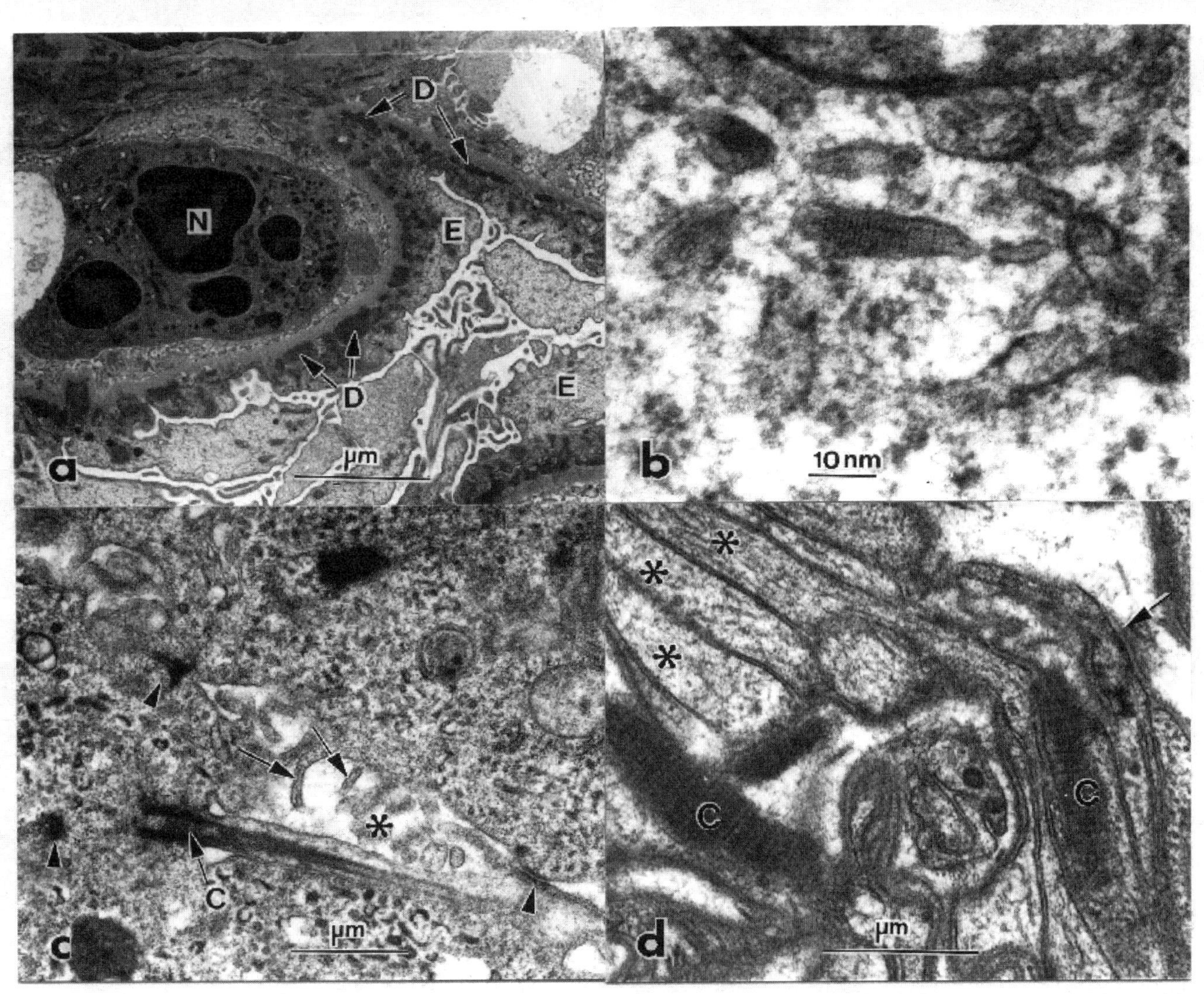

Figure 1. Ultrastructure of human biopsy specimens prepared with microwave processing from (a) renal glomerulus from lupus nephritis patient showing subepithelial dense deposits (D), effacement of foot processes in epithelial layer (E) and neutrophil (N) in capillary; (b) melanoma, showing striated structure of a stage 2 melanosome essential for diagnosis; (c) ovarian adenocarcinoma, showing microvilli (arrows) and a cilium (C) extending into acinar space (*) bounded by junctional complexes (arrowheads); (d) schwannoma, showing typical layered cell processes (*), fragmented external lamina (arrow), microtubules and cytoplasmic filaments and fibrous long spacing collagen.

Microsc. Microanal. 8 (Suppl. 2), 2002
DOI 10.1017.S1431927602102169

The Use of Microwave Technology in a Clinical E.M. Laboratory

R. L. Austin

Louisiana State University Medical Center, Dept. of Pathology, Electron Microscopy Laboratory, Shreveport, LA. 71130

Since the first modified microwave oven, to process tissue for E.M., was marketed by Ted Pella in 1994, the microwave has advanced into a more sophisticated instrument allowing for even faster processing time than the present literature has suggested [1]. In a previous publication by this author clinical specimens were processed with an average processing time of about 4 hours from wet tissue to polymerized block, with another 1 to 1½ hours to get the tissue to the E.M. for diagnostic viewing (Fig. 1b). This produced a processing time of about 5 to 6 hours depending on the number of specimens to be processed, sectioned, and stained. At that time this was a great improvement in turn around time from previous methods. The microwave of just a few years ago had a fixed wattage setting of approximately 750 watts, and the load cooler temperature was set at a fixed 40^{0}C before it would start cooling the circulating water load (Fig. 1a). The additional water loads had to be manually reloaded and readjusted to eliminate hot spots (Fig.1a). Specimens and fixative had to be pre-cooled in ice baths before they could be safely microwaved [2].

Microwave processing has been simplified through the development of a water re-circulation device called the ColdSpot™. This device brings about the virtual elimination of hot and cold spots in the microwave cavity. It thus provides a uniform surface area (8.5"x11.0") to process tissue, eliminating the need for ice baths and additional water loads. The load cooler has an adjustable temperature control so that a temperature lower than 40^{0}C can be maintained in the ColdSpot™ cooler processing conditions. The wattage setting is now adjustable from a fixed 750 watts in the old Model 3450 to as low as 150 watts in the Model 3451 and lower still in the newest microwave device with the incorporation of a wattage control mechanism. In addition, a vacuum system can be easily used in combination with the ColdSpot™. These features, in combination with the variable temperature load cooler, (Fig. 2a) enhance fixation and infiltration and reduce the processing time even further [3]. We are now looking at a processing time of 2 hours from wet to polymerized tissue. This is twice as fast as the earlier microwave capabilities. Considering an added 1 to 1½ hours for sectioning and staining we now have a system that can get newly received tissue to the electron microscope for diagnostic viewing in roughly 3 to 3½ hours (Fig. 2b).

References

[1] R.T. Giberson et al., J. Vet. Diagn. Invest 9 (1997) 61.

[2] R. L. Austin., Microwave, Techniques and Protocols. Humana Press Inc. (2001) 38.

[3] Ted Pella et al., Pelco Inc. Technical Notes. (2001) 1.

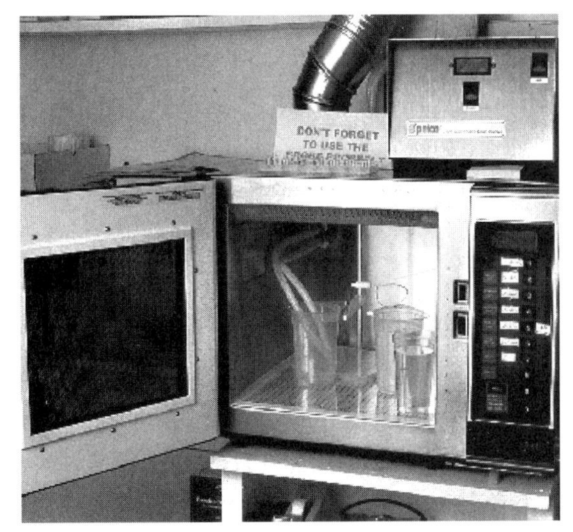

Figure 1a. Original microwave model 3450 series without wattage control or cold spot.

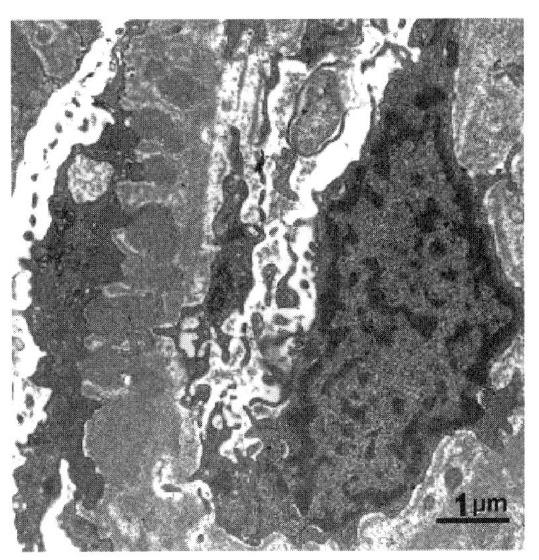

Figure 1b. Results demonstrate oven capabilities to process tissue.

Figure 2a. The newer version of the model 3450 series show variable wattage control unit and cold spot, with vacuum chamber in place.

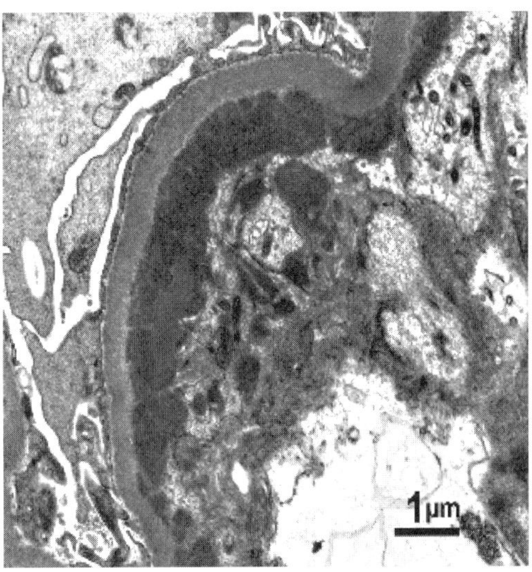

Figure 2b. Results from the improvements made to the microwave oven.

Microsc. Microanal. 8 (Suppl. 2), 2002
DOI 10.1017.S1431927602102170

Immunocytochemistry: A New Microwave Application

Jonathan R. Day*, Richard Demaree*, Richard Giberson**, and Teresa Elena Munoz*

*Department of Biological Sciences
California State University at Chico
Chico, California

**Ted Pella Inc.
Redding, California

We use Immunocytochemistry to examine proteins within the hippocampus of the rat*in vivo* and *in vitro*. This includes determining cell specific and subcellular expression patterns. We also use this tool to monitor changes in the expression after experimental manipulation? Conventional immunocytochemical protocols require primary and secondary incubations of several hours to several days. Reduced signal to noise ratios associated with high background and epitope masking vary with antibody preparations. The advent of variable power microwave ovens with effective temperature control have dramatically reduced these inconsistencies and produced a standardized protocol that can be used with a variety of monoclonal and polyclonal antibodies. It works equally well with fluorescence and peroxidase-linked detection methods. Finally, there is an astonishing 10-fold reduction in the time required to achieve consistent, reproducible results using this microwave protocol. Simple fluorescence protocols can be completed in less than an hour.

The basic protocol is not fundamentally different from conventional protocols that require overnight antibody incubations, nor are any special solutions required. The protocol we use for peroxidase linked immunocytochemistry on fixed brain sections follows.

4C fix overnight in 4% paraform
30% sucrose equilibration
35 micrometer sliding microtome floating sections

Step		Conventional	Microwave
3 rinses PBS @ RT 5 min/ea		15 min	120 s
0.3% H202 @ RT agitate		20 min	60 s
3 rinses PBS @ RT 5 min/ea		15 min	120 s
10% horse serum/PBS/ 1% BSA/ 0.3% triton X-100 (block) @ RT		30 min	60 s
Primary antibody in PBS or block @ RT (or overnight@ 4C)		3 h or 16 h	6 min
3 rinses 1% block @ RT 15 min/ea		45 min	120 s
Secondary antibody @ RT		1 h	6 min
3 rinses PBS @ RT 15 min/ea		45 min	120 s
ABC		1 h	6 min
3 rinses PBS @ RT 10 min/ea		30 min	120 s
2 rinses Imidazole 5 min/ea		10 min	120 s
DAB for up to		7 min	6 min
Rinse and mount	total time	8.5 - 21.5 h	38 min

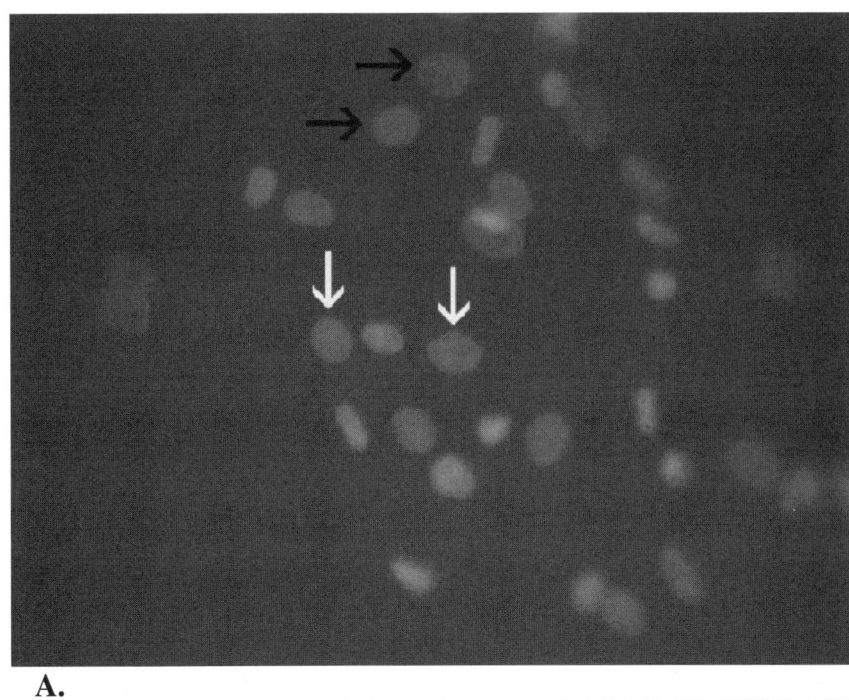

A.

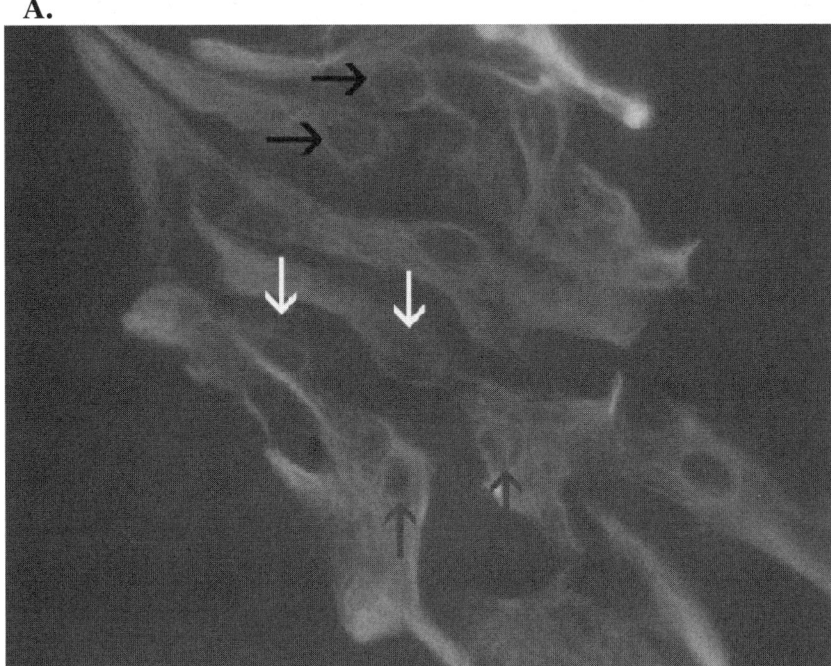

B.

Figure1. shows nuclear profiles of primary astrocytes isolated fromE18 rat cortex after 7 days *in vitro* revealed by *bis*benzamide and UV epifluorescence illumination(A). Panel B shows the same field under blue epifluorescence illumination. Glial fibrillary acidic protein (GFAP) immunoreactivity is revealed using an FITC conjugated secondary antibody to a monoclonal primary antibody directed against a human GFAP epitope. The cells were labeled on the glass coverslips where they were grown using the basic microwave protocol described in the abstract. Arrows of the same color and direction point to the same nuclei in each panel.

DOI 10.1017.S1431927602102182

Recent Advances in Microwave Assisted Specimen Processing: Keeping it Cool.

Mark A. Sanders

Imaging Center, College of Biological Sciences, University of Minnesota, St. Paul, MN, 55108

Microwave energy as a method of rapid tissue processing is gaining increasing acceptance as an alternative to routine chemical processing in laboratories where rapid and accurate specimen processing is required. Conventional chemical fixatives and protocols used for standard preservation of biological specimens can result in alterations in morphology as a consequence of solubility and conformational changes of cellular constituents. These changes often result in morphological artifacts and compromised antigenicity of many tissue proteins. Conventional chemical fixatives in combination with temperature controlled microwave processing can distinctly shorten the processing time while maintaining better specimen preservation. In addition to tissue fixation, very low wattage in a dedicated laboratory microwave (<250 watts, Pella 3450) can also be used to facilitate access of vital stains into developing embryonic organisms. The present study will have two objectives: to rapidly prepare specimens for fluorescent in situ hybridization and to utilize low wattage, temperature controlled microwave energy in *in vivo* experiments in plant and animal tissue. Low power microwave energy was used in conjunction with a recently developed microwave accessory called the ColdSpot™ (Ted Pella, Inc., Redding, CA) that creates a uniform environment of microwave irradiation to the samples.

This report demonstrates the use of low power microwave energy (250W or less) in conjunction with the ColdSpot™ in an attempt to overcome the problems reported in the past (Boon and Kok[1], Choi et al.[2], Chicoine and Webster[3]). A number of different nucleic acid and protein probes were tested on known systems. Tissue culture cells, tissue sections and living organisms were labeled using the same basic protocol to examine the viability of a standardized approach. The reported results indicate low power microwave energy combined with better control of the microwave environment produce uniform results across a wide spectrum of ultrasturtural labeling techniques.

Additionally in order to attempt to improve co-localization studies of DNA and protein distributions, we are to developing a routine technique to co-localize specific nucleic acid sequences and multiple antigens simultaneously. We have applied this technique to a variety of specimens including nematode embryos (see Figure 1).

We have also found that under reduced wattage conditions (~150 watts), living plant and animal tissue can be made "permeable" to fluorescent vital dyes such as the Syto series of nucleic acid stains (Molecular Probes, Eugene OR). Brief microwave treatment in the presence of the stain will greatly increase the depth and rate of penetration of the dye. Time-lapse fluorescence or confocal microscopy has shown promising results on several developing model systems including *Arabidopsis* trichomes, root tips, *Drosophila* and nematode embryos.

In conclusion, this method allows for the detection of multiple DNA sequences and antigens at the same tissue level with high sensitivity and good morphological preservation. Given the versatility of the system, it should be useful wherever simultaneous detection of multiple antigens and nuclear acid sequences is required. These procedures can have broad reaching applications in the preparation of specimens in clinical and research laboratory environments.

References:
[1] Boon, M.E., and Kok, L.P., 1994. Microwaves for immunocytochemistry. Micron., 25:151-170.

[2] Choi, T.-S., Whittlesey, M.M., Slap, S.E., Anderson, V.M., Gu, J., 1995. Advances in temperature control of microwave immunohistochemistry. Cell Vision., 2:151-164.

[3] Chicoine, L., and Webster, P., 1998. Effect of microwave irradiation on antibody labeling efficiency when applied to ultrathin cryosections through fixed biological material. Microsc. Res. Tech., 42:24-32.

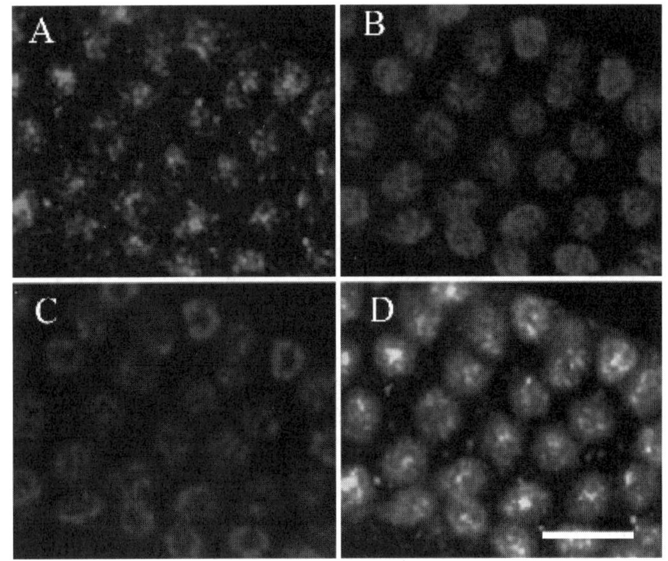

FIG. 1. LSCM image of *c. elegans* embryo triple labeled (D) with (A)Alexa 488 labeled D.v.TIM (RNAi), (B) Texas Red conjugate to anti-histone and (C) Syto 63 nucleic acid stain following microwave specimen processing utilizing a power level of 200W and the ColdSpot™ (Ted Pella, Inc., Redding, CA). Scale bar =20 microns.

Microsc. Microanal. 8 (Suppl. 2), 2002
DOI 10.1017.S1431927602100341

Microwave Processing and Pre-embedding Nanogold Immunolabeling for Electron Microscopy

JoAnn Buchanan,* Kristina D. Micheva* and Stephen J Smith*

*Department of Molecular and Cellular Physiology, Stanford University School of Medicine, Beckman Center, Stanford, CA 94305

The introduction of microwave assisted processing for electron microscopy (EM) Login and Dvorak [1], Giberson et al. [2], Giberson and Demaree [3] has dramatically reduced the amount of time required for sample preparation from days to hours. In addition, the introduction of the cold spot and variable wattage controller by Giberson (Ted Pella, Inc.) has provided a means of regulating specimen temperature during processing steps. We have applied the microwave technology to immuno-labeling of the green fluorescent protein (GFP).

Transiently transfecting cells with GFP-fusion proteins and subsequent imaging by live confocal microscopy is a powerful method for studying the function of proteins *in vivo*. However, because many GFP tagged structures are below the resolution limit of the confocal microscope, we expanded our observation to the Electron Microscope (EM) level. Paupard et al. [4] combined microwave fixation with LR White embedding and post-immunolabeling with GFP antibodies on *C. elegans*. In this study we used commercially available anti-GFP antibodies (Roche, Indianapolis, IN) and Nanogold labeled secondary antibodies (Nanoprobes, Stony Brook, N.Y) for pre-embedding immuno-labeling of hippocampal neurons in culture. We have found that microwave processing dramatically shortens all steps of the immuno-labeling procedure, including incubation in both in primary and secondary antibodies. It also speeds up the permeabilization step to allow better access of antibodies while maintaining structural integrity. This method combines the use of the latest in microwave technology combined with the use of ultra small gold labels to obtain excellent ultrastructural preservation and good labeling density in 3 hours.

Hippocampal neurons (div 7-9) were transfected with the PLC-δ PH domain fused to GFP (PH-GFP). This PH domain binds specifically and with high affinity to phosphatidylinositol 4,5-biphosphate (PIP2). As observed by confocal microscopy, PH-GFP labeled the plasma membrane of cell bodies and neuronal processes. Weaker diffuse cytoplasmic staining was observed as well. Accordingly, at the electron microscope level, the silver enhanced gold label was seen on the neuronal plasma membrane and, at a lower density, in the cytoplasm. Heavier labeling was observed in many small processes, at neuronal branch points and synaptic contact areas. Interestingly, no labeling was seen on the synaptic vesicles. Control (untransfected cells) cells showed no labeling. As an additional control, neurons were also transfected with a GFP fusion to a PH domain having a point mutation in the binding site to PIP2 (PHM-GFP). In this case, at the light level the GFP fluorescence appeared to be uniformly distributed throughout the neurons. At the electron microscope level, the immunolabeling was dispersed throughout the cell cytoplasm, with a heavier concentration of silver enhanced gold in the cell nucleus. This highly specific immunostaining of GFP-fusion proteins and the well-preserved cellular ultrastructure, combined with live imaging of transfected cells, can provide essential information for deciphering the function of biologically important molecules.

[1] R.T.Giberson and R.S. Demaree, *Microwave Technique and Protocols*, Humana Press, 2000
[2] G.R.Login and A.M. Dvorak, *The Microwave Tool Book A Practical Guide for Microscopists*, Beth Israel Hospital, 1994.
[3] R.T. Giberson and R.S. Demaree, in *Electron Microscopy Methods and Protocols*, Humana Press, 1999.
[4] M. Paupard et al., *J of Histochem. and Cytochem.*, 49 (2001) 949.
[5]K.D. Micheva et al., *J. Cell Biol.*, 154 (2001) 355.

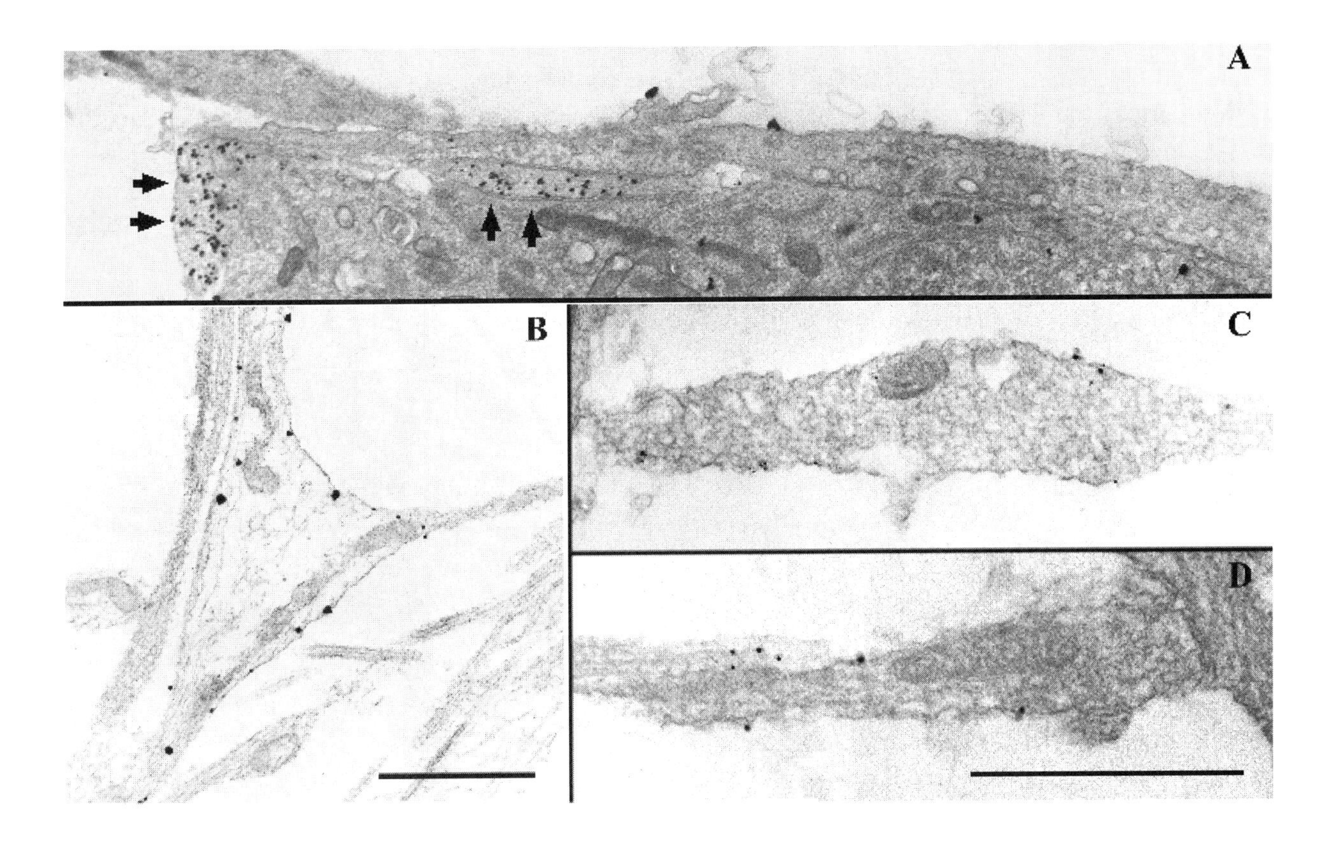

A. Small process of a hippocampal neuron 14 days *div* immuno-labeled with anti-GFP (PHM-GFP) primary and Nanogold conjugated secondary antibodies. Silver intensification.
B. Hippocampal neuron showing specific membrane labeling for anti-GFP (PH-GFP), unlike the PHM-GFP staining pattern. Scale bar 1 micron. Same magnification in A and B.
C. Membrane of an axonal varicosity of hippocampal neuron immuno-stained for anti-GFP (PH-GFP). Synaptic vesicles fill the process.
D. Synaptic varicosity of hippocampal neuron immuno-stained for anti-GFP (PH-GFP) shows silver enhanced gold particles along the membrane. Scale bar 1 micron. Same magnification in C and D.

Microsc. Microanal. 8 (Suppl. 2), 2002
DOI 10.1017.S1431927602102200

The Use of New Microwave Techniques to Facilitate the Immunostaining of Paraffin Sections on Glass Slides

R.T. Giberson

Research and Development, Ted Pella, Inc., Redding, CA 96003

Microwave-assisted processing is gaining acceptance as a routine laboratory technique due to improved technology and better standardization of processing methods.[1] This evolution is primarily due to better control of the microwave environment during processing.[2] Recent development of the ColdSpot™ (Patent US6329645) has normalized the microwave environment and eliminated the necessity to employ a microwave calibration scheme as part of a protocol. A second component of the evolution has been the incorporation of continuous power control, from 50 to 750W, as an integral part of microwave technology (PELCO BioWave™, Ted Pella, Inc.).

The combination of these elements has made sample processing a simple process for electron microscopy. The efficacy of this combination as a viable tandem for immunocytochemistry was elegantly demonstrated by the work of Sanders and Gartner detailing the in vivo labeling of *Drosophila* embryos and *Allium* sp. root tips with nuclear vital stains.[3]

Previous reports using microwave methods to accelerate immunostainingapplications for light or electron microscopy have reported variable results attributable to either differences between antibodies, length of microwave exposure or an incorrect temperature maximum.[4,5] To address the problem of the inconsistancies described by others a protocol for paraffin sections on glass slides was developed. The foundations were based on low power (250W-PELCO BioWave™), the ColdSpot™ and a slide rack with coverplates from ThermoShandon (Pittsburgh, PA). Control slides from DAKO® (Carpinteria, CA) were stained with IMMUNON™ pre-diluted primary antibodies (ThermoShandon, Pittsburgh, PA). An IMMUNON™ Universal Streptavidin/Biotin Immunoperoxidase Detection System with DAB (Kwik™ Kits, ThermoSandon) was used for the demonstration of each antigen in question.

The slide rack was used for all steps except the first two. When used, it was placed on top of the ColdSpot™ (Table 1). Steps 8, 10 and 12 (Table 1) were programmed as 2 minutes on, 2 minutes off and 2 minutes on. The addition of 0.1% Triton X-100 to all buffer rinse steps greatly improved the results of this method. Microwave heating was minimal due to the ColdSpot™ and a low microwave power setting (250W). Immunolabeling results were consistent run to run with little or no background staining. Figure 1 is indicative of the results obtained when using this technique. The problems and/or criteria ascribed to microwave-assisted labeling, as described by other authors, were not apparent using this new system.[4,5] Successful results did not depend on changing microwave parameters for any of the antibodies tested and all reagents were used at their supplied dilutions. When the protocol was not microwave-assisted, the results were inconsistent.

References

[1] Microwave Techniques and Protocols, Humana Press, Totowa, NJ, 2001.
[2] R.T. Giberson, Microsc. Microanal. 7 (Suppl. 2:Proceedings) (2001) 1192.

[3] M.A. Sanders and D.M. Gartner, in R.T. Giberson and R.S. Demaree, Eds., Microwave Techniques and Protocols, Humana Press, Totowa, NJ (2001) 155.
[4] J. Gu et al., Cell Vision. 2 (1995) 257.
[5] L. Chicoine and P. Webster, Microsc. Res. Tech. 42 (1998) 24.

TABLE 1. Protocol outline for microwave-assisted labeling of paraffin sections on glass slides.

Microwave Step	Time	Slide Holding Device in Microwave
1. Deparaffinization-xylene	4 min.	Coplin jar
2. 100% Ethanol	1 min.	Coplin jar
3. Isopropanol	1 min.	Slide rack
4. 70% Ethanol	1 min.	Slide rack
5. Proxidase block-3%H_2O_2	1 min.	Slide rack
6. Buffer rinse	1 min.	Slide rack
7. Blocking	1 min.	Slide rack
8. Primary antibody	6 min.	Slide rack
9. Buffer rinse	1 min.	Slide rack
10. Biotinylated secondary	6 min.	Slide rack
11. Buffer rinse	1 min.	Slide rack
12. Streptavidin peroxidase	6 min.	Slide rack
13. Buffer rinse	1 min.	Slide rack
14. DAB chromogen	4 min.	Slide rack
15. Water rinse	1 min.	Slide rack
16. Hematoxylin counterstain	3 min.	Slide rack

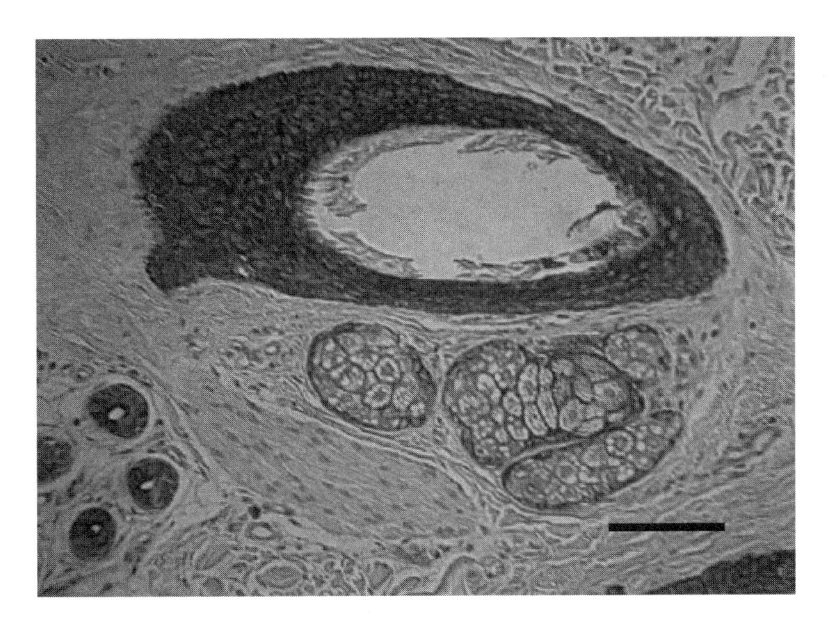

FIG. 1. DAKO® control slide S-100, Human Skin (Code T1072) stained by microwave-assisted methods with pre-diluted IMMUNON™ polyclonal Cytokeratin, wide spectrum and antigen detection with DAB using an IMMUNON™ Universal Streptavidin/Biotin Imunoperoxidase Detection System. Bar = 100μm.

Microsc. Microanal. 8 (Suppl. 2), 2002
DOI 10.1017.S1431927602102212

A Basic Strategy for Biomineralization: Taking Advantage of Disorder

S. Weiner, Y. Levi-Kalisman, S. Raz, I.M. Weiss and L. Addadi

Department of Structural Biology, Weizmann Institute of Science, Rehovot, Israel 76100

Biologically formed minerals are renowned for their unique shapes, preferred orientations, organisation into ordered mineral-matrix composites, their remarkable mechanical properties and their beauty [1]. Organisms seem to have an ability to control mineral formation that far exceeds our own in vitro capabilities. Have organisms evolved a really novel underlying strategy for controlling mineral formation? Possibly, at least in the case of calcium carbonate deposition.

Three crystalline polymorphs of calcium carbonate exist (calcite, aragonite and vaterite), and 3 hydrated forms (monohydrocalcite, hexahydrate and amorphous calcium carbonate (ACC)) [2]. ACC is traditionally regarded as being amorphous because it is isotropic under crossed polarizers and it does not diffract X-rays. It does however have short range order that differs between ACC deposits formed by different organisms [3]. It is highly unstable and soluble in vitro. It is thus surprising that 16 species from 7 animal phyla are known to form ACC, in addition to quite a few plant taxa [4]. In most known cases, the ACC is stable at least during the lifetime of the organism. In 2 cases, the larval spicules of the sea urchin [5] and the larval shells of molluscan bivalves [6], the ACC is a precursor phase that transforms into one of the crystalline polymorphs. In this presentation I will focus on larval and adult mollusk shell formation processes, and in particular the possible role of ACC.

Since the beginning of the 20th Century it has been thought that all mollusk larval shells are composed of only aragonite crystals, with their c-axes perpendicular to the shell surface [7]. Using a high resolution ESEM we compared the ultrastructures of the larval shells of 2 bivalve species, *Mercenaria mercenaria* and *Crassostrea gigas*. We confirmed earlier observations that showed that they are both composed of an outer organic layer (periostracum), a relatively thick granular-homogeneous middle layer and an inner prismatic layer [8]. We noted that the larval shells are much more sensitive to radiation damage than the adult shells, and that the granular layer is composed of spherical sub-units; properties reminiscent of an amorphous phase. Using Raman and infrared spectroscopy, we demonstrated that the first-formed mineral phase is indeed ACC and that with time it converts into aragonite. This aragonite phase is much less crystalline than geological aragonite. Interestingly, there is no obvious correspondence between the observed changes in ultrastructure as the larval shell forms and the mineral transformation; apparently the amorphous mineral phase is initially moulded into specific shapes (for example prisms), and only subsequently does it transform into aragonite. We also noted that the aragonite in the adult *Mercenaria* shell is most surprisingly less crystalline than two different samples of inorganic aragonite we analysed. This may be a hint that the adult shell also contains some ACC and/or poorly crystalline aragonite, and if so, an indication that the adult shell also forms via an ACC precursor.

We used cryo-TEM to investigate the structure of the organic matrix of the nacreous layer of the adult bivalve *Atrina* [9]. We showed that the interlamellar sheets viewed in vitrified ice are composed mainly of highly ordered chitin that display lattice fringes. Aspartic acid-rich proteins are located on the surfaces of these sheets. The other major matrix component, silk fibroin, could not be imaged in the extracted organic matrices, or even when reconstituted on chitin in vitro. This suggests to us that it adopts a gel-like phase. Vitrified ice cross-sections of the interlamellar sheets showed that the spaces between sheets even after demineralization, contain organic material. We therefore proposed that this is the site of the silk fibroin. If so, this implies that the aragonite forms inside a hydrated silk fibroin gel. If the larval shell formation processes are any guide to the adult shell formation processes, it is conceivable that the silk is somehow involved in the formation and/or stabilization of a precursor ACC phase - an exciting prospect to be investigated.

Note that the two proven examples of ACC acting as a transient precursor phase, the larvae of the sea urchin and the larvae of bivalves, are from phyla that are on 2 quite different branches of the phylogenetic tree of the animal kingdom. If this strategy evolved divergently, this could imply that it is widespread in the animal kingdom. If so, we would need to understand the benefits of adopting this novel approach to crystallization, whereby the crystals nucleate and grow from a dense colloidal phase, possibly with its own tailor-made short range structure.

References

[1] H.A. Lowenstam and S. Weiner, *On Biomineralization*, Oxford University Press, New York, 1989.
[2] F. Lippman, *Sedimentary Carbonate Minerals*, Springer-Verlag, Berlin, (1973).
[3] Y. Levi-Kalisman et al., *Adv. Funct. Mat.* 12 (2002) 43.
[4] S. Weiner et al., *Conn. Tissue Res.* (in press).
[5] E. Beniash et al., *Proc. R. Soc. Lond. Ser. B.* 264 (1997) 461.
[6] I.M. Weiss, N. Tuross, L. Addadi and S.Weiner (in preparation).
[7] W.J. Schmidt, Die Bausteine des Tierkörpers in Polarisiertem Lichte, Friedrich Cohen Publisher, Berlin (1924).
[8] M.R.Carriker and R.E. Palmer, *Proc. Natl Shellfisheries Assoc.* 69 (1979) 103.
[9] Y. Levi-Kalisman et al., *J. Struct. Biol.* 135 (2001) 8.
[10] This research was supported by grants from the U.S. Israel Binational Science Foundation and the Minerva Foundation, as well as a grant to I.M. Weiss from the Deutsche Forschings gemeinschaft WE 2629/2-1.

Microsc. Microanal. 8 (Suppl. 2), 2002
DOI 10.1017.S1431927602102224

Hydroxyapatie Formation and Its Interaction with Osteoblastic Cells

H. Vali,* P. Ghiabi,* J. Henderson,** M.D. McKee,*** E. Chevet,* and S.K. Sears*[1]

* Electron Microscopy Centre, Department of Anatomy & Cell Biology, McGill University,3640 University Street, Montreal, QC H3A 2B2
** Lady Davis Institute, Jewish General Hospital, Montreal, Quebec H3T 1E2
*** Faculty of Dentistry/Department of Anatomy & Cell Biology, McGill University, Montreal, QC H3A 2B2

Uncertainty remains over the interpretation of nano-phenomena from various biological and geological environments as living organisms. A characteristic feature of nano-phenomena is their association with a variety of mineral phases. One of the most controversial issues involving nano-phenomena is the existence of *Nanobacterium sanguineum* isolated from human and cow sera [1]. A unique trait associated with this nano-organism is the formation of hydroxyapatite (HA). The nature and formation of the HA phase(s), however, is still not well understood in this system. It is clear that HA is associated with many pathological calcifications, including kidney stones, atherosclerosis, urolithiasis, calcification in hemodialysis patients, dental plaque, and dysfunctional calcification occurring in implanted cardiovascular devices. Although our results do not confirm bacteria controlled or induced mineralization leading to formation of thesemineralized nanoforms, the association of HA with serum protein suggests a biogenic origin.[2] To investigate the interaction of nanoforms with MC3T3-E1 cells and the mechanism(s) of calcification, several sets of experiments were conducted under variable cell culture conditions. Extensive variation in morphology occurs between HA in nanoforms obtained from 10% FBS/DMEM (Fig 1a) and HA present in MC3T3-E1 cell cultures (Figs. 1b-f). Aggregates of altered nanoforms were present within a membranous organelle (Figure 1c, d). Occasionally, vacuole-like structures containing isolated crystals were observed that are likely a result of the disintegration of nanoforms. The original structure of nanoforms was not preserved either outside or inside of the cells. The distribution and concentration of mineralized structures within the cells differed extensively depending on the composition of the growth medium. The mineralized structures in the experiments conducted with nanoforms grown in EMEM show high concentrations of HA crystals within the cells. Larger well-ordered crystals, showing evidence of newly formed HA, are observed within membrane-bound regions (Fig. 1f). It is likely that the changes in size and morphology of the HA crystals are the result of the dissolution of the fine-grained HA crystals associated with the nanoforms obtained from serum (Fig. 1a) and the neoformation of distinct HA crystals observed in MC3T3-E1 cells (Fig. 1f). The mechanism of transformation and the involvement of MC3T3-E1 cells in this reaction are not understood. It may be that within the membrane-bound phagocytic compartments observed in the MC3T3-E1 cells, different regulatory controls are exerted over crystal growth that otherwise would occur extracellularly, thus resulting in the formation of large crystals. Further study mayreveal the significance of these experiments for pathological calcification and bone formation. However, the approach used in this study provides a unique opportunity to understand the mechanisms of HA formation in both normal and pathological calcification. In addition, this process may eventually be used for bone tissue engineering where mineralization is desirable with synthetic bone substitutes.

[1] E.O. Kajander & N. Ciftcioglu *Proc. Natl. Acad. Sci. USA* 95 (1998) 8279.
[2] H. Vali et al., *Geochim. Cosmochim. Acta* 65 (2001) 63.

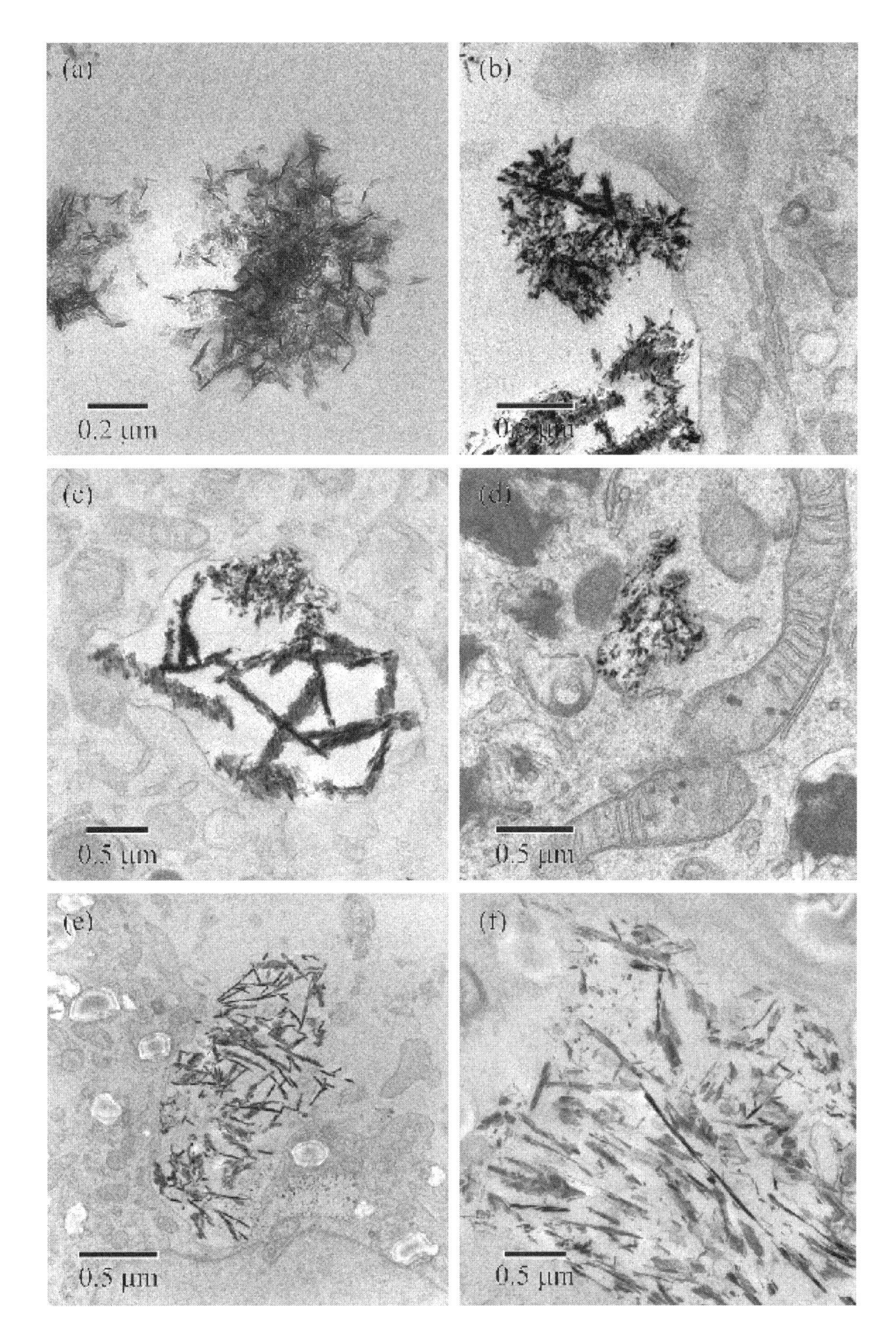

FIG. 1. TEM images of ultrathin sections obtained from nanoforms: (a) nanoforms obtained from 10% FBS/DMEM; (b) nanoforms exposed to MC3T3-E1 cells showing the stage of internalization; (c) internalized aggregates of HA showing a unique distribution; (d)dispersed HA crystals with dissolution features; (e) a large aggregate of HA engulfed by the cell; (f) large neoformed HA crystals present within a MC3T3-E1 cell.

Microsc. Microanal. 8 (Suppl. 2), 2002
DOI 10.1017.S1431927602102236

MAGNETIC AND STRUCTURAL CHARACTERIZATION OF BIOGENIC MAGNETITE

M.R. McCartney* and R. E. Dunin-Borkowski**

*Center for Solid State Science, Arizona State University, Tempe, AZ 85287-1704.
**Department of Materials Science, Cambridge University, Cambridge, UK

Magnetic crystals below 100 nm in size occur in organisms in many biological phyla[1]. For example, magnetotactic bacteria contain magnetosomes, which are intracellular, ferrimagnetic crystals of magnetite (Fe_3O_4) or greigite (Fe_3S_4). The magnetosomes are usually arranged in one or more linear chains within each bacterium and impart a permanent magnetic moment to the cell that results in its alignment and motion parallel to geo-magnetic field lines[2]. This behavior is thought to increase the efficiency with which such bacteria find their optimal oxygen concentrations or redox potentials at sediment-water interfaces or in water columns[3]. This fascinating phenomenon is of interest to both lay people and to scientists across many disciplines. Recently, evidence of similar magnetite crystals found in the Martian meteorite ALH84001 have been used to posit the possibility of life on Mars[4]. This identification relied on a comparison of the morphology of the meteoritic magnetite crystals with the magnetosomes from the magnetotactic bacteria, MV-1.

Clearly, what is needed is a comprehensive investigation into the morphology of biogenic magnetite. Transmission electron microscopy (TEM) bright field imaging is a projection technique and is relatively insensitive to the thickness of the sample. However, instead of obtaining bright-field images of crystals and analyzing their outlines to determine morphologies, it is possible to measure the projected thickness of a nanocrystal in the TEM in one of three ways: (i) by using energy selected imaging to form three-window, background-subtracted chemical maps [5], (ii) by using high-angle annular dark-field (HAADF) imaging with the microscope in scanning TEM mode[6]; or (iii) by using electron holography[7].

Figure 1 shows a TEM image of a chain of magnetosomes from Itaipu, Brazil. The magnetosomes are regular in size and the outline of the crystals reveals a symmetry which repeats every other crystal. Fig.2 is a high-resolution image of a corner of one of the crystals and shows the relation between the magnetite lattice and some of the facets of the crystal. Electron holography can be used in conjunction with careful tilting of the crystals to reveal the projected thickness and thus the three-dimensional morphology. Figure 3 show such an example where the electrostatic and magnetic phase shifts have been separated to show the projected thickness of several of the crystallites. Tiltiing experiments show that the crystallites are elongated along the (111) direction and have an hexagonal cross-section bounded by [110] planes. On the contrary, magnetite crystals from aquatic bacteria MS-1 have an octagonal shape bounded by [111] planes, as shown in Fig. 4. Perhaps, the best way to determine the morphology of these magnetosomes, is to use HAADF images as a series of tilt images for tomography. Figure 5 shows a processed projection of a magnetosome from a tomography series of 60 images taken at tilts of +/- 60°. A catalogue of 3-D images of biogenic magnetite would be an invaluable guide in evaluating extraterrestrial material for signs of life.

REFERENCES

[1] H. A. Lowenstam and S. Weiner, On Biomineralization, Oxford Univ. Press, New York, 1989.
[2] R. B. Frankel, Annu. Rev. Biophy. Bioeng. 13 (1984) 85.
[3] R. B. Frankel, et al., Biophys. J. 73 (1997) 994.
[4] K.L. Thomas-Keprta, et al., Proc. Natl. Acad. Sci. USA 98 (2001) 2164.
[5] R.E. Dunin-Borkowski, et al., Science 282 (1998) 1868,
[6] P.R. Buseck, et al, Proc. Natl. Acad. Sci. USA 98 (2001) 13490.
[7] M. R. McCartney, et al., Eur. J. Mineral. 13 (2001) 685.

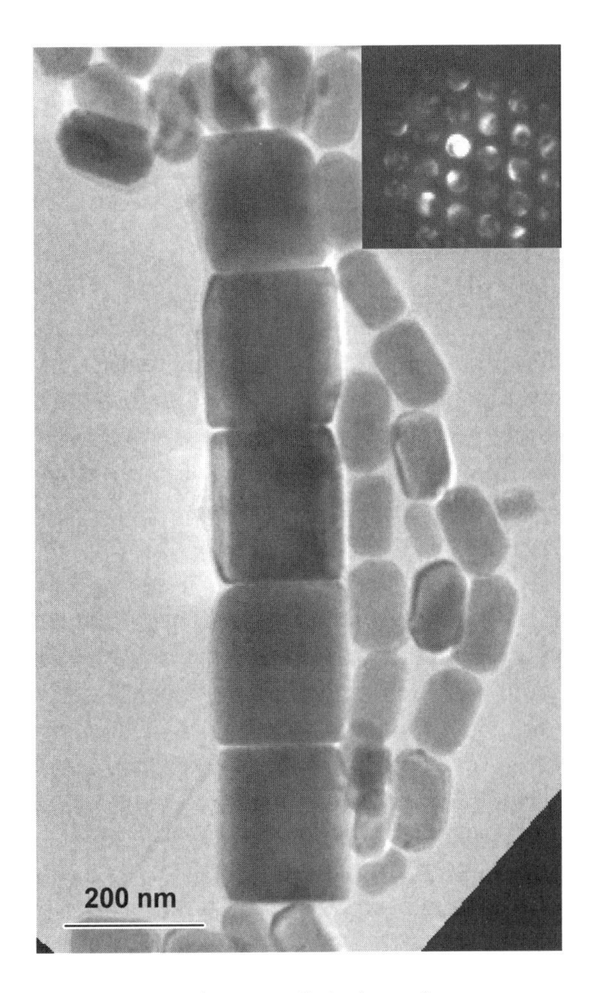

FIG. 1 TEM image of chains of magnetosomes from two different magnetotactic bacteria from Itaipu Brazil. Inset shows nano-diffraction pattern of one of the large crystals.

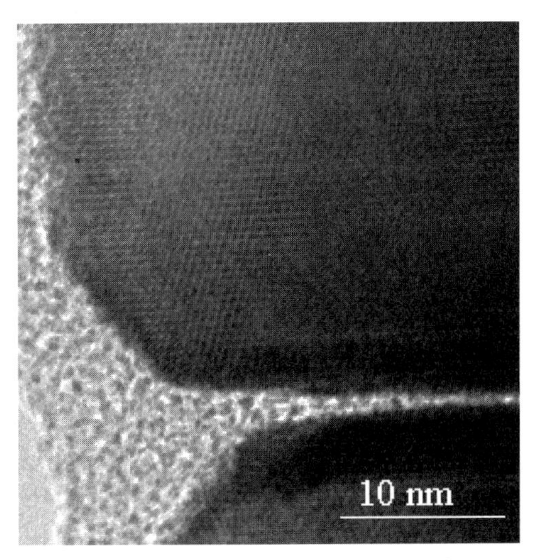

FIG. 2. High resolution image of (111) and (211) lattice planes from one of the larger crystals in Fig. 1.

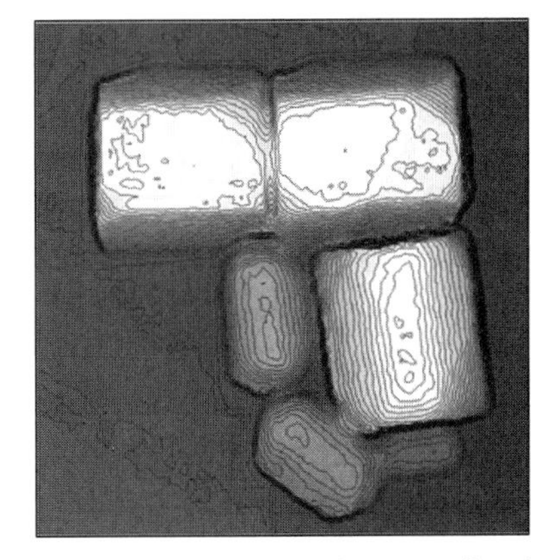

FIG. 3. Electrostatic contribution to the holographic phase shift for large Itaipu magnetosomes reveals projected thickness.

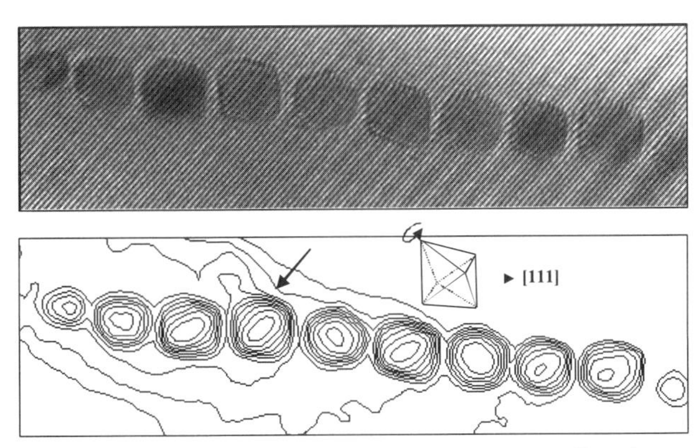

FIG. 4 Hologram and electrostatic contribution to the phase of a chain of magetite crystal from bacteria MS-1. The crystals are octahedral with (111) facets.

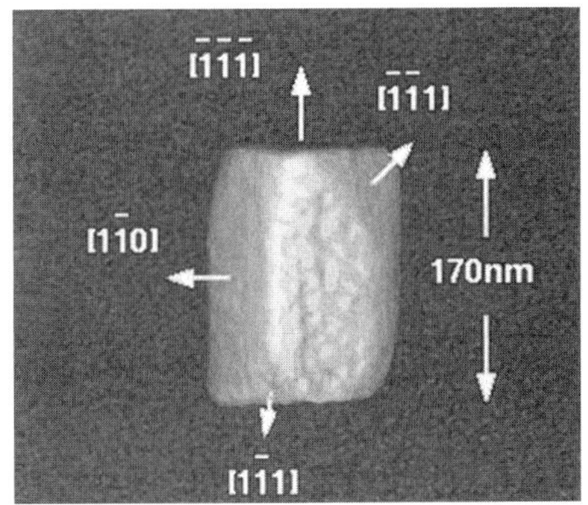

FIG. 5. Tomographic image of biogenic magnetite.

Microsc. Microanal. 8 (Suppl. 2), 2002
DOI 10.1017.S1431927602102248

Global and Local DNA Structure and Dynamics. Single molecule studies with AFM

Y. Lyubchenko*, L. Shlyakhtenko*, V. Potaman** and R. Sinden**

*Department of Microbiology, Arizona State University, Tempe AZ 85287
**Institute of Biosciences and Technology, TAMU, 2121 West Holcombe Blvd. Houston, TX 77030.

The interaction between specific sites along a DNA molecule is often crucial for the regulation of genetic processes. However, mechanisms regulating the interaction of specific sites are unknown. We show in this paper that the single molecule observations performed with the time-lapse atomic force microscopy (AFM) provide an important information on DNA dynamics. In our recent work [1] we took advantage of AP-mica to bind DNA molecules in a broad range of ionic strengths to observe directly the effect of ionic conditions on the global structure of supercoiled DNA. Continuous observations over the same scan area in aqueous solutions (time-lapse imaging mode) allowed us to observe the mobility of DNA at the surface-liquid interface. This approach was very useful for studies of the structure and dynamics of cruciforms [2]. Two families of the cruciform conformations were found and characterized: an X-type and an extended conformations. Statistical analysis of the interim measurements led to the conclusion that compact X-type cruciforms are very dynamic allowing the arms to move in a very broad range. Unlike the X-type conformation, the extended cruciform geometry is less dynamic. These conclusions were tested by direct observation of the cruciform dynamics with the time-lapse AFM. The results showed clearly a high mobility of the X-type cruciform conformation and a relatively static conformation of the extended cruciform geometry. It is remarkable that AFM has the capability to reveal structural dynamics of cruciforms that were not amenable to any current structural technique. Further analysis of global DNA conformation led to unexpected discovery that the structural transition between cruciform conformations can act as a molecular switch to facilitate or prevent communication between distant regions in DNA. It is important to note that the cruciform conformation exists in vivo and such local structures and their conformational transitions during gene expression could have dramatic structural effects on chromatin architecture and function.

We have recently succeeded in visualization of intramolecular triplexes [3]. Plasmid samples were prepared at acidic pH and control samples were prepared at neutral pH. The image is shown in Fig. 1A. A distinct feature of the molecules prepared at acidic pH is the formation of a clear kink with a short protrusion indicated with arrows. Such features are not present in a control that have inserts different than purine-pyrimidine repeat. The formation of a sharp kink is fully consistent with the model of intramolecular DNA triplex. To study directly the effect of pH on H-DNA stability and to follow the transition of H-DNA into B-conformation, the procedure for reproducible and gentle change of the buffer solution without interruption of the scanning has been developed. The buffer change procedure injection procedure itself does not influence the position of the molecule and its overall shape. This is illustrated by the data in Fig. 1 in which the images of the same DNA molecule obtained at initial conditions (pH 5) and after the buffer injection (pH 7.6) are shown.

H-DNA dynamics

The protrusions indicated with the arrows are observed and stably exist at conditions that stabilize H-DNA conformation (pH 5). However, these features become dynamic after replacing of the buffer. A series of images illustrating the dynamics of the part of the plasmid indicated with arrows is shown in Fig. 2. A complete set of 21 consecutive images suggests that overall changes of the molecule are accompanied with the local DNA conformation transition. However, the same DNA regions undergoes a series of structural changes after the disappearing of the protrusion suggesting that the H-to-B form transition is not a simple two-state conformational transition. There is a number of conformational states including unwound regions that are capable of visualizing with the time-lapse AFM.

We believe that further development of the single molecule AFM technique is extremely important for understanding the DNA dynamics and for direct observation of the role of local DNA structures in numerous biological functions.

References:

1. Y. L Lyubchenko. & L. S. Shlyakhtenko (1997). *Proc Natl Acad Sci U S A* 94, 496.
2. L. S Shlyakhtenko et al. (1998). *J Mol Biol* 280, 61.
3. W. Tiner, et al. (2001) *J. Mol. Biol.* 314: 353.

4. This work was supported by the grants GM62235 (NIH) and DBI-0100828 (NSF) to YL.

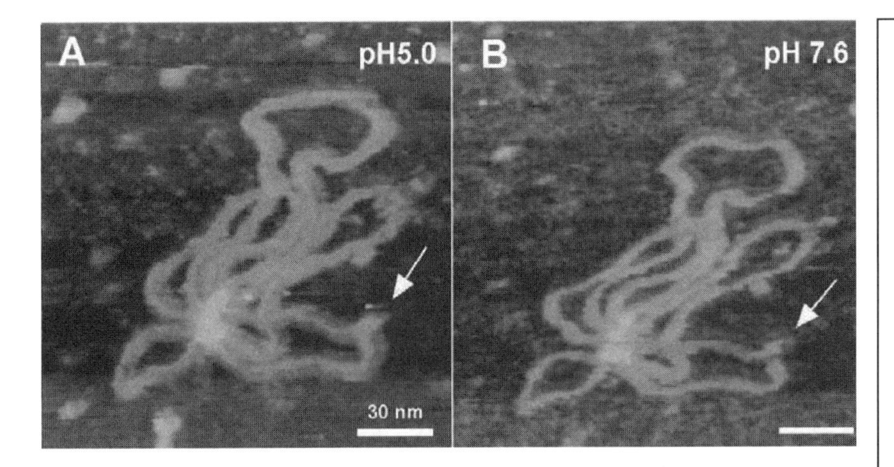

Figure 1. AFM images of the same plasmid obtained in different buffers. Image A was the last image acquired at in acetate buffer (pH5) and the image B is the first in the series after replacing the acidic buffer with a neutral one (TE –buffer, pH 7.6). The position of the protrusion that stably existed at acidic buffer (presumably H-DNA) is indicated with arrows.

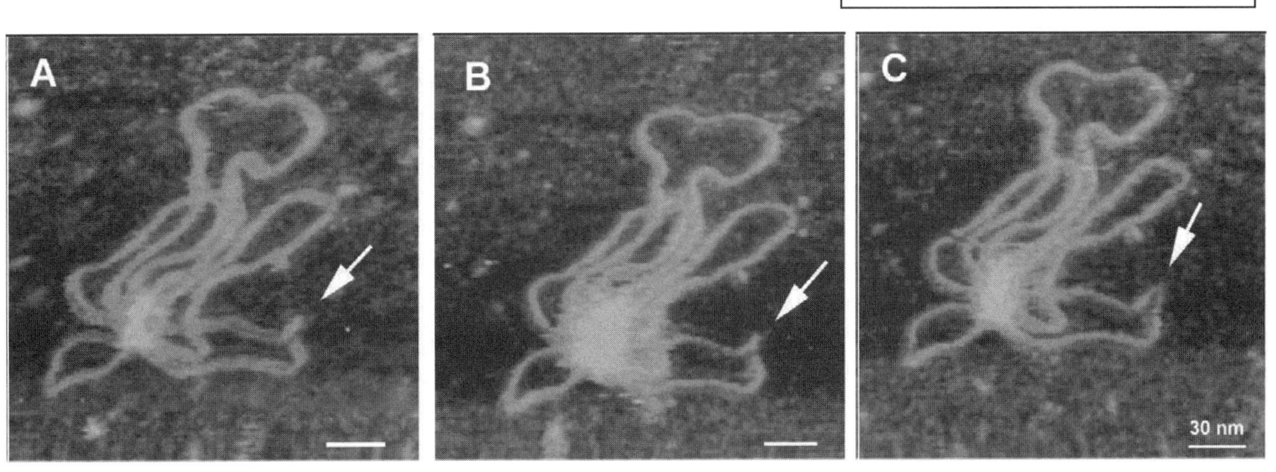

Figure 2. Time-lapse AFM images illustrating the transition of H-DNA into B-helix conformation.

Microsc. Microanal. 8 (Suppl. 2), 2002
DOI 10.1017.S143192760210225X

Microelastic Mapping of Living Cells: Changes in Relative Elasticity Between Nuclear and Cytoplasmic Regions of Mitotic MDCK Cells

Emad A-Hassan and Jan H. Hoh

Department of Physiology, Johns Hopkins University School of Medicine, Baltimore, MD 21205

Mechanical forces play a major role in the physiology of eukaryotic cells. Studies of the mechanical properties of cells are important for understanding their function in many physiological and pathological processes. Cytomechanics is an important factor in morphogenesis, in control of gene expression and protein synthesis and several other cellular processes especially in cellular division and proliferation. Progress has been made toward understanding the mechanics of cell division in a wide variety of organisms and cell types using several methods including molecular biology, light and electron microscopy, as well as various force transducers, methods including microneedles, laser and magnetic tweezers, and the atomic force microscope (AFM). Measurement of cellular viscoelasticity by such techniques can provide important functional information about the physiological role and mechanical properties of cellular components (in particular the cytoskeleton). Mitosis is a highly complex physiological and biochemical process. Although cellular structures and molecules involved in mitosis have been extensively characterized, the biomechanics of cell division remains poorly understood.

The AFM is emerging as a very valuable tool for studying biological samples. Besides its ability to acquire detailed surface images of various samples under physiological conditions, it also provides the means for examining sample-sample interactions and mechanical properties of samples [1]. These types of measurements are mainly performed by acquiring forces versus distant curves over the sample under consideration and then extracting elastic moduli [2]. Alternatively, we have developed an approach for relative microelastic mapping, referred to as force integration to equal limits (FIEL mapping), which is based on the fact that work done by the AFM cantilever during a force curve is related to the elasticity of the sample [3]. FIEL mapping has a number of advantages over determining local elastic moduli; there is no need to calibrate the cantilever, determine the tip radius, or know the precise tip-sample contact point in each curve.

Here we have used AFM and FIEL mapping to investigate the mechanical properties of mitotically active Madine Darby Canine Kidney (MDCK) cells. Microelastic maps were produced by collecting arrays of force curves, typically 64x64 curves, over subconfluent MDCK cells. Individual force measurements were made on a time scale, about 1 second, were viscous contributions were small. Low resolution FIEL maps of mature monolayers of MKCK cells typically appear similar to the topographic images with minor variations. Several distinct regions of the cell can be identified, predominantly the cell boundaries and the nucleus being the large structure in the center of the cell. Sub-confluent, mitotically active, MDCK cells have an elongated morphology. FIEL maps on these cells, in general, show similar features in elasticity to those obtained on mature monolayers. In the mitotically active MDCK cells, two distinct types of mechanical states can be seen (Figure 1). The first of which, described earlier, the FIEL maps show a soft intact nuclear region. In the second state, the nucleus, which appears to be divided in two parts by a soft band, is considerably stiffer than the cell body. These differences of relative elasticity appear to be related to the cell cycle, and are in agreement with point measurements made on bovine embryo skin and muscle cells [4].

References

[1] A. Vinckier and G. Semenza, *FEBS Lett.* (1998) 12.
[2] M. Radmacher et al., *Biophys. J.* (1996) 556.
[3] A-Hassan et al., *Biophys. J.* (1998) 1564.
[4] J.A. Dvorak and E. Nagao, *Exp. Cell Res.* (1998) 69.

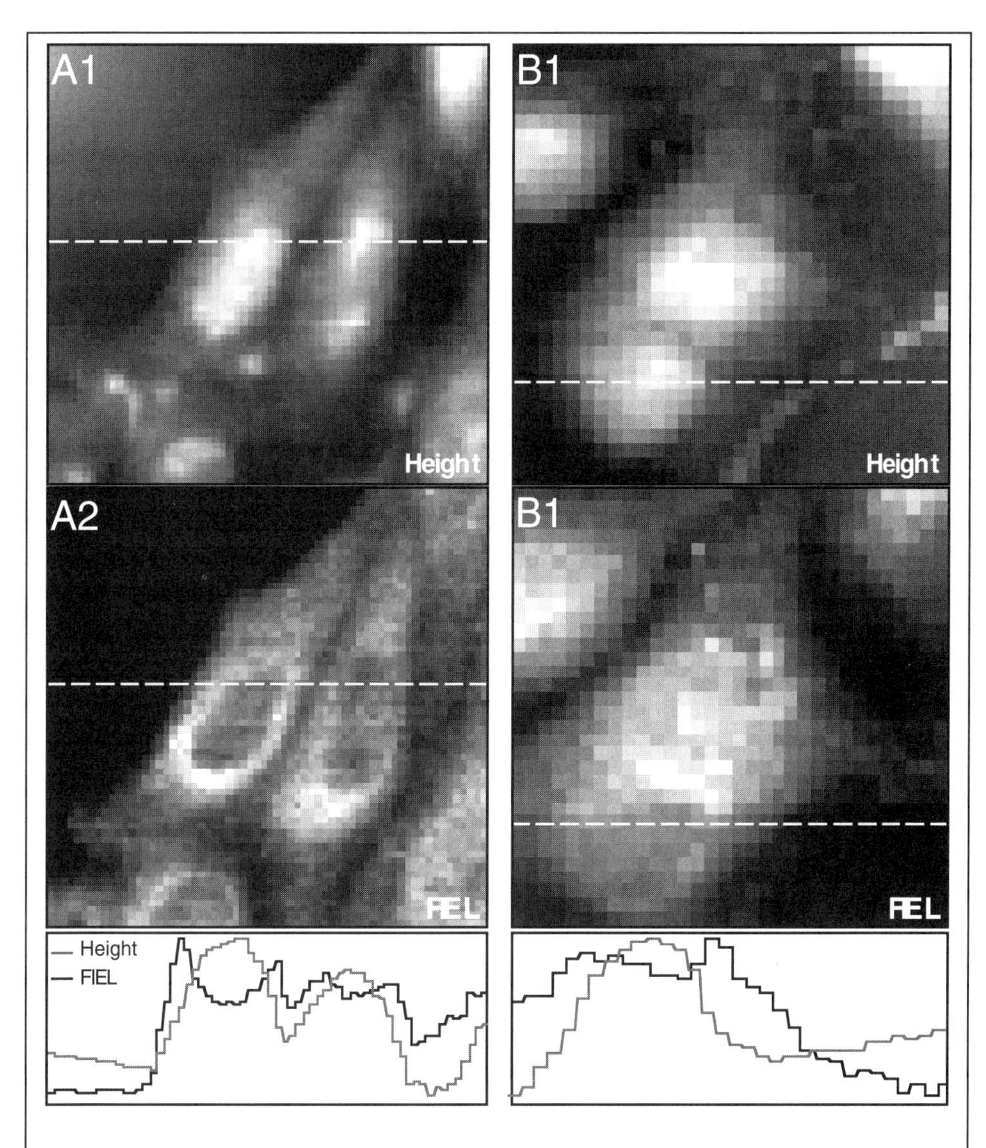

Figure 1. Topographic (A1 and B1) and elasticity maps (A2 and B2) of mitotically active MDCK cells. Note that the nuclear region of these cells is significantly stiffer than the surrounding cytoplasm. The convention adopted here for topographic AFM images is that the light areas are higher than the dark ones, while the gray scales in the FIEL maps is from relatively stiff dark areas to the soft light areas.

Microsc. Microanal. 8 (Suppl. 2), 2002
DOI 10.1017.S1431927602102261

Touching In Biological Systems: A 3D Force Microscope

[1,2]Richard Superfine*, [2]G. Bishop, [3]J. Cummings, [3]J. Fisher, [2]K. Keller, [4]G. Matthews, [1]D. Sill, [1,2]R. M. Taylor II, [2]L. Vicci, [2]C. Weigle, [2]G. Welch, [2]B. Wilde.

[1]Department of Physics and Astronomy and Curriculum in Applied and Materials Sciences*, [2]Computer Science Department, [3]Biomedical Engineering, [4]Cystic Fibrosis Center, University of North Carolina-Chapel Hill, Chapel Hill, NC 27599, USA

We are creating a microscopy system that allows a scientist to be an interactive actor in biological systems. Scanning probe microscopy (SPM) is being applied in a wide range of problems in biology, including the imaging of cell membranes and single molecules, and the measurement of the forces that bind ligands and maintain molecular conformations. Over the past decade we have developed user interfaces for SPM that allow the scientist to be an actor in the experiment. By moving a pen in held in hand, the user can control the motion of the probe tip on the surface, and through the force display capabilities of the pen, feel the surface and the biological sample. Applications of this interface include viruses, fibrin and nanotubes. However, performing force measurements inside cells and other biological systems remains a difficult challenge. The SPM tip, due to its overall dimensions, will strongly perturb a biological system such as a live cell or the extracellular matrix. The trick to solving this dilemma is to free the tip of the SPM probe from the cantilever.

We have developed a 3D force microscope by manipulating a freely suspended bead in liquid. By holding a pen in hand, the user can control the forces applied to and the resulting motion of a bead suspended in a biological system. The 3D force microscope uses magnetic fields to apply forces to ferromagnetic and paramagnetic spheres, and forward light scattering to track the particle. We have designed a tetrahedral electromagnetic pole geometry using finite element analysis to model the fields and forces. With pole spacing of 350 microns, we are able to generate over 100 pN of force with air-cooled coils. A tracking algorithm allows the particle to be tracked within the 100 micron x 100 micron x 15 micron range of the sample stage while the forward light scattering signal allows position measurements with high spatial resolution (<10nm) and bandwidth exceeding 10kHz. The microscope has been interfaced to a haptic (force feedback) interface that allows the positioning of the particle using a hand held pen. Motion of the pen in 3D controls the coil currents, and hence the applied force, and the force is displayed to the user hand through motors that drive the pen. We present results from our initial applications in the viscoelasticity of mucus and cell studies.

Taylor II, R. M. and R. Superfine, Advanced Interfaces to Scanning Probe Microscopes. Handbook of Nanostructured Materials and Nanotechnology. H. S. Nalwa. New York, Academic Press. **2:** 271-308. (1999).

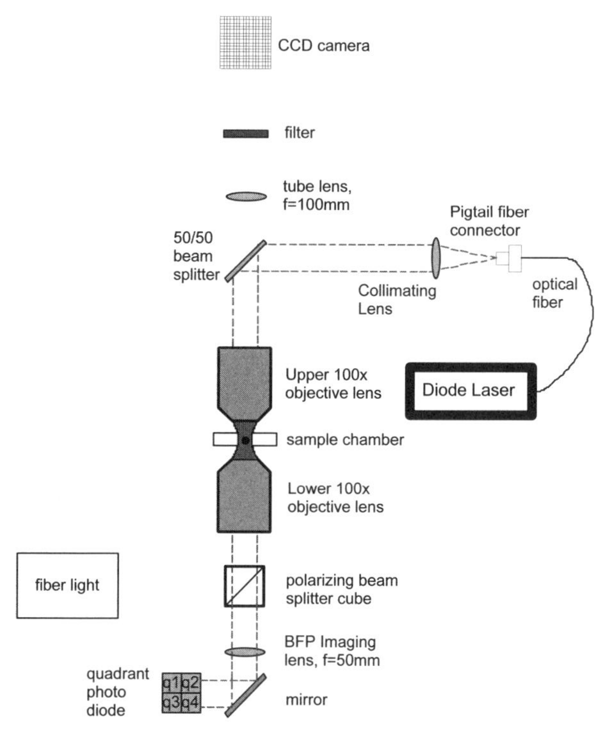

Fig. 1. Schematic of Optical tracking system for the 3DFM. This initial implementation uses 0.7 NA objectives (Mitutoyo) to allow room for the tetrahedral pole geometry.

Fig. 2. Side and Top views of the tetrahedral magnetic pole geometry. Two poles project from the top, and two poles, rotated 90 degrees from the top, project from the bottom. The sample slides between the poles.

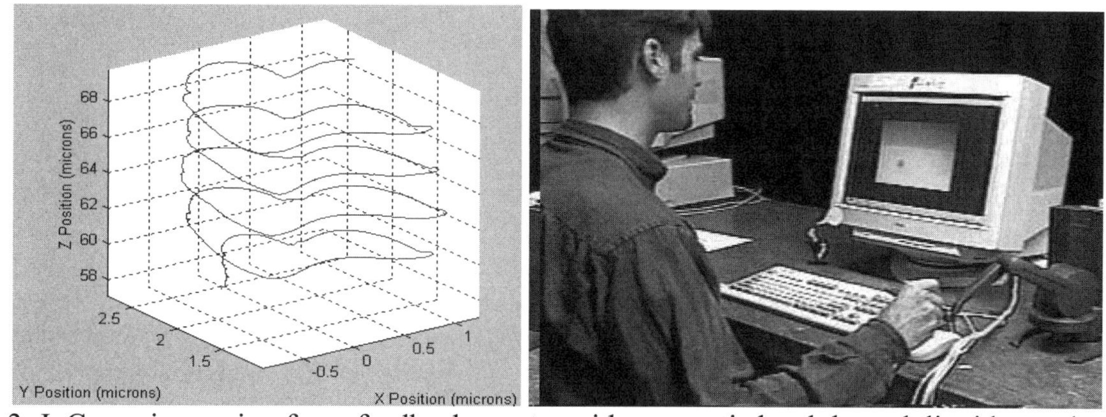

Fig. 3. J. Cummings using force feedback pen to guide magnetic bead through liquid sample.

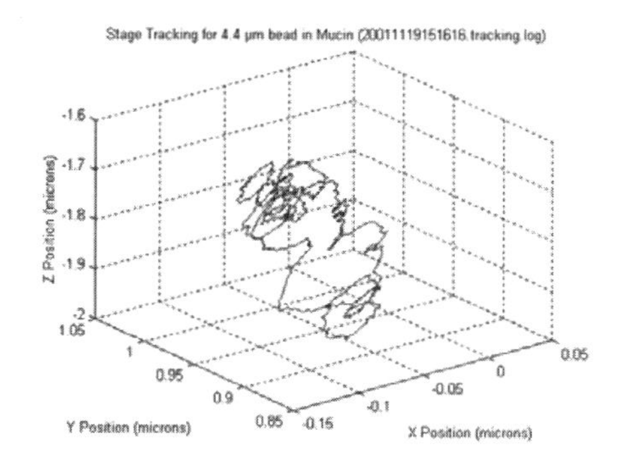

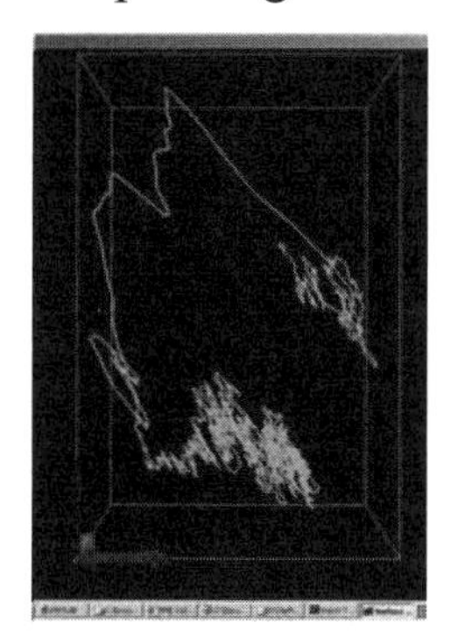

Fig. 4. Freely diffusing bead in mucus tracked in 3d (far left). Bead tracked in agarose gel show signs of slipping between “cages” that confine the bead for long times.

Microsc. Microanal. 8 (Suppl. 2), 2002
DOI 10.1017.S1431927602101681

Microstructural Characterization of Automotive Materials

W.T. Donlon, A.E. Chen, L. Gonzalez, J.Hangas, E. Lee and M.C. Paputa Peck

Ford Motor Company, Scientific Research Laboratories, P.O. Box 2053, Dearborn MI 48121-2053

Implementation of new materials or processes in the automotive industry are typically driven by the desire to improving fuel economy and customer satisfaction while reducing cost. Microstructural characterization is typically employed to assist in failure analysis of proto-type components or to evaluate the effect of process changes on microstructure. The types of materials under investigation include steels, aluminum, magnesium and polymers used primarily in body panel applications, steels and cast aluminum for powertrain components, and ceramics utilized in exhaust gas catalysts and various sensors. Three examples of materials characterization are presented below.

Aging studies of three-way exhaust gas catalysts.
Automotive three-way catalysts (TWCs) are most often comprised of a cordierite monolith coated with a washcoat containing Al_2O_3, CeO_2 and ZrO_2 which is impregnated with finely dispersed noble metal particles. Thermal aging and chemical poisoning are two of the primary deactivation mechanisms in TWCs. Post-mortem studies on aged catalysts can assist in understanding these failure mechanisms. Figure 1 shows WDS elemental maps of a cross section of a catalyst that was vehicle aged for 50k miles. Decomposition products of the oil additive zinc dialkyldithiophosphate (ZDDP) have deposited on the surface of the washcoat. The maps show that zinc and calcium are localized on the surface of the washcoat, while the phosphorus has migrated into the washcoat. The contamination layer on the surface blocks the washcoat pores and prevents the exhaust gas from reaching the catalytic material. TEM analysis has identified regions of $AlPO_4$ formed in the washcoat which can also block pores and cause a loss of surface area necessary for catalytic activity.

Development of polyolefin-based nanocomposites.
The development of a new generation of polyolefin-silicate clay nanocomposites requires a compatibilization process that can effectively disperse the clay throughout the polypropylene (PP) matrix in an economical way for large-scale automotive applications. Both supercritical fluid (SCF) processing and ultrasonic melt processing are currently being investigated to improve the dispersion, exfoliation and intercalation of the clay platelets. Figure 2a shows a bright-field (BF) TEM micrograph of a conventionally processed silicate-clay nanocomposite showing tightly layered tactoids. Clay platelets in materials which have been SCF processed (Figure 2b) show evidence of increased intercalation, while ultrasonication of the clay-PP mixtures during the melt state shows an increasing amount of dispersion and exfoliation (Figure 2c).

Heat-treatment optimization of a cast 319 aluminum (Al) alloy.
The substitution of 319 Al for cast iron in engine block and cylinder head applications results in significant weight savings. Since this alloy is heat-treatable (primary age hardening precipitate is θ-Al_2Cu), care must be taken to insure that material properties do not degrade during the life of the vehicle. This requires a thorough knowledge of temperatures encountered during operation and precipitation kinetics. The "as-cast" 319 alloy consist of aluminum dendrites surrounded by Si eutectic which also contains θ-Al_2Cu particles (Figure 3a). The remaining Cu in the Al dendrites is concentrated near the eutectic due to coring (Figure 3b). Solution treatment at 495°C for 1hr significantly increases and homogenizes the Cu in solid solution. Natural aging of solution treated

material for approximately 24hr increases the hardness of the alloy due to the formation of G.P. zones (Figure 3c). However, the alloy undergoes significant volume changes during aging until the G.P. zones are replaced by the metastable θ'-Al_2Cu (Figure 3d). θ'-Al_2Cu remain the predominate precipitate for agings of at least 1000h at 305°C.

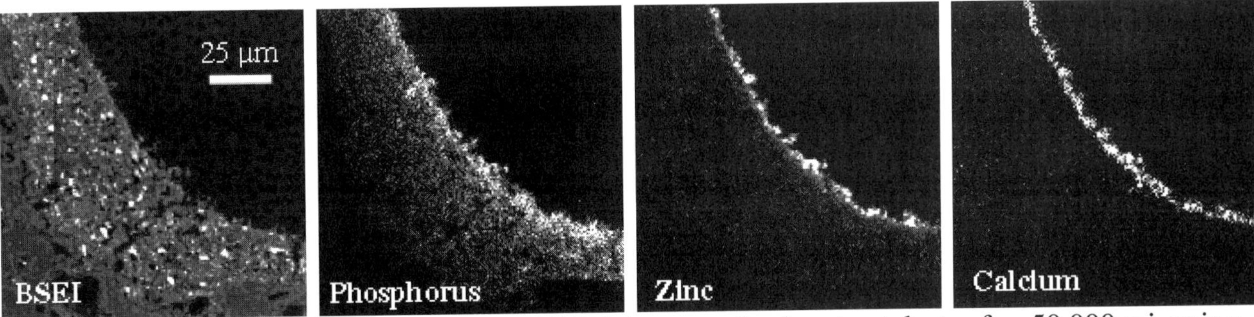

Figure 1. Electron probe micro-analyzer (EPMA) images of exhaust catalysts after 50,000 mi. aging. The presence of phosphorus throughout the washcoat and zinc and calcium on surface are due to the oil additive zinc dialkyldithiophosphate (ZDDP).

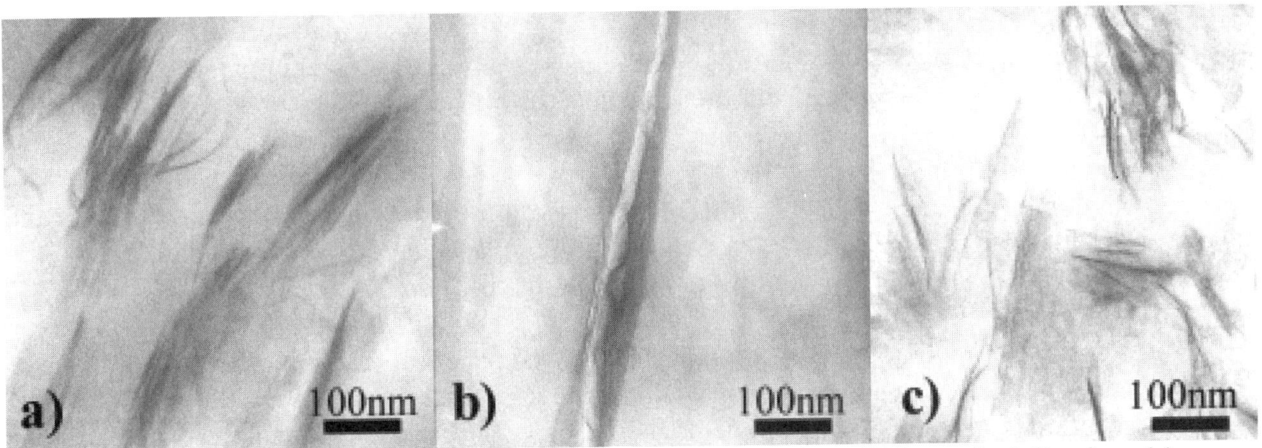

Figure 2. BF micrographs of clay nano-particles in a polypropylene (PP) matrix following; (a) conventional processing, (b) SCF processing, and (c) ultrasonic processing. Samples were cryo-microtomed and subsequently stained with ruthenium tetraoxide vapor.

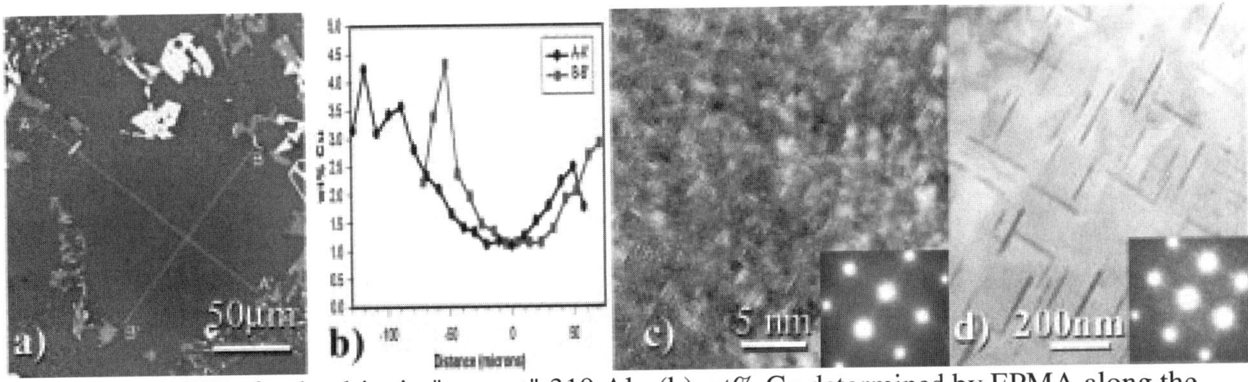

Figure 3. (a) BSEI of a dendrite in "as-cast" 319 Al. (b) wt% Cu determined by EPMA along the lines A-A' and B-B' shown in 3a. (c) BF micrograph of solution treated and naturally aged 319 Al. (d) BF micrograph of 319Al aged to a T7 condition (solution treated and aged for 5h at 260°C).

Microsc. Microanal. 8 (Suppl. 2), 2002
DOI 10.1017.S1431927602101693

Applied Microscopy for the Paper Industry

D.R. Rothbard

Institute of Paper Science and Technology, 500 10th Street, NW, Atlanta, GA 30318

Papermaking technology spans the sciences of biology, chemistry, engineering, and physics and often requires process understanding only possible through microscopy. Biotechnology is now applied to improve tree reproduction, disease resistance, and wood quality. Pulp and paper mills alter the physical and chemical properties of fiber, and involve energy intensive processes that can result in corrosion, scaling, and paper contamination. Coating and printing of paper and board demand additional material performance properties that are developed at the microscopic level. Paper product development, production, and quality control have all benefited from microscopic analysis.

Various forms of compound light microscopy and stereoscope imaging are essential to paper evaluation laboratories. Identification of pulp types and fiber species is accomplished by light microscopy using anatomic features with the aid of specialized stains. Microfibril angle in wood fibers can be determined with DIC microscopy and image analysis [1]. Contaminant particles, off color spots, and other appearance defects are investigated with a combined approach. This typically begins with light microscopy and may require FTIR microscopy or SEM/EDS. Sources of contamination can include additive packaging, improperly screened raw materials, and environmental dusts. Deposits from components of virginor recycled fiber can build up on rolls and roll fabrics in the natural course of mill operation. Printing plants have their own picking and deposit problems that vary with types of paper and printing processes.

SEM and TEM both have extensive applications in pulp and paper science [2]. SEM is more commonly used than TEM due to the nature of paper industry needs and, perhaps, the ease of sample preparation and instrument operation. SEM images of paper surfaces may reveal the degree of fiber bonding and the amount of external fibrillation, a consequence of the pulping, refining, and sheet forming processes. Tissue papers are formed with high bulk and porosity, and a controlled fiber furnish gradient, to optimize tactile smoothness and absorbency. Office copying and printing papers are heavily loaded with mineral fillers for a flatter surface with higher brightness. The distribution of fillers at the surfaces or in cross-section can be mapped with backscattered electron imaging. EDS is used to obtain elemental spectra that, along with morphology, often delineate filler types. At higher magnifications, polymeric sizing agents and coating binders can be visualized by TEM.

Cross sections of paper and board expose important thickness (Z) dimensional properties. These include the density and porosity of the sheet, the thickness and uniformity of coatings, and the retention of filler. Microtome sections for light microscopy are prepared to determine the penetration of starch sizing. For SEM examination oflarge sections, ground and polished cross sections have become more common [3]. The composition and microstructure of coatings can be determined in this way. The collapse of fiber walls and, sometimes, internal fibrillation are revealed. Figure 1 shows a polished cross section of a lightweight coated paper.

The availability of variable pressure and environmental scanning electron microscopes (VPSEM, ESEM) has made possible the study of paper fiber with its natural moisture content, and the evaluation of drying effects [4]. Coating components, like latex and plastic pigment, can be imaged without damage [5]. Additional applications are investigations of wood structure and mill corrosion products without the need for conductive coatings.

Operational problems in pulp and paper mills are sometimes resolved with the aid of SEM. Scales in pipes or vessels can develop in the evaporators of a pulp mill or the bleach plant of a paper mill. Shutdowns for maintenance cause expensive interruptions in production Light microscopy, SEM/EDS, and x-ray diffraction are commonly used to determine the texture and composition of mill scales. With this information, mill operations can be modified to minimize scale formation and expedite removal. Pulping liquors used in the kraft process may contain solids derived from feed chemicals, wood chips, or corrosion. Black liquor from the recovery boiler or green liquor from the washing of boiler residue contains fine solids that can be measured and chemically analyzed by SEM/EDS. The texture of fume particles that are produced in recovery boilers, as well as chars that are generated in black liquor gasification, have been measured by SEM to understand the impact of adjusting control variables. [6]

References:

[1] G.F. Peter et al., *J. Pulp & Paper Sci.,* 28 (2002) in press.
[2] D.R. Rothbard, in Z.R. Li (ed.), *Industrial Applications of Electron Microscopy*, Marcel Dekker, New York, (2002) in press.
[3] G.J. Williams and J.G. Drummond, *J. Pulp & Paper Sci.,* 26 (2000) 188.
[4] T. Enomae and P. LePoutre, *Nordic Pulp & Paper Res. J.*, 13 (1998) 280.
[5] Y. Xiang and D.W. Bousfield, *J. Pulp & Paper Sci.,* 26 (2000) 221.
[6] This work was made possible by the support of the member companies of IPST.

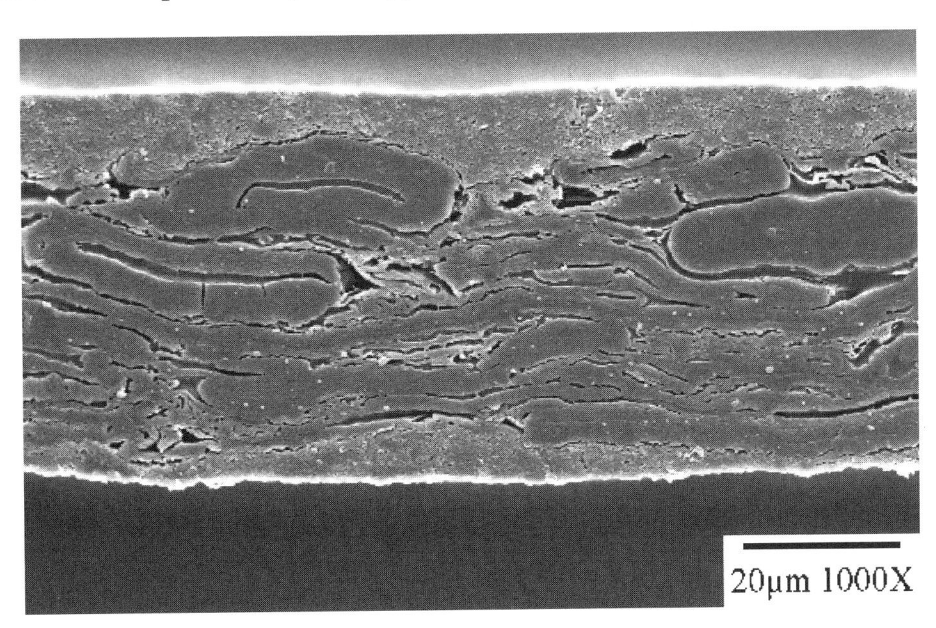

Figure 1. SEM secondary electron image of polished and etched cross section of lightweight coated paper. Bar = 20 μm. (Photo by G. Maghiari, IPST. Reproduced from [2] by courtesy of Marcel Dekker, Inc., www.dekker.com)

Microsc. Microanal. 8 (Suppl. 2), 2002
DOI 10.1017.S143192760210170X

Transmission Electron Microscopy Applications in the Semiconductor Industry - Challenges and Solutions for Specimen Preparation

Youren Xu and Chris Schwappach

Intel Corporation, New Mexico Materials Lab, 4100 Sara Road, Rio Rancho, NM 87124

Driven by Moore's law, transmission electron microscopy (TEM) has become an increasingly powerful technique supporting the development and manufacturing of deep sub-micron integrated circuits (ICs) due to its high resolution and nano-spot analysis capabilities [1]. The run rate of TEM analysis for major memory or logic chip manufacturers is about several thousand specimens per year. High precision and fast through-put time (TPT) are the two key factors that will determine if TEM can meet the growing demand for problem solving. The demand is driven by the increasing needs from new process development as well as from high volume manufacturing environments where a quick response to process excursion/line down situations is required. The introduction of the focused ion beam (FIB) technique and the emergence of the wedge (tripod polisher) method [2] achieved a breakthrough in TEM specimen preparation to meet these challenges. From a practical standpoint, overall specimen quality (thickness) and lack of end-point precision are two major issues associated with the conventional single beam FIB technique. These issues are primarily related to ion beam damage and endpoint control encountered during the final stages of specimen thinning.

The dual beam -focused ion beam (DB -FIB) system demonstrates superior performance over the single beam system including reduced TPT, ability to precisely section features in the sub micron range, high accuracy for endpoint control, and higher success rates. The DB-FIB technique can also be used to prepare high-precision plan-view TEM specimens [3]. Although the use of DB-FIB systems has greatly reduced specimen damage caused by the ion beam due to the added electron beam viewing capability, improper operating procedures can still induce unwanted ion beam damage. Fig. 1 shows a case where the Si substrate underneath the Ti/TiN adhesion layer was damaged due to missing e-beam Pt deposition prior to ion beam deposition. Our preliminary studies concluded that a minimum of 660 angstroms e-beam assisted Pt deposition is required for complete protection from a subsequent 350 pA ion beam assisted Pt deposition (bulk layer).

The mechanical grinding/polishing operation is always one of the most critical steps for TEM specimen preparation. It is a slow and labor-intensive process. Sela's "TEMstation" (now called "TEMpro") system offers a solution to this problem. This newly emerged automatic TEM specimen prep system eliminates manual grinding/polishing for the entire FIB sliver preparation, which is normally performed by a highly skilled expert technician. Our evaluation data showed that the average TPT for sliver preparation is 18 minutes per specimen and the success rate is ~90%. The "TEMstation" can produce standard slivers (~30 μm thickness) from both inline and end of line wafers (0.4 mm and 0.7 mm thickness). The total sliver thickness (Si + grid) is ~ 200 μm which fits both Philips and JEOL TEM specimen holders. The training needed for tool operation is less than a week for a person who does not have any grinding/polishing skills. This added capability will significantly improve the lab efficiency by better utilizing resources across functional areas.

The wedge method is a low cost approach offering a large transparent area with minimal or no ion milling. It is also a fast technique suitable for preparing specimens with repeatable structures.

Currently, 30% of specimens in our lab are prepared using this method. However, extensive training and exercise is needed to become proficient with this technique. The nearly 100% manual preparation and personnel-dependent operation procedure seems to prevent the adoption of wedge as the primary technique by TEM labs as compared with the DB-FIB cross-section method which offers standard and easy procedures and the added benefit of SEM for monitoring the thinning process.

The FIB "lift-out" method has become increasingly attractive to the TEM community due to the advantage of no mechanical grinding/polishing [4]. This technique can be used to prepare multiple specimens from within a very small area offering a powerful solution for failure analysis where multi-site sampling is required due to imprecise fault isolation [5]. We have successfully prepared two samples that were 5.5 μm apart from an FA unit using this technique (Fig. 2).

References

1. J Mardinly, Microscopy and Microanalysis, vol.7, supplement 2 (2001) 510.
2. S.J. Klepeis et al., Mater. Res. Soc. Symp. Proc., vol. 115 (1988) 179.
3. Z. Ma et al., Microscopy and Microanalysis, vol.5, supplement 2 (1999) 904.
4. L. Giannuzzi et al., Mat. Res. Soc. Symp. Proc., vol.480, (1997) 19.
5. Y. Xu, C. Schwappach, and R. Cervantes, Microscopy and Microanalysis, vol.6, supplement 2 (2000) 516.

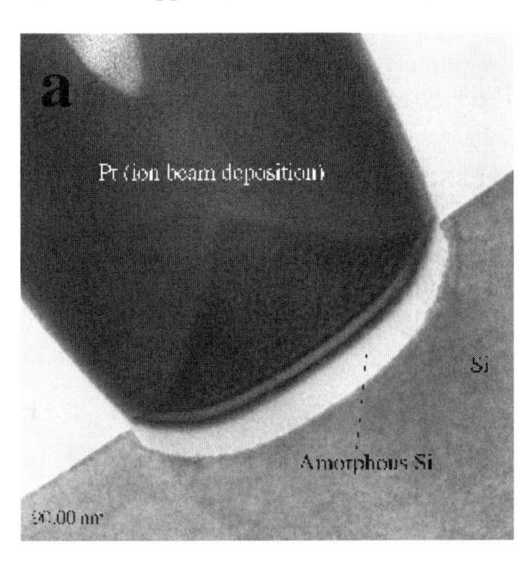

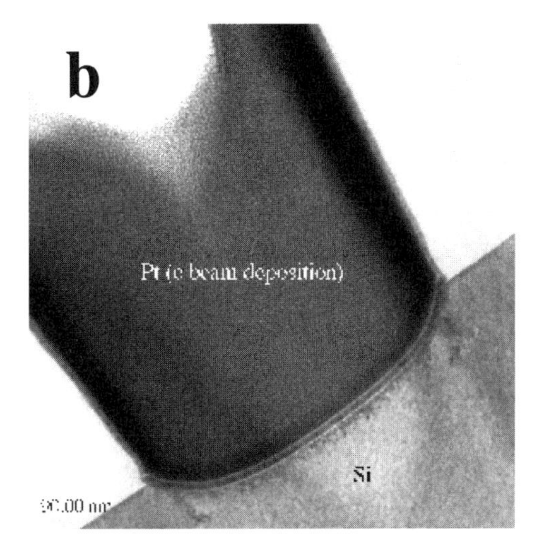

Figure 1. Use the proper Pt protective layer deposition: (a) ion beam Pt deposition caused Si amorphization; and (b) add a layer of e-beam deposited Pt prior to ion beam Pt deposition protecting the Si from ion damage.

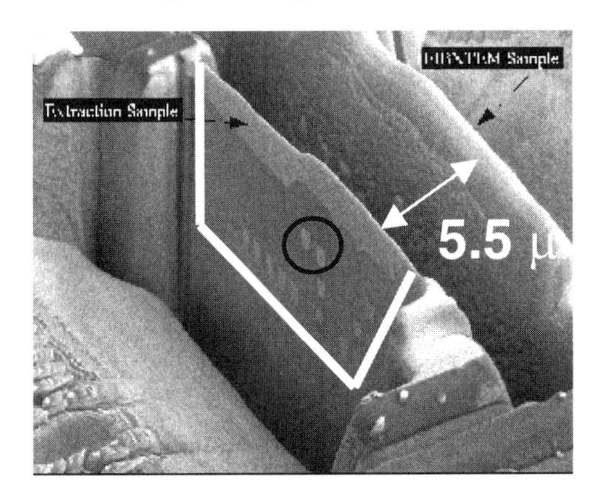

Figure 2. A case of multi-site sampling using the "lift-out" method. The distance between the extracted sample and the conventional FIB cross-section sample is about 5.5 microns.

Microsc. Microanal. 8 (Suppl. 2), 2002
DOI 10.1017.S1431927602101711

The Unique Diversity of Electron and Confocal Imaging Applications in a Natural History Museum Setting

ANGELA V. KLAUS

Core Imaging Facility, American Museum of Natural History, New York, NY, USA

Large public science institutions such as the American Museum of Natural History (AMNH) generally have two faces: the public exhibit halls and the "behind-the-scenes" research effort. The collective expertise of the curatorial staff ultimately contributes to the scientific content of the public exhibits. For the most part, however, day-to-day scientific research at AMNH is conducted as it would be in any conventional academic institution.

The AMNH Core Imaging Facility is a shared resource that maintains a state-of-the-art cold field-emission scanning electron microscope (FE-SEM) equipped for energy dispersive x-ray microanalysis (EDS) and cathodoluminescence spectroscopy (CLS; late 2002). The facility also houses a confocal laser scanning microscope (CLSM) and an image-processing lab where we maintain a number of 3-D reconstruction software packages, multiplatform (Windows NT/2000, Mac OS, and SGI/IRIX) computers, and peripheral devices such as a publication quality digital printer and a large format printer.

The imaging and microanalytical environment at AMNH is unique because applications come from the many diverse disciplines pursued in a natural history museum setting.[1] Broadly, these areas of research include anthropology, biological sciences, and geological sciences. Specifically, typical sample types examined include fossil bone and ammonites (paleontology); skulls, bone and teeth from extant vertebrate species (vertebrate zoology); mollusk shells, pearls, insects, and spiders (invertebrate zoology); meteorites and magmas (earth and planetary sciences); and pigments, metals, and hairs from cultural artifacts (anthropology).

Examples of typical confocal and electron imaging applications are shown in Figs. 1– 6. Figs. 1 and 2 are front and back views of the genitalia of a male mosquito (Family Culicidae). These images are 3-D reconstructions from 2-D confocal image stacks. The data were collected on a Zeiss 510 CLSM at 512 x 512 image resolution and reconstructed using Bitplane Surpass surface rendering software. Fig. 3 shows a close-up view of several types of sensory structures found on a wasp antenna (*Dolichovespula sylvestris*). This specimen is part of the permanent insect collection and was imaged uncoated at 1 kV accelerating voltage, 7 μA emission current. A fractured cross-section of the nacreous material making up the outer layers of a cultured freshwater pearl is shown in Fig. 4. This image became part of the public exhibition entitled "Pearls".[2] Fig. 5 is an image of part of a pigment fragment removed from a Zapotec urn. The rod-like structure of these crystals (palygorskite) and the absence of Cu and Co (data not shown) helped identify this material as Maya Blue, an unusual pigment used throughout Mesoamerica.[3,4] Fig. 6 is a backscattered electron (BSE) image of a porphyritic olivine chondrule within a thin section of the Allende meteorite. Chondritic meteorites represent one of the oldest (~4.6 billion years) forms of undifferentiated material available for laboratory study.[5] All electron images were collected on an Hitachi S-4700 cold FE-SEM.

References:

1. Klaus, A.V. Museum applications for SEM and x-ray microanalysis. *In* Industrial Applications of SEM (ed. Z. Li). Marcel Dekker: New York. In press.
2. Landman, N.H., et al. (2001) Pearls: A Natural History. Harry N. Abrams: New York.
3. Jose-Yacaman, M., et al. (1996) Maya blue paint: an ancient nanostructured material. Science 273: 223 – 225.
4. S. Alderson, AMNH Anthropology, personal communication.
5. McSween, H.Y. (1999) Meteorites and Their Parent Bodies, 2nd ed. Cambridge University Press: Cambridge.

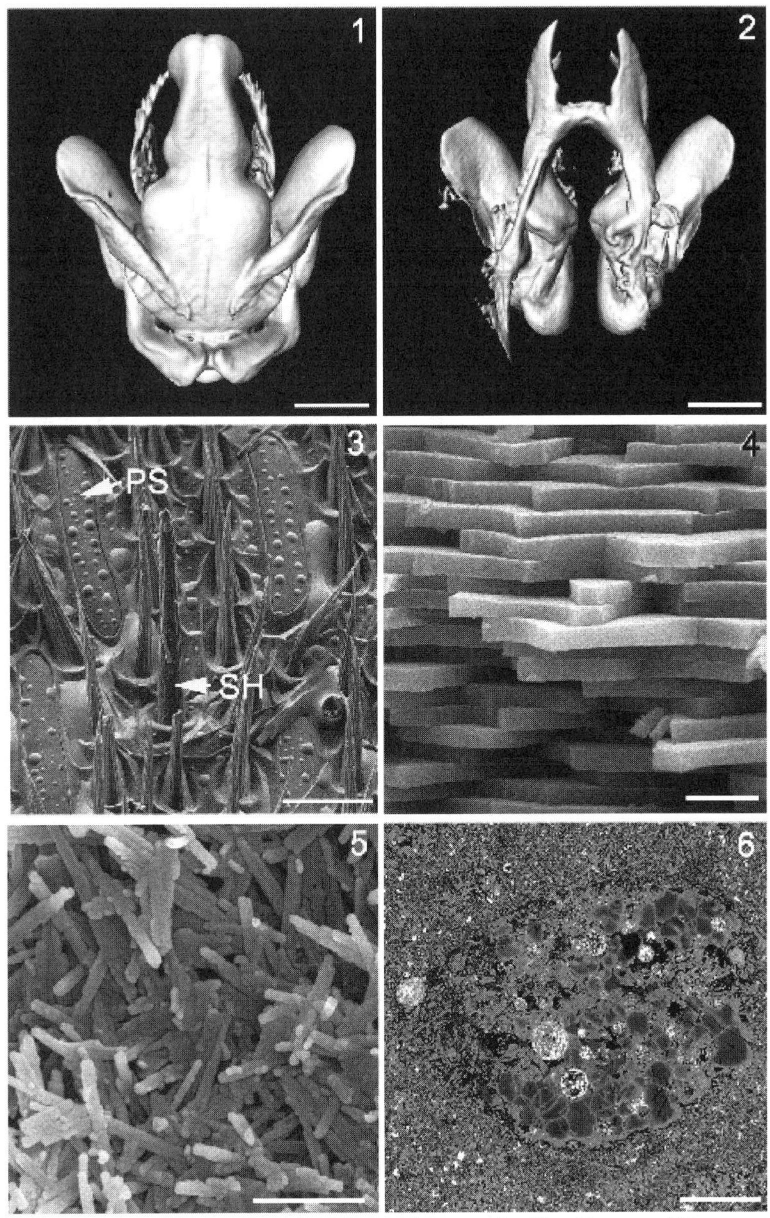

Figs. 1 and 2: 3-D reconstructions of male mosquito genitalia. Bars = 50 μm. Fig. 3: Wasp antennal sensory structures (PS = placoid sensillum, SH = sensory hair). Bar = 10 μm. Fig. 4: Fracture surface of cultured freshwater pearl. Bar = 2 μm. Fig. 5: Palygorskite crystals in Maya Blue pigment. Bar = 1 μm. Fig. 6: BSE image of Allende meteorite chondrule. Bar = 100 μm.

Microsc. Microanal. 8 (Suppl. 2), 2002
DOI 10.1017.S1431927602101723

3D Imaging of Polymer-based Materials by Laser Scanning Confocal Microscopy

L. Liang

DuPont Central Research and Development, P. O. Box 80228, Wilmington, DE 19880-0228

One of the greatest advances in microscopy in the past two decades is the development of laser scanning confocal microscopy (LSCM). [1] The technique built upon many new technologies offers non-destructive three-dimensional imaging capability. In conventional fluorescence microscopy much of the structural detail that could otherwise be resolved is obscured due to the contribution of fluorescence from regions outside of plane of focus. By using a pinhole restricting the collection of light from out-of-focus regions, confocal microscopy provides both significantly higher lateral resolution and possibility to produce 3-D images of the sample without destruction. By scanning in x-y plane and moving the microscope stage along the z direction step by step, a series of optical "sectioning" images can be generated. Towards the end of last century, new multi-photon excitation without pinhole has been added to LSCM to reach deeper penetration, lower UV radiation damage. Therefore, LSCM has been widely used in biological and medical research and development. However, the industrial applications of laser scanning confocal microscopy in material science and technology have been rather limited. [2-3] Recently, we have used a Zeiss-510 NLO two-photon scanning laser confocal microscope (TPLSCM) to investigate some industrially important polymer-based materials and report here.

Zeiss 510 NLO is composed of Axiovert 100 M inverse microscope, three visible laser lines, and one tunable near-infrared (NIR) Ti:Sapphire laser. The maximum scan resolution is 2048x2048 pixels in digital images. The stage can move a step of as small as 25 nm. Many long-pass or band-pass filters are available for the selection of the interested emission spectrum.

Electron microscopy including transmission and scanning electron microscopy (TEM, SEM), and conventional optical microscopy are very important microstructure characterization techniques to image multi-phase materials, such as TiO2 particles in a polymer matrix. However, the images obtained directly with these microscopes are all 2-D images since they are taken either from the surface or from two-dimensional projection of the samples. In some special cases, even 3D SEM images of multi-phase materials can be generated after time-consuming sample preparation process, many artifacts may be introduced due to destructive nature of these sample preparation techniques. If matrix materials are transparent, LSCM may be a very useful microstructure characterization technique. 3D images of multi-phase materials can be obtained without or with little sample preparation in short time. The second phase added in a polymer matrix may be organic (another polymer) or inorganic particles. The size distribution and dispersion of these particles in the polymer are usually important to the physical and chemical properties of the final products. In the past, to study these microstructure parameters, the samples were usually polished and then observed by scanning electron microscopy. However, due to polishing, the remaining parts of particles appear in the images may no longer reflect real size of the particles. Using LSCM, samples can be directly imaged as deep as several hundreds microns from the top surface without sample preparation. Based on digital LSCM 3D images, particle size distribution and dispersion should be adequately obtained.

This technique can also be used for the 3D topographic reconstruction of surface and the measurements of surface roughness of materials. Although there are other surface measurement techniques, one of important advantages of LSCM technique is its physically non-touching nature, which are important to surface microstructural characterization of many soft materials.

In industry, quantitative measurements are often needed such as size distribution of second phase particles in polymer matrix materials. LSCM images of the samples are all in digital form. Thus, quantitative measurement can be very conveniently conducted based on these 3D images through image analysis techniques. Further developments of 3D image analysis will over pass many limitations that was imposed by 2D image analysis.

Like many other microstructural characterization techniques, LSCM has its own limitations. First, the resolution is rather still low, just slightly higher than that of optical microscope. Second, LSCM is not suitable for the materials that are non-transparent or contain too many dark pigments. It should be noticed that if loading of the second phase materials are too high, the upper particles would block the light coming from the particles below and they would be invisible in the final 3D images. Thus, it is crucial to use multiple microscopy techniques to solve real industrially important problems.

References:

[1]. J. Pawley, Editor, *The handbook of biological confocal microscopy*, IMR Press, Madison, 1989

[2]. Y. Song et al., *Macromolecules*, 33 (2000) 4478.

[3]. A. E. Ribbe, *Trends in Polymer Science*, 5 (1997) 333.

Microsc. Microanal. 8 (Suppl. 2), 2002
DOI 10.1017.S1431927602101735

Application of the UHREM Technique in Atomic Modelling of Growth Defects in CVD-Diamond Films

D. Dorignac

CNRS, Centre d'Elaboration de Matériaux et d'Etudes Structurales, 31055 Toulouse, France

Diamond is of course known not only as the stuff of women's dreams - since at least two thousand years before Christ, "Diamonds have been a girl's best friend" - but also for its unique physico-chemical properties. When considering any characteristic property of a material, the value associated with diamond almost always corresponds to an extremum among all materials considered for that property. Diamond is effectively the hardest known material, with the highest Young's modulus, the highest sound velocity, a high wear resistance and a very low coefficient of friction. It is the best thermal conductor of any material, with a superior thermal shock resistance and a very low coefficient of expansion. It is transparent over a very broad spectral range, from X-rays to the far IR, with a low refractive index in the visible. It is chemically inert and highly resistant to acids. Diamond is also a very good electrical insulator, with wide bandgap, very high electrical breakdown strength, near zero workfunction, negative electron affinity, UV photoconductivity and it has the lowest dielectric constant of all the ceramic materials. And, finally, it can be doped to form p- and n-type semiconductors, which present the highest hole mobility and the highest saturated electron velocity, leading to the highest figures of merit in electronics, which exceed by far those of other semiconductors.

The above long-recognized combination of superlative properties with the recent chemical-vapour-deposition (CVD) ability to grow diamond as a film, which began in the 1980s and only really emerged in the 1990s, makes diamond an exceptional material for a wide variety of industrial uses and a vast range of potential applications. Much of the relevant literature can be found in references [1-5]. Typical applications can be classified, based on the range of extremely desirable properties of diamond, as follows: high hardness $\Rightarrow$ tooling; low chemical reactivity $\Rightarrow$ high-performance structural materials. But, noting that diamond can very efficiently transport, transmit or convert energy, and that its interactions in various ways with the entire electromagnetic spectrum can be valuable, it is much more instructive to look at these applications from the perspective of the energy spectrum. The main application areas and product classes are then found to be, in order of increasing energy: (i) phonons $\Rightarrow$ thermal management structures, sensors, surface-acoustic-wave devices; (ii) microwaves $\Rightarrow$ radomes, waveguides; (iii) photons $\Rightarrow$ IR and laser windows, optical switches and emitters, displays, UV detectors and focal plane arrays, X-ray windows and masks; and finally (iv) electrons $\Rightarrow$ high speed-, high temperature-, high power-, radiation hard-electronics (diodes, transistors and integrated circuits), and also capacitors and vacuum microelectronics (cold cathode components). It can be concluded that when ultimate properties are required, CVD-diamond is the ultimate answer, particularly in the semiconductor field because no other substance performs under duress as well as diamond. It is thus thought that diamond will become a widely used material in this 21st century, resourcefully engineered into high-value-added components and systems.

This highlights the importance of technology for controlling the purity, crystallinity and defect structure in diamond as required for each particular performance criterion. During the past decade, the growth defects present in this promising engineering material have thus been studied by high resolution electron microscopy with a view to understanding their crucial role in influencing its practical properties. Several investigations of the defect structure have been performed using microscopes with point-resolutions down to 0.17-0.16 nm [6-18] and down to 0.12-0.10 nm [19-24], since detailed atomic resolution images can effectively provide local crystallographic information on the defect core structures. The most striking feature observed in such films is the presence of multiple twinning on {111} planes. The twins are frequently present as large domains, the simplest interfaces corresponding to coherent first order (Σ=3) twinning. Higher order interfaces ($\Sigma=3^n$, n=2 to 4), pentagonal arrangements of twins, twinning dislocations, stacking faults and different types of associated dislocations have also been identified. My emphasis will thus be on the use of ultra-high-resolution electron microscopy (UHREM) at 0.12 nm to characterize, at the atomic level, the defects present in CVD-diamond. Plausible 3D atomic-scale models can be proposed, based on image-matching with computer simulations and built with reasonable compression-dilation of the bond lengths and distortions of the tetrahedral angles, according to recent self-consistent *ab initio* calculations of minimum-energy defect structures. From these models, evidence for unusual structural units, which are believed to be quite characteristic of the CVD-diamond growth, has sometimes emerged. Since CVD-diamond is expected to find gradually increasing acceptance in industrial applications during the 21st century, I hope that the UHREM technique will help researchers in this field in their endeavours to create deposition areas with a high degree of uniformity in terms of growth rate and material quality.

References

[1] K. E. Spear and J. P. Dismukes, Synthetic Diamond, Wiley, New York, 1994.
[2] D. M. Gruen and I. Buckley-Golder, Special Issue of MRS Bulletin 23 (9) (1998).
[3] B. Dischler and C. Wild, Low-pressure Synthetic Diamond, Springer, Berlin, 1998.
[4] M.A. Prelas et al., Handbook of Industrial Diamonds, Marcel Dekker, New York, 1998.
[5] P.W. May, Phil. Trans. R. Soc. Lond. A 358 (2000) 473.
[6] W. Luyten et al., Phil. Mag. 66 (1992) 899.
[7] D. Shechtman et al., J. Mater. Res. 8 (1993) 473.
[8] M. Joksch et al., Diamond Relat. Mater. 3 (1994) 681.
[9] D. Shechtman, Mat. Sci. Engn. A 184 (1994) 113.
[10] C. J. Chen et al., J. Mater. Res. 10 (1995) 3041.
[11] X. Jiang and C. L. Jia, Appl. Phys. Lett. 67 (1995) 1197.
[12] C. L. Jia et al., Phys. Rev. B 52 (1995) 5164.
[13] C. J. Chen et al., J. Mater. Res. 11 (1996) 1002.
[14] H. Verhoeven et al., Appl. Phys. Lett. 71 (1997) 1329.
[15] D. Wittorf et al., Diamond Relat. Mater. 6 (1997) 649; 9 (2000) 1696.
[16] H. Ichinose and M. Nakanose, Thin Solid Films 319 (1998) 87.
[17] Y. Zhang et al., J. Electron. Microsc. 48 (1999) 245.
[18] L.C. Nistor et al., Phys. Stat. Sol. 174 (1999) 5; 186 (2001) 207.
[19] D. Dorignac et al., Diamond Relat. Mater. 6 (1997) 758.
[20] S. Delclos et al., Diamond Relat. Mater. 7 (1998) 222; 8 (1999) 682; 9 (2000) 346.
[21] H. Sawada et al., Diamond Relat. Mater. 10 (2001) 2030.
[22] H. Sawada and H. Ichinose, Scripta Mater. 44 (2001) 2327.
[23] D. Dorignac et al., J. Phys. IV France 11 (2001) 971.
[24] D. Dorignac et al., Phil. Mag. B 81 (2001) 1879.

Microsc. Microanal. 8 (Suppl. 2), 2002
DOI 10.1017.S1431927602101747

Contributions of Microscopy to Advanced Industrial Materials and Processing

S. Dionne*, G.J.C. Carpenter*, G. A. Botton*, T. Malis*, and M. W. Phaneuf**

*CANMET-Materials Technology Laboratory, 568 Booth St., Ottawa, Canada K1A 0G1
**Fibics Inc., 556 Booth St., Ottawa, Canada K1A 0G1

The Canadian federal Materials Technology Laboratory supplies microstructure characterization services to projects involving a broad range of materials and product forms. TEM usage in industrial issues is often contingent on a preliminary evaluation suggesting that data is needed on a finer scale. Overviews of the microstructure are obtained via optical microscopy or SEM imaging so as to decide upon the key regions for TEM analysis. Recently, our TEM capability was augmented greatly by usage of focused ion beam (FIB) methodologies provided by a private sector partner. In this paper, we hope to convey a generic sense of industrial usage of analytical TEM via our own experience with characterization of advanced coatings and materials. Additional examples are presented in a recent compilation [1].

The hot-dipped iron-zinc coating known as galvanneal has seen expanded use for automotive applications. The in-service properties of galvanneal are very sensitive to the coating microstructure. TEM characterization of galvannealed steel is challenging since the coating is relatively thin and is sensitive to both beam heating and beam damage. Several phases in the Al-Fe-Zn system may be present which have overlapping composition limits and extensive solid solubility. A sequential use of SEM, FIB and analytical TEM has provided the most comprehensive set of results for galvanized and galvannealed steels. FIB images were useful for observing the relationship between the coating phases and substrate grains (FIG. 1) and for the selection, in-situ, of the best regions for TEM specimen preparation using the FIB lift-out technique. Elemental mapping of the uniform thickness cross-sections revealed unexpected features such as the diffusion of Zn along a ferrite grain boundary (FIG. 2). CBED and EDXS point analyses allowed phase determinations and higher resolution measurements on selected regions. Finally, plan view specimens prepared in the form of extraction replicas using a selective etching technique were invaluable for identifying the coating phases without the problem of Zn-depletion that sometimes accompanied ion preparation (FIG. 3).

Metal-matrix composites (MMCs) offer potential for improved specific properties in comparison with metal alloys. The strength and cleanliness of the reinforcement/matrix interfaces play a major role in determining the failure mode of MMCs. The effectiveness of fabrication procedures for Mg-based composites by the squeeze-casting route were continually evaluated using SEM and TEM. Coarse reaction products were detected in the SEM, while small interface features were best studied using analytical TEM. STEM Dark Field imaging proved to be a particularly effective means for quickly assessing the degree of MgO formation in a particular melt, as shown in FIG. 4. Ultimately, a proprietary binder was produced that, together with careful control of the processing parameters, virtually eliminated the undesired interface reactions.

References

[1] T. Malis et al., *Contributions of Microscopy to Advanced Industrial Materials and Processing* to be published in Industrial Electron Microscopy, ed. Z. R. Li, Marcel Dekker Inc., 2002.

[2] We thank Dr. Frank Goodwin of ILZRO for financial assistance and provision of material as well as Mr. M. Charest and Dr. J. Lo of MTL for valuable contributions of expertise and specimens.

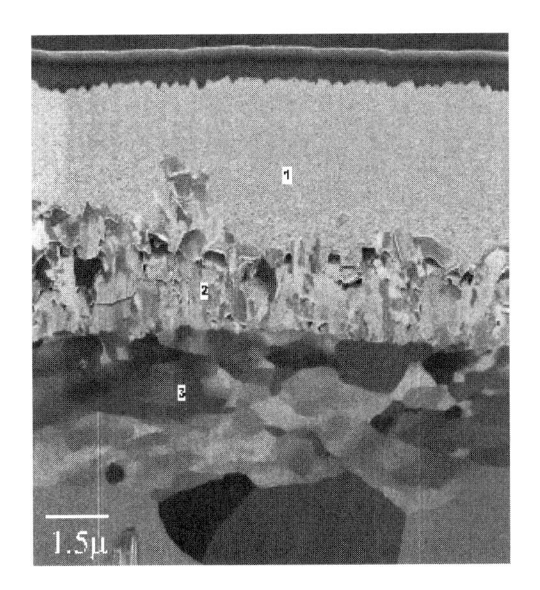

FIG. 1. FIB SE image of a cross-section of galvannealed steel showing (1) Zn overlay, (2) ζ and δ crystals resulting from the reaction and (3) steel substrate.

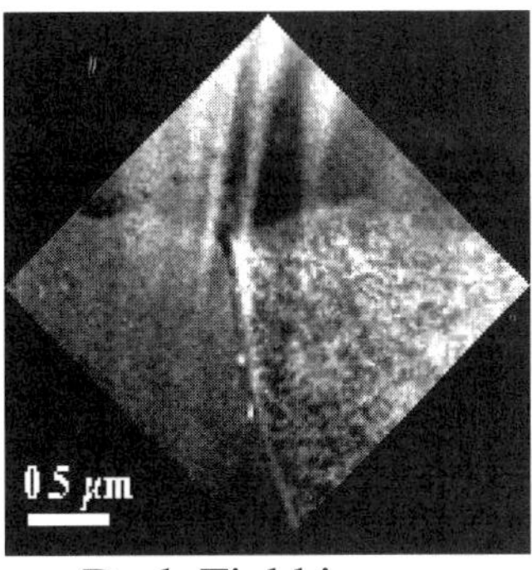

Dark Field image. Zinc map.

FIG. 2. STEM-DF image and zinc elemental map of a FIB lift-out specimen of galvanized low carbon steel showing zinc diffusion along an emerging ferrite grain boundary.

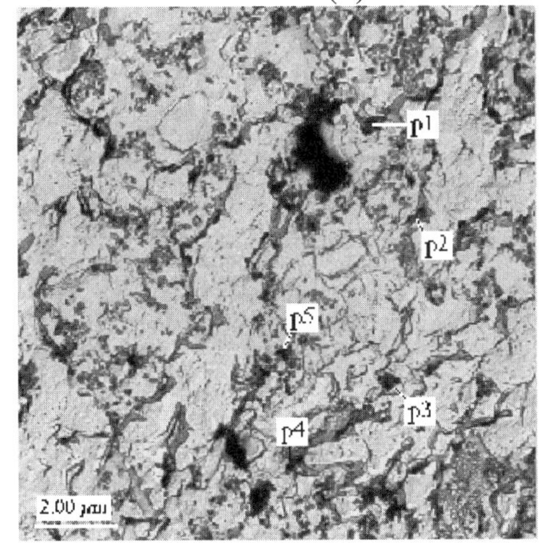

FIG. 3. Cluster of extracted inhibition layer particles on a replica of partially galvannealed Ti-IF steel. Particle 5 was identified as Fe_2Al_5 with the zone axis $[01\bar{1}]$ using CBED.

FIG. 4. STEM-DF image of a Mg-based squeeze cast composite, showing the strong visibility of the MgO films (arrowed) in this imaging mode.

Microsc. Microanal. 8 (Suppl. 2), 2002
DOI 10.1017.S1431927602102273

The Staining of Polymers

R.W. Smith

2450 Clarke Drive, Lake Havasu City, AZ 86403

A fascinating feature of polymer staining is that for some pliable networks (those that contain >C=C<) staining not only highlights target sites, but also provides rigidity to the polymer and enables room temperature microtomy. This means two staining protocols: 1) Bulk staining hardening of trimmed pyramids in preparation for ultramicrotomy, latexes and dispersions; and 2) Ultrathin section and cast thin film staining.

1) For bulk staining, the diffusivity of the staining media into the polymer network is critical and the media should be optimized for diffusion. It requires several days for *aqueous* osmium tetroxide to diffuse deeply enough into styrenic matrices (HIPS, ABS) to provide adequate staining-hardening depths but diethyl ether or tetrahydrofuran (THF) solutions of osmium tetroxide can provide adequate depths in matters of minutes [1]. Epoxy/Rubber matrices, especially crosslinked epoxies, require organic solvent media for effective staining. The use of organic solvent media must be carefully monitored to insure that the residence time of the solvent is not so long that it alters morphology. An intermediate between purely aqueous or nonaqueous staining media can sometimes be best. Rubber particles in PVC do not stain well (or at all) when using aqueous media and organic media tend to overly swell or partially dissolve the PVC matrix. Thus, a suitable intermediate for PVC/rubber is a staining medium of 5% v/v THF in water.

Bulk staining can present artifact. Figure 1 is a TEM cross section of poly (styreneacrylonitrile) modified with discrete poly (butadiene) particles that was bulk stained by immersion of a trimmed pyramid in 2% aqueous osmium tetroxide for four days. The stained-hardened layer is five microns deep and presents two cautionary layers. First is the layer of rubber particle distortion near the surface S where the pyramid was trimmed prior to staining. Microtoming the middle layer, G, yields good results. The deepest layer P, between stained and unstained particles, is where partially stained particles appear as donuts and could be wrongly interpreted as core-shell particles. A glance at the unstained matrix shows how effective the staining was for preserving morphology. A special case of bulk staining is the staining of polymer latexes (or dispersions) that is done by adding the staining media to a small volume of diluted latex. Here it is important to determine the proper dilution of the latex so that particles remain isolated upon drying. Generally this is in the order of fractional total solids.

2) Ultrathin sections and cast thin films can be stained using liquids or vapors in a straightforward manner but vapor staining is generally preferred due to ease of handling. Cryo sectioning, if available, can be used to circumvent bulk staining and has the advantage that there is less chance of artifact.

An awareness of what is being stained is essential. Mechanical deformation of some polymer systems may produce crazed matter and osmium can become trapped within the microfibrils of this matter and be confused with chemical staining. A different type of mechanical staining occurs for crystalline and amorphous regions that react differently to stains. This is a useful tool for differentiating between them since crystalline regions resist stains and amorphous regions "soak up" a telltale differential amount. A blend of two amorphous polymers, each with no stainablesites, may be differentiated based upon rates of stain perfusion- a phenomenon used to advantage by staining polymer blends at temperatures chosen so that one of them is near or at the glass transition temperature.

Negative staining can be useful other than for outlining particles sitting on flat surfaces. Figure 2 shows how protein layers normally present on surfaces of rubber latex particles delineate the particles in dipped products such as catheters. Aqueous silver nitrate accomplished this [2].

As the number and complexities of polymers has increased (blends, complex copolymers, blocks, grafts, core-shell particles, interpenetrating networks, amphilics, hydrogenated olefinics, etc.) so have the tactics for staining. Chemical alteration of the polymer to make it stainable, such as alkaline hydrolysis to enable osmium staining has been successful [3]. Most staining with ruthenium tetroxide is via vapor phase but aqueous immersion staining has been introduced [4]. Nonaqueous immersion stains are being used such as ruthenium tetroxide in carbon tetrachloride [5]; phosphotungstic acid in benzyl alcohol [6]; and osmium tetroxide in formaldehyde [7]. Preferential imbibition of styrene by one polymer of a blend and then staining the styrene with ruthenium tetroxide [8] introduces a new tactic that could be adapted for other systems. Multiple staining may be required, exemplified by a ruthenium tetroxide-uranyl acetate-lead citrate system [9].

Stains are aggressive chemicals and proper safety procedures must be followed.

References

[1] Riew and Smith, Journal of Polymer Science Part A 9 (1971) 2739
[2] Smith and Folt, Rubber Chemistry and Technology 50 (1977) 835
[3] Kanig and Neff, Journal of Colloid and Interface Science 29 (1975) 253
[4] Cao, Polymer Communications, 29 (1988) 66 and 67
[5] Nau, Journal of Material Science, 32 (1997) 5335
[6] Machedo, Journal of Polymer Science Part A 37 (1999) 1311
[7] Dutta, Rubber Chemistry and Technology, 65 (1992) 932
[8] Cudby, Journal of Natural Rubber, 12 (1997) 102
[9] Holsti-Miettia, Journal of Polymer Engineering and Science, 34 (1994) 395

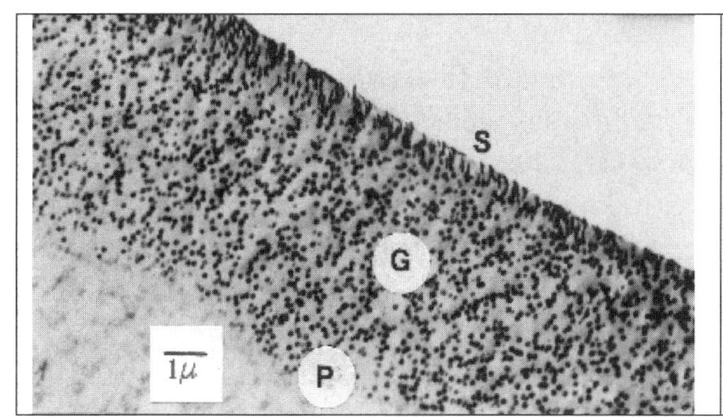

FIG. 1. Bulk Staining of ABS Type Resin Using Aqueous Osmium Tetroxide

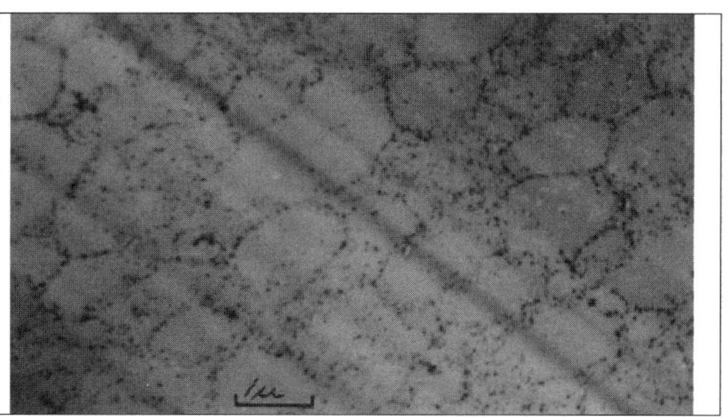

FIG. 2. Neg. Staining of NR Dipped Product Using Aqueous Silver Nitrate

Microsc. Microanal. 8 (Suppl. 2), 2002
DOI 10.1017.S1431927602102285

STAINING AND OTHER MICROSCOPIC TECHNIQUES FOR TEXTILES

E. K. Boylston

Southern Regional Research Center, P. O. Box 19687, ARS, USDA, New Orleans, LA 70179

Microscopical procedures for fibers are important to the textile industry in evaluating natural fiber maturity, mixed fiber blends, dyes and chemical finishes on fabrics. Staining (color for light microscopy and electron dense staining for TEM) of textile fibers as well as solubility and swelling tests along with special embedding methods for image analysis and Fourier Transform Infrared microscopy/spectroscopy are all essential in determining fiber properties and finishes. The Goldthwait test (ASTM D-1461) for determining maturity of cotton fibers is a differential dye test which reflects differences in effective pore size by using a combination of two dyes, Direct Green 26 (C.I. 34045) and Direct Red 81 (C.I. 28160). Immature, thin-walled fibers dye green, while mature, thick-walled fibers dye red. There are a number of "universal stains" composed of dye mixtures which have been developed for the identification of textile fibers. Testfabrics Inc. has several stain identification products that will stain different fibers (acetate or viscose rayon, acrilan, arnel, cotton, creslan, dacrons, nylons, orlon, silk, wood, wool, etc.) distinctive colors which can be identified from the supplied standard fiber strip. All natural fibers have a distinctive cross-sectional shape which also helps in identification by light microscopy. In many cases evidence obtained microscopically may be useful in the study of fiber damage and its causes. Types of suspected damage are abrasion, compression, tensile break and heat damage, as well as microbial damage due to enzymatic attack of microorganisms, or chemical damage. Electron microscopy of nearly all organic material requires some type of method to improve contrast because of low atomic masses present. This is especially true in the study of the substructure of cellulosic fibers such as cotton. The fiber is made up of concentric cellulose layers composed of an aggregate of fibrils, further divided into even smaller elementary fibrils. Exact measurement of fibrillar size is extremely difficult because of the inherent lack of contrast within the cellulose polymer due to low atomic weight. Electron staining encompasses two general approaches: a) negative staining with a substance such as phosphotungstic acid, where the electron-dense substance is deposited around the material, creating a negative contrast (Figure 1.), and b) positive staining, where the heavy metal is chemically bound to the material being stained. With protein fibers, positive staining is not a problem, but in cellulosic fibers, normally the cellulose polymer must be modified by attaching acidic, basic or unsaturated groups (reactive to heavy metal compounds) to the cellulose chain (Figure 2).

The expansion method is a microscopical procedure whereby cotton fibers are embedded wet in a low molecular weight methacrylate polymer[1]. Water in the fibers is replaced by the liquid plastic which is then further polymerized to a solid, which forces the layers of cellulose within the fiber apart. This procedure is useful in determining a) inter-fibrillar cross-linking (cross-linking between cellulose layers) - no swelling, b) fiber mercerization (swelling pattern of the fiber), and c) finishing agent deposition within the fiber, and particularly those finishes which do not contain elements easily detected by EDX.

Cuene, a cellulose solvent, is used to determine cross-linking of cellulose in cotton fabric treated for durable press, flame retardant finishing, etc.[2]. Cross-sections are mounted on TEM grids, covered with a drop of 0.1M cuene solution for 30 min., rinsed and examined in the TEM. Cotton fibers that are crosslinked will not dissolve in cuene solutions. Intra-fibrillar cross-linking (cross-linking within a cellulose layer) is apparent when fiber cross-sections have expanded but do not dissolve in cuene.

Three special embedding methods have been developed for cotton and cotton fiber blends[3]. 1) A rapid embedding method employing UV polymerization reactions has been devised for embedding fibers in acrylic and methacrylate media. After 10 min., the resultant thin, flat embeddings are suitable for both light and electron microscopy. 2) A process for the evaluation of yarns by light microscopy/image analysis has been developed. Approximately 2000 fibers before spinning, 50 yarn segments after spinning, or yarns removed from fabric after processing, can be encased in a tube, embedded in methacrylate plastic, quickly UV polymerized, and sectioned. All yarns are discretely separated and individual fibers are distinct so that image analysis and statistical evaluations are possible. Fiber maturities, yarn evenness, dye penetration, and position of different types of fibers in blend yarns are all visible and readily quantified. 3) A procedure for embedding cellulosic textiles has been developed for FT-IR microscopy whereby fibers are embedded in polystyrene. This polymer does not absorb in the same regions of the infrared spectrum as cellulose or traditional acrylate and epoxy resins that contain chemical groups in common with cellulose. Additionally, use of cross-sections mounted on a KBr disk has the advantage of better resolution than grinding and pressing fibers in a KBr disk.

References

1. E. K. Boylston and L. L. Muller, Journal of Applied Polymer Science, 19(1975)1079.
2. E. K. Boylston, et al., Textile Research Journal, 45(1975)790.
3. E. K. Boylston, et al., Biotechnic and Histochemistry, (1991)122.

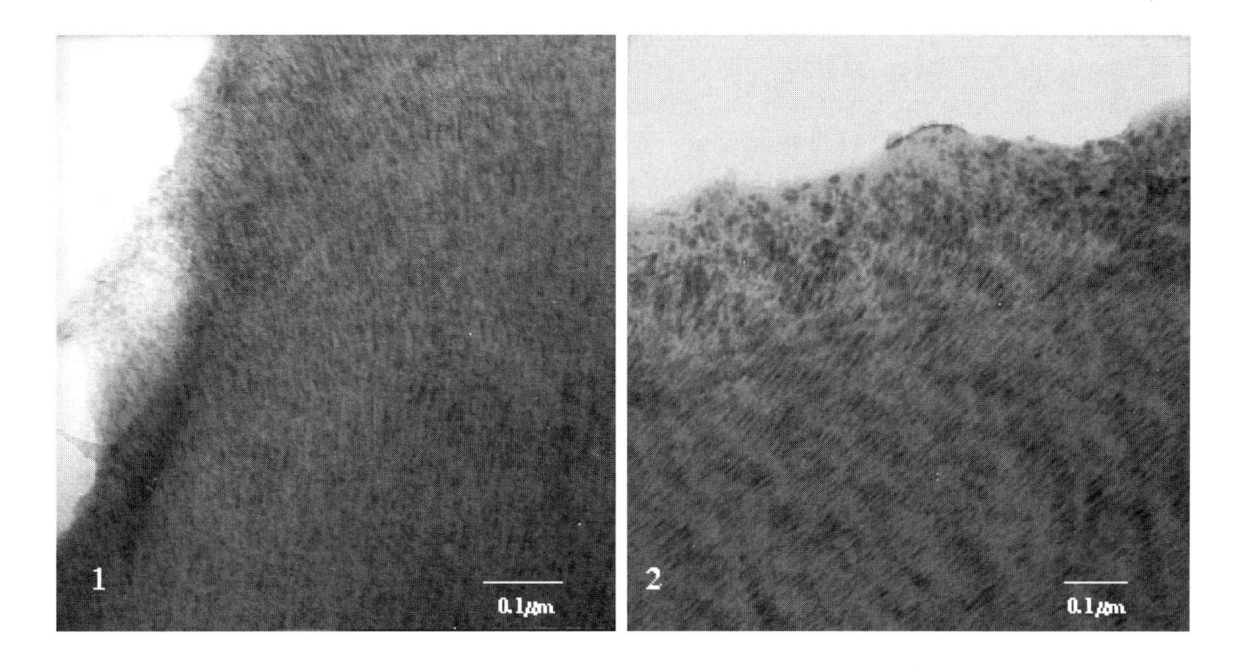

Fig. 1. Cotton fiber longitudinal section negatively stained with 2% phosphotungstic acid.
Fig. 2. Cotton fiber longitudinal section positively stained with 2% OsO_4 after treatment with sorbyl chloride.

Microsc. Microanal. 8 (Suppl. 2), 2002
DOI 10.1017.S1431927602102297

ALL THAT GLITTERS IS NOT GOLD: APPROACHES TO LABELING FOR EM

R.M. Albrecht and D.A. Meyer
Department of Animal Sciences, University of Wisconsin, Madison, WI 53706

A large number of natural and synthetic species ranging from virus particles to small polymer beads have been employed over the years as labels in electron microscopic applications. Perhaps the most useful and versatile of these labels are the colloidal heavy metals.[1] The colloidal labels can be synthesized as round regular spheres in sizes from 1 to 150 nanometers in diameter and thus can be used where molecular and sub-molecular ranges of spatial resolution are required. They are electron "dense" and good emitters of secondary and backscattered electrons and hence are readily detectable in both scanning electron microscopic imaging systems and transmission electron imaging systems. Larger particles, 10nm and above, can be identified via their shape in force based imaging systems. Each particle is basically a small sphere containing a number of atoms of the particular element in a relatively small volume. This improves detectability over less dense metal-organic complexes or proteins that are bound to specific elements such as boron or iron. In the latter case the metal density is relatively low and substantial numbers of the labels are often required for detection.

The bulk concentration of the colloidal particles in a tissue volume can be determined at very high sensitivity via Instrumental Neutron Activation Analysis, INAA, with minimal preparation and without significant specimen damage.[2] INAA analysis can be performed prior to or after imaging. INAA is very valuable in determining bulk concentrations of label to levels that would be difficult to detect by sectioning and counting due to insufficient particle numbers. INAA can be used with gold and other colloids to further improve the quantitative aspects of multiple labeling for EM. Label density in bulk, unsectioned tissue can be determined using INAA This permits a prediction of the number of labels in, for example, thin sections or whole mounts prepared from the labeled tissue. If particle numbers, as measured by INAA of bulk tissue, are too low or too high, labeling parameters or experimental design can be adjusted prior to tissue fixation, dehydration, embedding, sectioning, and imaging.[3]

The colloidal particles are also useful for correlative microscopic investigations since individual particles as small as 12nm can be detected (but not resolved) and tracked via their inflated diffraction image using interference based light microscopy.[1] Individual particles can also serve as nuclei for further addition of silver or gold in a process that enlarges each particle to a size that is resolvable in conventional light microscopy. Relatively small numbers particles can darken or color areas they stain due to their extremely high extinction coefficients compared to other dyes.

For use in labeling, colloidal particles can be readily conjugated, principally via strong hydrophobic interactions, to a variety of biologically active molecular species. The conjugates are stable and very little loss of ligand/antibody activity or specificity is encountered. A thin shell of bound ligand or antibody surrounds larger particles while the smaller particles intercalate within the structure of individual ligand or antibody molecule. Very small particles in the 2 to 5 nanometer range can be conjugated to active fragments of ligands or antibodies to produce labels with dimensions that readily permit identification and localization at sub-molecular levels of spatial resolution and with minimal steric interference. One small colloidal particle, conjugated to an active ligand fragment or a single Fab of an antibody molecule, results in labels having a single valence and hence one particle represents one binding site or epitope. This is advantageous for quantitative studies.

One of the critical factors in understanding cellular function is knowledge of the complex and dynamic cellular architecture and the interactions of individual molecular species within that architecture. Co-localization studies employing simultaneous multiple labels are often an extremely effective way in which questions of this type can be addressed. Photon based imaging systems use fluorescent dyes having different excitation and emission wavelengths. For studies at molecular and sub-molecular levels of resolution the standard approach is to use colloidal gold particles of different sizes. Unfortunately, to ensure that overlap in particle size is not a problem, generally only one or two particle sizes within the range that provide the desired level of resolution can be employed at once. Attendant steric problems arise because of the substantially greater mass of the larger particle. Substantial differences in ease of detection of the two particles also occur due to the larger size and greater mass of the larger gold particle. Therefore accurate localization/co-localization of multiple molecular species requires a labeling system that uses small particles of nearly identical size that can be conjugated to different identifier molecules (antibodies or ligands), or active fragments of antibodies or ligands. The different small conjugates permit simultaneous labeling of multiple epitopes or binding sites. However, the different particles must be easily and reproducibly identified from one another. We are pursuing two approaches to achieving this end.[4,5,6] The first is the synthesis of uniformly sized colloidal metal particles of differing geometry.[4] Unique shapes (spheres, asterisks, geometric, irregular) based on differences in particle nucleation and growth parameters, can be detected using standard TEM and SEM (and, within limitations, by Force Microscopy) This is assuming the shape variation is within the resolving limits of the particular instrumentation available. The second approach is the synthesis of uniformly sized colloidal metal particles of differing elemental composition such as Au, Ag, Pd, Pt, Ru, Rh.[5] Energy Filtering Electron Microscopy, EF-TEM, is used to identify and differentiate particles that are of the same size but of different elemental composition.[5,6] EF-TEM relies on the electron energy loss spectra, EELS, of different elements to selectively image materials. Colloidal particles of similar size but of differing composition will produce different, generally non-overlapping, energy losses in the beam electrons. Thus a typical EM image demonstrating ultrastructural features can be collected using zero loss electrons (zero loss also minimizes chromatic aberration seen with conventional EM imaging). Subsequently, labels of similar sizes but composed of different colloidal metals can be differentiated via EF-TEM imaging. Both shape and composition can be used in concert to differentiate particles and permit simultaneous identification and co-localization of multiple molecular species at molecular levels of resolution for in vitro and in vivo applications.

References

[1] R.M. Albrecht et. al., Immunochemistry, A Practical Approach, Oxford Univ. Press. Oxford (1993)151.
[2] J.F. Hillyer and R.M. Albrecht, Microsc. and Microanal. 4 (1999) 481.
[3] J.F. Hillyer and R.M. Albrecht, J. Pharm. Sci. 90 (2001) 1927.
[4] R.M. Albrecht and D.A. Meyer, Microsc. and Microanal.6 (Suppl. 2) (2000) 318.
[5] D.A. Meyer and R.M. Albrecht, Microsc. and Microanal. 6 (Suppl. 2) (2000) 322.
[6] D.A. Meyer and R.M. Albrecht, Microsc. and Microanal. 7 (Suppl. 2) (2001) 1032.

Microsc. Microanal. 8 (Suppl. 2), 2002
DOI 10.1017.S1431927602102303

Stains for the Determination of Paper Components and Paper Defects

J. H. Woodward

Buckman Laboratories, Inc., 1256 N. McLean Blvd., Memphis, TN 38108

Stains for fiber analysis have been utilized since the 1930's with the first color monograph being published in 1940 [1]. Throughout the years, stains and indicators have been routinely used to determine paper components and paper defects.

Interpretation of staining reactions requires a basic knowledge of the papermaking process. Wood fiber can be from either softwood (conifers) or hardwood (deciduous). Softwood fibers are generally characterized by long, thin-walled cells whereas hardwoods have relatively shorter, narrow cells and very short, wider cells. All wood fiber must be pulped prior to the papermaking process. The pulping process can be mechanical where the wood is basically ground. In this process, the lignin component of the fiber is not removed. Semi-chemical and chemical pulping, such as sulfite and sulfate (Kraft), remove various levels of lignin. The brightness of the resulting pulps ranges from low (e.g. brown paper bag) to relatively high (e.g. newsprint). Pulp can then be bleached by a variety of chemical means to achieve a higher brightness. Paper or paperboard is made at acid, neutral, or alkaline pH. Wet-end additives can include retention aids, starch, sizing, dyes, optical brighteners, and microbicides. Depending on the pH of the process, the fillers consist of calcium carbonate ($CaCO_3$), titanium dioxide (TiO_2), or clay. Coating components include binders, dispersants, preservatives, defoamers, clay, $CaCO_3$, and TiO_2.

For basic fiber analysis, the paper or paperboard must first be disintegrated. A sample is torn into pieces and boiled in distilled water. The pieces are then rolled into pellets, placed into a test tube of water, and shaken vigorously. Additional water is added and the tube is shaken again. This is repeated until the sample is completely defibered. The resulting suspension is further diluted for staining procedures. If the paper is not disintegrated by this method, a sample is boiled in a 1% NaOH solution, washed several times with distilled water, allowed to stand in a 0.05N HCl for several minutes, and again washed several times with water. The paper is then rolled into small pellets, placed into test tubes with water, and shaken as previously described. Disintegration methods for specially treated papers are described in TAPPI Official Test Method T 401 om-93 [2].

The most commonly used stain for fiber analysis is the Graff "C" stain which reacts with lignin to produce a yellow color. Unbleached chemical and semi-chemical pulps will stain yellow with the depth of color dependent upon the type and degree of cooking. When the pulp is bleached, the fiber will stain a red color. As seen in FIG.1, thermo-mechanical pulp stains a bright yellow due to the high lignin content. Another useful stain for fiber analysis is the Herzberg stain because it differentiates chemical pulp from mechanical pulp. Both bleached and unbleached chemical pulps stain blue. Fibers from a paper towel stained blue, yellow, and light red (FIG. 2) indicating the presence of chemical and mechanical pulp.

Starch, commonly found in and on paper, reacts with iodide-potassium iodide to form a characteristic blue-purple color. The procedure can be as simple as placing a drop of the indicator on the paper sample. A ninhydrin solution can be used to detect protein in a coating as well as microorganisms in a paper defect.

Basic microbiological stains prove useful in the identification of paper defects. For example, an unknown yellow defect appeared to be a fiber lump when viewed by a stereomicroscope. Lactofuchsin was applied and portions of the defect turned pink (FIG.3). Under oil immersion, the stain revealed filamentous bacteria wrapped around fibers (FIG. 4), indicating that the defect was due to the presence of actinomycetes growing in one of the furnishes.

Both traditional and non-traditional stains can be useful in the identification of paper components and defects.

References

[1] J.H. Graff, *Color Atlas for Fiber Identification*, Institute of Paper Chemistry, Appleton, WI, 1940.
[2] *2000-2001 TAPPI Test Methods*, TAPPI, Atlanta, 2000.

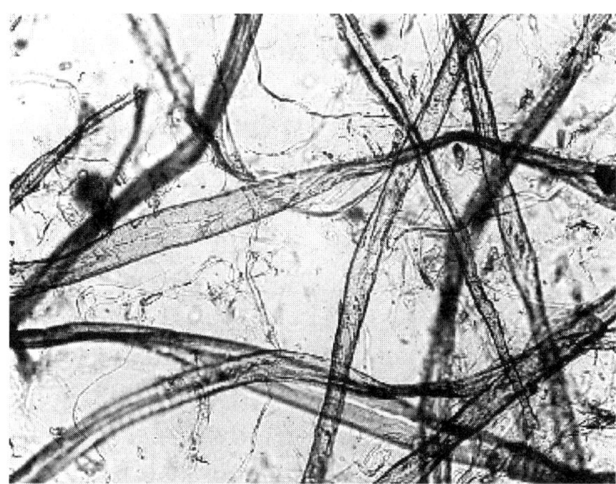

FIG. 1. Graff "C" stain of thermo-mechanical pulp with high lignin content. 40X

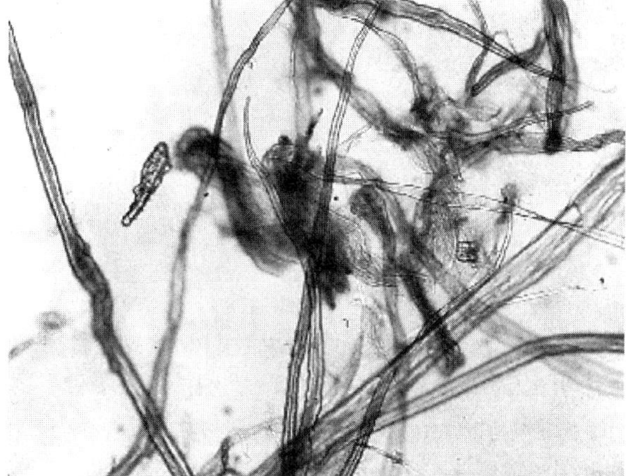

FIG. 2. Herzberg stain of paper towel fibers indicating chemical and mechanical pulp. 40 X

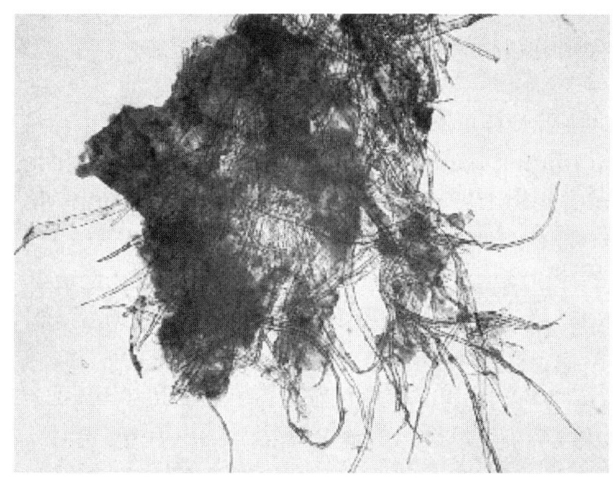

FIG. 3. Lactofuchsin stain of a paper defect. 10X

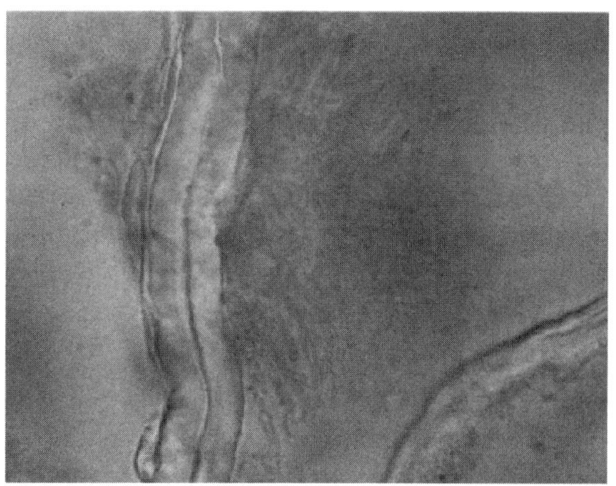

FIG. 4. Higher magnification of the defect revealing filamentous bacteria. 1000X

Microsc. Microanal. 8 (Suppl. 2), 2002
DOI 10.1017.S1431927602102315

Fluorescent Specimen Preparation Techniques for Confocal Microscopy

Judith A. Drazba

Imaging Facility, The Cleveland Clinic Foundation, 9500 Euclid Avenue, Cleveland, OH 44195

Every good microscopist knows that specimen preparation is critical for obtaining good images. Unfortunately some people think that a confocal microscope can make poorly stained specimens look good. A confocal makes images sharper by rejecting out-of-focus light but it cannot discriminate between specific and non-specific staining or between a specific probe and an autofluorescent background. It takes a picture of all the light coming from a selected focal plane. If anything, a confocal will render artifacts sharper and more beautiful. Therefore, good staining protocols and adequate controls are more important than ever.

Confocal microscopes require specimens to be either reflective or fluorescent. Since biological cells and tissues are not very reflective we use fluorescent molecules that can either be tagged onto biological macromolecules to show their specific cellular location (i.e. antibodies) or bind to specific cellular components (i.e. dyes). Until recently the most frequently used fluorophores have been FITC, which appears green when excited, and TRITC, which appears red. Despite their utility, these molecules come with problems such as only getting to see them for seconds to minutes before they are irretrievably bleached. Today's fluorescent molecules come in "designer" colors that can be better separated based on their emission wavelengths for multiple labeling and are much brighter and more stable than their older cousins. The advent of improved antiphotobleach mounting media now allows us to view specimens for hours and keep them viable in storage for years.

Dye selection The first decision in developing a staining protocol is what colors to use. If your specimen has considerable autofluorescence you'll need to determine what wavelength range is the most offensive and then use a fluorophore as far away from that range as possible. If you're labeling multiple molecules in the same sample you'll want to choose fluorophores whose emission profiles are sufficiently non-overlapping that you can separate them well in the microscope, but also have the best excitation profile for the available confocal light sources, usually lasers. For multiple labeling you also need to minimize cross-reactivity by carefully choosing the animal sources of the antibodies to which the probes are conjugated. Organelle specific dyes in combination with an antibody can tell you if your molecule of interest is associated with a particular cellular component.

Fixation and Permeabilization Appropriate fixation is critical to good staining. Avoid glutaraldehyde if possible. Even small quantities produce intense autofluorescence that will interfere with visualizing your probe. Quenching protocols can help, but it's best avoided. Paraformaldehyde is the better alternative. If you're using fluorescent dyes be sure to buy those that are "fixable" because, unlike antibodies, some can only be used on live cells and tissues. Most antibody staining is done on fixed tissue, which often requires permeabilization. A frequently asked question in confocal microscopy is: "How thick can my sample be?" The appropriate question in response is: "How far can you get your antibody to penetrate into the tissue?" In most instances that distance is less than the working distance of most objective lenses. The porosity of the tissue, the length of staining and the concentration of detergent will all play a role. Collection of confocal optical sections through the specimen will quickly reveal the success or failure of the staining (Fig 1). Analysis of an "XZ" section, available on some microscopes, is an even better indicator of stain penetration (Fig 2).

Quantity vs. Quality Titering of antibody or dye concentration is very important. In the case of antibodies, too much is as bad as too little. Over saturation can actually decrease signal due to steric hindrance and will substantially increase non-specific binding even if there is adequate blocking. Cellular components can appear very convincingly stained this way. To have any chance of interpreting the results it is very important to have controls that have not seen any antibody as well as controls that have only been exposed to secondary antibodies. Signal strength will also be determined by the method of antibody application and should be matched to the prevalence of the molecule being probed. Antibodies that are directly conjugated to a fluorochrome may produce less background, but since there's little amplification you may not be able to see molecules that exist in low quantity. Sandwich techniques work better in these cases either with fluorescently labeled secondary antibodies or with biotinylated secondaries and fluorescently labeled avidin.

Future Trends The advent of fluorescent proteins is a recent solution to the antibody penetration problem. Through the magic of molecular biology specific molecules can be tagged and then endogenously expressed by cells so that the specimen comes virtually "pre-labeled." Other techniques such as fluorescent in-situ hybridization are providing valuable information, but also require careful preparation for reliable results.

References:

1. R. Bacallao et al.,Handbook of biological Confocal Microscopy, Plenum, New York, 1995, J.B. Pawley (ed.)
2. R. Bacallao, EH Stelzer, Preservation of biological specimens for observation in a confocal fluorescence microscope. Methods Cell Biol 1989; 31:437-52

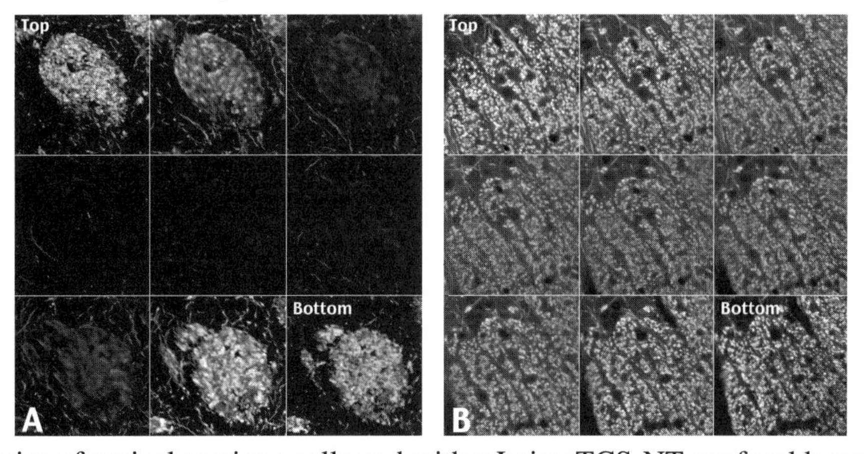

Figure 1. Series of optical sections collected with a Leica-TCS-NT confocal laser scanning microscope comparing antibody penetration into 30 μm thick free-floating sections of rat brain permeabilized with either (A) 0.5% or (B) 10% Triton-X 100 detergent.

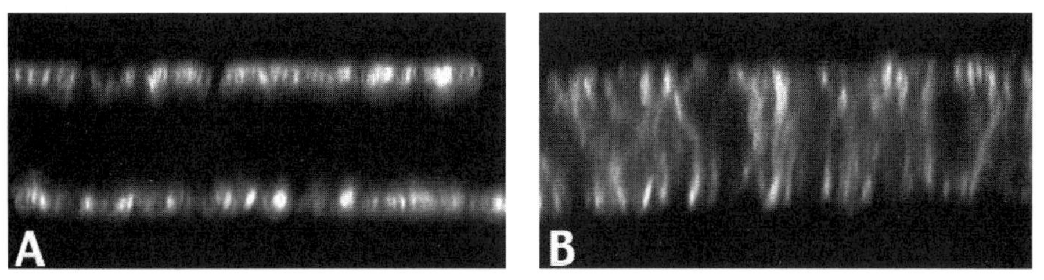

Figure 2. Same specimen shown in Figure 1 imaged in cross-section (XZ slice).

Microsc. Microanal. 8 (Suppl. 2), 2002
DOI 10.1017.S1431927602100006

Transmembrane Signalling of the Insulin Receptor: 3D Reconstruction from STEM Imaging, Crystallography and NMR Spectroscopy.

F.P. Ottensmeyer, A.C.H. Oh, R.Z.T. Luo*, A. Fernandes, D.R. Beniac and C.C. Yip*

Ontario Cancer Institute, Toronto, M5G 2M9; *Banting and Best Department of Medical Research University of Toronto, Toronto, M5G 1L6, Canada

One of the major goals in molecular biology is the understanding of the relationship between the 3D structure of macromolecules, alone or in complex assemblies, and the biological function or mechanism of action that ensues as the result of that structure. To achieve that goal, the need for structural detail at atomic resolution has been a tacit assumption. Almost by definition such a requirement eliminates the use of electron microscopy (EM) of biological specimens, with a concentration instead on x-ray crystallography and, more recently, nuclear magnetic resonance (NMR)spectroscopy.

Nevertheless, EM and electron diffraction of thin monomolecular crystals, although not yet providing atomic resolution directly, have resulted in 3D structures with about 3.5 Å detail [1,2]. However, the *forte* of EM is its capability to image individual macromolecules and complexes which have been refractory to crystallization and which are too large to be solved by NMR spectroscopy. The spatial resolution for 3D reconstructions of large macromolecules or complexes from such single particle imaging is generally between 10 and 20 Å. However, when subdomains of such macromolecules have been solved by crystallography or NMR, such domains can be docked or fitted into the EM reconstruction to much greater accuracy using the constraints of the centre of mass of the subdomain and its EM counterpart, the envelope of that part of the EM reconstruction constraints, and the limitations to possible rotation due to the asymmetry of the structure.

This combination of techniques has been used by us to result in the virtually complete atomic structure of the 480 kDa insulin receptor in the presence of insulin [3,4], resulting in an understanding of the binding of insulin ligand at the level of amino acid side chain interactions. Moreover, the connectivity of the structural domains and their three-dimensional spatial juxtaposition led to the prediction of the mechanics of transmembrane signalling and activation of this tyrosine kinase membrane receptor. This prediction has been substantiated by our current reconstruction of the insulin-free receptor (Fig. 1).

Technically, most single particle EM approaches use bright field phase contrast imaging. In contrast, we have developed and exploited low-dose, low-temperature dark field techniques using scanning transmission electron microscopy, to take advantage of the high contrast of that procedure which permits easy visualization of freeze-dried macromolecules as small as a few tens of kilodaltens. Imaging doses as low as 2 e/Å^2 were made possible via the simultaneous signal acquisition of elastic and inelastic electron scatter signals from the specimen at each picture point delineated by the 3 Å beam. Three-dimension reconstruction was carried out using by filtered back-projection after deriving the relative 3D orientations of individual molecular images by reference-free iterative quaternion-assisted angle determination [5,6].

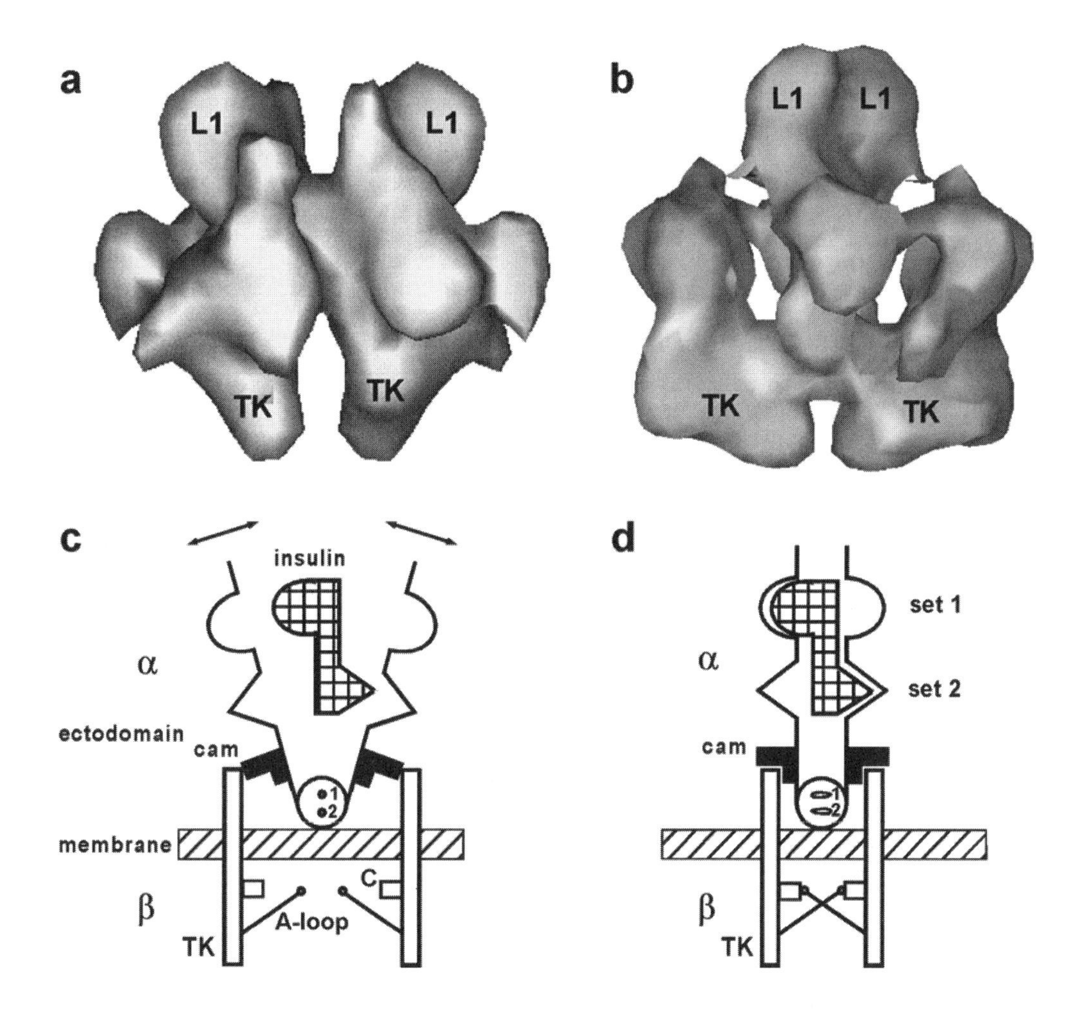

Fig. 1 Structure and Mechanism of Transmembrane Signalling and Activation of the Intrinsically Dimeric Insulin Receptor
(a) 3D reconstruction of the insulin-free receptor. (b) 3D reconstruction of the insulin receptor with insulin bound. (c) Schematic of the structure of the insulin-free receptor showing the structural relationship between the extracellular α-subunits and the transmembrane β-subunits of the receptor. (d) Schematic of the structure of the insulin-bound receptor. L1 = large domain 1; TK = tyrosine kinase domain; C = catalytic site; A-loop = activation loop; 1,2 = covalent cys-cys bonds between receptor monomers; cam = specific structural feature; set 1,2 = specific sets of binding side chains on α-subunits.

References

[1] R. Henderson et al., J. Molec. Biol. 213 (1990) 899.
[2] W. Kühlbrandt, D.N. Wang, Y. Fujiyoshi, Nature 367 (1994) 614.
[3] R.Z.T. Luo et al., Science 285 (1999) 1077.
[4] F.P. Ottensmeyer et al., Biochemistry 39 (2000) 12103.
[5] N.A. Farrow, F.P. Ottensmeyer, J. Opt. Soc. Am. A9 (1992) 1794.
[6] N.A. Farrow, F.P. Ottensmeyer, Ultramicroscopy 52 (1993) 141.

Microsc. Microanal. 8 (Suppl. 2), 2002
DOI 10.1017.S1431927602100018

Structure of the Eukaryotic Transcription Machinery: Insights into the Mechanism of Transcription Initiation and Regulation

Francisco J. Asturias*, Roger D. Kornberg**, John L. Craighead*, Joshua A. Davis*, and Weihau Chang**

*Dept. of Cell Biology, The Scripps Research Institute, La Jolla CA 92037
**Dept. of Structural Biology, Stanford University Med. School, Stanford CA 94305

Precise regulation of gene expression underlies development, oncogenesis, and the constant reshaping of the cell in response to a variety of metabolic and environmental cues. Regulation is mediated by a large number of factors that control the action of RNA polymerase II (RNAPII), the enzyme responsible for synthesis of all mRNA in eukaryotes. Structural information about the complex molecular machinery responsible for transcription and its regulation is essential to advance our understanding of the mechanism of gene expression. Macromolecular single particle cryo-electron microscopy (cryo-EM) is ideally suited to the structural study of such large systems, and can provide moderate (10-50Å) resolution structural information. Combined with high-resolution structures of individual components obtained by x-ray crystallography or NMR spectroscopy, cryo-EM structures of large assemblies will lead to an understanding of the entire transcription apparatus at the atomic level. Single particle cryo-EM analysis does not require the use of crystalline samples. Comparatively small amounts of material are sufficient to study macromolecular structure under physiologically relevant conditions, and without the artificial constraints imposed on molecular conformation by the formation of crystals.

We have used single particle cryo-EM to study the conformation in solution of RNAPII, and its implications for interaction of the enzyme with promoter DNA. Our analysis indicates that the conformation of RNAPII under physiologically relevant conditions precludes it from interacting with double-stranded promoter DNA (Fig. 1). Our 18Å resolution structure of RNAPII also reveals the location of subunits Rpb4/Rpb7, which was not determined in previous x-ray studies of the polymerase [1]. The essential Rpb7 subunit contains two single-stranded nucleic acid binding sites [2], and our structure shows that it is ideally positioned to determine the path of the nascent RNA transcript exiting the active site cleft of RNAPII (Fig. 1). Transcription is closely coupled to processing of the RNA transcript, and the essential role of Rpb7 is most likely related to the efficient delivery of the nascent RNA to the RNA processing machinery.

RNAPII is unresponsive to regulation without the intervention of a coactivator complex, that functions as an essential interface between polymerase and the activator and repressor proteins that bind to enhancer and repressor DNA control sequences. We have pursued structural characterization of the holoenzyme formed by RNAPII and the Mediator coactivator complex. Mediator was first identified in the yeast *Sacharomyces cerevisiae*, but homologs of the complex have now been characterized in all eukaryotes, from yeast to man. A 35Å resolution structure of the RNAPII holoenzyme (>35 polypeptide components, and >1.5 MDa MW) shows Mediator organized in three distinct structural modules, wrapped around RNAPII (Fig. 2A). Although multiple contacts are established between Mediator and RNAPII, must of the polymerase surface remains accessible for interaction with additional components of the transcription machinery. Comparison with the x-ray structure of RNAPII has identified RNAPII subunits involved in contacts with Mediator. The interaction is centered on RNAPII subunits Rpb3 and Rpb11 (Fig. 2B). These subunits are located opposite to the entrance to the active site cleft of the enzyme, implying that interaction of RNAPII with the coactivator complex does not interfere with association of the enzyme with its promoter DNA substrate. Factors involved in transcription regulation in prokaryotic organisms have been shown to interact with the aa' homodimer, the bacterial homolog of the eukaryotic Rpb3/Rpb11 complex, and mutations in the Rpb3 subunit of RNAPII have been shown to interfere with activated transcription [3]. Therefore, it appears that a similar mechanism is involved in regulation of transcription from prokaryotes to man.

References

[1] P. Cramer, D. A. Bushnell, R. D. Kornberg, Science 292 (2001) 1863.
[2] F. Todone, P. Brick, F. Werner, R. O. Weinzierl, S. Onesti, Molecular Cell 8 (2001), 1137.
[3] O. Tan, et al., Genes Dev.;14 (2000) 339.

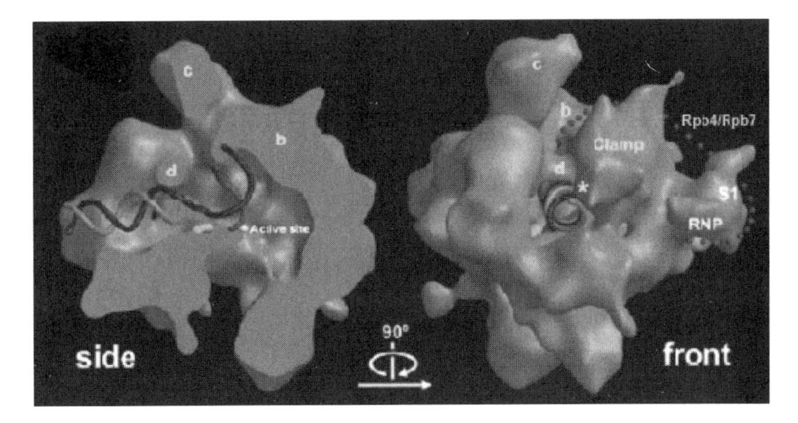

Fig. 1. Structure of wild-type RNAPII in solution and location of subunits Rpb4/Rpb7.

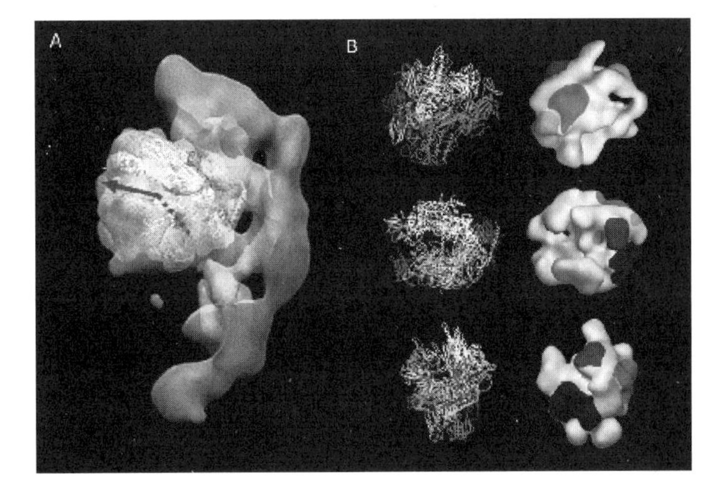

Fig. 2. Structure of the RNAPII holoenzyme and analysis of the RNAPII/Mediator interface.

Microsc. Microanal. 8 (Suppl. 2), 2002
DOI 10.1017.S143192760210002X

Functional Architecture of a Protein-Degradation Machine

Takashi Ishikawa*, Michael R. Maurizi**, Alasdair C. Steven*

* Laboratory of Structural Biology, NIAMS; ** Laboratory of Cell Biology, NCI,
National Institutes of Health, Bethesda, MD 20892, U.S.A.

Protein quality control protects cells from adverse effects of defective or foreign proteins. The intracellular population is monitored for such molecules which are then refolded or eliminated by proteolysis. The latter function is carried out by energy-dependent proteases, enzymes with rings of chaperone-like ATPase subunits stacked on rings of peptidase subunits. The ATPases recognize and unfold substrates, and deliver them to the peptidases for degradation. To elucidate their mode of action, we have studied the ClpAP protease of *E. coli* and its interactions with a model substrate, phage protein RepA. ClpA ATPase consists of two copies of the AAA ATPase domain (D1 and D2), preceded by an N-terminal domain (Fig. 1) and forms a hexamer. ClpP peptidase is a double heptamer: ClpA binds to one or both ends of ClpP. RepA substrates bind initially to sites on the distal surface of ClpA and upon ATPhydrolysis is then translocated ~ 15 nm axially into the digestion chamber inside ClpP ([1], Fig. 2a,b,c,e). Our current work is concerned with the disposition of the various protein substrates and enzyme domains during the reaction cycle.

We have calculated a 3-dimensional model of ClpA at 1.8 nm resolution (FSC = 0.5) from 1200 tilt pairs of ClpA side-views extracted from the image of ClpAP (Fig. 3). Data were recorded on a CM120 (FEI) with Gatan 626 cryoholder. The reconstruction was done with EMAN [2], using an earlier reconstruction at 2.8 nm [3] as starting model.

The resulting model has two tiers, corresponding to D1 and D2 respectively (Fig. 4). Each tier is hexagonal with 15 degrees relative axial rotation. The distal face of D2 is in contact with ClpP [4]. Adjacent D2 domains are loosely associated in the ring and have lateral protrusions that may represent the small domains in the distal portion of D2. There is no sign of the N-domains although they are fully half the size of D1 or D2. Similarly, the N-domains are not evident in averaged side projections (Fig 2d) and only show up as blurred density in a difference image between ClpAP and a mutant complex with deleted N-domains (Fig 2f). From these results, we infer that the N-domain is connected to D1 by a flexible linker. Fig. 4 illustrates two possible modes, in which RepA binds respectively to the N-domains or to a site on the distal face of D1, laterally displacing the N-domains which assemble on the axis in the absence of RepA. Both can account for the broad crescent of terminal density associated with RepA binding ([1]; Fig. 2b, e) and the stoichiometry of one RepA dimer per ClpA hexamer.

References

[1] T. Ishikawa et al., Proc. Natl. Acad. Sci. 98 (2001) 4328.
[2] S. Ludtke et al., J. Struct. Biol. 128 (1999) 82.
[3] F. Beuron et al., J. Struct. Biol. 123 (1998) 248.
[4] S. K. Singh et al., J. Biol. Chem. 276 (2001) 29429.

FIG. 1 Schematic diagram of domain structure of ClpA

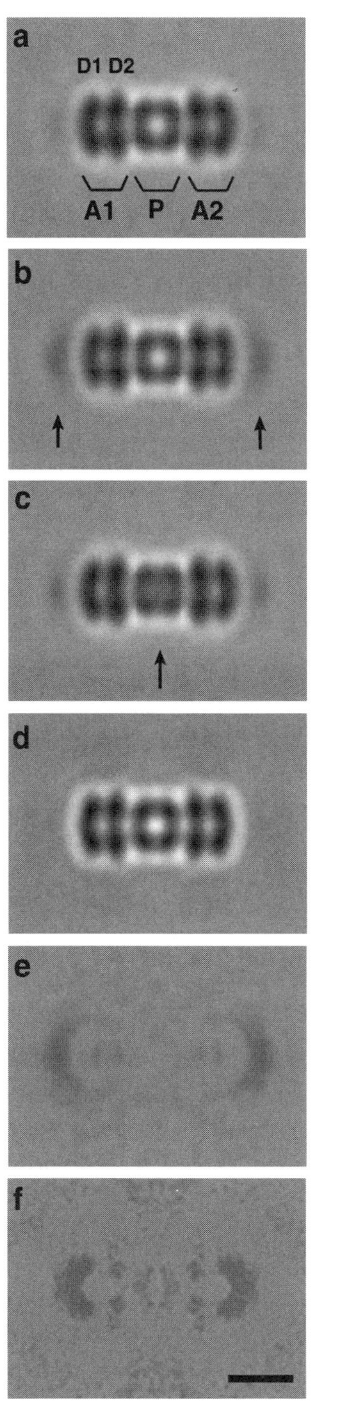

FIG. 2 (a-d) Averaged sideviews: (a) ClpAP+ATPγS; (b) ClpAP/RepA+ ATPγS; (c) ClpAP/RepA+ATP. The density due to RepA is shown by arrows. (d) ClpAP with ClpA lacking its N-domain. (c, f) Difference maps: (e) between (b) and (a); (f) between (a) and (d). Bar=10nm

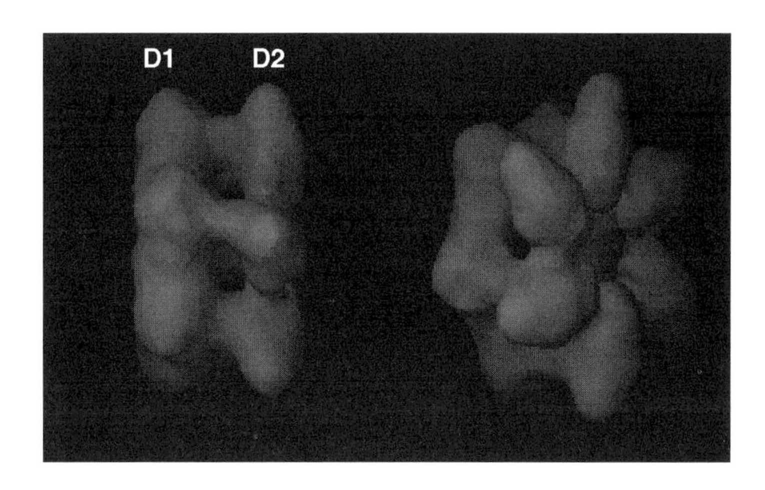

FIG.3 Three-dimensional reconstruction of ClpA hexamer from cryo-electron micrographs. On the left is the same view shown as A1 in FIG.2 On the right is a tilted view.

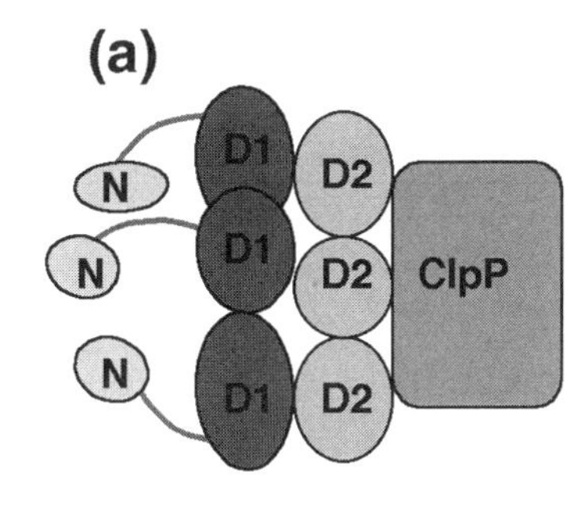

FIG. 4
Schematic representation of the alignment of domains in ClpA (a). Two models of substrate binding are shown (b, c).

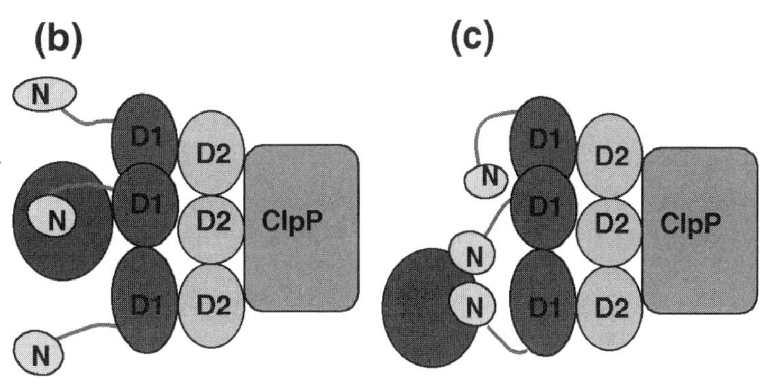

Microsc. Microanal. 8 (Suppl. 2), 2002
DOI 10.1017.S1431927602100031

The Ribosome – Ligand Interactions and Dynamics as Inferred by Cryo-EM

J. Frank*,***, J. Sengupta**, M. Valle*, and R.K.Agrawal**,***

*Howard Hughes Medical Institute, Health Research, Inc. at the **Wadsworth Center and ***Dept. of Biomedical Sciences, State University of New York at Albany, Empire State Plaza, Albany, NY 12201-0509

Protein synthesis involves the processive interaction of multiple ligands (mRNA, tRNA, EF-Tu, EF-G) with the ribosome, coupled with conformational changes of both ribosome and ligand molecules. Cryo-electron microscopy combined with three-dimensional reconstruction of single particles is ideally suited to study these processes during the main phases of translation: initiation, elongation, and termination. Thus far, the elongation cycle, in the course of which a new amino acid is added to the nascent polypeptide chain, has received the most attention. It is driven by the alternate binding of (i) the ternary complex consisting of elongation factor Tu (EF-Tu), aminoacyl-tRNA, and GTP, and (ii) elongation factor G (EF-G). The first of these binding interactions catalyzes the incorporation of a new, cognate tRNA into the A site, while the second catalyzes tRNA translocation (A->P, P->E, E->out). Both interactions require GTP hydrolysis.

Cryo-EM has been used to investigate both interactions, taking advantage of antibiotics that prevent the release of the factors. As expected from the close similarity of the two ligands ("molecular mimicry" [1]), they bind at the same location on the ribosome [2-5], but this immediately poses a problem in explaining the alternate binding modes. However, closer analysis using new cryo-EM data on the binding of ternary complex [6] and EF-G in the GTP form [7] shows that the residues contacting the ribosome form different constellations for the two factors. This coincides with the observation that EF-G binding (in the presence of a noncleavable GTP analog) triggers a large conformational change, described as a rotational ratchet motion between the two ribosomal subunits [8]. Tentatively, then, the elongation cycle might involve a sequence of the following kind [9]: factor A binds to the ribosome, performs its work, triggering a conformational change that destabilizes its own binding and leads to a conformation that invites factor B; factor B binds to the ribosome, etc. This concept suggests further experiments, but requires that cryo-EM advances toward higher resolution.

[1] P. Nissen et al., *Science* 70 (1995) 1464.
[2] H. Stark et al., *Nature* 389 (1997) 403.
[3] R.K. Agrawal et al., *Proc. Natl. Acad. Sci. USA*. 95 (1998) 6134.
[4] R.K. Agrawal et al., *J. Cell Biol.* 150 (2000) 447.
[6] M. Valle et al., submitted
[7] M. Valle et al., in preparation.
[8] J. Frank and R.K. Agrawal, *Nature* 406 (2000) 318.
[9] J. Frank and R.K. Agrawal, *Cold Spring Harbor Symposia on Quantitative Biology*, Cold Spring Harbor Laboratory Press, New York, 2001
[10] Supported by Howard Hughes Medical Institute, NIH R37 GM29169 (to JF) and R01 GM61576 (to RKA)

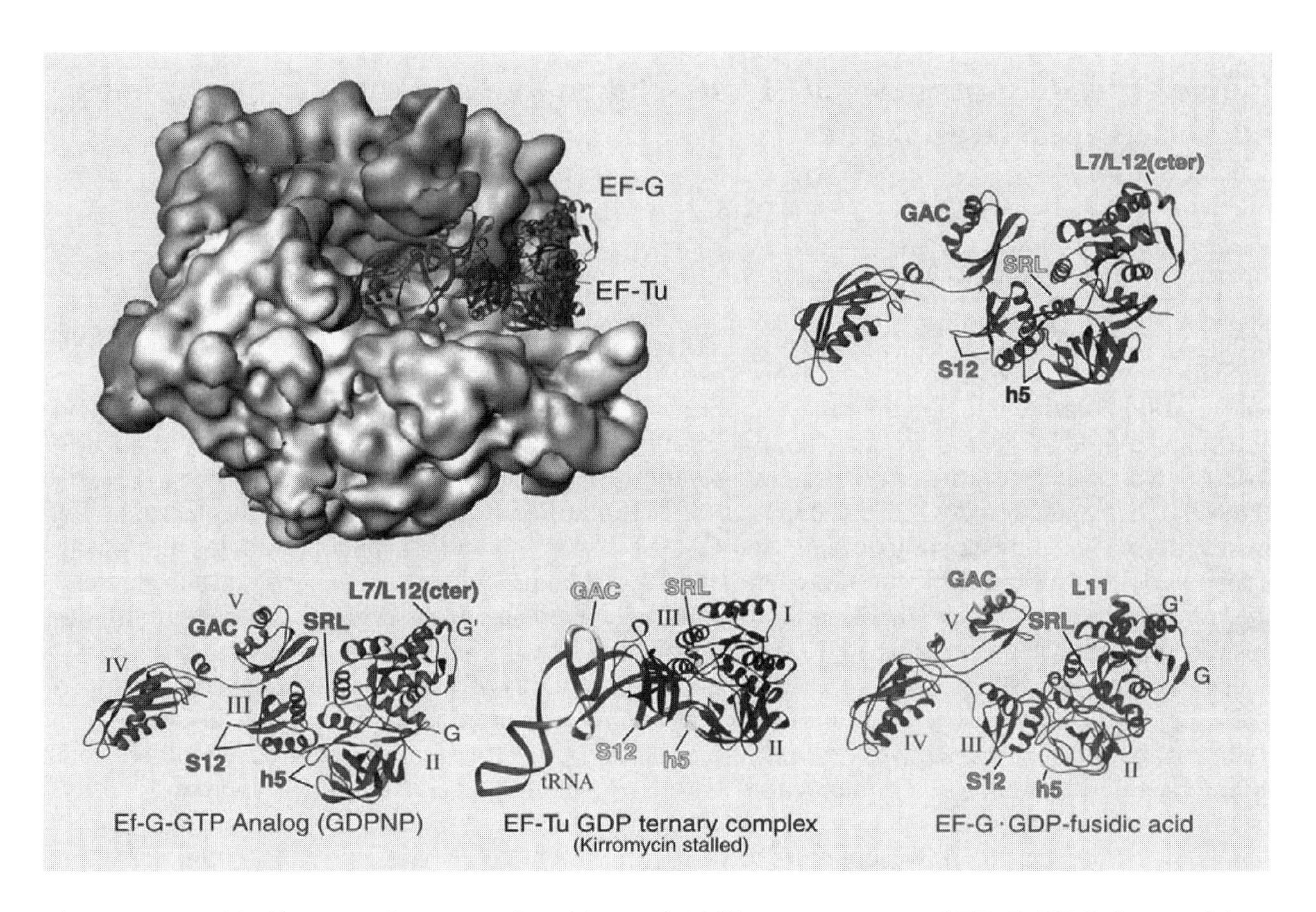

Fig. 1. Top: 70S ribosome from *E. coli*, with overlaid X-ray structures of EF-G (GDP form, with fusidic acid) and aa-tRNA·EF-Tu·GDP·kirromycin. Bottom: comparison of ternary complex with EF-G in both GDP and GTP forms. The comparison shows that the residues known to contact the ribosome at equivalent sites (indicated by labels) form different constellations for all three structures, implying that the ribosome *must* change its conformation to accommodate the binding of the factors in succession.

Microsc. Microanal. 8 (Suppl. 2), 2002
DOI 10.1017.S1431927602100043

Sindbis Virus Reconstruction at 11Å Resolution Reveals Details of Glycoprotein and Nucleocapsid Organization

W. Zhang, S. Mukhopadhyay, S. V. Pletnev, R. J. Kuhn, M. G. Rossmann and T. S. Baker

Department of Biological Sciences, Purdue University, West Lafayette, IN 47907-1392

Sindbis virus (SINV) is one of 26 members of the *alphavirus* genus (family *Togaviridae*), all of which are enveloped, positive strand RNA viruses.[1] The 52MDa SINV contains an 11.8kb genome, a host-derived lipid membrane, and three structural proteins: glycoproteins E1 and E2, and nucleocapsid protein (NCP). Because no high resolution crystallographic analysis of any alphavirus virion has yet been successful, structural investigations of alphaviruses have primarily been performed using electron cryo-microscopy (cryoEM) and three-dimensional image reconstruction of viruses [2, 3] and X-ray crystallographic analysis of the NCP [4] and E1 glycoprotein [5] along with atomic modeling of these molecules into cryoEM density maps [2, 5]. Based on such studies, *alphaviruses* have been shown to be spherical, multi-layered structures of ~720Å diameter with the capsid and glycoproteins organized with T=4 icosahedral quasi-symmetry. The viral genome is encapsidated by 240 NCPs, which extend to a maximum radius of ~220Å where they are enveloped by the viral lipid bilayer. The exterior of the virus contains 80 prominent, trimeric spikes, each of which consists of three E1-E2 heterodimers. E1 and E2 both have C-terminal transmembrane tails and the E2 tail forms one-to-one interactions with NCP which are believed to be important in virus assembly and stability.[6] Recent studies of SINV deglycosylation mutants have shown that *alphavirus* E1 and E2 proteins adopt markedly different orientations (E1, more tangential; E2, more radial) that are consistent with the known functions (E1, fusogenic; E2, cell attachment) of these glycoproteins.[7]

In this study, aimed at defining alphavirus components with improved clarity, we computed a three-dimensional reconstruction of SINV deglycosylation mutant E2-N318Q at 11Å resolution (FIG.1). Virus sample was prepared for cryoEM and images were recorded in a Philips CM200-FEG electron microscope at 38000 × nominal magnification.[8] Micrographs were digitized at a 7 μm step size and subsequently were bin-averaged to 14 μm pixels, corresponding to 3.68 Å intervals in the specimen. The final reconstruction was computed from 4931 particle images extracted from 27 micrographs recorded at defocus levels ranging between 1.1 and 2.1 μm.

The crystal structure of the Semliki Forest virus (SFV) E1 glycoprotein (~52% sequence identity with SINV E1) was fitted into the SINV 3D reconstruction making use of the known E1 glycosylation sites as constraints. A difference map, computed by subtracting the modeled E1 densities from the SINV reconstruction, provide clues about the location and structure of E2, which appears to have an extended structure quite similar to E1. The E1 and E2 glycoproteins wind around each other and form one petal of a trimeric spike. The SINV reconstruction clearly shows pairs of rod-like features with a left-handed twist that tranverse the bilayer. These features are attributed to the E1 and E2 transmembrane helices and were modeled as an α-helical coiled-coil (FIG.1C). The crystal structure of SINV NCP [4] fits nicely into the corresponding density of the SINV reconstruction (FIG.1C), but in an orientation quite distinct from that originally determined from a 25Å resolution reconstruction of Ross River virus [2, 9].

References:

[1] J. H. Strauss and E. G. Strauss, *Microbiol. Rev.* 58(1994) 491.
[2] R. H. Cheng et al., *Cell*, 80(1995) 621.
[3] E. J. Mancini et al., *Mol. Cell* 5(2000)255.

[4] H. K. Choi et al., *Nature* 354(1991)37.
[5] J. Lescar et al., *Cell* 105(2001)137.
[6] T. L. Tellinghuisen et al., *Genetic Engineering* 23(2001)83.
[7] S. V. Pletnev et al., *Cell* 105(2001)127.
[8] T. S. Baker et al., *Microbiol. Molec. Biol. Reviews* 63(1999)862.
[9] We thank R. Ashmore, P. Chipman, and C. Xiao for help and discus sions. Work supported in part by the NI H Program Project Grant to MGR and TSB (AI45976), the NIH grant to RJK (GM56279), a shared equipment grant from NSF to TSB (BIR 9112921), and an instrumentation reinvestment grant from Purdue University to the Purdue Structural Biology faculty.

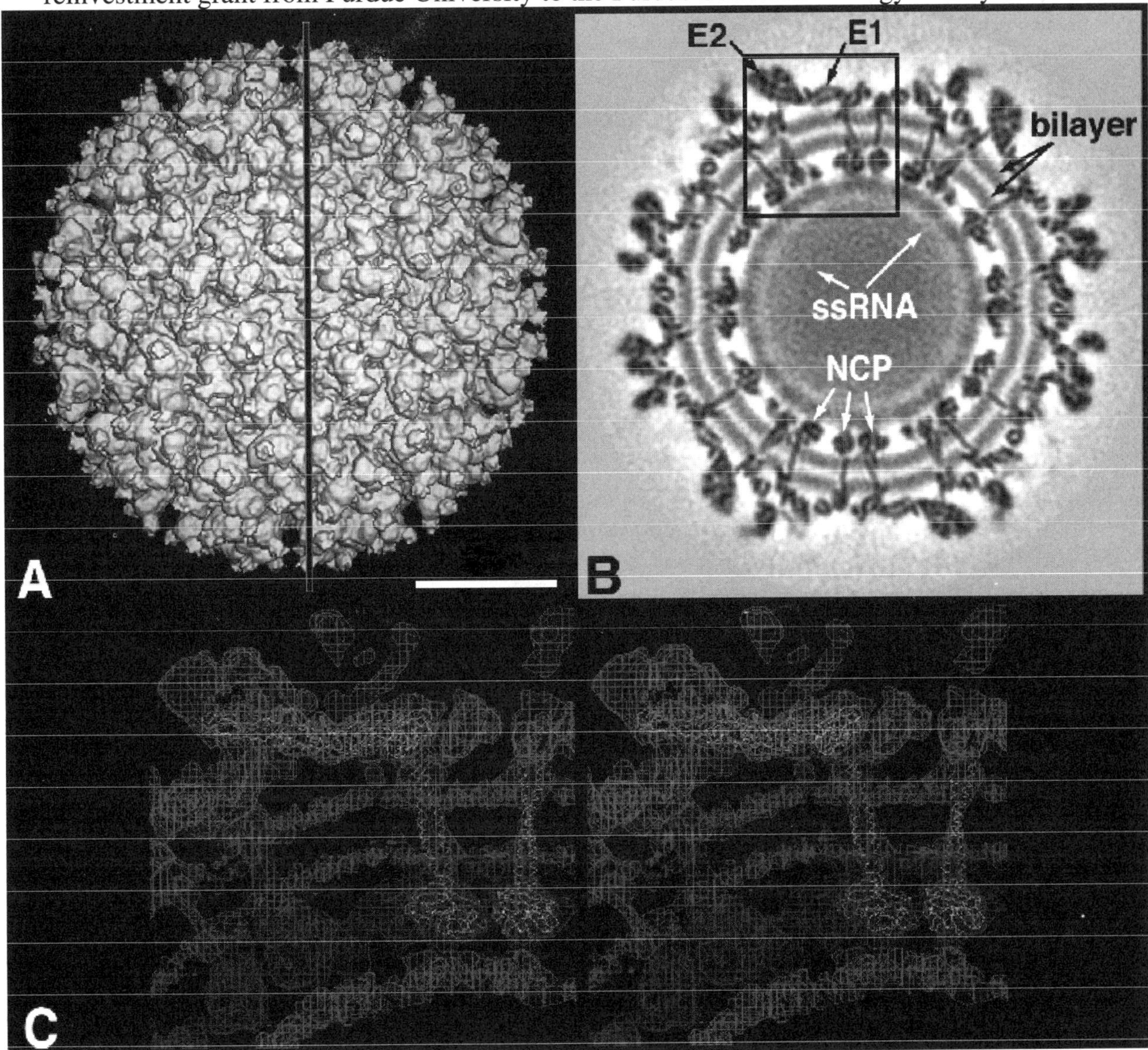

FIG.1 A. Surface-shaded representation of the Sindbis virus N318Q mutant viewed along a 2-fold symmetry axis. B. Cross-section from reconstructed density map (dark line in A) depicts the multilayer organization of SINV (highest density features are darkest). C. Magnified, stereoscopic view of the region outlined by the black square in (B). C_α backbone models of the X-ray structures of SFV E1, general control protein, GCN4, coiled-coil helices, and SINV NCP were fitted into the reconstruction (grey contours). Scale bar = 200Å for (A,B).

Microsc. Microanal. 8 (Suppl. 2), 2002
DOI 10.1017.S1431927602100316

Cryo-EM and X-ray crystallographic studies on the monomeric kinesin motor KIF1A

Masahide Kikkawa*,**, Elena P. Sablin***, Yasushi Okada**, Hiroaki Yajima**, Robert J.Fletterick*** & Nobutaka Hirokawa**

* University of Texas, Southwestern Medical Center, Dept. Cell Biology, Dallas, TX, 75390-9039 (current address)
** University of Tokyo, Dept. Cell Biology and Anatomy, Graduate School of Medicine, 7-3-1 Hongo Bunkyo-ku, Tokyo 113-0033, Japan
*** Department of Biochemistry/Biophysics, University of California, San Francisco, CA 94143-0448

Kinesin motors are specialized enzymes that use hydrolysis of ATP to generate force and movement along their cellular tracks, the microtubules. Although numerous biochemical and biophysical studies have accumulated much data that link microtubule assisted ATP hydrolysis to kinesin motion, the structural view of kinesin movement has remained unclear.

This study of the monomeric kinesin motor KIF1A combines X-ray crystallography and cryo-electron microscopy, and allows analysis of force-generating conformational changes at atomic resolution.

By using X-ray crystallography, the motor is revealed in its two functionally critical states: complexed with ADP and with a non-hydrolysable analog of ATP [FIG 3]. The conformational change observed between the ADP-bound and the ATP-like structures of the KIF1A catalytic core is modular, extends to all kinesins and is similar to the conformational change used by myosin motors and G proteins.

In order to understand the interaction between the motor and the microtubule, the structures of KIF1A-microtubule complex were studied by high resolution cryo-EM and helical image analysis [FIG 1,2]. Docking of the ADP-bound and ATP-like crystallographic models of KIF1A into the corresponding cryo-electron microscopy maps revealed detailed picture of the interaction between the motor and microtubule and suggests a rationale for the plus-end directional bias motion associated with the kinesin catalytic core [FIG 3].

References

[1] Kikkawa, M., et al. *Nature*, 411 (2001) 439-445
[2] Kikkawa, M., et. al *Cell*, 100 (2000) 241-252
[3] This research was supported by a Center of Excellence Grant-in-aid from the Ministry of Education, Science, Sports, and Culture of Japan

Microsc. Microanal. 8 (Suppl. 2), 2002
DOI 10.1017.S1431927602100328

Electron Cryo-microscopy and Image Reconstruction of Adeno-Associated Virus Type 2 empty Capsids

B. Böttcher[*], S. Kronenberg[**] and J. Kleinschmidt[**]

[*]EMBL, Meyerhofstrasse 1, D-69117 Heidelberg, Germany
[**]DKFZ, Im Neuenheimerfeld 242, D-69120 Heidelberg, Germany

Adeno-associated virus type 2 (AAV-2) is a human parvovirus. The virus capsid is composed of three different structural proteins, VP1, VP2 and VP3 which have molecular masses of 87 kDa, 72 kDa and 62 kDa respectively and differ in their N-terminal amino-acid sequences. The likely molar ratio of VP1, VP2 and VP3 is 1:1:8. In total sixty copies of these structural proteins occupy symmetrically equivalent positions in an icosahedrally arranged protein shell.
We have investigated empty capsids of AAV-2 by electron cryo-microscopy and icosahedral image reconstruction[1]. The three-dimensional map at 10.5 Å resolution showed sets of three elongated spikes surrounding the three-fold symmetry axes and narrow empty channels at the five-fold axes (Figure 1A, 1B). The inside of the capsid superimposed with the previously determined structure of the canine parvovirus[2], whereas the outer surface showed clear discrepancies. Globular structures at the inner surface of the capsid at the two-fold symmetry axes were identified as possible positions for the N-terminal extensions of VP1 and VP2 (Figure 1B, arrows).
The data was collected on a Philips CM 120 Biotwin at 100 kV. This microscope is designed for high contrast at low spatial frequencies by using a lens with a high C_s value (C_s=6.4 mm). The payoff of this design together with the use of a standard LaB_6-filament was a loss in high resolution information, which depended on the defocus and the illumination conditions used for taking a particular micrograph. The dampening of higher spatial frequencies is described by an envelope function K_s which modulates the contrast-transfer-function ctf (Figure 1C). K_s is different for each micrograph taken at a different defocus or under different illumination conditions. To allow for these variations of K_s, K_s was included in the previously described ctf-correction[3] of the observed values of a particular frequency component F. For evaluating the altered ctf-correction-algorithm, the Fourier-shell-correlation between two maps calculated from half of the particle images each was determined. Figure 1D shows the Fourier-shell-correlation after correction for the ctf including the calculated K_s for each micrograph in trace 1 and after correction for the ctf neglecting K_s (by setting it to one) in trace 2. The Fourier-shell-correlation values were clearly higher when K_s was included in the correction-algorithm, highlighting the importance of the K_s-correction.
The overall dampening of the high resolution information, caused by the lack of spatial coherence of the microscope was evident in the fall off of $\sum Ctf_i^2 w_i$ with increasing spatial frequencies (trace 3). The value of $\sum Ctf_i^2 w_i$ is a direct measure for expected information transfer of the microscope for the average of the observed data and can range from zero for no transfer up to one for transfer without loss due to the microscope settings. A considerable loss in the transfer of information was already apparent at frequencies of 1/25 $Å^{-1}$. However, this drop was not reflected in the fall-off of the Fourier-shell-correlation (trace 1), illustrating that information can be recovered even though it is only weakly transferred.

References:

[1] Kronenberg, S., Kleinschmidt, J.A. and Böttcher, B. (2001) EMBO Rep 2, 997-1002.

[2] Xie, Q. and Chapman, M.S. (1996) J Mol Biol 264, 497-520.
[3] Böttcher, B. and Crowther, R.A. (1996) Structure 4, 387-394.
[4] Reimer, L. (1993) Transmission Electron Microscopy. Berlin, Heidelberg, Springer Verlag.

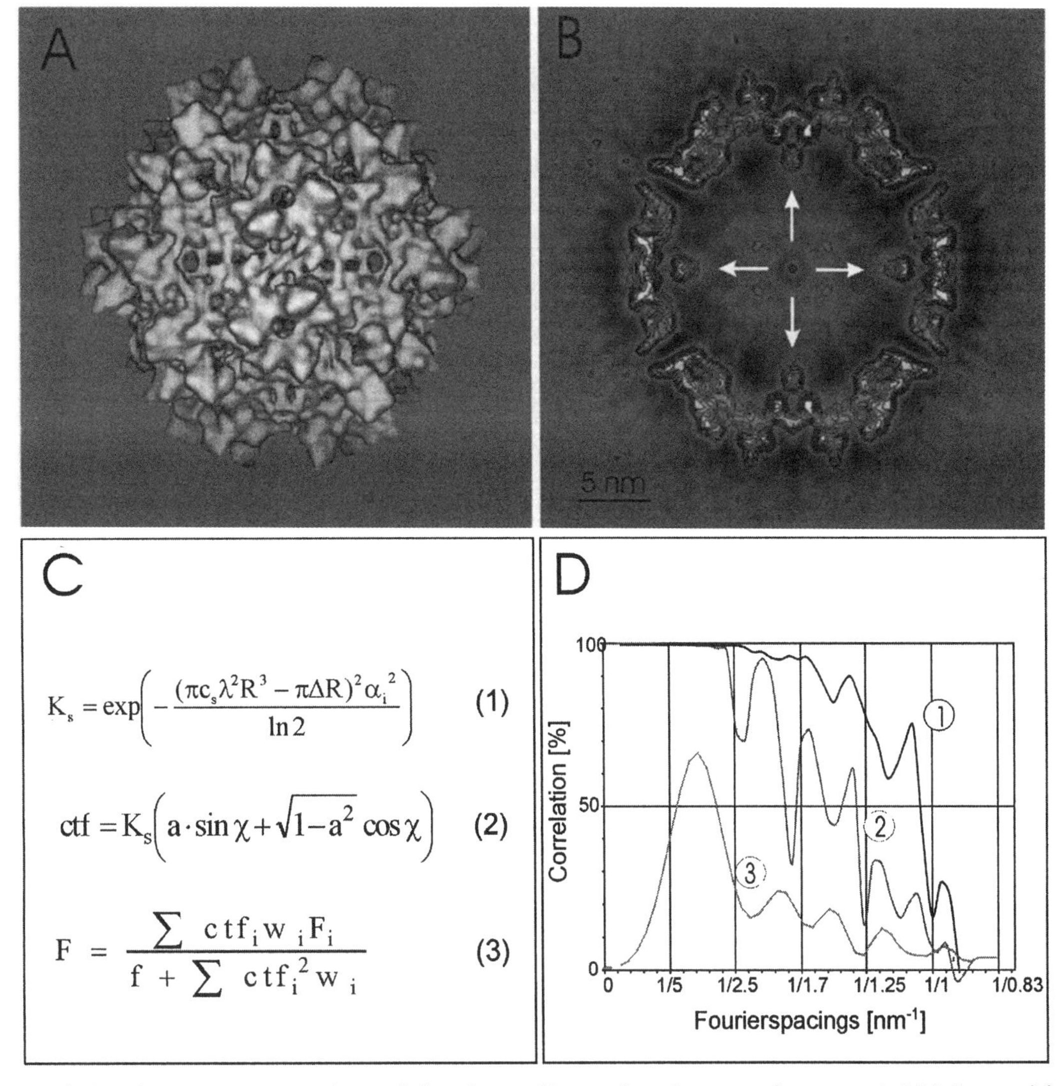

Figure 1: A) Surface representation of the three-dimensional map of empty AAV-2 capsids at 10.5 Å resolution. B) Equatorial slice of the map. C) 1) Envelope function K_s caused by the spatial coherence of the electron source[4] (C_s spherical aberration coefficient, λ wavelength of electrons, R spatial frequency, Δ defocus, α illumination half angle). 2) contrast transfer function *ctf* modulated by the envelope function K_s (χ phase-shift; a fractional amplitude contrast). 3) Combination of the observed Frequency components F_i in i reconstructions of particle images from i different micrographs to a *ctf* corrected Frequency component F (w fractional weight of each map, f factor akin to signal to noise ratio in a Wiener filter)[3]. D) Fourier-Shell-Correlation of the ctf-corrected data using the calculated value for K_s (trace 1) and with K_s set to 1 (trace 2). Trace 3 shows $\sum Ctf_i^2 w_i$, a direct measure for the expected information transfer of the microscope for the average of the observed data.

Microsc. Microanal. 8 (Suppl. 2), 2002
DOI 10.1017.S143192760210033X

Molecular Structure of an Icosahedral Pyruvate Dehydrogenase Complex

Sriram Subramaniam*, Dan Shi*, Richard N. Perham**, and Jacqueline L. S. Milne***,

* Laboratory of Biochemistry, National Cancer Institute, NIH, Bethesda, MD USA 20892
** University of Cambridge, Cambridge CB2 1GA, UK.
***Laboratory of Cell Biology, National Cancer Institute, NIH, Bethesda, MD USA 20892

Complex cellular processes such as signal transduction, gene expression, motility and energy metabolism are often implemented using multi-component molecular assemblies. Knowledge of the architecture of these molecular machines is central to an understanding of their mechanisms. Some large complexes such as the nucleosome, the ribosome and the proteasome have been successfully analyzed by X-ray crystallographic techniques. Structures of other assemblies, such as the family of 2-oxo acid dehydrogenase multienzyme complexes, have proved impossible to obtain by crystallographic methods, in part because of their large size but also because of the intrinsic structural heterogeneity introduced by the mobile domains and "swinging arms" that are essential to transfer catalytic intermediates between the active sites of their constituent enzymes.

Like the PDH complexes of eukaryotes and of other Gram-positive bacteria, the *B. stearothermophilus* PDH is assembled around a core of 60 dihydrolipoyl acetyltransferase (E2) chains arranged with icosahedral symmetry. Each E2 chain consists of three domains: (i) an N-terminal 9 kDa lipoyl domain, which visits the active sites of the pyruvate decarboxylase (E1) and then those of E2 and dihydrolipoyl dehydrogenase (E3), (ii) a 4 kDa peripheral subunit-binding domain (PSBD) to which E1 and E3 bind tightly and mutually exclusively, and (iii) a C-terminal 28 kDa catalytic (acetyltransferase) domain, 60 copies of assemble to form the icosahedral inner core. These domains are linked by stretches of extended, conformationally flexible polypeptide chain.

We have recently arrived at three-dimensional models for (i) the 1.8 megadalton icosahedral complex composed of sixty copies of the E2 catalytic domain (at a resolution of 14.5 Å), and (ii) an 11 megadalton, icosahedral PDH complex composed of sixty copies each of E1 and E2 (at a resolution of 28 Å). Atomic interpretations of each of these complexes were obtained by combining the electron microscopy-derived structures with previously determined atomic coordinates of the individual components of the complexes. Analysis of the models provides a number of novel insights into the design and function of this molecular machine. A key feature is that the E1 molecules are located on the periphery in an orientation that allows each of the 60 mobile lipoyl domains tethered to the inner E2 enzyme to access multiple E1 active sites from inside the icosahedral complex. This unanticipated architecture provides a highly efficient mechanism for active site coupling and catalytic rate enhancement, which we propose is achieved by the motion of the lipoyl domain in the restricted annular region between the inner and outer cores of the complex.

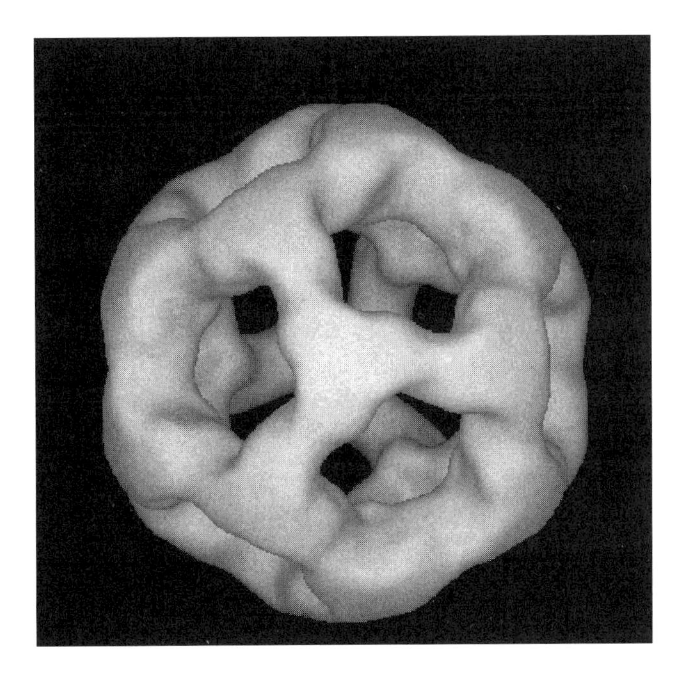

Fig. 1. Surface representation of the icosahedral complex formed by the E2 catalytic domain. The complex has a diameter of about 225 Å.

QuickTime™ and a
SGI decompressor
are needed to see this picture.

Fig. 2. A cross-sectional view of the E1E2 complex displaying the probable locations of the three domains of the E2 chain and the E1 tetramer. The complex has a diameter of about 475 Å. The E1 tetramers and the E2 binding domain (in red) are on the periphery, the E2 catalytic domain is at the core, and E2 lipoyl domain is in the annular region.

Microsc. Microanal. 8 (Suppl. 2), 2002
DOI 10.1017.S1431927602100341

Merging Focal Pairs for Improved Particle Selection and Orientation Determination

Steven J. Ludtke and Wah Chiu

* - National Center for Macromolecular Imaging, Verna and Marrs McLean Department of Biochemistry and Molecular Biology, Baylor College of Medicine, Houston, TX 77030

In single particle electron cryomicroscopy, radiation damage is virtually always the factor limiting the resolution of the final reconstruction. When attempting to achieve resolutions close to the limits of the available instrument, both dose and defocus must be carefully controlled to insure both that the specimen is not damaged at the desired resolution and that the envelope function of the microscope does not fall off too rapidly. Even on microscopes equipped with field emission guns, very far from focus micrographs do not extend to as high a resolution as those close to focus. However, the low overall contrast of close to focus micrographs often prevents accurate particle orientation determination, and in some cases even locating the particle within the micrograph or CCD frame becomes impossible.

Historically a popular approach for dealing with this issue is to collect a focal pair. The initial, low damage, micrograph is collected sufficiently close to focus as not to compromise resolution. A second micrograph of the identical field is then collected much further from focus. Generally this second micrograph is used to locate particles, and in some cases, particularly icosahedral particles, used to determine initial estimates of the 3D particle orientation.

We present the next logical step in the evolution of this technique. For projects pushing the frontiers of resolution, it is still often desirable to collect such focal pairs, since they allow recording of images with the best possible high-resolution information, and still obtaining some amount of high contrast low-resolution information. Rather than treating the micrographs separately or performing a simple average of the micrographs (which has been proposed previously), we apply the robust CTF correction procedure used by EMAN [1,2] to the process of combining the two images optimally. The combined image then replaces the far from focus image in standard reconstruction techniques.

The merging process occurs in three phases. First, the second micrograph is aligned to the first using a routine which can rapidly align two very large (sometimes over 12,000x10,000 pixels in size) images to 1 pixel accuracy. Second, both micrographs are analyzed to determine CTF, envelope function and Noise paramaters within EMAN's standard model. Finally, the images are combined with CTF correction and optional Wiener filtration.

When Wiener filtration is applied as part of the merging process, the resulting image has optimized contrast. Such images are useful for eliminating false positives often generated by single particle autoboxing techniques. The individual particles can be distinguished more easily, and in many cases, it is possible to distinguish damaged or otherwise unsuitable particles.

For orientation determination, Wiener filtration is not used, since it suppresses much of the high resolution information required for accurate orientations. While the incorporation of the damaged second image may be unsuitable for use directly in high-resolution studies, it can, at least, provide initial values for particle orientation, which can then be refined using image data from the first micrograph alone. The merging process allows one to take advantage of both micrographs rather than just using the low resolution information from the second microgrpah.

This technique is available in the standard EMAN (http://ncmi.bcm.tmc.edu/~stevel/EMAN/doc) distribution. Future work will include more automation of the merging process, and additional testing of the reliability of the orientations determined from the merged particles.

[1] S.J. Ludtke, P.R. Baldwin, and W. Chiu, "EMAN: Semiautomated Software for High-Resolution Single-Particle Reconstructions," J. Struct. Biol. 128: 82-97 (1999).
[2] S.J. Ludtke, J. Jakana, J. Song, D.T. Chuang and W. Chiu. "An 11.5 Å Single Particle Reconstruction of GroEL Using EMAN". J. Mol. Biol., 314:241-250 (2001).
[3] This research is supported by NCRR (RR02250), NSF through NPACI, and the Agouron Institute

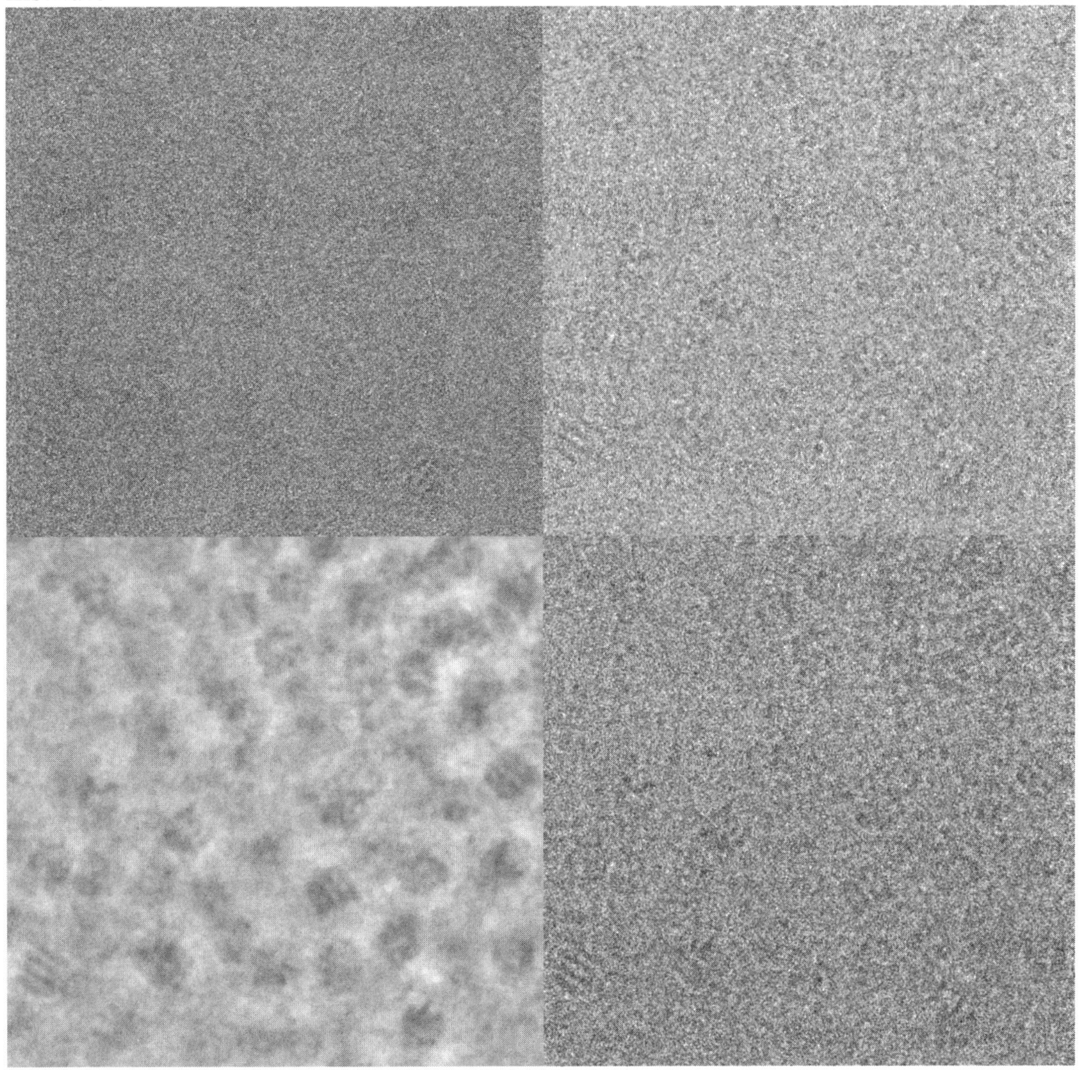

Figure 1- An example of focal pair merging. All four frames represent the same field from a pair of micrographs. In the upper left and right are the original close and far from focus images. In the lower left is the merged image with Wiener filter and in the lower right is the merged image without Wiener filter.

Microsc. Microanal. 8 (Suppl. 2), 2002
DOI 10.1017.S1431927602100353

Challenges in the Automation of Cryo-microscopy of Macromolecular Structure

C.S. Potter, D. Fellmann, R. Milligan, J. Pulokas, C. Suloway, Y. Zhu and B. Carragher

Department of Cell Biology, The Scripps Research Institute, La Jolla, CA, 92037

Although the methodology of molecular microscopy has enormous potential, it is time consuming and labor intensive. The techniques required to produce a three dimensional (3D) electron density map of a macromolecular structure normally require manual operation of the electron microscope by a skilled operator and manual supervision of the sometimes complex software needed for analysis and calculation of 3D maps. Typically it will take weeks to months to collect and analyze a dataset in order to reconstruct a map at 10-20Å resolution. It is generally agreed that in order to increase the resolution to a range where secondary structure will be discernable in the map (~7Å) will require an order of magnitude increase in the amount of data collected and analyzed. It is clear that this will only be practical as a mainstream technique if we can automate the imaging and analysis processes and greatly improve the overall throughput.

We will report on our system, called Leginon [1], which provides a high level of automation by locating a specimen on the grid using a multi-scale image analysis strategy. During a 24 hour experiment, Leginon can typically collect 500-1000 image defocus pairs without the necessity of operator intervention. Further automation is achieved by integrating Leginon with automated filament selection algorithms and a helical reconstruction software package so that a map is computed in parallel with data collection. We have shown the feasibility of this approach by automatically calculating an electron density map of tobacco mosaic virus to a resolution of ~10Å within 20 hours of inserting the grid in the microscope. The map is shown in figure 1. To produce the map, 167 grid squares were analyzed and 411 high magnification image pairs were collected. From these images, 725 TMV filaments were extracted and 100 of these were used in the final map. We estimate that approximately 130,000 subunits (particles) from the 100 filaments were used in constructing the average. We propose that the throughput of a data collection system be defined by (i) the *rate* at which data can be collected (e.g. particles/hour), (ii) *yield* - the fraction of the data that is useful and (iii) *sustainability* - a measure of the duration over which data can be continuously acquired. For our TMV experiment, data was acquired at a rate of approximately 48,000 particles/hour, with an overall yield of 14%, sustained through 20 hours of continuous operation. The total throughput was 130,000 particles for the entire experiment. Since minimal operator intervention is required for collecting the data and creating the map, we could feasibly perform 3-5 such experiments/week or collect approximately 0.5 million particles/week.

One of the principal factors that currently limits data throughput is the geometry and resolution of the CCD cameras that are available. We collect data using a 2Kx2Kx12bit CCD camera that has a bin size of 24x24μm^2. At the voltages that we typically use to collect images (120kV) the camera performs satisfactorily at about 1/3 of the Nyquist frequency, i.e. if the pixel size is 2Å at the specimen, we are able to collect data to a resolution of about 6Å. To achieve this resolution we need to work at a magnification of about 88,000x. In contrast a piece of film can be digitized to yield approximately 7Kx7K pixels at an equivalent resolution. Thus there is a factor of ~12 between the effective collection area of film vs. the 2Kx2K CCD camera when acquiring a single image. Since CCD cameras in cryoEM have been mostly used for manual data collection, this factor of 12 has clearly had a major dampening effect on the enthusiasm of the community for CCD's and they have consequently have not been generally adopted for data collection in spite of the advantage they offer in negating the need to develop and scan films.

If the TMV experiment we describe above had been performed using film and traditional manual data collection methods, an experienced microscopist could have collected perhaps 3 cassettes of film (~150 micrographs) in a reasonable day's work (~8 hours). Given the factor of 12 in effective collection area for film over CCD's, the corresponding data collection rate would be ~576,000 particles/hour. If we assume the same overall yield of 14%, the total throughput for the 8 hour data collection is on the order of 650,000 particles/experiment. However at the end of the data collection session, the films must be developed and scanned and then processed further. Given the time consuming nature of these tasks it might be reasonable to expect that one such experiment of this type could be performed each week on a sustained basis by a single investigator, which would make the overall throughput of manual and automated data collection very comparable (~0.5 million particles/week).

Our challenge is to substantially improve both the sustained throughput and the yield of automated data collection. We will report on the current performance and status of the Leginon system.

[1] Carragher, B., et al. (2000) Journal of Structural Biology, 132, 33-45.
[2] This work was supported by the National Science Foundation (DBI-0296063) and the National Institutes of Health (GM61939).

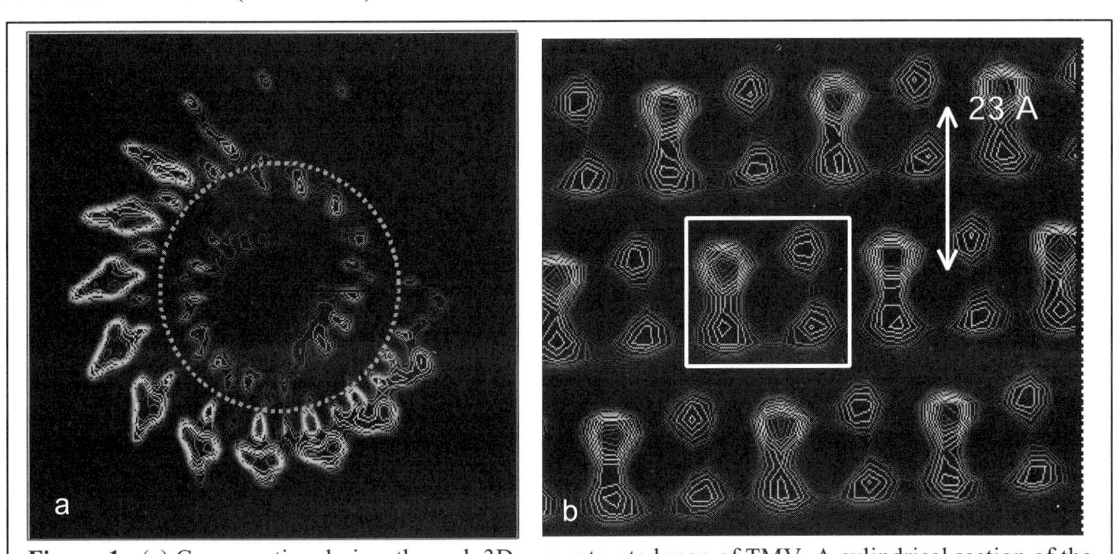

Figure 1: (a) Cross sectional view through 3D reconstructed map of TMV. A cylindrical section of the map, is shown in (b). The area within the box is consistent with an interpretation that this structure represents a group of 4 alpha helices that are expected to be in this position and orientation from an analysis of the known x-ray structure.

DOI 10.1017.S143192760210016X

THE MICROVASCULATURE OF THE BRAIN OF THE STERLET, *ACIPENSER RUTHENUS L.* A SCANNING ELECTRON MICROSCOPE (SEM) STUDY OF VASCULAR CORROSION CASTS

M. Klein,* B. Stöttinger,* B. Minnich,* W.D. Krautgartner,** and A. Lametschwandtner*

University of Salzburg, * Department of Vascular- & Performance Biology, ** Department of Light-, Electron Microscopy & Digital Image Acquisition,
A-5020 Salzburg, Austria

Introduction

The brain cannot store oxygen and glucose and thus depends on a continuous blood supply via the cerebral circulation. While the neuroanatomy of the vertebrate brain is well documented [1] its angioarchitecture is studied to a much lesser extent. Existing knowledge on the brain microvascularization is mainly derived from India-ink injected, cleared and sectioned brains and a few SEM studies of microvascular corrosion casts [for a bibliography see 2]. In bony fish the vascularization of the pineal gland [3] and in cartilaginous fish those of the rhombencephalic choroid plexus [4] and the hypothalamo-hypophyseal area [5] only are studied by the later technique. The aim of the present study is to describe gross arterial supply, venous drainage and the capillary bed of the ganoid brain.

Materials and Methods

Vascular corrosion casts of eleven brains of the sterlet, *Acipenser ruthenus L* (total lengths: 10-46 cm) were studied by scanning electron microscopy (SEM). Briefly, fish were heparinized by an intraperitoneal injection of 5000 I.E heparin and anesthetized after 10 minutes by immersion in MS 222 (0.5%; Sandoz, Basle, Switzerland). Animals then were decapitated, the lower jaw was removed by a horizontal section at the midlevel of the gills and a glass cannula was introduced via the first efferent branchial artery into the ipsilateral internal carotid which arises from the rostral portion of the lateral dorsal aorta. After fixing the cannula by a fine ligature and after tieing off the contralateral dorsal aorta between first und second afferent gill arteries the circulatory system of the brain was rinsed with Amphibian Ringer solution and 5 ml Mercox-Cl-2B diluted with monomeric methylmethacrylate (4+1; v+v) were injected manually. After hardening of the injected resin specimens were processed as described elsewhere [2].

Results and Discussion

Inspection of casts with low magnification revealed great individual variations in course, dimension and appearance of cerebral vessels, especially in olfactory arteries and basilar prosencephalic arteries (Figs. 1, 2). In contrary to former reports [6] a single branch of the dorsal mesencephalic artery, which we named „central cerebellar artery" was found to supply the central portion of the cerebellum. Some individuals had two arterial circles at the base of the brain, some one (Figs. 3,4) or even no arterial circle. The very small calibre of the anastomoses connecting posterior cerebral arteries (Fig. 3) might have been overlooked in light microscopical studies. SEM of microvascular corrosion casts clearly allows to demonstrate even smallest anastomoses. The venous drainage was found to be by the unpaired main choroidal vein, the course of which can vary greatly, and by paired middle and posterior cerebral veins, which had a rather constant course.

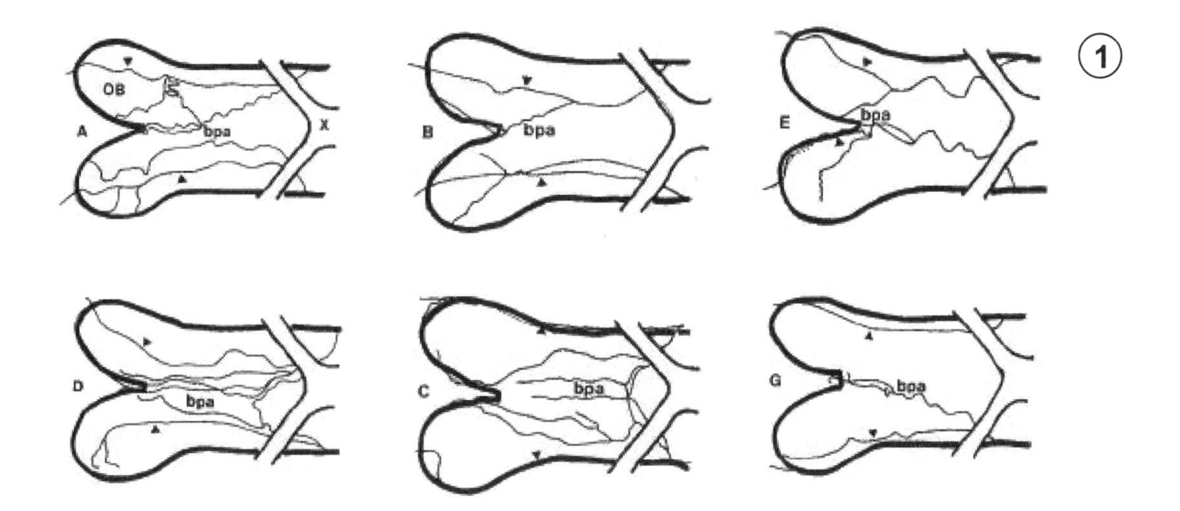

FIG. 1: Six variations of the course of the olfactory arteries (solid triangle) and the basilar prosencephalic arteries (bpa) as seen in vascular corrosion casts. OB olfactory bulb, X optic chiasma. Rostral is to the left.

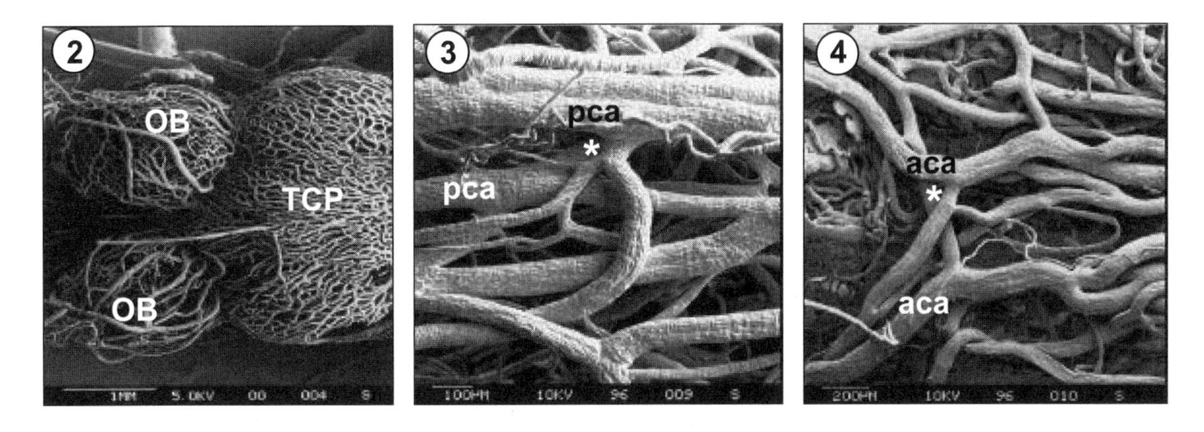

FIG. 2. Dorsal view at the olfactory bulbs (OB) and telencephalic choroid plexus (TCP).
FIG. 3. Anastomosis (*) between posterior cerebral arteries (pca) closing the posterior arterial circle of Willis.
FIG. 4. Anastomosis (*) between anterior cerebral arteries (aca) closing the anterior arterial circle of Willis.

References

[1] Nieuwenhuys, R., Donkelaar, H.J. ten, Nicholson, C.: The Central Nervous System of Vertebrates. Springer, Berlin – Heidelberg – New York. 1998., 2219 pp.
[2] Aharinejad, S. and Lametschwandtner, A.: Microvascular corrosion casting in scanning electron microscopy. Springer-Verlag, Wien 1992. 380 pp.
[3] S. Syed Ali et al., Cell and Tissue Research, 250 (1987) 425-429.
[4] T. Weiger et al., American Journal of Anatomy, 182 (1988) 33-41.
[5] K. Kotrschal et al., Journal fürHirnforschung, 26 (1985) 333-351.
[6] H. Splechtna, Gegenbaurs Morphologische Jahrbücher, 119 (1973) 401 – 421.

DOI 10.1017.S1431927602100171

Development of Subretinal Venous Sphincters in Spontaneously Hypertensive Rats (SHR)

S. Aharinejad*, U. Firbas

*Laboratory for Cardiovascular Research, Department of Anatomy, University of Vienna, Waehringerstr. 13, A-1090 Vienna, Austria

Venous sphincters are tufts of smooth muscle that have been first described in the pancreatic and pulmonary microvascular bed of rats [1,2]. Venous sphincters are capable of contraction and depending on their tone, they can add a substantial resistance to the vascular pressure [1,2]. SHR are a strain bred of Wistar Kyoto (WKY) rats that spontaneously develop systemic hypertension [3]. In this strain, the pulmonary venous sphincters can contract deeply and narrow the luminal diameter of veins up to 50% [4]. Existing reports indicate that SHR show hypertensive ocular fundus changes [5]. Other reports suggest that with ongoing systemic hypertension in SHR retinal vessels become tortuous, narrowed, and show localized constrictions [6]. However, it remains unknown whether venous sphincters are present in the ocular vasculature of SHR. We hypothesized that SHR might develop venous sphincters in their eyes and that the depth of sphincters´ contraction might correlate with systemic blood pressure.

To examine our hypothesis, we used 16-18 week-old male SHR and age-matched WKY rats (Charles River, Germany). SHRs were untreated or treated with carvedilol, a combined α and β receptor antagonist for 4 weeks. The systemic blood pressure was measured via a catheter in the abdominal aorta and recorded online using a telemetry system (Data Sciences International, Rochester, MN). The ocular vasculature was cast using Mercox CL-2B (Ladd Research, Williston, VT). For histology, the eyes were perfusion-fixed with paraformaldehyde. Cast specimens were prepared for scanning electron microscopy (SEM) as described earlier [1,4]. Histological specimens were sectioned serially and stained with H & E.

In WKY rats the mean systemic blood pressure was 96 ± 7; in untreated SHR it was 148 ± 15; in carvedilol-treated SHR it was 131 ± 10 ($P < 0.001$ vs. untreated). SEM of the posterior eye pole in untreated SHR showed multiple focal constrictions on the surface of cast veins, narrowing the luminal diameter up to 40% (FIG. 1). In tissue sections, muscular tufts in the venous walls were readily identified (FIG. 2), corresponding to the venous sphincters [4]. These venous sphincters were located in the subretinal vascular network. In carvedilol-treated SHR only shallow constrictions were found on the surface of subretinal venous casts. The luminal surfaces of corresponding veins in WKY rats were smoothly outlined; venous sphincters were not found. This study suggests that venous sphincters are related and develop reactive to elevated systemic blood pressure in the ocular vasculature of SHRs. Venous sphincters might be of significance in regulating blood flow in the ocular vascular system.

References

[1] S. Aharinejad et al., Scanning 12 (1990) 280.
[2] D.E. Schraufnagel et al., Am. Rev. Respir. Dis. 141 (1990) 721.

[3] K. Okamoto et al., Clin. Sci. Mol. Med. 45 (Suppl. 1) (1973) 11.
[4] S. Aharinejad et al., Am. J. Pathol. 148 (1996) 281.
[5] Y. Hamada et al., Clin. Exp. Pharmacol. Physiol. 22 (Suppl.) (1995) S132.
[6] I.A. Bhutto, T. Amemiya, Ophthalmic Res. 29 (1997) 12.

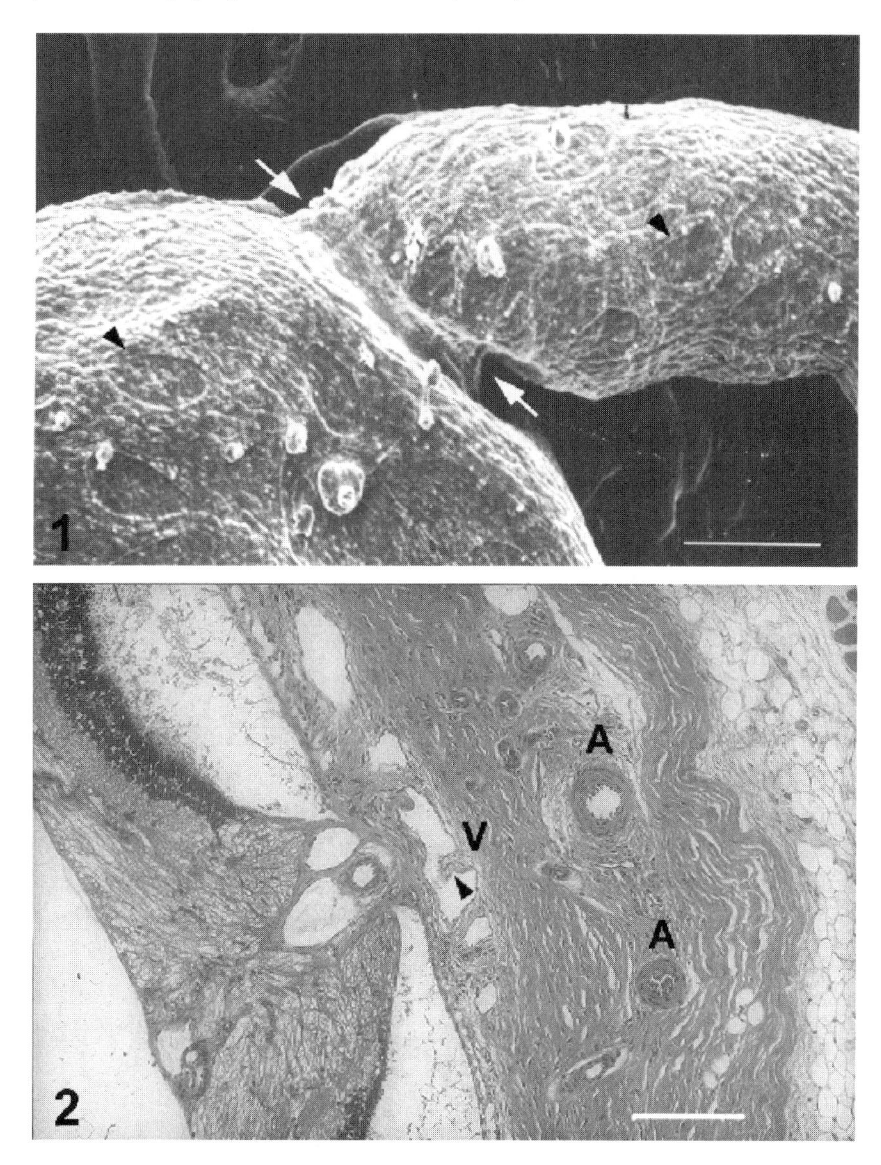

FIG. 1. SEM of a cast vein in the subretinal vascular network of an untreated SHR. White arrows mark a deeply contracted venous sphincter. Arrowheads mark characteristic endothelial cell nuclear imprints. Bar = 25 μm.

FIG. 2. Light microscopy image of the ocular posterior pole of an untreated SHR. A smooth muscle sphincter (arrow head) bulges into the lumen of a longitudinally sectioned vein (V) in the subretinal vascular network. Two arteries (A) are seen close to the vein; note the hypertrophy of the media in the lower artery. Bar = 100 μm.

Microsc. Microanal. 8 (Suppl. 2), 2002
DOI 10.1017.S1431927602100183

Microvasculature Of The Urinary Bladder Of The Dog Studied With Light Microscopy, Electron Microscopy And Vascular Corrosion Casts

C.W. Ridner*, R.L. Kao*, and F.E. Hossler**

East Tennessee State University, J.H. Quillen College of Medicine, Depts. of Surgery* and Anatomy & Cell Biology**, Johnson City, TN, 37614 USA.

The urinary bladder functions to store and expel urine, and the bladder wall acts as an excellent barrier to the passage of water and electrolytes. These functions require the delivery of nutrients via a rich blood supply to the bladder wall. Yet, ischemia may result from chronic distension or from interruption of the blood flow by inflammation due to injury or neoplasm [1]. Sustained ischemia often compromises the barrier function of the wall [2] and could lead to the loss of the urothelium and even necrosis [3]. Fortunately, complete recovery of the urothelial lining can ensue after relief from the ischemic insult, due partially to the rich blood supply [4]. The gross blood supply to the bladder is well known, but despite its importance, few investigations have been directed towards understanding the three dimensional anatomy of the microcirculation of this organ. In a previous study of the rabbit bladder, we described special features of the microcirculation which might enhance blood flow, despite the normal expansion and contraction of this organ [5]. In the present study we describe the microcirculatory anatomy of the dog bladder using light microscopy, electron microscopy and vascular corrosion casting.

Adult male dogs were anesthetized (pentobarbital and halothane) and anticoagulated (heparin, 500 U/kg). The abdominal aorta was exposed and cannulated, the abdominal vena cava was opened, the external iliac arteries as well as the internal pudendal and parietal iliac branches of the internal iliac artery were ligated, and the remaining caudal vasculature was flushed of blood using warm saline. In some cases the bladder volume was adjusted though a range of about 50 to 200 ml. The caudal vasculature was then perfused with buffered aldehyde fixative for microscopy or with Mercox resin for vascular corrosion casts. Fixed tissue was embedded in Epon-Araldyte resin for thin and thick sections. Casts were prepared by macerating the tissue in 5% KOH and warm water, and cleaned with formic acid and distilled water. Casts were dried by lyophilization, mounted on stubs with colloidal carbon, sputter coated with gold, and viewed by routine scanning electron microscopy. In some cases blood flow was observed by injecting 1% India ink.

The dorsal wall of the bladder is supplied by the cranial vesicular branches of the umbilical arteries. The basal and lateral walls are supplied by caudal vesicular branches of the urogenital arteries. These vessels spread apically and ventrally and anastomose freely to supply collateral circulation to the apical and ventral bladder walls. The serosal vasculature, especially near the apex, is characterized by large vessels with extensive coiling and abundant venous valves. Coiled arteriole branches ascend to the mucosal surface where they branched abruptly in a Ahydra-like≅ fashion to supply a dense, and freely anastomosing plexus of capillaries just deep to the urothelium. The vasculature of the wall appears to be separated into serosal and mucosal plates joined by the coiled arterioles. The mucosal capillary bed is continuous with similar capillary beds lining the ureters.

As in the rabbit, the vasculature of the wall of the dog bladder appears well designed to: (1) enhance unidirectional blood flow despite expansion and contraction of the organ (via venous valves, coiled vessels, and collateral circulation); (2) permit blood flow without excessive stress on vessels or vessel kinking during expansion and contraction (via coiled vessels connecting two distinct vascular plates); and (3) provide a rich vascular supply to the urothelium (via an extensive, freely-anastomosing, sub urothelial capillary bed).

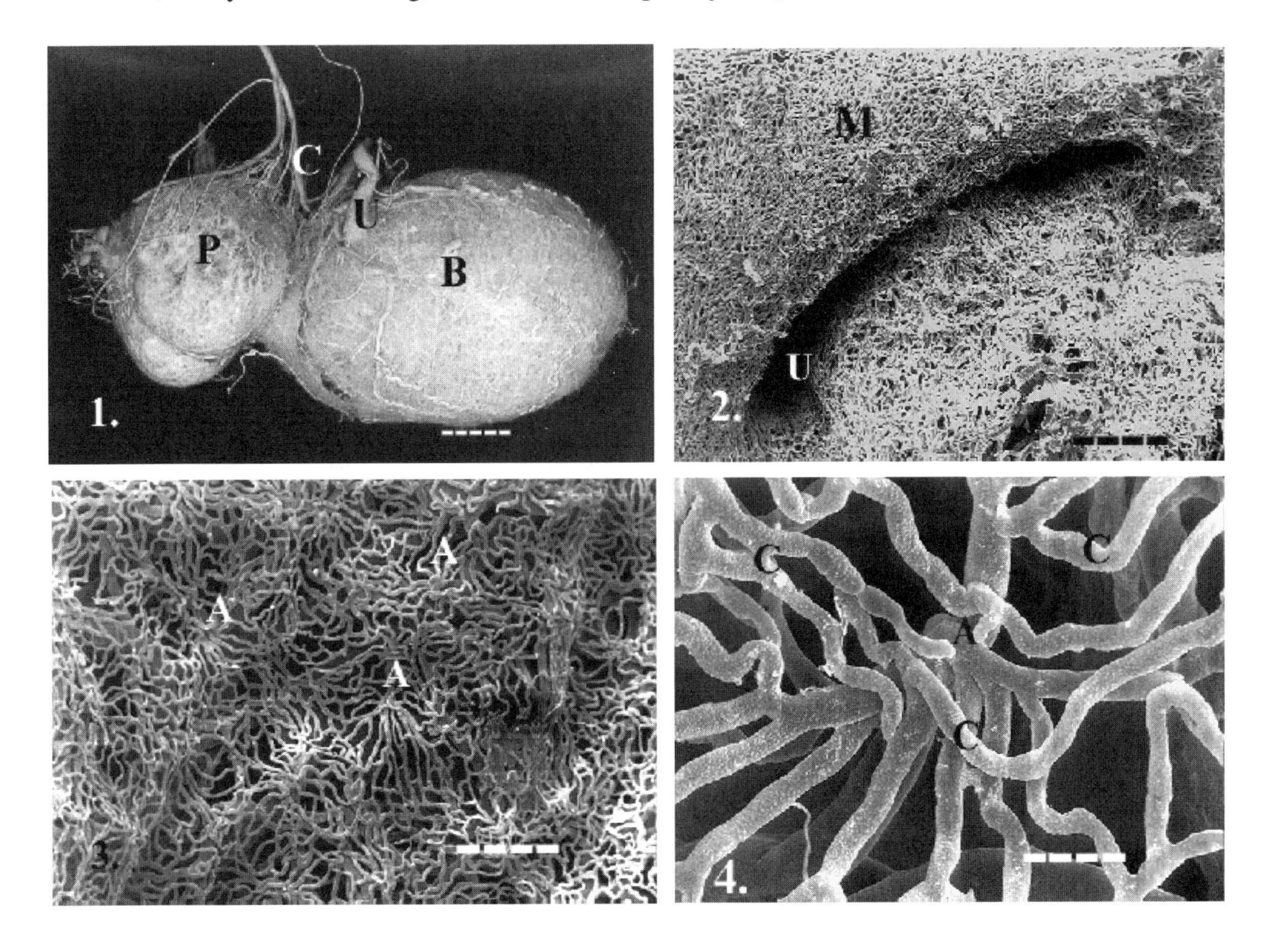

Fig. 1 Vascular corrosion casts of intact dog bladder. B, bladder vasculature; C, caudal vesicular artery; P, prostate vasculature; U, vasculature of the ureter. Bar = 10mm. **Fig. 2** Mucosal capillary bed (M) of bladder showing continuity with mucosal capillary bed of ureter; U, opening to ureter. Bar = 500μm. **Fig. 3** Mucosal capillary bed of bladder. Note multiple capillaries arising from the termini of supplying arterioles (A). Bar = 300μm. **Fig. 4** Origin of multiple mucosal capillaries (C) from a supplying arteriole (A). Bar = 45μm.

References

[1] M. Dunn, Br. J. Urol. 46 (1974) 67.
[2] R.M. Levin et al., Neurourol. Urodyn., 9 (1990) 269.
[3] F.F. Splann and W.H. Walker, J. Urol. 102 (1969) 429.
[4] H. Gill, et al., J. Urol. 139 (1988) 1350.
[5] F.E. Hossler and F.C. Monson, Anat. Rec. 243 (1995) 438.

Microsc. Microanal. 8 (Suppl. 2), 2002
DOI 10.1017.S1431927602100195

Blood Flow Regulating Structures in the Avian Kidney

H. Ditrich*

*Institute for Zoology, University of Vienna, Althanstrasse 14, A-1090 Vienna, Austria

The vascular supply of the kidney of birds is far more complicated than in mammals. In addition to arterial input and venous drainage, there is a prominent venous portal system, as in all non-mammalians (Fig. 1). The latter delivers blood from the leg (V. iliaca externa -*Ua*), the lateral pelvis (V. ischiadica – *Ub*) and the cloacal region (V. hypogastrica and V. pudenda – *Uc* and *Ud*). This blood then passes the peritubular capillaries of the kidney, providing the substrate for uric acid excretion, the main avian nitrogenous waste product. However, not all of the portal blood passes through the renal capillaries. A variable proportion bypasses the kidney via the common iliac / posterior caval vein (*VCP*) and the coccygo-mesenteric vein (*VCM*).

Vascular corrosion casting is the method of choice for studies on delicate vascular systems. As previously shown [e.g., 1], not only the course of the vessels but also their relative diameters and certain structures of the vessel walls can be replicated. In this study, vascular corrosion casting and scanning electron microscopy (SEM) were applied to study some of the structures responsible for flow control in the avian kidney.

Adult animals of the domestic chicken (*Gallus domesticus* – four specimens), the mallard (*Anas platyrhynchus* – eleven specimens) and the turtle dove (*Streptopelia roseogrisea* – nine specimens) were used for this study. The animals were either injected with Mercox® (Jap. Vilene Co.) diluted with 20% methyl-methacrylate monomer as described previously [2], or perfusion-fixed with 2% formaldehyde and 0.5% glutardialdehyde and then processed for conventional light- and transmission electron microscopy or critical point dried (CPD) for SEM.

The main shunt of portal blood from the kidney is regulated by the well studied valve at the external- / common iliac junction (Fig. 2 – *R3*). (Partial) closure of this valve forces blood into the anterior (*R2*) and middle lobes (*R1*) of the kidney. Physiological studies show that this valve is under neuronal (noradrenalin/acetylcholine) control [3]. The role of the coccygo-mesenteric vein (*VCM*) is less clear. Tracer studies indicate that this vessel may be able to reverse it's flow [4, 5]. I. e., instead of forming a bypass for the blood from the dorso-posterior region that does not enter the renal portal system, this vessel could serve as an additional renal portal vein. In any case, a flow-regulating structure (*R7*) has to be postulated at the junction of the right (*I7*) and left posterior renal portal veins with the coccygo-mesenteric vein. Despite minor anatomical differences in the three investigated species, preliminary data show that distinct surface impressions (Fig. 3) can be found on corrosion casts at the emergence of the coccygo-mesenteric vein. The variable depth of these imprints indicates, that these structures are at the site of constricting muscles in the vascular wall.

[1] K.C. Hodde et al., Scanning Microsc. 4 (1990) 693.
[2] H. Ditrich and H. Splechtna, Scanning Microsc. 1 (1987) 1339.
[3] A.R. Akester, The cardiovascular system. In: Physiology and biochemistry of the domestic fowl. (Eds.: D.J. Bell, B.M. Freeman), Academic Press, New York, 1984.
[4] K. Shimada and P.D. Sturkie, Jap. J. Vet. Sci. 35 (1973) 57.
[5] B.W. Oelofsen, Zool. Africana 8 (1973) 41.

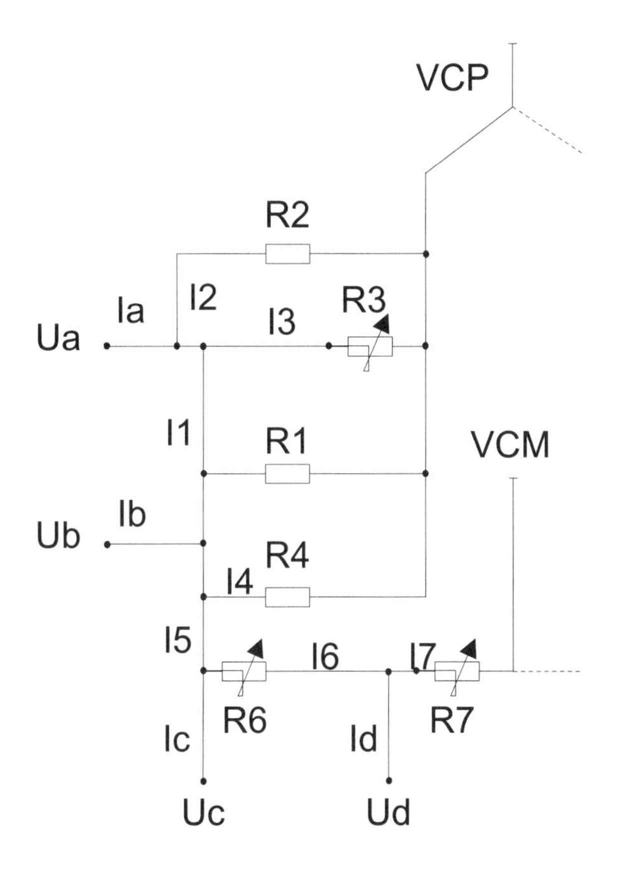

Fig. 1. Diagram of the venous system of the right kidney of birds. Note that the resistance at *R6* is based on theoretical considerations and has yet to be confirmed.

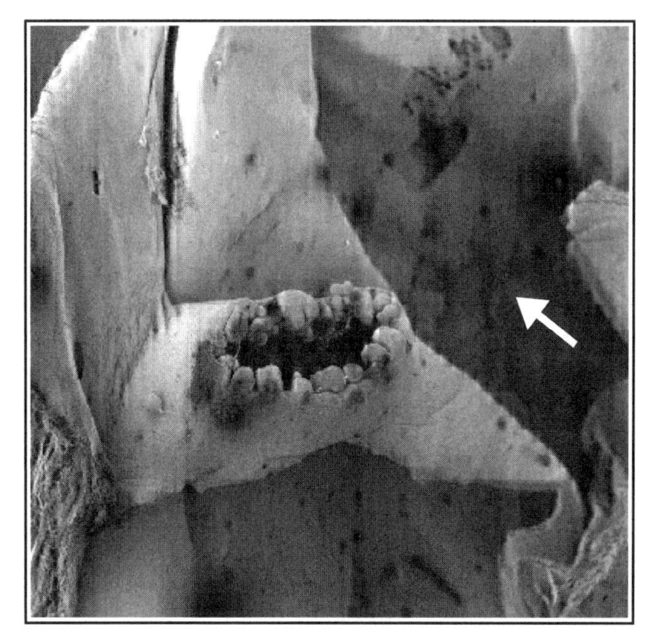

Fig. 2. Bypass valve (*R3*) at the external- (bottom) / common iliac (top) junction in the dove. The efferent renal vein is joining from the right (arrow). (CPD tissue in SEM,);

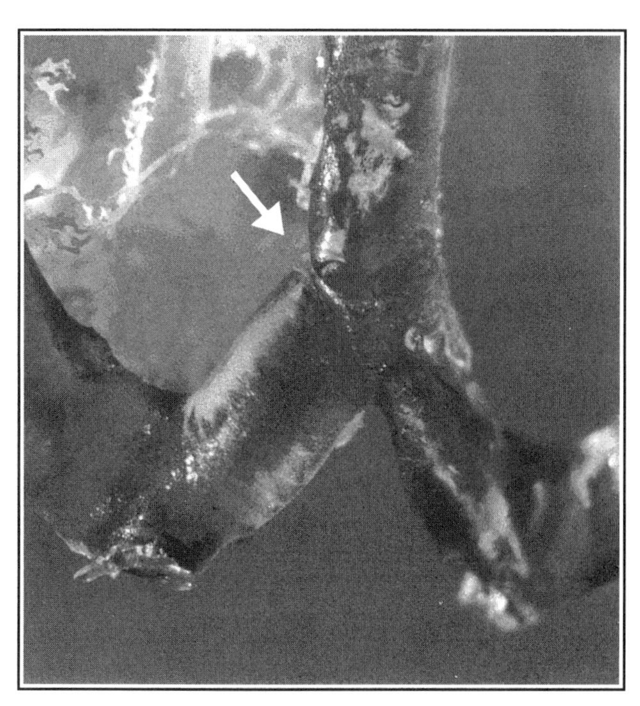

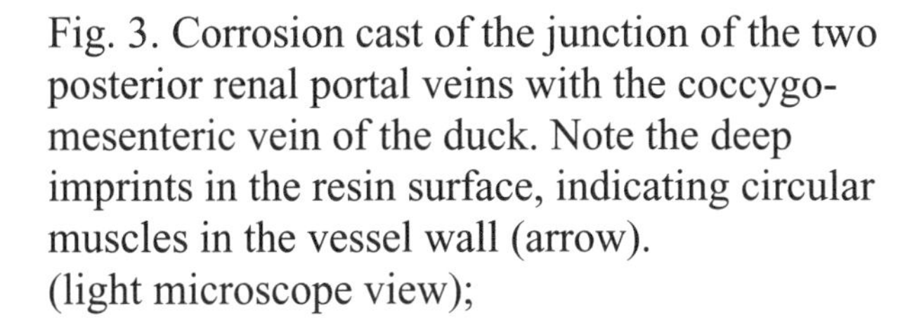

Fig. 3. Corrosion cast of the junction of the two posterior renal portal veins with the coccygo-mesenteric vein of the duck. Note the deep imprints in the resin surface, indicating circular muscles in the vessel wall (arrow). (light microscope view);

Microsc. Microanal. 8 (Suppl. 2), 2002
DOI 10.1017.S1431927602100201

DYNAMICS OF RAT ENDOMETRIUM AS STUDIED WITH SEM AND VASCULAR CORROSION CASTS.

Udo M. Spornitz and Irena Bartuskova

Dept. of Electron Microscopy, Institute of Anatomy, University of Basel, CH 4056 Basel, Switzerland,

The rat is known to have a relatively short endometrial cycle which lasts only about four days. The cycle comprises proestrus, estrus, diestrus I, and diestrus II. Every stage lasts for about 24 hours. Towards the end of the rats life span the cycle may increase to about 6 days. During each of the individual stages characteristic morphological features are present which allow easy identification [1]. One of the outstanding features of estrus e.g. is the presence of pseudoglands, i.e. numerous depressions (pits) of the surface epithelium which resemble glandular orifices (fig. 1). They are formed through necrosis and apoptosis of the endometrial epithelium, they do not however penetrate the underlying basallamina, thus forming a cul de sac. Apoptosis of the involved cells can clearly be detected on the basis of nuclear fragmentation and the marginalization of chromatin. In the necrotic cells giant lysosomal vesicles are formed (fig. 2). The digestion of necrotic and apoptotic cells causes the endometrial surface epithelium to sink below the level of the surrounding cells thus causing the pits or pseudoglands. - Scanning electron microscopic detection of alkaline phosphatase shows a maximum amount during estrus and a minimum during diestrus II [2]. During early pregnancy the alkaline phosphatase activity decreases to zero on day 5 i.e. the actual day of blastocyst adhesion [3]. During the immediate postpartum period again the alkaline phosphatase activity is very low and the rat could become pregnant again. This stage is called postpartum diestrus. Shortly after this stage has been passed the rat then enters postpartum estrus, during which the alkaline phosphatase activity becomes rather high again. This in turn goes along with the fact that the rat could not become pregnant. The absence or presence of alkaline phosphatase activity could thus be taken as a marker for the relative receptivity of the animal. - Despite of the short cycle of the rat, the vascular pattern of the endometrial vessels also changes considerably as investigated with corrosion cast methods (fig. 3). Particular the capillary network seems to be constantly remodeled from one cycle day to the next. One of the outstanding characteristics of the immediate subsurface capillaries of the rat endometrium is that they change their diameter during the course of the cycle. Thus the average diameter measures about 18 µm during diestrus II and less than 10 µm during proestrus. For this three possible mechanisms could be responsible: 1. The capillary network is regulated through precapillary sphincters. 2. The capillary network is in fact constantly remodeled. 3. The diameter is controlled through the action of pericytes.- On the corrosion casts we could only find impressions from precapillary sphincters in connection with deeper capillaries and not with the immediate subsurface capillaries. This would in fact imply that the immediate subsurface capillaries are either remodeled or that their diameter is perhaps controlled through a large number of pericytes.

[1] U .M. Spornitz, et al. 1999, Anat. Rec. 254: 116 - 126
[2] U. M. Spornitz, et al. 1997, Recent Advances in Microscopy of
Cells Tissues and Organs, Pietro Motta ed. Antonio Delfino Publisher, Rome 553 - 560
[3] A. Winkelmann and U.M. Spornitz 1997, Acta Anat. 158: 237 – 246

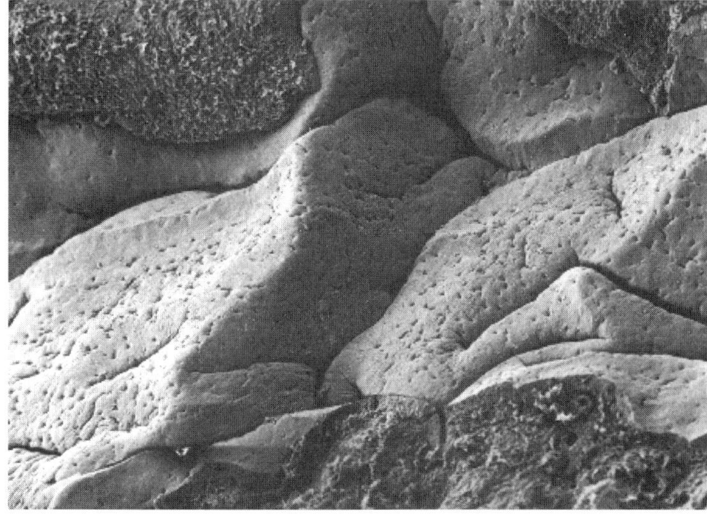

Fig. 1
Low power SEM micrograph of the luminal surface of the endometrium during estrus with folds and ridges like they are typical for this cycle stage. The many depressions are "pseudoglands", i.e. pits formed through necrosis and apoptosis of the underlying cells.

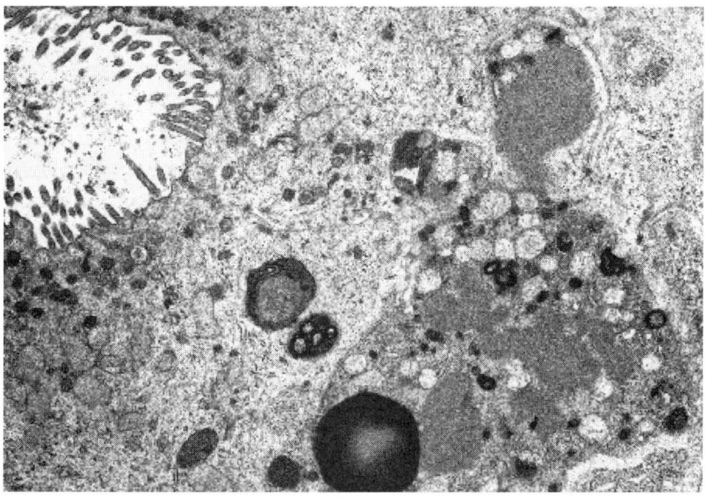

Fig. 2
Transmission electron micrograph which shows the deepest part of a depression in the upper left corner. These "pseudoglands" are formed through necrosis and apoptosis. Large parts of this micrograph is filled with an autophagic vacuole, which contains nuclear fragments among other more or less digested cellular organelles.

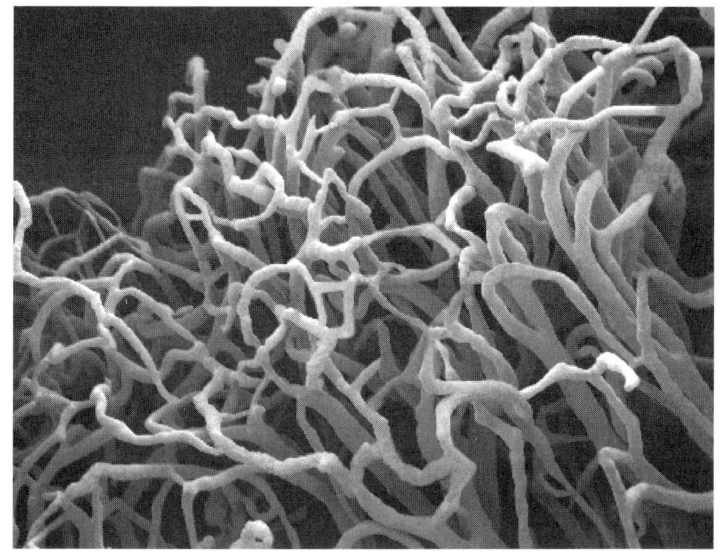

Fig. 3
Corrosion cast of the capillary network during diestrus II. Note the many interconnections between the immediate subsurface capillaries. These are much smaller in diameter than the capillaries of deeper layers.

Microsc. Microanal. 8 (Suppl. 2), 2002
DOI 10.1017.S1431927602100213

The Circulatory System Of Decapod Crustaceans And Its Functional Role In Cardiovascular Dynamics.

I. J. McGaw

Department of Biological Sciences, University of Nevada, Las Vegas, 4505 Maryland Parkway, Las Vegas, NV 89154-4004.

The decapod crustacean circulatory system has received extensive investigation during the last decade ♥1,2♠. Despite our increased understanding of physiological control mechanisms, recent work on the gross anatomy of this system is still scant. A number of classical papers describe the anatomy of the decapod crustacean system in detail♥3,4♠. However, most of these articles were written at the beginning of the last century or the end of the previous century before the advent of modern imaging techniques allowed the detailed analysis of the circulatory system to be resolved.

In the present study the circulatory system of adult decapod crustaceans, was mapped using corrosion cast techniques. Batsons Monomer (Polysciences Inc) was thinned with methyl acrylate methyl ester. It was injected into the pericardial sinus of live crustaceans using a syringe pump. A small cut was made in the arthrodial membrane of a pereiopod to allow hemolymph to escape during resin perfusion. Following injection, the animals were flash frozen in liquid nitrogen. The soft tissue was then macerated, over a 3 day period, in a saturated solution of potassium hydroxide. The remaining resin cast and chitinous exoskeleton, was washed and then dried at room temperature. The chitinous tissue was removed by soaking in 1M solution of hydrochloric acid for 6 hours. This left a corrosion cast of the circulatory system without the need for any manual removal of tissue, which could damage the fine vessels. In order to check the validity of this method, the circulatory system was also mapped by injecting a suspension of barium sulfate into the pericardial sinus, followed by radiography of the specimens.

Seven arteries (five arterial systems) arise from the heart. The anterior aorta exits from the anterior dorsal surface of the heart and branches into 3 arteries. The paired optic arteries arise at right angles and supply hemolymph to the eyestalks. The small cerebral artery, which branches laterally from the main artery, supplies hemolymph to the supraesophageal ganglion. Hemolymph flow into this structure may be aided, in part, by a small accessory pump, the cor frontale. The paired anterolateral arteries also exit from the anterior dorsal surface of the heart; these arteries and their sub-branches supply hemolymph to the gonads, hepatopancreas, stomach, antennal gland, mandibular muscles and the hypodermis of the anterior cephalothorax. The paired hepatic arteries exit the heart anteriorly and ventrally and branch profusely within the hepatopancreas. A smaller side branch, the pyloric hepatic artery supplies hemolymph to the pyloric stomach and midgut. The smallest artery, the posterior aorta, branches off the posterior ventral surface of the heart; it joins with the inferior abdominal artery in the region of the second abdominal segment and these arteries supply hemolymph to the hindgut and abdomen. The largest artery is the sternal artery, which exits from the ventral surface of the heart; the ventral thoracic artery branches off the sternal artery and supplies hemolymph to the chelae, the mouthparts and to each pereiopod.

The present study shows that the circulatory system is highly developed, with arteries dividing into smaller capillary-like vessels that ramify profusely within individual organs. The return vessels, the sinuses, are discrete channels rather than random open spaces as previously described. The presence of these sinuses, rather than a complete venous return system, defines the system as an open circulatory system. However, the terms open and closed are ambiguous. The systems of cladoceran, anostracan, calanoid, and cirriped crustaceans are definitely classed as open, because they lack the discrete vessels of the decapod crustaceans ♥5♠. The decapod crustacean system, however, is far more complex, with hydrostatic pressures that rival those of some lower vertebrate systems♥1♠ and neurohormonal control mechanisms that allow fine regulation of cardiac function and regional hemolymph flow ♥6♠. This study suggests that there may be room for a third category of circulatory system, one that would define the decapod crustacean circulatory system as Apartially closed@ rather than open. Thus the present study refines and advances descriptions of the circulatory system and is discussed in relation to recent work on hemolymph flow in crustaceans.

This study also invites further investigation into whether the decapod crustacean circulatory system has a two fluid-compartment system (intracellular and hemolymph), or a more vertebrate-like three-fluid-compartment system (intracellular, interstitial, and blood). Because the vascular system of decapods in general, is recognized for its complexities, further examinations of the local or central regubtory capabilities of regional blood flow should also support the regulatory comparisons made between decapods and lower vertebrates ♥7♠.

References

♥1♠. L.E. Burnett & B.R. McMahon. Physiol. Zool. 63 (1990) 35.
♥2♠. I.J. McGaw & C.L. Reiber. J. Morphol. 251. (2002) 1.
♥3♠. C. Claus. Ar Zool. Inst Univ. Wien U Zool. Sta. Triest. 5 (1884) 271.
♥4♠. J. Pearson. Livpl. Mar. Biol. Mem. 16 (1908) 1.
♥5♠. P.A. McLaughlin. Biology of Crustacea. Vol 5 (1983) 1.
♥6♠. I.J. McGaw & B.R. McMahon. J. Crust. Biol. 19(1999) 435.
♥7♠. This work was supported by the Office of Research at UNLV. The aid of Dr Carl Reiber, UNLV Biological Sciences, is gratefully acknowledged

Microsc. Microanal. 8 (Suppl. 2), 2002
DOI 10.1017.S1431927602100225

Blood Supply of the Symphysis Pubis

R.C.G. da Rocha*, R.P. Chopard*

*University of São Paulo, Department of Anatomy, Av.Prof. Lineu Prestes 2415, São Paulo, SP, 05508-900.

It has been inferred that fibrocartilagenous tissues are avascular structures and when blood vessels are present, they are restricted to the most peripherical portion of the tissue considered [1]. Several works were done to describe blood vessels in these structures. There are vessels in the intervertebral disc [2], knee menisci [3] and in the temporomandibular disc [4] among others. The vascularization of these structures is very rich in animals that are in development. With aging, there is a regression in the number of blood vessels, which leads to a rich vascularization confined tothe most peripherical portion of these structures. Since fibrocartilage has no pericondrium, these vessels are the main source of nutrients for fibrocartilage and the main pathway for metabolites drainage [5]. It is generally accepted that the symphysis pubis is an avascular structure. But in 1934, the human symphysis pubis was classified in to periods: one, vascular (development) in which the interpubic space shows a rich vascularization and a second, avascular (adult) where the symphysis shows a very poor vascularization limited to one or two vessels in its periphery [6].

The purpose of our study was to investigate the blood supply of the symphysis pubis during development and in adult age. For this we used Wistar rats as animal model with ages of 30 days (young) and 90 days (adult) of both sexes. Both groups were euthanized with pentobarbital 60mg/kg with heparin 700µl/kg and had their abdominal aorta canulated. Both were washed with saline solution until the caudal extremities appeared clean of blood. One group was injected with India ink and analyzed under light microscopy. The other was injected with Mercox® for corrosion casting (maceration with KOH 10%) and analyzed under scanning electron microscopy [7].

We were able to identify a very rich vascularization in the younger group in which blood vessels were present in whole interpubic space and also a rich vascularization restricted to peripherical portion of the symphysis in adult group. There were no differences between sexes. The vessels were originated from the pubic branches of the inferior epigastric and obturatory arteries, and formed an authentic articular arterial circle. We purpose the name: "Arterial Circle of the Symphysis Pubis". These observations may change not only the prognosis but also the treatment (manual therapy, ultra-sound and exercises) of pubic problems like osteitis pubis and pubalgia *post partum.*

References

[1] M. Benjamin, E.J. Evans, J. Anat. 171 (1990) 1-15.
[2]B. Grignon et al., Bull. L'Assoc. Anat. 243 (1994) 53-57.
[3]K. Messner, J. Gao, J. Anat. 193 (1998) 161-178.
[4]P. Etienne, A. Lametschwandtner, Archs.Oral Biol 40 (1995) nº6, 499-505.
[5]L.C. Junqueira, J. Carneiro, Histologia Básica 9th ed. Guanabara Koogan, Rio, 1999.
[6]R. Gatta, Arch. Ital. Anat. Embriol.
[7]F.E. Hossler, F.C. Monson, Anat Rec 252 (1998) nº3, 477-484.
[8]This research was supported by CAPES.

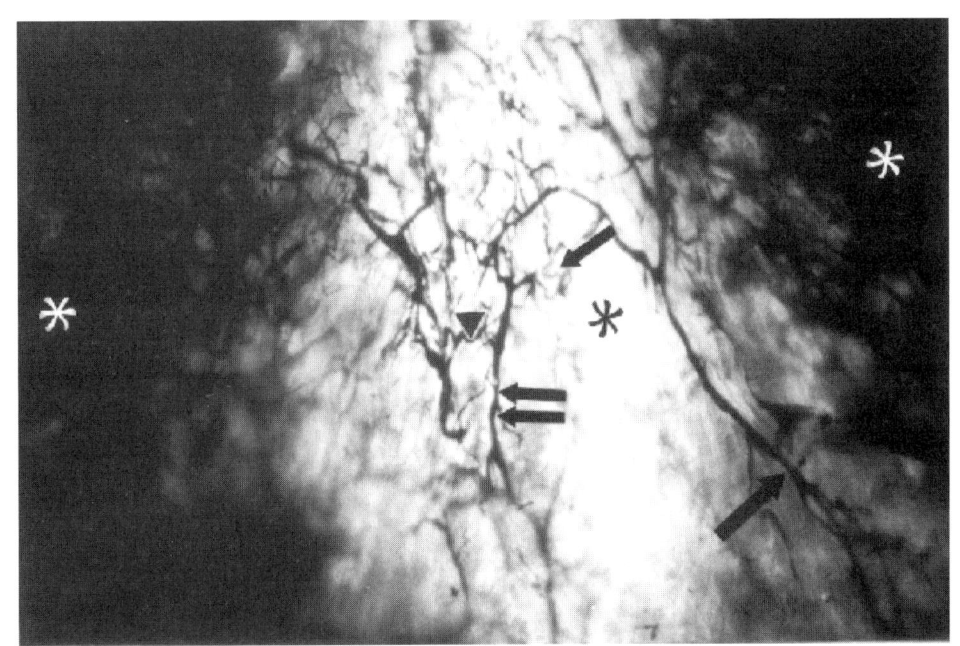

Figure 01: Anterior view of the symphysis pubis of a adult male Wistar rat injected with India Ink. We can observe a large vessel (large arrow) coming from the periphery giving smaler longitudinal branches (double arrows). We can also observe the presence of some anastomotic bridges (arrow head) between these longitudinal branches. The vascularization ends in capillary loops (small arrow) that face the central avascular portion (black *). Pubic Bones (white *). x16.

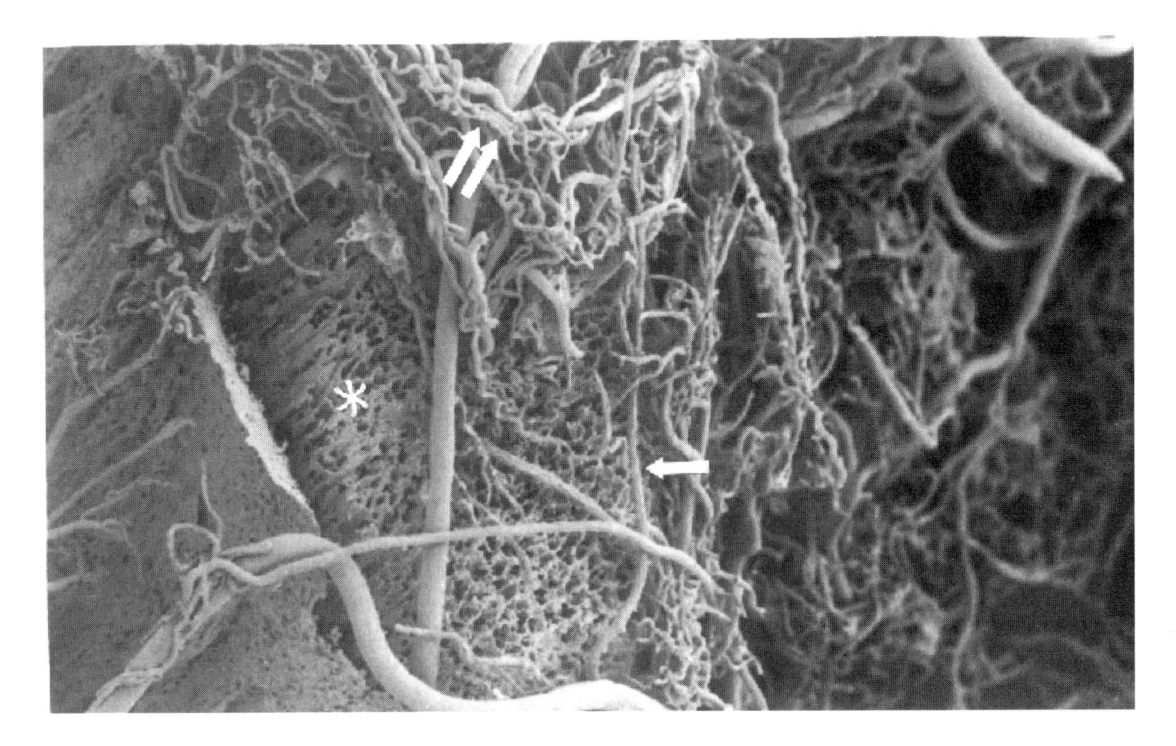

Figure 02: Anterior view of the symphysis pubis of a young male prepared for corrosion casting. The pubic branch of the obturatory artery (large arrow) gives longitudinal arterioles (arrows) wich finish in a spiraled capillary web (double arrows). Articular surface of the pubic bone (*). x60.

Microsc. Microanal. 8 (Suppl. 2), 2002
DOI 10.1017.S1431927602100237

Corrosion Casting of the Microvasculature in Normal Limbs and Limbs with Venous Ulceration.

M.N. Phillips,* ** A.M. van Rij,* M. Zhang,** and G.T. Jones*.

* Department of Medicine and Surgery, University of Otago, Dunedin 9001, New Zealand
** Department of Anatomy and Structural Biology, University of Otago, Dunedin 9001, New Zealand .

Controversy exists regarding the presence of valves in veins less than 2mm in diameter. Recently venous valves have been identified in the small superficial veins of the human lower limb using vascular corrosion casting. The role that these valves have in both normal limbs and limbs with venous ulceration is unclear. This study proposed to assess differences in valve morphology in the microcirculation of normal and diseased limbs, using microvascular corrosion casting.

Five amputated lower limbs with venous disease (CVI class 6; three males, two females; 61 -86 years) were retrogradely filled with resin via the dorsal vein of the foot. Superficial tissue (20x20mm) from the gaiter, lower, mid and upper calf were removed and macerated (15% NaOH). Resin casts were viewed by light and scanning electron microscopy (FIG 1), and compared with similar casts from six normal cadaver limbs (CVI class 0-1; four males, two females; 59-92 years).

Small venous valves were present in all regions from the normal limbs and limbs with venous ulceration (FIG 2), The valve number, density and size distribution between the normal and diseased limbs were not different. However, the limbs with venous ulceration had a greater density and tortuosity of veins filled with resin compared with normal limbs (153.8 vessels/cm^3 vs. 15.1 vessels/cm^3 respectively). Consequently, the valves/vein-density was lower in the diseased limbs when compared to the normal (0.02 vs. 0.16 respectively). This may reflect microvenous reflux in these diseased limbs.

This study shows, for the first time, that the number, density and size of microvenous valves does not differ between corresponding sites in normal and venous ulcerated limbs. Significantly, the microvalves of the venous ulcerated limbs were incompetent and allowed retrograde filling of the microvasculature. Consequently, the number of valves/vein-density is decreased in venous ulcerated limbs. This finding may aid in understanding factors influencing the impact of reflux on the microcirculation in venous hypertension.

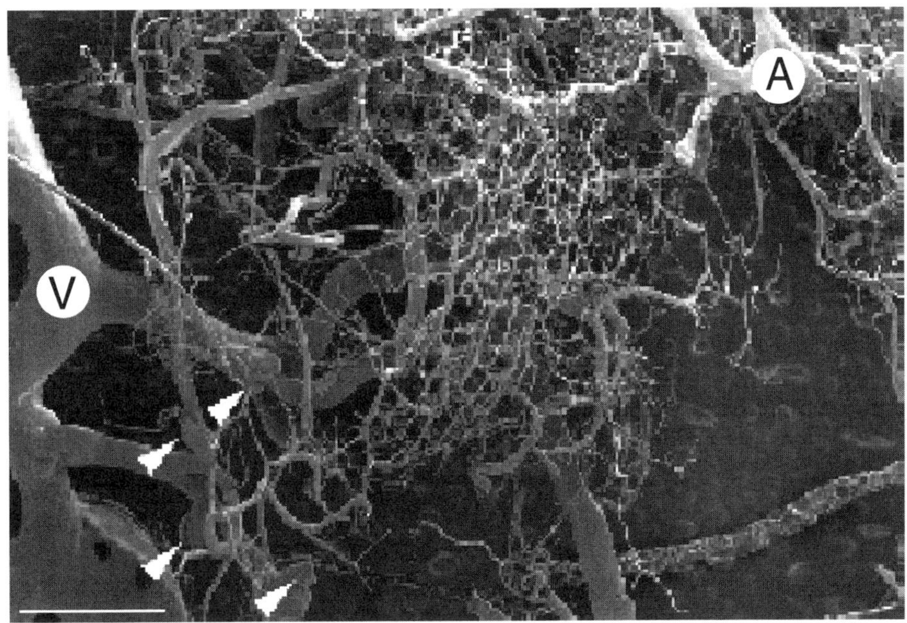

FIG 1. The capillary bed networks associated with an artery (A) and vein (V). Four venous valves (arrowheads) are easily visible. Bar = 2mm.

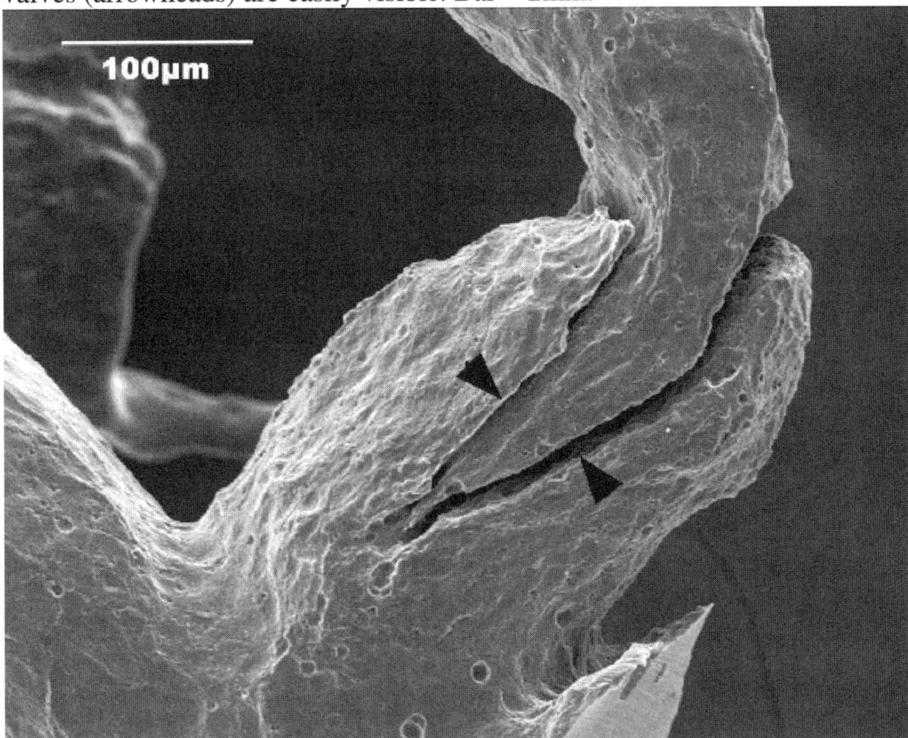

FIG 2. Example of the appearances of bicuspid venous valve identified by SEM of resin corrosion casts. A full venous valve cast, containing both the proximal and distal halves of the valve, joined by a small amount of resin passing between the cusps (arrow heads). Bar = 100μm.

Microsc. Microanal. 8 (Suppl. 2), 2002
DOI 10.1017.S1431927602100109

EXPERIMENTAL MICROSCOPY OF FOOD SYSTEMS

Eric Kolodziejczyk & Martin Michel
Food Science & Process Research,
Nestlé Research Center, CH- 1000 Lausanne 26.

Foods are complex products, which must fulfill at least two requirements: be stable and have acceptable sensory properties (flavor, texture), corresponding to the consumer' s acceptance.

Obviously, the stability and the textural properties of food products are strongly influenced by the functionality of the ingredients and their interactions during the process. It is also well known that these interactions result in a structuration of the food product, thereby creating their textural properties.

During mixing and along processing, ingredients interact at a nanoscale and, often due to thermo-dynamic incompatibilities leading to phase separations, build up micrometric structures whose mechanical properties have a critical impact onto the global mechanical properties of the bulk.

However, by creating these micro particles, numerous interfaces are also generated. The mechanical responses of these interfaces to mechanical factors, such as shearing, contribute also importantly to the global mechanical properties of the bulk. One can also easily deduce that these interfaces create more or less permeable barriers having also a profound influence onto the chemical and thermodynamic properties of the product and, its stability. The stability of these interfaces is critical for the product and is governed by thermodynamic compatibilities/imcompatiblities.

Subsequently, understanding the mechanisms sustaining the structuration/destructuration of these interfaces, is an essential issue for the food sciences and, hence, the food industry. This explains why food science is now moving and focusing more and more toward the science of interfaces.

Interfaces play an essential role in two important structural components of food products: foams and emulsions, both important partners of highly appreciated products such as ice creams. This explains the amount of work devoted to this subject in the food science literature. However, as food products are extraordinarily complex, a clear understanding of the basic phenomena grounding food structuration/de structuration cannot be achieved on finished products and simplifications are needed via the constructions of models of food and appropriate devices adapted to the microscopy of this models.

Numerous analytical techniques are available to investigate and understand food properties. Among these, imaging, at resolution ranging from the macrometric to the nanometric scale (macroscopy, light microscopy, Scanning electron microscopy, transmission electron microscopy, Atomic force microscopy, SNOM etc…), play a still growing key role in this issue, not only due to the resolution increase, but also because, more and more microscopy techniques are integrating highly specific tools guiding microscopy closer and closer to microchemistry and micro-physic.

Among these microscopy techniques, Confocal Scanning Laser Microscopy (CSLM) occupies a place of choice because it allows high resolutions in the 3 and 4 dimensions. Combined to the possibility of tagging ingredients with fluorescent probes, a very specific localization of the ingredients during the structuration process extends further the unique capabilities of this instrument. Moreover, applying these tools to simple models gives new insight into the basic phenomena which occur during the structuration/destrcuturation of the interfaces. By increasing the complexity of these models of foods, going closer and closer to the real target product, following a very structured approach, our understanding of food structuration/destructuration increases rapidly, resulting in a better control of food structure and texture.

The scope of this presentation is to review the place of the possible imaging approaches in the field of food processing.

Based on the literature and our own work, we will demonstrate the advantage of combining diffferent techniques (macroscopy, conventional epifluorescene, CSLM, micro-rheoloy) with different models of food and simplified versions of foos systems (flat films, bubbles) to understand better protein polysaccharides interactions leading to segregative or associative phase separations in bulk and their behavior at interfaces, leading to the creation of interfacial films. The correlation of the information provided by these models with the structures observed in complex systems close to real products, such as ice-creams using cryo-fluorescence techniques, will be presented and discussed.

Microsc. Microanal. 8 (Suppl. 2), 2002
DOI 10.1017.S1431927602100110

Microstructural Analyses to Study Ingredient Functionality, Interactions and Quality in Frozen Foods.

H. Douglas Goff

Dept. of Food Science, Univ. of Guelph, Guelph, ON N1G 2W1, Canada, dgoff@uoguelph.ca

During the freezing of foods, ice forms as the temperature is lowered below the equilibrium freezing temperature. With decreasing temperature, the quantity of ice increases and is dictated by the nature of the solutes in solution, as they are freeze-concentrated in a decreasing quantity of water (Fig. 1). The size, morphology and distribution of ice crystals is dependent in large part on the type of freezing process (quiescent versus under shear) and the rate of heat removal [1]. The unfrozen phase can undergo a glass transition due to high concentration and low temperature, and this gives the frozen product much greater storage stability.

During storage at temperatures above the glass transition, especially in fluctuating temperatures, ice crystals undergo recrystallization, in which they grow in size and decrease in number. Several phenomena are involved, including ripening, accretion, melt-diffuse, and shrink-regrow recrystallization. In most cases, these changes can detrimentally affect the quality of the frozen product [1,2].

Polysaccharide stabilizers have been used in frozen food systems for many years, to slow down the process of ice recrystallization and structural changes during frozen storage. Ice cream is one example where their use is of great importance, due to the detrimental effects of recrystallization on texture. Understanding how these polysaccharides function to control ice crystal growth, however, has been challenging, but is of importance to both finding and assessing new ingredients and or modifications to existing food products. Polysaccharides are well known viscosity-enhancers in solution, which occurs through their bulky structure and their ability to interact with each other and with other polymers [3]. As freeze-concentration occurs, viscosity within localized regions (so-called "micro-viscosity") can limit mobility and hence delay ice crystal growth [3]. Through microstructural analyses, two additional phenomena are evident in frozen systems, both of which are also relevant to polysaccharide functionality. The first is cryo-gelation, an inter-molecular interaction induced by freeze-concentration which sets up an element of 3-dimensional macromolecular structure, and the second is incompatibility and phase separation with proteins, if present in the food system, which is also enhanced by freeze-concentration and adds a further element of structure [4].

Various microscopic techniques can be used to explore functionality of polysaccharide stabilizers in frozen systems, including specific staining with cold-stage light or freeze-substitution transmission electron microscopy [5] and the use of labeled polysaccharides and cold-stage fluorescence microscopy (Fig. 2) [4]. These techniques augment structural analysis by cryo-scanning electron microscopy [6], to yield valuable information for frozen food formulations and processing.

References

[1] H. D. Goff. Food Research Internat. 25 (1992) 317.
[2] H. D. Goff in Rencontres AGORAL, TEC and DOC, Paris, 1999, 147.
[3] S. Bolliger et al. Internat. Dairy J. 10 (2000) 791.
[4] H. D. Goff et al. Food Hydrocoll., 13 (1999) 353.
[5] A. Regand. M.Sc. Thesis, Univ. of Guelph, 2001.
[6] K. B. Caldwell et al. Food Structure 11 (1992) 1.

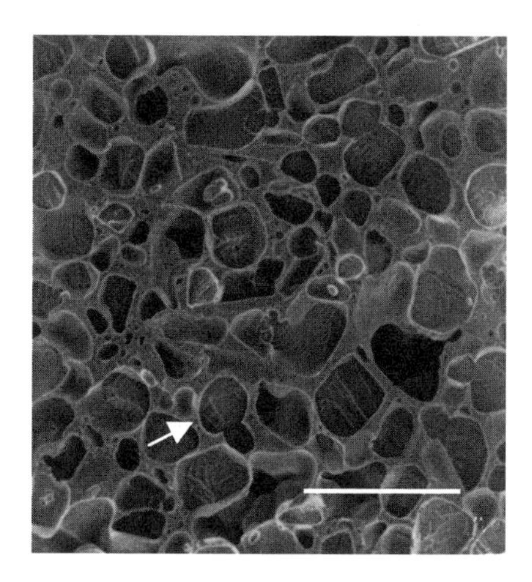

FIG. 1. Cryo-SEM micrograph of ice cream structure, showing ice crystals and the freeze-concentrated unfrozen phase (arrow). Bar = 50 μm.

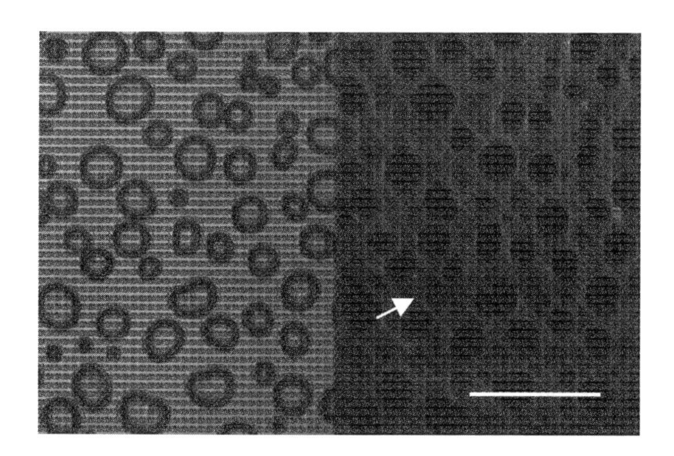

FIG. 2. Brightfield (left) and fluorescence (right) images from the same field of a frozen 40% sucrose + 0.42% rhodamine labelled-locust bean gum (LBG) solution after 5 cycles from −18°C to −10°C, at −10°C. Note the gel-like LBG structure (arrow) in the fluorescence image formed by freezing and temperature cycling. Bar = 50 μm.

Microsc. Microanal. 8 (Suppl. 2), 2002
DOI 10.1017.S1431927602100122

Effect of Acidity on Optical Properties of Isolated Skeletal Muscle Fibers

H.J. Swatland

University of Guelph, Dept. Food Science, Guelph, Ontario, N1G 2W1

Extremely pale meat is unattractive, looses fluid and is difficult to use for secondary processing. The problem is worst in pork, but also occurs in poultry. If anaerobic glycolysis is rapid or extended, low-angle x-ray diffraction shows fluid is released from the myofilament lattice, where lateral electrostatic repulsion is minimal at the isoelectric point of muscle proteins [1]. But what causes paleness at a low pH? There are two common explanations. Paleness originates from scattering by precipitated sarcoplasmic proteins [2], and shrinkage of the myofilament lattice at a low pH increases reflection at the myofibrillar surface [3]. A third possibility considered here is that muscle fiber refraction contributes to paleness. If refraction is weak, the original directionality of the light may be maintained so light passes deep into the meat rather than being scattered back to the surface to appear as paleness. If refraction is strong, directionality may be lost, refractive scattering may be increased, so meat appears pale. To test this working hypothesis, experiments were undertaken on individual muscle fibers in a microscope chamber perfused with 0.2 M phosphate buffer.

After washing out myoglobin, transmittance (T) spectra through 20 pork fibers were measured from 300 to 800 nm at high and low pH (pH 7.26 ± 0.15 and 5.49 ± 0.06), starting half the samples at low and half at high pH. As fibers were moved across the measuring aperture, light at short wavelengths was first to be refracted through the fiber and lost by scattering. Finally, with all wavelengths traversing the full depth of the fiber, T was higher at high than low pH (300 to 740 nm, $P < 0.05$; 310 to 510 nm, $P < 0.01$). The Becke line was detected in many fibers as a bright line caused by refraction at the edge of the fiber [4].

One would not necessarily expect the photometric laws of Lambert, Beer and Bouguer to hold true for microscopic distances through strongly scattering fibers, so pork fibers ($n = 50$) were examined at pH 5.5 or 7.0 with a mechanical scanning stage and photometer. The objective was to measure T in relation to fiber diameter, anticipating attenuation of T with increasing diameter. However, T across fibers was often asymmetrical, showing fibers retained flattened sides after being released from bulk pork. In bulk pork, the range in flattened sides is from 3 to 8, with a mode of 5 or 6 sides, depending on sample origin within a muscle [5]. A lateral measurement of fiber diameter (89.76 ± 17.07 µm at pH 5.5 and 97.43 ± 20.04 µm at pH 7.0, $P > 0.05$) was compared with minimum T at any point across a fiber. No T attenuation was detected at pH 7.0 ($r = 0.07$). At pH 5.5, instead of attenuation, the relationship of T with diameter was unexpectedly positive ($r = 0.44$, $P < 0.05$), possibly from differences in myofibrillar composition between small and large diameter fibers (fibrillenstruktur versus felderstruktur). However, attenuation at pH 5.5 was greater than at pH 7.0 (2.76 ± 1.19 vs 1.42 ± 0.65 ?T mm^{-1}, respectively, $P < 0.001$) and was strongly related to wavelength ($r = 0.97$, $P < 0.001$). Thus, fiber shape and physiological fiber-type composition may affect optical properties of bulk meat. Results agreed with myofibrillar refraction contributing to post-mortem development of normal paleness in pork.

To check these findings were applicable to bulk meat, bulk refractive index (*n*) was measured with an Abbe refractometer using a red laser for *T* and a green laser for reflectance. The critical angle, although obscured by scattering, was detected subjectively at the red-green boundary. The refractometer also was operated under computer control, detecting *n* photometrically. Pork longissimus thoracis ($n = 20$) had higher *n* than biceps femoris (1.357 ± 0.004 versus 1.352 ± 0.005, respectively, $P < 0.001$). Japanese pork colour scores (*JPCS*) are widely used in the meat industry and range from 1 (very pale) to 5 (very dark). Longissimus thoracis had lower *JPCS* than biceps femoris (2.92 ± 0.37 versus 3.87 ± 0.48, respectively, $P < 0.001$). In pooled samples ($n = 40$), *n* was correlated with *JPCS*, $r = -0.55$, $P < 0.001$. Thus, the negative slope showed pale pork had a higher refractive index than dark pork, as anticipated from microscopy of single fibers.

Deviations from the photometric laws in bulk meat were examined using optical fibers mounted in gauge-13 hypodermic needles with a 15° tip. The tip angle modified the cone of illumination produced by the optical fiber, but the elliptical window still responded in a logical manner to test objects. *T* spectra obtained as needles moved apart to lengthen the light path through meat were atypical, being almost flat. Pushing needles into the meat to shorten the light path, however, yielded typical spectra for meat, with a secondary absorbance band at 550 nm and low absorbance at 700 nm. The effect of optical path length through the meat was detectable ($P < 0.01$) for all visible wavelengths, but the magnitude of change was greater at 700 nm than at 400 nm. *T* measured perpendicularly through pork muscle fibers was a linear, arithmetic function of path length (slope -0.103 ± 0.038 T cm^{-1} at 700 nm, $r = -0.99$, $P < 0.001$) with no evidence of an exponential relationship predicted by the photometric laws.

Research in progress uses polarized light. In myofibrils, light splits into two components with different velocities, the ordinary ray (*O*) and the extraordinary ray (*E*), with *O* – *E*. Birefringence is given by $n_E - n_O$. Retardation, the decrease in velocity of light caused by interaction with muscle proteins, is detected as phase retardation, the interference caused by path difference $E \neq O$. The path difference of a depth of fiber (*G*) is measured by ellipsometry with a de Sénarmont compensator [6], **G nm = K? nm$^{-1°}$. u^{o}** where u is the angle in degrees required for compensation, and K? is the monochromatic de Sénarmont constant or path difference for 1° of rotation. Following the same logic as for measuring attenuation of *T* in relation to fiber diameter, if the optical path length can be found as well as *G*, it may be possible to quantify the effect of acidity on birefringence, giving us an appreciation of the refractive contribution to meat paleness from the molecular to the bulk state. Preliminary results confirm that *G* increases as pH decreases, but finding the exact relationship is difficult. Low molarity of the buffer makes it difficult to control fiber pH, whereas equilibration to 0.2 M buffer removes some actomyosin [7].

[1] H.J. Swatland et al., *J. Anim. Sci.* 67 (1989) 1465.
[2] J.R. Bendall and J. Wismer-Pedersen, *J. Food Sci.* 27 (1962) 144.
[3] R. Hamm, *Adv. Food Res.* 10 (1960) 355.
[4] R.C. Faust, Optical properties. In R. Meredith and J.W.S. Hearle, *Physical Methods of Investigating Textiles* (pp. 320-345). Textile Book Publishers, New York, 1959.
[5] H.J. Swatland, *J. Anim. Sci.* 41 (1975) 78.
[6] M. Pluta, M. *Advanced Light Microscopy.* Amsterdam: Elsevier. Vol. 1, p. 38. 1988.
[7] Research supported by a grant for basic research from the Danish Bacon and Meat Council.

Microsc. Microanal. 8 (Suppl. 2), 2002
DOI 10.1017.S1431927602100134

Fat crystal networks – structure and properties

Alejandro G. Marangoni
Dept. of Food Science, University of Guelph, Guelph, ON N1G2W1, Canada
Ph: (519) 824-4120, Fax: (519) 824-6631, e-mail: amarango@uoguelph.ca

When triglycerides are cooled from the melt to a temperature below their melting point, they undergo a liquid-solid transformation to form primary crystals with characteristic polymorphism. These crystals associate, or grow into each other, to form increasingly larger aggregates, which further interact, resulting in a continuous three-dimensional network. The macroscopic properties of a fat crystal network are affected by all these levels of structure, however, most directly by the level of structure closest to the macroscopic world. Most of the research in this area has been directed towards establishing a link between molecular structure and phase behavior, crystal habit, in particular crystal polymorphism, and the macroscopic properties of fats. Not including all levels of structure in the analysis is undoubtedly a major reason why these attempts have only been partially successful. In order to truly understand, and eventually predict, the macroscopic properties of fat-containing products, it is necessary to characterize and define the *different levels of structure* present in the material and their respective relationship to a macroscopic property. A macroscopic property is not always simply and directly related to molecular structure.

Work in our laboratory has shown that the rheological properties of edible fats are strongly dependent on the microstructure of their fat crystal networks using fractal scaling relationships and developed a particle counting method for the determination of the fractal dimension from polarized light micrographs (Narine and Marangoni, 1999. Physical Review E 59:1908). We have also developed a mechanical and structural model which relates the Young's modulus (E) of an isotropic material to particle properties, solids' volume fraction, and the spatial distribution of that mass (Marangoni, 2000. Phys. Rev B 62: 13951; Narine and Marangoni,1999. Physical Review E 60:6991), namely:

$$E \sim \frac{A}{\pi a \gamma d_o^2} \Phi^{\frac{1}{3-D}}$$

where A is Hamacker's constant, a corresponds to the diameter of primary crystal aggregates (<5 m), to the strain imposed during the mechanical test, d_o to the inter-cluster separation distance, to the volume fraction of solids, and D to the fractal dimension of the network. Processing conditions (heat and mass transfer) can affect some, or all, of these parameters, particularly at the level of liquid structure. The possibility of structuring the solid state via manipulation of liquid structure will be explored in this presentation.

Figure 1. Atomic force micrographs of cocoa butter crystallized at 22°C at different levels of magnification. Crystallite clustering into increasingly larger clusters is shown.

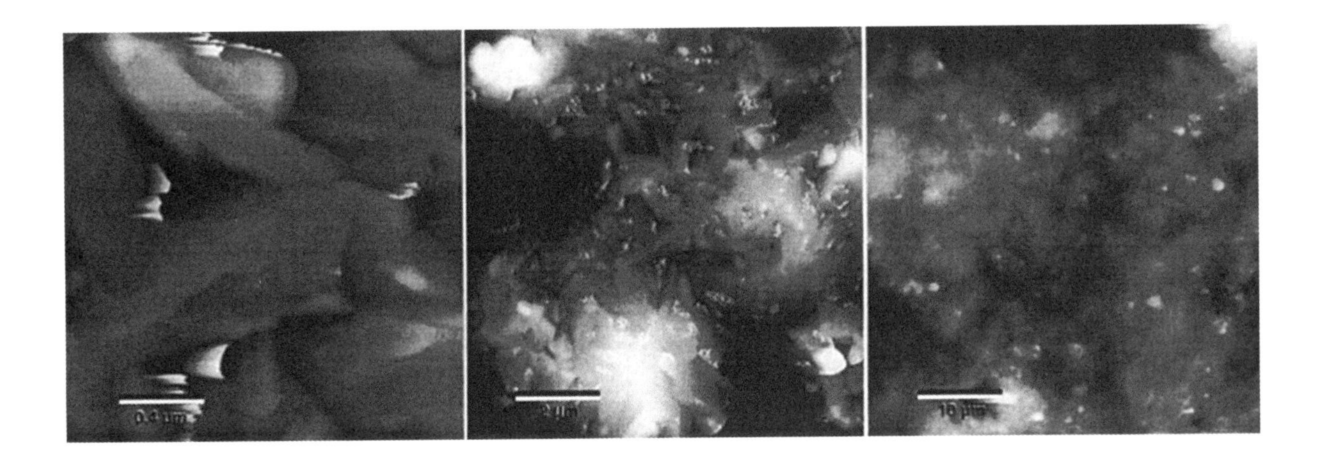

Microsc. Microanal. 8 (Suppl. 2), 2002
DOI 10.1017.S1431927602100146

The Effects of a Non-polar Surfactant on the Crystallization Behavior and Physical Properties of the High-Melting Triglyceride Fraction of Milkfat

J.W. Litwinenko, A.G. Marangoni*

University of Guelph, Department of Food Science, Guelph, ON, Canada, N1G 2W1
* Author to whom correspondence should be sent

The texture of foods that are high in fat, such as butter, can be dramatically altered by crystallization conditions and the presence of minor polar components [1]. The effects of non-polar surfactants on emulsion properties have been extensively studied, but little is known about their effects on single-phase fat systems. In this study, the effects of the surfactant Tween60 (at levels between 0% and 3% w/w) on the crystallization of a model plastic fat system were investigated. The model system was composed of milkfat's high-melting triglycerides diluted in triolein at a 30:70 (w/w) ratio. These high melting triglycerides solidify first upon supercooling, and hence play a major role in determining the resulting crystal network structure and rheological properties.

The effects of Tween addition on rheology were profound, including a sharp increase in breaking force at 0.1%, followed by a decrease at 0.5% Tween as shown in Fig. 1. Differential scanning calorimetry showed no differences in melting shape or peak melting temperatures, while NMR showed no differences in solid fat content (SFC). It was thus hypothesized that changes in rheological properties were due to changes in microstructure. Crystallization kinetics were characterized using the Avrami model. This analysis yields two parameters – a rate constant of crystallization (k) and a growth mode exponent (n). SFC curves at 28°C allowed for the computation of n and k, which predicted maximum and minimum rates of nucleation and crystal growth at 0.1% and 0.5% levels respectively. Lower Avrami exponents (n) were found at levels associated with higher rates of nucleation and growth, and suggested less spontaneous nucleation at 0.5%. Induction times and rates of crystallization determined by turbidimetry corresponded to the predictions of the Avrami model, with shorter induction times found at 0.1%. Polarized light micrographs and image analysis revealed that at the 0.1% level, the crystals that made up the network were about two times smaller, and four times more numerous than at 0.5% (Fig 2). Fractal dimensions remained constant (~2.0) indicating no change in the spatial distribution of the network and suggesting that differences in breaking force were primarily due to particle properties. This is consistent with the predictions of the Avrami model, trends in the storage modulus and induction times, and our knowledge of crystallization in general (with shorter induction times resulting in more spontaneous nucleation, smaller and more numerous particles, and a more rigid network). Time-lapsed microscopy allowed for the visualization of the kinetics of crystallization and the normalized percentage of crystal mass was plotted against time yielding crystallization curves that showed the same relative differences in both induction times and rate of crystal growth as found by NMR and turbidimetry. The role of PLM in corroborating the results of previous experiments vastly enhanced both the qualitative and quantitative findings of the study.

References

[1] A..J. Wright, *J. Am. Oil Chemists' Soc.* 7 (2000) 463.

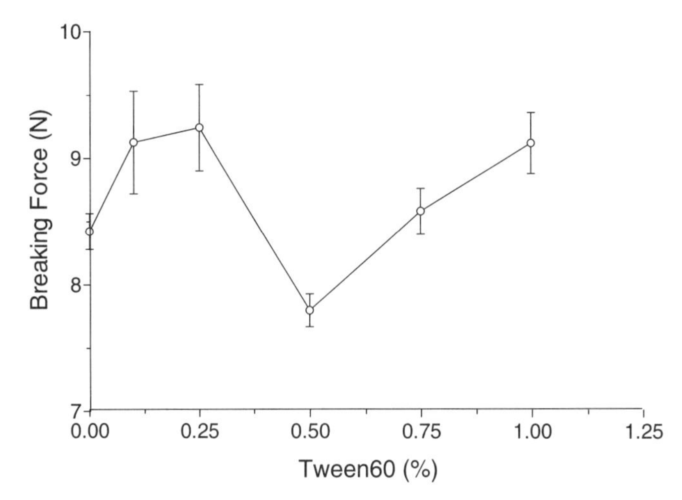

FIG. 1. Breaking force (N) as a function of Tween60 added (%) to 70:30 triolein:HMF following storage at 5°C for 24h. Samples were compressed beyond fracture by parallel plate compression at a crosshead speed of 10mm/sec using a 5kg load cell.

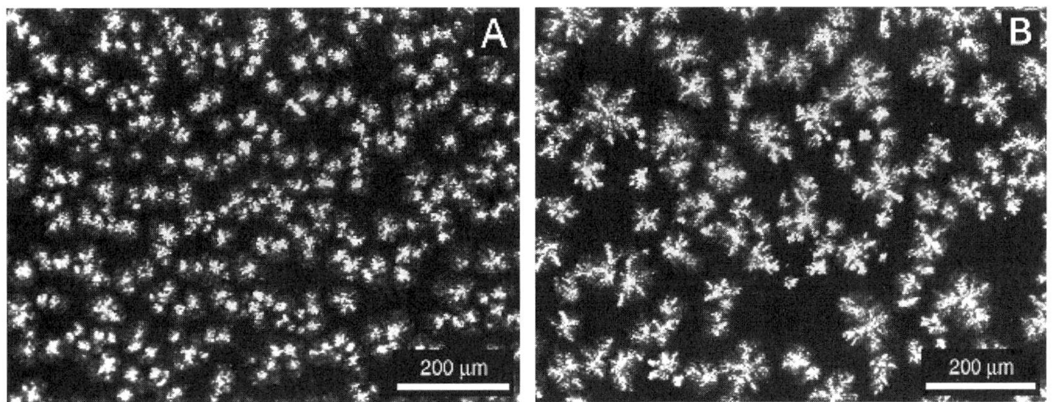

FIG. 2. A) Polarized light micrograph of 70:30 triolein:HMF with 0.1% Tween60 after 15 minutes at 28°C. B) Polarized light micrograph of 70:30 triolein:HMF with 0.5% Tween60 under the same crystallization and storage conditions as A.

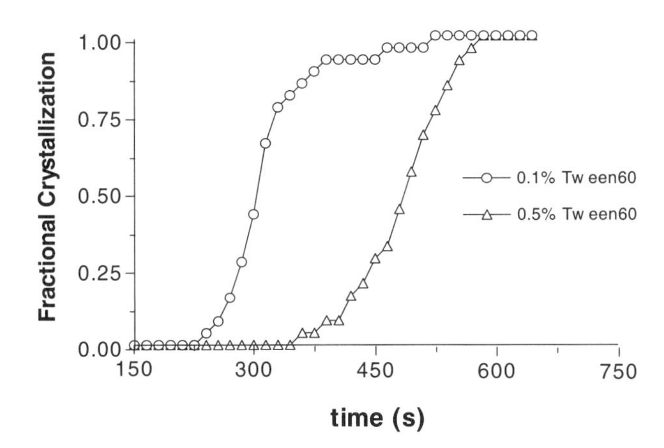

FIG. 3. Fractional crystallization as a function of time for samples of 70:30 triolein:HMF with 0.1% and 0.5% Tween60 added. Fractional crystallization was designated as the percentage of crystal mass relative to the final mass attained, determined by % black upon thresholding.

Microsc. Microanal. 8 (Suppl. 2), 2002
DOI 10.1017.S1431927602100158

Light Microscopy and TEM to Study the Effect of Biopolymers on Ice Recrystallization in Ice Cream

Alejandra Regand and H. Douglas Goff

Department of Food Science, University of Guelph, Guelph, ON, N1G 2W1, Canada

Hydrocolloid stabilizers are widely used in ice cream because they hinder ice crystal growth as temperature fluctuates during storage [1]. However, their mechanisms of action are still uncertain. Light microscopy (LM) and transmission electron microscopy (TEM) techniques have been applied to study microstructure and ice recrystallization in ice cream model solutions. Sucrose solutions with or without stabilizers (carboxymethyl cellulose (CMC), xanthan gum, locust bean gum (LBG), and gelatin) and with or without milk solid-non-fat (MSNF) were frozen in a scraped surface heat exchanger and temperature cycled (5 cycles from –6°C to –20°C). Brightfield images were acquired from samples before and after cycling. Measurements of ice crystal size were made by manually tracing the perimeter of the crystal with a computer mouse, the area of each crystal was automatically calculated by the software *Scion Image 1.62* [2]. Ice crystal size distributions were characterized by the logistic dose response model [1]. In the absence of milk proteins, xanthan and LBG were the most effective stabilizers at retarding recrystallization; while in their presence, only xanthan had an effect.

Cycled samples were freeze-substituted (3% v/v glutaraldehyde) and low-temperature-embedded (LR Gold). Resin blocks were sectioned at a thickness of 0.5μm for LM and 90nm for TEM. Stabilizer gel-like structures were observed in sections from LBG, gelatin and gelatin/MSNF solutions after being subjected to differential staining (leucobasic fuchsin for carbohydrates and amido-black for proteins [3]) in LM and uranyl acetate-lead citrate staining in TEM [4]. Representative LM (Fig. 1) and TEM (Fig. 2) micrographs show thermodynamic incompatibility and phase separation between biopolymers that promotes localized high concentrations of milk proteins located at the ice crystal interface, probably exerting a water-holding action that significantly enhances the stabilizer effect in retarding recrystallization. Phase separation was directly proportional to ice crystal growth inhibition [5].

References

[1] Flores, A. A. and H. D. Goff. J. Dairy Sci. 82 (1999) 1408.
[2] *Scion Image 1.62.*Scion Corporation, USA.
[3] Clark, G. Staining Procedures. Williams & Wilkins, Baltimore, MD, 1981.
[4] Lewis, P. R. and D. P. Knight. Staining methods for sectbning material. Elsevier, UK, 1977.
[5] Special gratitude to Sandy Smith, Ken Baker and Christine Epp for their contribution. This research was supported by NSERC Canada and CONACYT Mexico.

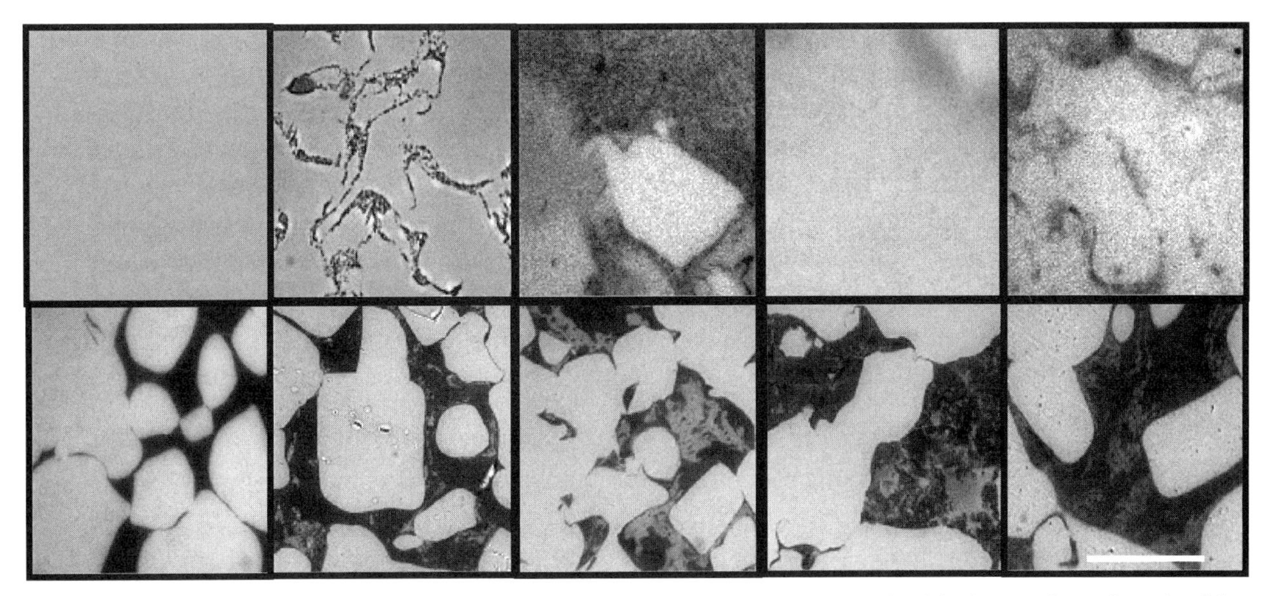

Fig. 1. LM micrographs of freeze substituted and low temperature embedded samples after double staining with leucobasic fuchsin and amido-black. Samples contain sucrose a) without MSNF, b) with MSNF, 1) control, 2) LBG, 3) xanthan 4) CMC, 5) gelatin. *No results.

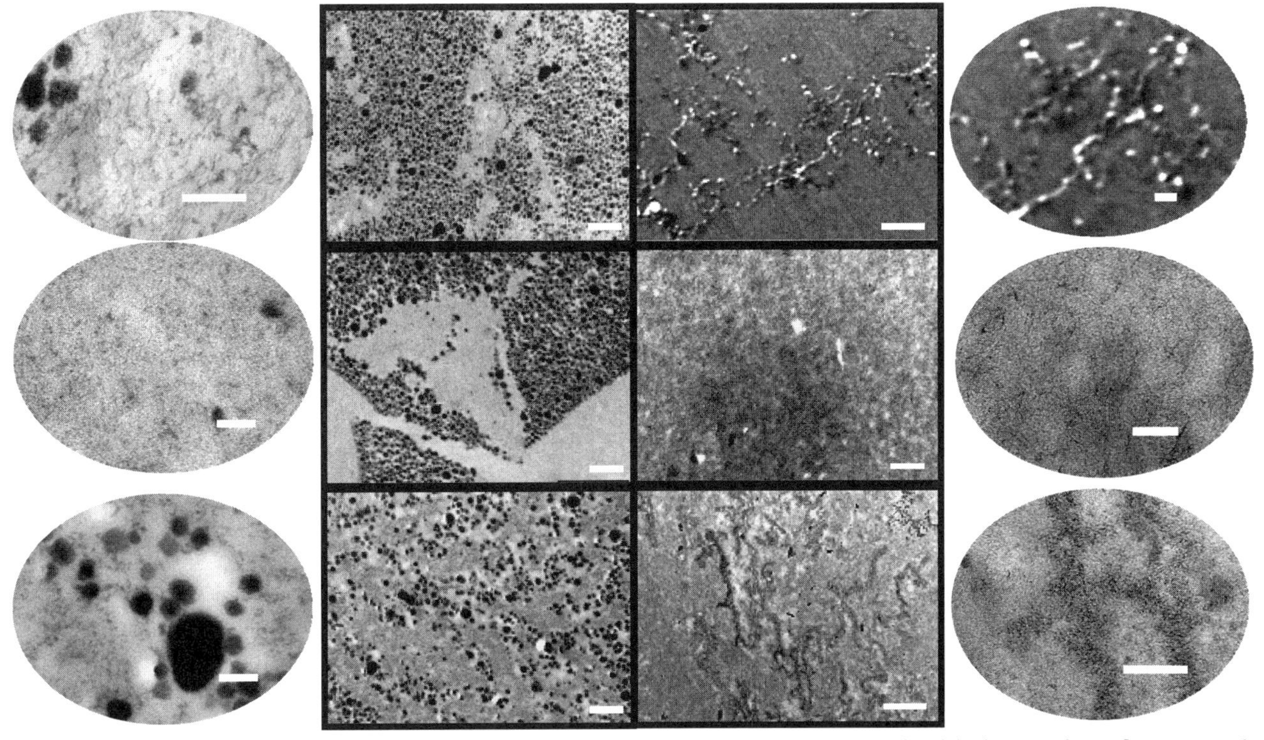

Fig. 2. TEM micrographs of freeze substituted and low temperature embedded samples after uranyl acetate-lead citrate staining. Samples contain sucrose a) with MSNF, b) without MSNF, 1) LBG, 2) xanthan, 3) gelatin. Calibration bar for square images=1μm, for oval images=0.2μm.

Microsc. Microanal. 8 (Suppl. 2), 2002
DOI 10.1017.S1431927602100055

Cytological features of programmed cell death in *Nicotiana tabacum* cells in relation to the expression and localization of death regulators.

Louise Brisson, Nathalie Bolduc, Mario Ouellet, Frédéric Pitre and Israel Fortin.

Department of Biochemistry and Microbiology, Life and Health Sciences Research Building, Laval University, Québec, Canada, G1K 7P4.

Plant cells could benefit from programmed cell death (PCD) in a number of developmental and physiological processes [1]. For example, during the establishment of the hypersensitive response (HR) PCD could play an important protecting role by limiting pathogen ingress. This death program is possibly regulated by a number of genes. To date, some regulators of plant PCD have been identified, but very few have been found to be homologous to animal PCD regulators [2]. Among these is the plant Bax Inhibitor-1 (BI-1) protein, which possesses, like its human counterpart, the ability to suppress the lethality induced by the proapoptotic regulator Bax [3-4]. Of interest, expression of BI-1 is rapidly upregulated in plant during wounding or pathogen challenge [4]. To increase our understanding of plant PCD and its mechanism of regulation, we have exploited genetic engineering to study the cytological features of Bax expression in mammalian and tobacco cells as well as to study the intracellular distribution of BI-1 protein in plant cells.

Cytological studies were conducted using transformation of human 293 cells, tobacco BY-2 cells or tobacco Xanthi leaves. Localization studies of Bax and BI-1 were investigated with the fusion proteins BaxGFP and BnBI-1GFP. Cells transformed with untargeted GFP, or GFP targeted to the ER were used as controls. Compared to non transformed cells, murine Bax expression was associated with an increased accumulation of mammalian and tobacco dead cells. This Bax-induced death was reduced by co-transforming mammalian cells with Bcl-2 or with BI-1 indicating that both plant BI-1 homologues and Bcl-2 can suppress Bax-induced apoptosis at a similar level (Figure 1). Production of Bax and Bcl-2 in transfected cells was attested by immunoblot analysis (Figure 1).

BY-2 cells expressing GFP alone or in fusion to Bax showed diffuse fluorescence distributed throughout the cytoplasm and in the nucleoplasm (Figure 2A). Bax-induced-cell death was frequently accompanied by typical “apoptosis”-related changes in nuclear morphology. In BY-2 cells expressing BnBI-1 GFP, the fluorescence was mainly observed in the vicinity of the nuclear envelope and within endoplasmic reticulum (data not shown). Of interest, fluorescence of BY-2 cells expressing BnBI-1-GFP or ER-GFP was localized to a number of spherical vesicles after an exposure to 125 mM SA, which was markedly different to the distribution observed in untreated cells (data not shown). Taken together, our observations suggest that regulators of cell death could operate in diverse organisms indicating that this process is functionally conserved throughout the evolution.

[1] A. Danon et al., *Plant Physiol. Biochem* 38 (2000) 647.
[2] R.I. Pennell and C. Lamb, *Plant Cell* 9 (1997) 1157.
[3] M. Kawai et al., *FEBS Lett.* 464 (1999) 143.
[4] P. Sanchez et al., *Plant J.* 21 (2000) 393.

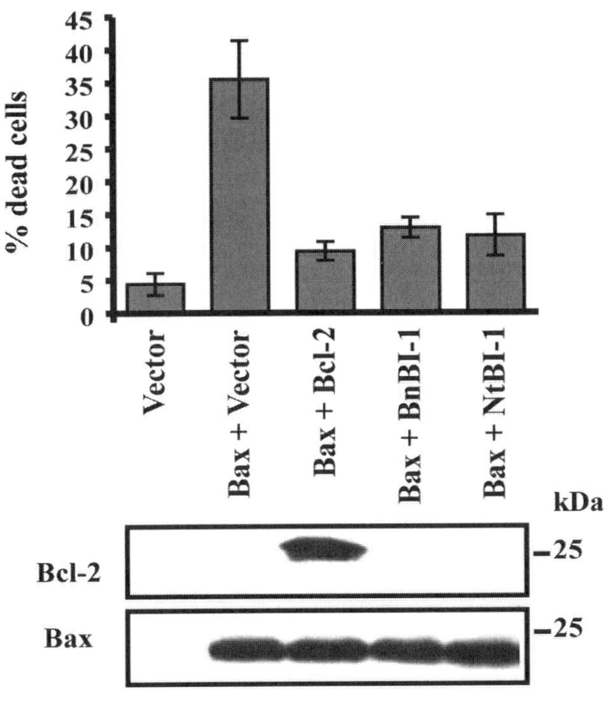

Figure1. Death assay in human 293 cells. Cells were either transfected with control vector, or cotransfected with Bax plasmid together with control vector or plasmids encoding Bcl-2, BnBI-1 or NtBI-1. Dead cells were determined using trypan blue dye exclusion assay (top). Protein extracts from a fraction of transfected cells were subjected to immunodetection using antibodies specific for Bcl-2 and murine Bax proteins (bottom).

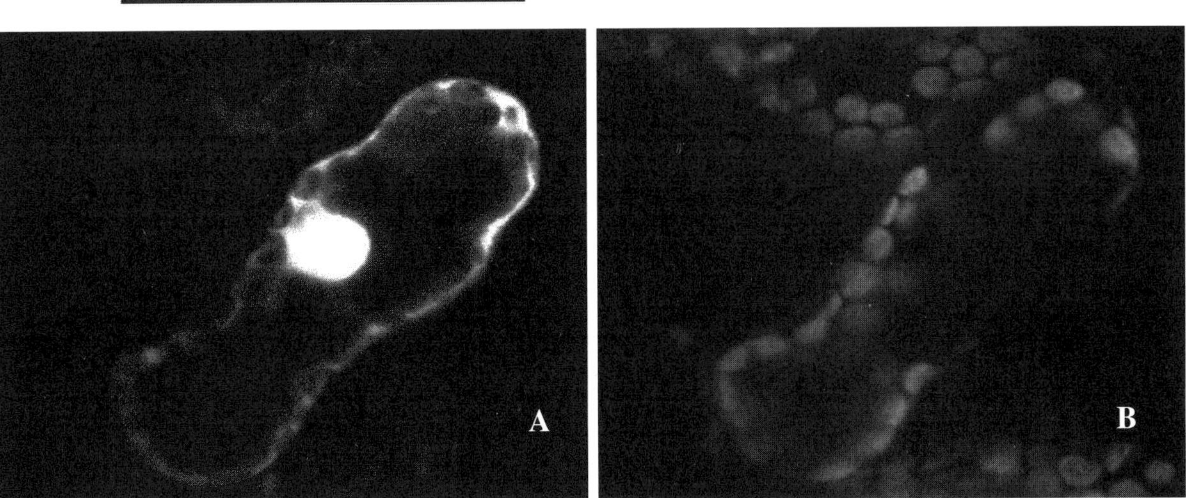

Figure 2. Transient Bax expression in palisade mesophyll tobacco cells. Transformed plant cells were viewed under a confocal microscope at an excitation wavelength of 488 nm (A) and 543 nm (B) to view the fluorescence of GFP and chlorophyll, respectively.

Microsc. Microanal. 8 (Suppl. 2), 2002
DOI 10.1017.S1431927602100067

In Situ Localization of AOS in Host-Pathogen Interactions

Klaus B. Tenberge ,* Marcus Beckedorf,* Britta Hoppe,* Alexander Schouten,** Martina Solf,* and Marcell von den Driesch*

* Institut für Botanik, Westfälische Wilhelms-Universität Münster, Schlossgarten 3, D-48149 Münster, Germany; email: tenberg@uni-muenster.de
** Wageningen University Plant Sciences, Laboratory of Phytopathology, PO Box 8025, 6700 EE Wageningen, The Netherlands

Active oxygen species (AOS) of host origin accumulate transiently in plant apoplasts in response to pathogen attack. This rapid reaction is called oxidative burst and probably functions in defense reactions directly or in signaling [1]. The aim of the presented work is to gain insight in role and source of AOS by visualizing the spatio-temporal occurrence of AOS during pathogenesis. As the role of AOS seems to vary with the life strategy of a pathogen, we investigate the interactions of a necrotrophic and of a biotrophic pathogen with their respective hosts.

Botrytis cinerea causes serious diseases, called grey mould, in at least 235 plant species, including important crops. The pathogen is a typical necrotroph, inducing host cell death before invasion. As the oxidative burst usually results in host cell death, the pathogen might utilize this induced suicide during the preparation of its primary infection court. In addition, the pathogen might produce AOS itself to intensify the effect.

The histopathology was investigated using light and electron microscopy. Spore adhesion, germination and swellings of germ tube tips attached to the surface by a matrix material was documented on tomato and bean. Germ tube tip swellings were demonstrated to develop functional appressoria that mediated cuticle penetration and infection hypha formation in the outer epidermal wall underneath (Fig. 1). Using LM, chloronaphthol and NBT staining were applied to visualize hydrogen peroxide and superoxide, respectively. The primary lesion area stained positive for AOS and most intense at its border zone. With time this concentric zone spread in the leaf mesophyll. The source of this wavelike AOS production was the host oxidative burst. This was visualized even by LM at higher magnification and further substantiated in TEM. The cerium chloride technique specific for hydrogen peroxide resulted in electron dense precipitate of cerium perhydroxide at the interface of *B. cinerea* and host cells [2]. This precipitate was present in the periplasmic space, in the host cell wall and at the outer surface of the host cell as well as at the outside of the fungal wall. Consequently, H_2O_2 was produced in host cells at the plasma membrane and diffuses through the host cell wall into the intercellular space.

Rectangular crystals were found to be an interaction-specific reaction product on lesion surfaces. Cytochemistry and ESEM with EDX revealed that they were formed of calcium oxalate. This could serve as substrate of fungal oxalate oxidases generating hydrogen peroxide. Notably, other cytological and biochemical evidence established the presence of an AOS secretion pathway in the pathogen. AOS specific staining was observed inside and outside of vital fungal cells in axenic culture and during pathogenesis. Activity staining and use of specific inhibitors substantiated that fungal superoxide dismutases were involved in this process. Using TEM, secreted hydrogen peroxide was visualized during penetration, indicating an oxidative attack upon the host cuticle and outer epidermal cell wall (Fig. 2). This seems to be a new and important discovery, since target mutagenesis ruled out an essential role of a cutinase gene in the pathogenesis of grey mould [3].

Claviceps purpurea causes ergot disease by infecting ovaries of grasses. This pathogen is a holobiotroph which obtains nutrients only from living host tissue while managing to maintain host cell viability for extended periods. Since *C. purpurea* produces compatible interactions with hundreds of host species, we hypothesized that the fungus might interfere with H_2O_2-mediated defense by means of secreted catalases and other AOS scavenging enzymes [4, 5, 6]. While investigating the spatio-temporal distribution of different AOS and related enzymes in this system, cytochemical evidence indicated that this pathogen also produces AOS itself [7].

References

[1] C.S. Bestwick et al., *Plant Cell* 9 (1997) 209-221.
[2] T.W. Prins et al., in *Fungal Pathology*, pp. 33–63, Kronstad J.W., ed., Kluwer Academic Publishers, Dordrecht, 2000.
[3] J.A.L. van Kan et al., *Mol. Plant-Microbe Interact.* 10 (1997) 30–38.
[4] V. Garre et al., *Phytopathology* 88 (1998) 744–753.
[5] V. Garre et al., *Mol. Plant-Microbe Interact.* 11 (1998) 772–783.
[6] P. Tudzynski and K.B. Tenberge, in *Clavicipitalean Fungi: Evolutionary Biology, Chemistry, Biocontrol and Cultural Impacts*, J.F. White et al., eds., Marcel Dekker, Inc., New York (in press).
[7] Our research was funded by the EU in the EU-FAIR project and by the DFG.

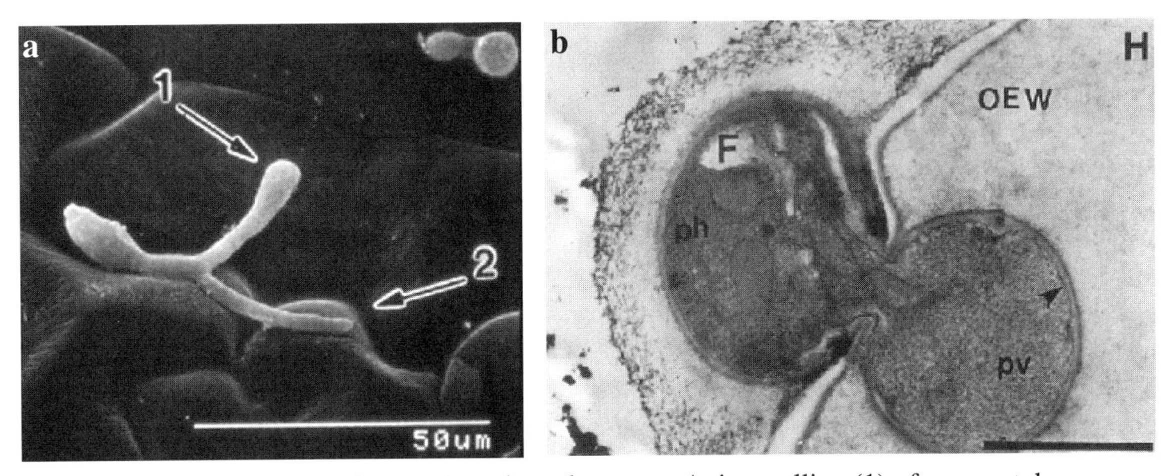

Fig. 1. Infection structures of *Botrytis* on bean leaves. a, A tip swelling (1) of a germ tube. b, Appressorium-mediated penetration of *B. cinerea* (F) into the outer epidermal wall (OEW) of a bean leaf (H) at 12 hpi. Scale bar, 1 μm.

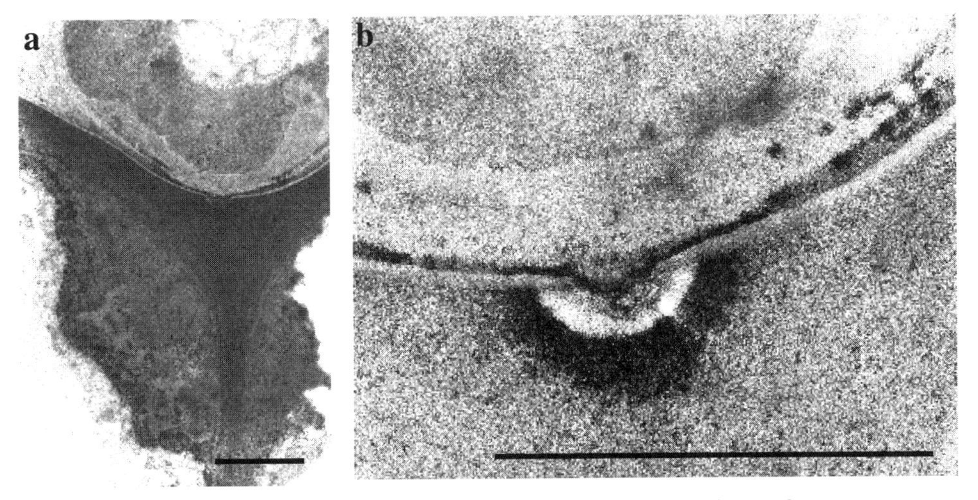

Fig. 2: Localization of H_2O_2 by cerium chloride at the interface of a *B. cinerea* appressorium and the outer epidermal wall of a tomato leaf at 12 hpi. Overview (a) and detail (b) from a section series.

Microsc. Microanal. 8 (Suppl. 2), 2002
DOI 10.1017.S1431927602100079

Use of High Pressure Freezing and Freeze Substitution to Study Host-Pathogen Interactions in Fungal Diseases of Plants

C. W. Mims,* and E. A. Richardson**

*Department of Plant Pathology, University of Georgia, Athens, GA 30602
**Department of Botany, University of Georgia, Athens, GA 30602

The value of high pressure freezing followed by freeze substitution (HPF/FS) for ultrastructural studies of host-pathogen interactions in fungal diseases of plants was first demonstrated in rust infections by Knauf et al. [1] Encouraged by these results, we began to explore the use of HPF/FS for the study of host-pathogen interactions in a variety of different types of plant diseases caused by biotrophic fungi. Here we discuss some of our results. Emphasis is placed upon the ultrastructure of haustoria and details of the haustorium-host cell interface.

Infected leaf samples approximately 1-mm^2 were placed in either brass or aluminum planchettes along with a drop of cryoprotectant (15% dextran in water) and frozen usinga Balzers 010 high pressure freezing machine. Frozen samples were transferred to liquid nitrogen and then to a substitution fluid consisting of 2% OsO_4 in anhydrous acetone containing 0.05 % uranyl acetate. Samples were substituted according to the procedures of Hoch [2] and then processed for study with transmission electron microscopy following the procedures of Mims et al. [3]

In the four host-pathogen systems discussed here we obtained many well-preserved host cells and fungal haustoria. Haustoria of *Exobasidium vaccinii* (Fig. 1), a parasite of *Rhododendron,* consisted of short, lobed branches that extended only a short distance into host cells. Each lobe possessed an electron dense cap-like structure and contained conspicuous branched, membranous inclusions. The haustoria of *Puccinia hemerocallis* (Fig. 2) and *Frommeëla mexicana* var. *indicae* (Fig. 3), two rust fungi we studied, were similar to those of other rust fungi. [1] Each possessed a slender neck region and an expanded body separated from the host cell cytoplasm by an apoplastic extrahaustorial matrix and an extrahaustorial membrane continuous with the host cell plasma membrane. The extrahaustorial membrane was well-preserved in cryofixed samples and exhibited a smooth profile (Fig.4). This membrane was continuous with tubular elements that extended into the host cell cytoplasm (Fig. 2). In *P. hemerocallis* many of these elements were beaded in appearance while others were connected to flattened cisternae that bore short beaded chains (Fig. 5). In powdery mildew of poinsettia, haustoria were produced only in leaf epidermal cells. Each haustorium consisted of a long slender neck and an expanded body with numerous finger-like branches (Fig. 6). Haustoria were separated from the host cell cytoplasm by an extrahaustorial matrix and an extrahaustorial membrane, portions of which were highly convoluted.

References

[1] G.M. Knauft et al., Physiol. Mol. Plant Path. 34 (1989) 519.
[2] H.C. Hoch, in Ultrastructure Techniques for Microorganisms, Eds. H.C. Aldrich and W. T. Todd, New York: Plenum (1986) 183.
[3] C.W. Mims, Can. J. Bot. 79 (2001) 383.

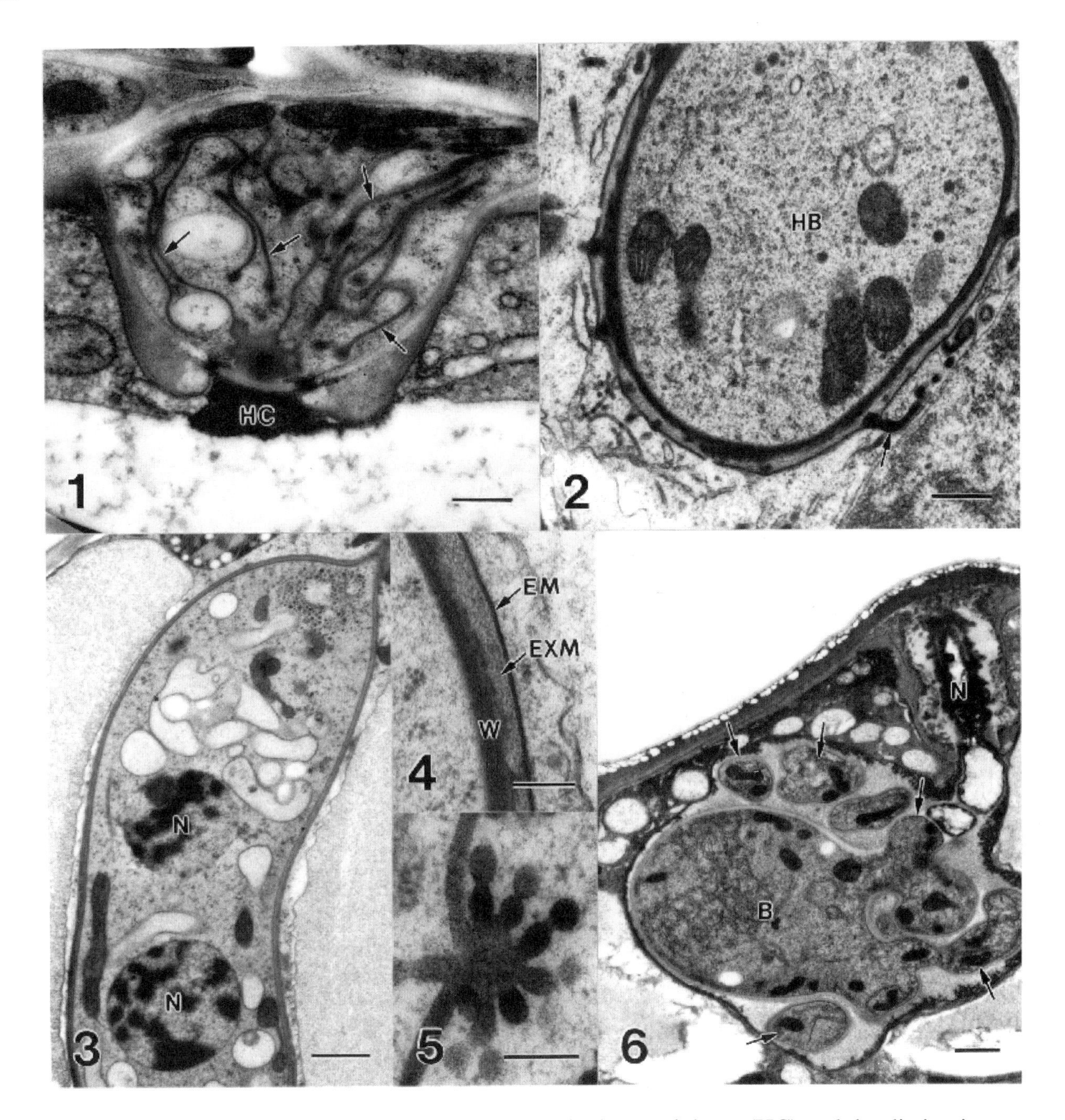

FIG. 1. Haustorium of *Exobasidiun vaccinii.* Note the haustorial cap (HC) and the distinctive membranous inclusions (arrows) present in the haustorium cytoplasm. Bar: 0.3 μm.
FIG. 2. Cross section of a haustorial body (HB)of *Puccinia hemerocallis* in a daylily leaf cell. Note the tubular element (arrow) protruding into the host cell cytoplasm. Bar: 0.3 μm.
FIG. 3. Longitudinal section of part of a haustorial body of *Frommeëla mexicana* var. *indicae* in a leaf cell of Indian strawberry. The nuclei of the haustorium are visible at N. Bar: 1 μm.
FIG. 4. The host-cell interface between *P. hemerocallis* and a daylily cell. The haustorial wall (W), extrahaustorial matrix (EXM) and extrahaustorial membrane (EM) are visible. Bar: 0.15 μm.
FIG. 5. A flattened cisterna bearing beaded chains in an infected daylily cell. Bar: 0.3 μm.
FIG. 6. Haustorium of *Oidium* sp. in an epidermal cell of a poinsettia leaf. Note theneck (N), body (B) and finger-like branches (arrows) that comprise the haustorium. Bar: 0.3 μm.

Microsc. Microanal. 8 (Suppl. 2), 2002
DOI 10.1017.S1431927602100080

Cytochemical Localization of Fungal Wall Components in Host-Pathogen Interactions: Particular Labeling with Gold-complexed Probes

G. B. Ouellette*, R. P. Baayen**, H. Chamberland***, M. Simard* and P. M. Charest***

*Canadian Forest Service, Laurentian Forestry Centre, Quebec, Canada
**Plant Protection Institute, Wageningen, The Netherlands
***Universite Laval, Health and Life Sciences Research Building, Quebec, QC, Canada

Gold-complexed lectins, mostly WGA, and enzymes are currently used to localize fungal wall components. Labeling for chitin is generally associated with the lucent wall layer, and erratic labeling is paralleled by irregularities in the wall layer [1]. Concentrations of gold particle occurrence close to fungal cells have also been observed suggesting that chitin components may be liberated into the surrounding medium. When using gold-complexed rabbit polyclonal antibodies to fungal compounds, control tests with pre-immune serum have often shown that the probe strongly attached to fungal cell walls, indicating that using concomitantly such a test is essential. Without it, however, the results obtained with the antibody may still be informative. For example, tests with a labeled polyclonal antibody to fungal fimbriae to identify extracellular structures produced by *Phaeotheca dimorphospora* showed that the probe attached to components in the medium at a distance from the fungal colony; these were similar to structures marginating the fungal cell (Figs 1, 2). As this fungus inhibits other fungal plant pathogens [2], and as a compound obtained from the agar medium distally from the fungal colony is strongly inhibitory as well to a fungal human pathogen (L. Giasson, Universite Laval, unpublished), these labeling results may be meaningful.

Another peculiar result was obtained using a gold-complexed monoclonal antibody (mab) against esterified pectin in histopathological studies of infection by *Fusarium oxysporum* f.sp. *dianthi*, causing a wilt in carnation. Walls of fungal cells, even of endocells, clearly labeled with the probe (Fig. 3), which was shown in the same tests, to bind specifically to host plant cell walls. As questioned by Chamberland [3] the labeling in such cases might not have been indicative of a fungal wall component. Indeed, subsequent similar tests with the fungus in culture gave negative results. However, by growing the fungus on an agar medium supplied with citrus pectin, this substrate was found to bind closely to the fungal wall (Fig. 4), indicating that in the infected plant, pectin residues might have been fully adsorbed by the fungus cell walls. As fungal walls in susceptible plants generally labeled less strongly than in resistant plants, these results may have a bearing on our understanding of the host-pathogen relationships in this disease.

References

[1] G.B. Ouellette *et al.* 2001. Chitin Enzymology. Atec Edizioni, Grottamare, Italy. 79-89.

[2] D. Yang *et al.* 1993. Can. J. Bot. 71:426.

[3] H. Chamberland. 1994. Host Wall Alterations by Parasitic Fungi. APS Press. pp 1-11.

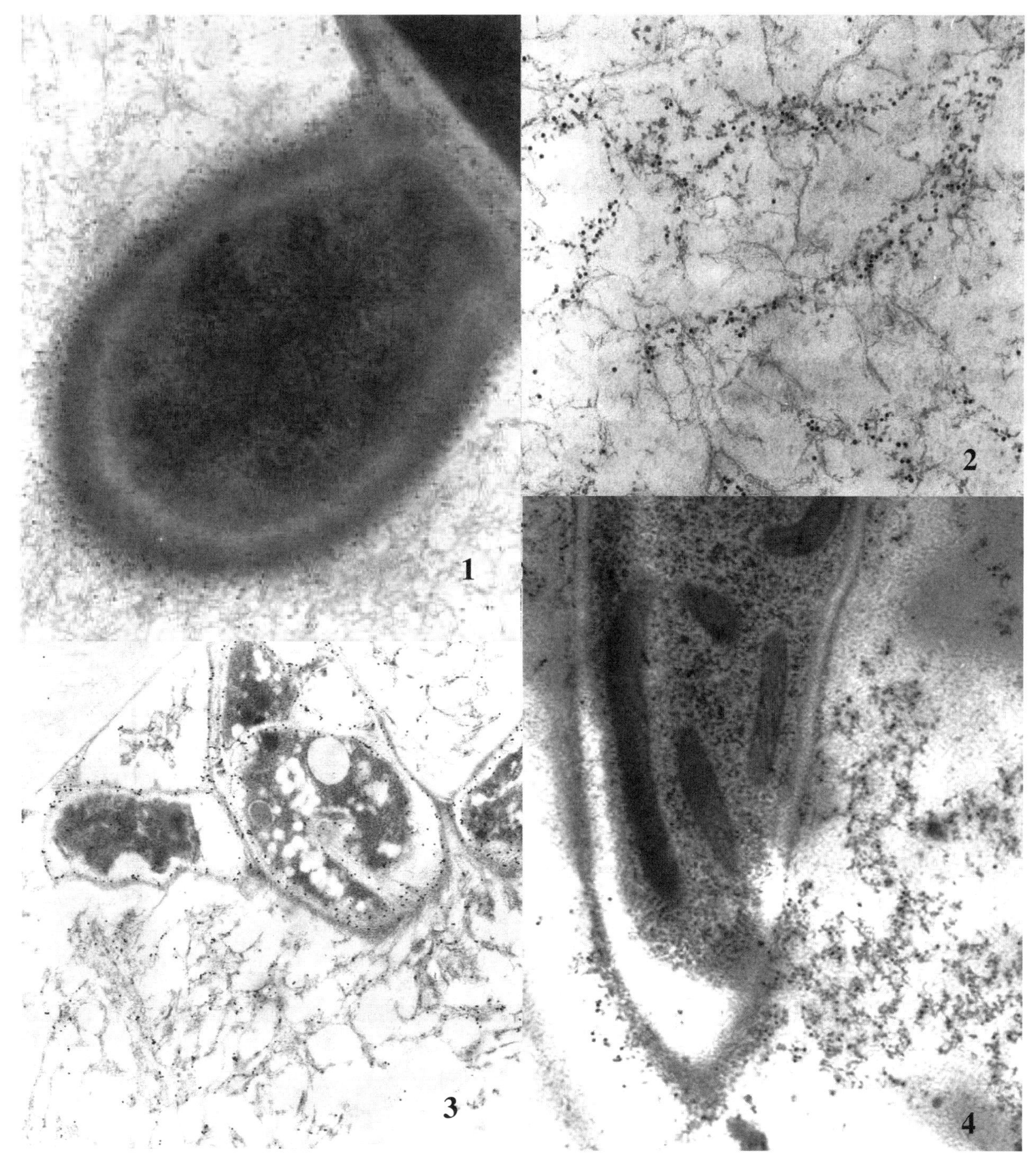

FIGS 1, 2. *Phaeotheca dimorphospora*. Labeling with polyclonal antifimbriae (kindly furnished by Dr. A Day, Univ. Western Ontario). FIG. 1 Outside layer of fungal cells contains labeled fimbriae extending into the surrounding medium. FIG. 2. Labeling is concentrated on the more compact patches of amorphous material, bordered by more fibrillar matter which extended some distance from the fungal cells into the medium. FIGS 3, 4. *Fusarium oxysporum* f. sp. *dianthi*. Labeling with mab to esterified pectin (JIM 5). FIG. 3. Walls of altered fungal cells and fibrillar material nearby, in vessel lumen, are well labeled. FIG. 4. Walls of fungal cell in culture containing pectin components, are not labeled, except sparsely where the pectin substrate contacts the fungus wall.

Microsc. Microanal. 8 (Suppl. 2), 2002
DOI 10.1017.S1431927602100092

Ultrastructural Investigation of the Mycoparasitic Interaction between *Stachybotrys elegans* and its host *Rhizoctonia solani*

P. M. Charest* G. Taylor* and S. H. Jabaji-Hare**

*Université Laval, Dept. de Phytologie, Québec, QC, Canada G1K 7P4
**McGill University, Dept. of Plant Science, Montreal, QC, Canada H9X 3V9

Stachybotrys elegans, a soil-borne fungus, has been described as a destructive mycoparasite of *Rhizoctonia solani* [1]. Detailed light and ultrastructural studies of *R. solani* hyphal and sclerotial cells parasitized with *S. elegans* revealed that parasitism of host cells was characterized by hyphal attachment and enhanced production of extracellular fibrillar material during contact, penetration of host cell walls and proliferation of trophic hyphae within the host cytoplasm [2].

S. elegans produces two exo- and one endo-acting chitinases when grown on chitin. We purified to homogeneity one of the exo-acting chitinases, β-*N*-acetylhexosaminidase, and partially characterized its physical and biochemical properties. The native enzyme has a molecular mass of 120 kDa when determined by gel filtration, and 68 kDa by SDS-PAGE which suggests that the protein may occur as a dimer in solution. The purified β-*N*-acetylhexosaminidase is most active at pH 5.0 and at 40° C; it hydrolyzed the ρNP-*N*-acetyl-β-D-glucosaminide with apparent K_m of 84.6 μM. Polyclonal antibodies raised against the 68 kDa β-*N*-acetylhexosaminidase (NAG-68) indicated that the antibody is highly specific and recognizes the protein in crude filtrate preparation (Fig. 1) [3].

Transmission electron microscopy observations of areas sampled from the interaction zone of both fungi in dual culture revealed that *S. elegans* penetrated *R. solani* thick hyphal cell walls and the cytoplasm of the host cells was in an advanced state of disintegration (Fig. 2). Interestingly, in sections sampled from areas where both fungi were in close vicinity but not in direct interaction, *R. solani* cytoplasm was highly altered (data not shown). This indicates that the mycoparasite can affect its host in advance to penetration. At penetration sites (Fig. 3), the extent of the extracellular fibrillar material surrounding S. elegans cells was often abundant. In order to demonstrate that an exo-acting chitinase of *S. elegans* is involved in the degradation process of *R. solani*, we carried out immunocytochemical investigation using the characterized antibody described here. These studies are complemented with other cytochemical tests for the localization of chitin.

References

[1] Benyagoub M. *et al.* 1994. Mycological Research 98:493-505.

[2] Benyagoub M. *et al.* 1996. Mycological Research 100:79-86.

[3] Taylor et al. 2002. Purification and characterization of an extracellular exochitinase, β-*N*-acetylhexosaminidase from the fungal mycoparasite *Stachybotrys elegans*. Can. J. Microbiol. (in press).

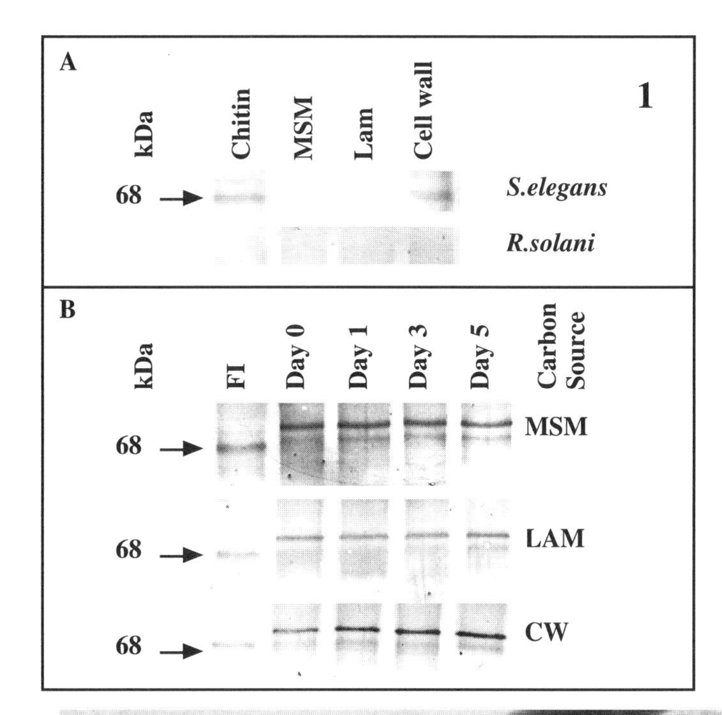

Figure 1: Western blot analysis of *S. elegans* and *R. solani* interactions using a polyclonal antibodies raised against the purified 68-kDa exochitinase. (A) Intercellular proteins (250 ng/lane) extracted from *S. elegans* and *R. solani* grown on MSM supplemented with different carbon sources ($0.5 g.L^{-1}$). Chitin (CHT), laminarin (LAM), *R. solani* cell wall fragments (CW), or no carbon source (MSM). (B) Intercellular proteins extracted from the interaction zone of *S. elegans* and *R. solani* in dual culture plates 0,24, 72 and 120 h after contact. NAG-68kDa in fraction F I (240 ng /lane)

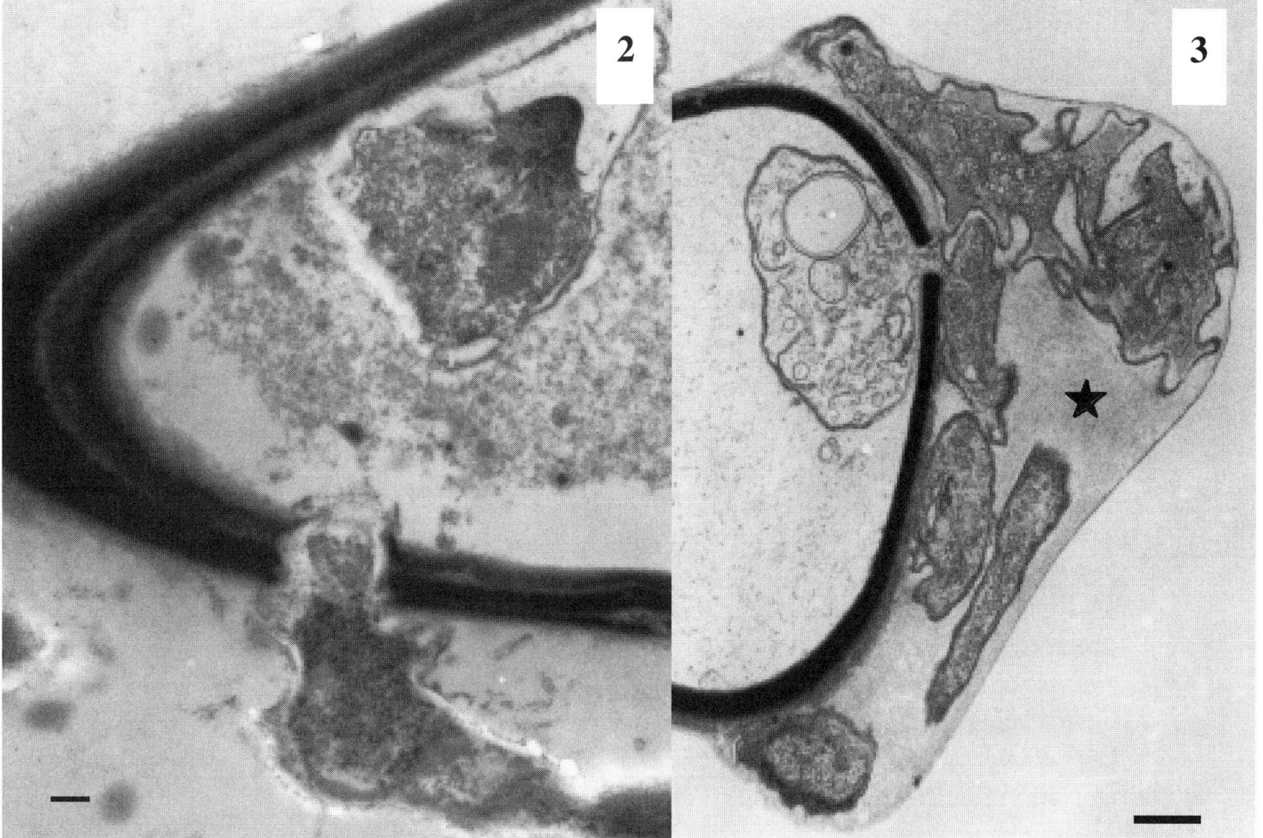

Figure 2: Penetration of *R. solani* hypha by *S. elegans*. The host cytoplasm is severely altered, no organelle is discernible. Scale bar = 500 nm. Figure 3: Moribund hypha or *R. solani* that has been penetrated by *S. elegans*. The cytoplasm of the host cell is desintegrated. The *S. elegans* hyphae of unusual form are included in a fibrillar matrix (*) outside the hypha of *R. solani*. Scale bar = 1 µm.

Microsc. Microanal. 8 (Suppl. 2), 2002
DOI 10.1017.S1431927602100432

Directing Protein Assembly At Interfaces: Balancing Electrostatic And Hydrophobic Forces and the Role of Epitaxy.

Christopher M. Yip*

*Departments of Chemical Engineering and Applied Chemistry, and Biochemistry, Institute of Biomaterials and Biomedical Engineering. University of Toronto 407 – 4 Taddle Creek Rd Toronto, Ontario, Canada M5S 3G9

The rational design of protein-based supramolecular architectures requires careful consideration of not only intramolecular structure but also the intermolecular interactions that control their self-association into higher order structures. We are particularly interested in the role of interfacial structure and chemistry in defining the nucleation and growth of these systems and specifically the synthesis of extended two-dimensional protein arrays. Extended-duration in situ tapping mode atomic force microscopy was employed to investigate the role of electrostatic and hydrophobic forces on the assembly of β-helix and β-sheet forming peptides at ordered interfaces.

While a 33-residue fragment of the hexapeptide LPX-A rapidly self-associated on mica to form well-defined rectangular rod structures in epitaxy with the (001) surface of mica, on highly oriented pyrolytic graphite, we only observed an amorphous 8 Å thick film. This is in stark contrast to the results obtained for a model hexapeptide repeat (NAKIGD) that was found to assembly epitaxially on both mica and HOPG. Similar studies performed with insulin revealed epitaxial growth of two-dimensional fibril domains on HOPG but only amorphous aggregation on mica.

Protein assembly is critically dependent not only on the amino acid sequence but also the conformation and relative exposure of specific functional domains. This is of particular importance for understanding the mechanisms by which proteins form coherent supramolecular architectures, such as extended beta-sheet or beta-helical fibres. Such a motif has been proposed to underlie the unique mechanical and physical properties of the extracellular protein, elastin, which can undergo reversible gelation, or coacervation, at elevated temperatures [1]. We report here the results of a variable temperature in situ atomic force microscopy study of the assembly of recombinant human tropoelastin proteins EP I and EP II at ordered interfaces. On mica, both of these peptides assemble as discrete stable globular aggregates. However, on highly ordered pyrolytic graphite (HOPG), the assembly process proceeds via the initial formation of an two-dimensional film and subsequent nucleation of high-aspect ratio rods on this prelayer [Figure 1] Remarkably these aggregates appear to form in an ordered epitaxial relationship with the underlying hexagonal lattice of the (0001) cleavage plane of HOPG. For EP I and II, the surface transition temperature was higher than the known solution coacervation temperatures suggesting that surface confinement of the peptide presents an additional energetic barrier to fibril nucleation that is not present in solution [Figure 2].

These unique characteristics provide an intriguing design strategy for the creation of novel peptide-based nanostructures. Our results suggest that careful manipulation of surface electrostatics and hydrophobics can be used to control protein assembly at interfaces [2]

[1] C.M. Bellingham et al., Biochim Biophys Acta 1(2001), 6
[2] This work was supported by the Canadian Institutes of Health Research (MT-14769), NSERC (194435-99), the Canada Foundation for Innovation, the Ontario Research and Development Challenge Fund, and the Ontario Innovation Trust. C. M. Y. is a Canada Research Chairs Program Chairholder.

Figure 1a. In situ tapping mode AFM image of EPI peptide on HOPG acquired in Tris buffer solution at 50°C 30 minutes after peptide addition. Note that fibril formation has begun and is in alignment with underlying substrate. Image size: 4 μm x 4 μm

Figure 1b. In situ tapping mode AFM image of EPI peptide on HOPG acquired in Tris buffer solution at 50°C approximately 2 hrs after the initial addition of the peptide. Note the high density of ~ 1 nm thick fibres on the surface. Image size: 4 μm x 4 μm

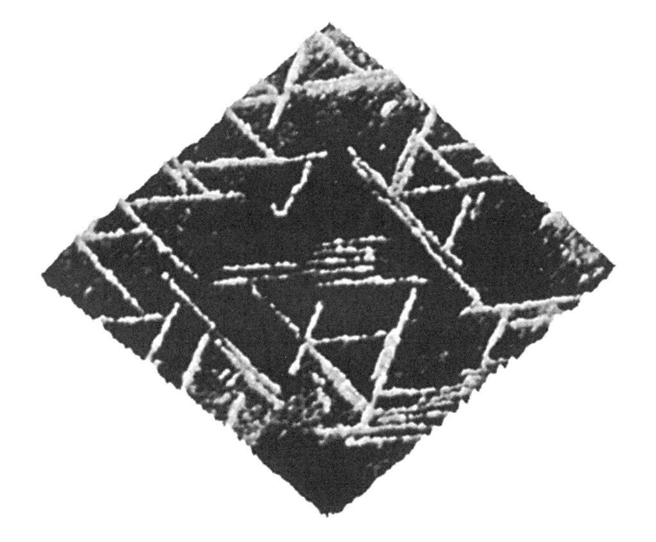

Figure 2a. In situ tapping mode AFM image of EPI peptide on HOPG acquired in Tris buffer solution at 50°C. This image clearly reveals the complex nature of the EP I multilayer assembly Image size: 1 μm x 1 μm

Microsc. Microanal. 8 (Suppl. 2), 2002
DOI 10.1017.S1431927602100444

Spreading of Fibrinogen at Model Surfaces Studied by AFM

C. A. Siedlecki[a,b] and A. Agnihotri[b]

Departments of [a]Surgery and [b]Bioengineering, The Milton S. Hershey Medical Center,
The Pennsylvania State University, Hershey, PA 17033

Fibrinogen is a crucial protein in surface-induced thrombus formation due to its importance in mediating platelet adhesion to biomaterials by serving as an adhesive ligand for the $\alpha_{IIb}\beta_{III}$ integrin receptor. The adsorption and function of fibrinogen have been widely studied by a variety of techniques, and have demonstrated both time- and material-dependent properties of the protein[1], and have contributed to our understanding of fibrinogen's interesting properties on biomaterial substrates. In this study, we have utilized atomic force microscopy (AFM) to examine the changes in the protein structure as a function of both time and material as a step towards an ultimate goal of understanding molecular level structure/function relationships. These studies demonstrate that fibrinogen denatures much more rapidly and to a greater degree on a hydrophobic model surface than on a hydrophilic surface, as might be expected. Furthermore, analysis of the individual D and E domains of the molecule indicate differences in the rate of structural change within the molecule itself, a phenomenon which may be associated with submolecular properties of the protein

Fibrinogen structure was examined on two substrates of varying surface wettability, highly ordered pyrolytic graphite (HOPG), a hydrophobic material with an advancing water contact angle over 100°, and muscovite mica, a hydrophilic material with an advancing water contact angle of <10°. These model materials were chosen as they not only possess vastly different surface properties, but also are smooth enough to permit nonambiguous imaging of individual proteins. Fibrinogen was added at concentrations ranging from 500 ng/ml to 2 μg/ml in 1 mM phosphate buffer for 15 minutes in order to yield submonolayer amounts of protein. Imaging was initiated and continued for 2 to 3 hrs. Cross-sectional analysis was used to determine the dimensions of individual fibrinogen molecules. Due to the complexities associated with image enlargement, the height of the molecules was used as an initial indicator of molecular spreading.

Figure 1 illustrates a typical image for fibrinogen adsorbed on muscovite mica. A 1μm x 1μm image (Figure 1A, z-range = 8 nm) illustrates a field of both individual fibrinogen monomers and aggregates. The inset (1B) shows a single fibrinogen molecule with a characteristic trinodular structure. A cross-section (1C) clearly shows the three domains, labeled as D-E-D, and the measured heights of each of the domains. Figure 2 shows fibrinogen on HOPG in a similar manner. Again, a trinodular structure is observed, and cross-sectional analysis shows the height of the domains. Analysis of the molecules at sequential time points for each material results in a curve illustrating the time-dependence of the heights. This is seen in figure 3. For mica, the heights of both the D and E domains remain relatively constant over the time points studied. However, on HOPG, not only are the values of the measured heights ~½ that seen on mica, but they continue to decrease over the 2 hr time range. This is particularly notable for the D domains, and is consistent with what others have reported concerning the role of increased hydrophobicity of the D domain in fibrinogen film formation[2]. In summary, these results are consistent with extensive unfolding of adsorbed fibrinogen molecules on nonpolar hydrophobic surfaces as compared to polar hydrophilic materials, particularly in the relatively hydrophobic D domains[3].

References
1. T. A. Horbett. *Cardiovascular Pathology* **2,** (1993)137S.
2. T. C. Ta et al. *Langmuir* **14,** (1998) 2435.
3. The authors would like to acknowledge financial support from the Whitaker Foundation.

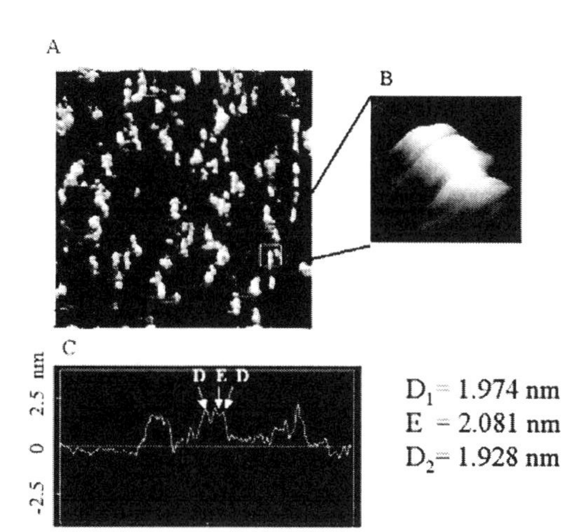

Figure 1: Fibrinogen molecules adsorbed on a hydrophilic mica substrate. Figure 1A is a 1 μm x 1 μm image, with a z-range of 8 nm. Figure 1B is a zoom of a single fibrinogen molecule as shown in the box in 1A. The trinodular structure is clearly visible. In 1C, a cross sectional analysis shows the three individual domains, labeled as D-E-D, as well as the measured heights of each.

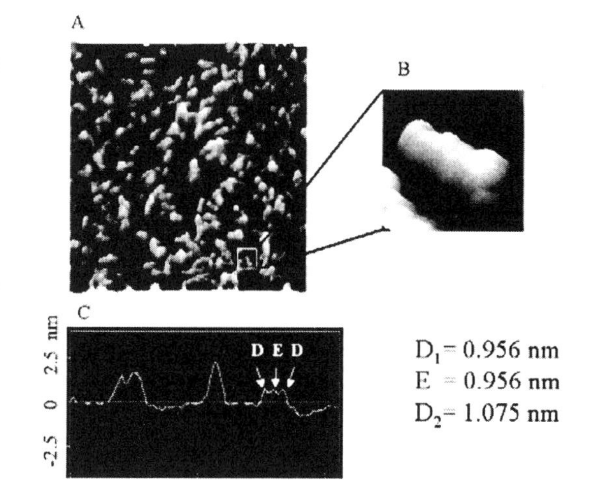

Figure 2: Fibrinogen molecules adsorbed on hydrophobic HOPG. Figure 2A shows a 1 μm x 1 μm image, while a higher magnification zoom of an individual fibrinogen molecule is shown in figure 2B. As with the mica substrate, the trinodular structure is clearly visible in this protein. In figure 2C, the cross-section shows these three domains, and indicates that the height of the molecules on graphite is significantly less than that seen on mica for each of the three domains.

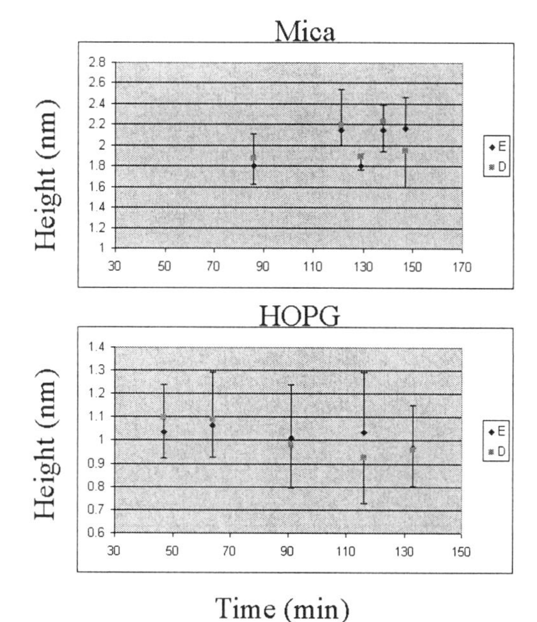

Figure 3: Time dependence of adsorbed fibrinogen height on muscovite mica and HOPG. On mica, the heights of both the D and E domains remain relatively constant and similar, while on HOPG, the height of the E domain remains relatively constant, but the height of the E domain decreases with time to a final value less than 1 nm. Note that both E and D domains are nearly ½ the height on HOPG as they are on mica. These results suggest that the relatively more hydrophobic D domain undergoes spreading to a greater degree than does the E domain, which may also be stabilized by the disulfide knot holding the six polypeptide chains together.

Microsc. Microanal. 8 (Suppl. 2), 2002
DOI 10.1017.S1431927602100456

Linking Atomic Force Microscope Images of Proteins to Their Genetic Sequence

Steven J. Eppell and Brian A. Todd

Nanoscale Orthopedic Biomaterials Laboratory, Dept. of Biomedical Engineering, Case Western Reserve University, Cleveland, OH 44106-7207

Efforts to link protein structure and biochemical function have focused mostly on primary structure, largely because of the abundance of information at this level; all of the amino acid sequences of the ~30,000 proteins in the human proteonome are known, whereas the three-dimensional crystal structures of only ~1000 proteins have been solved. Clearly, techniques that provide details concerning three-dimensional protein structure are needed to discover mechanisms governing biological function. Here, we have identified a motif in protein tertiary structure at sites of biochemical activity using scanning force microscopy (SFM). We show that sites on aggrecan, a cartilage proteoglycan, that are susceptible to catabolic enzymes are more flexible than other regions of the molecule. The results demonstrate a powerful new technique for investigating molecular scale structure-function relationships and suggest the role of flexibility in aggrecan degradation. This model system will be used to show how tip asymmetries can be accounted for using morphological processing coupled with AFM imaging simulation.

In Fig. 1A, we start with a model of a surface adsorbed aggrecan molecule. The backbone of the model is a 6 element 3rd order spline. Using an independently measured tip shape image (Fig. 1B), we then dilate the model to produce a simulated AFM image, Fig. 1C. The simulated image is then subtracted from a real AFM image, Fig. 1E, yielding an error image. This process is iterated using a method previously published to find the best fit of the model to the AFM data.[1] The result is an analytical function representing the surface adsorbed molecule.

Using intrinsic asymmetries in the models, we were able to register several such functions to obtain an average model. We looked for kinks in this model by evaluating it's second derivative (inversely proportional to the local radius of curvature). The values of these derivatives were then plotted as a function of contour length down the model backbone, Fig. 2A. Integrating previously published TEM data,[2] we mapped the backbone to the known genetic sequence[3] of the aggrecan molecule. The three highest sites of curvature overlap with three of the known aggrecanse cleavage sites.

1. W. Huyer and A. Neumaier, Global optimization by multilevel coordinate search, J. Global Optimization **14,** (1999), 331-355

2. M. Morgelin, M. Paulsson, A. Malmstrom, D. Heinegard, *J Biol Chem* **264**, (1989) 12080-90.

3. T. M. Hering, J. Kollar, T. D. Huynh, *Arch Biochem Biophys* **345**, 259-70. (1997).
We thank the NIH AR45664-02 and the Whitaker Foundation for their generous support.

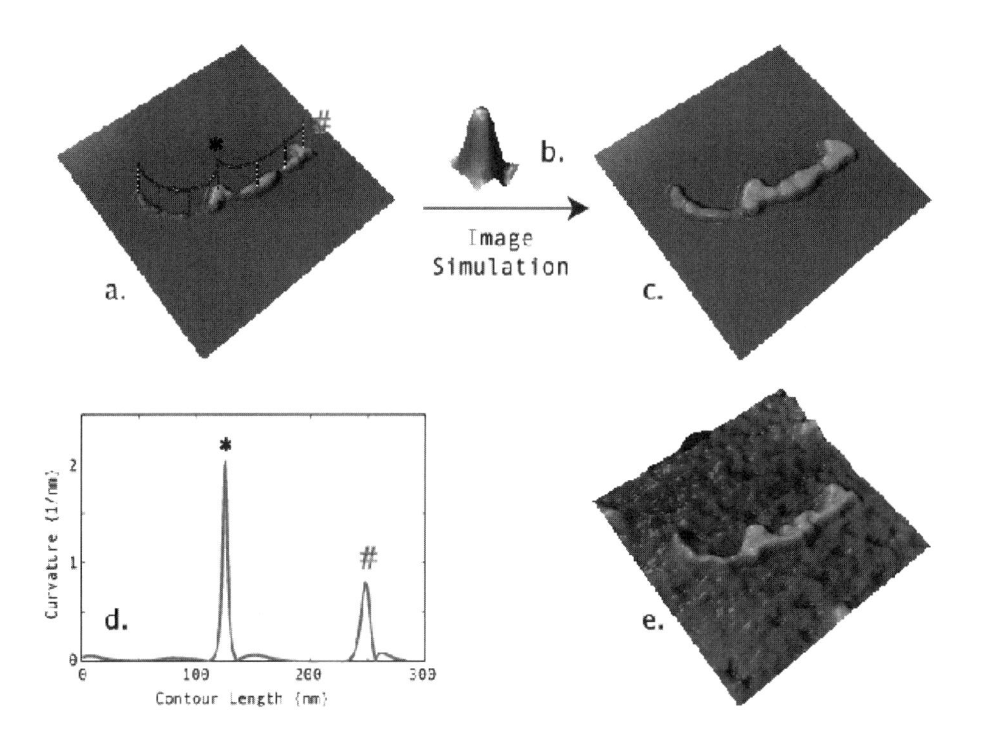

Figure 1

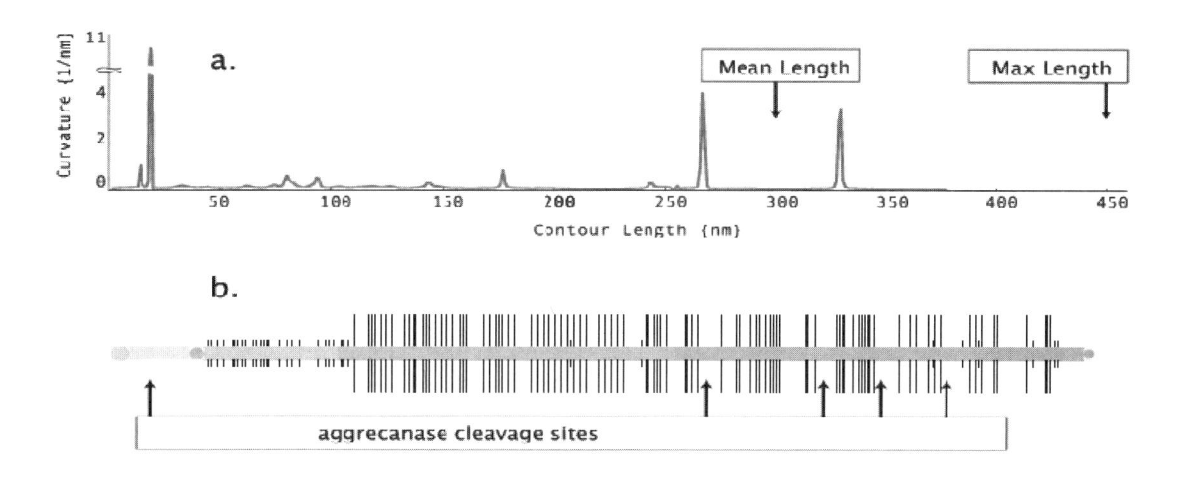

Figure 2

Microsc. Microanal. 8 (Suppl. 2), 2002
DOI 10.1017.S1431927602100365

Laser scanning microscopy: seeing more by imaging less.

N.S. White

Oxford University,
Sir William Dunn School of Pathology (Microscopy Lab), Oxford. UK. OX1 3RE
nick.white@path.ox.ac.uk

Conventional fluorescence microscopy results in all points of a sample being imaged simultaneously to a complete representation in a wide field of view. Illumination is not focused but bathes the whole specimen. The laser scanning microscope (LSM), has advantages by not adhering to this model.

In a non-confocal single-photon LSM, continuous laser illumination is focused (imaged) into a specimen. In-focus and out-of-focus fluorescence are not imaged but integrated by a large detector into a single measurement. The optical probe is scanned through the sample to buildup a 2-D picture. The imaging components determine the instrument resolution. Thus the shorter wavelength of the simple LSM illumination results in superior resolution to the longer fluorescence emission image of the conventional system. This has important consequences for multi-photon configurations.

In the Confocal (C)LSM, fluorescence from the focused laser at each sequential point in the scan is imaged via a small aperture by a 'point' detector. The aperture blocks out-of-focus light, resulting in optical sectioning. The detection is no longer wide-field, and the imaged spot means that the fluorescence wavelength must now be considered in addition to the excitation when determining resolution. Fortunately, the combination of two imaging components(in this case excitation and detection optics) gives an additional 1.4x increase in resolution (approx) and so the CLSM can deliver superior resolution to the conventional and simple LSM cases. This result also has implications for multi-photon configurations.

The MPLSM uses an optical probe confined to a small volume at the focus of a lens. This 3D probe is derived entirely from near-infrared laser illumination, in contrast to confocal (C-) LSM that combines illumination and detection optics for a similar result.

By increasing the laser power and operating at two (or three) times the usual excitation wavelength, multiple photons may be absorbed simultaneously by each fluorescent molecule. Excitation efficiency then increases by the 2nd or 3rd power, respectively, of the laser intensity and so only happens near the focus. An ultra-fast pulsed laser (~100fs @ 100 MHz) is needed for a high peak intensity but low time-averaged power to prevent specimen damage. With this multi-photon excitation no out-of-focus fluorescence is generated, so no confocal aperture is needed for optical sectioning. The simple LSM configuration described above can then provide maximum detection efficiency. The 2-photon process provides the equivalent of the dual-focussed excitation and detection of confocal microscopy in delivering a $\sqrt{2}$ increase in resolution compared to a conventional microscope operating at the same wavelength. 3-photon excitation can show a cube-root improvement. Unfortunately the resolution, as in the simple non-confocal case, is dependent on the excitation laser wavelength – i.e. in the near IR for 2- and 3-photon excitation, and is intrinsically worse than both the previous LSM examples under ideal conditions. The major benefits of MPLSM can, however, be realized when collecting data under non-ideal conditions.

Although the ultra-fast laser pulses used for multi-photon excitation have a wider bandwidth than continuous single-photon excitation, chromatic aberrations are substantially reduced with largearea detectors. Monochromatic corrections (such as spherical) can be incorporated more easily (even automated) in the excitation optics.

By exciting the fluorescence only in a localized volume around the focus, photo-damage away from the optical section is avoided. This is in contrast to the single-photon cases where unwanted out-of-focus planes are always excited (and thus subject to photo-damage) but the resultant fluorescence just degrades the image or is discarded in the confocal case by the detectoraperture.

For thick specimens that scatter light significantly we can further exploit the multi-photon configuration. As scattering losses decrease at longer wavelengths, near-IR illumination can penetrate significantly further than the shorter wavelengths used for single-photon excitation. However effective the IR light is at penetrating the specimen it still remains to collect the emerging fluorescence that will also be scattered by the sample. By removing the requirement to image the fluorescing spot onto a confocal aperture, the detection optics can be made to collect scattered light more effectively when focused deep within the specimen.
.
In order to assess the relative merits of these (and other) LSM configurations for different specimens, it is desirable to separate the many image-degrading processes into simple reproducible measurements. These can readily be achieved using some simple reflection and fluorescence test samples. SPR (surface plasmon resonance) signal from small gold particles offers an additional well-defined non-bleachable test specimen for some critical measurements.

Microsc. Microanal. 8 (Suppl. 2), 2002
DOI 10.1017.S1431927602100377

Characterization and Use of Wide-Field Fluorescence Microscopy and Image Restoration in Quantitative Live Cell Imaging

Melpomeni Platani*, Angus I Lamond*, and Jason R. Swedlow *

*Division of Gene Regulation and Expression, MSI/WTB Complex, University of Dundee, Dundee DD1 5EH Scotland

Digital fluorescence microscopy is now a standard tool for determining the localization of cellular components in fixed and living cells. Two fundamentally different imaging technologies are available for imaging fluorescently labelled cells and tissues, in either the fixed or living state. The laser scanning microscope uses a diffraction-limited focused beam to scan the sample and develop an image point by point. In addition, a pinhole placed in a plane confocal to the specimen prevents emitted out-of focus fluorescence from reaching the photomultiplier tube (PMT) detector. By combining spot illumination and selection of in-focus fluorescence signal, the laser scanning confocal microscope (LSCM) creates an image of the specimen largely free of out-of-focus blur [1]. By contrast, a wide-field microscope (WFM) illuminates the whole specimen simultaneously and detects the signal with a spatial array of point detectors, usually a charge-coupled device camera (CCD) [2]. This approach collects an image of all points of the specimen simultaneously and includes all the out-of-focus blurred light. Subsequent restoration by iterative deconvolution generates an estimate of the specimen, largely free of out-of-focus blur [3]. While many other fluorescence imaging modalities exist, these two methods represent the majority of the fluorescence imaging systems currently in use in biomedical research.

We (JRS) have recently completed a quantitative comparison of LSCM and WFM [4], focussing on commercially available versions of these microscopes. This analysis revealed that the most important parameter for comparing the performance of the two systems was the signal-to-noise ratio (SNR). Using both live cells and fluorescent beads standards, we showed that the intrinsic noise levels in commercial LSCMs are higher than in WFM, largely due to illumination power fluctuations. For low contrast specimens, especially live ones where illumination levels are kept low to preserve cell health, WFM gives superior performance. However, WFM did not perform well when thick tissues were examined. Again, the SNR dictated the choice. In thick tissue, out of focus noise in the WFM subsumes the in-focus signal, removing all contrast. For these specimens, the LSCM has a higher SNR and performs better. We are currently undertaking an analysis of a spinning disk microscope to assess how this system performs.

Based on this information, we have used a WFM to study the dynamics of the Cajal bodies in living cells. Cajal bodies (also known as coiled bodies), are subnuclear organelles that contain specific nuclear antigens, including small nuclear ribonuclear proteins (snRNPs) and a subset of nucleolar proteins [5, 6]. Cajal bodies are localised in the nucleoplasm and are often found at the periphery of the nucleolus. We have constructed a stable HeLa cell line, HeLa$^{GFP\text{-}coilin}$, that expresses the Cajal body marker protein, p80 coilin, fused to GFP [7]. The localisation pattern and biochemical properties of the GFP-coilin fusion protein are identical to the endogenous p80 coilin. Time lapse recordings on 63 nuclei of HeLa$^{GFP\text{-}coilin}$ cells showed that all Cajal bodies move within the nucleoplasm. Movements included translocations through the nucleoplasm, joining of bodies to form larger structures and separation of smaller bodies from larger Cajal bodies. The GFP-coilin protein is dynamically associated with Cajal bodies as shown by changes in their fluorescence intensity over time.

To determine the mechanism of Cajal body mobility, we tracked the trajectories of 350 CBs in the nuclei of 50 living cells and calculated mean square displacements (MSD) of individual CBs [8]. Brownian diffusion is the primary mode of CB motility, although we also noted some examples of diffusion with flow. Plots of MSD vs Δt revealed that many CBs diffuse within a confined volume. To identify the nature of this confinement, we transfected $HeLa^{GFP\text{-}coilin}$ cells with a plasmid encoding YFP-histone H2B and recorded the dynamics of CBs and chromatin by time-lapse fluorescence imaging. CBs that diffused within confined volumes were either immediately next to chromatin or surrounded by chromatin, suggesting that CBs can either be tethered to or confined by chromatin. Furthermore, calculation of the diffusion constant (D) of individual CBs through a time-lapse sequence showed large increases in D when chromatin constraints were relieved. We conclude that a single CB can switch between dynamic states, apparently because of changes in the physical association of the CB with chromatin.

This type of mobility is similar to what has been described previously as anomalous subdiffusion [9]. Indeed, using statistical analyses, we were able to show that Cajal body mobility is described by anomalous subdiffusion. Most interestingly, treatments that decrease the association of CBs with chromatin like ATP depletion and transcriptional inhibition also decrease the amount of subdiffusion.

Our results show that WFM combined with image restoration can produce novel analysis of cellular dynamics. In our case, we have used the low SNR levels of the WFM and the improvement in contrast obtained by deconvolution to characterize the dynamics of the nucleus in vivo.

References

[1] S. Inoué., Handbook of Biological Confocal Microscopy, J. Pawley, Plenum Press, New York, 1995.
[2] D. A. Agard, Y. Hiraoka, P. Shaw and J. W. Sedat., Methods Cell Biol. 30 (1989) 353-377.
[3] J. R. Swedlow, J. W. Sedat and D. A. Agard., Deconvolution of Images and Spectra, P. A. Jansson, Academic Press, New York, 1997.
[4] J. R. Swedlow, K. Hu, P. D. Andrews, D. S. Roos and J. M. Murray., Proc. Soc. Natl. Acad. USA. in press (2002)
[5] A. G. Matera., Trends Cell Biol 9 (8) (1999) 302-309.
[6] J. G. Gall., Annu Rev Cell Dev Biol 16 (2000) 273-300.
[7] M. Platani, I. Goldberg, J. R. Swedlow and A. I. Lamond., J. Cell. Biol. 151 (2000) 1561-1574.
[8] M. Platani, A. I. Lamond and J. R. Swedlow., Submitted
[9] M. J. Saxton., Biophys J 70 (3) (1996) 1250-62.
[10] J.R.S. is a Wellcome Trust Career Development Fellow (054333).

Microsc. Microanal. 8 (Suppl. 2), 2002
DOI 10.1017.S1431927602100389

Biological Photonic Crystals – Revealed by Multi-photon Nonlinear Microscopy

P. C. Cheng[1], C. K. Sun[2], B. L. Lin[3], S. W. Chu[2], I. S Chen[2], T. M Liu[2], S. P. Lee[3], H. L. Liu[4], M. X. Kuo[4] and D. J. Lin[4]

[1]Department of Electrical Engineering, State University of New York, Buffalo, NY 14260 USA
[2]Department of Electrical Eng., Natl. Taiwan Univ., Taipei, Taiwan, 10617, Republic of China
[3]Molecular & Cell Biology Div., Development Ctr for Biotechnology, Taipei, Taiwan, 10659, ROC.
[4]Department of Physics, Natl. Taiwan Normal University, Taipei, Taiwan, 116, Republic of China

Highly optically active nonlinear bio-photonic crystal structures in living cells are studied by a novel multi-modal nonlinear microscopy. Numerous biological structures, including stacked membranes and arranged protein structures are highly organized in nano scale and are found to exhibit strong optical activities through second-harmonic-generation (SHG) interactions, behaving similar to man-made nonlinear photonic crystals. The microscopic technology used in this study is based on a combination of different imaging modalities including SHG, third-harmonic-generation (THG), and multi-photon-induced fluorescence. With no energy deposition during harmonic generation processes, the demonstrated nonlinear-photonic-crystal-like SHG activity in highly organized biological nano-structrues is useful for investigating the dynamics of structure-function relationship at subcellular level and is ideal for studying living cells. In order to allow both SHG and THG fall within the visible spectrum, and also achieve low illumination attenuation [1, 2, 3, 4], we move the excitation wavelength to the 1200-1350nm regime by using a Cr:forsterite laser with center wavelength around 1230 nm, resulting not only the visible but also the near infrared (NIR) spectrum open for simultaneous recording of 2PF, three-photon fluorescence (3PF), SHG and THG. Our previous studies using 1230nm light indicated efficient excitation of 2PF and 3PF in common bio-probes [3, 5]. Figure 4 shows the experimental setup of a multi-modal nonlinear laser scanning microscope. (Figure 4: Laser: Cr:forsterite femtosecond laser centered at 1230nm; BS: beam splitter; L: mercury lamp; S: sample; TS: 3-dimensional translation stage; F: color filter; SM: spectrometer; C: computer, CA: camera).

Figure 1 is a set of x-y-λ images of the adaxial surface of rice leaf showing dumb-bell-shaped silica cells (s, profiled by dashed line). The illumination polarization is shown as white arrows in each image. The orientation of SHG images obviously follows the illumination polarization while no correlation can be found in THG images. Scale bar =15μm. Figure 2 shows a set of x-y-λ images of a live mesophyll cell from *Commelina communis* L. revealing the distribution of chloroplasts as imaged by THG, SHG and 2-photon fluorescent modes. The enlarged images show the individual chloroplasts (Fig. 3). By comparing the SHG image with TEM images of similar samples, the crescent-shaped SHG image was identified to be the grana and the oval-shaped structures are starch granules. (Scale bar: 15 μm). Figure 5 demonstrates the starch granules in a piece of fresh potato can generate strong frequency doubled output. An objective lens was used to focus the 1230nm illumination onto a small piece of potato. Our results reveal photonic crystal -like optical activity in living cells using a multi -modal microscopy. Highly organized nano -period structures in biological samples exhibit strong SHG activities resembling nonlinear photonic crystals and thus can be treated as "nonlinear bio -photonic crystals". N umbers of biological structures exhibit strong bio -photonic effect, such as cell wall, starch granule, silica deposition, cuticular papillae on the epidermal cell of rice leaf, crystalline myosin and actin nano-filaments in the myofibrils of skeletal muscles, and grana in chloroplast.

[1] Lin, B. L. *et al.*, (2001): Maize Genetics Cooperation News Letters **75**, 61-62.

[2] Cheng, P. C. *et al.*, (1998): J. Microsc. 189:199-212.
[3] Cheng, P. C. *et al.*, (2000): Scanning, 22, 187–188.
[4] Cheng, P. C. *et al.*, (2001): SPIE Proceedings, 4262, 98-103.
[5] Liu T. M. *et al.*, (2001): Scanning 23, 249-254.
Supported by grants from the National Science Council of ROC [NSC89-2215 -E-002-064, NSC89-2112-M-002-082 (CKS)], NSC89-2811-E-002- 0058 (PCC). [NSC89-2311-B-001-134, NSC89-2311-B-001-137 (BLL)], Academia Sinica (BLL). National Taiwan University (CKS).

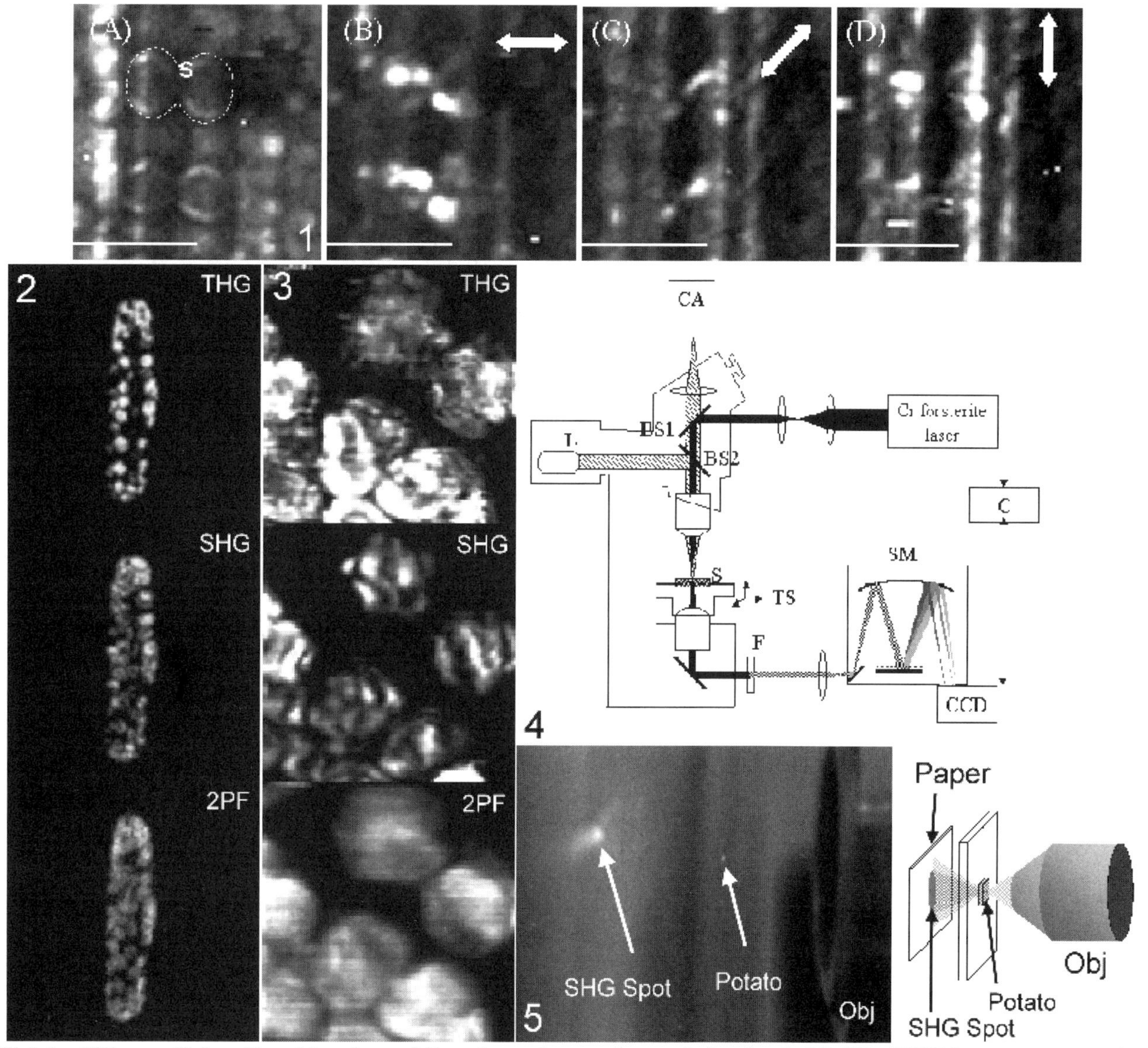

FIG 1A: Silica cell (outlined by dotted lines) on the surface of rice leaf. 1B-1D: SHG images at varying polarization planes (white double-head arrows). Scale bar= 15μm.
FIG 2: Live mesophyll cell of *C. communis* imaged by THG, SHG and autofluorescent modes.
FIG 3: Chloroplasts in live mesophyll cell of *C. communis* imaged by THG, SHG and autofluorescent.
FIG 4: Experimental setup. L: light source; BS1, BS2: beam splitter; S: sample; SM: spectrometer.
FIG 5: strong frequency doubled output generated by a piece of fresh potato. An objective lens was used to focus the 1230nm illumination onto a small piece of potato as shown in the diagram.

Microsc. Microanal. 8 (Suppl. 2), 2002
DOI 10.1017.S1431927602100390

Fluorescent Probes for Ultrasensitive Cytochemical and Histochemical Imaging

Iain Johnson, Molecular Probes, Inc., 4849 Pitchford Avenue, Eugene, OR, 97405

Immunofluorescence techniques using multicolor labeling and analysis of colocalization are essential for unraveling the structural and functional complexity of cells and tissues. Improvements in the capabilities of these techniques are dependent on the development of new dyes and methods for attaching them to antibodies and other markers.

The Alexa Fluor series of dyes have optimized spectroscopic and physical properties that enable preparation of bioconjugates with consistently strong and photostable fluorescence output. The series consists of fourteen dyes with fluorescence excitation peaks ranging from 350 nm to 750 nm. Functionally important characteristics of Alexa Fluor dyes include strong absorption at the output wavelengths of common excitation sources, resistance to photobleaching and self-quenching, and water solubility to facilitate coupling reactions with proteins and other biomolecules. The well-differentiated spectra of the fourteen Alexa Fluor dyes provide many options for multicolor detection (FIG 1) and fluorescence resonance energy transfer (FRET). In addition to their use for immunofluorescent labeling, Alexa Fluor dyes have been incorporated into a wide range of specialized probes including neuronal tracers, fluorescent lipopolysaccharides and cholera toxin B conjugates for detection of lipid rafts.

Simultaneous multicolor immunofluorescence detection requires the use of primary antibodies from unrelated species of animals in conjunction with species-specific secondary antibodies, or combinations of biotin–avidin reagents with directly labeled primary and secondary antibodies. The straightforward approach of directly labeling primary antibodies is often overlooked, primarily because chemical coupling of dyes, haptens and enzyme labels to antibodies is a moderately difficult and relatively low-yield process that is impractical for very small quantities of protein. Direct labeling via formation of binary antibody complexes circumvents many of these limitations. Sub-microgram quantities of a primary antibody can be labeled, and the reactions are usually quantitative with respect to the primary antibody. Labeling and purification of the complexes can be completed in a matter of minutes. In this way, customized sets of direct conjugates can be readily prepared for multicolor immunofluorescence applications.

Enzyme-mediated detection methods permit ultrasensitive detection of low abundance targets that are below the detection threshold of other fluorescence techniques. Tyramide Signal Amplification (TSA) and Enzyme-Labeled Fluorescence (ELF) are two complementary techniques that generate signal amplification based on *in situ* deposition of multiple fluorescent substrates per enzyme label. TSA uses horseradish peroxidase (HRP) enzyme labels and tyramide substrates, which can be coupled to a wide assortment of dyes and haptens. Higher levels of signal amplification can be achieved by detection of Oregon Green 488 tyramide with anti-fluorescein/Oregon Green antibody conjugates labeled with HRP, followed by a second round of fluorescent tyramide deposition. The dye–anti-dye primary antibody system also provides a useful alternative to biotin–streptavidin detection in cells and tissues where the presence of endogenous biotin produces substantial background signals. ELF is based on alkaline phosphatase enzyme labels and a fluorogenic substrate that forms a yellow-green–fluorescent precipitate at the site of phosphatase activity. The precipitate is extremely photostable and exhibits an exceptionally large fluorescence Stokes shift, allowing signals to be readily discriminated from background autofluorescence and other fluorescent labels.

Investigation of physiological responses by fluorescence microscopy requires probes that can be introduced, retained and detected in living cells. Some recent developments in this area include phospholipase substrates used for *in vivo* imaging of lipid metabolism and fluorescent analogs of phoshointositides that allow the intracellular distributions of these important lipid second messengers to visualized. In addition, new fluorescent indicators for intracellular zinc have been developed that can be used to detect zinc secretion associated with insulin release from pancreatic islets.

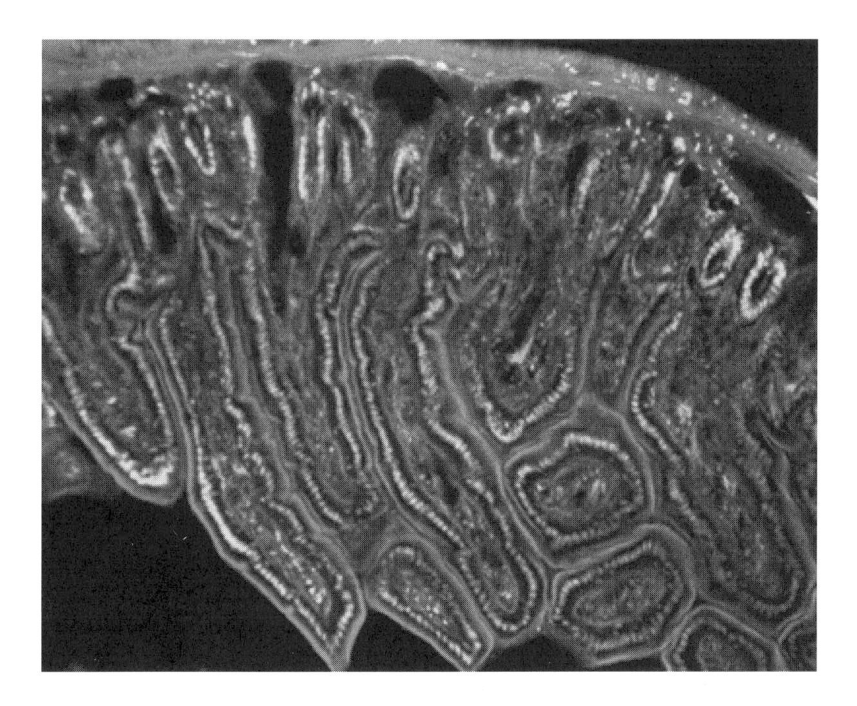

FIG 1. Mouse intestine cryosection showing basement membranes labeled with an anti-fibronectin antibody and Alexa Fluor 488 goat anti-chicken secondary antibody (green fluorescence). Goblet cells and crypt cells are labeled with Alexa Fluor 594 wheat germ agglutinin (red fluorescence). The microvillar brush border and smooth muscle layer are visualized with Alexa Fluor 680 phalloidin (pseudocolored purple). The section was counterstained with DAPI (blue fluorescence).

Microsc. Microanal. 8 (Suppl. 2), 2002
DOI 10.1017.S1431927602100407

Confocal Microscopy System Performance: Foundations for Measurements, Quantitation and Deconvolution.

Robert M. Zucker and Tammy E. Stoker

Reproductive Toxicology Division, National Health and Environmental Effects Research Laboratory, Office of Research Development, U.S. Environmental Protection Agency, Research Triangle Park, North Carolina 27711

The confocal laser-scanning microscope (CLSM) has enormous potential in many biological fields. The reliability of the CLSM to obtain specific measurements and quantify fluorescence data is dependent on using a correctly aligned machine that contains a stable laser power. For many applications it is useful to know the CLSM system's performance prior to acquiring data images so the necessary resolution, sensitivity and precision can be obtained. Applications of deconvolution, FRET and quantification necessitate that the confocal is correctly configured and operating at the highest performance levels.

The most common method in many laboratories to measure system performance involves the use of a histological slide to create a "pretty picture". Although this test evaluates many parameters in a crude manner (laser power, field illumination and lateral resolution) that can influence a CLSM image, the interpretation of this histological image is subjective and range of acceptability is variable. In fact, many confocal microscopes can indeed obtain "pretty pictures" even when they are sub-optimally functioning. Furthermore, it is impossible to compare similar machines for proper functionality when using only an image as a reference standard. This study involves methods that can be used to QA a CLSM. Without the use of these various performance tests, it cannot be absolutely determined whether CLSM machines are working at appropriate performance levels.

Tests methods have been devised on the Leica TCS-SP1 and TCS-4D confocal microscope systems to ensure that these machines are operating correctly. The tests measure the following: field illumination, lens functionality and lens clarity, field illumination, spectral registration, total laser power, laser stability, dichroic reflectance, spectral registration of the beams, axial resolution, scanning stability, overall machine stability, and system noise (1, 2). It is anticipated by using this type of test data, performance standards for confocal microscopes will be determined and the current subjectivity in evaluating CLSM performance will be eliminated. These tests will help serve as guidelines for other investigators to assess both the performance of their machines and the quality of data derived from their machines. These tests have been applied in a similar manner to Zeiss 510 confocal system.

One of our research objectives was to try to quantify fluorescence using a CLSM. We have used reproductive tissues (oocytes, embryos and fetal limbs) as model systems to quantify fluorescence. The distribution of cortical granules has been measured in normal and hormonally delayed rat zygotes. The pattern of fluorescence from these two oocyte groups suggests that extra sperm may enter an egg if there is incomplete release and uneven distribution of the cortical granule materials (illustrated below). We have used Bitplane (Imaris/Surpass) programs to quantify the amount of fluorescence at different gray scale values (GSV) in the zygotes with 3D visualization after using a water immersion 63x objective (Na 1.2). We have previously developed a technique that uses LysoTracker Red to measure late stage apoptosis in embryos (3). This dye accumulates in acidic regions associated with increased lysosomes and phagocytosis. Optimization of sample preparation techniques of fixation, MEOH dehydration and BABB clearing increased the transmission of laser light to enable the visualization of tissues as thick as 500uM using a 10x objective (NA=0.40). We have used Bitplane software (Imaris/Surpass) to measure the number of particles and the volume of these particles at different gray scale values (GSV). This quantification technique will be discussed in terms of cell death using the following model systems: 1) somites in normal GD10 rat embryos; 2) GD14 developing limbs from dams exposed to different concentrations of 5-Flurouracil (5-FU); 3) GD 8-9 day developing embryos treated with different concentrations of toxicants and 4) cortical granule release in delayed zygotes.

References

1. Zucker, RM. Price OT. Cytometry 44:273-294 2001
2. Zucker, RM, Price OT. Cytometry 44: 295-308 2001
3. Zucker, RM. Hunter S. Rogers JM: Cytometry 33:348-354 1998

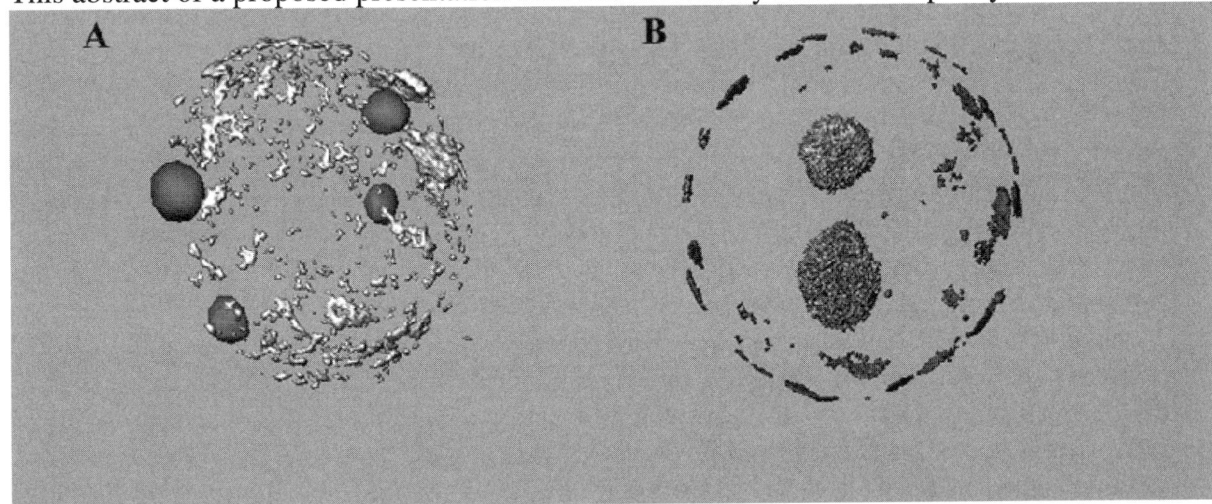

Polyspermic rat zygote (A,left) containing 3 sperm pronuclei and one-egg pronuclei stained with DAPI for the DNA and Lens Culinaris Aggluntin (LCA) Rhodamine labeled lectin for the cortical granules. The cortical granule distribution is uneven in this abnormal zygote. The light gray fluorescence represents cortical granules and can be quantified using Imaris/Surpass software. The normal zygote (B,right) contains one sperm pronuclei and one egg pronuclei. The cortical granule staining pattern is more symmetric and has less fluorescent intensity.

Microsc. Microanal. 8 (Suppl. 2), 2002
DOI 10.1017.S1431927602100419

Spatial and temporal assays to determine the dynamics of protein localisation and organelle movement in single living cells

R.J. Errington*, P.J. Smith**, S.C. Chappell**, W.H. Evans*, A. Fajardo-Bermudez* and P.E.M. Martin***

Department of Medical Biochemistry*, Department of Pathology** and Department of Diagnostic Radiology***, University of Wales College of Medicine, Heath Park, CF14 4XN, Cardiff, UK.

Visualisation and monitoring of protein localisation and function using Green Fluorescent Protein (GFP) tagged proteins is a well-established approach. We have sought to develop robust assays for quantifying the dynamic localisation of proteins at the single cell level. This approach allows us to determine properties of trafficking wild-type proteins and their mutated counterparts. It provides a functional genomics approach to screening drug effects and investigating the consequence of protein mutations. This opens the door to using combined GFP-tagging and laser scanning microscopy in a high-content medium-through-put arena.

A simple application of spatial localisation has been used to screen the targetting and delivery of chimeric proteins in sub-cellular compartments identified using vital cell markers to, for instance, endoplasmic reticulum, mitotracker and DNA.

A more advanced co-localisation approach has been developed for studying protein movement during gap junction assembly and collapse. Gap junctions are ubiquitous intercellular channels underpinning cell-cell communication. The sub-unit protein of gap junctions, connexins, are thought to be processed along the secretory pathway during which oligomerisation occurs into hemichannels. Many connexin proteins have been identified and also connexin mutations have been shown to be associated with key pathologies in which connexin (Cx) trafficking, gap junction assembly and also function are compromised. We have examined, using single and dual channel confocal microscopy, the delivery of tagged connexins to gap junctions in control conditions and after disruption of Golgi or microtubule function. To determine the kinetics of intracellular trafficking of different connexins we have acquired image sequences consisting of single optical sections collected over time(x,y,t). To quantify Cx-GFP movement multiple time points were compared and the temporal colocalisation coefficients or mobility coefficients were extracted. This works on a pixel-by-pixel basis that can be best represented in a scatter plot. However if we assign a colour red, green and blue to three chosen time points the combined or merged image provides a single view depicting the local movement of vesicles. The coefficient describes the amount of fluorescent pixels (at 30 seconds) overlapping with fluorescent pixels of a subsequent time point (at 15 minutes). If the mobility coefficient tended towards 1 then the extent of overlap was 100%, indicating no movement. If the value tended towards 0 then the extent of overlap was minimal and maximal vesicular movement had occurred. The approach is analogous to that previously described to measure spatial localisation [1]. A marked disruption of connexin movement was demonstrated in all connexin types with a typical shift in mobility coefficient from 0.7 to approximately 0.96 using nocodazole, a microtubule disrupting agent. However Brefeldin A (a drug which disrupts Golgi) had differential effects on connexin types, high-lighting alternative (or multiple) mechanisms for connexin trafficking [2]. Gap junctions are responsible for intercellular communication this is promarily via the transfer of calcium ions, however ironically intracellular calcium levels act to modulate connexin trafficking such that sustained levels of calcium prevents gap junction assembly. We have been able to track and characterise connexin mobility and ER movement [3], using the colocalisation approach, at the same time as monitoring calcium levels in single cells. The potential implications on the regulation and modulation of gap junction communication during apoptosis where cells experience high levels of intracellular calcium will be discussed.

References

[1] EM Manders, FJ Verbeek and JA Aten (1993) Measurement of colocalisation of objects in dual-colour confocal microscopy. J Microsc. 169: 375-382.
[2] PEM Martin, G Blundell, S Ahmad, RJ Errington and WH Evans (2001). Multiple pathways in the trafficking and assembly of connexin 26,32 and 43 into gap junction intercellular communication channels. J Cell Sci 114, 3845-38555.
[3] HL Baker, RJ Errington, SC Davies and AK Campbell (2002). A mathematical model predicts that calreticulum interacts with the endoplasmic reticulum Ca^{2+}-ATPase. Biophys J 82(2): 582-590.

Microsc. Microanal. 8 (Suppl. 2), 2002
DOI 10.1017.S1431927602100420

Medical Diagnosis Using Miniaturised Confocal Microscopes

Alan R. Hibbs, BIOCON, 7 Walhalla Drive, Ringwood East VIC 3135 Australia.

Confocal microscopy offers a number of important advantages for imaging live tissue samples. These include the ability to image below the tissue surface, to collect images at varying depths, to image relatively deeply into the tissue and to remove out of focus noise. These advantages have been exploited by Optiscan Pty Ltd (Melbourne, Australia), for the development of miniaturised confocal microscopes for aiding medical diagnosis of important tissue abnormalities.

Optiscan was one of the earliest developers of the confocal microscope for the biological research community. Their original research grade confocal microscope (the F900e) was developed using single mode fibre delivery and pickup of the fluorescent signal [1]. The use of fibre optics allowed the scan head to be small and robust, with most of the optics and electronics being remote from the scan head. In the mid 1990's Optiscan embarked on the development of a number of confocal microscopes for clinical examination using their own fibre optic technology.

The Stratum (Figure 1) hand held confocal scan head is designed for imaging skin lesions in vivo. The probe consists of a small hand held device that contains the scanning mirrors, and the front objective lens. All other optical and electronic components are attached using a single mode fibre optic connection. The Stratum is capable of imaging below the surface of the skin, down even as far as the vascular layer. The Stratum is being developed as an imaging device for medical examination of important dermatological problems, including skin cancer diagnosis.

The flexible confocal endomicroscope (Figure 2) is a highly miniaturised scan head that fits within an available channel in a conventional endoscope. This microscope is designed for the examination of cell abnormalities, particularly those leading to cancer in less accessible locations within the gastrointestinal tract. The ability to fully utilise the conventional functions of the endoscope allow both the macroscopic localisation of possible areas of cell abnormality, and the ability to locally deliver relevant dyes for diagnostic purposes.s

The development of miniaturised confocal medical imaging devices will substantially aid in the diagnosis of tissue lesions without the necessity for tissue biopsy. An important part of the development of these instruments is to also provide a reliable and relatively easy to use interface for use by clinicians that are experts in their own area of medical diagnosis, but are not confocal or imaging experts. This technology also provides the future possibility of remote diagnosis using internet connection and image transfer to a central specialist clinic.

Optiscan Imaging Limited:
PO Box 1066, Mt. Waverley MDC, VIC 3149, Australia (www.optiscan.com.au)

Alan R. Hibbs is not directly associated with Optiscan, but has worked as a consultant for Martin Harris, the Director of fundamental research at Optiscan.

References:

1. Delaney, PM and MR Harris (1995). Fiberoptics in Confocal Microscopy. In "Handbook of Biological Confocal Microscopy" 2nd edition, edited by James B. Pawley. Kluwer/Academic Press/Plenum NY.
2. McLaren W, Anikijenko P, Barkla D, Delaney TP, King R. (2001). In vivo detection of experimental ulcerative colitis in rats using fiberoptic confocal imaging (FOCI). Dig Dis Sci 2001 46(10):2263-76

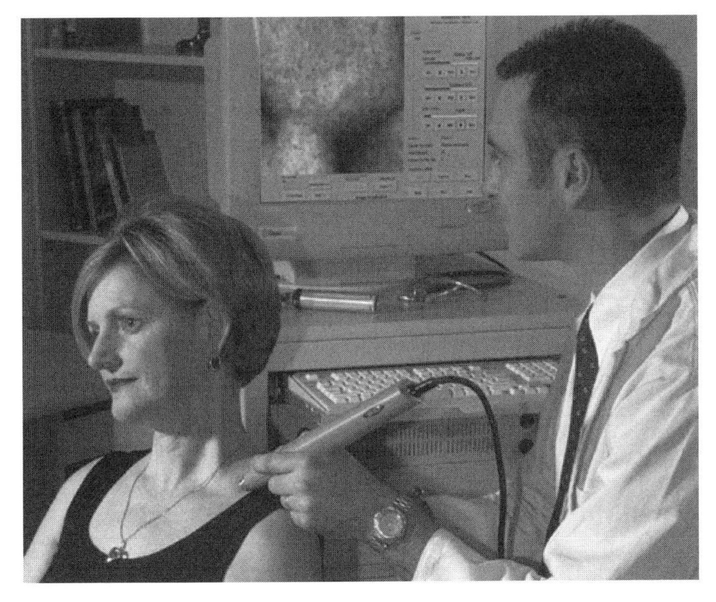

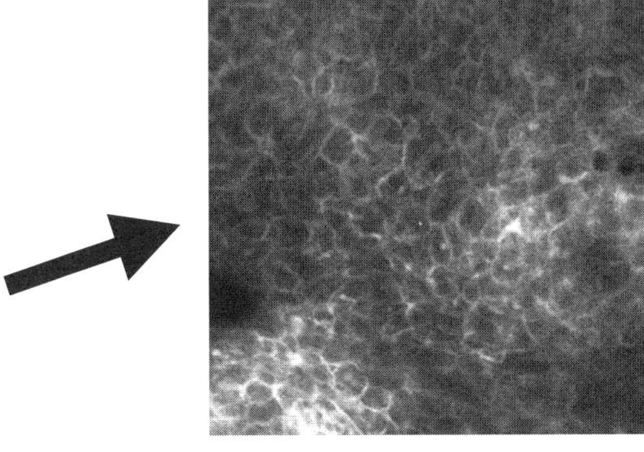

Figure 1: "Stratum" hand held confocal microscope imaging probe.
The Stratum hand held probe for in vivo confocal imaging of human skin has been developed for medical examination of skin lesions. The microscope provides non-invasive, below surface, imaging of human skin, allowing one to image the sub-cellular structure of skin in real time. The right hand panel shows confocal imaging of human stratum spinosum and basal skin cells in vivo, obtained using the Stratum confocal microscope probe. FOV = 250 um.

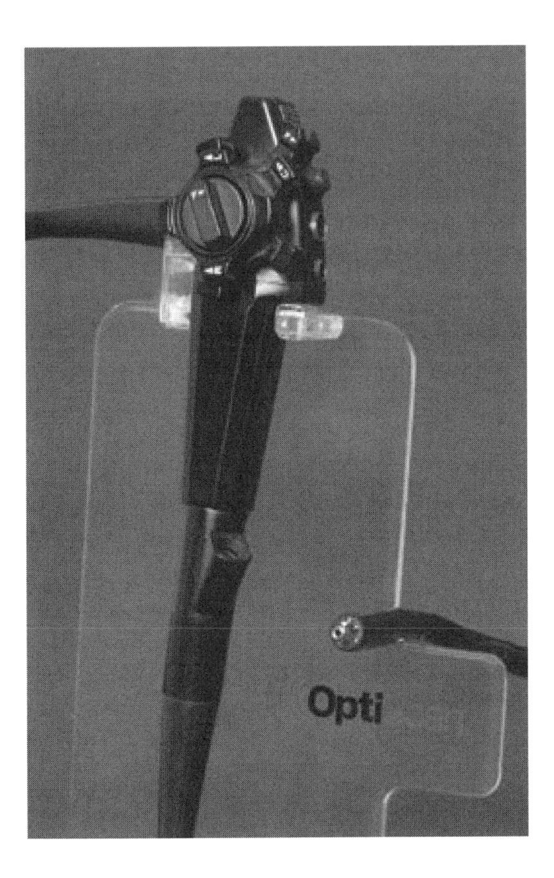

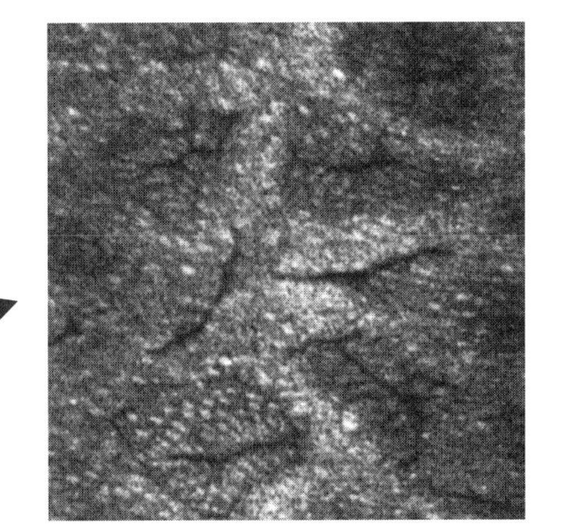

Figure 2: Live Micro Imaging (LMI) using a confocal endomicroscope.
Further miniaturisation of the Optiscan fibre optic based confocal scan head has allowed the production of a confocal microscope that can be inserted into an existing channel within a conventional endoscope. This instrument allows the collection of high resolution microscopic images of, for example, human colon in vivo, while at the same time allowing the macroscopic and dye delivery capabilities of the conventional endoscope to be available. The right panel shows human colonic crypts stained with acriflavine and imaged using a miniaturised confocal endomicroscope.

Microsc. Microanal. 8 (Suppl. 2), 2002
DOI 10.1017.S1431927602100249

New aspects of the skin barrier organisation assessed by diffraction and electron microscopic techniques

J. Bouwstra*, Y Grams*, G. Pilgram**, H. Koerten***

* Leiden/Amsterdam Center for Drug Research, Einsteinweg 55, Leiden The Netherlands

** Department of Molecular Cell Biology, Utrecht University, The Netherlands

***Center for Electron Microscopy, Leiden University Medical Center, The Netherlands

The skin barrier for most substances is found in the upper layer of the skin, the stratum corneum (SC). The SC consists of corneocytes, flattened dead cells, which are embedded in lamellar lipid regions. The composition of the SC lipids strongly differs from that of cell membranes of living cells in which phospholipids are the major lipid components. In the SC and the major lipid classes in the SC are ceramides, cholesterol and free fatty acids. The lipid-protein matrix of the SC not only restricts the passive diffusion of lipophilic molecules, but also severely limits the transport of hydrophylic compounds across the membrane. In order to study the lipid organization of healthy and diseased skin and to understand the mode of action of formulations used to increase the drug transport, several techniques have been explored. With these techniques we intend to elucidate either the lipid organization in the SC, the penetration pathways of the active agents, or the swelling mechanism of corneocytes. The latter is of interest, especially for moisturizers.

In this presentation first a brief description of the X-ray and electron diffraction method will be given. While the X-ray diffraction technique provides integral information on the lipid organization in SC, the electron diffraction can be used to study the local lipid lateral packing and provides detailed information on the orientation of lipid crystals in the SC. Furthermore by employing a stripping technique in combination with electron diffraction, structural information can be obtained as a function of depth in vivo. This makes the electron diffraction technique extremely powerful.

Using the X-ray diffraction technique it was found that the lipids in SC form two crystalline lamellar phases with periodicities of 6.4 and 13.4 nm. Furthermore, the 13.4 nm phase consist of one narrow central lipid layer with fluid domains with on both sides a broad layer with a crystalline structure. However, using X-ray diffraction it was not possible to detect whether besides an orthorhombic phase also a hexagonal lateral packing was present. For this reason the electron diffraction technique was explored. Electron diffraction revealed that only in the superficial layers of the SC a hexagonal lateral packing was coexisting with the orthorhombic packing. When the lateral packing is compared to the lipid organization in lamellar ichthyosis skin (a skin disease), it was found that in lamellar ichthyosis skin the hexagonal lateral packing was more pronounced present. This difference in phase behavior can explain at least partly the impaired barrier function in lamellar ichthyosis skin.

Electron diffraction does not provide information on the lamellar organization in SC in vivo. For this reason we explore the freeze fracture electron microscopy and combine it with the stripping technique. In this way freeze fracture electron microscopy can be used to study the lamellar phases in the SC as function of depth in humans in vivo in a similarly as with electron diffraction. This is illustrated by studies carried with healthy and lamellar ichthyosis skin. It was found that not only the lateral packing, but also the lamellar organization in lamellar ichthyosis skin is different from that in healthy skin. Information on the relationship between lipid organization in diseased skin is of great importance to unravel the mechanism controlling the skin barrier function. This is again demonstrated in lamellar ichthyosis skin, in which an impaired barrier function parallels an altered lipid composition and organization.

However, not only changes in SC lipid organization, but also the permeation pathways of the active agents need to be studied in order to understand the mode of action of delivery systems. For this reason methods have been developed to examine the diffusion into fresh unfixed human skin. This might be the permeation pathway in the SC and the targeting of fluorescent dyes to the hair follicles.

Microsc. Microanal. 8 (Suppl. 2), 2002
DOI 10.1017.S1431927602100250

NON-INVASIVE DIAGNOSIS OF SKIN STRUCTURE AND BIOCHEMISTRY BASED ON NON-LINEAR OPTICAL MICROSCOPY & SPECTROSCOPY

PETER T. C. SO*, KI HEAN KIM*, LILY H. LAIHO*, KARSTEN BAHLMANN*, CHRISTOF BUEHLER**, CHEN Y. DONG ***

* Massachusetts Institute of Technology, Dept. of Mechanical Engineering and Biological Engineering, 77 Massachusetts Avenue, Room 3-461A, Cambridge, Massachusetts 02139
** Paul Scherrer Institute, CH-5232 Villigen PSI, Switzerland
***National Taiwan University, Dept. of Physics, Taipei, Taiwan\

The study of skin physiology and pathology can be benefit from non-invasive optical imaging of tissue structure and biochemistry. Optical diagnostic of skin is hindered by its high scattering coefficient and its spatially heterogeneous index of refraction. These difficulties have been partially overcome by recent development of in non-linear imaging and spectroscopic technologies. This presentation describes a number of important advances in this field.

On the theoretical area, the degradation of microscope imaging point spread function due to tissue scattering and index of refraction heterogeneity has been carefully quantified (Figure 1). An understanding of image degradation mechanisms is critical to further improve non-invasive imaging techniques for skin structures.

On the instrumentation area, significant progresses have been made in many fronts. One of the most important advances lies in the development of video rate tissue imaging (Figure 2). High speed imaging allows physiologically relevant tissue area to be efficiently sampled allowing statistically significant measurement of skin properties. Another major advance involves the further development of non-linear microscopy based on second harmonic generation. Second harmonic microscopy has been shown to be a powerful method to visualize collagen structures in the dermis.

Another major advance in the instrument area involves the incorporation of spectroscopic measurement into high resolution microscopy. Powerful new instrument has been developed that allows spectral resolved measurement of tissue properties based on fluorescence emission wavelength, lifetime, and polarization.

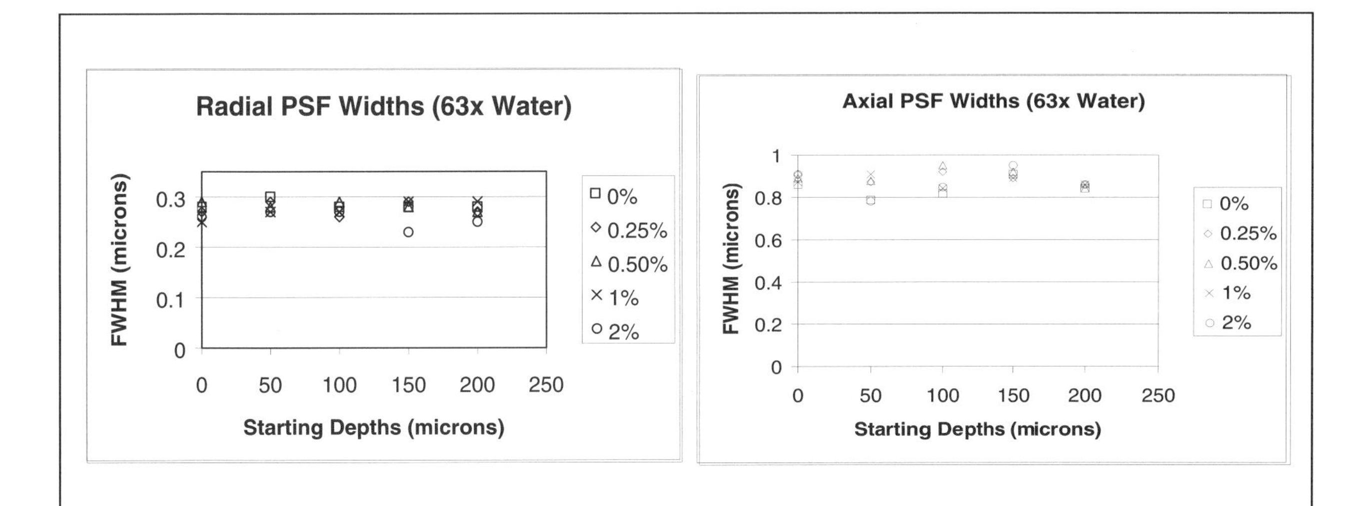

Fig. 1: Averaged widths of the radial and axial PSF from latex spheres at 4 depths and 5 Liposyn III concentrations (0, 0.25, 0.5, 1, and 2%), taken with 63x, water immersion, C-Apochromat objective (NA 1.2).

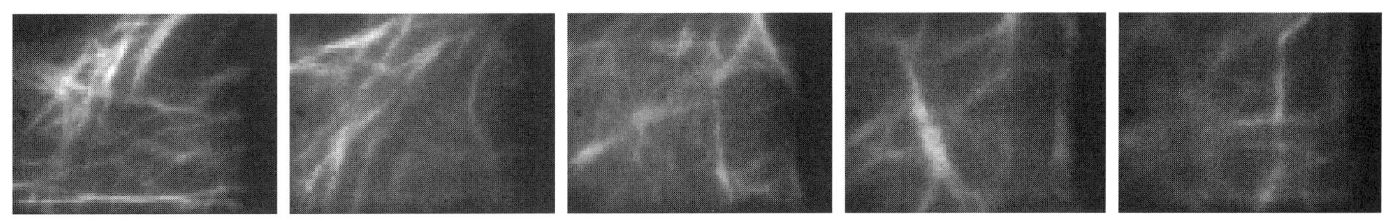

Fig. 2: Images of collagen/elastin fibers in the dermis of frozenhuman skin acquired at successive depths using two-photon video-rate microscopy. The image contrast is based on collagen/elastin autofluorescence. The five panels (from left to right) are images acquired at successive depth of 80, 100, 120, 140, 160μm below the skin surface. The size of each image is $80x100\mu m^2$.

Microsc. Microanal. 8 (Suppl. 2), 2002
DOI 10.1017.S1431927602100262

Detectability of reflectance and fluorescent contrast agents for real-time *in vivo* confocal microscopy

Milind Rajadhyaksha,[1,2] Salvador González[1]

[1]Massachusetts General Hospital, Dermatology - Bartlett Hall Ext 630, Boston, MA 02114
[2]Lucid Inc., Rochester, NY.

Abstract: Reflectance agents (liposomes, polystyrene microparticles, aluminum chloride salts, acetic acid) and fluorescent agents (polymer- and cosmetic actives-tagged fluorescein and rhodamine compounds, green fluorescent protein) enhance contrast of real-time confocal images of skin and microcirculation in vivo. Quantitative analysis of signal detectability versus contrast agent properties, and experimental images are presented. These results provide a basis for optimizing confocal microscope design and imaging parameters for real-time in vivo applications.

Introduction: Confocal microscopy is a non-invasive optical imaging modality with typical lateral resolution of 0.5-1.0 μm and axial resolution (section thickness) 2-5 μm within living tissues. This resolution allows histologic detail to be imaged and visualized in human and animal tissues in vivo. Nuclear and cellular detail, organelles, circulating blood cells, collagen, and connective tissues have been imaged to maximum depth of 200-500 μm in human skin [1-3] and human oral mucosa in vivo [4], and surgically exposed animal tissues in situ [5, 6]. However, the lack of structure-specific reflectance or fluorescent contrast limits the usefulness of confocal microscopy to morphologic investigations at the cellular- and nuclear-level. Morphologic and functional imaging at specific organelle- and ultrastructure-levels will be possible only if we develop contrast agents that may be used and detected in vivo.

We carried out a preliminary quantitative analysis and imaging experiments of the detectability of reflectance and fluorescent contrast agents for visualizing human skin and animal microcirculation in vivo. The goal was to understand signal detectability and contrast as a function of the optical properties of contrast agents and instrumentation parameters that would be necessary for real-time confocal imaging in skin in vivo.

Detectability of reflectance contrast agents: Based on Mie's optical scattering theory, we expect strong back-scatter and high contrast from organelles and microstructures that have size (d) on the order of the illumination wavelength (λ), and refractive index (n) greater than that of the surrounding epidermis (n=1.34) or dermis (n=1.38). When imaging at the infrared wavelength of 1064 nm, we detect typically 100-10000 photons/pixel (experimentally measured) from the epidermis to dermis, relative to a background of 100 photons, which then gives a signal-to-noise ratio of about 3-40 and signal-to-background of 1-100 [3].

Liposomes (n=1.41, d=0.7 μm) could thus be detected and they strongly enhance the contrast of microcirculation in the dermis of Sprague-Dawley rats. Topically applied aluminum chloride salts and polystyrene microspheres (diameter 0.1-0.5 μm) were found to enhance the contrast of sweat ducts and hair follicles in human skin; brighter and deeper imaging is possible with these agents. Topically applied 5% acetic acid causes the intra-nuclear chromatin (which normally exists as 30-

100 nm-thin diffuse filaments) to condense into convoluted 1-5 μm-thick strands. The condensed chromatin strongly back-scatters light, making the nuclei appear bright and easily detectable; this contrast-enhancing technique allowed detection of non-melanoma cancers in skin [7].

Detectability of fluorescent contrast agents: In fluorescence, detection of contrast agents in vivo and at video-rate (i. e., pixel integration time of 100 nsec) is much more challenging [8]. In general, our analysis predicts $\sim 10^5$ molecules in a high-resolution confocal probe volume of about 6×10^{-12} ml (measured lateral resolution 0.5 μm and section thickness 2 μm) when we use a 0.9 NA water immersion objective lens, a detection pinhole of diameter 10 resels and blue illumination at 488 nm. After accounting for losses due to scattering through tissue, collection efficiency of the objective lens and transmission through the confocal optics, 5000-10000 photons per pixel will typically be detected from the dermis, relative to a background of 500 (experimentally estimated). Thus the signal-to-noise ratio will be 50-80 and signal-to-background 10-20.

With non-toxic dosage of ~1 mg/kg, we could image polymeric particles labelled with either FITC or rhodamine B sulfonyl chloride in the microcirculation of Sprague-Dawley rats. In another pilot study, the penetration of topically-applied hydrophilic and hydrophobic compounds through human stratum corneum in vivo was investigated. The compounds were labelled with either fluorescein or octadecyl fluorescein ester. For the various compounds, we could visualize either the adsorption in the superficial corneocytes or the inter-cellular penetration to a depth of 10-20 μm in the stratum corneum. Topically applied fluorescein has also proven useful to characterize the distribution of corneocytes in healthy versus unhealthy skin. In another experiment, we have been able to image the expression of green fluorescent protein in mouse epidermis in vivo.

Summary: The analysis and experiments provide an understanding of confocal detectability of exogenous reflectance and fluorescent contrast agents within human and animal skin in vivo. Such information will provide a basis for developing optimum optical properties of contrast agents and optimum confocal microscope design and imaging parameters for real-time in vivo applications.

References

1. M. Rajadhyaksha et al., J. Invest. Dermatol. 104, 946-952 (1995).
2. M. Rajadhyaksha et al., J. Invest. Dermatol. 113, 293-303 (1999).
3. M. Rajadhyaksha et al., Appl. Opt. 38, 2105-2115 (1999).
4. W. M. White et al., Laryngoscope 109, 1709-1717 (1999).
5. F. Koenig et al., Urology 53, 853-857 (1999).
6. T. Keck et al., Pancreatology 1, 48-57 (2001).
7. M. Rajadhyaksha et al., J. Invest. Dermatol. 117, 1137-1143 (2001).
8. M. Rajadhyaksha and S. Gonzalez, Fluorescence in Biomedicine, Marcel-Dekker, in press.

Microsc. Microanal. 8 (Suppl. 2), 2002
DOI 10.1017.S1431927602100274

Mapping Inter-Cellular Water in Skin

A. Aitouchen*, S, Shi**, M. Libera* and M. Misra**

*Stevens Institute of Technology, Hoboken, NJ 07030
**Unilever Research, 45 River Road, Edgewater, NJ 07020

The main functions of skin's outer skin, stratum corneum (SC), are to prevent the water loss from underlying tissue, limit penetration of extraneous chemicals and save the viable cells of skin from environmental insults. Normal skin cells embark upon a well-orchestrated differentiation pathway that eventually results in cell desquamation. Proper desquamation depends upon the degradation of intercellular plugs, desmosomes, which in turn is believed to depend upon skin hydration [1]. Water, therefore, plays a critical role in maintaining skin's function and is also important for skin's physical and even optical properties. Earlier *in vitro* measurements of water distribution in skin [2] have relied on indirect measurement of water requiring dehydration of the sample in the microscope column. Here we report the application of electron energy-loss spectroscopy (EELS) and spectrum imaging (SI) [3] to measure cellular water in skin maintained in its native state. The main advantage of the EELS based measurements is that the specimen is maintained in frozen state during analysis, and therefore dehydration-induced artefacts are avoided and it also provides a direct measure of water in hydrated skin.

Thin slices (~200 μm) of freshly excised porcine skin (a by-product) were high pressure frozen using Leica Impact. Vitrified tissue blocks were trimmed and cut into 150 nm thick sections using Leica Ultracut EM FCS (at -150 C). Frozen samples were cryo-transferred at –170C to a Philips CM20 FEG STEM, operating at beam energy of 200 kV and equipped with a Gatan spectrometer Model CCD *ENFINA*. Spectrum-images were acquired using the Emispec data acquisition system. Data was collected at 5-15 kX using a probe size of about 40 nm, dwell time 20 ms, at 0.1 eV /channel and using a collection angle 10 mrad. The electron dose was about 2000 electrons/nm^2. The images were transferred to a Macintosh G4 500MHZ dual processor for further processing. Pixel wise spectral contribution in spectrum images was computed after spectral pre-processing and by multiple least squares fitting of the reference spectra to SI data set [3].

The frozen-hydrated skin, as expected, appears predominantly featureless (Fig. 1a) but the computed water map distinctly shows water-rich SC cells (Fig. 1b). White in this map corresponds to high water content. Water was found to reach a plateau of about 60% concentration at about 24 micron from the skin surface. The presence of a water gradient is seen in the outer layers of skin (Fig. 1c). Outermost region of the water map displays alternating, ~ 0.5 micron wide, regions of high water content, corresponding to intracellular regions of SC, separated by dark regions corresponding to intercellular lipid rich regions. Desmosomes that are clearly visualized in conventionally fixed tissue (Fig. 2a) appeared as water-rich (60-65%) dense pockets punctuating the lipid-rich intercellular regions of skin (Fig. 2b, ellipses). Such regions were predominantly present in the compactum, immediately above the viable cells.

1. A. Watkinson et al., Arch. Dermatol. Res. 293 (2001) 470
2. R. Warner et al., J. Invest. Derm. 100 (1993) 528.
3. S. Sun et al., J. Micros., 177 (1995) 18.

Fig. 1 Frozen hydrated section (a) water map (b) and computed water gradient (c) from SC to outer viable skin cells.

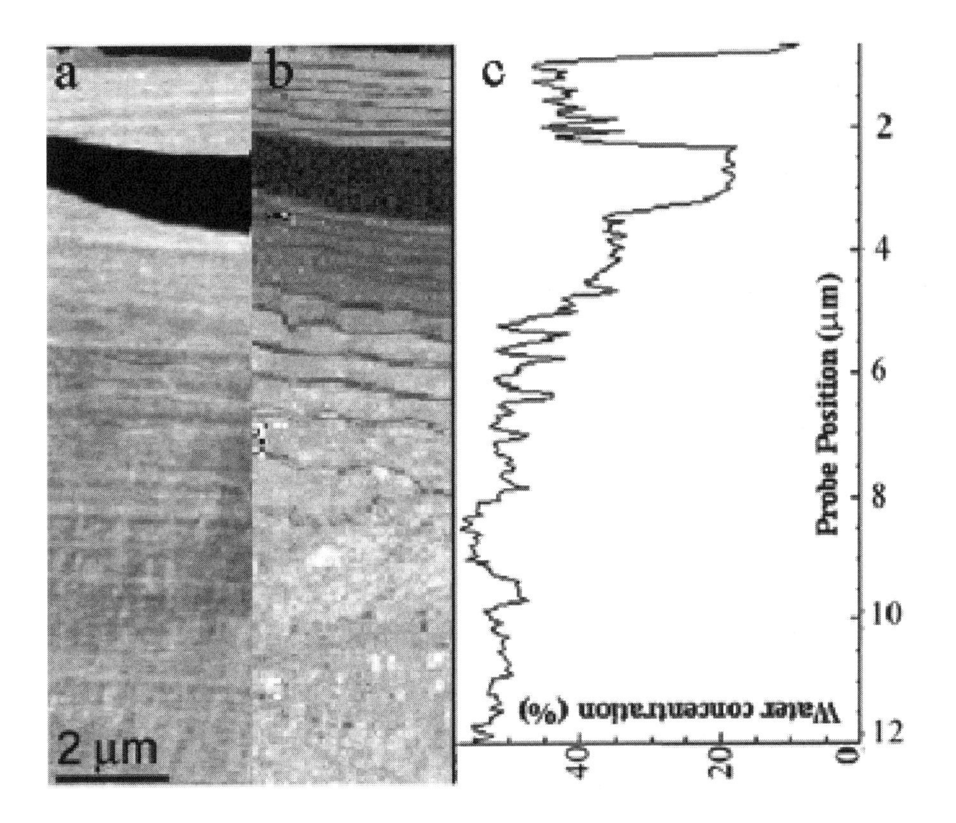

Fig. 2 Conventionally prepared micrograph of skin (a) showing inter (arrows) and intra-layer (arrowheads) desmosomes and water map of frozen-hydrated skin showing inter-layer (circles) desmosomes (b).

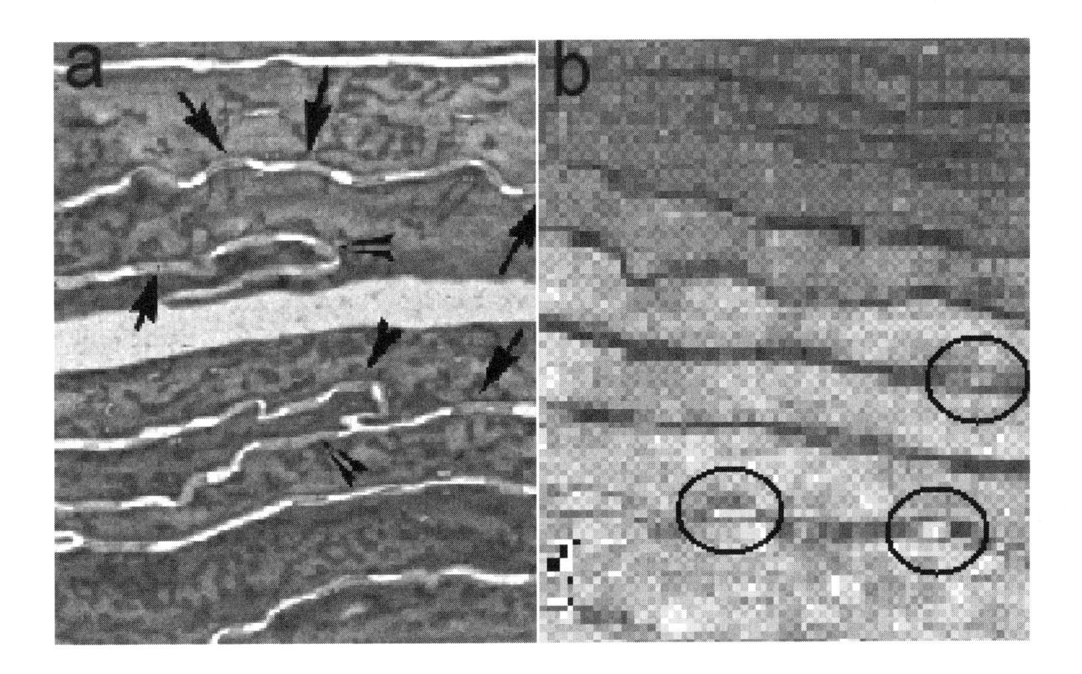

Microsc. Microanal. 8 (Suppl. 2), 2002
DOI 10.1017.S1431927602100286

Monitoring Skin Hydration and Product Induced Changes by Near-Infrared Spectroscopic Imaging

S.L. Zhang, T.M. Hancewicz, D.J. Palatini, P. Kaplan, M. Misra, and E.M. Attas[**]

Unilever Research U.S., Edgewater Laboratory, 45 River Road, Edgewater, NJ 07020
[**] Institute for Biodiagnostics, NRC Canada, 435 Ellice Avenue, Winnipeg, MB, Canada R3B 1Y6

Skin hydration is a key factor in skin health. Hydration measurements provide diagnostic information on the health of skin and can indicate the integrity of the skin barrier function. Near-infrared (NIR) methods measure the water content of living tissue by its effect on tissue reflectance at particular wavelengths. NIR imaging is a rapid, non-contact and non-invasive technique and has the important advantage of showing the degree of hydration as a function of location.

A near infrared spectroscopic imaging system has been developed [1]. It uses a liquid-crystal tunable filter (LCTF) in front of the objective lens and incorporates a 12-bit digital camera with a 320x240-pixel indium-gallium arsenide (InGaAs) array sensor. Reflectance images over the range of 960 to 1700 nm were used to monitor changes in skin hydration on delineated regions of the forearms and lower outer legs of volunteers following treatment with acetone, a commercial moisturizer, or other skin care products. Presented in this abstract are the results of the tissue water content obtained from the integrated intensity of the strong absorption band of water centered at 1450 nm. Other approaches to the data analysis of these NIR hydration images have also been attempted. They include cluster analysis, principal component analysis, and multiway methods.

Effectiveness of the new instrumentation for skin hydration measurement has been successfully demonstrated. A typical result from the previous study [2] is shown in Figure 1. NIR images were taken before product application (baseline) and 30, 90, 180, and 300 minutes after the forearm was treated with acetone and a moisturizer in the marked areas. One area was left untreated as control. Shown in the figure is the difference image between the 30-minutes image and the baseline image. Clearly, the acetone-treated left square close to the elbow treated with acetone shows the lowest water content, while the moisturizer-treated square on the right close to the wrist shows increased skin hydration, when compared with the middle, untreated area.

The NIR method has been extended to examine smaller but more realistic changes in skin hydration induced by daily use of personal care products and by seasonal effects from mild fall to cold winter. An example from the former is shown in Figure 2. In this study, the arms of eight volunteers were each marked with 4 circular areas. The baseline images were taken on the first day before product treatment. Three areas were treated twice a day following a standard protocol for 5 consecutive days with 3 different products, respectively. One area was left untreated as a control. NIR images were taken prior to the second treatment each day. Figure 2 shows typical results as difference images from the baseline. These images clearly indicate both positive and negative skin hydration changes induced by the products. More specifically, observed for this subject were increased hydration in the area treated with product 3 (far right circle), no change or slightly increased hydration in the product 2 area (2nd circle from right) and significantly decreased hydration in the product 1 area (far left circle). The control site is the 2nd circle from left.

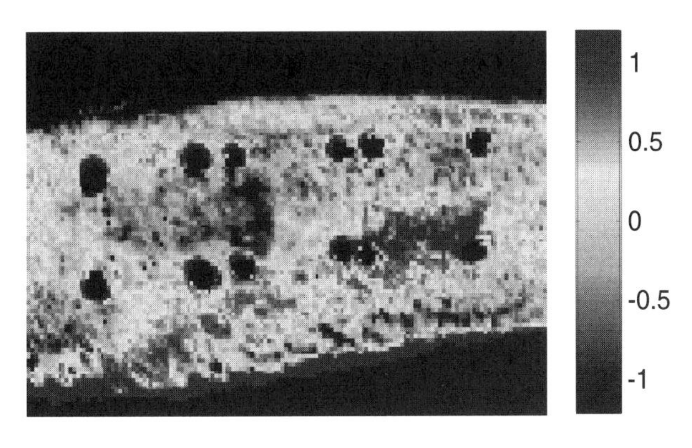

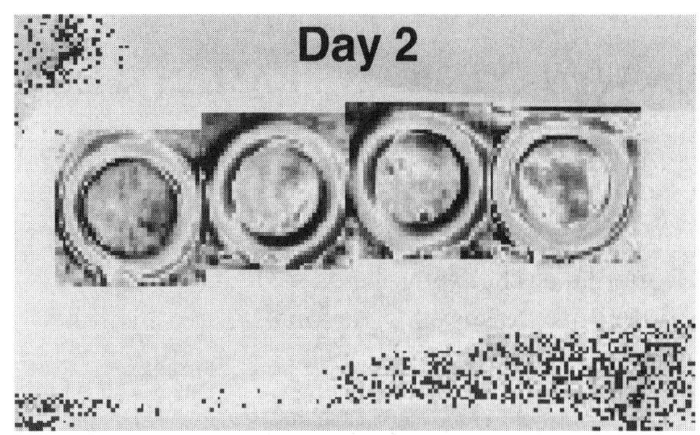

(Above)
Figure 1. NIR hydration image of volar forearm using the 1450 -nm band, shown as the difference between images taken before and 30 minutes after product application. The scale indicates area of the absorption band in NIR spectrum. The left square close to elbow was treated with acetone, the right square on the right close to wrist was treated with moisturizer, and the middle one was untreated.

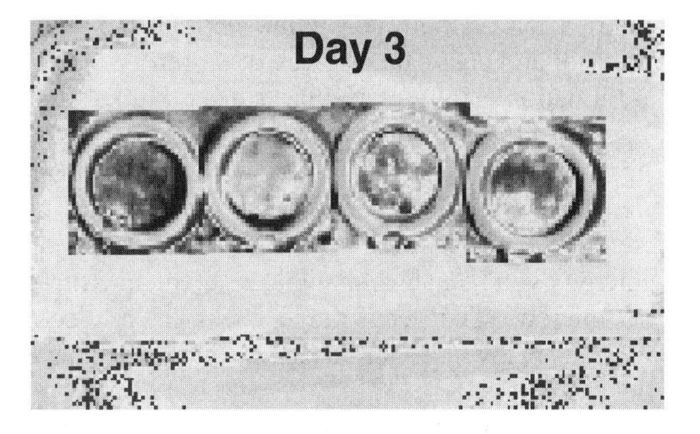

(Four panels on the right)
Figure 2. Changes in volar forearm hydration from baseline (day 1) during a 5 -day treatment with skin care products as measured by NIR imaging. Product application sites are (from left to right circles): Product 1, Untreated, Product 2, and Product 3. The color scale is defined similarly as in Figure 1.

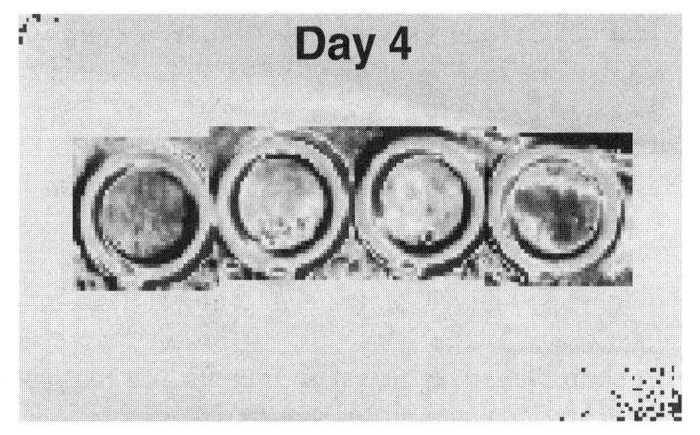

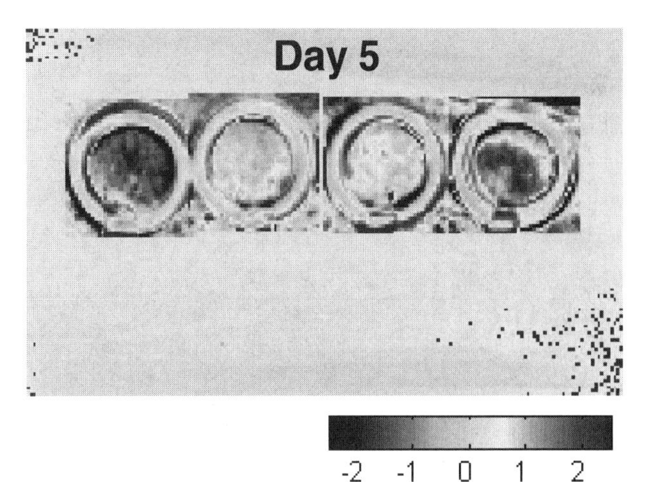

References

[1] M. Attas, T. Posthumus, B. Schattka, M. Sowa, H. Mantsch, and S. Zhang, "Skin hydration imaging using a long -wavelength near infrared digital camera," *Biomarkers and Biological Spectral Imaging,* eds. G.H. Bearman, D.J. Bornhop, and R.M. Levenson, *Proc. SPIE*: Bellingham, Washington, 2001; Vol. **4259**, pp. 75-84.

[2] E.M. Attas, M.G. Sowa, T.B. Posthumus, B.J. Schattka, H.H. Mantsch, S.L. Zhang, *Biospectroscopy* **67**(2), (2002, in press).

Microsc. Microanal. 8 (Suppl. 2), 2002
DOI 10.1017.S1431927602100298

IN VIVO CONFOCAL RAMAN SPECTROSCOPY OF THE SKIN

P. J. Caspers*, G. W. Lucassen**, H. A. Bruining*, G. J. Puppels*

* Erasmus University Rotterdam, Department of General Surgery, Dr. Molewaterplein 40, 3015 GD Rotterdam, The Netherlands
** Philips Research, Personal Care Institute, Prof. Holstlaan 4, 5656 AA Eindhoven, The Netherlands

In vivo methods that provide qualitative and quantitative information about the presence and depth distribution of molecular compounds in the skin are of great help in addressing many biological, medical and cosmetic research questions. Existing non-invasive *in vivo* methods provide little depth information about molecular concentrations across the epidermis [1]. *In vivo* confocal Raman microspectroscopy is a new method to determine the depth distribution of a range of molecular compounds in the skin [2-4]. We have established a confocal Raman microspectroscopic system dedicated to skin research. Measurements are performed directly on the skin and are fully non-invasive. The spatial resolution in the direction perpendicular to the skin surface is currently about 5 micron, which enables spatially resolved measurements within the stratum corneum (thickness about 15 micron). The depth range from which spectra can be obtained can be up to several hundred microns below the skin surface. A diagram of the Raman setup is shown in figure 1.

Skin hydration and skin moisturizers have received attention from both medicine and cosmetics. The skin itself produces a mixture of moisturizing substances in the lower part of the stratum corneum (SC). This mixture of mainly free amino acids and amino acid derivatives is referred to as natural moisturizing factor (NMF). The SC provides a highly efficient water barrier to the skin, which results in a steep water concentration gradient across the SC. Much of what is known about skin hydration, NMF and the skin barrier is based on in vitro information. We have used *in vivo* confocal Raman spectroscopy to determine concentration profiles across the SC of various compounds including water, and compounds of natural moisturizing factor and sweat. Figure 2 illustrates the unique possibility of *in vivo* confocal Raman microspectroscopy to study skin hydration non-invasively. Water concentration profiles were determined for the stratum corneum of the lower forearm. This was repeated after intense hydration for 45 minutes with a water soaked bandage. Dramatic changes in the shape of the *in vivo* water concentration profile is observed. Clearly the water concentration in the SC (0-15 micron) has largely increased to about 60%.

References:

[1] J. Serup and G.B.E. Jemec, Non-invasive methods and the skin,CRC Press, Boca Raton,1995.
[2] P.J. Caspers et al., Biospectr. 4, (1998) S31.
[3] P.J. Caspers et al., J. Raman Spectrosc., 31 (2000) 813.
[4] P.J. Caspers et al., J. Invest. Dermatol. 116 No. 3 (2001) 434.

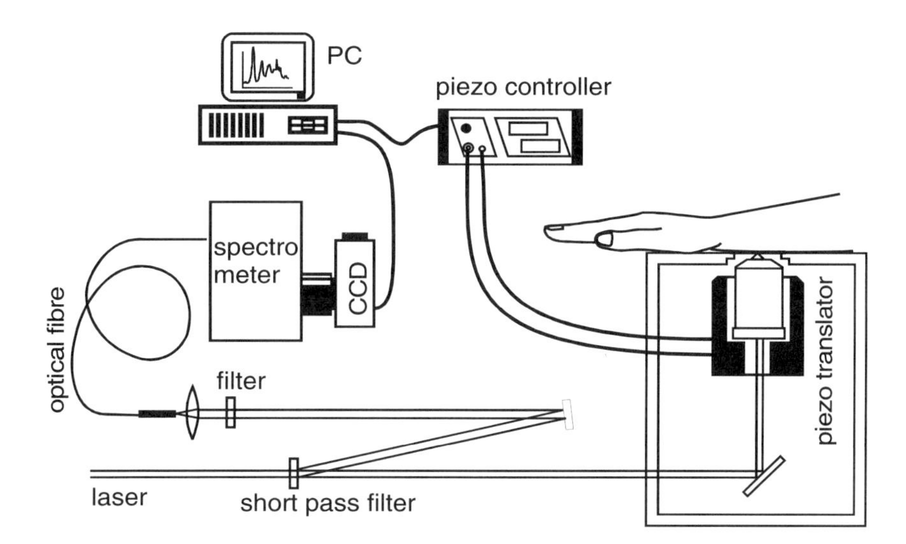

FIG. 1. Diagram of the *in vivo* confocal Raman microspectrometer. A laser beam is focused in the skin by a microscope objective. The Raman scattered light is reflected by a short-pass filter and coupled into an optical fiber, which is connected to the spectrometer. Vertical translation of the microscope objective changes the position – i.e. the distance to the skin surface - of the laser focus. During measurement the skin rests on a fused silica or CaF_2 window, which prevents movement of the skin surface.

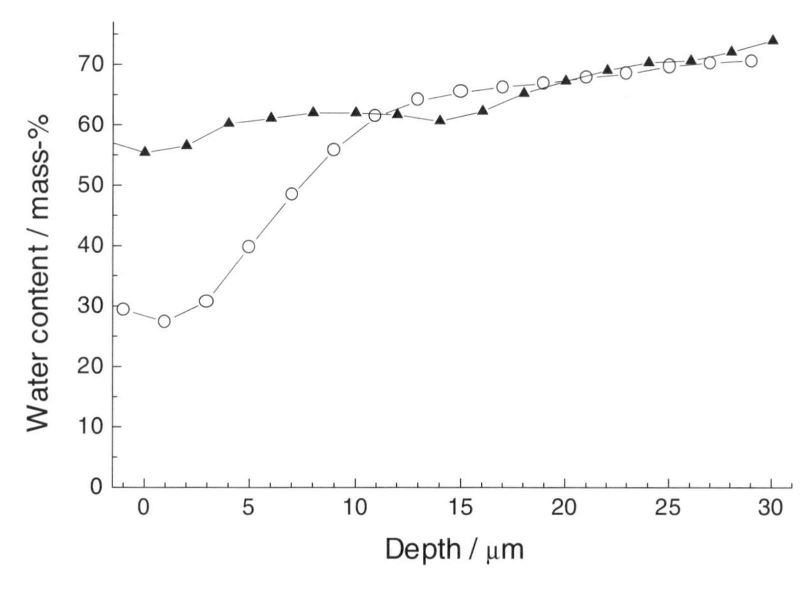

FIG. 2. *In vivo* water concentration profiles for the forearm before (circles) and after (triangles) hydration for 45 minutes.

Microsc. Microanal. 8 (Suppl. 2), 2002
DOI 10.1017.S1431927602100304

Imaging Of Cellular Trafficking In Skin Using Multiphoton And Handheld Confocal Microscopy Techniques.

Simon C. Watkins, Glenn D. Papworth
Center for Biologic Imaging, University of Pittsburgh, Pittsburgh PA, 15261.

We will present data showing the utility of both multiphoton microscopy and handheld confocal microscopy techniques for minimally invasive investigation of cell trafficking in human and murine skin.

One alternative to the conventional cytotoxic agents used in conventional cancer threrapies is a more defined molecular approach for cancer immune treatment; promotion of the immune system specifically to target and eliminate tumor cells on the basis of expression of tumor-associated antigens (TAA). It has been well documented that certain antigen-presenting cells (APC) are capable of recognizing, processing and presenting TAA, in turn initiating a specific antitumor immune response (1). Results from several laboratories and clinical trials (2,3) suggested significant efficacy of TAA-pulsed APC to generate a more efficient and effective anti-tumor immune response, following administration to tumor-bearing hosts.

It is fundamentally necessary to dynamically assess cell abundance within the microenvironment of the tumor in the presence of the appropriate therapeutic agent. Multiphoton microscopy and hand-held confocal microscopy were used to assess the trafficking of antigen presenting cells in mice, both in vivo and ex vivo. These imaging techniques are particularly suited to such investigations, because of their deep-tissue and minimally-invasive cellular imaging capabilities.

1. Shurin MR. Dendritic cells presenting tumor antigen. Cancer Immunol Immunother 1996;43(3):158-64.
2. Hamblin TJ. From Dendritic cells to tumor vaccines. Lancet 1996;347:705-706.
3. Girolomoni G, Ricciardi-Castagnoli P. Dendritic cells hold promise for immunotherapy. Immunol. Today 1997;18:1021-1024.

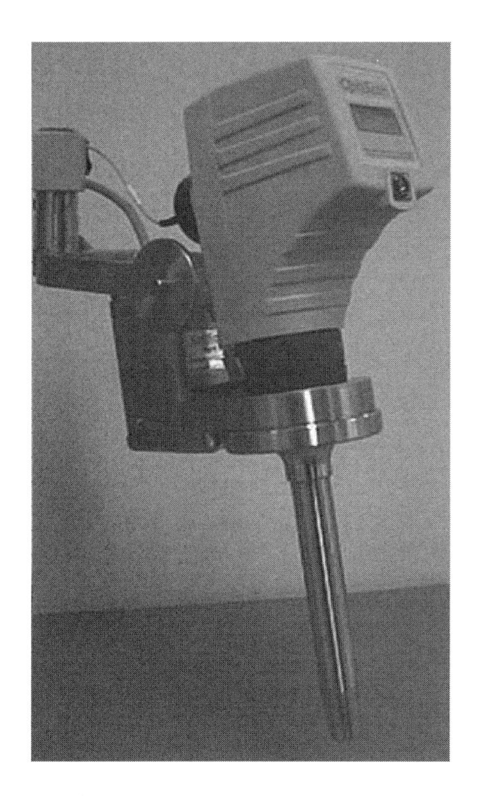

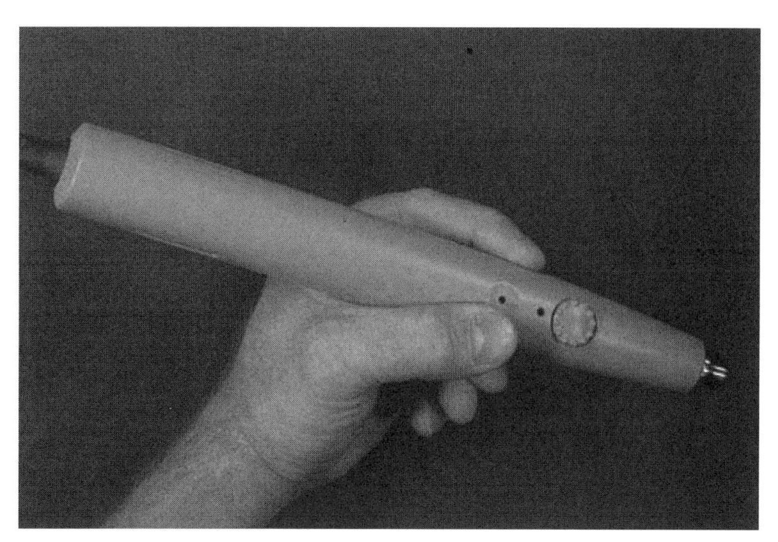

FIG. 1. Hand-held confocal imaging probes

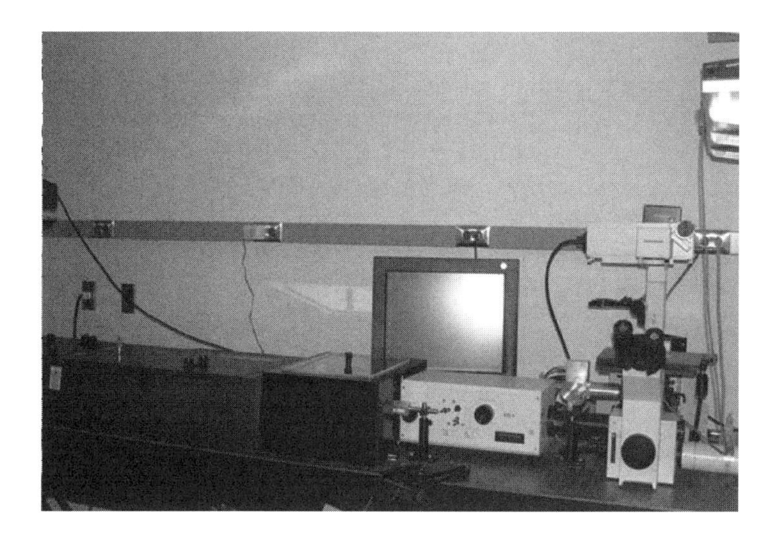

FIG.2. Multiphoton microscope system

Microsc. Microanal. 8 (Suppl. 2), 2002
DOI 10.1017.S143192760210078X

Irradiation-Induced Development of Nanoscale Features in Steel: Complementary 3D-APFIM and FEG-STEM Characterization

M. G. Burke, M. Watanabe, D.B. Williams and J.M. Hyde

Bettis Atomic Power Laboratory, West Mifflin, PA 15122 USA
Lehigh University, Bethlehem, PA 18015 USA
AEA Technology, Harwell, Oxon., UK

Neutron irradiation can promote significant changes in both microstructure and properties of steels. These changes are particularly sensitive to composition, specifically the Cu, Mn, and Ni content of the low alloy steels. Irradiation causes a hardening of the steel, which may lead to a reduction in toughness of the material. The ultra-fine features that are responsible for the change in mechanical properties require analytical techniques with very high spatial and chemical resolution. Atom probe field-ion microscopy (APFIM) has been used with great success to identify and quantify these nanoscale irradiation-induced features. Diffuse solute-enriched "clusters" or "precipitates" containing significant quantities of Fe, as well as Cu, Mn, Ni, and Si, have been documented in a wide variety of low alloy steels and welds. (1-3) Conventional transmission electron microscopy has been unable to identify these chemically complex features, which are likely to exhibit the same body-centered cubic structure as the Fe matrix. In this investigation, the techniques of 3D-APFIM (using the Oxford University Energy-Compensated Optical Position Sensitive Atom Probe) and field emission gun – scanning transmission electron microscopy (FEG-STEM) energy dispersive x-ray (EDX) microanalysis (using the Lehigh University VG HB603) have been used to provide independent analyses of the irradiation-induced nanoscale structure responsible for the changes in mechanical behavior of the material.

The steel used in this investigation was a bainitic low alloy A508 Gr4N forging steel, containing 3.7 wt.% Ni – 0.3% Mn – 1.8% Cr – 0.5% Mo – 0.08% Cu – 0.05% Si – 0.2% C (bal Fe). The steel was irradiated to a dose of 68 milli-displacements per atom (mdpa) at ~250°C. Specimens for analytical electron microscopy and 3D-APFIM characterization were prepared using conventional techniques. Quantitative FEG-STEM x-ray mapping was performed using experimentally generated Cliff-Lorimer "k" factors and the Zeta factor technique (4,5). This permitted the generation of local thickness maps in addition to the composition maps.

The APFIM analyses provided direct evidence of discrete, well-defined ~2 to 4 nm "precipitates" distributed within the matrix, an example of which is shown in Figure 1. The average "precipitate" composition (at%) was ~33Ni -~15 Mn-~6 Cu-~5 Si, despite the low levels of Mn, Cu and Si within the alloy. The APFIM estimated number density was ~5 X $10^{23}/m^3$. The composition (wt.%) of the matrix (precipitate-free) was ~3Ni–1.2Cr–0.2Mn– 0.4Mo–0.08Cu–0.03Si (bal Fe). Quantitative EDX maps showed the presence of very fine regions (~3 nm) enriched in Ni and Mn (Figure 2) and depleted in Fe. The number density of these local zones from the Ni-map and thickness map was estimated as ~2 X $10^{23}/m^3$. This value is remarkably consistent with the ~5 X $10^{23}/m^3$ number density determined by 3D-APFIM of a relatively small volume of material (~16 nm X ~16 nm X ~85 nm). Discrete spot analyses provided a measure of the matrix composition, which was generally consistent with the 3D-APFIM analysis. The maximum Ni level measured via mapping, ~8%, is consistent with the incorporation of an irradiation-induced solute-enriched "precipitate" within the analyzed volume. Thus, APFIM and FEG-STEM EDX mapping have successfully provided independent confirmation of the nanoscale features responsible for irradiation-induced hardening in these alloys.

References

1. M. G. Burke and S. S. Brenner, **J. de Physique 34:C2** (1986) 239.
2. M. K. Miller and M. G. Burke, **J. Nuc. Materials, 195** (1992) 68.
3. M. G. Burke et al., **Proc. 10th Intl. Symposium on Environmental Degradation of Materials in Nuclear Power Systems**, ed. G. S. Was (NACE, 2001) in press.
4. M. Watanabe and D. B. Williams, **Ultramicroscopy, 78**(1999) 89-101.
5. M. Watanabe, Z. Horita and M. Nemoto, **Ultramicroscopy, 65**(1996) 187-198.

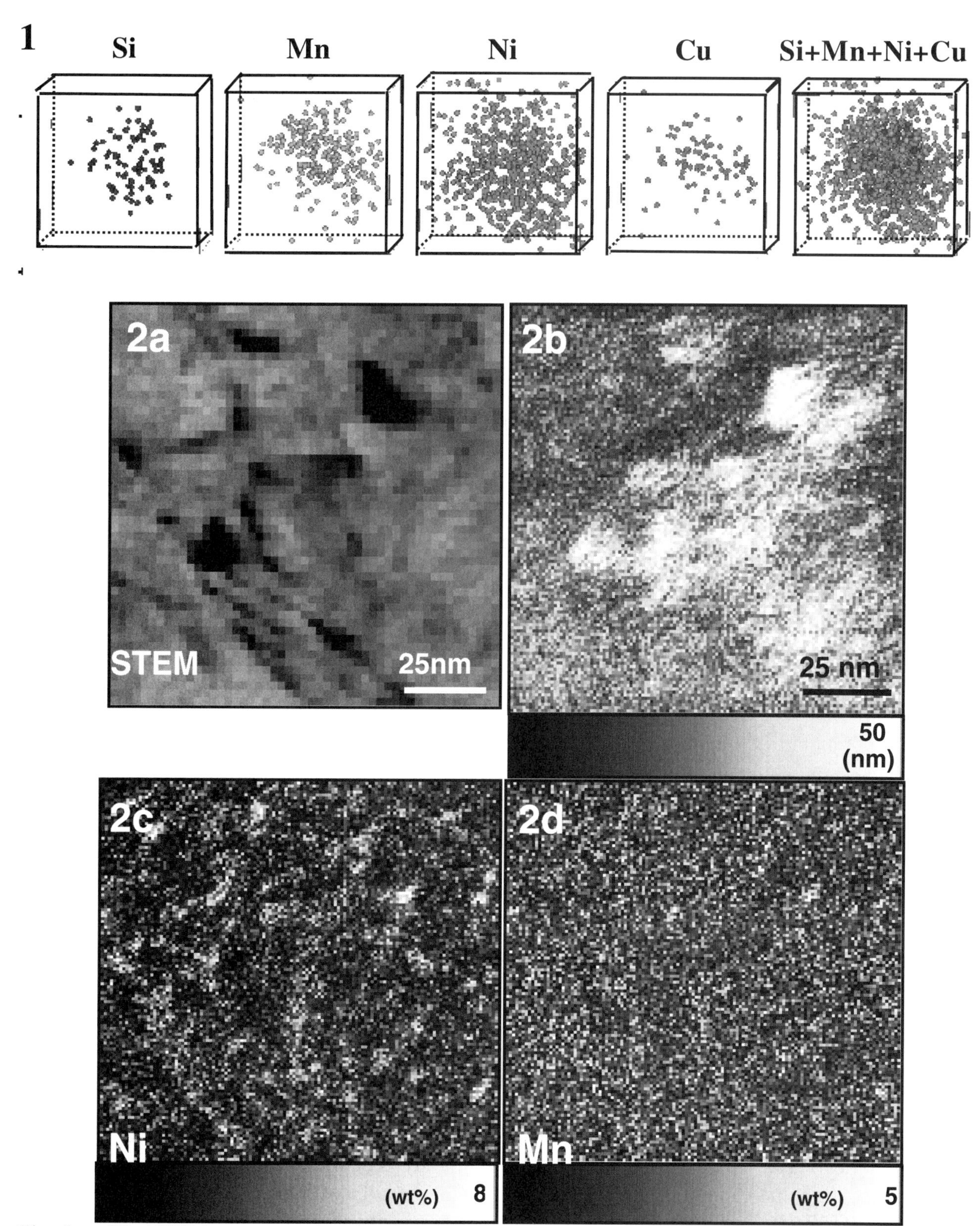

Fig. 1. 3D-APFIM reconstruction for Si, Mn, Ni, and Cu in a typical irradiation-induced solute-rich "precipitate" formed in A508 Gr4N steel. vol: 6 nm X 6 nm X 3 nm.

Fig. 2. (a) STEM image, (b)foil thickness and (c,d) quantitative x-ray maps showing the presence of ~3 to 4 nm irradiation-induced "precipitates". The nanoscale "precipitates" are not visible in the STEM image.

Microsc. Microanal. 8 (Suppl. 2), 2002
DOI 10.1017.S1431927602100791

On the Control of Atomic Clustering, Segregation and Partitioning: Nanoscale Materials Technology

S.P. Ringer

Australian Key Centre for Microscopy & Microanalysis, The University of Sydney, NSW, 2006, Australia

This work summarises the application of microscopy and microanalysis at high resolution to study atomic clustering, segregation and partitioning in selected materials developments. Such studies are increasingly necessary in order to understand structure-property relationships in nanoscale materials technology. Whilst the structure, orientation and, in many cases, the composition of primary and secondary phases including nanoscale precipitation of GP zones and other phases are readily observed and measured using analytical transmission electron microscopy (TEM), clear observation of clustering, segregation and partitioning requires more advanced approaches. In this work, TEM, STEM microanalysis and three-dimensional atom probe (3DAP) to observe these effects in a selection of scientific alloys.

A series of Al-2.5Cu-xMg alloys have been studied, where 0.1<x<1.7 Mg (wt. %). It was shown that the well-known rapid hardening effect in this system requires a threshold of ~0.5 Mg. The behaviour of the quenched-in defect structure upon ageing is shown to be distinctly different for alloys containing x = 0.5 Mg in that the rate and extent of growth of dislocation loops was much higher in these higher Mg containing alloys (Fig. 1). We have correlated this to the vacancy content and solute-vacancy interactions. On the basis of measurements of the vacancy-loop characteristics from TEM images such as Fig. 1, we have calculated that concentration of Mg atoms per nm on the average dislocation loop may be as much as 8-10 times higher in the 0.5Mg alloy compared to the 0.2 alloy.

The effect of trace additions (~0.01 at. %) of Cd on the enhanced precipitation of? ' (Al_2Cu, *I4/mcm*, *a*=0.404 nm and *c*=0.580 nm) has been studied in a set of Al-4Cu (-0.3Mg-0.5Cd) (wt. %) alloys. The effect was confirmed in all Cd-containing alloys and in the Al-Cu-Cd alloy, elemental Cd precipitates were detected uniformly throughout the matrix and attached to θ'. On the other hand, co-clustering of Cd-Mg atoms was detected using 3DAP in the Al-Cu-Mg-Cd alloy during the early stages of ageing. This clustering was followed by the precipitation of a Cd-Mg rich precipitate, which was observed in association with all strengthening phases. Furthermore, it was shown that all of the Cd-bearing phases are able to nucleate the θ' and concomitantly, that the θ' is able to nucleate the Cd-bearing phases.

Finally, the crystallisation of nanocrystalline α-Fe in $Fe_{89-x}Zr_7B_3Cu_1Ge_x$ alloys has been studied using a dedicated VG STEM. Although the T_C of the Ge-free alloy is below room temperature, it is enhanced to 360 K at $x = 5$, indicating that Ge is effective in enhancing the exchange stiffness of Fe-rich amorphous Fe-Zr-B. These results indicate that Ge-induced magnetic softening in $Fe_{89-x}Zr_7B_3Cu_1Ge_x$ is due to the preferential enrichment of Ge into the residual amorphous phase which results in an enhancement of the exchange stiffness in the intergranular region.

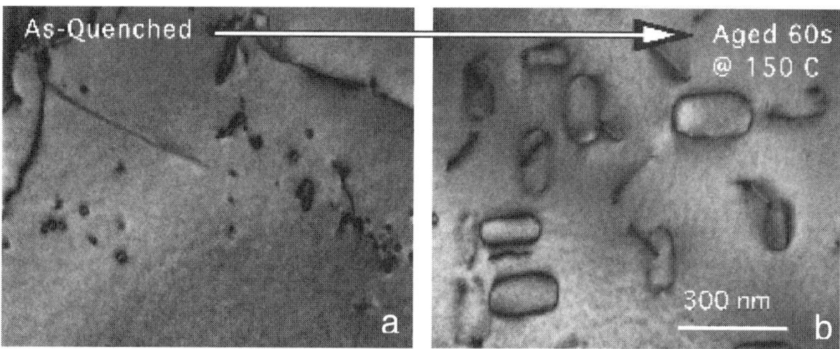

FIG. 1. BF TEM images recorded near <001>$_\alpha$ of the defect structures of Al-2.5Cu-0.5Mg (wt. %) after quenching (a) and after ageing for 60 sec at 150°C (b). Approximately 60% of the hardening reaction occurs in the first few seconds of ageing.

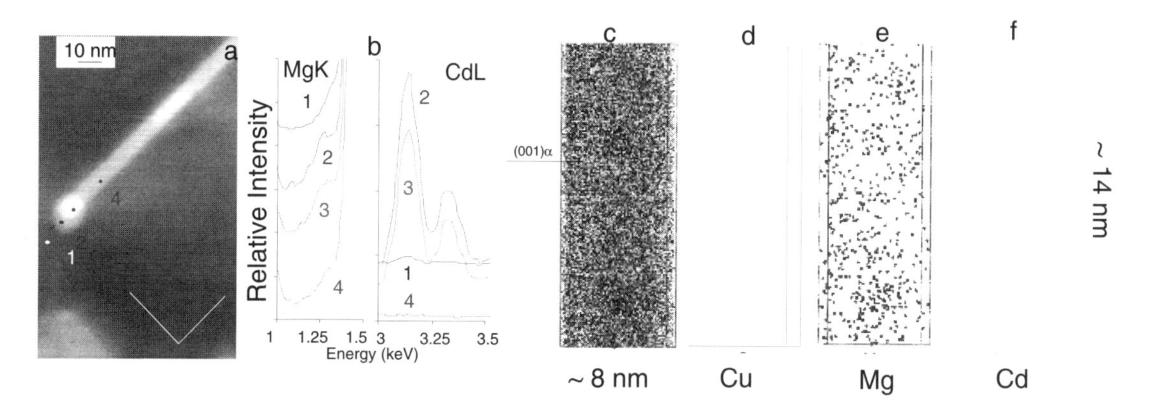

FIG. 2. (a) <001>$_\alpha$ DF STEM image of typical θ' in the peak hardness microstructure of the Al-Cu-Mg-Cd alloy with the Mg and Cd EDXS profile from the points 1-4, indicated. The cube traces are marked in (a). (c-f): 3DAP elemental mapping of Al-Cu-Mg-Cd alloy aged at 200 °C, 15 min showing the presence of two clusters. Maps for (c) all atoms (d) Cu (e) Mg and (f) Cd atoms.

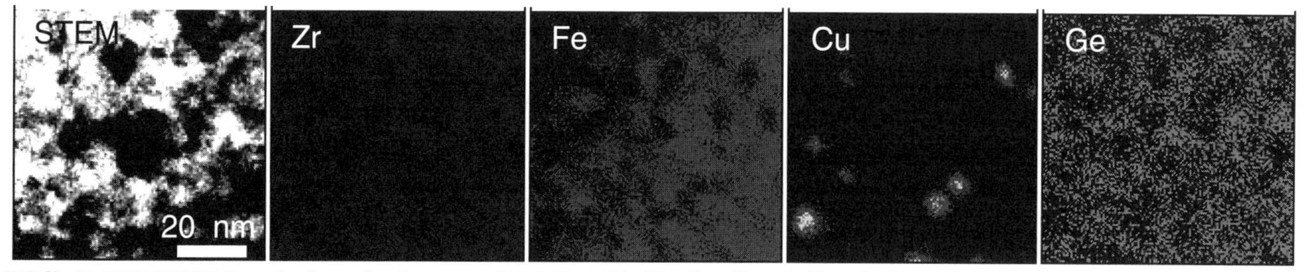

FIG. 3. BF STEM and chemical maps from $Fe_{86}Zr_7B_3Ge_3Cu_1$ following annealing for 1 h at 550 °C. The Ge is partitioned into the residual amorphous and is rejected from the nanocrystalline α-Fe grains. Clusters of Cu were also detected. A probe size of ca. 1 nm was used.

This research was supported by the Australian Research Council. Contributions and valuable discussions from B. T. Sofyan, G. B. Winkelman and K. Suzuki from Monash University are gratefully acknowledged.

Microsc. Microanal. 8 (Suppl. 2), 2002
DOI 10.1017.S1431927602100808

Nanoscale materials for information storage

A K Petford-Long*, P Shang**, Y G Wang* and N Owen*

*Department of Materials, University of Oxford, Parks Road, Oxford OX1 3PH, UK.
**IBM Storage Division , 5600 Cottle Road, San Jose, CA 95193, USA.

The novel materials being developed to address the need in information storage systems for smaller bit size and faster memory tend to be very complex and consist of many nanometre thick layers with microstructure that is inhomogeneous on a fine scale. Examples of such materials are spin-valve (SV) and spin tunnel junction (STJ) structures which rely on the giant magnetoresistance (GMR) effect [1] and have applications as magnetoresistive read-heads [2] and as memory elements [3]. In their simplest form, SV and STJ structures consist of a ferromagnetic (FM) layer whose magnetisation direction is fixed by an adjacent antiferromagnetic (AF) pinning layer through exchange coupling, separated by a nonmagnetic spacer layer from a second FM layer (sense layer) whose magnetisation can be rotated in a low external magnetic field. The spacer is a metal in an SV and an insulator in an STJ. A further example is that of an array of magnetic dots or antidots [4], which are being developed for potential use as patterned magnetic media, with each dot (or magnetic region between adjacent antidots) supporting a magnetic bit.

The behaviour of these materials relies on the local magnetic domain structure and magnetisation reversal mechanism, and one of the techniques enabling micromagnetic studies at the sub-micron scale is Lorentz transmission electron microscopy (LTEM) which allows the magnetic domain structure and magnetisation reversal mechanism of a FM material to be investigated dynamically in real-time with a resolution of a few nm. This presentation will highlight some recent research in this field. For example, Figure 1 shows a series of LTEM Foucault mode images of the magnetisation reversal of the sense layer in an STJ element (width 2 μm) of structure: Ta/NiFe/MnFe/NiFe/Al_2O_3/NiFe/Ta (5/6/10/4/2.5/7/5 nm). The arrows indicate the in-situ applied magnetic field direction and the numbers indicate the field value. After applying a positive field, a 360° domain wall formed by the combination of two 180° domain walls (imaged as adjacent bright and dark lines). The wall is pinned at a structural defect at the element surface. The field was then increased to +60 Oe and decreased (centre row of images). The 360° wall defect acted as a domain nucleation site. If the magnetisation cycle was then repeated but with the positive field increased to +80 Oe, the 360° wall was removed and reversal of the sense layer magnetisation occurred at a higher negative field. This shows the difficulty of producing large arrays of STJ elements for memory applications, in which the magnetisation reversal field of all the elements must be the same.

Figure 2 shows a Foucault mode LTEM image of an antidot array with 0.1 μm width holes. The array was formed using a focused ion beam (FIB) system to mill a series o f holes in a thin $Ni_{80}Fe_{20}$ film. Competition occurs between the spatially varying demagnetising field arising because of the free poles induced on the anti-dot edges and the intrinsic uniaxial anisotropy. Micromagnetic simulations have predicted remanent states should exist when the demagnetising and anisotropy fields are of the same order. Regular, well defined remanent domains (one is arowed) can be seen between the anti-dots in the image; these may be suitable for information storage applications

References
[1] R L White, *IEEE Trans. Mag.* 30(2) (1994) 346.
[2] J A Brug, T C Anthony and J H Nickel, *MRS Bulletin* 21 (1996) 23.
[3] J M Daughton, *Thin Solid Films* 216 (1992) 162.
[4] R P Cowburn, A Adeyeye and J A C Bland, *J. Magn. Magn. Mater.* 173 (1997) 193.
[5] This research was supported by the Engineering and Physical Sciences Research Council and Hewlett-Packard Labs, Palo Alto.

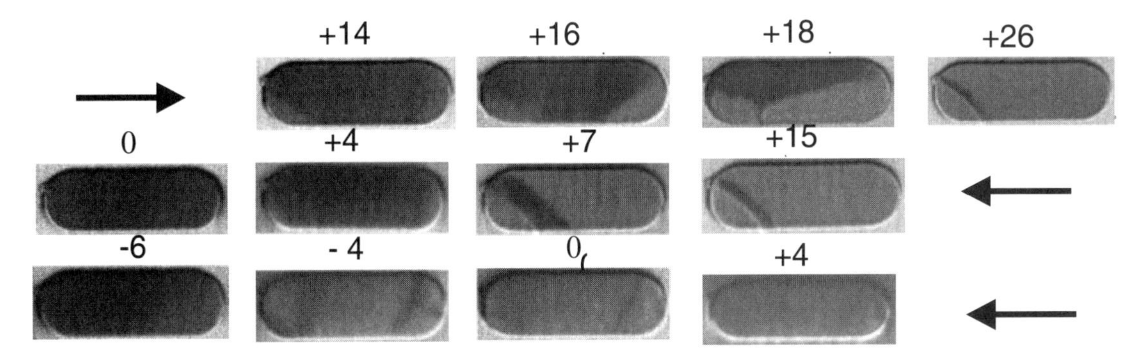

Fig. 1. Foucault mode LTEM images showing in-situ magnetisation reversal of the sense layer in a MnFe/NiFe/Al_2O_3/NiFe STJ element (width 2 μm). Arrows indicate applied field direction.

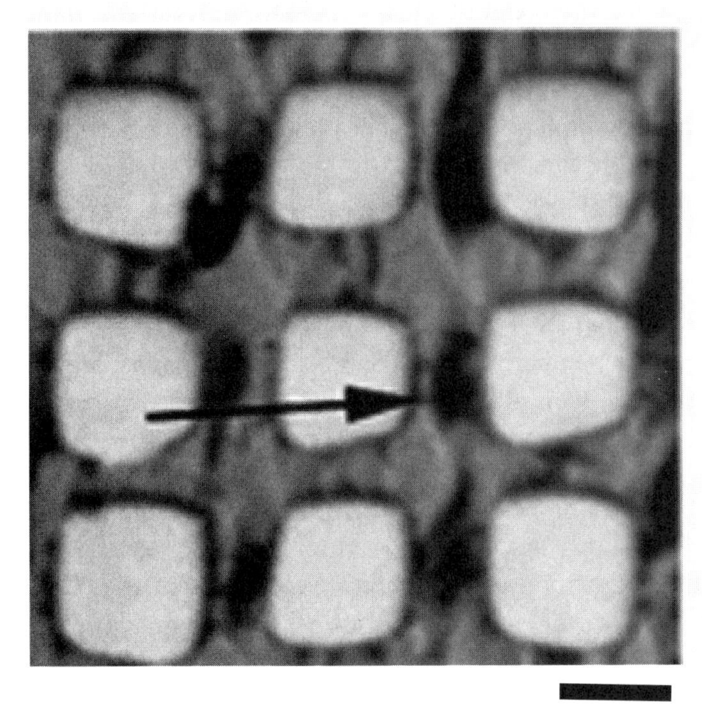

Fig. 2. Foucault mode LTEM image of antidot array cut into a thin Ni80Fe20 film. The magnetic field has been applied along the hard axis of the film. A remanent state domain between two antidots is arrowed.

Microsc. Microanal. 8 (Suppl. 2), 2002
DOI 10.1017.S143192760210081X

Future Hard X-ray Nanoprobe at the Advanced Photon Source

Derrick C. Mancini, Jörg Maser, G. Brian Stephenson

Argonne National Laboratory, 9700 S. Cass Avenue, Argonne, IL 60546

We are designing and constructing a new hard x-ray nanoprobe at the Advanced Photon Source. The nanoprobe will allow real space mapping of density and elemental composition through transmission; crystallographic phase, strain, and texture through diffraction; trace element analysis through fluorescence; chemical states through spectroscopy; magnetic domain structure through linear and circular dichroism; and morphology through x-ray tomography. The nanoprobe will be a versatile tool that can be applied to research in nanoscience and nanotechnology. Examples of application to research thrusts at Argonne National Laboratory include to image and track domain evolution in ferroelectric and magnetic nanostructures, to observe strains in microelectronic interconnects, to measure composition and phase distributions in layered nanoparticles for catalysis, or to determine the location, elemental composition, and chemical state of hybrid inorganic/organic nanoparticles interacting with biological systems.

We aim with the design of the x -ray nanoprobe to advance the state of the art of hard x-ray microscopy by providing the highest spatial resolution achievable from the brilliance of a third generation synchrotron radiation source such as the Advanced Photon Source. A dedicated source, beamline, and optics will be used to optimizing the capab ilities of the nanoprobe. This unique instrument will not only be key to the specific research thrusts in nanoscience at Argonne National Laboratory, but will be of general utility to the broader nanoscience community. It will offer diverse capabilities in studying nanomaterials and nanostructures, particularly for embedded structures. The combination of diffraction, fluorescence, and phase contrast in a single tool will provide unique characterization capabilities for nanoscience. Current x -ray microprobes based on Fresnel zone plate optics have demonstrated a spatial resolution of 150 nm. With advances in the fabrication of zone plate optics, and using an optimized beamline design, we expect to be able to achieve 30 nm resolution. The nanoprobe will cover the x -ray energy range of 3 -30 keV with a working distance between the focusing optics and the sample typically of 10-30 mm.

The x -ray nanoprobe will complement other microscopy techniques such as optical microscopy, scanning electron microscopy, transmis sion electron microscopy, scanning probe microscopies, and soft x -ray microscopy. In particular, the x-ray nanoprobe provides advantages such as being nondestructive, non-invasive, quantitative, requiring minimal sample preparation, providing sub-optical spatial resolution, having the ability to penetrate inside a sample and study its internal structure, and having enhanced capability to study processes *in situ*. Another important distinction compared to charged-particle probes is that x-rays are non-interacting with applied electric or magnetic fields, which is a significant advantage for in-field studies of dielectric and magnetic materials.

The nanoprobe will operate in two modes: a primary scanning probe mode, in which an x-ray spot of 30 nm FWHM is used to probe the specimen, and a full-field transmission mode, in which the full undulator beam will be used for transmission imaging with 30 nm spatial resolution. The scanning probe mode (figure 1) makes use of the high brilliance of an APS insertion device by focusing the spatially coherent fraction of the undulator beam into a diffraction-limited spot. It is suited for spectroscopy with high trace element sensitivity and strain contrast imaging. The full-field transmission mode, which will use the full photon flux provided by the undulator, will allow fast acquisition of a larger object field at high spatial resolution in amplitude or phase contrast. It can also be used as an alignment tool to allow quick, *in situ* identification of a small specimen area of interest for scanning probe studies. In full-field transmission mode, the nanoprobe will be invaluable for applications such as high-throughput 2D imaging and high-resolution tomography. The combination of both scanning probe and full-field transmission mode in a single x-ray microscope will be a uniquely powerful tool.

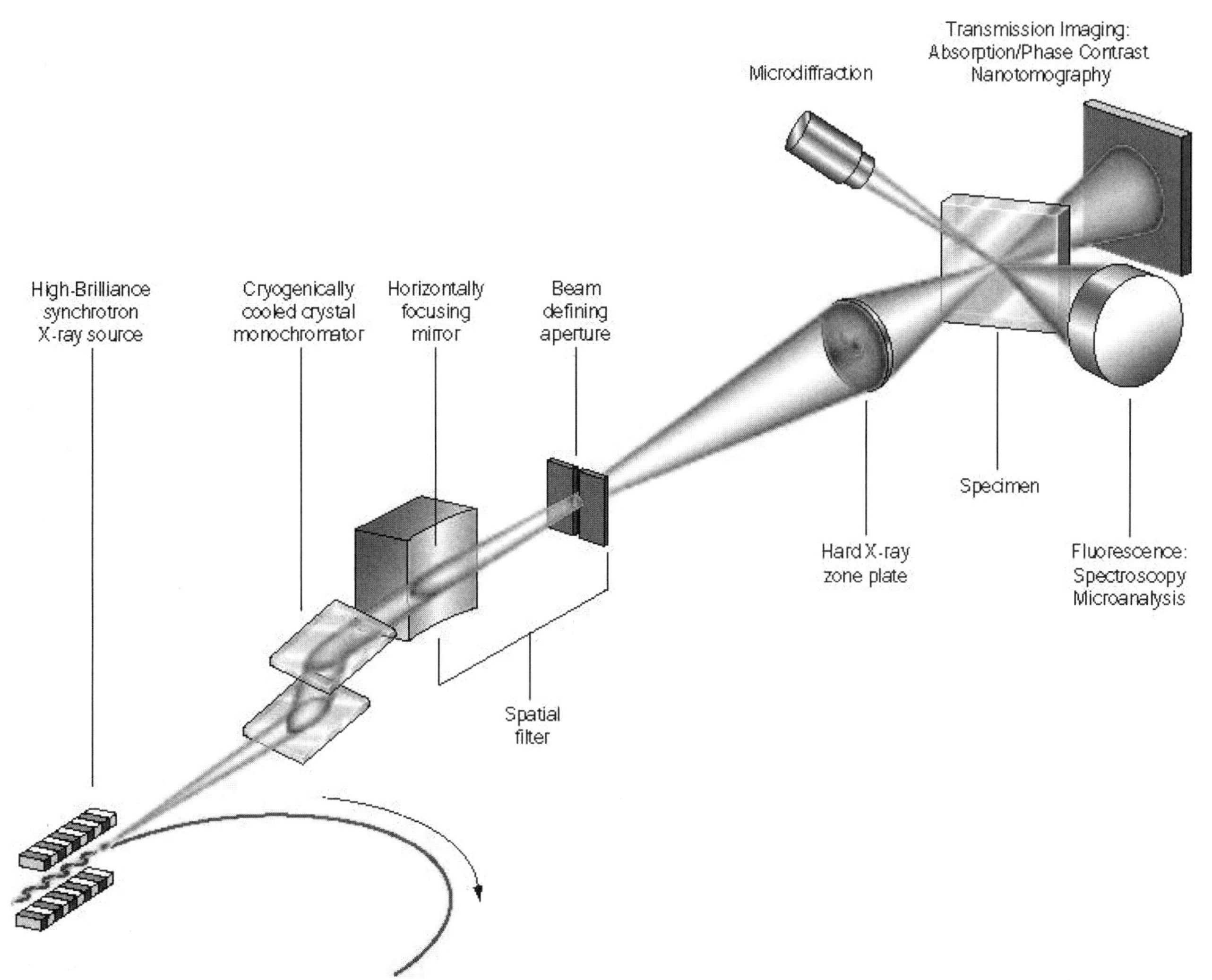

Figure 1. Conceptual layout of the planned hard x-ray nanoprobe at the Advanced Photon Source.

Microsc. Microanal. 8 (Suppl. 2), 2002
DOI 10.1017.S1431927602100821

Scanning Probe Microscopy in TEM : an In-situ Approach for Nano-scale Property Measurements

Zhong Lin (ZL) Wang*

Center for Nanoscience and Nanotechnology, and School of Materials Science and Engineering, Georgia Institute of Technology, Atlanta, GA 30332-0245.
* e-mail: zhong.wang@mse.gatech.edu

Property characterization of nanomaterials is challenged by the small size of the structure because of the difficulties in manipulation. Here we demonstrate a novel approach that allows a direct measurement of the mechanical properties, the electron field emission and the ballistic quantum conductance in individual nanotubes and nanowires by an conjunction operation of scanning probe microscopy in transmission electron microscopy (TEM). The technique is powerful in a way that it can directly correlate the atomic-scale microstructure of the carbon nanotube with its physical properties, providing a one-to-one correspondence in structure-property characterization.

Bending modulus of a carbon nanotube: To carry out the property measurement of a nanotube, a specimen holder for an TEM was built for applying a voltage across a nanotube and its counter electrode [1]. Static and dynamic properties of the nanotubes can be obtained by applying controllable static and alternating electric fields. To measure the bending modulus of a carbon nanotube, an oscillating voltage is applied on the nanotube with ability to tune the frequency of the applied voltage. Resonance can be induced in carbon nanotubes by tuning the frequency (Figure 1), from which the bending modulus can be derived. For nanotubes produced by arc-discharge, which are believed to be defect-free, the bending modulus is as high as 1.2 TPa (as strong as diamond) for nanotubes with diameters smaller than 8 nm, and it drops to as low as 0.2 TPa for those with diameters larger than 30 nm [1]. Nanotubes produced by catalyst-assisted pyrolysis containing a high density of point defects have bending modulus of 7 times smaller than that of a defect-free nanotube [2].

Electron field emission from individual carbon nanotubes: For carbon nanotube emitters, most of the electrons are emitted from the tips of the tubes, and it is the local work function that matters to the properties of the tube field emission. We present a novel approach for measuring the tip work functions of individual carbon nanotubes [3,4]. Our results indicate that the tip work function show no significant dependence on the diameter of the nanotubes in the range of 14-55 nm. Majority of the nanotubes (~75%) have a work function of 4.6-4.8 V at the tips, which is 0.2-0.4 V lower than that of carbon. A small fraction of the tubes (~25%) have a work function of ~5.6 V, about 0.6 V higher than that of carbon. Field emission induced structural damage has also been observed and the results explain the current fluctuation in field emission [5].

Ballistic conductance in carbon nanotubes: The conductance of a carbon nanotube was measured as a function of the depth with which the tube was inserted into the mercury (Figure 3) [6]. Surprisingly, the nanotube displays quantum conductance No heat dissipation was observed in the nanotube. This is the ballistic conductance, and it is believed to be a result of single graphite layer conductance.

[1] P. Poncharal, Z.L. Wang, D. Ugarte, and W.A. de Heer, *Science* 283 (1999) 1516;
Z.L. Wang, P. Poncharal and W.A. De Heer, *Pure Appl. Chem.*, 72, Nos. 1-2 (2000) 209;
Z.L. Wang, P. Poncharal, and W.A. De Heer, *J. Physics & Chemistry of Solids*, 61 (2000) 1025.
[2] R.P. Gao, Z.L. Wang, Z.G. Bai, W. de Heer, L. Dai and M. Gao, *Phys. Rev. Letts.*, 85 (2000) 622.
[3] Z.L. Wang, P. Poncharal, and W.A. De Heer, *Microscopy and Microanalysis*, 6 (2000) 224.
[4] R.P. Gao, Z.W. Pan and Z. L. Wang *Appl. Phys. Letts.*, , 78 (2001) 1757.
[5] Z.L. Wang, R.P. Gao, W. de Heer and P. Poncharal, *Appl. Phys. Letts.*, in press (2002).
[6] S. Frank, P. Poncharal, Z.L. Wang, and W.A. de Heer, *Science* 280 (1998) 1744.
[7] Thanks to the support from the NSF grant DMR-9733160 and Georgia Tech. This work was collaboratively carried out with P. Poncharal, W. de Heer and R.P. Gao.

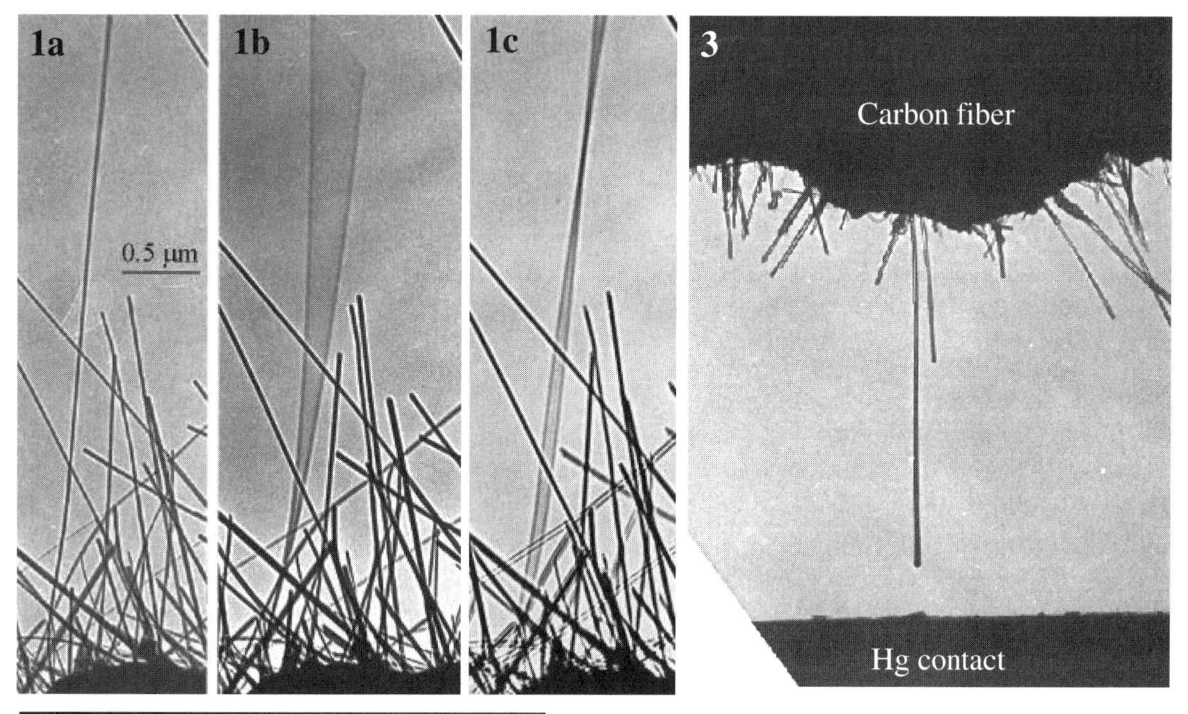

Fig. 1. Electric field induced mechanical resonance. A selected carbon nanotube at (a) stationary, (b) the first harmonic resonance, and (c) the second harmonic resonance.

Fig. 2. Electron field emission from the tips of carbon nanotubes.

Fig. 3. Ballistic transport measurement from a single carbon nanotube in TEM. Electric current can be passed through the nanotube.

Microsc. Microanal. 8 (Suppl. 2), 2002
DOI 10.1017.S1431927602100833

TEM Characterization of Thin, Epitactic Ni_2MnGa films on GaAs.

S. McKernan, J. W. Dong and C. J. Palmstrøm

Department of Chemical Engineering and Materials Science, Amundson Hall, University of Minnesota, Minneapolis, Minnesota 55455.

Spintronics is a concept based on the intimate interaction between magnetics and semiconductors that is being applied to the fabrication of advanced devices. Advances in this field require an understanding of three principal components: a suitable spin injector material, the detection of electron spin once it has been injected into a semiconductor, and the fundamental limitations on spin transport across interfaces.

Several materials have been proposed for the spin injector layer, which is required to be a high quality single crystal ferromagnet with a Curie Temperature greater than ~300K for room-temperature operation of a device. Most choices of ferromagnetic metals have large lattice mismatches with any semiconductor substrate, resulting in a defect-rich interface. Recently we have reported the MBE growth of the ferromagnetic Heusler Alloy Ni_2MnGa on GaAs [1]. This alloy has a 3% lattice mismatch with the substrate, and can be grown pseudomorphically with respect to the substrate using suitable intermediate layers. It exhibits a number of different polymorphs depending on the temperature and stress of the alloy. At high temperature the structure is the cubic $L2_1$ ordered structure. At low temperature it undergoes a martensitic phase transformation to either a non-modulated $L1_0$ structure or one of several modulated variants [2]. In order to determine the suitability of the structure for spintronic device fabrication the magnetic and structural properties have been determined by a number of techniques.

Samples of the thin Ni_2MnGa films on GaAs substrates were prepared for TEM by traditional dimpling and ion-milling techniques in both cross-section and plan-view geometries. The plan-view samples were back-thinned to remove the substrate material, leaving the epilayer film able to relax without the constraint of the substrate.

Images obtained from the plan-view samples show that the microstructure is evidently not one of a continuous single crystal. In Fig 1 it can be seen that there are several domains present, (these are elongated in shape 100-200 nm wide and ~1μm in length) and within these domains twin-bands are observed . The twins are irregularly spaced down to the 5 nm scale. The diffraction pattern from the central twinned domain is shown in figure 4. Dark-field images from this region are shown in figures 2 and 3, corresponding to the reflections indicated in figure 4. It can be seen that the minor component of the twins in the central domain is strongly present in the bottom domain. The diffraction pattern from this domain is shown in figure 5, and is indexed in terms of the "double" $L1_0$ lattice according to Pons et. al. [2]. Using this indexing scheme, the close packed planes in the structure are {111} planes, and it is apparent that both the twin boundaries and the domain boundaries are edge-on in this orientation, and parallel to {111}. In none of the domains observed here were any additional diffraction spots observed that could not be accounted for by twinning, indicating that this is a non-modulated martensite. This observation is borne out by the fact that the structure remained in this state to well above 400C whereas the modulated martensite structures would have transformed to the parent cubic structure below that temperature.

The structures of the relaxed epilayer film are consistent with those described in bulk Ni_2MnGa [2], although our TEM and XRD measurements show that the film is strained in order to be lattice matched to the substrate. The strain relief may be the reason that the crack in the epilayer visible in figure 1 has occurred. The crack edges appear to follow the {220} planes in the differently oriented domains.

The presence of these defects in the released and unreleased epilayer, and the quality of the semiconductor-ferromagnetic interface will be discussed. [3]

References

1. J. W. Dong, L. C. Chen, J. Q, Xie, T. A. R. Muller, D. M. Carr, C. J. Palmstrøm, S. McKernan, Q. Pan, R. D. James, J. Appl. Phys. **88**, 7357-7359 (2000)
2. J. Pons, V. A. Chernenko, R. Santamarta and E. Cesari, Acta mater, **48**, 3027-3038 (2000)

3. This research is supported in part by the Characterization Facility in the University of Minnesota Institute for Technology, and by ONR/DARPA Contract No. N/NN00014-99-1005.

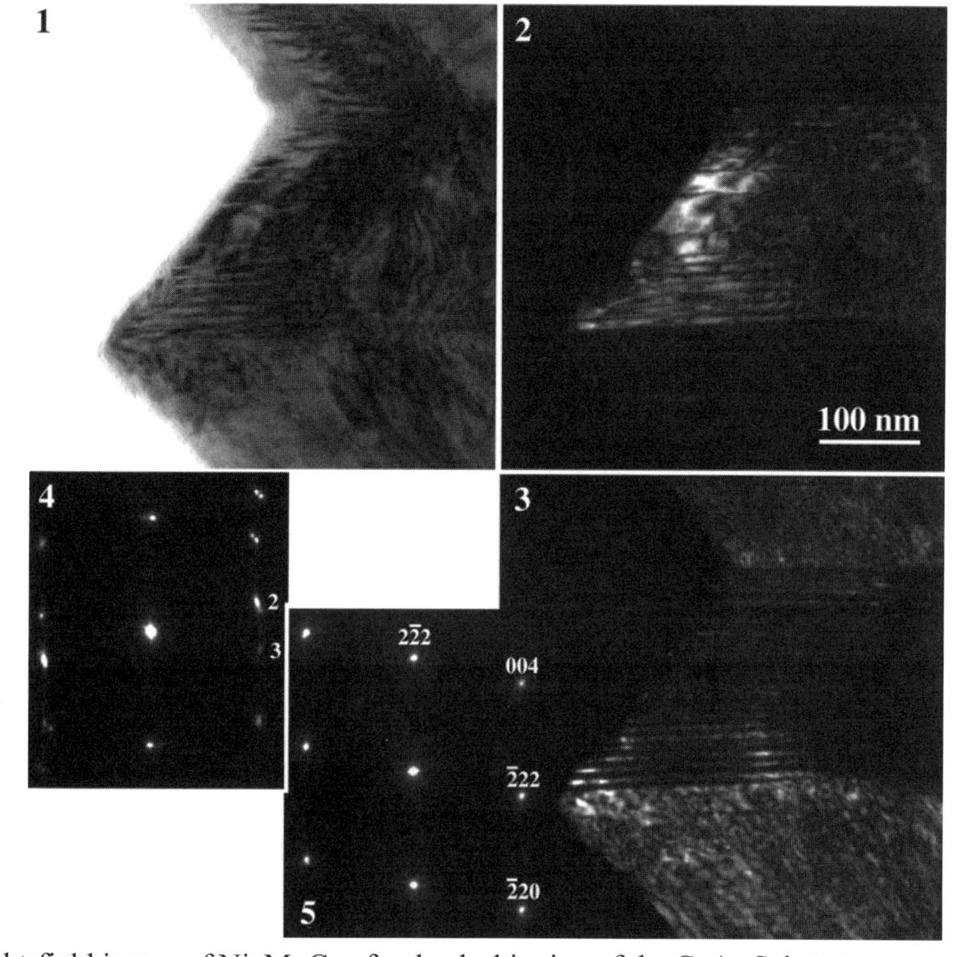

FIG 1. Bright-field image of Ni_2MnGa after back-thinning of the GaAs Substrate.
FIG 2 and 3. Dark field images of the region shown in fig 1 using the reflections indicated in fig 4.
FIG 4. Diffraction pattern of central domain in fig 1.
FIG 5. Diffraction pattern of lower domain in fig 1 indexed as a "double" $L1_0$ lattice

Microsc. Microanal. 8 (Suppl. 2), 2002
DOI 10.1017.S1431927602100845

Mechanics of Nanowires

Rodney S. Ruoff*, Xinqi Chen, Dmitriy Dikin, Weiqiang Ding, Ming-Feng Yu[a], Gregory J. Wagner

Department of Mechanical Engineering, Northwestern University, 2145 Sheridan Road, Evanston, IL 60208
[a]Zyvex Corporation, Advanced Technologies Group, 1321 North Plano Road, Richardson, TX 75081

Carbon nanotubes (NTs) and boron and SiO_2 nanowires (NWs) have many potential applications in scientific and industrial fields due to their size and expected excellent mechanical and electrical properties. Here we present the further development of the mechanical resonance method for characterizing the mechanical properties of such nanowires.

A homemade nanomanipulation testing stage (shown in Fig. 1), having X, Y, Z and rotational degrees of freedom, was used in these measurements of the resonance response of C NTs and SiO2 NWs. The measurements were made in an SEM chamber (Hitachi S-4500 FEG-SEM). The NTs or NWs were attached on a Pt/Ir wire with conductive carbon tape. A tungsten tip was brought into proximity and an ac electric field was applied between the Pt/Ir wire and the tungsten tip. The NT or NW could be driven into resonance by varying the frequency of the electric field. The bending modulus of multi-wall C NTs and SiO_2 NWs were obtained by fitting the measured resonate frequencies. [1,2] In addition to this method of electromechanical resonance excitation, the resonance could be mechanically excited by a piezoelectric oscillator (without the presence of an applied electric field). Figure 2 shows a SiO_2 NW on resonance.

Parametric resonance, a resonance phenomenon resulting from an oscillation system with a time-varying coefficient, has found wide applications in nonlinear optics and electronics for detecting and amplifying small signals. We also realize such resonance with boron nanowires. Resonance at drive frequencies near $2f_0/n$, where f_0 is the fundamental resonance of the nanowire, for n from 1 to 4 were observed inside a scanning electron microscope and analyzed. These results, of fundamental interest because they are likely representative of a ubiquitous response of nanoscale systems that intrinsically have smaller damping than macroscale systems, are also of practical interest for applications in the sensor field. Thus we will devote some time developing the theory, and method of modeling, for the audience. *ONR MIS and NASA support for current work, NSF support for building tool.*

References:

[1] P. Poncharal, Z. L. Wang, D. Ugarte, W. A. de Heer, *Science*, **283** (1999) 1513.
[2] Z. L. Wang, R. P. Gao, P. Poncharal, W. A. de Heer, Z. R. Dai, Z. W. Pan, *Materials Science and Engineering C*, **16** (2001) 3.

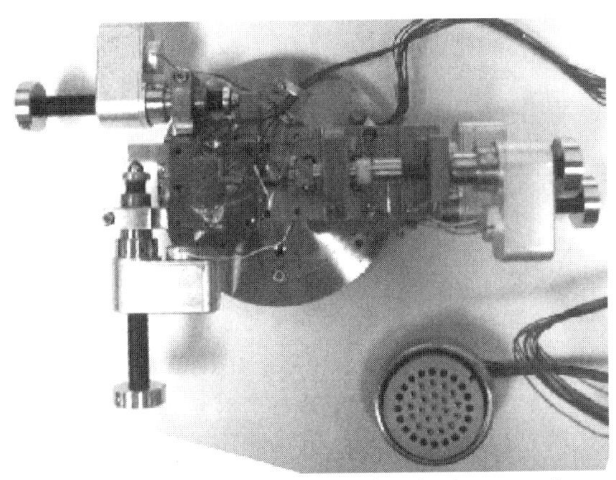

Figure 1. Homemade nanomanipulation testing stage.

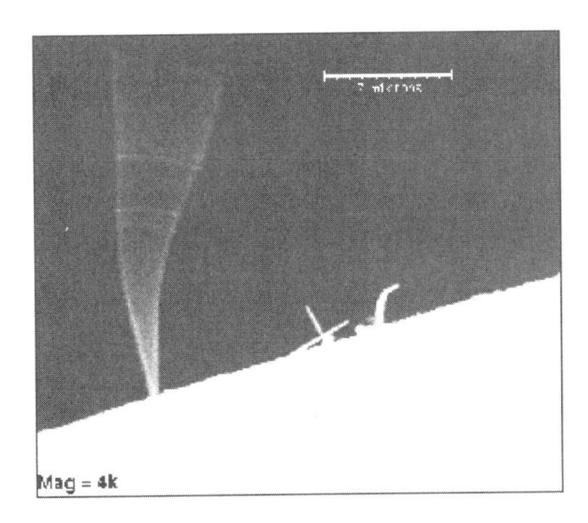

Figure 2. Resonate vibration of a SiO2 nanowire excited by anac electric field; a similar resonance response, but not identical, is achieved by mechanically driving via direct physical contact with an oscillating piezoelectric.

Microsc. Microanal. 8 (Suppl. 2), 2002
DOI 10.1017.S143192760210105X

Electron Microscopy and its Application to the Study of Incommensurately Modulated Compositionally and/or Displacively Flexible Phases

R.L.Withers, L.Norén, Y.Liu and F.Brink

Research School of Chemistry, Australian National University, Canberra, A.C.T, 0200, Australia

Many crystalline phases behave as the text-books say they should *i.e* as three-dimensionally periodic objects characterized by well-defined stoichiometries, unit cells and space group symmetries. An ever-increasing number of compositionally and/or displacively flexible phases, however, do not fit into this category and are modulated in one form or another [1]. The nature of these modulation/s can be purely displacive in character, of either static or dynamic origin, or of mixed compositional and displacive character. An understanding of the local crystal chemistry of such flexible phases can not be had until these modulations are properly characterized. The ability of electron diffraction to reveal weak subtle features of reciprocal space from small local regions in conjunction with the capacity to image in various modes in real time with excellent spatial resolution and over a considerable range of temperature make the modern TEM an extremely powerful instrument for the study of many such modulated flexible phases.

Fig.1a, for example, shows a $<-1,1,3>_p$ zone axis electron diffraction pattern (edp) of the ferroelectric α-polymorph of $K_3MoO_3F_3$. Indexation is with respect to the underlying elpasolite (or ordered perovskite) type parent structure. In addition to the existence of supercell reflections, note the presence of a spectacular and highly structured, previously unreported, diffuse intensity distribution related to local O/F ordering and associated structural relaxation. Likewise careful TEM investigation has revealed that there exist four quite distinct phase regions within the one previously reported $Co_{2-x}Se_2$, $0 \leq x \leq 0.5$, solid solution field. The second of these phase regions is a Co/vacancy ordered, (3+2)-d incommensurately modulated phase region, see, for example, the $<110>_p$ (p for the underlying *NiAs*/CdI_2 type parent structure) zone axis micro-diffraction pattern shown in Fig.1b. The presence of a multitude of higher order harmonic satellite reflections in the [001] zone axis edp of the incommensurately modulated $Ni_{6-x}Se_5$ solid solution phase (see Fig.1c) suggests a non-conservative interface modulated structure mechanism for the accommodation of non-stoichiometry in this system as is confirmed by the corresponding [001] HREM image (see Fig.1d).

Non-stoichiometry is not, however, the only known cause of structural flexibility. Many framework structures built out of corner-connected, essentially rigid polyhedral units are also inherently (displacively) flexible, particularly in their high temperature polymorphic forms. Fig.1e, for example, shows a $<-1,1,0>$ zone axis edp of the high temperature polymorph of SiO_2-tridymite. The curved diffuse distribution in this edp traces out thermally excited, low energy RUM (Rigid Unit Mode) modes of distortion of this tetrahedral framework structure. On cooling, one or other of these RUM modes of distortion often condense out giving rise to a statically distorted, lower temperature polymorphic form. Fig.1f, for example, shows a $[-2,0,\sim6.8]_p$ zone axis edp of the (3+2)-d incommensurately modulated, room temperature polymorph of the mineral fresnoite, $Ba_2TiSi_2O_8$. The weak incommensurate satellite reflections in Fig.1f arise from two such condensed RUM modes of distortion of the parent fresnoite framework structure.

[1] R.L.Withers, S.Schmid & J.G.Thompson, Prog. Solid St. Chem. **26** (1998) 1.

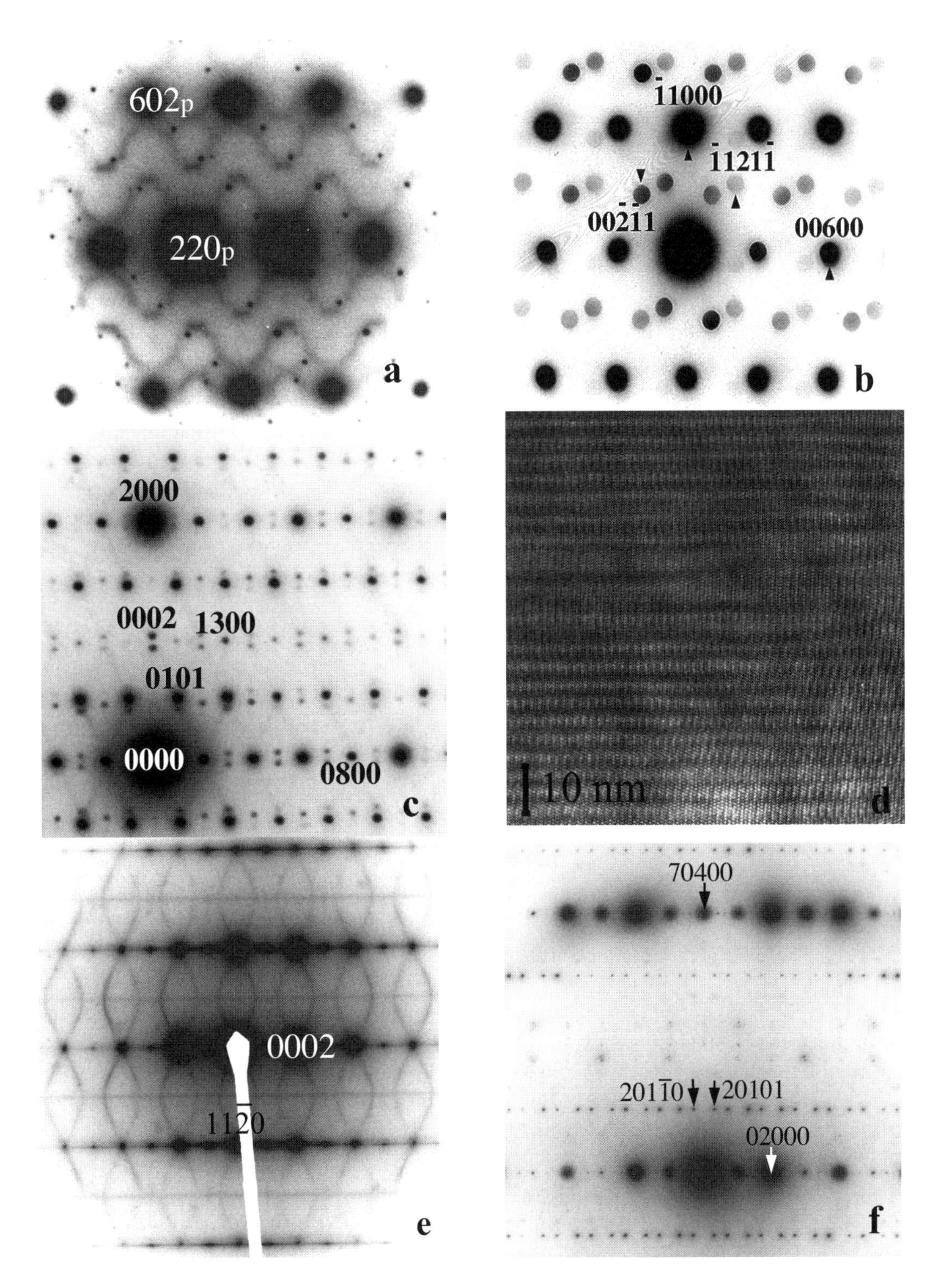

Fig.1 (a) a <–1,1,3>$_p$ zone axis edp of $K_3MoO_3F_3$. Indexation is with respect to the underlying elpasolite type parent structure, (b) a <110>$_p$ micro-diffraction pattern of an incommensurately modulated phase in the $Co_{2-x}Se_2$ system, (c) an [001] zone axis edp of the incommensurately modulated $Ni_{6-x}Se_5$ solid solution phase along with the corresponding HREM image in (d), (e) a <–1,1,0> zone axis edp of the high temperature polymorph of SiO_2-tridymite and (f) a [-2,0,~6.8]$_p$ zone axis edp of the incommensurately modulated, room temperature polymorph of the mineral fresnoite, $Ba_2TiSi_2O_8$.

Microsc. Microanal. 8 (Suppl. 2), 2002
DOI 10.1017.S1431927602101061

Electron Crystallographic Study of Incommensurate Modulated Structures

Fan Hai-fu, Li Yang, Wan Zheng-hua, Fu Zheng-qing, Mo You-de, Cheng Ting-zhu & Li Fang-hua

Institute of Physics, Chinese Academy of Sciences, Beijing 100080, China

As a tool of structure analysis, electron crystallography has well-known advantages in comparison with X-ray techniques. For example, it enables the combination of microscopic imaging and diffraction analysis making the phase problem much easier to solve. Apart from this and some others, electron crystallography has additional advantages in studying incommensurate modulated structures. First, crystals having incommensurate modulation are often such imperfect that they are too poor and too small to carry out an X-ray single-crystal analysis. However those crystals are suitable for electron microscopic observation. Secondly, diffraction patterns of incommensurate modulated structures consist of main reflections and satellites. The latter is the main carrier of modulation information. Electron diffraction shows much stronger satellites enabling observation of weaker structural modulation in smaller area. A big problem of electron crystallography is the strong dynamical-diffraction effect. However, this effect is considerably weakened by the imperfection of periodicity due to incommensurate structural modulation.

Strictly speaking, incommensurate modulated structures do not have 3-dimensional periodicity. However, they can be considered as a 4- or higher-dimensional periodic object cut with the 3-dimensional physical space. Direct methods have been extended to use in multi-dimensional space enabling *ab-initio* solution of incommensurate modulated structures [1]. The main points are as follows. (i) The basic/average structure is solved by conventional direct methods in 3-dimensional space using only the main reflections. (ii) Phases of satellite reflections are derived by multi-dimensional direct methods based on the known phases of main reflections. (iii) A multi-dimensional Fourier map is calculated with experimental structure-factor magnitudes and direct-method phases derived from the previous steps. The actual incommensurate modulated structure will then revealed objectively on the 3-dimensional hyper-section of the multi-dimensional Fourier map. (iv) The modulated structure model is constructed according to the multi-dimensional Fourier map without any preliminary assumptions on the property of modulation.

The above method has been applied to study incommensurate modulation in bismuth based high Tc superconductors [2 - 4]. Results are shown in Figures 1 - 3. Influence of the dynamical effect of electron diffraction to the results has been extensively examined by a series of simulating calculation. Some of the results are given in Figure 4. It turns out that the method may be applied to samples as thick as 300Å. An MS Windows program *VEC* (Visual computing in Electron Crystallography) [5] has been written and includes the direct method for solving incommensurate modulated structures. The program is freely available for academy use and has the following features: (i) searching defocus value from a single electron microscopy image; (ii) resolution enhancement of electron microscopy images using direct methods; (iii) simulation of dynamical/kinematical electron diffraction patterns and electron microscopy images for conventional and incommensurate modulated crystals; (iv) 2-, 3- and 4-dimensional FFT; (v) 2-dimensional half-tone-graph display of 2-, 3- and 4-dimensional Fourier maps; (vi) direct-method solution of incommensurate one-dimensionally modulated structures and composite structures of two subsystems.

References

[1] Hao Quan, Liu Yi-wei & Fan Hai-fu, *Acta Cryst.* A**43**, 820-824 (1987).

[2] Mo Y.D., Cheng T.Z., Fan H.F., Li J.Q., Sha B.D., Zheng C.D., Li F.H. & Zhao Z.X., *Supercond. Sci. Technol.* **5**, 69-72 (1992).

[3] Z.Q. Fu, D.X. Huang, F.H. Li, J.Q. Li, Z.X. Zhao, T.Z. Cheng & H.F. Fan, *Ultramicroscopy*, **54**, 229-236 (1994).

[4] Fan Hai-fu, *Microscopy Research and Technique,* **46**, 104-116 (1999).

[5] Wan Zheng-hua, Liu Yu-dong, Fu Zheng-qing, Li Yang, Cheng Ting-zhu, Li Fang-hua & Fan Hai-fu, *VEC, a program for Visual computing in Electron Crystallography*, Institute of Physics, Chinese Academy of Sciences, Beijing, China (2000).

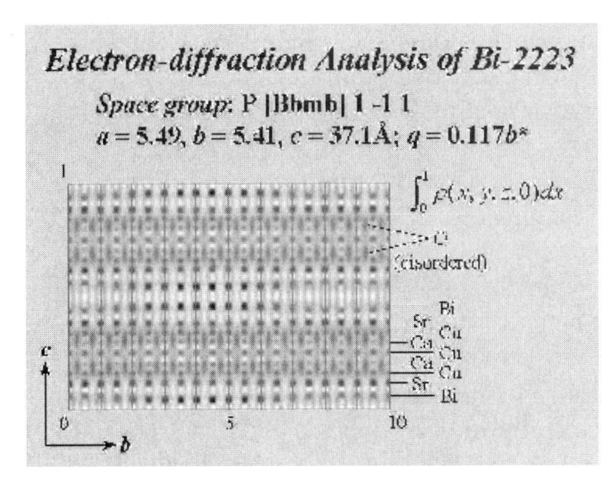

Figure 1. Potential distribution of the high Tc superconductor Bi-2223 projected down the ***a*** axis

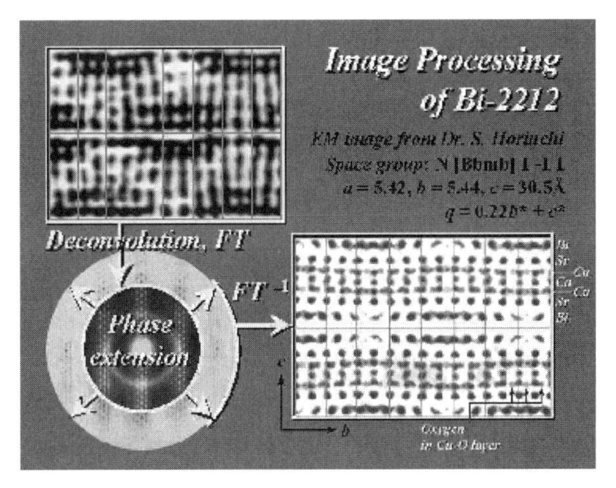

Figure 2. Electron micrograph (upper left), electron diffraction pattern (lower left) and potential distribution (right) of Bi-2212

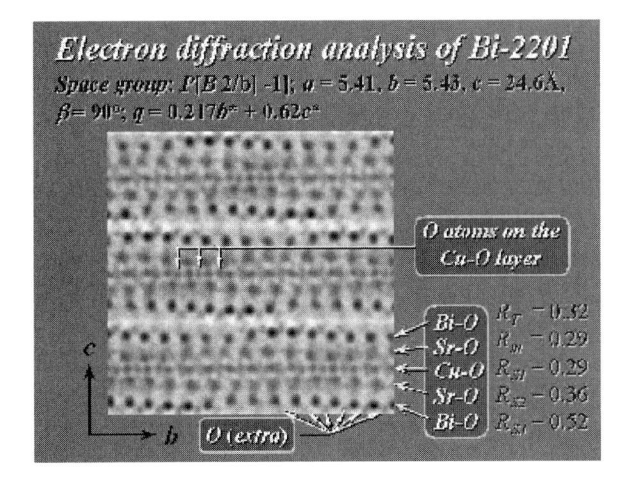

Figure 3. Potential distribution of Bi-2201 projected down the ***a*** axis

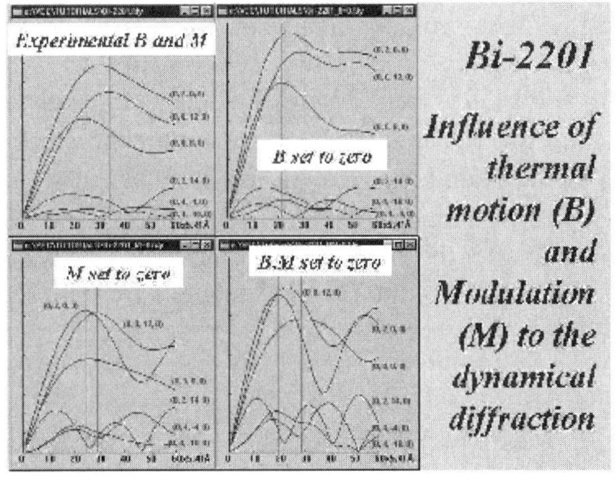

Figure 4. Curves of dynamical electron-diffraction intensity versus sample thickness

Microsc. Microanal. 8 (Suppl. 2), 2002
DOI 10.1017.S1431927602101073

Electron Microscopy Study Of Misfit Layer Structures In The Sb -Nb-S And Bi-Nb-S Systems

L. C. Otero-Díaz

Dpto. Quím. Inorgánica, F. CC. Químicas and CME, Univ. Complutense, E-28040, Madrid, Spain

The misfit layer chalcogenides constitute a broad group of minerals (as cylindrite, cannizzarite,...) and synthetic inorganic materials [1,2]. Their ideal stoichiometry is $(AX)_{1+\delta}$ $(BX_2)_n$ (A=rare earth, Sn, Pb, Sb, Bi; B=Ti, V, Cr, Nb, Ta; X=S, Se) and they are built up by alternate stacking of two sub-structural units with their own underlying average lattices: (a) a so-called H sublattice, with pseudo-orthohexagonal symmetry, formed by n adjacent (BX_2) sandwiched three-atom-thick layers with a structure similar to the one found in the parent transition metal chalcogenide and; (b) a so called Q sublattice, with pseudo-tetragonal symmetry, formed by one (AX) two-atom-thick layer with a distorted rock-salt structure. $(1+\delta)$ is given by the interlayer misfit.

Samples with nominal composition ANb_2S_5 (A=Sb, Bi) were prepared from the elements [3,4]. Electron microscopy observations were performed in scanning (30 kV) and transmission (200 and 400 kV) microscopes. The SAEDP's taken along the main orientations have allowed us to characterize seven new misfit layer structures in both systems. So, figure 1 (a, b, c, and d) corresponds to the Fourier-filtered high resolution images and diffraction patterns of the monolayer $(SbS)_{1.195}NbS_2$ along the indicated directions. Note the modulations bands marked by arrows in a, and the stacking planes in c. Figure 2-a shows a SEM micrograph of an agglomerate of $(BiS)_{1.166}(NbS_2)_2$ misfit layer crystals. The SAEDP of one of these crystals, taken along [001], is shown in 2-b. The main reflections, marked with arrowheads, are indexed as 200_Q, 200_H and $020_{H,Q}$ (without labelling). The arrows indicate the modulation q vector (~0.4 nm period) that defines the misfit modulation. The small arrows denote and additional modulation, which seems to be beam sensitive. Thus in the figure 2-c the high resolution processed image contains only information coming from the primary modulation, see the arrows.
We have also observed big crystals (mm-sized) with variable tubular morphology (see the scanning micrograph given in figures 3-a,b). BSE imaging and WEDS analysis of single tubular crystals and TEM of their cross sections (Figures 3-c,d) revealed a strong compositional and inter-laminar stacking disorder along the tube radius. Slabs with stacking sequences corresponding to binary 3 R-NbS_2 (~0.6 nm) and $BiNb_2S_5$ (~1.74 nm) which dominate in the crystal have been found, see figure 3-e. Besides the disordered areas, a new related phase ~$BiNb_4S_9$ with a stacking sequence ..Q, H, H, H, H.. and ~2.92 nm periodicity was observed. In table 1 we present the crystal unit cell refined parameters for two phases, a monolayer ~$SbNbS_3$ (..Q,H..) and a bilayer ~$BiNbS_5$ (..Q,H,H..). Those values were obtained after the approximated ones were measured from the SAEDP's along the main orientations. The distortions of the Q and H subcells from ideal tetragonality and hexagonality, **b/a** values, are also included.

References

[1] E. Makovicky and B. G. Hyde, Mater. Sc. Forum 100-101 (1992) 1.
[2] G. A. Wiegers, Prog. Solid State Chem, 24 (1996) 1.
[3] L. C. Otero-Díaz et al, J. Solid State Chem, 115 (1995) 274.
[4] A. R. Landa-Cánovas et al, Micron 32 (2001) 481.
[5] This work was supported by the Spanish CYCIT project MAT2000-0753-C02-01. The help of Mr. Gómez-Herrero and Dr. Landa-Cánovas from Univ. Complutense is gratefully acknowledged.

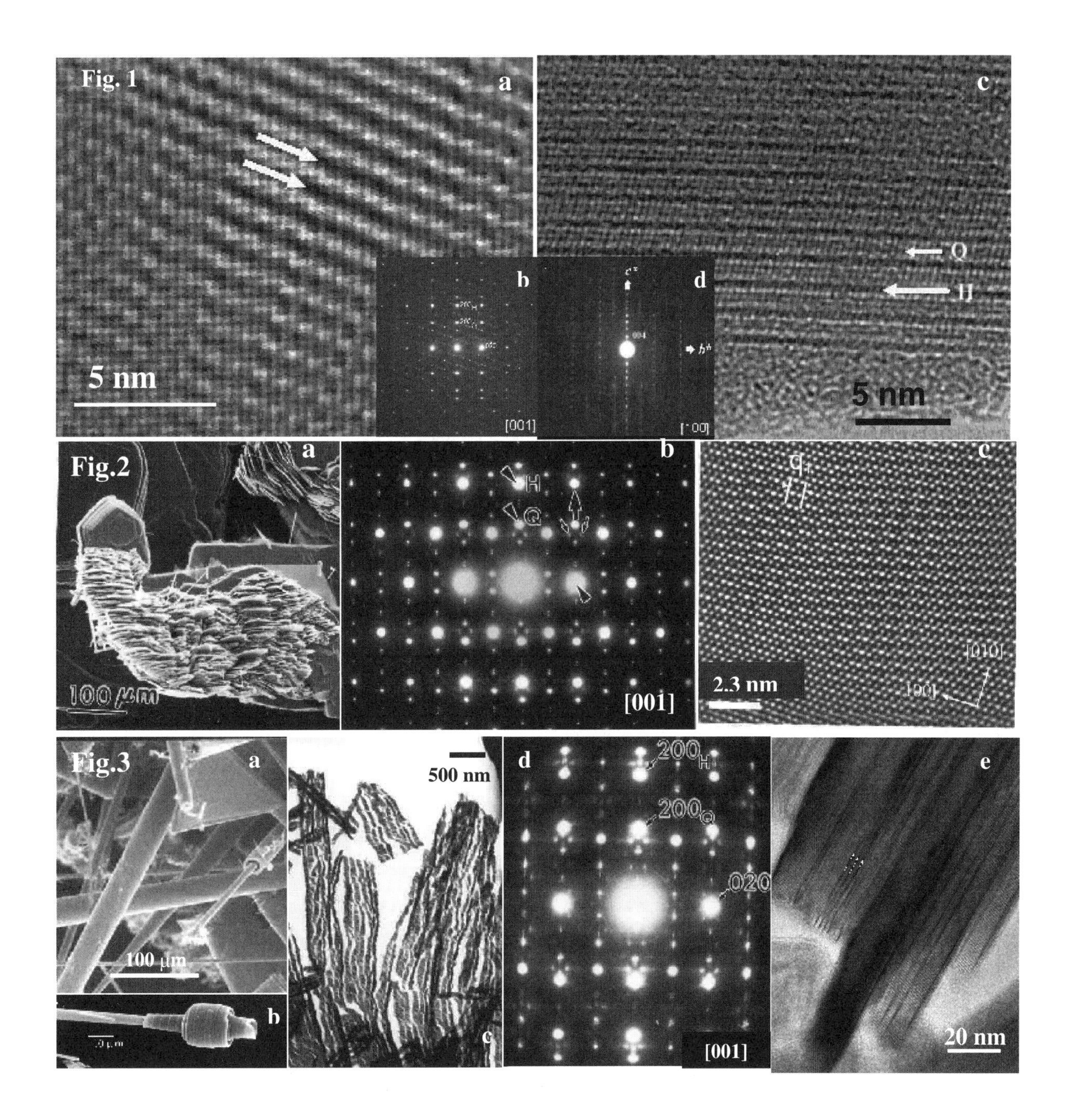

Table 1: Refined unit cell parameters from X-ray powder diffraction data.

Phase	Subcell	**a** (nm)	**b** (nm)	**c** (nm)	α (°)	β (°)	γ (°)	**b/a**
$(SbS)_{1.195}NbS_2$	Q (SbS)	0.5560(1)	0.5693(4)	2.2834(3)	90	90	90	1.024
	H (NbS_2)	0.3321(6)	0.5701(8)	1.1411(4)	90	90	90	1.717
$(BiS)_{1.166}(NbS_2)_2$	Q (BiS)	0.5682(2)	0.5769(1)	1.7416(5)	93.26(2)	90	90	1.015
	H (NbS_2)	0.3313(1)	0.5774(1)	1.7419(3)	93.15(2)	90	90	1.743

Microsc. Microanal. 8 (Suppl. 2), 2002
DOI 10.1017.S1431927602101085

Scanning Tunneling Microscopy of Modulated Surface Structures

A. Prodan[*], H. J. P. van Midden[*], N. Jug[*], F. W. Boswell[**], J. C. Bennett[***], and H. Böhm[****]

[*]Institute Jožef Stefan, Jamova 39, SI-1000 Ljubljana, Slovenia
[**]Department of Physics, University of Waterloo, Waterloo, Ontario, Canada N2L 3G1
[***]Department of Physics, Acadia University, Wolfville, Nova Scotia, Canada B0P 1X0
[****]Geosciences, Johannes Gutenberg University, D-55099 Mainz, Germany

Modulated structures are commonly found in transition-metal chalcogenides with compositions between MX_4 and M_3X_4. Their origin is in the low-dimensional character of these compounds, which supports phenomena like charge density waves (CDW). These can be further stabilized at surfaces, where the bulk dimensionality is additionally reduced. Modern surface methods, combined with calculations of the electronic properties, can give a new insight into such phenomena.

Two weak incommensurate modes, superimposed onto the main breathing mode, were discovered during the initial TEM studies of $Nb_{1-x}Ta_xTe_4$. They were detected at the $NbTe_4$ side of the phase diagram as weak and diffuse satellites in overexposed electron diffraction patterns [1]. Although a bulk phenomenon, these modes were for the first time successfully resolved by STM [2]. This was possible due to the preferential cleavage of the $NbTe_4$ crystals, which exposed complete tetragonal anti-prismatic columns. The weak transversal modes were observed as a surface phenomenon, superimposed onto the much stronger breathing mode (Fig.1).

The observed sliding of two apparently independent low-temperature CDWs in $NbSe_3$ was recently explained on basis of a new model [3], which took into account statistically distributed nano-domains of both CDW modes. The results of earlier low-temperature STM [4] and satellite dark field TEM studies [5] were shown to be in a good accord with the model. It was further shown that sub-monolayer gold deposits on the $NbSe_3$ van der Waals surface exhibit poorly correlated modulated regions, whose periodicities are comparable to those of the two CDWs in pure $NbSe_3$. In another trichalcogenide, $ZrTe_3$, a surface compositional modulation [6] was shown to be a result of ordered, surplus tellurium atoms, which occupied the interstitial positions on the (001) $ZrTe_3$ surface. Calculations within the extended Hückel tight binding (EHTB) approximation show that these tellurium atoms also account for the relatively high metallic conductivity of this compound.

CDW domains of two orientational variants were detected at the edges of disk-like surface defects in $NbSe_2$, which was an indication that the deformed regions of the Se-Nb-Se sandwiches assumed an octahedral rather than trigonal-prismatic coordination [7]. It was further shown that surface domain boundaries between orientational variants in the monoclinic $NbTe_2$ were often strongly modulated (Fig.2) with periodicities identical to those of the adjacent domains [8]. In contrast to the modulation in $NbSe_2$, the $NbTe_2$ modulation was interpreted on basis of a misfit between the domain boundaries across the van der Waals gaps.

Finally, the Nb_3X_4 (X = S, Se, Te) compounds, characterized by large hexagonal tunnels [9], were studied by means of TEM and STM. From the three compounds forming the family only Nb_3Te_4 shows bulk conductivity anomalies at 95 K, which were attributed to a CDW with a modulation vector $\mathbf{q} = \pm(1/3\mathbf{a}^* + 1/3\mathbf{b}^* + 3/7\mathbf{c}^*)$. Intercalation with indium and thallium raises the CDW onset temperature in Nb_3Te_4 and supports CDW formation in the other two compounds [10,11], without significantly deforming the host structure. The electronic properties, calculated within the EHTB method, confirmed that intercalation flattens the Fermi surfaces and consequently improves the one-dimensional character of these compounds. Poorly correlated modulation along the hexagonal columns was consistently observed in room-temperature STM images of these compounds, indicating that precursor effects to full CDW formation are surface supported phenomena.

References:

[1] F. W. Boswell and A. Prodan, Phys. Rev. B 34 (1986) 2979.
[2] A. Prodan et al., Phys. Rev. B 57 (1998) 6235.
[3] A. Prodan et al., Phys. Rev. B 64 (2001) 115423-1.
[4] Z. Dai et al., Phys. Rev. Lett. 66 (1991) 1318.
[5] K. K. Fung and J. W. Steeds, Phys. Rev. Lett., 45 (1980) 1696.
[6] A. Prodan et al., Surf. Sci., 482-5 (2001), 1368.
[7] N. Ramšak et al., Phys. Rev. B 60 (1999), 4513.
[8] D. Cukjati et al., to appear in J. Cryst. Growth (2002).
[9] K. Selte and A. Kjekshus, Acta Cryst. 17 (1964) 1568.
[10] F. W. Boswell and J. C. Bennett, Mat. Res. Bull. 31 (1996) 1083.
[11] F. W. Boswell et al., J. Sol. St. Chem. 144 (1999) 454.

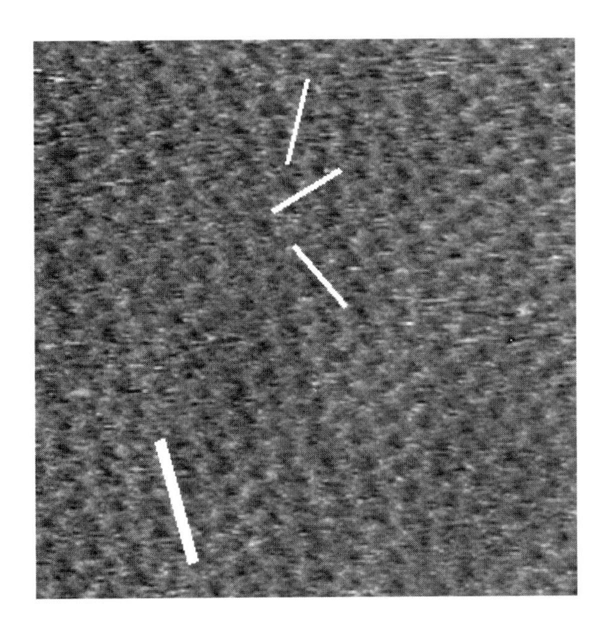

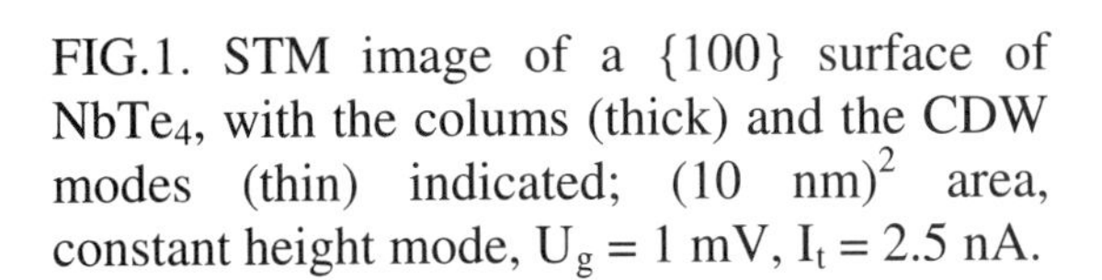

FIG.1. STM image of a {100} surface of $NbTe_4$, with the colums (thick) and the CDW modes (thin) indicated; $(10\ nm)^2$ area, constant height mode, $U_g = 1$ mV, $I_t = 2.5$ nA.

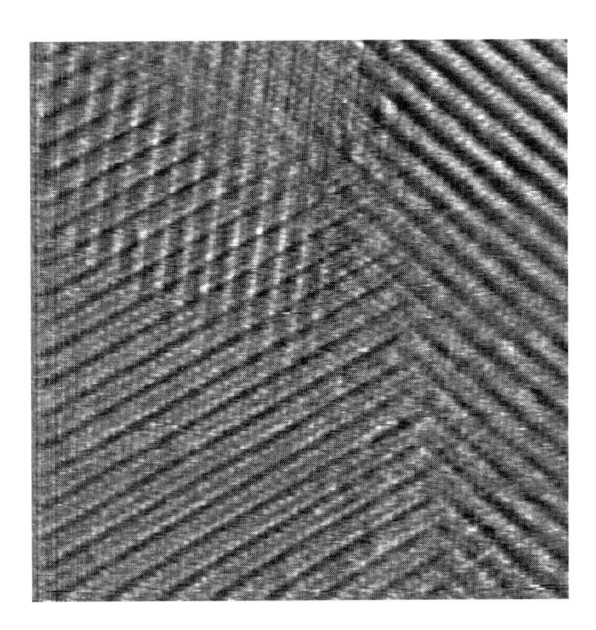

FIG.2. STM image of three domains in $NbTe_2$, separated by strongly modulated boundaries; $(20\ nm)^2$ area, constant height mode, $U_g = 0.2$ mV, $I_t = 2$ nA.

Microsc. Microanal. 8 (Suppl. 2), 2002
DOI 10.1017.S1431927602101097

Quasi-periodic Materials – Crystal Redefined

Dan Shechtman

Technion, Haifa, Israel

Crystallography has been one of the mature sciences. Over the years, the modern science of crystallography that started by experimenting with x-ray diffraction from crystals in 1912 has developed a major paradigm – that all crystals are ordered and periodic. Indeed, this was the basis for the definition of "crystal" in textbooks on crystallography and x-ray diffraction. Based upon a vast number of experimental data, constantly improving research tools, and deepening theoretical understanding of the structure of crystalline materials no revolution was anticipated in our understanding the atomic order of solids.

However, such revolution did happen with the discovery of the Icosahedral phase, the first quasi-periodic crystal (QC) in 1982, and its announcement in 1984 [1,2]. The discovery created deep cracks in this paradigm, but the acceptance by the community of crystallographers of the new class of ordered crystals did not happen in one day. In fact it took almost a decade for QC order to be accepted by most of the community of crystallographers. The official stamp of approval came in a form of a new definition of "Crystal" by the International Union of Crystallographers. The last objection to the existence of quasi-periodic order in crystals faded in about 1994. The paradigm that all crystals are periodic has thus been changed. It is clear now that although most crystals are ordered and periodic, a good number of them are ordered and quasi-periodic.

While believers and nonbelievers were debating, a vast number of experimental and theoretical studies were published. This was created by a relentless effort of many groups around the world. Quasi-periodic materials have developed into an exciting interdisciplinary and international science.

The atomic order of the Icosahedral phase contains 5-fold 3-fold and 2-fold axes. The lack of periodicity is clearly observed in high resolution TEM images, and well manifested in electron diffraction and in single crystal x-ray diffraction patterns.

Soon after the discovery of the Icosahedral phase in rapidly solidified Al-Mn, the Decagonal phase was discovered in rapidly solidified Al-Cr alloys. This phase contains a 10-fold rotation axis, and periodic stacking perpendicular to it. In addition, other rotation axes observed in quasi-periodic materials are 8-fold and 12-fold

All the quasi-periodic structures observed during the first years were metastable, and transformed into periodic stable phases upon heating. This supported the false notion that quasi-periodic structures can not be the low energy phase of a given composition. The discovery of several stable quasi-periodic phases not only proved that quasi-periodic phases can be stable, but also confirmed early calculation regarding possible stability of QC structures. In addition, it enabled growing large QCs, and serious x-ray diffraction experiments. This was important to the community of crystallographers that depend mainly on precise x-ray diffraction measurements. The early QCs were discovered and studied intensively by TEM. This was necessary, since the QC grain size was only several microns. Electron diffraction in the TEM is a powerful qualitative tool for small crystals, but its precision is limited compared to x-ray diffraction. The community of crystallographers waited until high quality x-ray diffraction patterns of large nearly perfect QCs became available, and then adopted QCs into the family of crystals.

Quasi-periodic materials have several characteristic properties. Many of them have low electrical and heat conductivity; in fact at low temperature the electrical conductivity of some of them drops to a very low value. They are hard and some of them have low friction coefficient and resist wear. These properties may indicate potential uses and several applications have been tried, and reached the markets. For example, one of the potential uses is strengthening of alloys with quasi-periodic precipitates. This has been commercialized in stainless steel, and experimented in aluminum and magnesium alloys with very good results.

This talk will outline the discovery of QCs and discuss their structure and some of their properties.

1. D. Shechtman and Il Blech, "The Microstructure of Rapidly-Solidified Al6Mn", Met. Trans. **16**A (June 1985) 1005-1012.
2. D. Shechtman, I. Blech, D. Gratias and J.W. Cahn, "A Metallic Phase with Long-Ranged Orientational Order and Broken Translational Symmetry", Phys. Rev. Letters, Vd **53**, No. 20 (1984) 1951-1953.

Microsc. Microanal. 8 (Suppl. 2), 2002
DOI 10.1017.S1431927602100857

Imaging of self-assembly and self-assembled materials

P. V. Braun

Department of Materials Science and Engineering, University of Illinois at Urbana-Champaign, Urbana, IL 61801

Self-assembled materials have recently been of great interest to a wide range of scientists for multiple, and indeed quite disparate applications. Self-assembly has been proposed as a route to photonic band gap materials, semiconductor devices, molecular sieves, biomaterials, and molecular monolayers. For many of these processes, there are few very good routes to image the self-assembly process, or even the structure of the final product. Because we are generally interested in three dimensional systems, visualization of the process becomes even more difficult. Furthermore, we often desire to obtain structural information on the nanoscale, and finally, many of our systems are solution based, and thus not amenable to conventional high vacuum imaging procedures.

We utilize a range of techniques to obtain morphological information on our systems. These include single photon confocal microscopy and multiphoton microscopy, and conventional transmission electron microscopy. Depending on the system, three dimensional structural information can be obtained from either holography or reconstruction of serial sections. We utilize confocal microscopy to investigate two different material systems, and for real time visualization of the assembly process. The first use we have for confocal microscopy is for self-assembled membrane materials, where we are determining the three dimensional pore structure of the membrane (figure. 1), and the second is for photonic band gap materials (figure 2), where the key aspects are to determine the assembly process and resulting three-dimensional structure, and to create defined defects into the structure. It is these defined defects that allows for the fabrication of waveguides and optical cavities (figure 3).

Both the membrane and photonic band gap material studies are performed on a Leica confocal microscope. The core unit of the self-assembled photonic band gap materials are monodisperse colloidal particles. Under the appropriate conditions, these particles can be crystallized into three-dimensional materials with long range periodicity with a lattice constant similar to the wavelength of the light to be modulated. The assembly process takes between minutes and hours, which is much slower than the frame rate of the microscope. Thus real time imaging is possible. The multiphoton imaging is performed using a mode locked Ti-sapphire laser, this same laser is used to write defects into the self-assembled colloidal crystals (figure 3).

Most of the electron microscopy is performed on a Philips CM12 or CM20. In the electron microscopy experiments, we are imaging the structure of semiconductor nanostructures formed through growth in a self-assembled matrix (figure 4). Here, a soft template, specifically a non-ionic hexagonal lyotropic liquid crystal, is used to control the growth of a hard, inorganic phase, in this case CdS. The growth process in some aspects resembles the biomineralization process, as biominerals such as bone or shell are also formed in three dimensional spaces defined by self-assembled biomolecules [1].

[1] This work supported in part by the US DoE and the NSF.

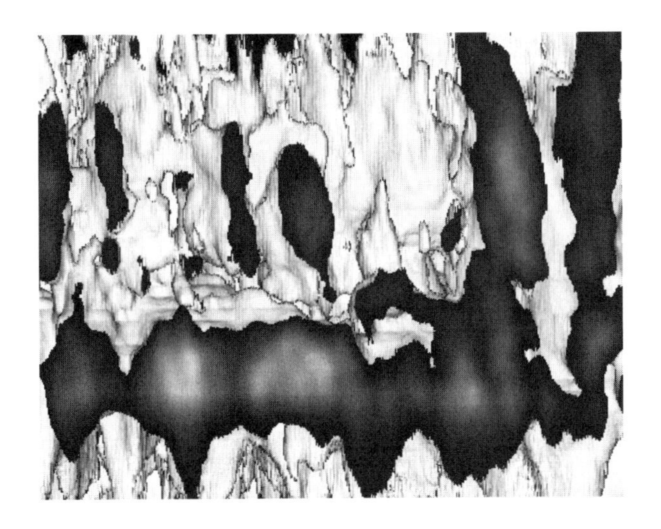

Figure 1. Three dimensional reconstruction of a membrane generated from 60 slices.

Figure 2. Silica colloidal crystal formed from 1.6 μm diameter silica colloidal spheres. Space between spheres filled with $1x10^{-4}$ molar Rhodamine-6G in DMF, Excitation wavelength: 488nm, fluorescence widow: 520 to 580nm.

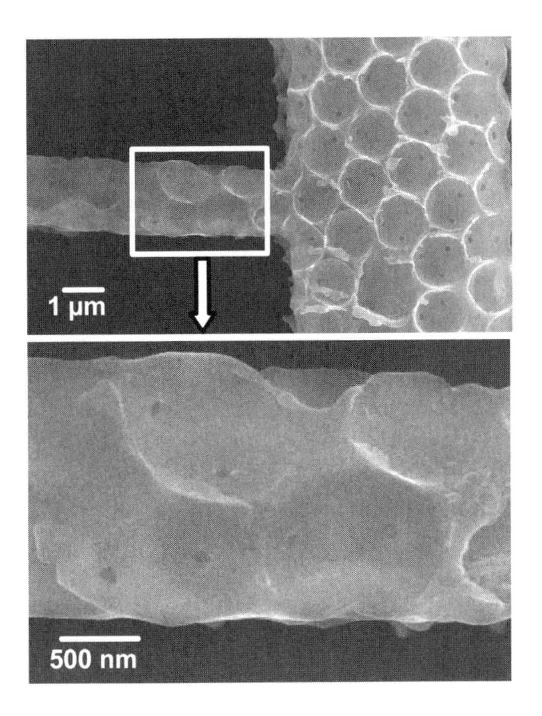

Figure 3. SEM micrograph of polymer feature formed through three-photon polymerization within a silica colloidal crystal. The silica has been removed with HF, exposing the polymer.

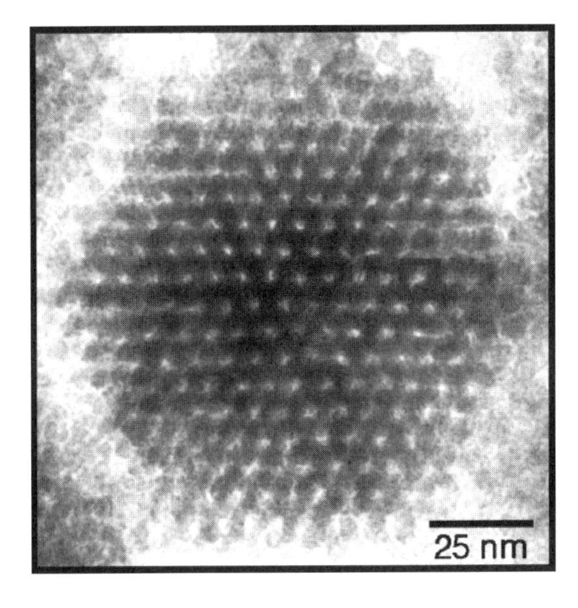

Figure 4. CdS nanostructure formed through templating with a hexagonal lyotropic liquid crystal.

Microsc. Microanal. 8 (Suppl. 2), 2002
DOI 10.1017.S1431927602100869

Self Assembled Phenylene Vinylene Materials

Martin U. Pralle*, Gregory N. Tew,* Mehmet Sayar,*** Leiming Li,***, and Samuel I. Stupp***

* Ion Optics, Inc., Waltham, MA
** Department of Polymer Science and Engineering, University of Massachusetts at Amherst, Amherst, MA
*** Departments of Materials Science and Engineering, Chemistry and Medical School, Northwestern University, Evanston, IL

A new class of self assembling rodcoil molecules was recently synthesized and their solid state structure was rigorously characterized. These molecules have a triblock architecture with a rigid rod molecular compound coupled to a diblock coil composed of an oligomeric flexible spacer and a bulky coil. The molecular structure has integrated photonic properties resulting from phenylene vinylene segments in the rigid backbone of the rod segment. It has been found that these molecules aggregate spontaneously when cast from solution into discrete objects with dimensions on the order of a few nanometers. The driving force for this aggregation is the crystallization of rod segments observed by electron diffraction. The aggregates were imaged by transmission electron microscopy using the difference in scattering intensity of the crystallized rod segments and the amorphous coils to generate contrast. From this data we suggest that these aggregate are mushroom like with the rods arranged into a crystalline stem and coils randomly splaying to form the cap. These supramolecular building blocks are packed into nanosheets which stack on top of one another to form macroscopic materials as observed by small angle x-ray scattering and TEM.

Detailed cross sectional electron microscopy together with selective staining techniques has documented the nature of the z direction packing. The mushroom aggregates pack in a parallel rather than anti-parallel arrangement within each layer and furthermore these layers stack in a polar fashion. To probe the structure-property relationship piezoelectric measurements were carried out on these self assembled films and spontaneous piezoelectric activity was demonstrated without prior poling procedures. Upon poling the net polarization was enhanced and piezoelectric activities equal that of quartz were observed. By applying both positive and negative fields a hysteresis was observed suggesting that these materials exhibit ferroelectric character.

Through novel synthesis a supramolecular diode structure was assembled in this way with emissive phenylene vinylene in the rod segment and hole transporting triphenylamine (TPA) groups in the coil. This molecular architecture afforded strong energy transfer as can be seen in figure 1. The TPA absorbs strongly at 302 nm but then transfers the energy to the phenylene vinylene segment from which light is emitted at visible wavelengths. Through subsequent characterization it was shown that the formation of nanoaggregates is governed by the molecular architecture rather than the specific chemical functions of the 3 molecular segments.

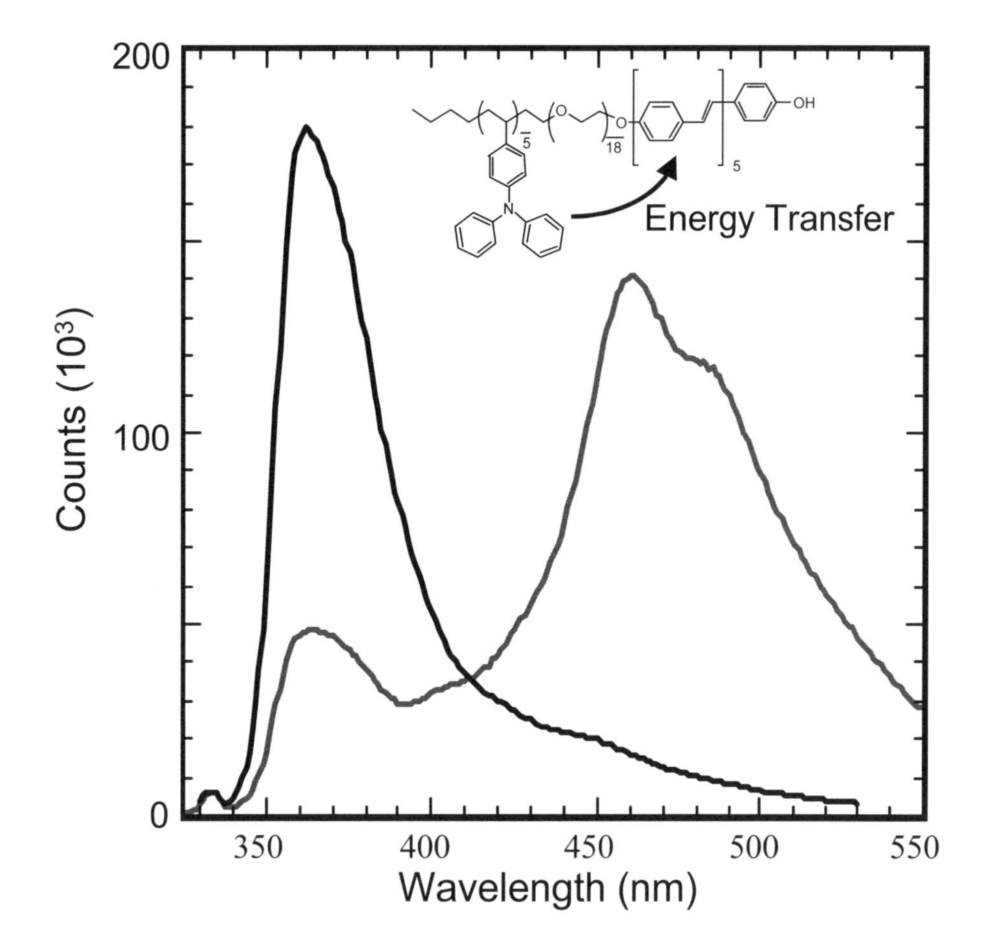

Figure 1. Fluorescence spectra of the TPA coil alone (black) and of the rodcoil molecule (grey) showing a drastic reduction in the TPA fluorescence intensity. The samples were excited at 302 nm where the phenylene vinylene is nearly transparent and TPA has an absorption maximum.

Microsc. Microanal. 8 (Suppl. 2), 2002
DOI 10.1017.S1431927602100870

SELF-ORDERED COLLOIDAL ARRAYS AS PHOTONIC CRYSTAL HYDROGELS FOR TRAINABLE METAL ION SENSORS AND AS SUPERPARAMAGNETIC MATERIALS

ANJAL C. SHARMA*, XIANGLING XU*, MICHELLE S. WARD*, LONG LI**, JUDITH C. YANG** AND SANFORD A. ASHER*

*Department of Chemistry, 234 Chevron Science Center, University of Pittsburgh, Pittsburgh, PA 15260.
**Department of Materials Science and Engineering, 848 Benedum Hall, University of Pittsburgh, Pittsburgh, PA 15260.

We have developed novel materials based on colloidal self-assembly, which are ordered arrays of colloidal particles formed in a liquid. We have polymerized these arrays in solid films that change dimension in response to chemical, electrical, and thermal environmental changes. Here, we present two applications of these self-ordered colloids: (1). Photonic crystal hydrogels as trainable metal ion sensors and (2) Superparamagnetic materials.

We have utilized a soft hydrogel matrix, which contains an embedded photonic crystalline colloidal array [1], to control the microenvironments around metal ion binding sites [2]. This sensor material consists of a ca. 90% aqueous polyacrylamide network hydrogel, which is lightly crosslinked with methylenebisacrylamide, and has pendant 8-hydroxyquinoline ligand groups (Figure 1). When this material is treated with aqueous solutions of a metal ion, the hydrogel undergoes a volume collapse, resulting in a blue shift of the Bragg diffracted wavelength of light from the embedded photonic crystal. This is due to the formation of 2:1 ligand:metal ion complex sites,[3] and a rearrangement of the matrix, to create cavities around these sites. These sites are extremely robust, with very large metal:ligand affinities, as evidenced by the inability to remove the residual metal content from the gel. In addition, it is believed that the ligand bound metal ion is reduced and produces colloidal metal clusters, which stay embedded within the matrix. This hydrogel then functions as a selective optical sensor for aqueous solutions of the same metal ion, and shows only red shifts in the diffracted wavelengths upon increasing the concentration of the metal ion, due to the formation of 1:1 ligand:metal ion complexes [3]. Aqueous solutions of a different metal ion produce much smaller red shifts in this material. This effect of conservation of site architecture and hydrogel trainability to selectively bind the original metal ion is a general one, and selective sensors for Zn^{2+}, and Cu^{2+}, have been produced in this way. Hence this technique is an effective route to create hydrogel based sites for the selective recognition and sensing of metal ions, *via* control of the microenvironments around these metal ion binding sites.

To provide structural insights into the metal ion sensor mechanisms of this novel material, scanning (SEM) and transmission (TEM), including analytical, electron microscopy methods will be used. Specifically, we will investigate the position and structure of the metal (e.g. Cu) atoms within the matrix, possible oxidation states of the metal and metal interactions with the binding site, as well as the structure of the self-ordered colloidal array in the hydrogel matrix prior and after metal ion exposure.

Superparamagnetic colloid particles can also be synthesized by emulsion polymerization of styrene in the presence of nanoscale ferrite (Fe_3O_4). The ferrites were incubated inside the ~ 150 nm polystyrene particles, and appeared as black aggregates in the TEM image (Figure 2). These particles would self assemble into superparamagnetic photonic crystals which could be utilized as optical switch controlled by magnetic fields. Further TEM studies can provide guidelines to new synthesis methods of superparamagnetic colloidal arrays with higher magnetic moments and greater homogenous distribution.

References

[1]. (a). Asher, S. A.; Holtz, J.; Liu, L.; Wu, Z. *J. Am. Chem. Soc.* **1994** *116*, 4997-4998; (b). Holtz, J.H.; Asher, S. A. *Nature* **1997** *389*, 829-832.; (c). Holtz, J. S.; Holtz, W.; Munro, C.H.; Asher, S. A. *Anal. Chem.* **1998** *70*, 780-791.
[2]. Polymeric materials have been utilized to control the microenvironments around metal ion binding sites. See for example: Sharma, A. C.; Borovik, A. S. *J. Am. Chem. Soc.* **2000**, *122*, 8946-8955.
[3]. (a). Kim, Y-S.; Shin, J-H.; Choi, Y-S.; Lee, W.; Lee, Y-III. *Microchem. J.* **2001**, *68*, 99-107; (b). Lee, S. C.; Izzat, R. M.; Zhang, X. X.; Nelson, E. G.; Lamb, J. D.; Savage, P. B.; Bradshaw, J. S. *Inorg. Chim. Acta* **2001**, *317*, 174-180.
[4]. This research program is supported by the Department of Energy (DE-FG07-98ER62708) and Office of Naval Research (ONR) (N00014-94-1-0592).

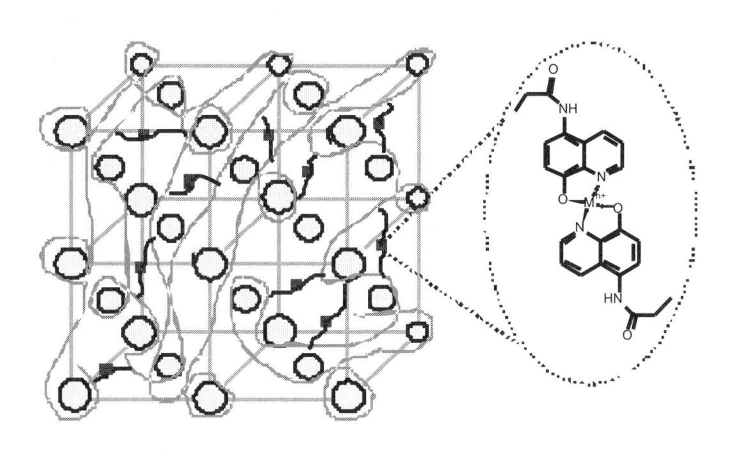

Figure 1: Illustration of the metal ion sensor material, which is a hydrogel matrix with an embedded photonic crystalline colloidal array. The enlarged diagram shows the pendant 8-hydroxyquinoline ligand group.

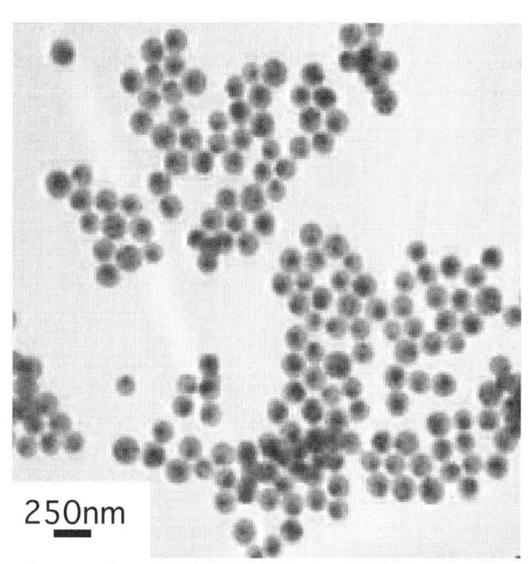

Figure 2: Bright-field image of Fe_3O_4 embedded in polystyrene as a paramagnetic material.

Microsc. Microanal. 8 (Suppl. 2), 2002
DOI 10.1017.S1431927602100882

Near-Field Optical Imaging of Microphase Separated and Semi-Crystalline Polymer Systems

Michael J. Fasolka*, Lori S. Goldner***, Augustine M. Urbas**, Jeeseong Hwang***, Kathryn Beers*, Peter DeRege**, Edwin L. Thomas**

*Polymers Division, NIST, Gaithersburg, MD 20899
**Dept. of Materials Science and Engineering, MIT, Cambridge, MA 02139
***Optical Technology Division, NIST, Gaithersburg, MD 20899

Polymer self-assembly presents an attractive means of creating the micro- and nano-patterned spatial arrays required for many opto-electronic and coatings technologies. Two of these ordering processes are microphase separation (MS), exhibited by block copolymers, and crystallization, common in many polymer species. In this work, we present optical micrographs of block copolymer MS morphology and thin-film polymer crystallites having sub-diffraction-limit resolution (≈100nm) as afforded by Near-Field Scanning Optical Microscopy (NSOM). Images obtained via transmission aperture NSOM and polarization-modulated (PM) NSOM, which yields the local dichroism and birefringence, will be discussed. These images provide insights into the structure and local optical properties of these specimens, resolved at the level of single microphase domains and defects.

A detailed review of aperture transmission NSOM can be found in reference [1]. Our PM-NSOM uses Al-coated aperture probes, fabricated from drawn optical fibers, and a 488nm light source. Measurements of polarimetric quantities were made via the PM technique [2], adapted for NSOM in a manner similar to McDaniel and Hsu [3], but improved upon here by employing Fourier analysis. Modulated (50kHz) source-light polarization is created via a photoelastic modulator and complimentary optics. Fourier analysis of the transmitted signal yields the sample dichroism and birefringence (if a post-sample analyzer is in place) at each point. Our measurement techniques and analysis rectify probe-fiber birefringence and any inherent dichroism of the probe-aperture.

Microphase separation is driven by the immiscibility of the end-linked polymer chains, or blocks, constituent to a block copolymer (BC). A variety of pattern motifs and equilibrium periodicities (L_0) are achieved by controlling the BC composition and molecular weight (MW), respectively [4]. Recent synthetic efforts have created ultrahigh MW BCs, with L_0 of 150-300nm, that exhibit tunable photonic band gaps in the visible range [5,6]. Figs. 1-3 shows NSOM and PM-NSOM data collected from a 100nm-thick polystyrene-b-polyisoprene (PS-b-PI) photonic BC specimen. OsO_4 was used to impart a light stain to PI domains. Z-ranges are supplied in brackets, [ ], in each figure caption. Transmission NSOM (Fig. 1) illuminates single lamellar domains, and defects exhibiting enhanced optical contrast. PM-NSOM data gives the local dichroism (Fig. 2), which marks domain interfaces, and birefringence (Fig. 3), which reflects the local domain/chain orientation.

Polymer crystallites consist of layers (lamella) of folded chains and intermediate amorphous domains. While the structure of bulk polymer crystallites (spherulites) is well established [7], a variety of less-understood forms are found in ultra-thin (<100nm) films [8]. The sensitivity and resolution of PM-NSOM can illuminate the structure of these "2D" crystallites, where traditional techniques may fail due to low resolution and the sparse signal inherent to thin samples. This capability is demonstrated in Figure 4, which shows PM NSOM micrographs of a 100nm-thick isotactic PS spherulite. These images show the radial arrangement of crystallite lamella, defect structures located near the crystal nucleus, and possible chain alignment in the "depletion zone" at the spherulite periphery.

[1] R. C. Dunn, Chem. Rev. 99 (1999) 2891 (and references therein).
[2] J.W.P. Hsu, Mat. Sci. and Eng. R. 33 (2001) 1 (and references therein).
[3] E.B. McDaniel, J.W.P. Hsu, J. Appl. Phys. 80 (1996) 1085.
[4] F.S. Bates and G.H. Fredrickson, Annu. Rev. Phys. Chem. 41 (1990) 525.
[5] Y. Fink et al., J. Lightwave Technol., 17 (1999) 1963.
[6] A. Urbas et al., Adv. Mater. 12 (2000) 82.
[7] B. Wunderlich, Macromolecular Physics, vol 1, Academic Press, NY, 1973.
[8] K. L. Beers, J. F. Douglas, E. J. Amis, A. Karim, Macromolecules, submitted (2002).

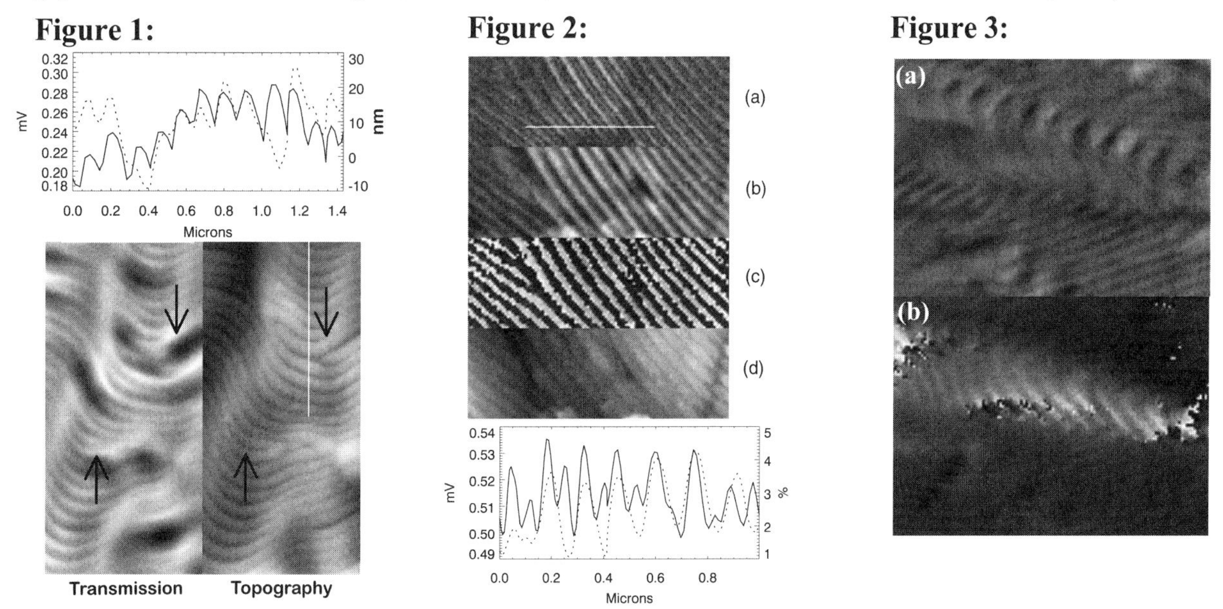

Fig. 1. NSOM micrograph of PS-b-PI. The top plot shows topography (dotted) and transmission (solid) along the white line (1.4 μm) in the topography image. Arrows indicate edge defects.

Fig. 2. PM-NSOM of PS-b-PI. (a) Dichroism [5%], (b) Transmission [0.2-0.5mV (a.u.)], (c) Dichroism angle [180°], (d) Topography [25nm]. The bottom plot shows transmission (dotted) and dichroism (solid) along white line (1 μm) (a).

Fig. 3. PM-NSOM of PS-b-PI twin boundary. (a) Transmission [0.2-0.25mV (a.u)] (b) Birefringence angle (fast axis) [180°] reflecting lamellar orientation. Images are 4 μm wide.

Figure 4:

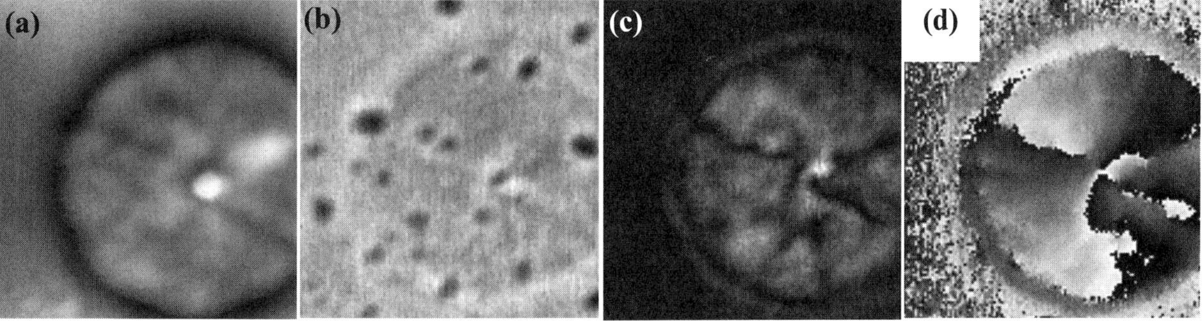

Fig. 4. PM NSOM of PS spherulite. Images are 6x6μm. (a) topography [43nm], (b) transmission [0.2-0.25mV (a.u.)], (c) Birefringence (retardance) [6°], (d) Birefringence angle (fast axis) [180°].

Microsc. Microanal. 8 (Suppl. 2), 2002
DOI 10.1017.S1431927602100560

Ultramicrotomy of Polymers Using an Oscillating Diamond Knife; Improving Polymer Morphology.

J.S.J. Vastenhout* and H. Gnägi**.

*Dow Benelux B.V., Analytical Sciences, MC-SMX Discipline, P.O. Box 48, 4530 AA Terneuzen, The Netherlands.
** Diatome Ltd Switzerland, P.O. Box 557, CH-2501 Biel, Switzerland.

Sample preparation is one of the crucial steps for successful transmission electron microscopy (TEM) and atomic force microscopy (AFM) investigations of polymers. Some of the typically encountered complications include ability to obtain very thin sections (50 nm or less), compression of ultra thin sections, induced surface roughness, need of working at cryogenic temperatures etc. It is generally known that the ultramicrotomy procedure results in compres sion of the ultra thin sections. Ultramicrotomy of polymer materials done at room temperature using a diamond knife with a 45° angle and the ultra thin sections floating on water results in an average compression of 20 -40 % depending the material. Reducing the knife angle results in lower compression [1]. Therefore nowadays it is common practice in materials ultramicrotomy of to use a diamond knife with a 35° angle instead of a 45° angle, which reduces the original compression with 33 – 50 %. The compressi on of ultra thin sections increases when these sections are cryoultramicrotomed [2] due to the fact that there's no lubricating effect anymore of the water. Because a lot of polymeric materials need to be cryoultramicrotomed it is essential to lower the co mpression as much as possible. The lower the compression, the better the morphology will be.

Recent developments in reducing the compression of ultra thin sections have lead to the invention of an oscillating diamond knife [3]. The oscillating diamond kni fe reduced the compression almost completely without additional cutting artifacts observed. A prototype of this oscillating diamond knife was evaluated and tested to prepare ultra thin sections of polymer materials both at room temperature and at cryo tem peratures. The ultra thin sections made with the oscillating diamond knife were compared with sections of the same sample but prepared with a 35° angle diamond knife. A variety of polymer materials and blends were studied and for all materials an improveme nt of the morphology was observed. The overall morphology was better preserved and in some materials more detailed structure was visible that was normally destroyed by the sectioning procedure. An additional, but important advantage of the oscillating knif e is the possibility to section materials at room temperature that normally need to be sectioned at cryo temperatures (Fig.1,2). The elimination of the compression also results in flatter surfaces which are essential in the preparation of samples for AFM as well as the reduced section thickness, down to approximately 20 nm, is.

References

[1] Jesior, J.-C., *Scanning Microscopy* (Suppl.3) (1989) 147-153.
[2] Richter, K., *Micron* 25 (1994) 297-308.
[3] Studer, D. and Gnägi, H., *Journal of Microscopy* 197 (2000) 94-100.
[4] We would like to thank Dr. Daniel Studer of the Institute of Anatomy of the University of Berne, Switzerland for his editorial comments and contribution in the discussion of the results.

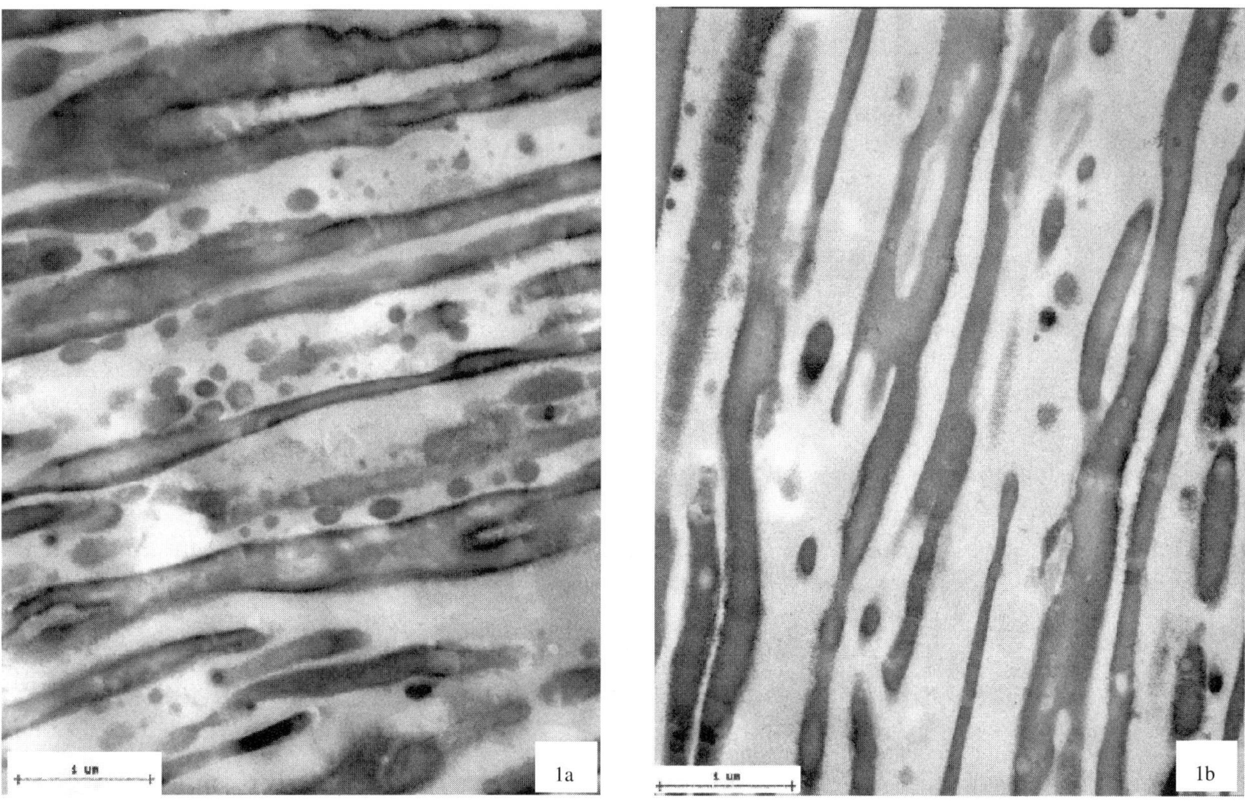

FIG. 1 Blend of polypropylene and polyethylene (a) sectioned at room temperature using a 35° angle diamond knife, (b) sectioned at room temperature using an oscillating diamond knife.

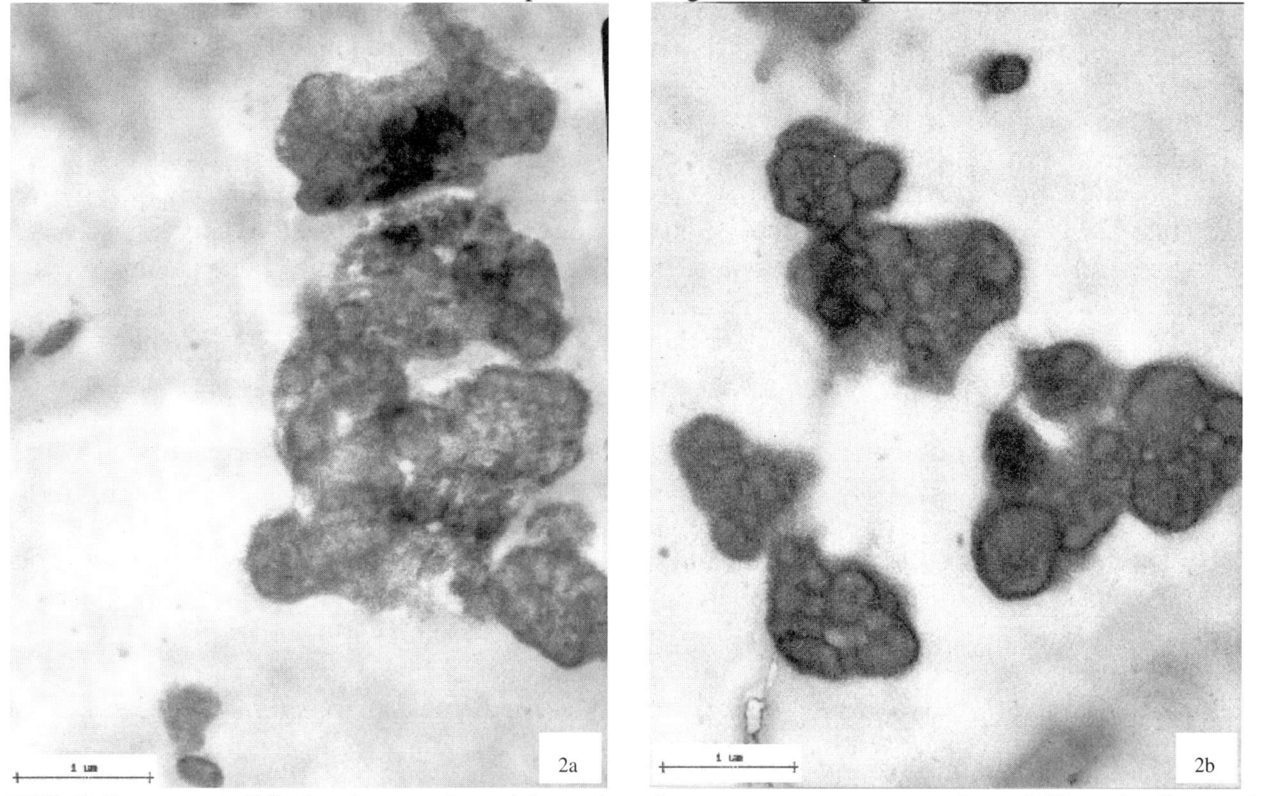

FIG. 2. Impact modified polypropylene (a) sectioned at room temperature using a 35° angle diamond knife, (b) sectioned at room temperature using an oscillating diamond knife.

Microsc. Microanal. 8 (Suppl. 2), 2002
DOI 10.1017.S1431927602100572

Total Microscopy of a Tire

R.W. Smith

2450 Clarke Drive, Lake Havasu City, AZ 86403

Every aspect of an automotive passenger tire can be probed by some sort of microbeam instrument and Figure 1 identifies the aspects that are common to today's wire belted radial passenger tires. The salient features of each aspect are discussed with respect to the microscopical information obtainable from the area, how this information impacts on understanding tire performance, and the microscopical activity used to extract the information.

Tread Surfaces **1** –The type of abrasion patterns generated on tread surfaces during tire operation reflect the mechanism of tread wear [1]. Three possible mechanisms for tread wear have been identified: random cutting and tearing, pattern abrasion, and roll formation. A confirmation of the prevailing mechanism reflects the tire usage history. In some instances it may be important to know the direction of rotation of the tire and this also can be learned from abrasion patterns. Methods used are optical microscopy (OM) and scanning electron microscopy (SEM) on tread surfaces, SEM on cross sections cut normal to tread surfaces, and transmission electron microscopy (TEM) of ultrathin sections cut normal to tread surfaces [2].

Tread Wear Debris Particles **2** - The internal morphology of tread wear debris particles can provide additional evidence for establishing the mechanism of tread wear since the roll formation mechanism can be easily verified by OM examination of microtomed sections. Energy dispersive spectroscopy (EDS) fingerprints of the rubber compounds in the roll add to this evidence [2].

Bulk Rubber Compounds **3** - The degree of carbon black dispersion in tire compounds is one of several indicators that directly affect tire performance. The referee method for evaluating this is by quantitative OM on microtomed sections [3]. This method also provides critical information of tire durability, i.e. assessment of cement lines, splices, junctions, flow patterns and/or bulk grain effects. Morphological changes in bulk compounds caused by tire operation dynamics can be monitored by OM, SEM, and EDS [4], [5]. Investigations that reveal the bonding interaction between rubber and carbon black are conducted by straining ultrathin sections up to 300% prior to TEM examination [6].

Steel Wire Cable Belts Under the Tread **4** – Vulcanization pressure and cable twist patterns affect the depth of penetration of rubber into cable void spaces. Trapped air that promotes wire corrosion must be avoided. These aspects can be seen by OM and SEM. The complex adhesive system of rubber to brass plated steel wires is reflected by several metal oxide and metal sulfide layers that form the bond between the rubber and metal. SEM/EDS on surfaces and TEM selectedarea electron diffraction (SAED) on ultrathin sections provides critical information [7].

Steel Wire Belt Cut Ends **5** – This is a critical region that needs constant monitoring of adhesion between rubber and cut ends and is best done by OM and SEM. Thisis also the area of "belt edge separation" which may be a principle failure mode in some catastrophic events. The shape of individual cut wire filaments must be such that they do not act as stress raisers. OM and/or SEM can also reveal polishing of the cut ends that occurs as failures progress.

Textile Cord **6 –** Adequate adhesion requires optimum penetration of adhesive dip into cord interstices and this can be evaluated by OM, SEM on single step systems. For multiple dip adhesive systems the distribution of adhesives is best done by TEM on ultrathin sections [9].

Ply Turnup Ends **7** - Penetration of rubber into cut cord ends and the absence of porosity are diagnostics of cure pressure and can be observed best by SEM.

Bead Area **8** - Compaction of rubber into and around bead wires and the absence of trapped air are essentials for proper bead integrity and easily observed by OM and SEM.

Other Surfaces **9** - Molding processes can be evaluated by checking for: insufficient mold filling; presence of blemishes; existence of mold flow problems; and "health" of sidewall vents. OM and SEM/EDS methods are effective here.

The total microscopy of a tire must include the microscopy of failures. All techniques useful for the above aspects 1 through 9 can be applied to investigations of failure incidents. A necessary adjunct is that of applying intentional damage to the structure to see what morphological changes can be useful for diagnostic purposes. This would include the microscopy of punctures, road hazards, abuse, and other assaults common or uncommon to tires.

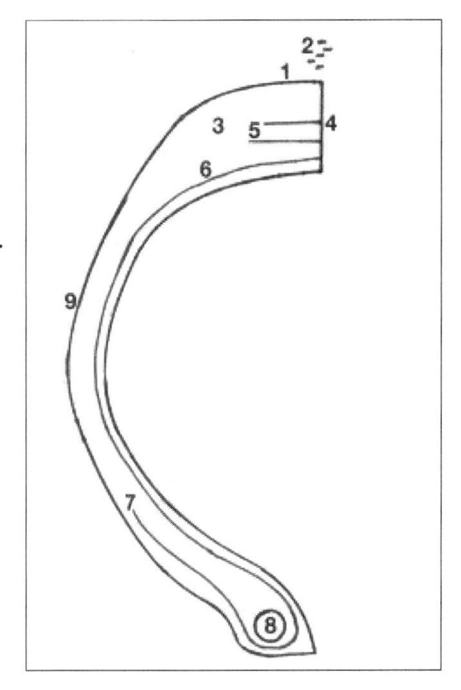

FI G. 1. Generic Tire

1. Tread Surface
2. Tread Wear Debris
3. Bulk Rubber Compound
4. Steel Belt Plies
5. Steel Belt Cut Ends
6. Carcass Ply
7. Carcass Ply Turnup Ends
8. Bead Package
9. Other Surfaces

References

[1] A. Schallamach, Rubber Chem.& Tech. 26, (1953) 230
[2] R.W. Smith, ibid. 55 (1982) 469
[3] C.H. Liegh-Dugmore, ibid. 29 (1956) 1303
[4] W.M. Hess, ibid. 36 (1963) 754
[5] R. W. Smith, ibid. 37, (1964) 338
[6] R.W. Smith, ibid. 40 (1967) 350
[7] W.J. van Ooij, Kautschuk Gummi Kunst. 44 (1991) 345
[8] R.W. Smith, Rubber Chem. & Tech., 70 (1997) 283
[9] R.W. Smith, Polymer Sci. & Tech., 9A (1975) 289

Microsc. Microanal. 8 (Suppl. 2), 2002
DOI 10.1017.S1431927602100894

Self-assembled Nanostructures: from Nanocrystals to Mesopores and to Nanobelts

Zhong Lin (ZL) Wang*

Center for Nanoscience and Nanotechnology, and School of Materials Science and Engineering, Georgia Institute of Technology, Atlanta GA 30332-0245.
* e-mail: zhong.wang@mse.gatech.edu

Size and shape selected nanocrystals behave like molecular matter that can be used as fundamental building blocks for constructing nanocrystal assembled superlattices. The nanocrystals form a new class of materials that have orders in both atomic and nanocrystal length-scales. The nanocrystals are passivated with organic molecules (called thiolates) that not only protect them from coalescence but also act as the molecular bonds for forming the superlattice structure. The interparticle distance is adjustable, possibly resulting in tunable electric, optical and transport properties. This paper reviews our current progress in self-assembly of shape-controlled nanocrystals, photonic crystals and mesoporous materials.

Soft materials made of self-assembled nanocrystals: Self-assembled arrays involve self-organization into monolayers, thin films, and superlattices of size-selected nanocrystals encapsulated in a protective compact organic coating [1]. Nanocrystals are the hard cores that preserve the ordering at the atomic scale; the organic molecules adsorbed on their surfaces serve as the interparticle molecular bonds and as protection for the particles in order to avoid direct core contact with a consequence of coalescing. The interparticle interaction can be changed via control over the length of the molecular chains. Quantum transitions and insulator to conductor transition could be introduced, resulting in tunable electronic, optical and transport properties. We have reported the preparation and structure characterization of three-dimensional (3-D) packing of Au and Ag nanocrystal super-structures, with an emphasis on the roles played by crystal shapes in forming the nanocrystal superlattice (NCS) [2]. In NCS of Ag nanocrystals with truncated octahedral shape, the type of orientation ordering is determined by the nanocrystal's faceted morphology, as mediated by the interactions of surfactant groups tethered to the facets on neighboring nanocrystals. The particles tend to be packed face-to-face to lower the energy, the surface adsorbed molecules are distributed on the facets of the particles and are self-assembled into bundled and interdigitative molecule bonds [3], which have been found to be extraordinarily strong, stabilizing the NCS to temperatures as high as 500 $^{\circ}$C [4]. The passivation molecules form bundles on the surface, which have been directly observed by energy-filtered TEM [9]. NCS's formed by spherical-like (e.g., icosahedral and decahedral) Ag nanocrystals form *hcp* superlattices [5]. Defect structures present in NCS's, including: twins, slip planes, multiply-twins and dislocations, have been studied in detail [1,6]. The purified CoO nanocrystals can also form NCS with ordered structure (Figure 3) [7]. CoO particles have a dominant tetrahedral shape of sizes 4 nm. In-situ TEM studies of CoO and Ag NCS's have proved the structural stability up to temperatures of ~ 500 $^{\circ}$C [8].

Mesoporous materials as photonic crystals and for low-loss dielectrics and catalysis: There are two main groups of photonic crystals. One is the self-assembly of silica spheres. The other is the ordered porous structure. The ordered hollow structure was made through a template-assisted technique, in which self-assembly of size selective polystyrene spheres creates an ordered template. Metal-organic solution containing the ions for synthesizing the required structure is infiltrated between the spheres. Heating the template at a higher temperature gets rid of the spheres, while the skeleton of the metal oxide still preserves [9]. Ordered self-assembly of hollow structures are potential candidates for low-dielectric-constant materials and catalysis of high surface areas. We have synthesized structures that have ordering and porosity on two length-scales, one is at the scale of hollow spheres created by a template of polystyrene (PS) spheres, and the other is the nanocavities created by self-assembled molecular co-polymers [10].

Semiconducting oxide nanobelts – a new family of nanomaterials: Ultra-long belt-like, quasi-one-dimensional nanostructures (so called nanobelts or nanoribbons) have been successfully synthesized for semiconducting oxides of zinc, tin, indium, cadmium, gallium and lead, by simply evaporating the desired commercial metal oxide powders at high temperatures [11-14]. The as-synthesized oxide nanobelts are pure, structurally uniform, single crystalline and most of them free from defects and dislocations; they have a rectangular-like cross-section with typical widths of 30-300 nm, width-to-thickness ratios of 5-10 and lengths of up to a few millimeters. The belt-like morphology appears to be a unique and common structural characteristic for the family of semiconducting oxides with cations of different valence states and materials of distinct crystallographic structures. The nanobelts are an ideal system for fully understanding dimensionally confined transport phenomena in functional oxides and building functional devices along individual nanobelts [15].

[1] Z.L. Wang, Adv. Mater. 10 (1998) 13.
[2] S.A. Harfenist, Z.L.Wang, et al., J. Phys. Chem. B 100 (1996) 13904.
[3] Z.L. Wang, S.A. Harfenist, et al., J. Phys. Chem. B, 102 (1998) 3068.
[4] S.A. Harfenist and Z.L.Wang, J. Phys. Chem. B, 103 (1999) 4342.
[5] S.A. Harfenist, Z.L.Wang, M.M.Alvarez, I.Vezmar, and R.L.Whetten, Adv. Mater. 9 (1997) 817.
[6] Z.L. Wang, Mater. Charact. 42 (1999) 101.
[7] J.S. Yin and Z.L. Wang, Phys. Rev. Lett. 79 (1997) 2570.
[8] J.S. Yin and Z.L. Wang, J. Phys. Chem. B, 101 (1997) 8979.
[9] J.S. Yin and Z.L. Wang, Adv. Mater. 11 (1999) 469.
[10] J.S. Yin and Z.L. Wang, Appl. Phys. Lett. 74 (1999) 2629.
[11] Z.W. Pan, Z.R. Dai and Z.L. Wang, *Science*, 291 (2001) 1947.
[12] Z.W. Pan, Z.R. Dai, and Z.L. Wang, *Appl. Phys. Letts.*, 80 (2001) 309.
[13] Z. L. Wang, R. P. Gao, Z. W. Pan and Z. R. Dai, *Adv. Eng. Mater.*, 3, (2001) 657.
[14] Z.R. Dai, Z.W. Pan, Z.L. Wang, Solid State Comm.,118 (2001) 351.

[15] Thanks to the contribution made by my group members: J.S. Yin, S.A. Harfenist, R.L.Whetten, Z.W. Pan, Z.R. Dai, C. Ma and Y. Berta. Thanks for the financial support from NSF grants DMR-9733160.

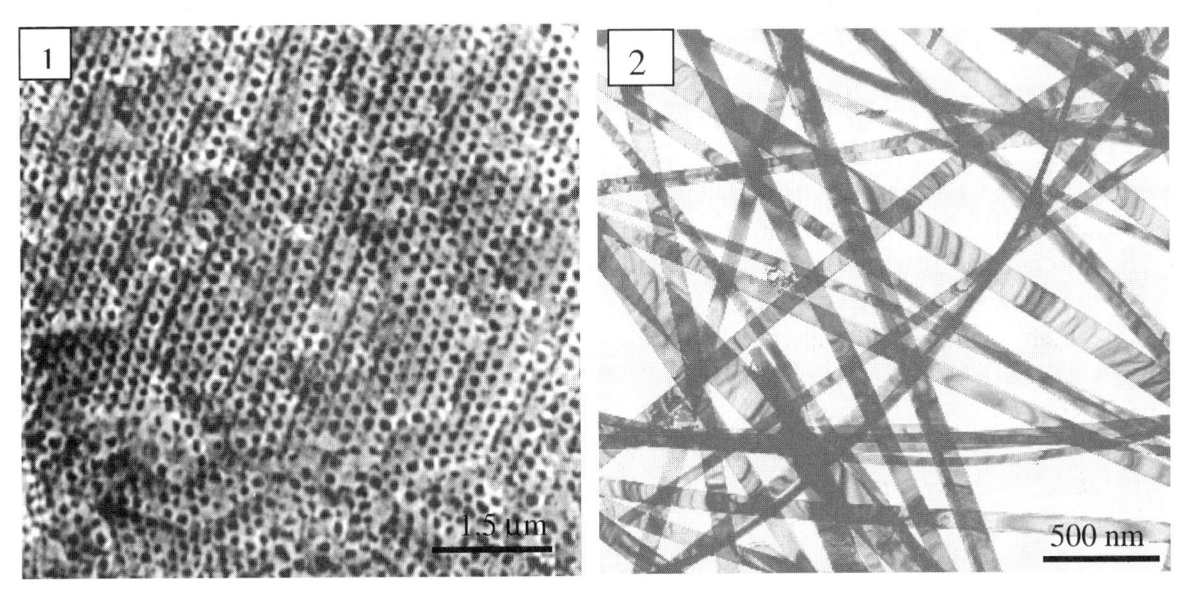

Fig. 1 Scanning electron microscopy images of the porous titania nanostructures.
Fig. 2 Semiconducting SnO_2 nanobelts – a new family of nanomaterials.

Microsc. Microanal. 8 (Suppl. 2), 2002
DOI 10.1017.S1431927602100900

THREE DIMENSIONALLY ORDERED MACROPOROUS BIOACTIVE GLASSES

David C. Bell, Kai Zhang[&], Hongwei Yan[*], Lorraine F. Francis[&], Andreas Stein[*]

Institute of Technology Characterization Facility, [&]Department of Chemical Engineering and Materials Science, [*]Department of Chemistry
The University of Minnesota, Minneapolis, MN 55455

Bioactive ceramics including glasses have been used to repair and reconstruct damaged parts of the skeletal system [1]. A three dimensionally ordered macroporous (3DOM) structure can be formed using the sol-gel technique to produce a glass that is bioactive [2]. Bioactive 3DOM materials develop a biologically active, hydroxycarbonate apatite layer on their surfaces within minutes of exposure to body fluids either *in vivo* or *in vitro*. The composition and morphology of sol-gel derived glass materials have direct impact on the design of the new generation of biomaterials for use in bone regeneration, membranes and possible drug delivery systems.

The composition, processing temperature, surface chemistry and structural morphology of the sol-gel bioactive glasses affect the apatite formation and growth [3]. The bioactivity is enhanced by increasing the surface pore size and developing bulk porous structure. The 3DOM sol-gel bioactive glasses were synthesized for this research using monodispersed poly- methyl methacrylate, (PMMA), spheres of different sizes (100-1000 nm), followed by a sol gel process as described in [4, 5]. The resultant 3DOM glass is characterized using SEM as shown in figure 1 and is shown to have an fcc arrangement of pores. The preparation and microstructure of sol-gel bioactive glasses with 3DOM structure will be presented.

To test *in vitro* bioactivity the 3DOM sol-gel glass is soaked at body temperature in a simulated body fluid (SBF) to observe the apatite formation. The composition and preparation of SBF is after Abe et al. [6]. The 3DOM structure, after a sufficient soaking time, completely converts to a flake like material, which was demonstrated to be crystalline hydroxycarbonate apatite (HCA) by XRD, FTIR and TEM characterization. Figure 2, shows CaO-SiO_2-P_2O_5 3DOM sol-gel bioactive glass after soaking in SBF for 4 days. The images show that the surface is covered with HCA. The development of HCA is faster for 3DOM sol-gel bioactive glasses than for traditional sol-gel bioactive glasses [2]. Differences in bioactive behaviors are directly related to the microstructural morphology; the SBF can readily penetrate the larger macropores in the 3DOM bioactive glass compared to the mesopores in traditional sol-gel processed bioactive glasses, while minor differences in chemical composition of 3DOM glass have little effect. Pore size can be adjusted by using different sizes of PMMA spheres. Factors determining apatite growth such as pore size and ordering, which is critical for skeletal reconstruction applications, are discussed.

References:

[1] Hench L.L., *J. Am. Ceram. Soc.,* 81, 1705-1728, 1998
[2] Yan H., Zhang K., Blanford C.F., Francis L.F., Stein A., *Chem. Mat.*, 13, 1374-1382, 2001
[3] Pereira M.M. et al. *J. Am. Ceram. Soc.,* 78, 2463-2468, 1995.
[4] Zhang K., Yan H., Bell D.C., Stein A., Francis L.F., *Trans. of Soc. for Biomaterials,* 28 2002
[5] Holland B.T., Blanford C.F., Stein A., *Science,* 281, 538-540, 1998

[6] Abe Y., et al., *Chem. Mat.*, 12, 1134-1141, 2000
[7] Support of the University of Minnesota, IT Characterization Facility is gratefully acknowledged.

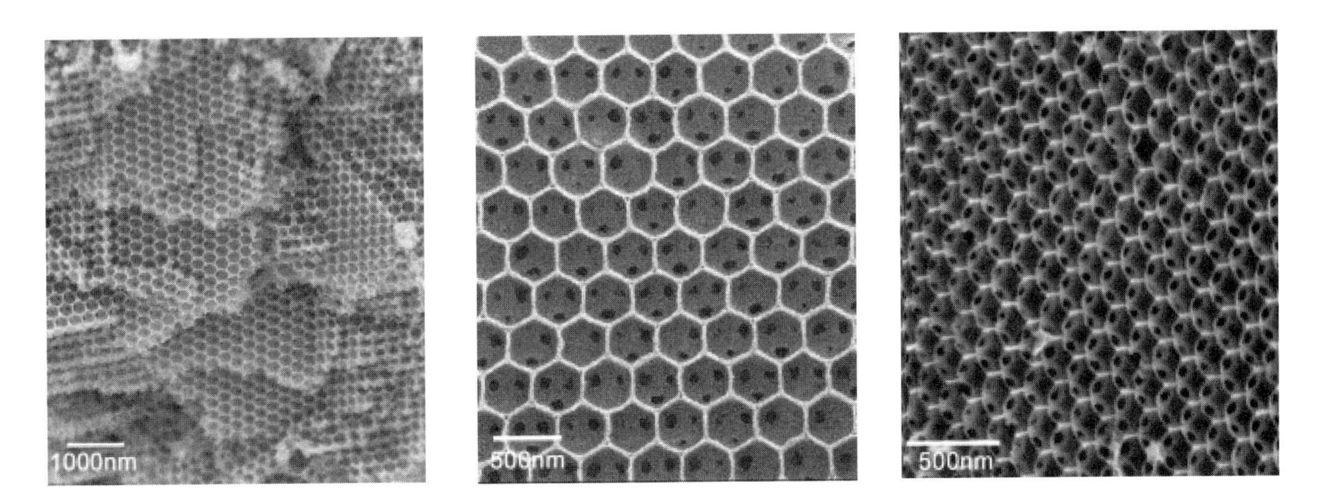

Figure 1. SEM Images of 3 DOM bioactive glass formed using the sol-gel process described, indicating the macro-sized pore structure.

Figure 2. (Left) TEM image of apatite formation on CaO-SiO_2-P_2O_5 3DOM sol-gel bioactive glass after soaking in SBF for 4 days. (Right) SEM image of surface of CaO-SiO_2-P_2O_5 3DOM sol-gel bioactive glass after soaking in SBF for 4 days showing flake like HCA.

Microsc. Microanal. 8 (Suppl. 2), 2002
DOI 10.1017.S1431927602100912

Energy Loss Spectroscopy and Electron Microscopy of Photoluminescent p-type Porous Silicon Treated with NaOH and NH_3 Solutions

Minghui Song, Yingda Yu, Yoshio Fukuda, and Kazuo Furuya

National Institute for Materials Science, 3-13 Sakura, Tsukuba, Ibaraki 305-0003 JAPAN

Porous silicon (PS) formed electrochemically on crystalline silicon at room temperature is known to be visibly photoluminescent [1], and is thought to be a prospective material for optoelectronic devices. Several models have been proposed for the mechanism of photoluminescence. However, little agreement has been attained yet. It is known that the PL is sensitive to both chemical components and microstructures of PS layer. To clarify the effect of microstructure and chemical components of PS on PL is definitely helpful to understand the mechanism of PL of PS. In the present work, electrochemically formed PS on p-type silicon wafers were etched by NaOH and NH_3 hydro-solutions, in order to study the relation between PL and Si-oxide layer on the surface of the PS layer. The change in PL of the treated specimens has been correlated with the changes in local chemical states and microstructure of PS layer with EELS and TEM.

PS samples were prepared by anodizing p-type Si (100) wafers with resistivity of 0.04-0.06 Ωcm. Anodization was carried out in a solution of 50wt%HF:C_2H_5OH=1:1 at room temperature for 10 seconds by applying a current density of 350 mA cm^{-2}. The specimens were then etched using 0.05% NaOH hydro-solution from 5 to 60 seconds and 0.14% NH_3 hydro-solution from 1 to 45 minutes at room temperature. The PL spectra were obtained using a photoluminescence-lifetime measurement system with a streak camera (Hama-photo C4780). The local chemical states and microstructure of as-prepared and solution treated PS were analyzed and observed using a JEM-ARM1000 TEM, to which a post-column parallel detection EELS system is attached under the camera chamber.

As-prepared PS layers were shown to be photoluminescent with a peak wavelength around 640nm. The NaOH solution etching resulted in a decrease of PL intensity as shown in figure 1. For NH_3 treated specimens, the PL around 640 nm decreased after etching for 1 minute. Further etching up to 10 minutes increased the PL but in a shorter wavelength range. The increased PL then decreased again after 45 minutes etching. The spectra of the PL are shown in figure 2. TEM observation revealed that the PS layer structurally consists of so called a sponge-like part near the surface of the PS and a tree-like structure region in the inner part of the PS layer [2]. The sponge-like layer was apparently reduced by the NaOH etching, while the tree-like structure remained almost unchanged. For the NH_3 treated specimens, TEM observation revealed that the sponge-like layer on surface of PS was resolved by 1-minute etching, and that further treatment up to 10 minutes formed the new sponge-like layer on the surface as shown in figure 3. The 45-minutes treatment resolved the sponge-like layer again. The decrease and increase of the sponge-like layer correlated to the decrease and increase of PL, respectively. The near edge EELS spectra (ELNES) of Si-$L_{2,3}$ of PS layers were obtained. The Si^{4+} edge at 108 eV corresponding to the chemical bond of Si in Si-O tetragonal and Si^{0+} edge at 100 eV corresponding to Si in silicon crystals were analyzed. These edges changed in its shape and relative intensity after the etching. The correlation of PL intensity in wavelength of 640 ± 20nm with the counts ratio Si^{4+}/Si^{0+} in EELS spectra for a NaOH treated specimen is shown in figure 4, which reveals that the decrease of PL after NaOH etching is consistent with the decrease of Si^{4+}/Si^{0+}. The present work suggests that the sponge-like structure near the surface of PS contributes to PL much more than the inner structure of PS layer and that the Si-O chemical structure relates to the PL strongly.

References

[1] L.A.Balagurov et al, *Appl. Phys. Lett.*, 69 (1996) 2852.
[2] M. Song et al , *Micron*, 31 (2000) 429.

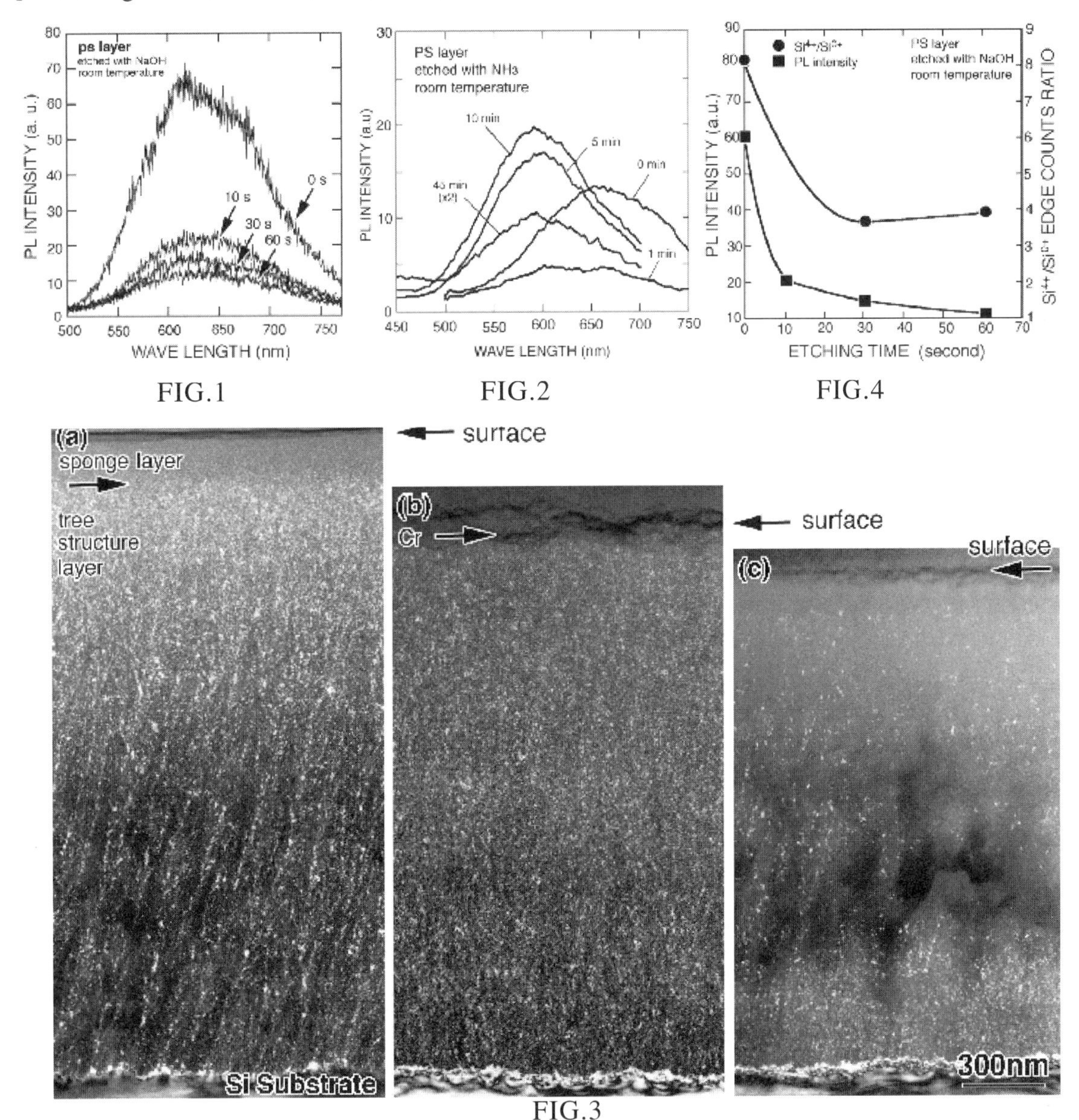

FIG. 1 PL of NaOH solution treated PS layer formed on low resistivity p-type Si wafers in a current of 350mA cm^{-2} at room temperature.

FIG. 2 PL of NH_3 solution treated PS layer.

FIG. 3. Cross sectional TEM dark field micrographs of PS layer etched with NH_3 solution. a) As-prepared, b) etched for 1 minute. c) etched for 5 minutes. Cr layer was coated on the surface to protect PS.

FIG. 4. Correlation of PL intensity with the count ratio of edges Si^{4+}/Si^{0+} in EELS spectra of PS layer.

Microsc. Microanal. 8 (Suppl. 2), 2002
DOI 10.1017.S1431927602100924

Applications of SEM and Con-focal Laser Microscopy in Developments of Macro-Porous Materials Produced by Sintering

K. Ishizaki, M. Ohyagi, K. Jodan, K. Matsumaru and M. Nanko
Nagaoka University of Technology, Nagaoka, Niigata, 940-2188, JAPAN

Our group has developed many kinds of new macro-porous materials by applying a hot isostatic process (HIP) [1-3] and pulsed electric current sintering (PECS) [3,4]. In order to develop macro-porous materials, evaluation of microstructure is important to consider production parameters and properties of porous materials.

A scanning electro microscope (SEM) is used to observe microstructure of macro-porous materials. For example, bonding parts between grains are important for mechanical properties of porous materials. Figure 1 shows the SEM images of bonding parts of SiC grains in a grinding wheel with a vitrified bonding agent. Cracks located in the bonding part disappear by HIPing. Applications of HIP for this materials promise to increase mechanical strength of porous materials [1,2].

A con-focal laser microscope has also great potentials for observing microstructure of porous materials. The laser microscope enables to observe microstructure with relatively high resolution and large focus depth in large area. Since the laser microscope is available in air and requires no particular treatment to sample surfaces, it leads to reduce an operation time compared with SEM. Figure 2 shows the comparison of an SEM image and con-focal laser microscopic one of porous Bi-system superconducting oxides. The laser microscope provides excellent images to understand microstructure of porous materials.

Furthermore the con-focal laser microscope provides surface profiles, which is useful for studying surface treatments of porous materials. For example, to refresh a surface of grinding wheels, a dressing is applied to a grinding wheel when its grindability degrades during machining. Dressing is one of key processes to increase efficiency of precise grinding advanced ceramics and semiconductors. Our group has developed a new dressing technique with laser, which realizes in-process dressing [5, 6]. Figure 3 shows the surface image and surface profiles of a porous cast-iron matrix diamond grinding wheel treated by an Nd:YAG laser-dressing technique. Applying the laser dressing, fresh diamond grains appear on the grinding wheel surface. Surface roughness, which influences grinding performance, can be evaluated quantitatively and simultaneously when the surface morphology is observed.

REFERENCES

[1].K. Ishizaki, A. Takata and S. Okada, J. of Ceram. Soc. Jpn., 98, 533-40 (1990).
[2] K. Ishizaki and M. Nanko, J. Porous Mater., 1 (1995) 19-27.
[3] M. Nanko et al., Ceramic Industry, Jan. (1996) 31-37.
[4] H. Onishi et al., J. Porous Mater., 4 (1997) 187-198.
[5] H. Funakoshi et. al., Interceram, 50 (2001) 466-69.
[6] K. Jodan et. al., ATM, 2 (2000) 117-123

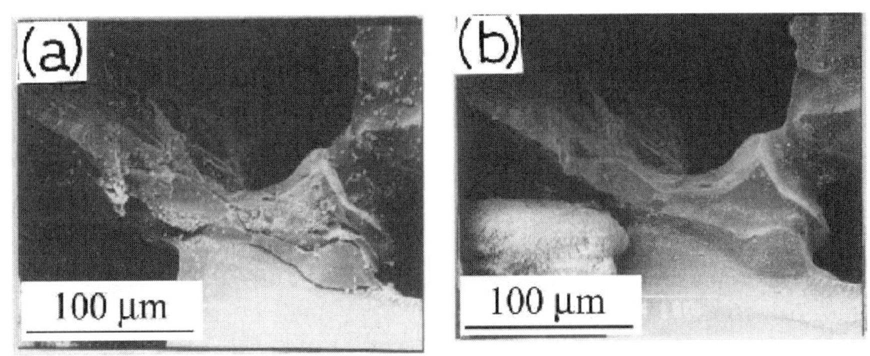

Fig. 1 SEM images of bridges of vitrified bond and SiC grains in grinding wheels sintered in a conventional sintering process (a) and by a hot isostatic pressing method (b).

Fig. 2 Microstructures of a Bi-phase porous superconducting material in an SEM image (a) and optical image taken by a con-focal laser microscope (b).

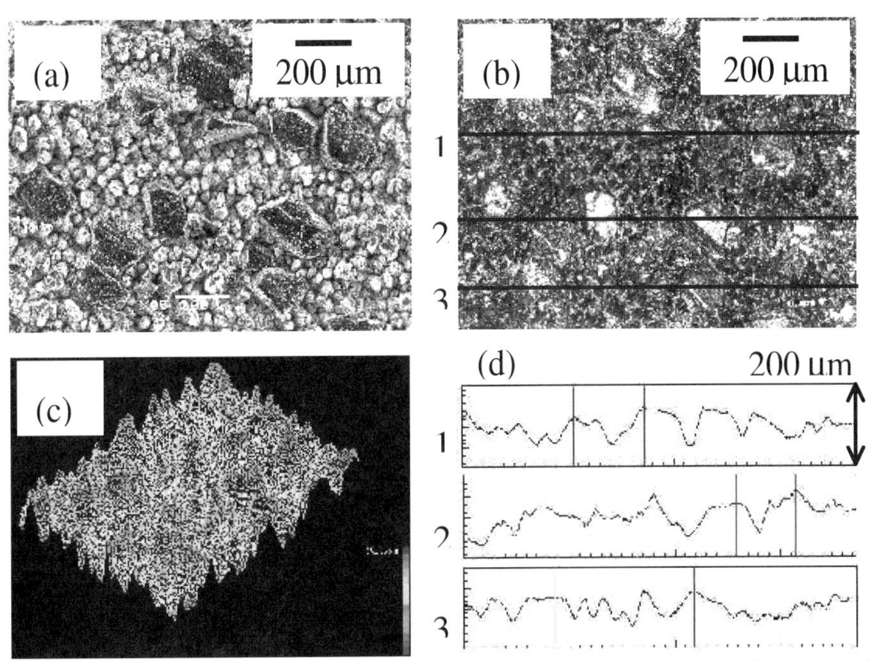

Fig. 3 Surface of a laser-dressed porous cast-iron bonded diamond grinding wheel. (a) is an SEM image. (b), (c), and (d) are an optical image, 3D, and line profiles given by a con-focal laser microscope, respectively. Lines 1, 2, and 3 in (b) correspond to profile lines 1, 2, and 3 in (d).

Microsc. Microanal. 8 (Suppl. 2), 2002
DOI 10.1017.S1431927602100936

Pore Hierarchies in High-Temperature Composite Refractories

W. E. Lee, S. Zhang and S. Hashimoto

University of Sheffield, Dept. of Engineering Materials, Mappin St., Sheffield, S1 3JD, UK

Refractories are high-temperature, porous, composite materials containing from 5-90 volume % porosity. They are enabling materials used to line furnaces for production of other materials including metals, ceramics, glasses, cements, semiconductors and single crystals [1]. Massive volumes of them are used successfully in some of the most severe environments seen by materials but researchers often neglect them as being too "traditional" and low tech. This is far from the case and their microstructural design is often complex and beautiful. Pores are present because the firing step in their fabrication, unlike in other ceramics, induces little if any shrinkage and densification; it is simply a bonding operation to improve properties [2]. Pores occur within and between the solid materials in the microstructure after the shape forming operation. While they are not removed on firing their size, shape and distribution may be altered by a range of processes including thermal expansion/contraction of phases present, phase transformations involving volume change, local solid state and liquid phase sintering, vitrification and reactions between phases and/or the atmosphere.

The porosity in refractories ranges from macroscopic (mm scale) to microscopic (μm to nm scale), may be open or closed and may be located in the grain/aggregate phases or the bond/matrix system or both. While refractories may be made as pre-fired bricks or shapes they are more often made as monolithics or unshaped mixtures, which can be installed and shaped *in situ*. Such monolithics include castables which are dry powder mixes which when water is added can be shaped by casting or vibrating, a hydraulic cement chemical bond forms at room temperature and converts to a physical ceramic bond on firing. The pore distribution is a strong function of installation technique. Typically, thermo-mechanical and chemical properties of refractories are also a function of the total pore system so its quantitative characterisation is important. Pores occur over a range of length scales (hierarchies) in the refractories microstructure and it is useful to perform interrupted firing experiments to catch the evolution of the pore structure on firing.

The microstructural evolution on firing and quenching a commercial vibratable ultralow cement castable has been fully characterized and related to high-temperature properties [3]. Figure 1 shows that after firing at 1200°C low melting calcium aluminium silicate (CAS) liquid formed in the matrix leaving bright CAS-rich regions around the cement agglomerates, while dispersed microsilica in the bond started to react with alumina from hydratable alumina agglomerates forming an aluminosilicate (AS) rim around them. Pores are black in this image and occur predominantly in the matrix system. They are seen e.g. as:

- rims around the calcium dialuminate (CA_2) agglomerates arising from its formation in the cement via reaction of calcium aluminate with alumina and shrinkage contraction,
- fine (μm) intragranular pores in the matrix from incomplete sintering.
- nm scale layer pores within the hydratable alumina agglomerates (labelled AS).

The origins of the layer pores can be traced back to the production of the hydratable alumina from Gibbsite ($Al(OH)_3$) using the Bayer process. The calcined product of the Bayer process is made up of transition alumina which is pseudomorphic with the original Gibbsite (Figure 2). The volume change associated with the calcination of Gibbsite to alumina and the orientation relation between the Gibbsite and alumina lead to this pore morphology. The emergence of refractory phases such as calcium hexaluminate, magnesium aluminate spinel and mullite in commercial castable refractories has been observed to have beneficial effects on hot bend strength and refractoriness under load due to development of morphologies which interlocked other phases and to pore filling from expansile reactions.

References

[1] WE Lee, Chapter 4.12 pp.363-385 "Refractories" in *Comprehensive Composite Materials* (edited by A Kelly and C Zweben) Volume 4. *Ceramic, Carbon and Cement Matrix Composites* (Elsevier 2000).
[2] WE Lee and WM Rainforth, *Ceramic Microstructures: Property Control by Processing*. (Chapman and Hall 1994).
[3] H Sarpoolaky, KG Ahari and WE Lee, "Influence of *in-situ* Phase Formation on Microstructural Evolution and Properties of Castable Refractories," *Ceramics International* 2002.

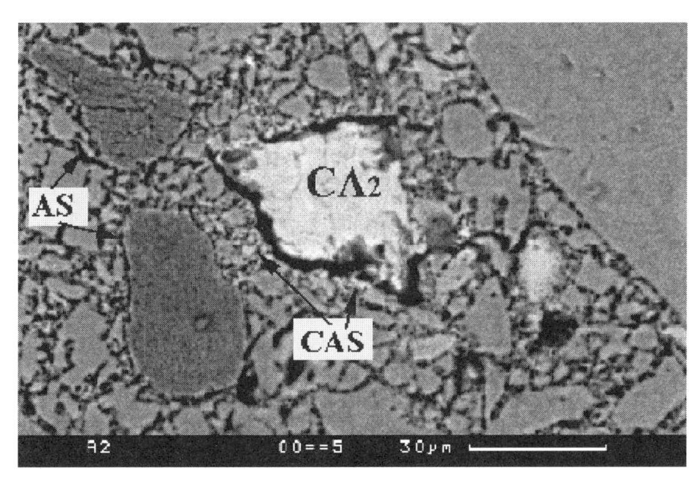

FIG 1. Backscattered electron SEM image of an ultralow cement castable after 3h at 1200 °C.

FIG 2. Secondary electron SEM image of pore fissures in alumina formed by dehydroxylation of Gibbsite at 1150°C.

Microsc. Microanal. 8 (Suppl. 2), 2002
DOI 10.1017.S1431927602101206

Understanding Complex Microstructures With High-Resolution Microanalysis in the Transmission Electron Microscope.

G.A. Botton*, J. A. Gianetto**, C. V. Hyatt# and M.W. Phaneuf§.

*Dept of Materials Science and Engineering, McMaster University, Hamilton, ON, L8S 4M1, Canada; ** Materials Technology Laboratory-CANMET, Natural Resources Canada, Ottawa, ON, Canada; #Defense Research and Development Canada, DREA, Halifax, NS, Canada; § Fibics Incorporated, 556 Booth, Ottawa, ON, K1A 0G1, Canada.

The development of intermediate voltage transmission electron microscopes equipped with field-emission sources, have made analytical transmission electron microscopy (TEM) a widely used tool in the characterization of fine microstructures. In the general area of metallography, however, TEM is often considered as a tool of last resort to characterize complex materials because of the difficulty associated with sample preparation and limited sampling of the microstructures. With access to new sample preparation tools such as the focused ion beam (FIB), metallographic analysis of samples by TEM is likely to become a routine tool to elucidate not only complex materials but also microstructural and chemical nature of materials subjected to environmental degradation. Specific regions in welded microstructures or areas showing preferential corrosion can be selected and analyzed at high spatial resolution. In this short abstract some examples illustrate the possibilities of identifying features in such complex microstructures that would not have been possible without making use of the high spatial resolution analysis available in analytical TEM. Previous work(1) has shown the potential of TEM and FIB to identify the response of various phases to electrochemical noise tests in saline water.

To further demonstrate this approach we discuss an example of analysis taken from on-going work related to the study of laser clad nickel aluminum bronzes (NAB) that were subjected to marine corrosion testing. In this work, laser clad coupons were subjected to alternately flowing and stagnant long-term seawater immersion testing for 1 to 4 years. To obtain a more fundamental understanding of the corrosion behaviour of the laser clad welds the thin corrosion product and the first few microns from the sample surface were analyzed(2). In the cross-sections of the sample, the corrosion product is clearly visible and fully preserved (Fig. 1). Measurements of the energy loss fine structures at the oxygen K edge reveal that the corrosion product contains hydroxides fully preserved in spite of the sample preparation used to obtain the thin sections (Fig. 2). Another area of interest along the laser-clad surface included a region where less obvious attack had occurred in the proximity of a grain boundary (Fig. 3a). This area was investigated to assess whether preferential corrosion occurs where boundaries intersect the surface of the sample. Although such effect was not observed, the high resolution energy dispersive x-ray spectrometry (EDXS) elemental mapping (Fig. 3b, c, d) demonstrated very clearly the depletion of Fe containing κ_{iv} precipitates (20-50nm in size) in proximity of the grain boundaries and near the larger precipitates. Such effect was not detected in the conventional TEM or in the annular dark-field micrographs. Al was detected at the sample surface.

This brief example shows the potential of the combination of analytical TEM with FIB to study complex microstructures and reveal features not visible by other analytical means typically used in conventional metallography techniques. Further examples related to corrosion studies of NAB and Ni based alloys will also be discussed.

(1) R.D. Klassen, et al. Can. Metal. Quart. 41, 1, p 121-132, 2002.
(2) J.A. Gianetto et al. Special Report DREA SR 2001-094, p.67-82, October 2001.

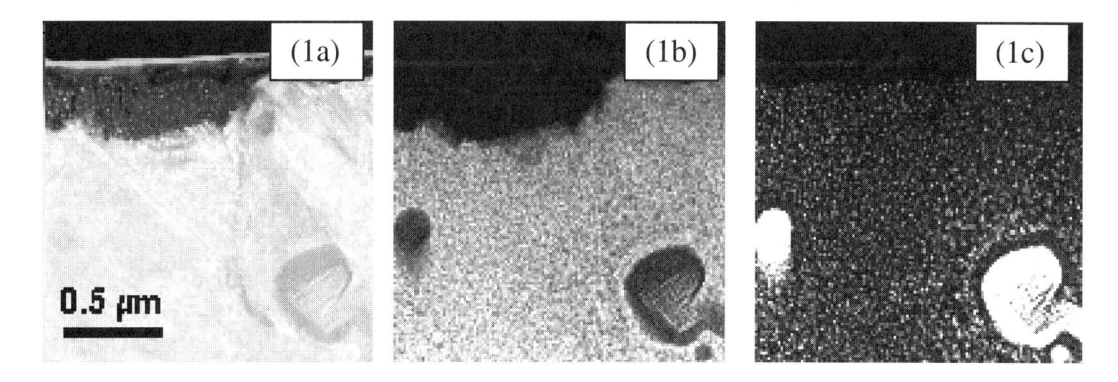

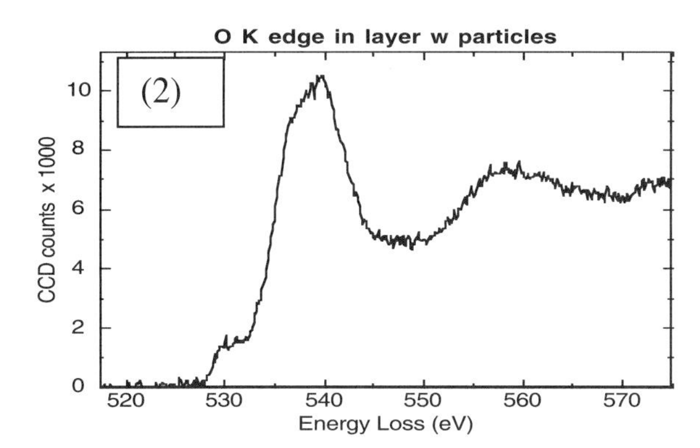

Figure 1. a) Dark-field image of an area with enhanced corrosion product in a NiAl bronze submitted to sea-water corrosion. EDXS elemental maps of Cu map (b) and Fe map (c).

Figure 2. Near-edge fine structure of the O K edge in the area with enhanced corrosion. The broad peak and the shoulder at 530eV suggest the presence of hydroxides

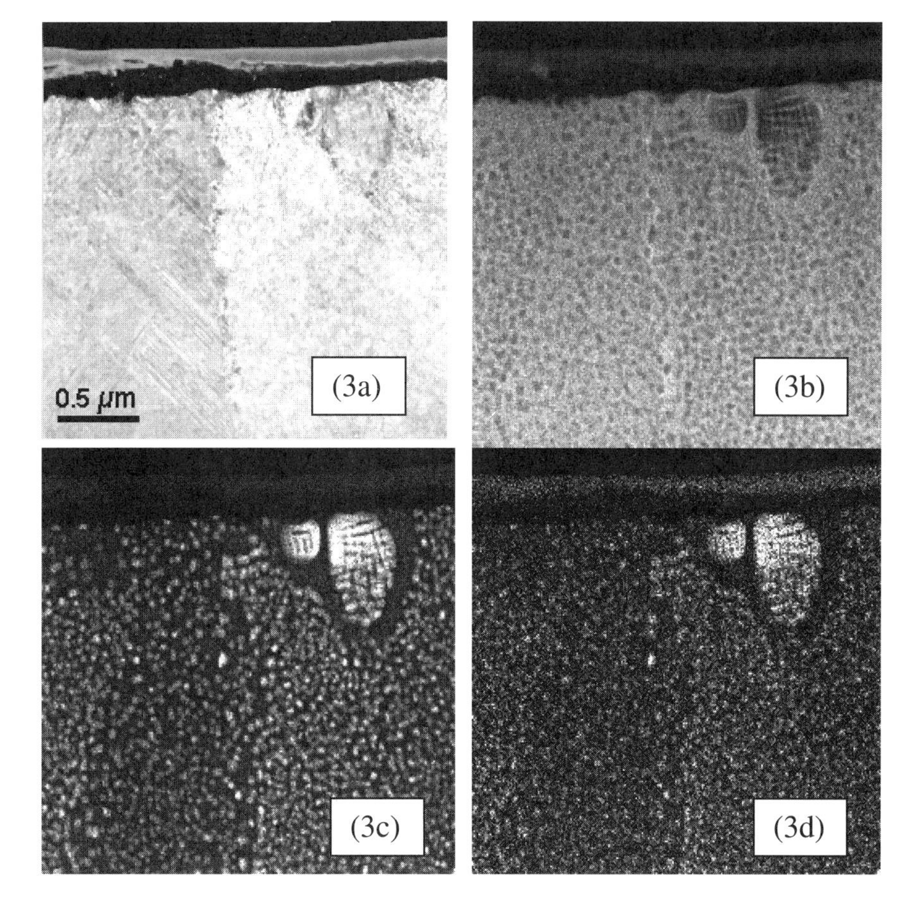

Figure 3. Microstructure and high-resolution elemental maps of the uniform layer formed on the surface of a Ni-Al bronze. a) Dark-field image, b) Cu elemental map, c) Fe elemental map, and d) Cr elemental map showing depletion of κ_{IV} precipitates in proximity of large precipitates and near a grain boundary in the centre of the field of view.

Microsc. Microanal. 8 (Suppl. 2), 2002
DOI 10.1017.S1431927602101218

Characterization of Dislocation Structures in Hexagonal Close-Packed Metals by X-ray Line-Broadening Analysis

M. Griffiths*, D. Sage* and D. Galindo**

* Atomic Energy of Canada, Chalk River Laboratories, Chalk River, Ontario, K0J 1P0.
** McGill University, Department of Mathematics, Montreal, Quebec.

In deformed hcp metals such as Zr, X-ray diffraction line profiles are determined by the convolution of a number of different broadening factors: (i) instrumental; (ii) intergranular strain distributions; (iii) coherent diffracting domain size; and (iv) lattice microstrains (strain distribution around a dislocation), [1]. Dislocation structures can be characterized by Fourier analysis methods, [2], provided that the only contributions to the line profile arise from the latter two factors, (iii) and (iv). The instrumental broadening effect, (i), can be extracted from the line profile using conventional deconvolution methods with an appropriate standard, [2,3]; the line-broadening is then compared relative to the standard (normally an annealed single crystal). The intergranular strain distributions, (ii), introduce an error into the determination of dislocation densities unless they too can be deconvoluted. Unfortunately, as the intergranular strains are dependent on the extent of deformation, there are no practical means by which a standard can be generated that is independent of the dislocation structure, that is also dependent on the extent of deformation. In principle, the intergranular strain distributions and lattice microstrains from dislocations are both dependent on the order of diffraction. This provides a means of separating them from the coherent domain size effect, which is independent of the order of diffraction, using the Fourier method of Warren and Averbach [2]. The residual peak profile will be comprised of a convolution of the two components, (ii) and (iv), and can only be separated if the functional form of each component can be defined or measured.

The aim of this work is to determine the functional form of dislocation microstrains and intergranular strain distributions. Experimental data for both types exist. For microstrains, the diffraction peaks from an irradiated single crystal containing dislocation loops are analysed. The single crystal peaks are, in principle, unaffected by intergranular strain distributions and can be processed using the method of Warren and Averbach [2] to obtain type (iv) profiles. For intergranular strain distributions, the diffraction peaks from a deformed sample are analysed. Those peaks corresponding with planes that are relatively undistorted by dislocations and are, in principle type (ii). Both peak shapes are compared with the calculated shape of the diffraction peak resulting from the strain field around a dislocation in order to verify the premise that the two effects originate from different sources and are separable. The dislocation loops are shown in Figure 1(a) and the corresponding prism plane line profile, with superimposed Gaussian and Lorentsian fits, for neutron irradiated single crystal Zr are shown in Figure 1(b). The line profile calculated from first principles, with superimposed Gaussian and Lorentsian approximations, is shown in Figure 2. These preliminary results indicate that the diffraction line profile for the strain around a dislocation is primarily Lorentsian in nature. Similar fits for intergranular strain distributions indicate that a Gaussian distribution is the better fit.

References

[1] M. Griffiths et al., *Trans. AIME, in press.*
[2] B.E. Warren and B.L. Averbach, *J. Appl. Phys.*, 21 (1950) 59
[3] S. Ergun, *J. Appl. Cryst.*, 1 (1968) 19.

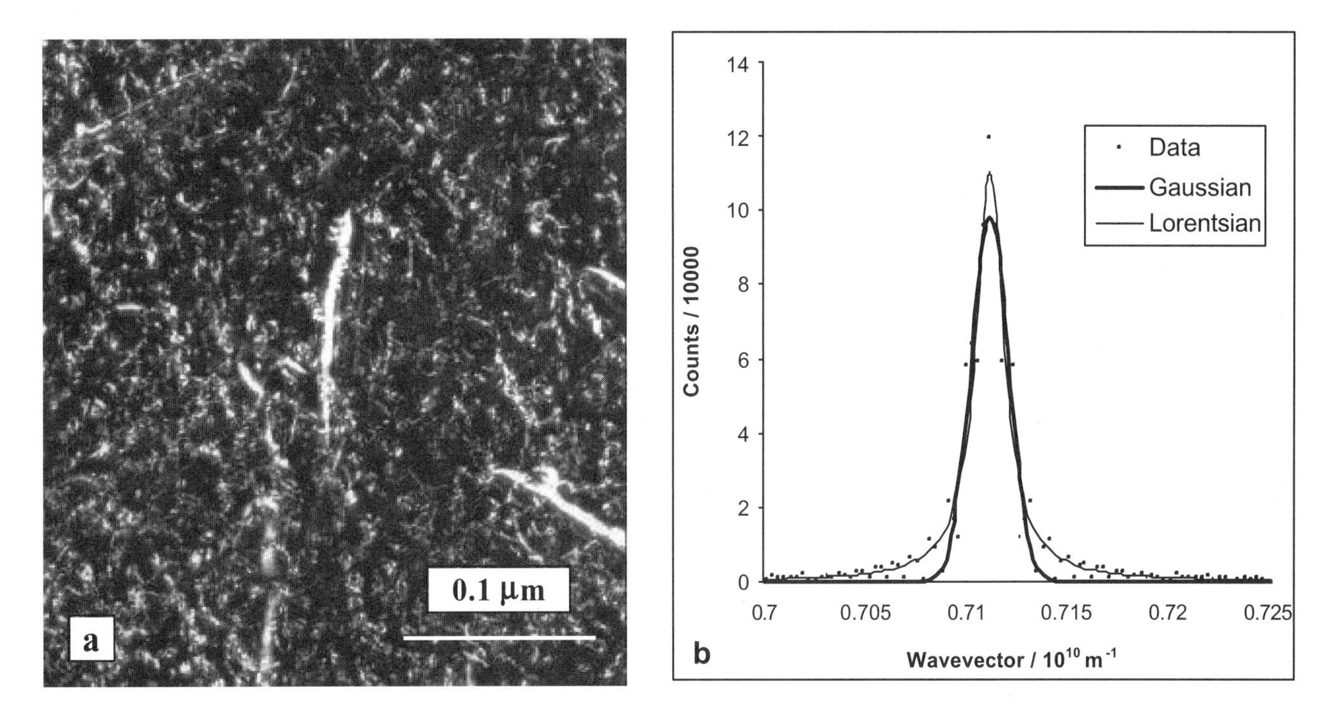

FIG. 1. (a) Micrograph showing dislocation loops in neutron irradiated single-crystal Zr; and (b) corresponding deconvoluted prism plane diffraction line profile.

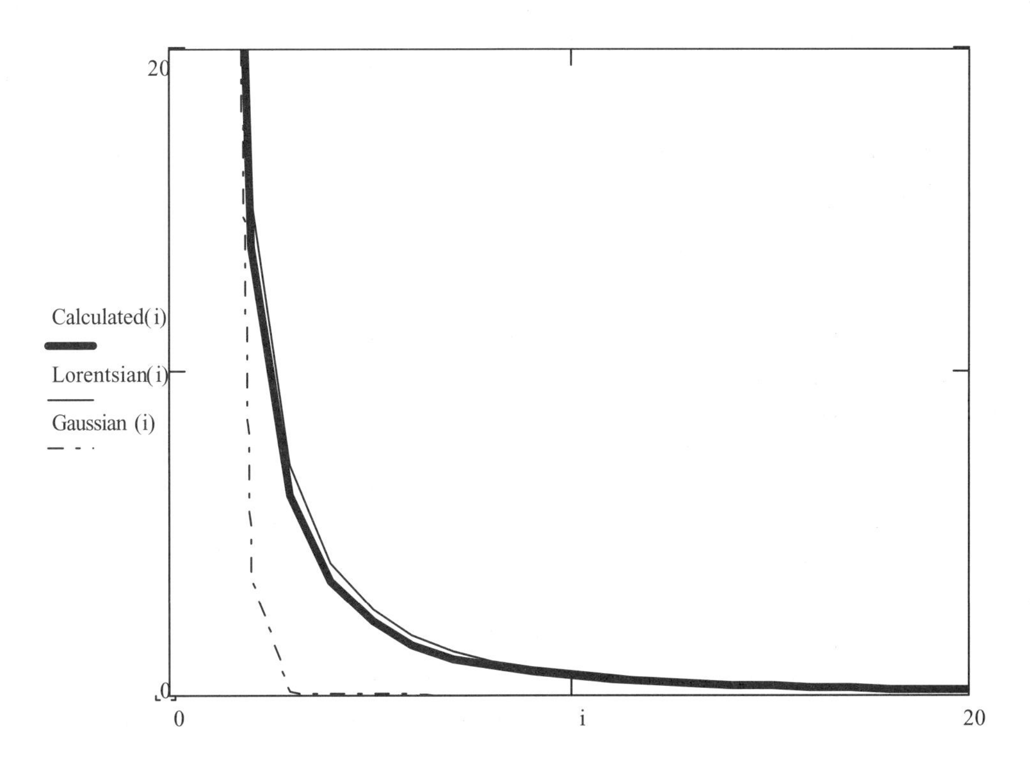

FIG. 2. Calculated diffraction line profile, based on the theoretical strain field around a single edge dislocation, compared with Gaussian and Lorentsian approximations.

Microsc. Microanal. 8 (Suppl. 2), 2002
DOI 10.1017.S1431927602101182

Neutron Diffraction as a Probe of Microstructure: Surveying the Forest Before Examining the Trees

John H. Root*

*National Research Council of Canada, Chalk River Laboratories, Chalk River, Canada K0J 1J0

With the advancement of microscopy technologies in recent years, and the reasonable availability of instruments, researchers can acquire an impressive amount of information at the micro and nano scales. Neutron diffraction is not suitable for imaging small details. However, it is relatively straightforward with neutron diffraction to obtain the volume fractions of phases [1], crystallite orientation distribution functions [2] and the evolution of microstructures under load [3,4] or at elevated temperatures, to complement the understanding of phenomena deduced from microscopy techniques. For specimens that require complex environments to exhibit their relevant responses (e.g. some combination of pressure, temperature, hydration or chemical stress), neutron diffraction may be the only practical method to obtain any microstructure information [5]. This paper will review two examples where neutron diffraction measurements suggested novel ideas about microstructure evolution, and pointed the way for subsequent investigation by microscopy methods.

In Ref. 6, a metal-matrix composite of titanium carbide (TiC) particles in a titanium matrix was made by sintering mixed-powder compacted specimens in vacuum at temperatures between 1000°C and 1500°C. By reaction with the metal matrix, TiC was converted to a carbon-depleted TiC_x, with a lattice parameter sufficiently reduced that its diffraction peaks could be resolved from those of TiC (Fig. 1a). The integrated intensities of these peaks were plotted versus time (Fig. 1b) to monitor the reaction between the carbide particles and the matrix during sintering. The observation of two distinct diffraction peaks during sintering (suggesting a single value of x) , rather than the gradual broadening of a single peak (continuous range of stoichiometry $1.0 > x > 0.5$), led the authors to surmise that the microstructure of the carbide particles might exhibit a distinctive outer shell, depleted in carbon content. Low-voltage, high-resolution field emission gun scanning electron microscopy (FEGSEM) provided sufficient contrast to confirm the sharp boundary between the inner and outer regions of carbide particles.

In Ref. 7, neutron diffraction was applied to study the kinetics of a phase transformation in zirconium hydrides precipitated by cooling a zirconium alloy specimen (containing about 200 mg deuterium per kg of zirconium) from 450°C to a holding temperature of 17°C. Over 60 hours, the (111) diffraction peak of δ-phase hydride, $ZrH_{1.6}$, decreased while the (111) peak from the γ-phase hydride, ZrH, increased as shown in Fig. 2. The authors noted that a transformation from the δ to the γ phase would require diffusion of hydrogen out of the prior δ-phase precipitates. They proposed that hydride precipitates might exhibit a core of untransformed δ-phase and that γ-phase hydrides would be found at the interface between the hydride precipitate and the zirconium matrix. The γ- and δ-hydride phases can be distinguished by electron energy loss spectroscopy (EELS) in a transmission electron microscope [8]. A TEM micrograph, Fig. 3 [9], was obtained from a specimen of material from the neutron diffraction study, and the EELS analysis identified γ-phase hydrides that appeared to grow like dendrites out of the boundary of a prior δ-phase precipitate.

References

[1] H. Abuluwefa et al., Metall. and Mater. Trans. 27B (1996) 993.
[2] H.J. Bunge, Textures and Microstructures 10 (1989) 265.
[3] M.A. Gharghouri et al., Phil. Mag. A79 (1999) 1671.
[4] P. Dawson et al., Mater. Sci. Engng. A 313 (2001) 123.
[5] I.P. Swainson and E.M. Shulson, Cement and Concrete Research 31 (2001), 1821.
[6] P. Wanjara et al., Acta mater. 48 (2000) 1443.
[7] M. Small et al., J. Nucl. Mater. 256 (1998), 102.
[8] O.T. Woo and G.J.C. Carpenter, Microsc. Microanal. Microstruct. 3 (1992) 35.
[9] The TEM micrograph in Fig. 3 was obtained by O.T. Woo, Atomic Energy of Canada Limited, Chalk River Laboratories, Canada.

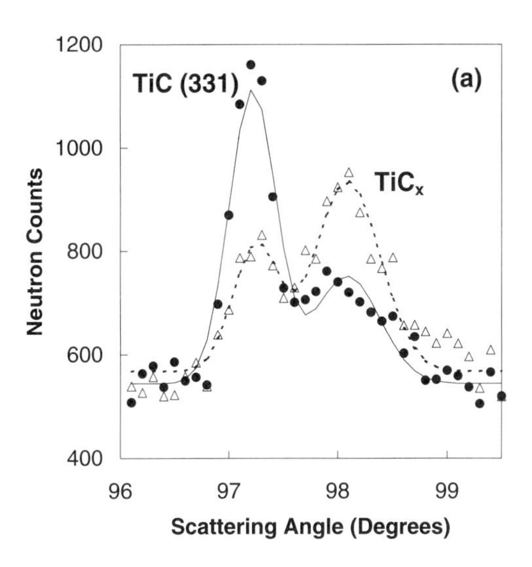

FIG. 1a. Distinct neutron diffraction peaks from two carbide phases, after 1h (+) and after 16 h of sintering (?). Lines indicate fitted gaussian functions.

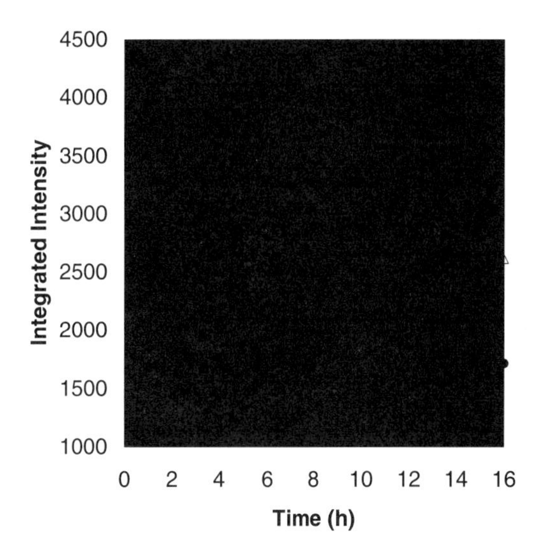

FIG. 1b. Erosion of TiC (+) with time and complementary growth of the carbon-deficient phase TiC_x (?).

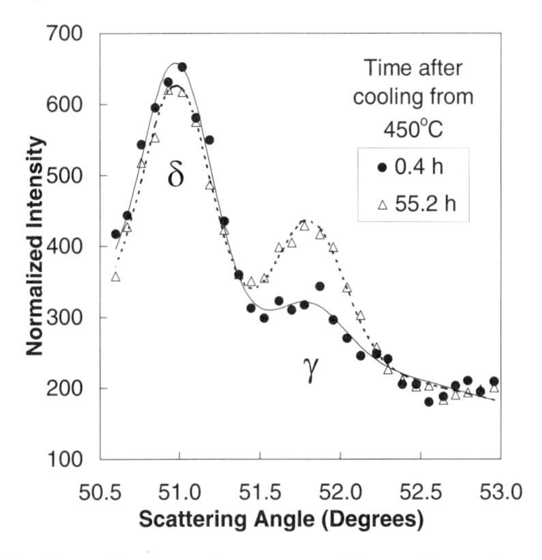

FIG. 2. Increasing quantity of γ-phase hydride over time, seen as growth of the (111) diffraction peak. The (111) δ-phase peak decreases slightly. T = 17°C.

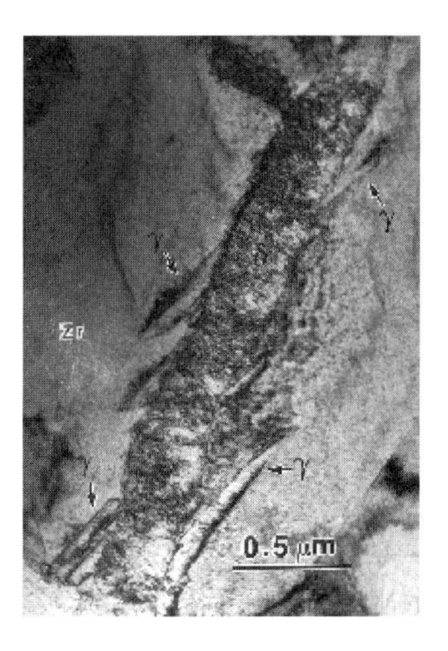

FIG. 3. TEM image of a δ-phase hydride precipitate with γ-phase dendrites protruding from the outer boundary [9].

Microsc. Microanal. 8 (Suppl. 2), 2002
DOI 10.1017.S1431927602101231

Failure Analyses of Three 6061-T6 Aluminum alloy Turbo-Expander Wheels Exposed to Natural Gas Environments,

Behzad Bavarian, PhD
Dept. of Manufacturing System Engineering and Management
California State University, Northridge 91330-8332

Failure analyses were performed on three fractured 6061-T6 Aluminum alloy turbo-expander wheels (#A, #B, & #C) that were exposed to mercury-bearing natural gas environment, using Scanning Electron Microscopy (SEM), Energy Dispersive Analysis(EDAX), optical microscopy, and mechanical tests to determine the cause of their premature failures.

This investigation showed that causes of these wheel failures were corrosion attacks by liquid metal embrittlement (LME) of the 6061 Al-alloy by mercury. Mercury corroded and embrittled the 6061-T6 Al-alloy resulting in several crack initiations at the edges of several blades, and inside expander wheel holes. LME also caused a loss of wheel toughness, which led to corrosion fatigue and stress corrosion cracking led to their premature failure of these wheels.

SEM fractographs of these wheels showed several fatigue crack initiation sites located at severely corroded regions on the edges of their blades, and EDAX analyses showed a presence of large amount of mercury, chloride and sulfide at these locations. The fracture surfaces were relatively flat, indicating the absence of an appreciable amount of gross plastic deformation, unusual for the ductile 6061-T6 Al-alloy. Clearly lack of toughness was observed for both wheels, brittle cleavage cracking modes, and the presence of a few droplets of mercury on the fracture surfaces strongly suggested that liquid mercury attacked the aluminum and initiated fatigue cracks; LME caused severe toughness loss of the alloy. Presence of some corrosion products and mercury droplets covering the wheel fracture surfaces were observed. These observations indicate that a few cracks were initiated by liquid metal attack, and then propagated until the final stage of fracture. The cleavage like faceted morphology of fatigue striations during the fatigue propagation stages indicated a low stress intensity level during these stages. These observations indicate that the wheel was designed properly for this application; however, severe mercury attacks shortened their lives. The last stage of failure for both wheels showed overloaded failure modes, demonstrating that these wheels possess proper toughness, but mercury embrittlement resulted in their toughness loss.

SEM/EDAX results showed some severe corrosion attacks on both wheels by mercury embrittlement. In addition, large amounts of sulfur and chloride were also detected in the corrosion products present on the fracture surfaces, which could also cause serious corrosion problems for aluminum alloys.

In conclusion, causes of these wheels' failure were corrosion attacks due to liquid mercury embrittlement of 6061 Al-alloy, which resulted in delayed corrosion assisted failures of these wheels. The source of mercury was possibly the flowing gas. The stress level was relatively low which indicates the design of the wheel was appropriate. The presence of some fatigue crack propagation prior to the final fracture of both wheels verified the sound design of these wheels.

Figure 1: Optical images of the fractured wheel showing a few of secondary cracks that were fractured to analyze their failure mode, Fracture surfaces were covered with Hg and corrosion attacks.

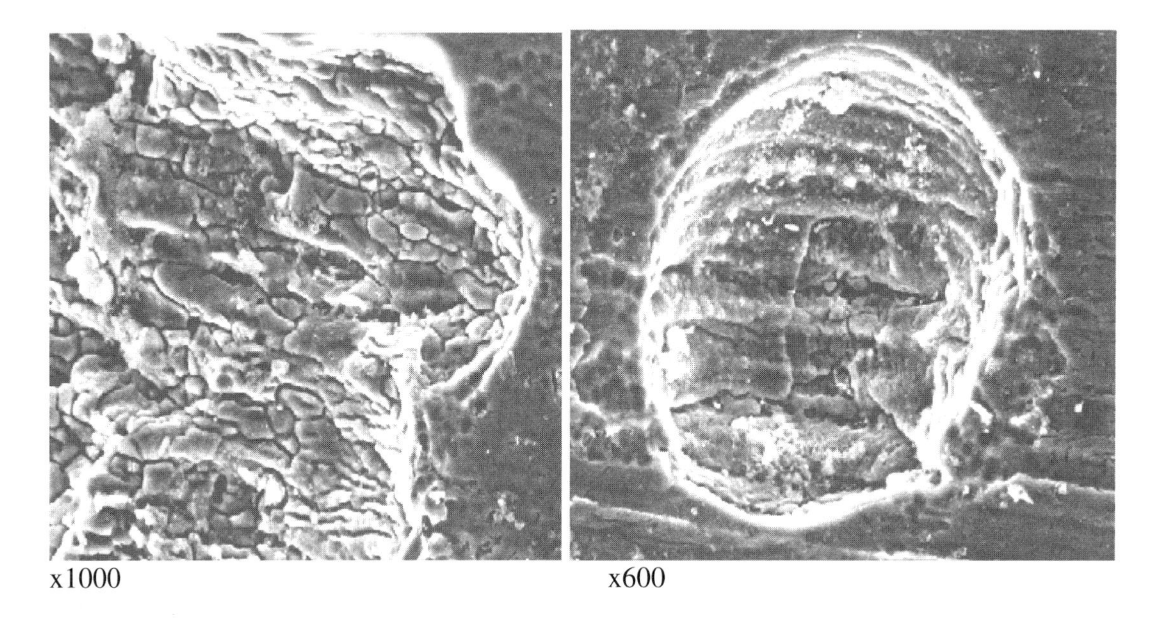

x1000 x600

Figure 2: SEM images of the fractured wheel, showing corrosion attacks inside a pit caused by Hg-attack.

DOI 10.1017.S1431927602101243

Failure Analysis of Small Gap Brazing of a Stainless Steel Heat Exchanger

Michael Neff, *Michael Neff Associates, Culver City, California*

ABSTRACT
A heat exchanger used to cool hot hydrogen in a high purity gas train failed in service. The heat exchanger failed in a brazed joint near the inlet (hot side) portion between the body and the tube sheet. The brazing joint was cross sectioned to open the fracture and determine the brazing joint geometry. It was found that the thickness of the brazing joint around the circumference of the joint varied from 0.001 minimum to 0.004" maximum. The use of a nickel-boron brazing alloy in this assembly corrected the problem.
Keywords: Brazing, Nicrobraz, Heat Exchanger

SUMMARY

A 316L Stainless Steel heat exchanger fabricated by furnace brazing failed at the braze joint in a high purity gas delivery application. Nicrobraz 51, a high chromium nickel alloy with phosphorus modification, was originally chosen for this heat exchanger because of the high corrosion resistance provided by the high chromium. The heat exchanger had been used successfully in previous applications with lower thermal stresses, but in the subject application in which it was used to cool hydrogen the thermal stresses were more severe. Hydrogen at 375°C entered the heat exchanger through openings in the tubesheet (Figures 1 and 2). Water at 20ºC entered the tube bundle cavity through an inlet just adjacent to the failed tubesheet. The high differential temperature between the hydrogen and the water coolant caused high radial stresses leading to fracture of the brazing alloy.

The recommended joint clearance for the Nicrobraz 51 alloy is very critical. While the flow characteristics of the alloy are good, the brazing joint failed in a brittle manner due to the high stresses introduced by the thermal differential. The brittle fracture was observed emanating from a groove in the center portion of the brazing joint, Figure 5.

The crack failed the entire cross section of the brazed joint in the upper quadrants of the circle, between 9 o'clock to 3 o'clock, (Figure 3). Toward the bottom quadrants, the crack terminated, so that the total separation of the gap was 0.004" at 12 o'clock and only 0.001" at 6 o'clock. Metallographic examination of the fracture showed that the brazing alloy was thicker near the groove, and this was the area where the brittle fracture appeared to originate, Figure 7.

In order to further evaluate the strength of the brazing alloy, a series of tensile tests were conducted in order to correlate the gap thickness with the strength of the brazed joint. It was found that the there was a sharp decrease in strength as the gap thickness increased for this alloy. The sensitivity of the brazing alloy strength to gap thickness was not as critical for standard boron modified brazing alloys. As a result of the testing, the brazing alloy was changed to a BNi-3 (3.1%B) which is much more forgiving with respect to embrittlement in thicker gaps, and the problem did not recur.

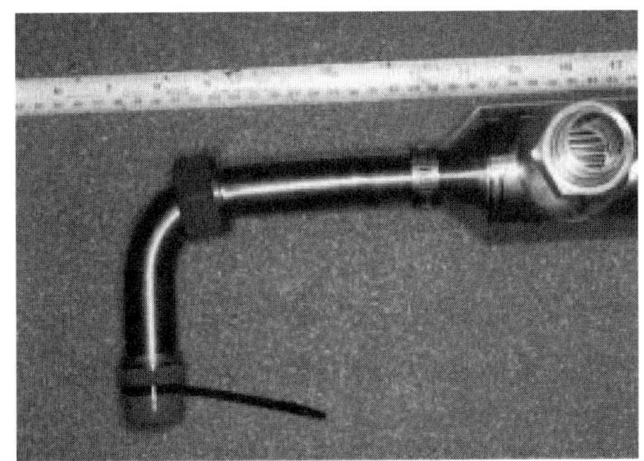

Figure 1. An optical photograph of the inlet side of the tube heat exchanger.

Figure 2. An optical photograph of the hydrogen inlet tubesheet braze assembly.

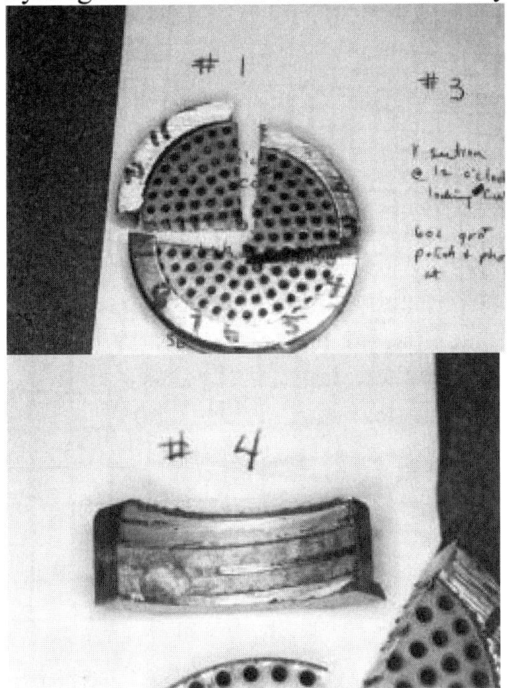

Figures 3 and 4. Optical photographs showing how the sections were removed for SEM examination and cross section. Also showing the brazing joint fracture.

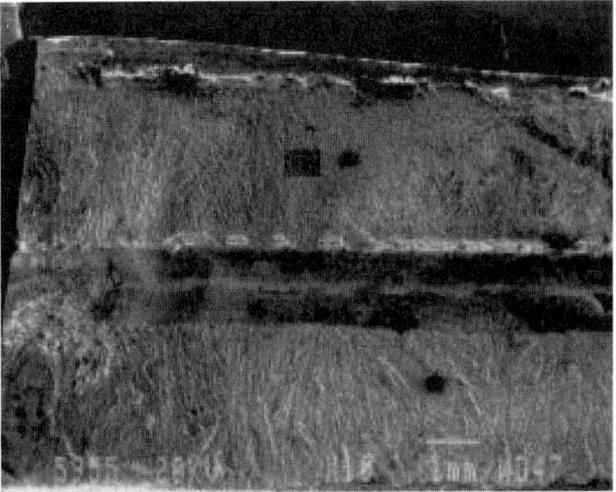

Figure 5. An SEM photograph showing the fracture initiating along a void in the center of the brazing joint.

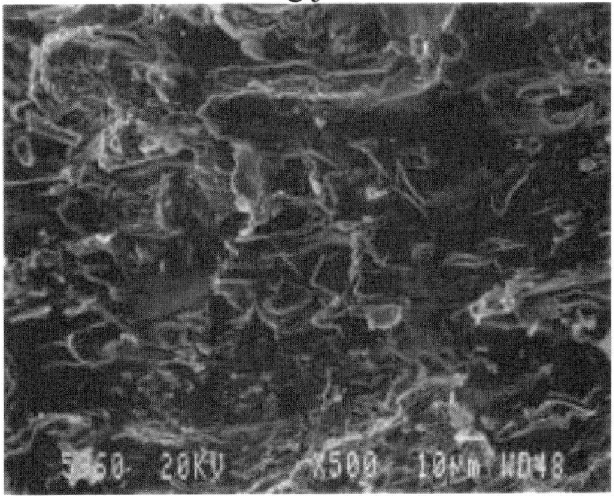

Figure 6. An SEM photograph showing the brittle fracture surface morphology.

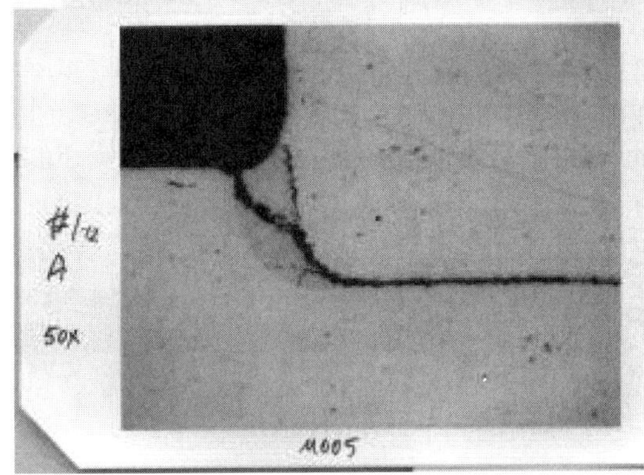

Figure 7. A metallographic section through the failed brazing joint showing the crack initiating in a thick portion of the braze joint next to the void.

Microsc. Microanal. 8 (Suppl. 2), 2002
DOI 10.1017.S1431927602101255

Grain Size Measurements – Variables to Consider

Janice Klansky

Buehler Ltd., 41 Waukegan Rd. Lake Bluff, IL 60044

In order to better understand the influence of the welding process on the mechanical properties of a finished component, it is useful to examine a metallographic cross-section. A typical weld will display three distinct regions: the solidified weld metal, grain growth and recrystallization due to the temperature gradient introduced by the welding process. Each region can have a dramatically different grain size and anisotropy. The goal of this study was to examine some of the variables surrounding automatic image analysis grain size measurements and determine some recommended practices when working with specimens with irregular microstructures.

Standardized approaches to measurement techniques have evolved with the technology available. For example, comparison charts and tables had determined the magnification requirements whereas the newer automated imaging standards reference the number of features per field of view or a set calibration factor. In addition, automation introduces new questions such as: just how well detected do all of the boundaries have to be to get an accurate answer and what are the consequences of having too few or too many grains in the field of view.

Two types of specimens were examined in order to understand the measurement process. The first was a plain carbon steel specimen with an equiaxed structure (Figure 1). Images from this specimen were used as a standard since most grain size methods assume an anisotropy value less than three. The second was a cross-section of a weld (Figure 2). This provided a variety of grain sizes and varying degrees of anisotropy.

Images of the equiaxed structure were captured at a resolution of 3072 x 3840 pixels. The grain size was measured across the entire field of view based on a count of the grains (planimetric method) and x and y intercepts [1,2]. Next the incomplete grains on the image boundaries were removed and grain size was determined based on the individual grain areas [2]. This procedure was repeated on subsections of the image to determine the influence of the relative number of grains on the boundaries compared to the total number of grains present in the image. Figure 3 demonstrates that most of the methods show an increase in the grain size number when fewer than fifty grains are present. The exception was the planimetric approach using a rectangle, which displayed no noticeable change in the grain size number.

Two of the measurement methods were then employed on a weld cross-section. In addition to calculating grain size, the x and y centroids of the grains were tracked to determine the relative positions of the grains. Figure 4 displays the grain size results relative to position using a constant magnification across the weld zone. Where the results from the two methods diverge there are either less than 50 grains present in the field of view or more than 300 grains. The difference in the average grain size is further amplified by the calculation method for average grain size. When individual areas were used to determine grain size the average was weighted by the number of grains present, not the actual area represented by the grains. The planimetric approach is more representative of an area weighted average.

[1] "Standard Test Methods for Determining Average Grain Size," E 112-96, *Annual Book of ASTM Standards*, ASTM, 2000.

[2] "Standard Test Methods for Determining Average Grain Size Using Semiautomatic and Automatic Image Analysis," E 1382-97, *Annual Book of ASTM Standards*, ASTM, 2000.

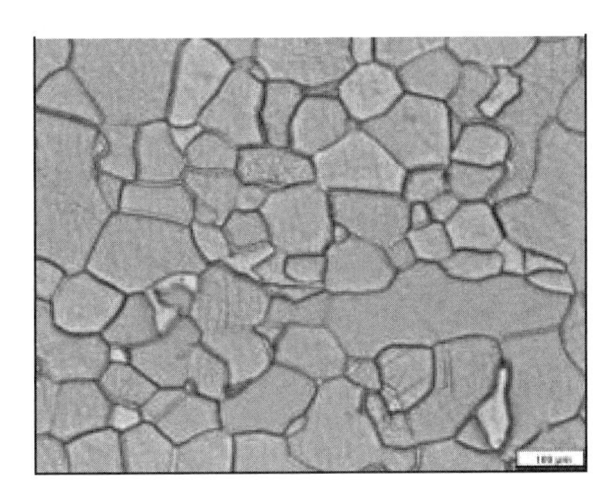

FIG. 1. A section of the equiaxed grain size image.

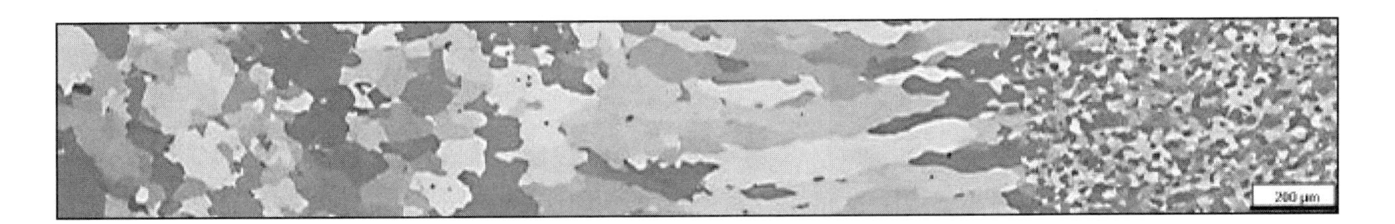

FIG. 2. A metallographic cross-section of a weld specimen.

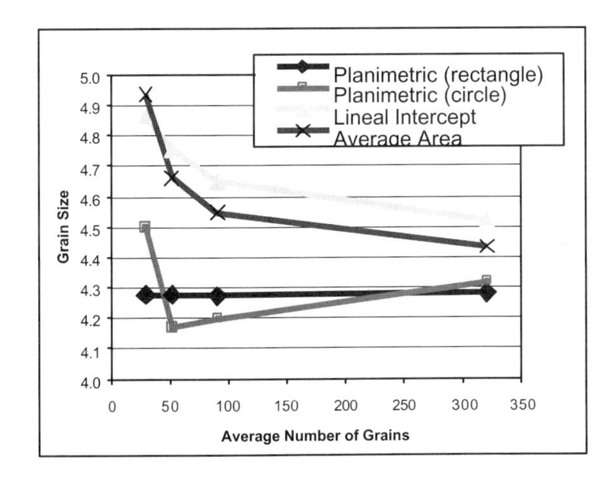

FIG. 3. A comparison of grain size measurement methods.

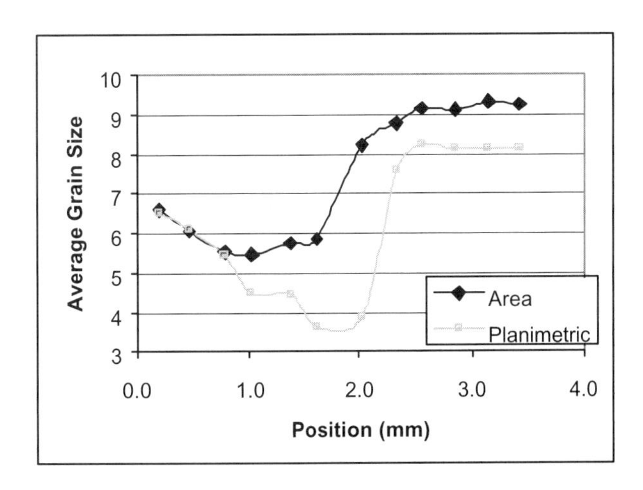

FIG. 4. A comparison of grain size measurement methods across a weld cross-section.

Microsc. Microanal. 8 (Suppl. 2), 2002
DOI 10.1017.S1431927602101267

Automatic Phase Segmentation of Spectrum Images

J. J. Friel* and R. Batcheler*
Princeton Gamma-Tech, C/N 863, Princeton, NJ 08542

Spectrum Images or Position-Tagged Files—depending on what technology was used to collect them—can be large datasets. A strategy to mine these datasets should consider, whether or not the search should take advantage of *a priori* knowledge. Statistical methods such as principal component analysis are called non-judgment methods and require no knowledge of the specimen composition or phase assemblage [1]. However, the method described here allows the user to specify which elements should be considered and is thus called a judgment method [2]. The segmentation of the dataset is based on the elemental maps chosen for display plus a sensitivity parameter. The user's knowledge is therefore incorporated into the segmentation strategy.

The algorithm is called recursive pixel allocation (RPA) and it consists of five passes through the dataset. These steps are: 1) Refine the average intensities and make preliminary pixel assignments. 2) Identify the phase boundaries, and reallocate indeterminate pixels. 3) Remove small phases. 4) Reduce the number of phases by combining those that are within sensitivity limits. 5) Recalculate average compositions. This process is not iterated, but the procedure can be rerun at different sensitivity levels or with different maps displayed to adjust the strategy. It should be noted that the RPA algorithm makes use of size information and elimintes small phases. If small phases are of interest, it is simple to increase the magnification to better define them. Although it is possible to generate false phases at high sensitivities, these are easily removed by lowering the sensitivity once their spectra have been examined. The output consists of a pseudocolored image of phases and a table showing intensities and area fractions. Spectra can be extracted by phase, and the compositions can be quantified based on the sum of all spectra from a phase. Position-tagged files with sufficient statistics can be collected in a minute or two at 128 x 100 resolution (~2mb), and very high quality datasets can be collected in about 10 mins. at 256x200 pixel resolution(<20 mb).

Fig. 1 shows an image and maps of an alloy consisting primarily of Fe, Mg, and Si, plus several minor constituents. The phases in this material are easily segmented, and the output of the AutoPhase routine is shown as an image in Fig 1. Reporting the output as an image enhances the ability of the mind to perceive spatial associations and contrasts, and tabular data are also available.

Fig. 2 shows a cast iron with graphite nodules. This material has only two "true" phases; namely, graphite and iron. However, in this sample, the silcon distribution was investigated by means of a tint etch, but the results were inconclusive. By using X rays and setting the AutoPhase sensitivity to the maximum, it was possible to distinguish two other distinct regions. One of these was high in silicon, and the other high in iron. In fact, the high-Fe regions are likely regions of less (thinner) etching, but the high-Si regions near the nodules appear to be real. Quantitative analysis in the bulk of the iron showed ~2% Si; whereas, 3.9 wt.% Si was found in the regions near the nodules.

The method of phase segmentation by recursive pixel allocation provides a measure of control of the automatic phasing process. It takes <1sec. to run and can sort the microstructure into phases based on the user's knowledge. A non-judgment method like Principal Component Analysis does not require this knowledge, but also cannot take advantage of it. Of course, identification of the elements present can also be automated. Complete manual control can be exercised by selecting representative regions and having the computer assign all pixels to one of the selected phases.

References

[1] P. G. Kotula, Spectral Imaging: Getting the Most from All that Data, this volume.
[2] F. Mosteller and J. W. Tukey, Data Analysis and Regression, Addison-Wesley, Reading, 1977

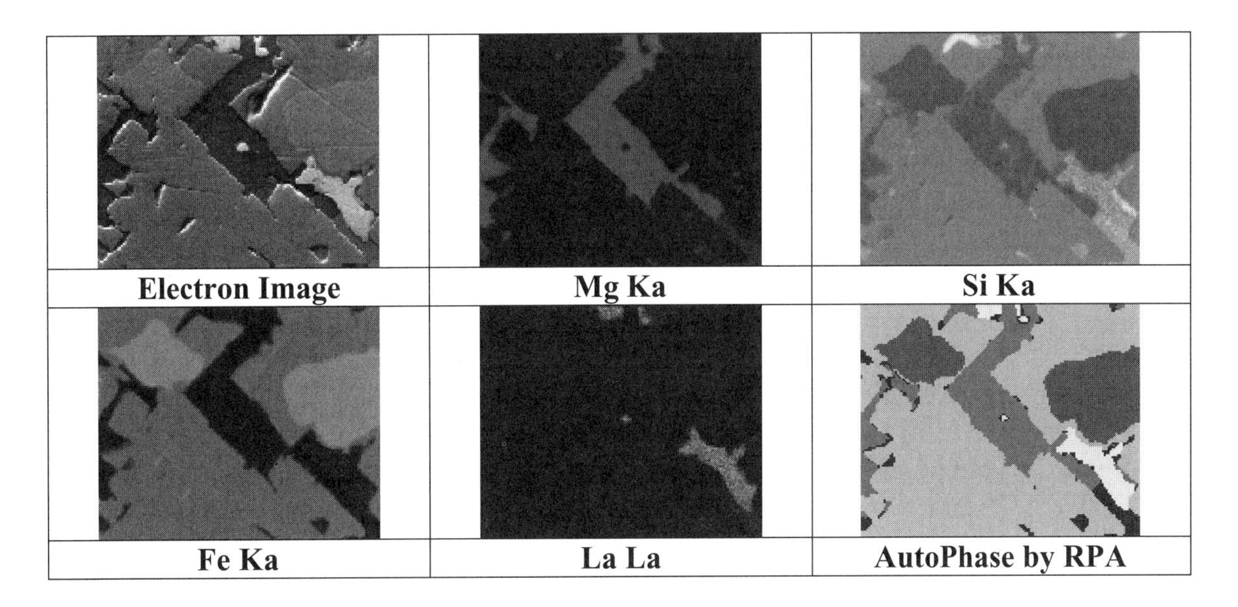

FIG. 1. Maps of an alloy of Fe, Mg, Si, and REE, showing a straightforward example of autophasing by recursive pixel allocation.

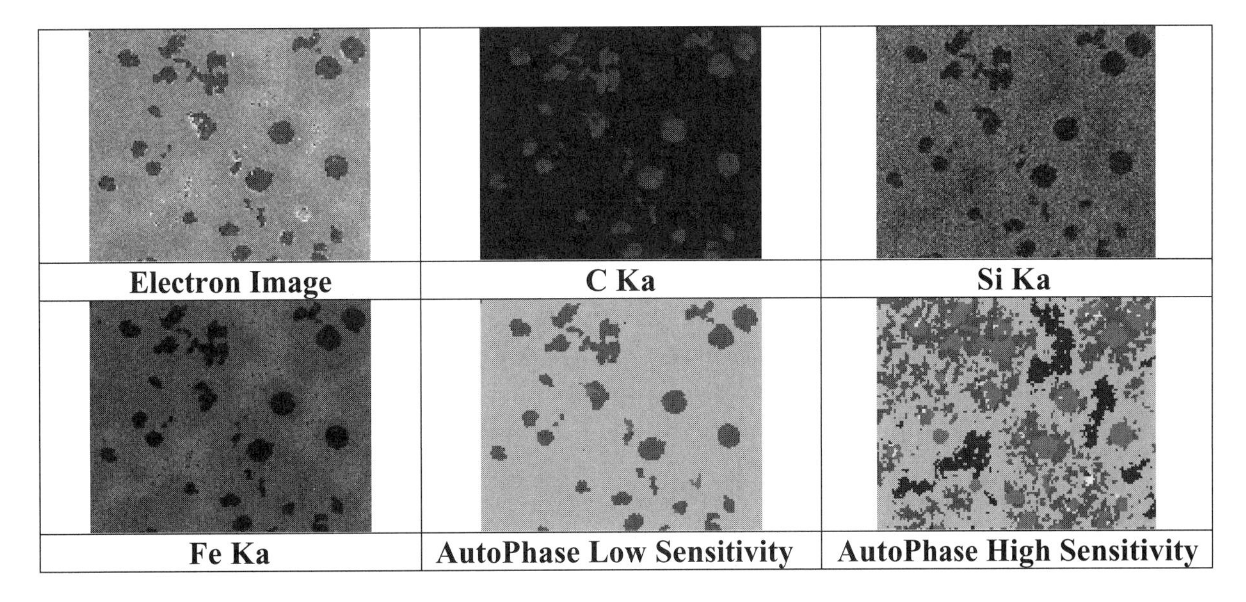

FIG. 2. Maps of a cast iron with graphite nodules. Autophase at low sensitivity shows only graphite and iron. At high sensitivity, it reveals a silicon-rich phase and an iron-rich one.

Microsc. Microanal. 8 (Suppl. 2), 2002
DOI 10.1017.S1431927602101279

A Transmission Electron Microscopy Study of Dual Phase High Strength Steels

I.A. Yakubtsov*, J.D. Boyd*, D. Emadi**

* **Department of Mechanical Engineering, Queen's University, Kingston, ON K7L 3N6, CANADA**
** **Materials Technology Laboratory, CANMET, Ottawa, ON K1A 0G1, CANADA**

The development of Grade 690 (σ_y = 690 MPa) linepipe steel having the best combination of strength, toughness and weldability requires a fine-scale bainitic microstructure. The bainitic ferrite laths nucleate intragranularly ("acicular ferrite" –AF) and grow with random orientations, and the remaining carbon-enriched austenite forms dispersed islands of martensite/austenite (M/A). The dual phase AF+M/A microstructure is achieved in low carbon microalloyed steels by a combination of thermomechanical processing and accelerated cooling. This paper presents the results of a study of AF+M/A microstructures by transmission electron microscopy.

For the current Grade 550 steel, an accelerated cooling rate (CR) of about 50 °C/s is required to obtain 100 % AF+M/A [1]. Fig.1b shows a typical microstructure comprising randomly oriented, high dislocation density bainitic ferrite laths and ~0.5 µm equiaxed M/A islands. The M/A islands are clearly distinguished by their light contrast in SEM-SE (Fig. 1a) and dark contrast in TEM (Fig. 1b). Carbon enrichment of the M/A relative to the bainitic ferrite matrix has been confirmed by SEM-EDS [2].

The current research investigates new steel compositions which produce the dual phase AF+M/A microstructure at CR $\leq$ 25 °C/s. The microstructures of 2 new steel compositions are shown in Fig. 1c-1f. The compositions of both steels are identical, except for carbon. Steel NC2 has 0.04wt% C and steel NC3 has 0.07wt% C. Both steels were produced as 15 mm thick plate, by the same processing schedule, with a CR of 20 °C/s. The microstructures for both steels are dual phase AF+M/A, with 13.7% M/A in NC2 and 17.9%M/A in NC3.

The M/A islands are completely martensite, with no evidence of retained austenite (Figs.1d, 1f). The detailed structure of the martensite is seen in Fig.1d to be a combination of blocky martensite and twinned martensite (Fig.1d@T). The martensite forms by the strain which develops in the C-enriched austenite adjacent to the growing bainitic ferrite laths [3,4]. The blocky martensite is due to accommodation strain which results in the formation of areas having different orientations. The twinned martensite is due to locally increased C concentration and/or a complex shear stress condition.

References.

[1] I.A. Yakubtsov et al., 43rd MWSP Conf. Proc., ISS, XXXIX (2001) 531.
[2] I.A. Yakubtsov and J.D. Boyd, Mater. Sci.& Technol., 17 (N3) (2001) 296
[3] E. Swallow and H.K.D.H. Bhadeshia, Mat. Sci. & Technol., 12 (1996) 121.
[4] G.B. Olson and M. Cohen, J. Less Common Met., 28 (1972) 107.

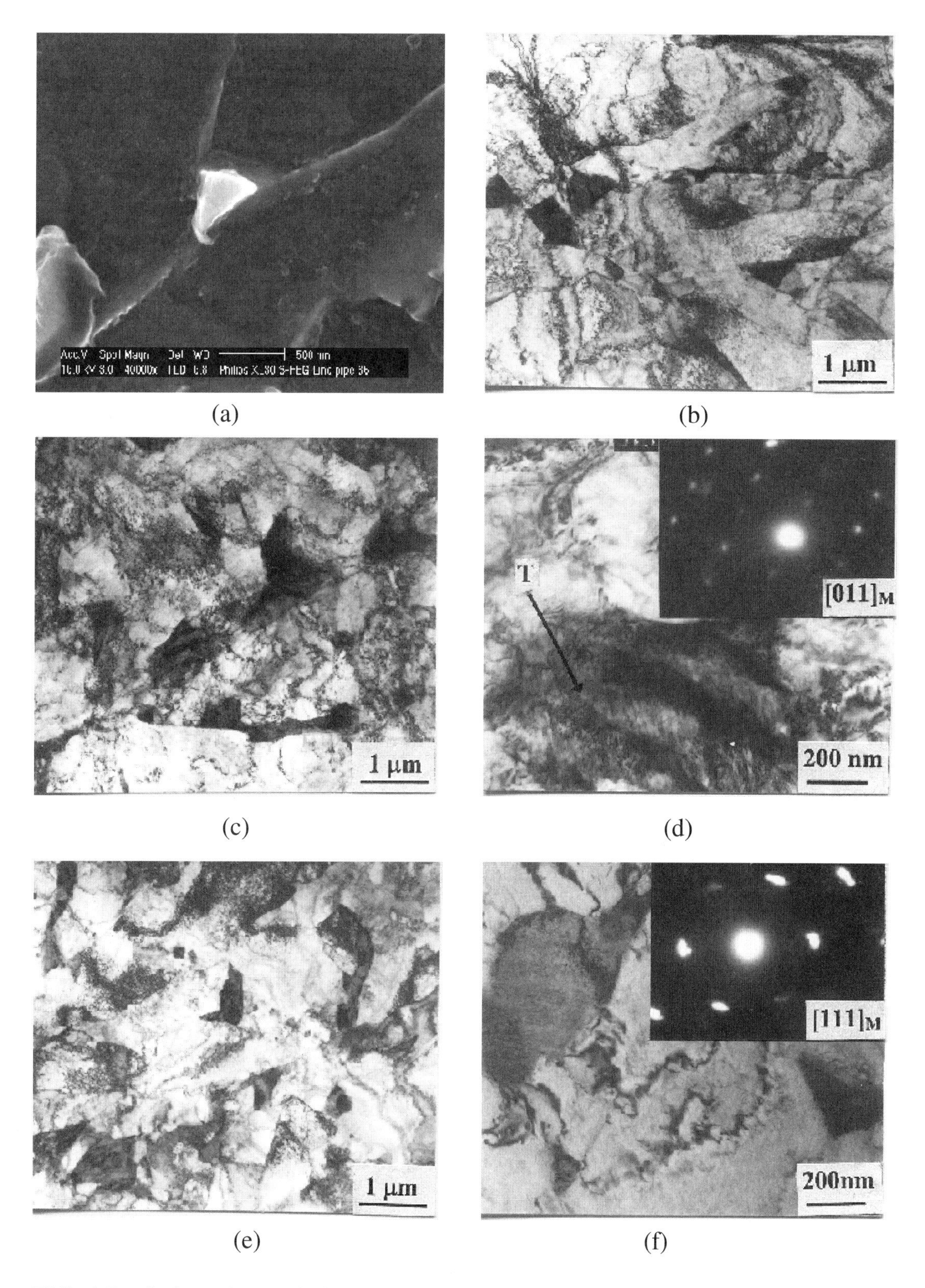

FIG. 1 Dual phasc AF+M/A in thermomechanically processed +accelerated cooled steels: (a, b) Grade 550 steel composition, CR = 50 °C/s; (c, d) NC2 steel, CR = 20 °C/s; (e, f) NC3 steel, CR = 20 °C/s.

Microsc. Microanal. 8 (Suppl. 2), 2002
DOI 10.1017.S1431927602101280

Fractographic Evaluation of Medium Carbon Steels with Low Hot Ductility.

O. Dremailova, D. Emadi, E. Essadiqi and J. R. Brown

Materials Technology Laboratory, CANMET, Ottawa, Ontario, K1A 0G1

Many steel companies incorporate electric arc furnace methodsin steel production, which involves the recycling of steel scrap. Of concern is the effect of increasing level of residual elements such as Cu and Sn on the properties of the steel at various stages of production. [1,2] This work examines the effect of these key residuals on the cracking susceptibility during continuous casting by conducting simulation tests using the Gleeble 2000.

Medium carbon (0.2%C) steels with various levels of Sn and constant Cu and Ni levels were cast in a vacuum induction furnace then reheated and rolled at CANMET. Samples were machined from these steels for subsequent hot ductility tests in the Gleeble. Testing consisted of performing insitu melting and solidification in the Gleeble followed by controlled cooling to varioustemperatures and subsequent tensile testing to fracture to generate hot ductility curves. Optical metallography, scanning electron microscopy with energy dispersive analysis and electron probe microanalysis were carried out to provide microstructural data to assist in the interpretation of the observed variations in ductility. A limited number of tests were interrupted at maximum load and quenched for subsequent Auger scanning electron microscopy. Figure 1 presents the hot ductility curves for each of the steels revealing minimum ductility at 750°C. The intergranular morphology observed for the plain carbon steel at this temperature is presented in Figure 2 with higher magnification image given in Figure 3. All steels fractured between 700 and 800°C were examined in the SEM and exhibited the same failure morphology, which was typically intergranular with a local ductile fracture appearance in the form of microvoid coalescence on the intergranular facets. SEM with energy dispersive X-ray analysis of polished sections close to the fracture revealed the presence of manganese sulfide particles on the prior austenite grain boundaries. These particles were also observed in the microvoids on the fracture surface as shown in Figure 4. The base steel samples quenched from 700 to 800°C contained proeutectoid ferrite, as presented in Figure 5.

The observed results for the plain carbon steel can be explained in terms of the presence of grain boundary ferrite and manganese sulfide precipitates. The deteriorationin ductility with increasing Sn was found to be associated with Sn segregation to grain boundaries in agreement with previously reported work. [3]

References

[1] S. Yue, J.J. Jonas and B. Mintz, "Relationship between hot ductility and cracking duringthe continuous casting of steel", *13th PTD Conference Proceedings*, Vol.13, Iron and Steel Society, 1995, pp. 45-52.
[2] C. Nagasaki and J. Kihara, *ISIJ International*, Vol. 37, No. 5, 1997, pp. 523-530.
[3] H. Matsuoka et al., "Influence of Cu and Snin hot ductility of steels with various carbon content", *ISIJ International*, Vol. 37, No. 3, 1997, pp. 255-262.
[4] This work was part of American Iron and Steel Institute Project 9705 jointly funded by US Department of Energy and AISI participating steelcompanies.

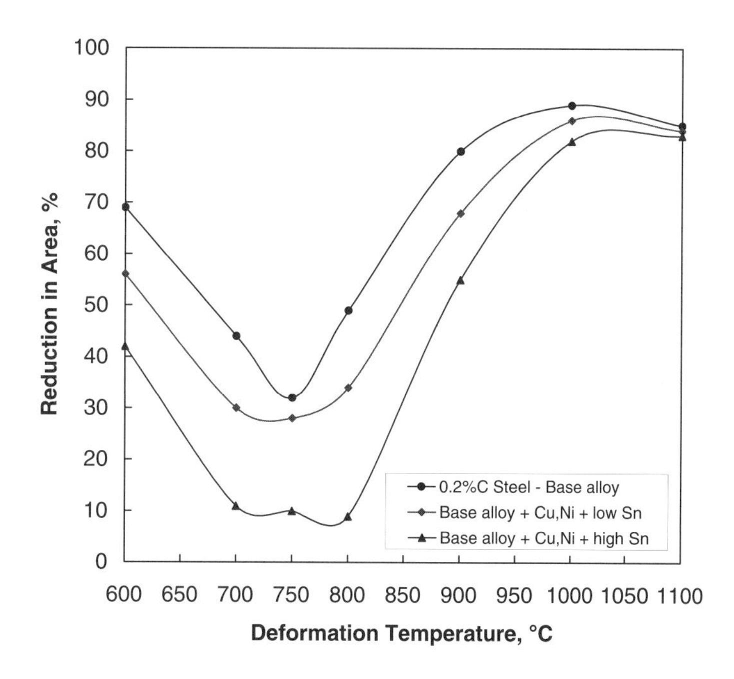

FIG. 1. Hot ductility curves of medium carbon steels.

FIG. 2. Fracture surface of plain carbon steel at minimum ductility temperature.

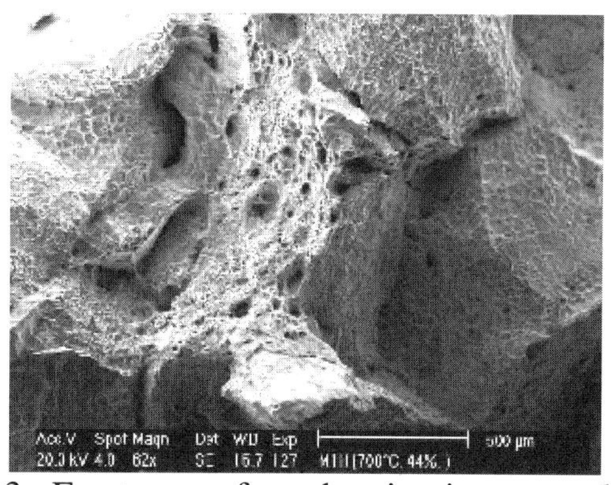

FIG. 3. Fracture surface showing intergranular facets with microvoid coalescence.

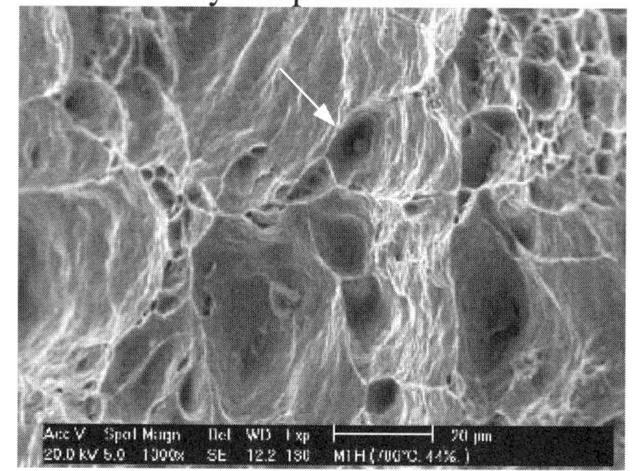

FIG. 4. SEM image of plain carbon steel showing a MnS particle in a void.

FIG. 5. Quenched sample of plain carbon steel revealing proeutectoid ferrite along the grain boundaries at minimum ductility.

Microsc. Microanal. 8 (Suppl. 2), 2002
DOI 10.1017.S1431927602101292

Recovery and Recrystallization of Ferrite in Warm Forging of a Medium Carbon Steel

P. Zhao and J.D. Boyd

Dept. of Mech. Eng., Queen's University, Kingston, ON K7L 3N6, Canada

Warm forging produces high dimensional accuracy and enhanced mechanical properties [1-3]. When forging is carried out at temperatures below Ar_1, the transformed ferrite + pearlite is deformed, and the ferrite can be refined by recovery and recrystallization. The current research investigates microstructural evolution in a Nb-microalloyed medium-C steel, during warm forging. 35-mm diam. bars were forged to 12-mm thick plates, following the schedule given in Fig. 1. Samples were quenched immediately following the first 2 deformations ('1' and '2'), and the final plate was air cooled ('3').

Sample 1 exhibits the elongated austenite grain structure produced by the first deformation (800°C). Sample 2 shows some transformed ferrite and pearlite. Grain boundary ferrite and a small amount of intragranular ferrite comprise 21% of the microstructure. The deformed ferrite recovers quickly and forms subgrains having a mean linear intercept of 0.46 ± 0.14 μm (Fig. 2). Some recrystallized ferrite grains are also observed in sample 2 (Fig. 3). The dimensions of these recrystallized grains are comparable to the subgrains in the recovered ferrite.

In sample 3, deformation at 650°C (below Ar_1) followed by air cooling produces elongated grain boundary ferrite and pearlite (Fig. 4). The mean dimensions of the elongated ferrite grains are length = 3.62 ± 2.01 μm and width = 1.33 ± 0.75 μm. Fig. 5 shows a typical area of recovered elongated grain boundary ferrite grains. The substructure comprises 0.73 ± 0.09 μm subgrains. It was confirmed by electron diffraction that the individual subgrains have small (< 10°) misorientations with respect to each other. Some areas of recrystallized ferrite are also observed in Sample 3 (Fig. 6). The mean linear intercept diameter of the recrystallized ferrite grains is 0.66 ± 0.27 μm, again comparable to the subgrain size.

It is concluded that warm forging can produce significant microstructural refinement through the mechanisms of ferrite recovery and recrystallization. Sub-micron ferrite subgrains and recrystallized ferrite grains can be obtained.

References

[1] S. Sheljaskov, *J. Mater. Processing Technol.* 46 (1994) 3.
[2] J. H. Reynolds et al, *Mater. Sci. Technol.* 4 (1988) 586.
[3] C. García-Mateo et al, *Iron & Steelmaker.* 27 (2000) 79.

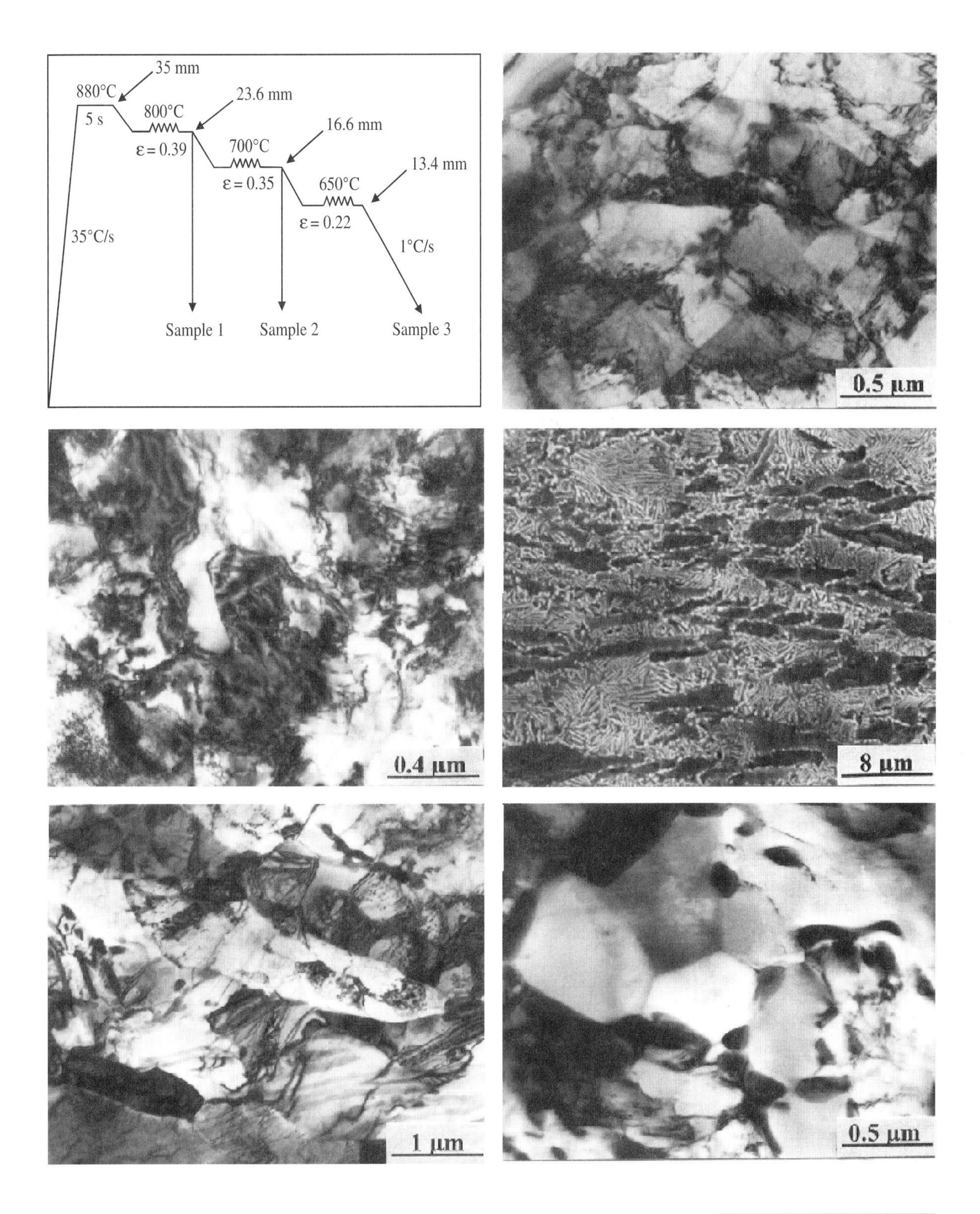

Figure 1 Experimental warm forging schedule.
Figure 2 Typical TEM microstructure of quenched sample 2.
Figure 3 Quenched sample 2, showing a recrystallized grain.
Figure 4 Typical SEM microstructure of air-cooled sample 3.
Figure 5 Typical TEM microstructure of air- cooled sample 3.
Figure 6 Air-cooled sample 3, showing a recrystallized area.

Fig. 1	Fig. 2
Fig. 3	Fig. 4
Fig. 5	Fig. 6

Microsc. Microanal. 8 (Suppl. 2), 2002
DOI 10.1017.S1431927602101309

Characterization of the Inhibition Layer on Galvanized Interstitial Free Steels

S. Dionne*, G. A. Botton*, M. Charest* and F. E. Goodwin**

* CANMET-Materials Technology Laboratory, 568 Booth St., Ottawa, Canada K1A 0G1
** ILZRO, 2525 Meridian Parkway, P.O. Box 12036, Research Triangle Park, NC 27709-2036 USA

Galvannealed steels have seen expanded use for automotive applications because of their superior properties over those of pure zinc coatings. During the galvannealing treatment, the iron aluminide inhibition layer that forms on the steel surface during galvanizing is broken down and interdiffusion of iron and zinc gives rise to a series of iron-zinc intermetallics. The formability and other in-service properties of galvannealed coatings are very sensitive to their composition and microstructure. Therefore, an understanding of the formation of the inhibition layer and its breakdown during the annealing treatment is required to optimize coating performance. In the present study, the microstructure of galvanized coatings produced on interstitial free steels under conditions typical of industrial continuous galvannealing (i.e. an effective Al content in the zinc bath of about 0.135% at 460°C) was examined using a combination of advanced characterizationtechniques.

The galvanized coatings formed on Ti-, Ti-Nb- and Ti-Nb-P- IF steels were not strongly influenced by the substrate composition. All of the samples had a continuous aluminum-rich inhibition layer with Al/Fe ratio in the Fe_2Al_5 range or above, as well as δ and ζ iron-zinc crystals with significantly higher Fe content (and Al content in the case of δ) than predicted from equilibrium considerations (FIG. 1 to 4). The δ crystals were located on top of the interfacial Fe_2Al_5 inhibition layer. There was no evidence of Zn or Al diffusion in the ferrite grains or emerging grain boundaries. Mn and/or Ti oxides were observed within the inhibition layer and at the interface between the substrate and the inhibition layer (FIG. 5). These surface oxides did not affect the thickness or composition of surrounding inhibition layer phases. Surface enrichment of alloying elements such as Mn, Si, P and Cr has been reported by several authors [1] [2]. The phenomenon is driven by selective oxidation of the alloying elements during annealing of the strip in a low dew point atmosphere.

The microstructure of the galvanized coatings is consistent with the following sequence of phase formation during galvanizing. When the strip initially makes contact with the zinc bath, dissolution of the substrate in the bath produces an increase of the Fe content in the melt close to the strip surface. The inhibition layer then precipitates on the strip surface from this Fe-supersaturated melt. With an effective Al content in the zinc bath at the knee point of the Fe-Zn-Al solubility curve and industrial processing conditions, Fe_2Al_5.was the first phase to form on the strip surface. Surface oxide particles within the inhibition layer suggest that precipitation of the Fe_2Al_5 crystals was concurrent with strip dissolution or that growth of the inhibition layer towards the ferrite substrate took place. Fe_2Al_5 formation was followed by the precipitation of δ crystals. Finally, Fe-rich ζ crystals were precipitated from the supersaturated zinc overlay during the solidification of the coating.

References

[1] M. Guttman et al., *Galvatech '95 : The Use and Manufacture of Zinc and Zinc Alloy Coated Sheet Steel Products into the 21st Century*, (Iron and Steel Society/AIME, 1995), pp. 295-307.

[2] W. van Koesveld et al., *Galvatech '95 : The Use and Manufacture of Zinc and Zinc Alloy Coated Sheet Steel Products into the 21st Century*, (Iron and Steel Society/AIME, 1995), pp.343-355.

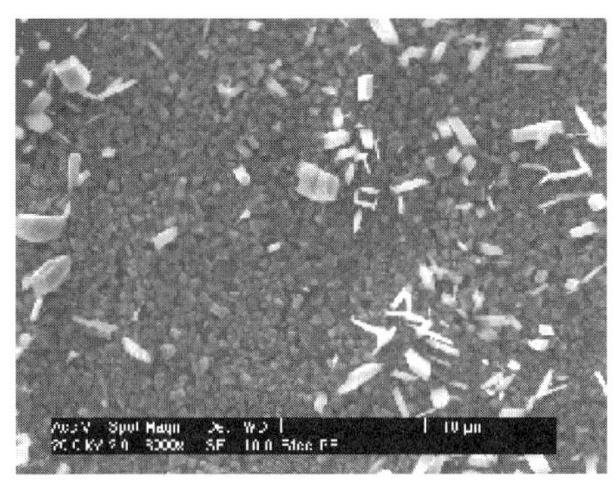

FIG. 1. SEM of a galvanized rephosphorized IF steel showing δ and ζ crystals (the zinc overlayer was removed by selective etching).

FIG. 2. SEM of a galvanized rephosphorized IF steel showing the δ and Fe_2Al_5 inhibition layer crystals (ζ crystals and Zn overlayer were removed by selective etching).

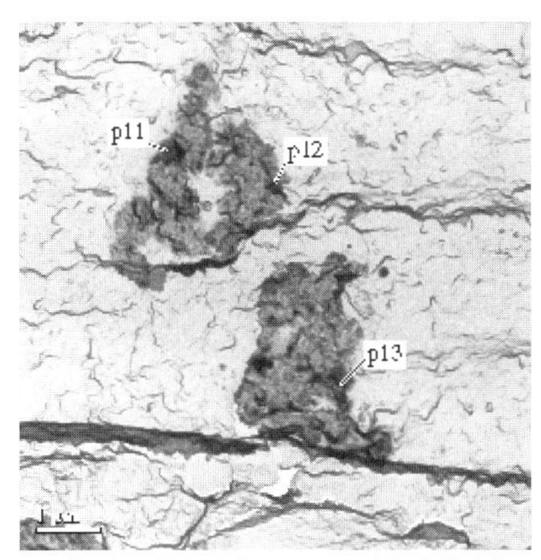

FIG. 3. TEM image of inhibition layer particles on a replica of a galvanized Ti-Nb-IF steel. Particle 11 was identified as Fe_2Al_5 with zone axis $[0\bar{1}2]$ using CBED.

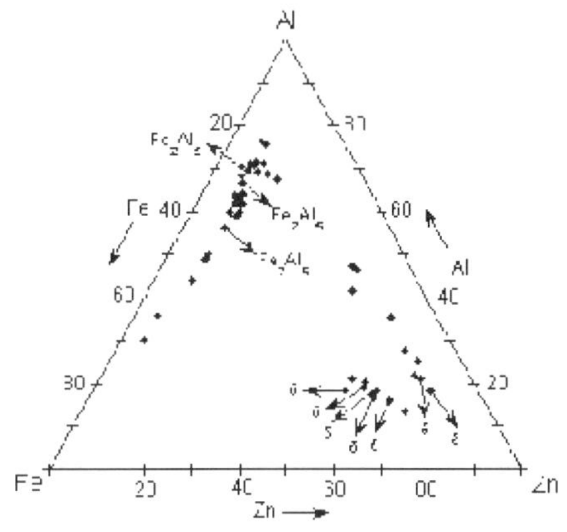

FIG. 4. Ternary plot showing the composition of inhibition layer crystals analyzed on extraction replicas prepared from a galvanized Ti-Nb-stabilized IF steel.

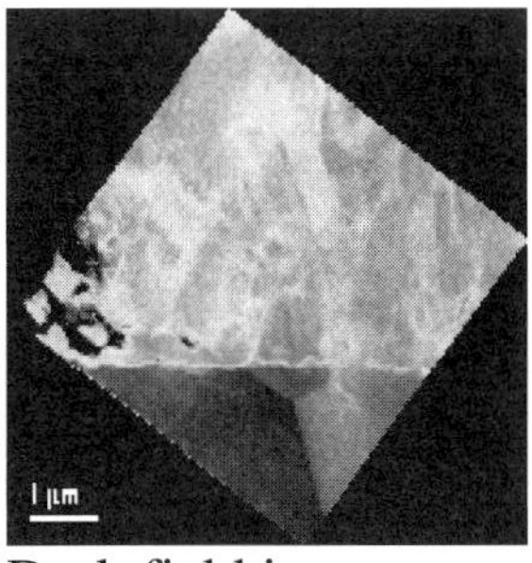

Dark field image.

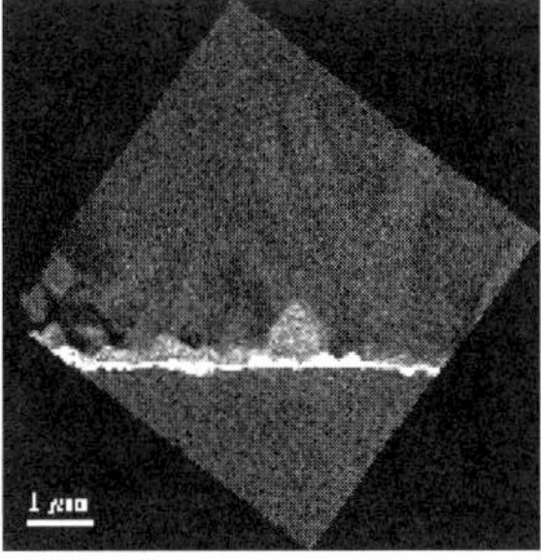

Aluminum map.

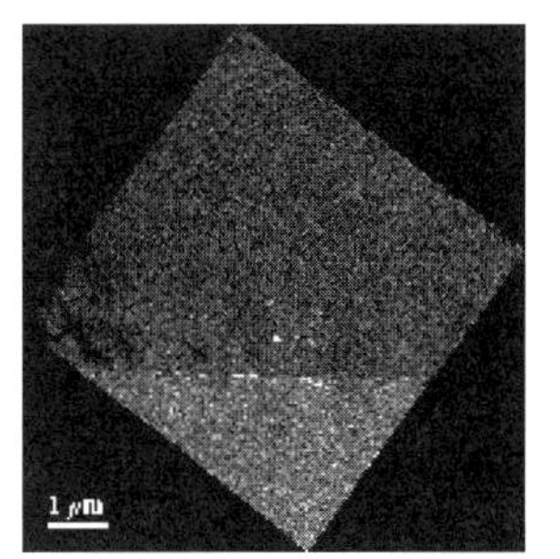

Manganese map.

FIG. 5. TEM image and elemental maps of a FIB lift-out cross-section of galvanized Ti-Nb-IF steel.

Microsc. Microanal. 8 (Suppl. 2), 2002
DOI 10.1017.S1431927602101310

Metallographic Methods for Troubleshooting of Roll Problems in the Finishing Train of a Ferrous Hot Strip Mill

Ron Webber*, Marian Lalik*

*Dofasco Inc., 1330 Burlington St East, Hamilton, Ontario, Canada, L8N 3J5

Rolls are a critical tool in the production of flat rolled steel. Being the only solid material that comes in contact with the product, they are essential to ensuring that the required product characteristics of dimension, flatness, profile and surface quality are achieved. In the last 10 years, revolutionary improvements in roll metallurgy have changed the status of the roll from a standard consumable to a technological tool. The drastic improvement in roll performance has also meant an increase in the price of advanced roll types. A better understanding by the user of the relationship between roll metallurgy, microstructure and roll performance is required.

In 1993, Dofasco was one of the first Companies to start using High Speed Steel (HSS) work rolls in the Finishing Stands of the Hot Mill. The initial attempt to relate microstructure to performance used standard metallographic techniques. Samples of the rolls were obtained by lathe cutting. The samples were mounted, polished, etched with 5% Nital (if required) and observed by either a Scanning Electron Microscope or optical microscopes. Other samples came from rolls that had catastrophically failed, a not too uncommon event in the early days of development. This old technique resulted in several key discoveries.

The superior performance of the HSS is in part due to the replacement of networked chromium carbides found in high chrome rolls with intergranular MC carbides and M_2C carbides (Figure 1a and b). It was determined that too high a volume of M_2C carbides resulted in an unfavourable structure. The brittle carbides created a pathway for cracks to propagate (Figure 2). The cracks would find their way back to the surface and cause a small spall that would lead to quality problems.

Typically, HSS rolls have a higher coefficient of friction than high chrome rolls. Using an SEM, it was shown that the matrix of the roll wears away preferentially and leaves the harder carbides protruding (Figure 3). The effect is much like the steel studs that were used on snow tires to improve traction.

Metallography was also used to confirm that thermal cracks are perpendicular to the roll surface and do not propagate after the initial creation. Mechanical cracks are not normal to the surface and will propagate until the roll fails.

For analysis of in-plant problems, another old technique is being employed. Using acetone tape, replicas can be obtained from rolls that are in service. The replica can be viewed under an optical microscope or, after coating with aluminum, observed with an SEM. This technique has resolved numerous problems and helped in the development of roll technology. A number of examples will be given of rolling problems that were swiftly and cheaply resolved.

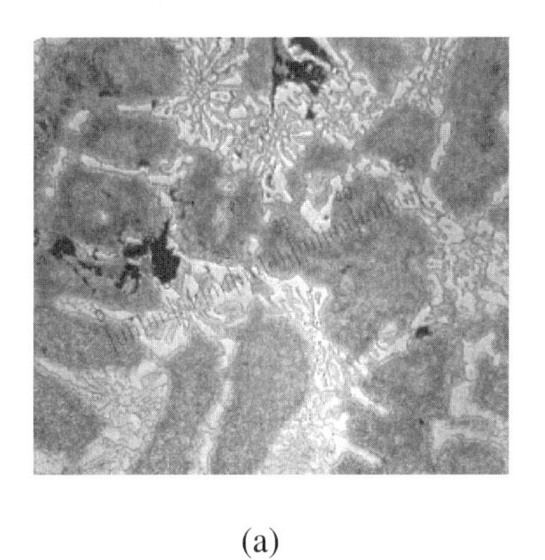

(a)

(b)

FIG. 1a. Microstructure of high chrome iron work roll(200 mag).
FIG. 1b. Microstructure of an HSS work roll (200 mag).

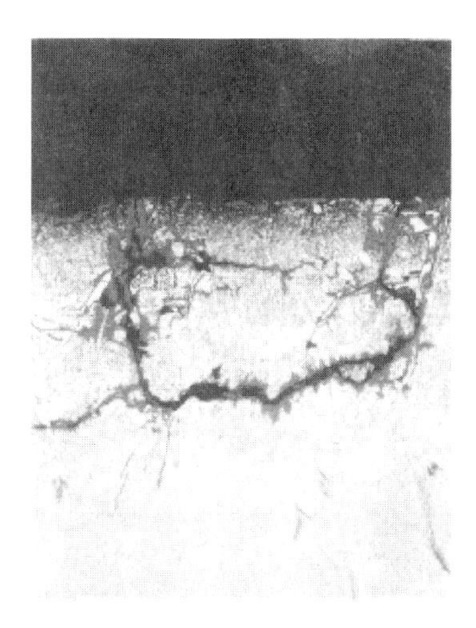

FIG 2. Microstructure of a crack following M_2C carbides (500 mag).

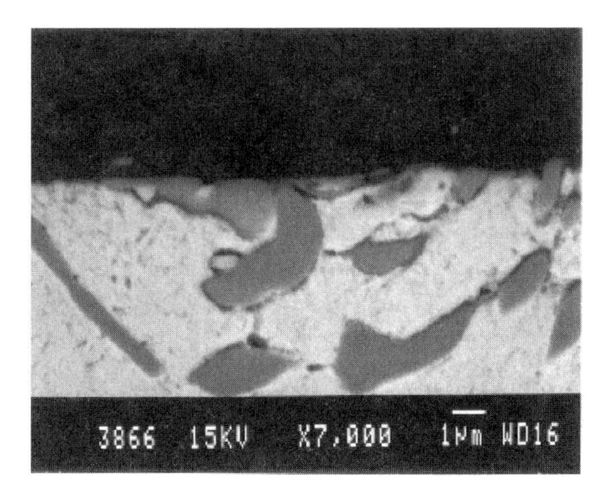

FIG. 3. SEM metallograph of hard carbides protruding from the HSS roll surface.

Microsc. Microanal. 8 (Suppl. 2), 2002
DOI 10.1017.S1431927602101322

Macro-etching of Continuous Cast Steel

Jim Casey, Dofasco Inc. *
* Box 2460, Hamilton, Ontario, Canada, L8N 3J5

Investigations on service failures in line pipe steels have underscored the importance of control of macro-segregation during slab casting. In some investigations, crack initiation in line pipe service failures has been traced to martensitic or hard segregation bands outcropping on the pipe surface. The formation of a brittle martensitic band is due to a hardenability increase from Mn and carbon enrichment resulting from macro-segregation in the casting operation. Oxide and sulfide inclusions occur in the macro-segregated regions. Hydrogen diffusion to these inclusion sites in sour-gas line pipe is known to cause catastrophic in-service failures. Segregation related problems are also known to affect the quality of the slab during the bending and unbending operation in the caster, resulting in problems with the seam welding of plates and hot rolled strip used to make pipes. Macro-segregation also affects subsequent field operations involving hot bending and field welding of pipes. Hence a method of preventing macro-segregation by fluid flow control in the caster, and having a method to monitor segregation have become an important quality assurance issue in the manufacture of line pipe and heavy plates. Since the base chemistry of the modern line pipe steel consists of ultra-low sulfur (< 0.005 wt % S) and low carbon (<0.06wt %C), the traditional Baumann sulfur printing method does not provide any useful information. Extensive research was carried out at Dofasco at the time of installation of their new caster to develop enhanced techniques for etching and documenting the macro-structure of low residual, as-cast slabs. Details of the techniques involving sample surface preparation, deep etching (*using three different techniques*) followed by ink highlighting will be outlined. The relative merits of documenting the resulting macrostructures using four different recording methods will be examined *(i.e., taping, paper printing, Intaglio printing and digitized imaging).*

The applications will be highlighted along with metallurgical case histories, to demonstrate the power of the techniques in problem solving. The first application is in characterization of macro-segregation and defects in as-cast slab hot-rolled and weld fabricated structures. The techniques were instrumental in establishing two underlying mechanisms, which initiated hook cracks in line pipe steel. The case histories will be presented. The second major application of the new techniques is in the identification of casting defects at the time of caster start-up. Figure-1 shows a typical example of midway cracking, as well as centre-line segregation in a slab of low residual chemistry. Figure-2 shows a typical Intaglio print of a conventional HIC resistant line pipe slab showing equiaxed crystals nucleated by a falling shower of dendrites through the remelting phenomenon. Based on the detailed information derived from the macro-structure, the base chemistry, and the casting machine variables used to control the fluid flow, such as casting speed, superheat and soft reduction ratios, effective steps can be taken to eliminate macro-segregation in the slab. Figure–3 shows crystal structure details, along with the center line segregation present in a thin slab structure having a low residual chemistry. The paper will show that the techniques can be used for off-line monitoring and control of macro-segregation in ultra-low residual chemistry steels.

With the emergence of thin strip casting, these techniques should prove valuable tools, in assisting metallurgists and production engineers to monitor and control the fluid flow and solidified structures encountered in the development of near net shape casting technology.

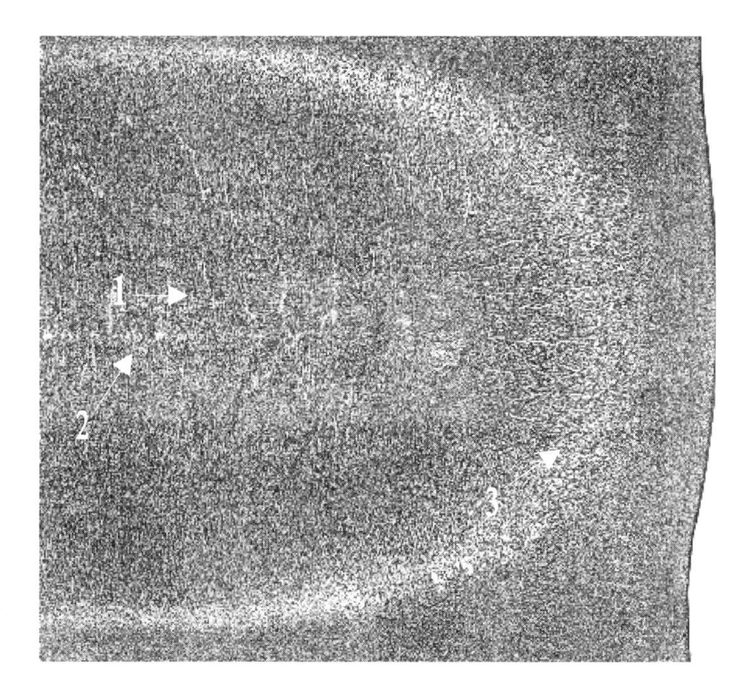

Figure 1: Relief Ink Print (X 0.35)
Continuous Cast Steel Slab, Transverse Section with moderately low residual chemistry*. Etched with hot HCL (*50% v/v at 90C*) and printed using a conventional relief ink printing method.
The structure reveals heavy mid way cracking (1), centerline (2) segregation and triple point cracks (3), not easily revealed by conventional sulphur prints.
* Analysis w% --- C (0.30), Mn (0.53), Si (0.233), S (0.003), P (0.012), Al (0.058, Ca (0.029)

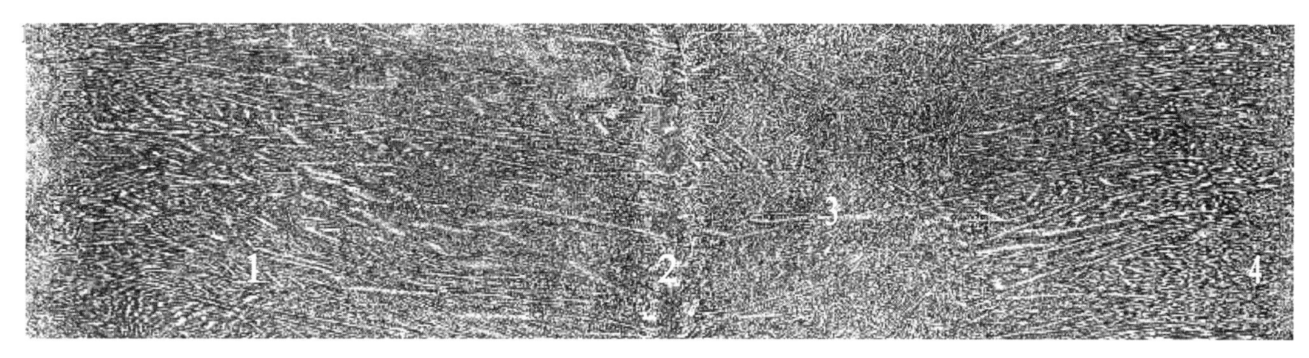

Figure 2: Intaglio Print (X 0.67)
Longitudinal section of a continuously cast, calcium treated, hydrogen induced cracking resistant line pipe steel, having a low residual chemistry. The steel is etched in two stages using Humfrey's Reagent. *Neutral, 12% Copper Ammonium Chloride, followed by subsequent immersion in acidified, (4% v/v HCL), 12% Copper Ammonium Chloride.* Etching is performed at room temperature. The image is an Intaglio ink print, made using printing ink and a high pressure, mechanical printing press.

This "Copper" etch produces excellent resolution of the primary solidification structures on what is a very low residual chemistry steel *, revealing primary dendrites (1), centerline (2), equiaxed zone (3) and the chill zone (4). A second advantage of this technique is that the inclusions remain intact and are not etched out, as in hot acid etching, therefore allowing for subsequent probe analysis and chemical characterization when doing a forensic analysis on failed or rejected steel sections.
* Analysis w% --- C (0.08), Mn (0.61), Si (0.214), S (0.0018), P (0.005), Al (0.075, Ca (0.0066)

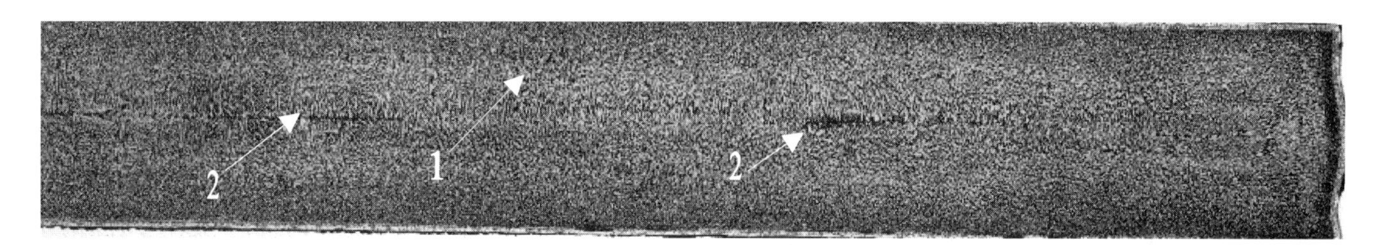

Figure 3: Intaglio Print (X 0.58)
Transverse section of a continuously cast thin slab with low residual chemistry *, etched using Humpfrey's Reagent. The fine primary dendrites (1) are easily revealed, as is solute segregation concentrated in patches along the centerline (2). Conventional etching in hot acid and the use of sulphur prints reveals nothing of the structure hidden within these thin as cast slabs.
* Analysis w% --- C (0.059), Mn (0.30), Si (0.030), S (0.002), P (0.009). Al (0.035)

Microsc. Microanal. 8 (Suppl. 2), 2002
DOI 10.1017.S1431927602100584

Metallic Magnetic Nanocrystals – Shapes, Self-assembly and Phase Transformation

Zhong Lin (ZL) Wang*, Zurong Dai* and Shouheng Sun**

* Center for Nanoscience and Nanotechnology, and School of Materials Science and Engineering, Georgia Institute of Technology, Atlanta, GA 30332-0245; *e-mail: zhong.wang@mse.gatech.edu
** IBM T. J. Watson Research Center, Yorktown Heights, New York 10598

Controlled assembly of magnetic nanocrystals (NCs) has been a key issue in fabricating functional nanodevices. Extensive investigations in NC assemblies have revealed that the symmetry of the observed superlattices is influenced by the NC size, NC shape and relative dimensions of the NC core and the organic capping. Our study has been focused on investigating the roles played by NC shape in self-assembly and the mechanism by which defects are created [1]. The key challenges in this self -assembly method are to control the defects of the self -assembly to ensure an ordered structure, and to prepare an assembly with uniform thic kness and large lateral dimension. For the 11 nm Co NCs and their superlattice assembly (Fig. 1), the Co NCs have anisotropic polyhedral shapes (Fig. 2) [2]. The self -assembly of the NCs and the defect structures in the NC arrays are governed by the anisot ropic shape. The self-assembly of nanocrystals with faceted shape is dominated by a surface-to-surface contact. This simple geometrical matching is the root for creating defects in the self-assembly. The structural transformation of shape-controlled nanocrystals has been studied by transmission electron microscopy.

Our second study is about the FePt nanocrystals, which have been demonstrated as a potential candidate for high-density magnetic transitions at room temperature [3]. Depending on the Fe to Pt elemental ratio, the Fe-Pt alloys can display chemically disordered face centered cubic (fcc) phase (A1, $Fm\bar{3}m$) or chemically ordered phases. High-resolution transmission electron microscopy (HRTEM) studies show that A1 to $L1_0$ phase transformation occurs at 530 °C [4,5]. The multilayered nanocrystal assemblies coalesce to form larger grains at 600 °C. The coalescent temperature of the nanocrystal monolayer assembly depends on the substrate used (Fig. 3). On SiO_2 surface, the FePt nanocrystal monolayer can stand up to 700 °C without any obvious aggregation. The coalesced nanocrystals show dominant {111} twin defect inside, while their surface and coalescent grain boundary consist of both {111} and (001) facets [6].

References

[1] Z.L. Wang (ed.) *Characterization of Nanophase Materials* (Wiley-VCH, 2000); Z.L. Wang, *Adv. Mater.* 10 (1998) 13; Z.L. Wang, *J. Phys. Chem. B*, 104 (2000) 1153; J.S. Yin and Z.L. Wang, *Phys. Rev. Lett.*, 79 (1997) 2570.
[2] Z.L. Wang, Z.R. Dai and S. Sun, *Adv. Mater.*, 12 (2000) 1944.
[3] S. Sun, C. B. Murray, D. Weller, L. Folks, A. Moser, *Science*, 287 (2000) 1989.
[4] Z. R. Dai, Z. L. Wang, and Shouheng Sun "Phase Transformation, Coalescence and Twinning of Monodisperse FePt Nanocrystals", Nanoletters. 1 (2001) 443-447.
[5] Z.R. Dai, Z. L. Wang, and S.H. Sun "Shapes, Multiply Twins and Surface Structures of Monodispersive FePt Magnetic Nanocrystals", Surface Science, submitted (2001).
[6] Research was supported by the NSF (DMR-9733160) and DARPA (DAAD19-01-1-0546).

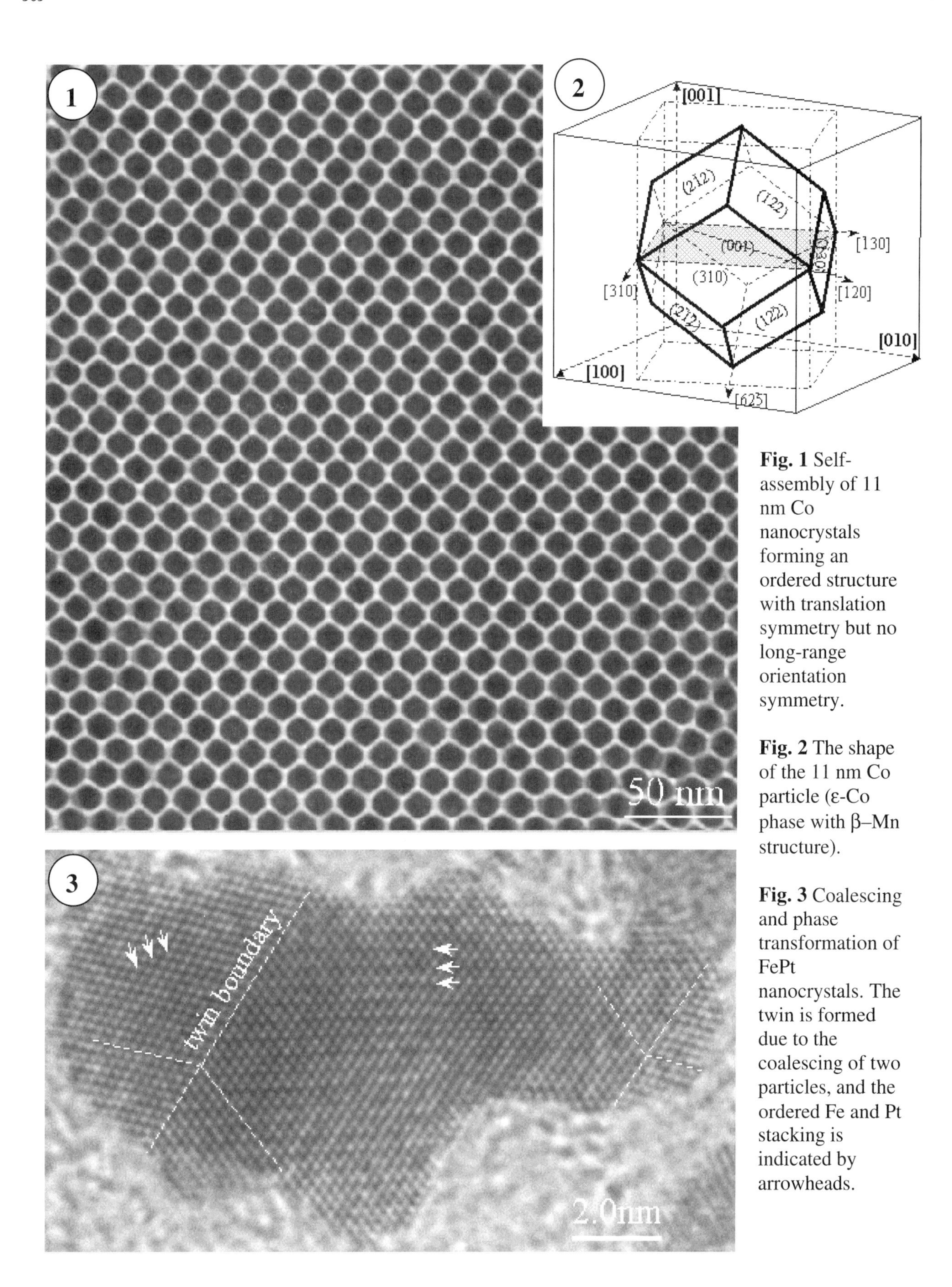

Fig. 1 Self-assembly of 11 nm Co nanocrystals forming an ordered structure with translation symmetry but no long-range orientation symmetry.

Fig. 2 The shape of the 11 nm Co particle (ε-Co phase with β–Mn structure).

Fig. 3 Coalescing and phase transformation of FePt nanocrystals. The twin is formed due to the coalescing of two particles, and the ordered Fe and Pt stacking is indicated by arrowheads.

Microsc. Microanal. 8 (Suppl. 2), 2002
DOI 10.1017.S1431927602100596

The Magnetism-Nanostructure Interface in Advanced Magnetic Materials

David J. Sellmyer

Department of Physics and Astronomy and Center for Materials Research and Analysis, University of Nebraska, Lincoln, NE 68588-0113

The critical relationship between microstructure and nanostructure of materials and their magnetic properties has been appreciated for decades. Electromagnetic machinery, permanent magnets, and data recording and electronic devices all have seen steady and sometimes spectacular advances over this period. At the present time the most interesting research in magnetism and magnetic materials arises from new developments in structuring materials on the nanometer length scale. This talk will present recent advances and challenges in furthering this work, with particular attention paid to extremely high density magnetic recording films, exchange-coupled high-energy-product permanent-magnet materials, high-temperature permanent-magnet materials, and self-organized and patterned magnetic nanoarrays.

Future High-Density Recording Media

Present longitudinal recording media are based mainly on CoCrPtX granular alloys and record at areal densities in the 20-50 gigabits per square inch (Gb/in^2) region [1]. Such films are expected to be limited by superparamagnetic and noise considerations to densities of about 100-200 Gb/in^2. Further advances to 1 $terabit/in^2$ (Tb/in^2) are likely to be provided by high-anisotropy, small-grained films with perpendicular anisotropy [2]. Candidates include $L1_0$ ordered phases as FePt or CoPt, or Co/Pt and related multilayers. Recent development of high-coercivity (~ 12 kOe), small-grain (6-8 nm) films with minimal exchange coupling, and their study with HRTEM will be outlined [3].

Exchange-Coupled Permanent Magnets

The development of significantly stronger permanent magnets, with energy products in the range 75-100 MGOe, through discovery of new compounds or alloys is problematic. However, the concept of nanostructuring hard-soft nanocomposites on the exchange-length scale of about 10 nm is of great interest [4]. The difficulties in this approach involve inventing or adapting synthetic methods to achieve strong exchange coupling between the hard and soft phases, without grain growth of either phase beyond about 10 nm. We have produced prototype FePt:$Fe_{1-x}Pt$ nanocomposites with energy products of about 53 MGOe, very close to those seen in the largest values known (~54 MGOe for $Nd_2Fe_{14}B$) [5]. New ideas and challenges in improving and studying nanostructures of this type will be discussed.

High-Temperature Permanent Magnets

$Sm_2(Co,Fe,Cu,Zr)_{17}$ [Sm_2TM_{17} or 2:17] magnets have been the best permanent magnets for use at temperatures up to about 300EC [6]. Their energy products of about 20 MGOe at 300EC are significantly degraded at T   500EC where new applications are sought for the all-electric airplane. The cellular nanostructure of the 2:17 magnets consists of 2:17 grains surrounded by Cu-containing $SmCo_5$-structure boundaries. Recent research has shown a record high temperature coercivity (12.3

kOe) at 500EC in a Sm-Co-Ti-Cu alloy [7]. A physical model for this behavior and relationship to nanostructures as determined by TEM will be discussed.

Self-Organized and Patterned Nanoarrays

A new topic of high interest is the study of magnetic nanoarrays by fabrication methods such as electrodeposition into self-assembled nanopore structures [8], laser-interference lithography [9], and focused-ion-beam synthesis [10]. Future applications have been suggested such as data storage, spin-logic devices, spin electronics, and quantum computing. If time permits one or more of these topics will be discussed.

This research is supported by DOE, NSF, ONR, ARO, AFOSR, NSIC and CMRA. It was performed in collaboration with R. Skomski, Y. Liu, M Yan, Y. Qiang, and J. Zhou.

References

[1] See: *Magnetic Storage Systems Beyond 2000*, Ed. G.C. Hadjipanayis Series II, Vol. 41 (Kluwer, Dordrecht, 2001).
[2] M. H. Kryder, Ref. 1., p. 559.
[3] D.J. Sellmyer *et al.*, *IEEE Trans. Mag.* 37 (2001) 1286.
[4] See: R. Skomksi and J.M.D. Coey, *Permanent Magnetism* (Inst. of Physics, Bristol, 1999), p. 298.
[5] D.J. Sellmyer, *et al.*, in *Handbook of Thin Films*, Ed. H.S. Nalwa, *Vol. 5: Nanomaterials and Magnetic Thin Films,* (Academic Press, New York, 2002), p. 337.
[6] G.C. Hadjipanayis in *Rare-Earth Iron Permanent Magnets*, Ed., J.M.D. Coey (Oxford University Press, Oxford, 1996).
[7] J. Zhou *et al.*, *IEEE Trans. Mag.* 37 (2001) 2518.
[8] D.J. Sellmyer *et al.*, J. Phys. Cond. Matt. 13 (2001) R433.
[9] M. Zheng *et al.*, *Appl. Phys. Lett.* 79 (2001) 2606.
[10] B.D. Terris *et al.*, *Appl. Phys. Lett.* 75 (1999) 403.

Microsc. Microanal. 8 (Suppl. 2), 2002
DOI 10.1017.S1431927602100602

TEM Microstructure Studies of Thin Film Magnetic Recording Media

R. Sinclair, U. Kwon and J. D. Risner

Department of Materials Science and Engineering, Stanford University, Stanford, CA 943052205

The remarkable increase in the magnetic recording density of computer hard disk systems (currently about 60% per annum) is achieved by microstructural control. Of particular importance are the grain size and orientation of the active cobalt alloy layer, the alloy content, and grain boundary segregation or separation effects. Since typical grain sizes are now close to 10 nm or so, it is clear that only transmission electron microscopy (TEM) has the capacity to study these effects in the necessary detail. This paper reviews some recent progress in this field.

TEM samples are prepared in both through-foil and cross-section orientations, although the former often contain most of the information for the magnetic technologists. High resolution (HREM) imaging is generally required for the grain size analysis [1,2] because of the lack of clarity in bright field images in identifying each individual grain. Segregation effects are studied with nanoprobe X-ray energy dispersive spectroscopy (EDS), combined with HREM for direct grain misorientation analysis, and collaborative work using electron energy filtered imaging has also been successful [3,4]. The basic alloys are cobalt-chromium based with additions of tantalum, platinum and/or boron. Tantalum promotes the grain boundary segregation of chromium, platinum improves the magnetic coercivity and boron refines the grain structure significantly.

Fig. 1 shows low and high magnification views of a CoCrPtB alloy and Fig. 2a shows its grain size analysis in the form of a cumulative percentage plot. Such curves allow immediate derivation of the median volume, and can easily distinguish grain size distributions of materials with apparently similar averages [1,2]. Because of the uniformity of the microstructure, and indeed the magnetic properties across the hard disk, we have found that sampling only 100-200 grains yields a satisfactory analysis, a result obtained elsewhere on a study of sputtered aluminum thin film grains [5]. Data such as these are related to the experimental deposition conditions and the thin film magnetic properties.

Fig. 2b shows EDS analysis of grain boundaries in a $Co_{80}Cr_{16}Pt_4$ alloy. As found previously [6], the degree of chromium segregation is significantly lower for platinum-bearing alloys compared to tantalum bearing. Indeed for the latter, up to twice the nominal chromium composition can be accommodated at the boundaries [3,4,7], which noticeably improves the magnetic recording signal-to-noise performance. We are currently correlating segregation effects to grain misorientation. An earlier report [8] indicates that an important structural relationship exists.

In summary, TEM is essential for microstructural studies of such fine-grain, high performance magnetic thin films.

References

[1] E.T. Yen et al., *IEEE Trans. Magn.* 35 (1999) 2730.

[2] D.W. Park et al., *J. Appl. Phys.* 87 (2000) 5687.

[3] J.E. Wittig et al., *IEEE Trans. Magn.* 34 (1998) 1564.

[4] J.E. Wittig et al., *Mat. Res. Soc. Symp. Proc.* 517 (1998) 211.

[5] D.T. Carpenter et al., *J. Appl. Phys.* 84 (1998) 5843.

[6] S.R. McKinlay, *Ph.D. Thesis*, Stanford University, in preparation.

[7] R. Sinclair et al., *Microsc. Microanal.* 3 (Supple. 2) (1997) 513.

[8] J. Wittig et al., *Microsc. Microanal.* 7 (Supple. 2) (2001) 298.

[9] This work was supported by Komag Corporation, and has also benefited from prior support from Seagate Corporation, HMT Technology and IBM. One of us (JDR) acknowledges receipt of a NSF fellowship.

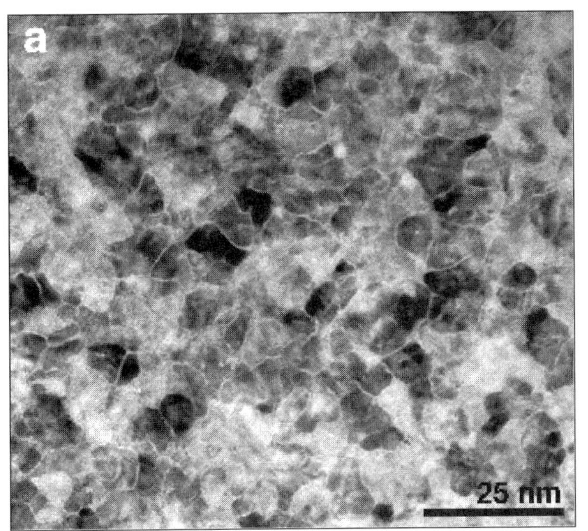

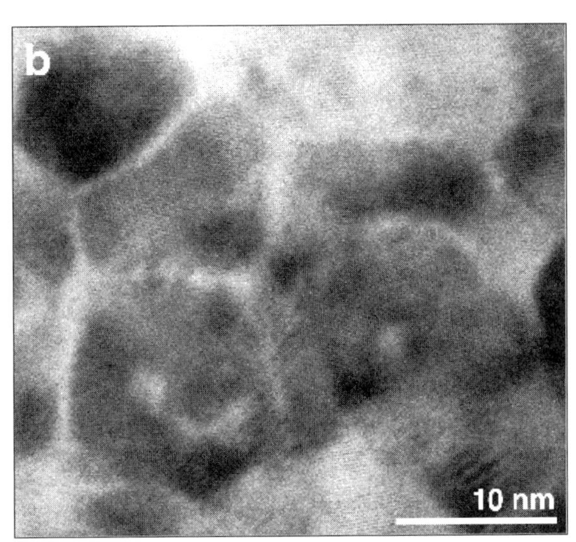

FIG. 1. (a) Bright field and (b) high resolution transmission electron microscopy images of a CoCrPtB alloy.

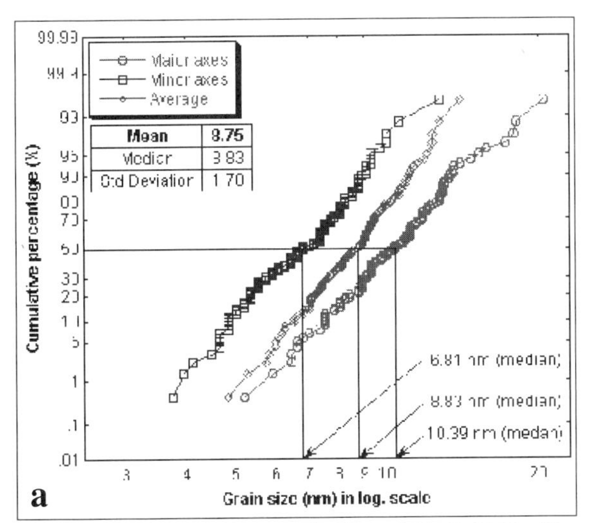

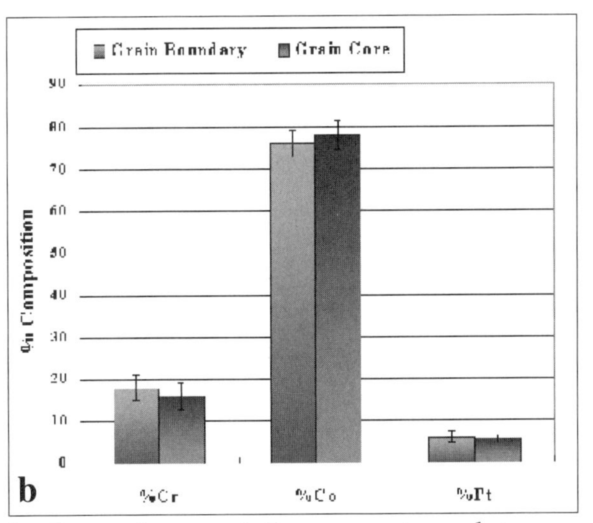

FIG. 2. (a) Grain size analysis of CoCrPtB alloy in the form of a cumulative percentage plot (b) Comparison of grain boundary and grain core Cr, Co, and Pt composition in a $Co_{80}Cr_{16}Pt_4$ alloy.

Microsc. Microanal. 8 (Suppl. 2), 2002
DOI 10.1017.S1431927602100614

Determination of Disordered Magnetic Structures in High-Coercivity Nd-Fe-Based Glassy Alloys

N. Lupu and H. Chiriac

National Institute of R&D for Technical Physics, 47 Mangeron Blvd., R-6600 Iasi, Romania

Binary rare earth - Fe amorphous alloys have been extensively studied in the past not only for their physical interest but also for technological importance for permanent magnets, magnetostrictive materials and magneto-optical recording media. These alloys form a variety of magnetic structures ranging from ferromagnetic to sperromagnetic and sperrimagnetic, depending on the relative magnitude of the random magnetic anisotropy (RMA) with respect to the exchange energy [1,2]. The high coercivities observed by Croat [3] in melt-spun rare earth – Fe (rare earth = Nd or Pr) alloys was the first evidence for the existence of an unknown metastable hard magnetic phase in the light rare-earth – transition metal binary amorphous system. More recently, it has been reported that Nd-Fe-(Si,Al) ternary amorphous alloys are formed in a wide range of compositions in different bulk shapes and exhibit large coercive fields at room temperature [4,5].

In this work, we focus on the determination of magnetic structures in high-coercivity $Nd_{90-x}Fe_xAl_{10}$ (x = 20 – 60 at. %) bulk amorphous alloys using RMC modeling of neutron diffraction data, Mössbauer spectra analysis and specific magnetic measurements. We measured both melt-spun ribbons with thicknesses between 25 and 120 μm and cast rods with diameters no larger than 2 mm.

Neutron diffraction measurements have been made at 12 temperatures between 15 and 773 K. The data have been modeled using the reverse Monte Carlo method (RMC) [6] under the assumption that the atomic structure is completely disordered (Figure 1). From the resulting configurations we find that the near-neighboring ordering is asperomagnetic, but the magnetic structure is predominantly ferromagnetic at temperatures below crystallization temperature (about 440^0C) in agreement with the static magnetic measurements [7] and Mössbauer spectra (Figure 2). The pair correlation function obtained by Fourier inversion of the diffraction patterns shows first neighbor peaks at 2.54, 2.85 and 3.36 Å. The peak at 2.54 Å is at the position expected for the nearest neighbor Fe atoms in the dense random packing (DRP) model; the other peaks cannot be correlated to combinations of the radii of Nd (1.82 Å), Al (1.43 Å) or Fe (1.27 Å) atoms as predicted by the DRP model. This disagreement could be ascribed to the development of a new type of disordered structure. Thus, one suggests that the structure consists in a dense random packing of nanometer-sized atomic clusters. Consequently, the magnetic structure will consist in non-collinear short-range magnetic ordered regions randomly distributed.

The granular magnetic structure consists in Fe-based nanosized clusters dispersed in amorphous Nd-rich matrix. Due to their very small dimensions they cannot be evidenced by X-ray diffraction measurements, but their presence is proved by the specific magnetic behavior, especially by the high coercive fields (up to 2 T) in the as-quenched state and the dependence on the cooling rate. The composition and size of the magnetic clusters as well as the composition of the matrix in which the clusters are embedded are very sensitive to the preparation conditions, mainly the cooling rate and the thermal history of the molten alloy.

Fe-based magnetic clusters are coupled between them through the amorphous matrix and the exchange coupling is strongly influenced by the magnitude of the Nd^{3+} single ion local anisotropy. Whereas at low temperatures the anisotropy plays the predominant role in the macroscopic magnetic response of the system, the increase of the temperature diminishes its effect and the main role is attributed to the ferromagnetic exchange energy. At low temperature, the local anisotropies of the Nd^{3+} ions are very strong and oriented randomly, and the local magnetic moments of Fe and Nd are "frozen" randomly. The anisotropy axis corresponding to the Fe-Nd pairs are oriented randomly, and a small external field is not enough to align them. The maximum of the coercive field is attained at that temperature at which the anisotropy energy and the exchange energy have similar values, i.e. the magnetic clusters have the size of the single magnetic domain (smaller than 100 nm). As larger the applied field as smaller the temperature at which the coercive field attains its maximum.

Despite of that, there are still many questions related to the microstructure of these materials and its interplay with magnetic properties The magnetic ground states of Nd-Fe-based cluster amorphous alloys and non-collinear structures existent in these materials are far from being fully characterized. These problems and others make the study of the magnetic properties of bulk amorphous alloys a fascinating field of research.

References

[1] J.M.D. Coey, *J. Appl. Phys.* 49 (1978) 1646.
[2] K. Moorjani and J.M.D. Coey, *Magnetic Glasses*, Elsevier, Amsterdam, 1984.
[3] J.J. Croat, *IEEE Trans. Magn.* MAG-11 (1982) 1442.
[4] A. Inoue et al., *IEEE Trans. Magn.* 33 (1997) 3814.
[5] H. Chiriac and N. Lupu, *J. Magn. Magn. Mater.* 196-197 (1999) 235.
[6] D.A. Keen et al., *Nuclear Instruments and Methods in Physics Research* A354 (1995) 48.
[7] H. Chiriac et al., *IEEE Trans. Magn.* 37 (2001) 2509.
[8] Access to facilities at the Studsvik Neutron Research Laboratory has been supported by the Access to Research Infrastructures activity, Improving Human Potential programme, of the European Commission under contract HPRI-CT-1999-00061. The aid of Prof. R.L. McGreevy is gratefully acknowledged.

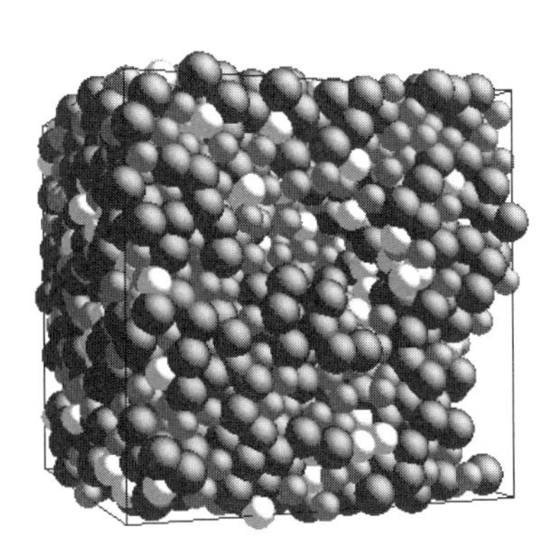

FIG. 1. Modeled atomic structure for amorphous $Nd_{50}Fe_{40}Al_{10}$ melt-spun ribbon, 25 μm in thickness

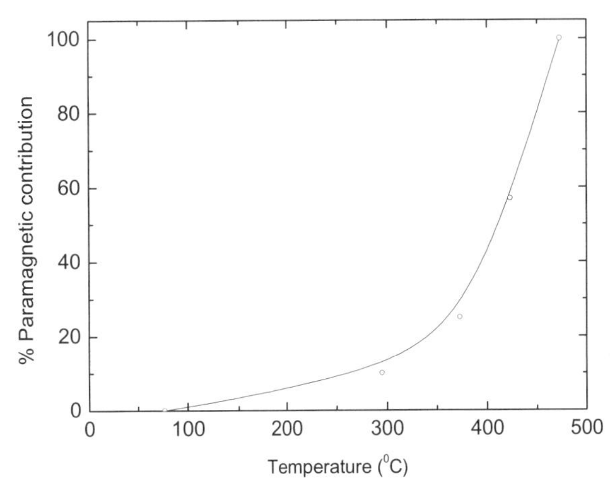

FIG. 2. The evolution of the paramagnetic fraction vs. temperature for $Nd_{50}Fe_{40}Al_{10}$ bulk amorphous alloys

Microsc. Microanal. 8 (Suppl. 2), 2002
DOI 10.1017.S1431927602100626

EELS Analysis of Magnetic Materials

C.G. Trevor, P.J. Thomas, R. Harmon, R. Alani and H.A. Brink

Gatan Research and Development, 5933 Coronado Lane, Pleasanton CA 94588

Modern mass storage devices often rely on ever-higher density magnetic storage to provide fast access to large amounts of data. As the scale of features in these devices decreases the analysis of their structure and composition presents greater challenges to current techniques. For these materials, electron energy-loss spectroscopy (EELS) is an attractive technique [1-2] due both to its high acquisition rate and high spatial resolution. We present elemental analyses of several such materials. In some examples the multiple overlapping edges make conventional power-law fitting infeasible. In this case the multiple linear-least-squares (MLLS) [3] approach enabled us to create elemental maps.

The first sample we studied was an $Nd_2Fe_{14}B$ based permanent magnet. The microscope was a J2010 TEM equipped with a Gatan Imaging Filter (GIF, model 860). Figure 1a shows a zero-loss image of the specimen. Application of the 3-window power-law technique to the Fe L (709eV) edge and the Nd M (978eV) edge yields the elemental distribution maps shown in Figures 1b and 1c. Acquisition of the three images took 12 seconds each for the Fe map and 10 seconds each for the Nd map. Simple systems like this are quickly and easily analyzed with the 3-window technique.

A more complex problem is presented by characterization of a magnetic recording head. The sample contained many elements including Al, O, Co, Fe, Ni, Cr, W and Ta. Some of these elements are easy to map using a conventional power law approach, whereas others are not straightforward because of overlapping edges. In this example the tantalum and tungsten M edges overlap. To map these elements a series of energy-filtered images were acquired covering the range from 1200eV to 2700eV, using a slit size of 30eV and stepping the energy loss by 30eV between images. The data was then analyzed as an EELS spectrum-image (SI).

Figure 2a shows the zero-loss image of part of this device. Figure 2b shows three spectra extracted from the spectrum image taken from locations marked on this figure. A map of the aluminum distribution was generated from the K edge (1560eV) by stripping off a power-law background based on a pre-edge window. For the tungsten and tantalum we applied the MLLS technique to generate compositional maps. Two reference spectra, the M edges of Ta and of W (the two lower spectra in Figure 2b) were fitted to each spectrum in the spectrum image. This resulted in a set of combination coefficients that specify the relative contribution of each input reference spectrum at that location in the spectrum image. The resulting elemental distribution maps are shown in Figures 2c and 2d.

For complex materials such as these, EELS coupled with the spectrum imaging and MLLS techniques provides a powerful tool with which to probe the elemental distribution.

References

[1] K. Kimoto, Y. Hirayama, M. Futamoto, *Journal of Magnetism and Magnetic Materials 159* (1996) 401-405.
[2] J. Bentley et al, *Microsc. Microanal.* 7 (Suppl. 2) (2001) 1140.
[3] R.D. Leapman and J.A. Hunt, *Microsc. Microanal. Microstruct.*, 2 (1991) 257

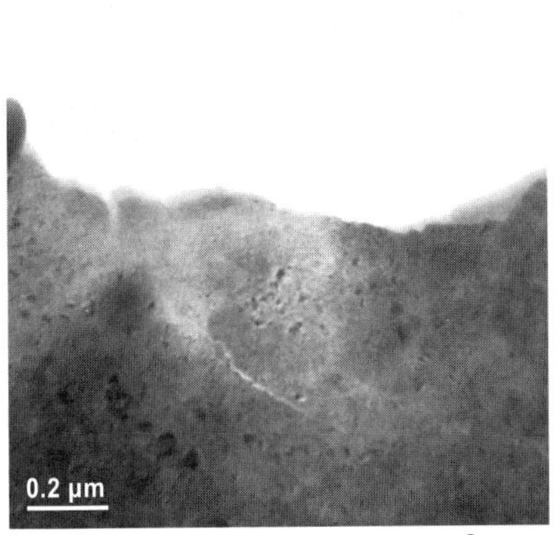

Fig. 1a. Zero Loss Image of permanent magnet

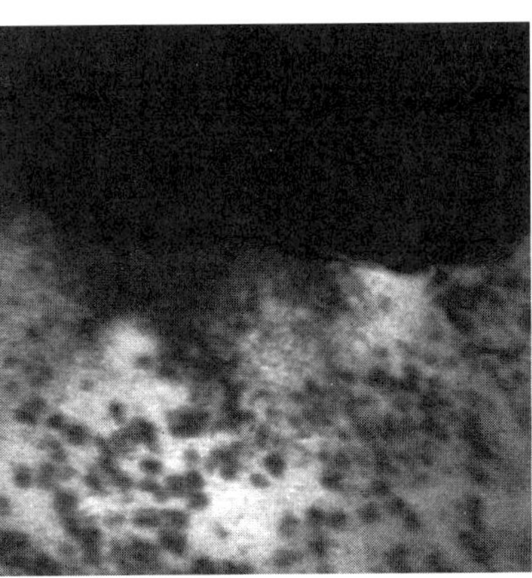

Fig. 1b. Fe map (L edge 709eV)

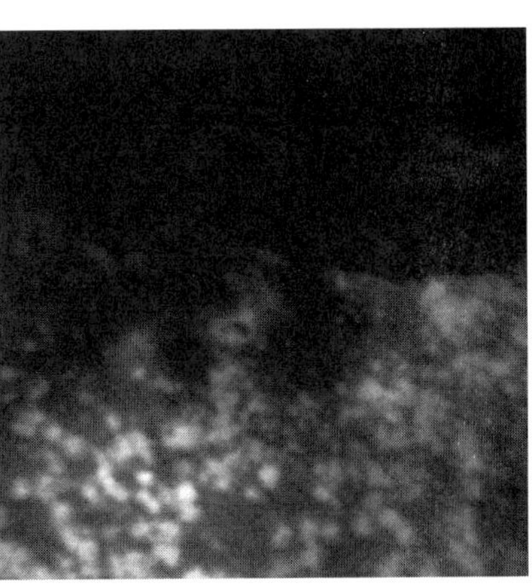

Fig 1c. Nd map (M edge 978eV)

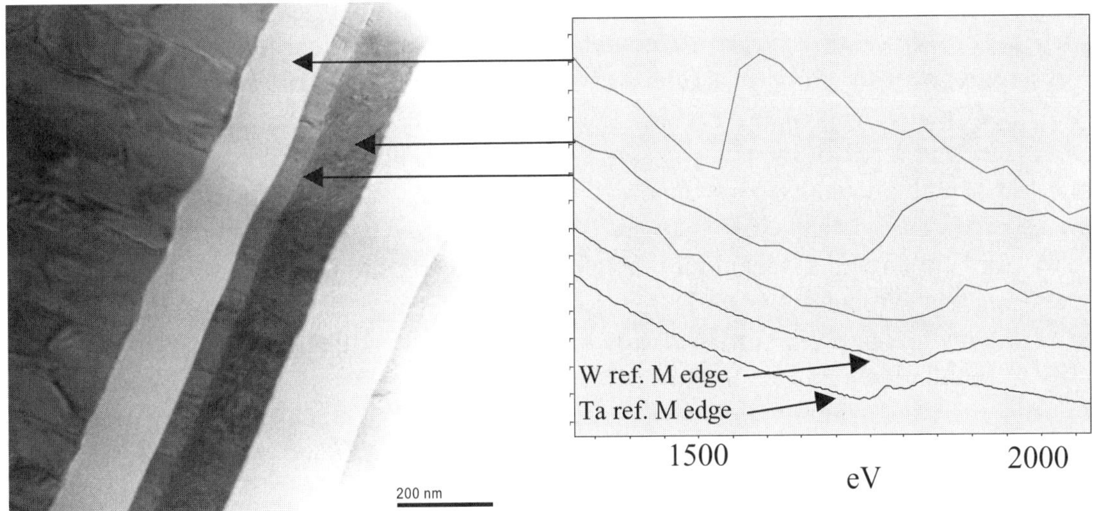

Fig. 2a. Zero Loss Image of MR head

Fig. 2b. Spectrum created from SI

Fig. 2c. W map (M edge 1809eV)

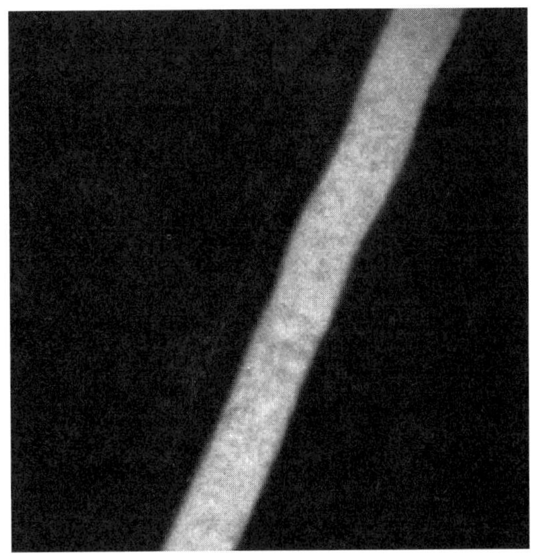

Fig. 2d. Ta Map (M edge 1735eV)

Microsc. Microanal. 8 (Suppl. 2), 2002
DOI 10.1017.S1431927602100638

Characterization of Magnetic Materials by Means of Neutron Scattering and Future Possibilities at a Next Generation Spallation Neutron Source

F. Klose* and G. Ehlers**

* Oak Ridge National Laboratory, Spallation Neutron Source Project, current address: Argonne National Laboratory, IPNS/SNS, 9700 South Cass Avenue, Argonne, IL 60439, USA
** Institute Laue-Langevin, 6, rue Jules Horowitz, BP 156, 38042 Grenoble Cedex 9, France

Due to its unique elementary characteristics, the neutron is especially suited for probing the magnetic properties of materials. It is not electrically charged, and therefore penetrates deeply into condensed matter. On the other hand, it possesses spin ½, and so interacts with atoms as well as with magnetic moments present in matt er. These two types of interaction are of comparable strength. Since the neutron has mass, its energy is inversely proportional to the square of its de Broglie wavelength. Neutrons with wavelengths in the Ångstroem range possess kinetic energies in the meV range, which are the typical energies of elementary excitations in condensed matter. Consequently, neutrons can simultaneously probe structural and magnetic *spatial* correlations on atomic to mesoscopic length scales, as well as structural and magnetic *temporal* correlations in the range of 10^{-14} to 10^{-7} s. By keeping track of the neutron spin and its change during the interaction with the sample, one can unambiguously separate nuclear and magnetic scattering processes.

Compared to other experimental techniques developed to investigate magnetic properties of matter, neutron scattering has particular merits: 1) transparent and easy experimental procedures that allow a straightforward conversion of experimental data into physical quantities, 2) a non -destructive nature, 3) sensitivity to both volume and surface properties, and 4) the capability to study a huge variety of magnetic phenomena and different classes of materials. The dependence of magnetic properties on temperature, pressure, or magnetic field can easily be explored since most sample environments can be made transparent to neutrons.

The applicability of neutron scattering techniques, however, is limited by the relatively low flux of useful neutrons generated by today's research reactors or pulsed sp allation sources, which is many orders of magnitude smaller than the flux of X -rays produced by contemporary photon sources. Recently, major efforts have been made to optimize existing sources and to develop new, more powerful sources.

A next -generation pulsed source, the Spallation Neutron Source (SNS), which has been under construction at Oak Ridge National Laboratory since 1999, will become operational in 2006. This facility will generate an effective neutron flux about one order of magnitude higher th an today's best neutron sources. Other approaches to gain intensity concern optimization of neutron optical components, development of new optical devices, and implementation of advanced instrument designs. Simulation calculations indicate that these approaches should further increase the usable flux by another order of magnitude for most SNS scattering instruments. The total intensity gain for SNS instruments, therefore, can be as high as two orders of magnitude, an improvement that will greatly enhance the quality of neutron scattering studies.

Figure 1 shows the current layout of the neutron scattering instruments in the SNS target hall [1]. Currently, seven instruments have been approved for construction (a backscattering spectrometer, a magnetism reflectometer, a liquids reflectometer, a small-angle scattering instrument, a powder diffractometer, a wide angle chopper spectrometer and a cold neutron chopper spectrometer). Seven more instruments are in the design phase (a high-pressure diffractometer, an engineering diffractometer, a single-crystal diffractometer, a disordered materials diffractometer, a high-resolution chopper spectrometer, a spin-echo spectrometer and a fundamental physics beamline). Construction will start as soon as funds become available. Further instruments will follow until eventually all 24 available SNS beam ports are occupied [2].

References
[1] For detailed information on particular instruments consult the SNS Instrument Systems web site at www.sns.anl.gov
[2] This work is supported by the Spallation Neutron Source Project (SNS). SNS is managed by UT-Battelle, LLC, under contract No. DE-AC05-000R22725 for the U.S. Department of Energy.

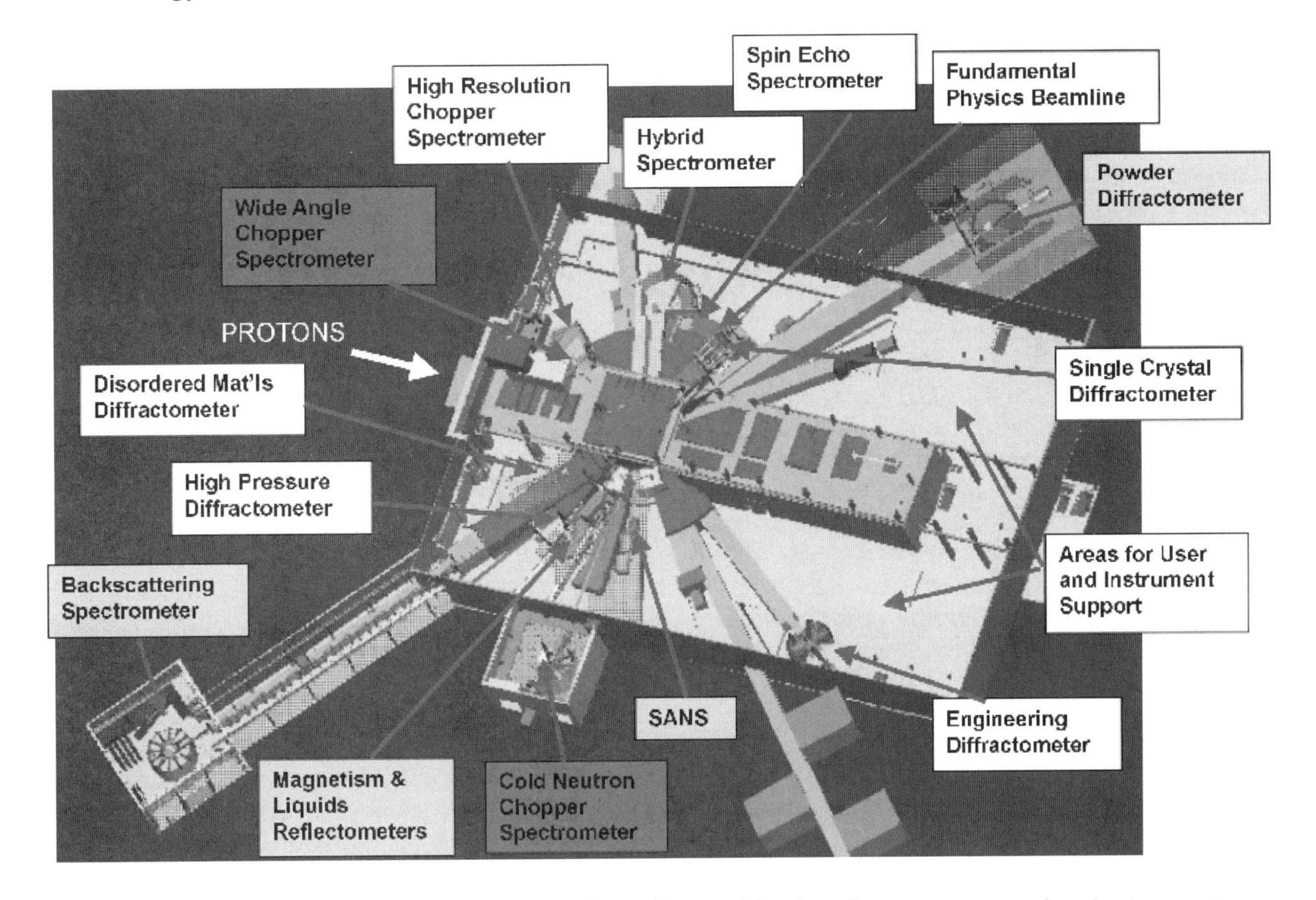

FIG. 1. Current layout of SNS target hall (size: 60 m x 90 m) and neutron scattering instruments. The yellow background characterizes instruments that are funded by the SNS Project, red background instruments are funded by (university) instrument development teams and white background characterizes instruments that are currently in the design phase.

Microsc. Microanal. 8 (Suppl. 2), 2002
DOI 10.1017.S143192760210064X

Quantitative Measurements of Magnetic Vortices using Position Resolved Diffraction in Lorentz STEM

Nestor J. Zaluzec

Materials Science Div., Argonne National Lab, Argonne, IL, USA

A number of electron column techniques have been developed over the last forty years to permit visualization of magnetic fields in specimens. These include: Fresnel imaging (Fuller and Hale[1]), Differential Phase Contrast (Chapman[2]), Electron Holography (Tonomura[3]) and Lorentz STEM (Zaluzec[4]). In this work we have extended the LSTEM methodology using Position Resolved Diffraction (PRD) to quantitatively measure the in-plane electromagnetic fields of thin film materials.

The experimental work reported herein has been carried out using the ANL AAEM HB603Z 300 kV FEG instrument [5]. In this instrument, the electron optical column was operated in a zero field mode (figure 1), at the specimen, where the objective lens is turned off and the probe forming lens functions were reallocated to the C1, C2, and C3 lenses. Post specimen lenses (P1, P2, P3, P4) were used to magnify the transmitted electrons to a YAG screen, which was then optically transferred to a Hamamatsu ORCA ER CCD array. This CCD was interfaced to an EmiSpec Data Acquisition System and the data was subsequently transferred to an external computer system for detailed quantitative analysis. In Position Resolved Diffraction mode, we digitally step a focused electron probe across the region of interest of the specimen while at the same time recording the complete diffraction pattern at each point in the scan.

Figure 2 shows a Bright Field LSTEM image of the lithographically produced array of ~ 100 nm thick Permalloy microdots on an ~ 75 nm thick SiN amorphous film, supported on a 1000 mesh Cu grid. No magnetic contrast is shown in this image, but the polycrystalline nature of the microdots is clearly visible. Independent confirmation by Magnetic Force Microscopy indicated that the out of plane magnetic fields is minimal for the microdot array. Quantitative analysis of the in-plane electromagnetic field distribution is obtained by measuring the vector components of Lorentz deflection (amplitude and orientation) of the transmitted electron diffraction patterns (EDPs) at each of the 900 points of a two dimensional 30x30 pixel LSTEM scan of selected areas of the SiN film and a individual microdots. In order to perform these measurements correctly it was essential that the instrument be precisely aligned and that all instrumentally induced displacements of the EDP be negligible. This was confirmed by operating the instrument in PRD mode and simply translating the stage until an empty grid square was present, essentially measuring “vaccum”. The results of such a null specimen measurement are shown in the polar plot of Figure 3a, where the total vector displacement can be discerned to be minimal (the group of 900 vectors is centered at the origin). In contrast, figure 3b shows the measurement of the Lorentz displacement vector of the EDPs measured from an ~ 0.5 μm^2 area of the SiN film (region A of fig. 2) , while in figure 4c, from a neighboring microdot (region B of fig. 2). In figure 3b, we see that the scattering is highly distorted and asymmetric due in part to charging of the SiN film, while in comparison, in 3c, we see that the scattering is symmetric distributed equally in nearly all amplitudes and orientations.

To more succinctly present the data of figures 3b and 3c, we replot the analysis as two-dimensional pseudo-images of the “Lorentz” amplitude and orientation for the SiN film (4a,b) and microdot (4c,d) respectively. In the SiN film we see that both pseudo-images show little systematic correlation as is expected from the polar plot (3b). On the other hand for the microdot, we see that the amplitude increases from near zero at the microdot center (black) to a maximum (bright) at the periphery and decreasing thereafter. In addition and equally importantly we observe that the orientation dependence of the displacement vector for the microdot rotates continuously around the center of the field of measurement (4d). These last two pseudo-images (4c,d) can therefore be interpreted to indicate that the microdot contains a simple magnetic vortex, the Lorentz field of which, is rotating continuously about the nominal center of the microdot. We also note, in passing, that the SiN film is charging and we can thus see that the PRD methodology is also applicable to measuring in-plane electrostatic fields.

References:

1.) H. Fuller and M. Hale, J Applied Phys., 31 No. 10, 1699, (1960).
2.) J. Chapman, J. Phys. D. 17, 623 (1984)
3.) J. Tonomura, J. Magn. Mag. Mater. 35, 963 (1983),
4.) N.J. Zaluzec, Microscopy & Microanalysis, 7, Number 5, Sup 1, (2001)
5.) N.J. Zaluzec, Proc. of 26th MAS Conf. SF Press. 137, (1991)
6.) This work was supported in part by the U.S. DoE under BES-MS W-31-109-Eng-38 at ANL.

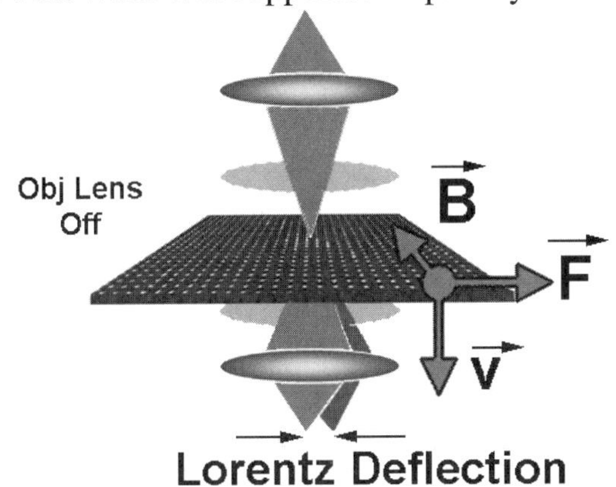

Figure 1. Electron Optical Configuration in Lorentz mode.

Figure 2. BF Lorentz STEM Image of Permalloy microdots on SiN film

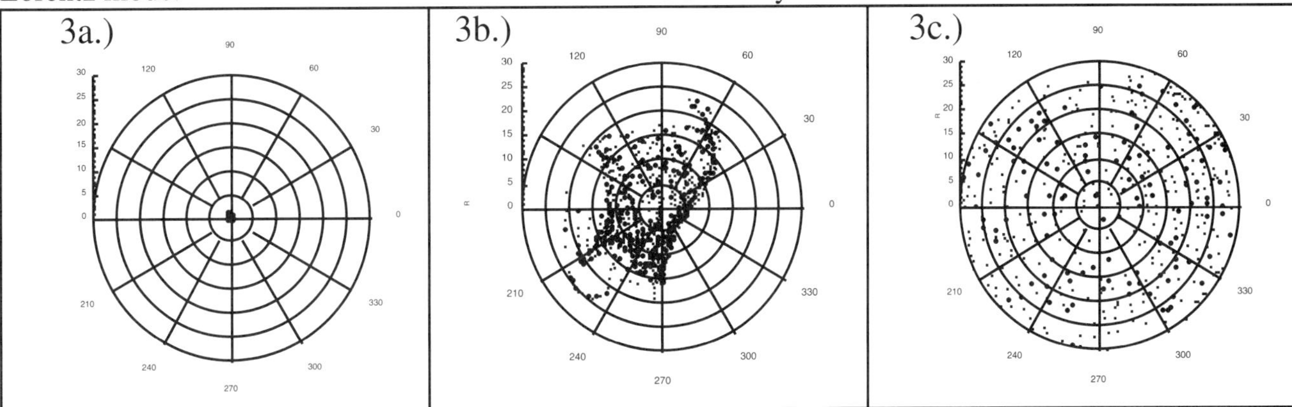

Figure 3 Polar plots of the Lorentz displacement field of 900 EDPs taken from: a) Vaccuun (no specimen), b) Area A of the SiN film [fig. 2], and c) Microdot B [fig.2]. Each point in these plots represents a single measurement of an EDP, in 3a the 900 data points are too closely spaced to be resolved.

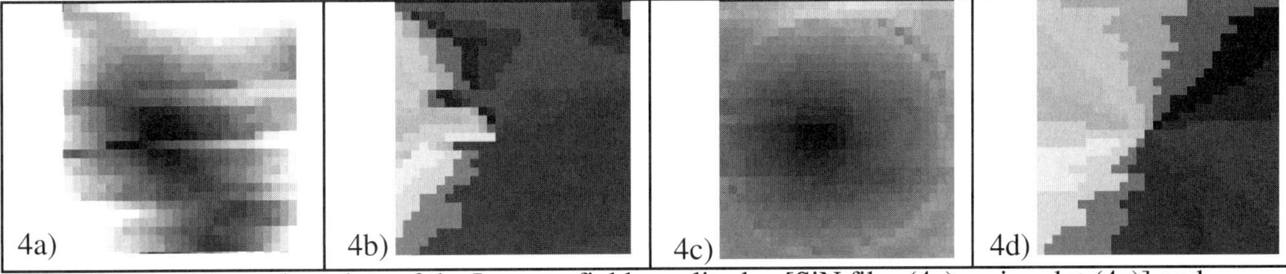

Figure 4.) Two dimension plots of the Lorentz field amplitude [SiN film (4a), microdot (4c)] and orientation [SiN film (4b), microdot (4d)]. All amplitudedata is directly grayscale coded and normalized from minimum (black) to maximum (white), while the rotation data is grayscale converted from a 360 degree full spectrum color wheel. For SiN film the orientation component is rotationally asymmetric and directional reflecting the data of fig 3b, while that of the microdot is clearly rotationally symmetric. Together, the pseudo-images of 4c,d characterizes the Lorentz field to be an in-plane vortex located near the center of the microdot.

Microsc. Microanal. 8 (Suppl. 2), 2002
DOI 10.1017.S1431927602100651

Advanced Magnetic Force Microscopy Tips for Domain Images of Soft Magnetic Materials under Magnetic Field

S. H. Liou

Department of Physics and Astronomy and Center for Materials Research and Analysis, University of Nebraska, Lincoln, NE 68588-0111

We have developed low stray field and high coercivity magnetic force microscopy (MFM) tips for the domain images of soft magnetic materials under magnetic field. A systematic domain evolution under magnetic field can clearly be observed by MFM in a garnet sample.

Magnetic domain structures of garnets have been studied by ferrofluid decoration methods and magneto-optic Faraday effect etc.[1, 2] The resolutions of these methods are limited by either the size of the magnetic particles used in the ferrofluid or the wavelength of the light used in the magneto-optic Faraday effect. The recently developed magnetic force microscopy has been proved to have a higher resolution than ferrofluid decoration methods and magneto-optic Faraday effect in magnetic domain imaging. There are a few investigations of the magnetic domain structures of garnets with or without external magnetic field using the MFM method. However, imaging magnetic domains of soft magnetic materials (low magnetic coercive field materials such as garnets) needs some special precautions due to the fact that the domain structures of soft magnetic materials are easily perturbed by the stray field from the magnetic force microscopy tips [3]. It is also noted that many studies use soft tips. The soft MFM tip is more difficult for magnetic domain images under magnetic field, due to the change of magnetization direction of the tip under magnetic field. The ideal magnetic force microscopy tip for imaging magnetic domains of soft magnetic materials should have a high coercive and low stray field.

In this paper, we have systematically studied magnetic domains of an epitaxial garnet film using magnetic force microscopy. The magnetic garnet is an epitaxial YGdTmGa/YSmTmGa film, grown on a nonmagnetic $Ga_3Ga_5O_{12}$ (GGG) substrate, with a zero-field stripe width 1.4 μm. To our knowledge, we have fabricated CoPt MFM tips with low stray field and high H_c for the first time. We demonstrated that these MFM tips can be used in the study of magnetic domain images of soft magnetic materials in the presence of an external magnetic field. An example of this type study is shown in Fig.1. In this case, the magnetic field was applied along the easy axis (perpendicular to the film surface). The magnetization curve was measured by a SQUID magnetometer, as shown in Fig. 2. That the domain structure of the garnet is sequentially varied with the applied magnetic field is clearly observed, showing that we are able to directly correlate the magnetic domain patterns with the magnetization curve. The MFM tip does not appear to disturb the domain structure of the garnet. The new developed MFM tips show great promise for magnetic domain images of soft magnetic materials under magnetic field. We show the direct correlation of the domain structure and the magnetic hysteresis curve. The moving of the domain wall structure that is affected by the stray field emanating from the tip is only noted in the high magnetic field range from 190 Oe to 290 Oe. In this field range the magnetic domain wall is easy to move. The MFM tip has been cycled through the positive and negative magnetic field. There are not any noticeable effects on the change of the magnetization direction of the MFM tip under the external magnetic field.

References

1. R. R. Katti, P. Rice, J. C. Wu, H. L. Stadler, IEEE Trans. Magn. 28, (1992) 2913.
2. A. Wadas, J. Moreland, P. Rice, R.R. Katti, Appl. Phys. Lett. 64, (1994) 156.
3. Fang Tian, Chen Wang, Guangyi Shang, Naixin Wang, Chunli Bai, J. Magn. Magn. Mater. 171, (1997) 135.

4. Research was supported by the Army Research Office under grant number DAAD19-00-1-0119 and Nebraska Research Initiative at the University of Nebraska.

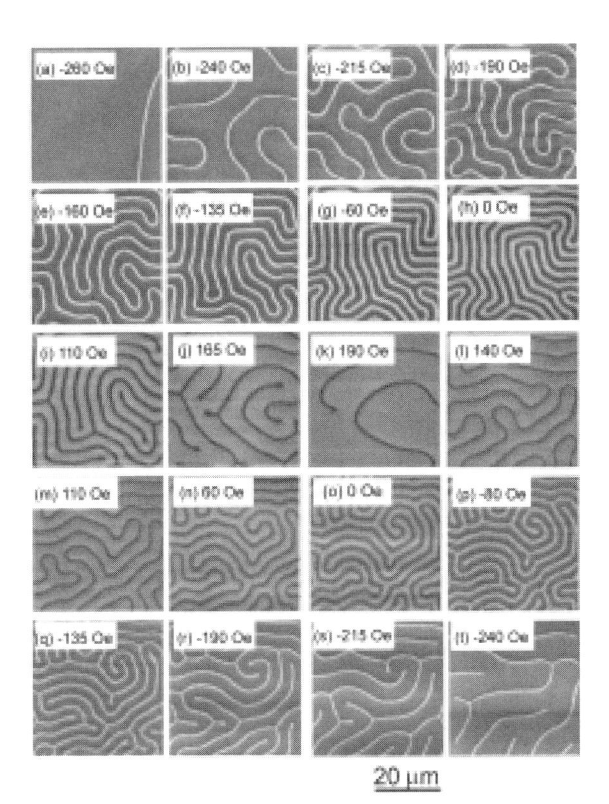

Fig 1 Magnetic domain images of a garnet film under magnetic field.

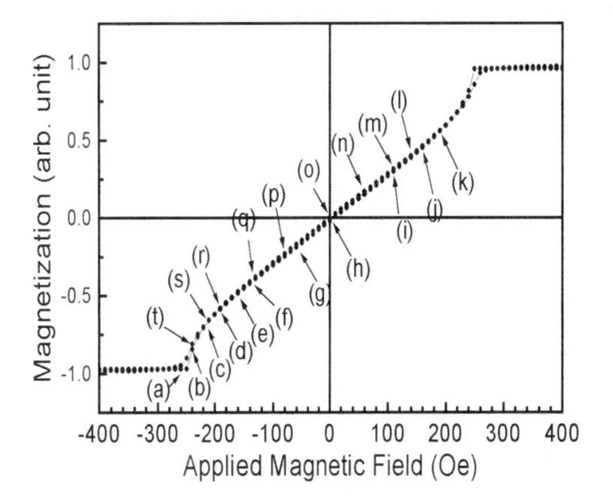

Fig.2 The magnetization curve of a garnet measured with the field perpendicular to the film surface. The labels correspond to these in Fig.1.

Microsc. Microanal. 8 (Suppl. 2), 2002
DOI 10.1017.S1431927602100663

Dynamic Observation of Vortices in High-T_c Superconductors by Lorentz Microscopy

A. Tonomura

Advanced Research Laboratory, Hitachi, Ltd. Hatoyama, Saitama, Japan
SORST, Japan Science and Technology Corporation (JST), Tokyo, Japan
Frontier Research System, The Institute of Chemical and Physical Research(RIKEN), Saitama, Japan

The behaviors of quantized vortices in superconductors affect the practical applications of superconductors. This is because the vortex motion due to the current-induced Lorentz force eventually leads to the breakdown of superconductivity. Therefore, vortex-pinning is a key problem for developing high-critical current materials, especially in high-T_c superconductors where vortices tend to move easily due to high-temperature operation and layered material structures.

Previously, we developed a method to directly and dynamically observe vortices in superconducting thin films by using the phase information of an electron beam transmitted through the films using our 350-kV field-emission electron microscope. Individual vortices can be observed as spots by Lorentz microscopy [1], and projected magnetic lines of force of the vortices were observed by holographic interference microscopy [2]. However, it is mainly effective for conventional metal superconductors such as Nb but not for high-T_c superconductors. This is because 350-kV electrons cannot pass through high-T_c films thicker than their magnetic penetration depths, which are one order of magnitude greater than that of Nb. The recently developed 1-MV electron microscope [3] has enabled to observe vortices in high-T_c superconducting films.

The observation method for Lorentz microscopy is illustrated in Fig. 1. A superconducting thin film is tilted at 30° - 45° to an incident electron beam and a magnetic field is applied from various directions. The phase distribution of an electron beam transmitted through the film is determined by the projected magnetic flux distribution along the direction of the electron beam due to the Aharonov-Bohm effect [5]. The phase distribution cannot be observed by an in-focus image but can be transformed into an intensity distribution by image defocusing as in a Lorentz micrograph.

This method of observing vortices is different from other methods, such as the Bitter method [6] or methods using scanning probe microscopes [7-9], in that it quantitatively detects not only the magnetic field leaking outside from the sample surface, but also the magnetic flux inside the superconductor. Therefore, we can obtain direct information about vortices and observe pinning centers simultaneously inside the sample though the images are blurred by image defocusing [10].

Our 1-MV microscope, which has the brightest beam ever obtained, is very suitable for observing fine magnetic structures for the following reason. The attainable precision in the phase measurement is determined by the collimation angle of the incident electron beam, and therefore a brighter electron beam is required. In addition, a higher accelerating voltage is desirable for the magnetic flux observation, since the magnetic flux sensitivity remains the same, *i. e.* a vortex magnetic flux of $h/2e$ always produces an electron phase shift of p, even when faster electrons are used. The other phase shifts of faster electrons such as those caused by thickness variations or by electric fields are decreased in general. Therefore, a 1-MV field-emission electron beam is suited for magnetic field observations.

In fact, we have obtained new results [4, 11] on magnetic vortices peculiar to high-T_c superconductors with this microscope that we could not achieve with our previous 300-kV microscope. One example is the observation of the chain-lattice state of vortices in Bi-2212 that was obtained when a magnetic field was applied obliquely to the layer plane (See Fig. 2). We found that chain vortices gradually began to disappear at high temperatures and attributed this phenomenon to the oscillation of vortices along the chain direction [11].

References

[1] K. Harada *et al.,* Nature 360 (1992) 51.
[2] J. E. Bonevich *et al.,* Phys. Rev. Lett. 70 (1993) 2952.
[3] T. Kawasaki *et al.,* Appl. Phys. Lett. 76 (2000) 1342.
[4] A. Tonomura *et al.,* Nature 412 (2001) 620.
[5] M. Peshkin and A. Tonomura, Lecture Notes in Physics 340 (SpringerVerlag, 1989)
[6] U. Essman and H. Träuble, Phys. Lett. 24A (1967) 526.
[7] H. F. Hess *et al.,* Phys. Rev. Lett. 64 (1990) 2711.
[8] A. M Chang *et al.,* Appl. Phys. Lett. 61 (1992).
[9] L. N. Vu *et al.,* Appl. Phys. Lett. 63 (1993) 1693.
[10] K. Harada *et al.,* Science 274 (1996) 1167.
[11] T. Matsuda *et al.,* Science 294 (2001) 2136.

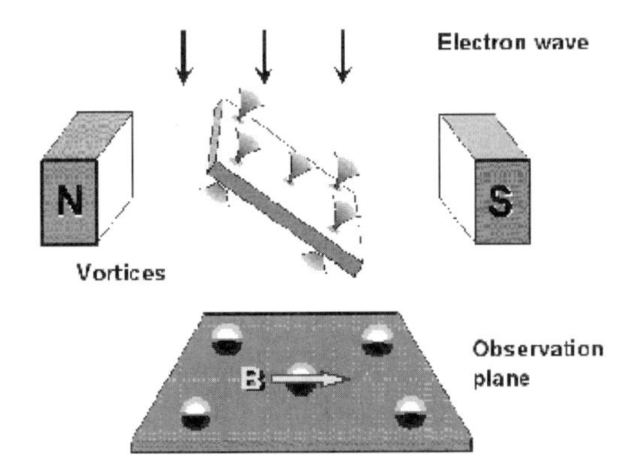

FIG. 1. Principle behind Lorentz microscopy of vortices.

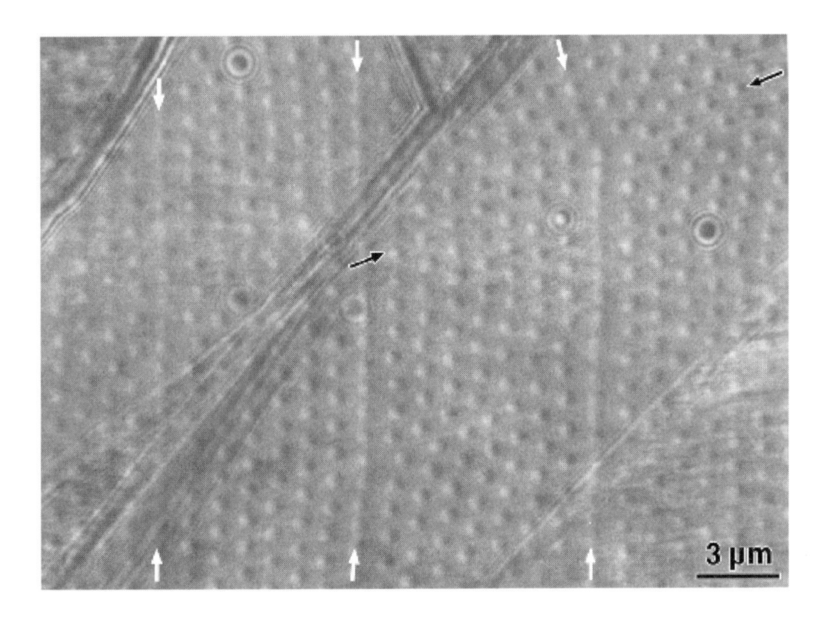

FIG. 2. Lorentz micrograph of vortices in the chain-lattice state (Sample temperature = 50 K, incidence angle of magnetic field = 80º)

Microsc. Microanal. 8 (Suppl. 2), 2002
DOI 10.1017.S1431927602100675

HREM CHARACTERIZATION OF MAGNETIC THIN FILMS AND MULTILAYERS

David J. Smith*

* Center for Solid State Science and Department of Physics and Astronomy, Arizona State University, Tempe, AZ 85287

The reduced vertical dimensions of magnetic thin films and multilayers lead to major and often unexpected changes in magnetic properties and behavior. In addition to their intrinsic scientific importance, these novel characteristics have obvious direct relevance to current and projected technological needs. In particular, future generations of magnetic storage devices and magnetic field sensors will rely on spin-dependent scattering in arrays of mesoscopic structures such as spin-valves, giant magnetoresistance (GMR) superlattices, and other spintronic devices that are scaled transversely and vertically on the nanometer scale. Implicit to the successful implementation of this technology is a detailed understanding of materials growth mechanisms at this level. Chemical and crystallographic structure must be correlated with micromagnetic structure and dynamic response before the fundamental limits of device performance can be firmly established. The transmission electron microscope and related techniques will continue to make a major contribution to the success of these ongoing investigations.

In the GMR effect, thin ferromagnetic (FM) layers are separated either by thin metallic spacer layers ("spin-valve" - SV) or thin insulating barriers ("magnetic tunnel junction" - MTJ), and the combination may display changes in resistance that can be as large as 40% or more at room temperature depending on whether the FM layers are coupled ferromagnetically (i.e., parallel spins) or antiferromagnetically (i.e., anti-parallel spins). The former SV configuration forms the basis for the GMR recording head, which is already in widespread commercial use. The latter MTJ configuration has been adopted in the more recent development of non-volatile magnetic random-access memory (MRAM) technology [1, 2]. The spin-dependent MTJ has attracted much recent interest, with research especially being directed towards optimizing the tunnel barrier thickness, and maximizing the GMR values. Some post-deposition annealing promotes improved GMR, which has been attributed to the enhanced abruptness of the FM-barrier interface, but a monotonic decrease in GMR values occurs for annealing temperatures above about 300°C, probably as a result of interlayer diffusion [3]. Atomic-level analysis by electron microscopy of structural and chemical changes induced by varying the growth temperature or by post-deposition annealing should play a major role in elucidating the reasons for the enhanced magnetic properties [4, 5]. Oxidized aluminum (Al_2O_3) has been the predominant barrier material but there is continuing interest in alternative oxides such as HfO_2 and CoO [6], as well as barriers of AlN and AlON [7], which have potential for band-gap and interface engineering. These materials could offer advantages for device applications as well as allowing additional insights to be obtained into the nature of the tunneling phenomena. For example, there has been ongoing debate about whether some part of the enhanced MR in MTJs results from ballistic transport through nanoscale defects in the barrier, but our recent comparisons with superconducting junctions of similar morphology suggest otherwise [8].

An important factor in the practical utilization of thin magnetic films has been the development of antiferromagnetic (AFM) pinning layers, usually Mn-containing binary alloys or oxides of Co or Ni

[9]. When cooled through their Néel or ordering temperature, in the presence of a strong magnetic field (~1-10 kOe), such AFM materials become ordered, and uncompensated spins at the AFM surface then cause "pinning" of the magnetization direction for any immediately adjacent FM layer. Hysteresis loops are also shifted sideways (i.e., exchange-biased) by substantial amounts (~50-200 Oe). Interfacial roughness and chemical effects are expected to play an important role in layer coupling and magnetization reversal, while further changes in behavior are also likely to occur for both thin films and nanostructures because individual grain size and orientation will have an enhanced influence on the magnetic response. Annealing of Mn-alloy AFMs at elevated temperature (~200-300°C) is usually required to achieve chemical ordering but significant interdiffusion across the FM/AFM interface then becomes inevitable. It was thus significant that spontaneous chemical ordering was observed in $Mn_{0.52}Pd_{0.48}$ films grown by molecular beam epitaxy on body-centered-cubic Fe(001) films at room temperature 10]. In our recent study of $Ni_{0.5}Co_{0.5}O$, it was found that deposition at below room temperature led to smaller columnar grain diameters and a significant increase in the exchange-bias field [11]. These latter two studies emphasize strongly the value of complementary TEM studies for reaching a better understanding of nanoscale magnetic behavior.

Discontinuous metal/insulator multilayers prepared by sputtering from two separate targets have recently started to attract attention because of several practical advantages compared with "traditional" MTJs which require oxidation treatment and some annealing [12]. The problem of incomplete Al oxidation and/or the presence of metallic pinholes through the insulator are greatly minimized in these discontinuous layers. Moreover, the multilayers are easily prepared, they are generally very stable both chemically and electrically, and, most importantly, they show high MR sensitivity at low fields [13]. Interesting, long-range magnetic correlations that depend on particle size have also been observed in non-percolated granular films [14]. Lorentz microscopy at low magnification is also useful in understanding these effects [15].

[1] S.S.P. Parkin, *et al.*, J. Appl. Phys. 85 (1999) 5828.
[2] S. Tehrani, *et al.*, J. Appl. Phys. 85 (2000) 5822.
[3] S.S.P. Parkin, *et al.*, Appl. Phys. Lett. 75 (1999) 543.
[4] R.E. Dunin-Borkowski, *et al.*, J. Appl. Phys. 85 (1999) 4815.
[5] M.J. Plisch, *et al.*, Appl. Phys. Lett., 79 (2001) 391.
[6] D.J. Smith, *et al.*, J. Appl. Phys. 83 (1998) 5154.
[7] M.M. Schwickert, *et al.*, J. Appl. Phys. 89 (2001) 6874.
[8] E.P. Price, *et al.*, Appl. Phys. Lett. 80 (2002) 285.
[9] K. Takano and A.E. Berkowitz, J. Magn. Magn. Mater. 200 (1999) 552
[10] R.F.C. Farrow, *et al.*, Appl. Phys. Lett. 80 (2002) 808.
[11] D. Martien, *et al.*, Appl. Phys. Lett. 74 (1999) 1314.
[12] B. Dieny, *et al.*, J. Magn. Magn. Mater. 185 (1998) 283.
[13] S. Sankar, *et al.*, Appl. Phys. Lett. 73 (1998) 535.
[14] S. Sankar, *et al.*, J. Magn. Magn. Mater. 221 (2000) 1.
[15] The author is pleased to acknowledge ongoing collaborations with Prof. Ami Berkowitz and his group at UC San Diego as well as Drs. Robin Farrow and Stuart Parkin at IBM Almaden.

Microsc. Microanal. 8 (Suppl. 2), 2002
DOI 10.1017.S1431927602100687

STEM Investigations of Defects and Interfaces In Complex Oxides

S. J. Pennycook[1], M. Varela[1], J. Santamaria[2], D. Kumar[3] and G. Duscher[1,4],

[1] Solid State Division, Oak Ridge National Laboratory, Oak Ridge, Tennessee.
[2] Departamento de Fisica Aplicada III, Universidad Complutense de Madrid, 28040 Madrid. Spain
[3] Center for Advanced Materials & Smart Structures, North Carolina A & T State University, Greensboro, NC
[4] Department of Materials Science and Engineering, North Carolina State University, Raleigh NC

The properties of complex oxides can be significantly affected or even dominated by the presence of extended defects or epitaxial strain. Evidence is accumulating that many effects may have a common origin due to non-stoichiometry in the highly stressed region of the dislocation cores. The combination of Z-contrast imaging, electron energy loss spectroscopy (EELS) and first-principles theory is a powerful means to unravel such phenomena. A good illustration is the prototypical perovskite $SrTiO_3$.[1] Z-contrast images revealed the location of the atomic columns, which provided starting models for theory. Calculations then revealed that the low energy structures that agreed with the images were non-stoichiometric, having a Ti/O ratio greater than the bulk. This in turn was confirmed by atomic-resolution EELS. The theory also showed that the excess d-electrons on the Ti atoms were in the conduction band (which is formed from Ti d orbitals). In an acceptor-doped bulk, these electrons would move off the boundary to neutralize nearby acceptors, leaving the boundary charged and setting up a space charge depletion layer. This is the origin of the electrical activity used for varistors, capacitors and other devices.

Dislocations in other complex oxides have similar core structures and are also expected to have strong effects on properties. Band bending due to non-stoichiometric grain boundaries in YBCO leads to hole depletion, and explains the exponential reduction in critical currents with misorientation[2]. In the manganites, grain boundaries have been linked to a useful low-field colossal magnetoresistance. Figure 2 shows the determination of interface structure and misfit dislocation core structure at the $LaAlO_3/La_{2/3}Ca_{1/3}MnO_3$ interface. The interfacial plane can be determined directly from intensity traces across the interface on the two cation sublattices, and the misfit dislocation core is seen to be located two unit cells into the film.

Epitaxial strain may also have a large effect on properties. Figure 3 shows a Z-contrast image of a $[YBCO_1/PBCO_5]$ superlattice with an EELS spectrum taken from the center of the single YBCO unit cell layer. The pre-edge peak is absent, indicating strong hole depletion. For comparison, the EELS spectrum from the center of the YBCO layer in a $[YBCO_8/PBCO_5]$ superlattice does show a substantial pre-edge feature. The reason for the difference is strain; the single layer superlattice is strained, the thicker layer is relaxed. Strain affects the charge in the CuO_2 planes and therefore Tc.

References

[1] .M. Kim et al., Phys. Rev. Letts. 86 (2001) 4056
[2] N D Browning et al Physica C 294 (1998) 183, also S. J. Pennycook et al., Mat. Res. Soc. Symp. Proc. 654 (2001) AA1.3.2
[3] This work was supported by the USDOE under contract DE-AC05-00OR22725 managed by UT-Battelle, LLC.

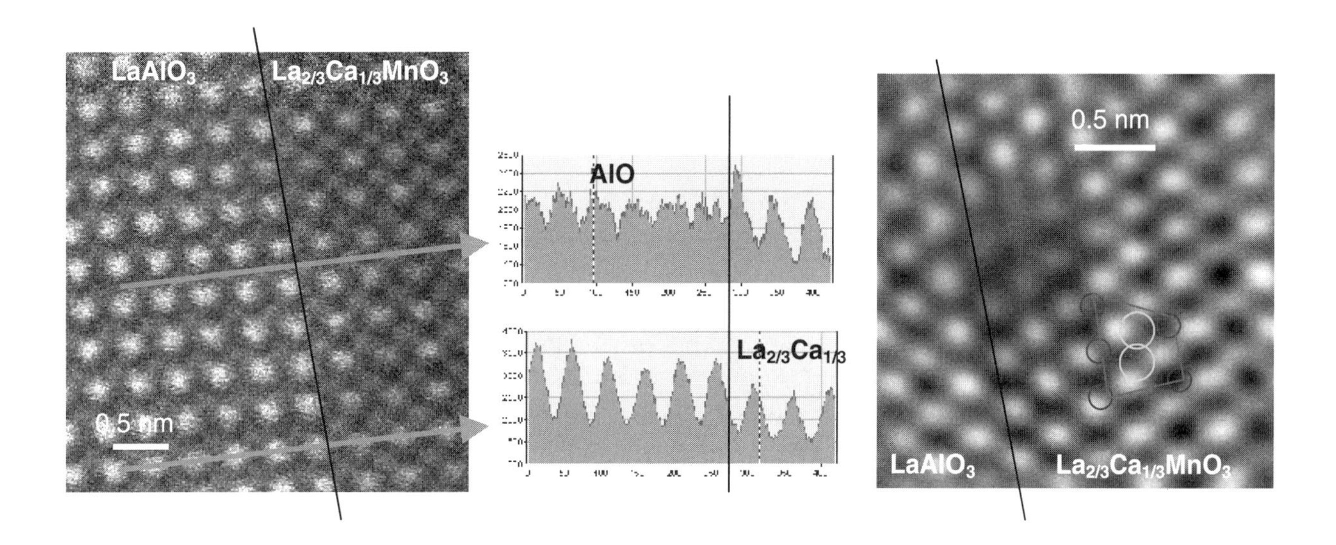

Fig. 1. (a) Z-contrast image of a $LaAlO_3/La_{2/3}Ca_{1/3}MnO_3$ interface showing the location of the interface from intensity traces across the two cation sublattices. (b) Higher magnificaton filtered image showing the location and core structure of a misfit dislocation.

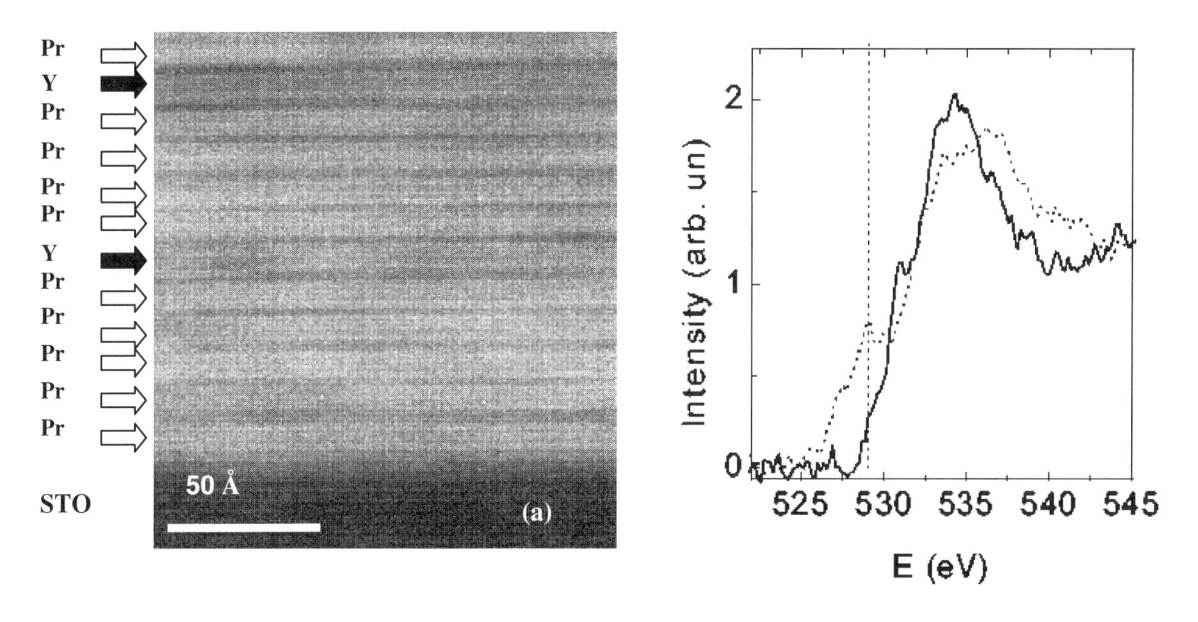

Fig. 2. Z-contrast image of a $[YBCO_1/PBCO_5]$ superlattice. The EELS spectrum of the O K edge shows no pre-peak (solid line), indicating strong hole depletion consistentwith the measured Tc of 35K. A $[YBCO_8/PBCO_5]$ superlattice (dotted line) does show a pre-peak and a Tc of 88K.

DOI 10.1017.S1431927602100699

A Challenging to Characterization of Superconductors: Accurate Measurements of Charge Distribution and Interfacial Displacement

Yimei Zhu,* Lijun Wu* and J. Tafto**

*Material Science Division, Brookhaven National Laboratory, Upton, NY 11973, USA
**also Department of Physics, University of Oslo, P.O. Box 1048 Blindern, 0316 Oslo, Norway

One of the most challenging tasks in solid-state physics and materials science is to understand electronic properties of crystals and interfaces. This is particular important for superconductors where electron/hole distribution in crystal and grain boundary determines the usefulness of the materials. To meet the challenge, we developed a novel electron-diffraction technique PARODI (Parallel Recording Of Dark-field Images) by focusing a small probe above the sample to measure charge density and lattice displacement. Since the method couples diffraction with images, it is thus suitable for studying crystals as well as defects. Combined with HREM and ELNES, the method allows us to address issues crucial to superconductivity mechanism.

To measure 1-D charge distribution, we record PARODI patterns of a systematic reflection row from a wedge crystal. Through quantitative intensity analysis, i.e., comparing the experiments with calculations via fitting and error analyses, we can accurately determine the structure factors of low-order reflections of crystals with a large unit cell, and thus valence electron distribution and charge transfer with very high accuracy [1]. In the case of $YBa_2Cu_3O_7$, a movement of 0.05 electron between CuO and CuO_2 planes that corresponds to rearranging 1 out of 5000 electrons in the crystal, changes the 001 structure factor of electron diffraction by 1 Å while we determine this structure factor with an accuracy of 0.5 Å. Fig.1 shows the charge-density distribution ρ, along the c-axis for $Bi_2Sr_2CaCu_2O_8$ (Bi/2212) together with the charge transfer Δρ, determined from the measurement and the first principle calculation, using formal valence as a reference. A 2-D charge density map of MgB_2 projected along the [001] direction is shown in Fig.2. It reveals not only bonding characteristics (B: covalent bonding, Mg: close to ionic bonding) but also significant charge transfer from Mg to B. Measurements of electrostatic potential variation across grain boundaries in Bi/2212 using PARODI and electron holography will also be presented.

A remarkable advantage of PARIDO is to simultaneously study charge distribution and lattice displacement from the same illuminated area because of a wide range of reflections available (electrons scattered at small-angles are sensitive to charge arrangement, while those scattered at large-angles are sensitive to crystal distortion). Fig.3 is a calculated PARODI pattern with a stacking fault in the middle. The interfacial interference fringes are out of contrast in reflections where **g·R** equals to an integer. We have measured grain boundary expansion of several (001) twist boundaries in Bi/2212. A typical HREM image of such a boundary [2] is shown in Fig.4. By examining the fringe contrast in very high-order reflections such as 0 0 30, we were able to determine the interfacial lattice displacement with an unprecedented accuracy down to 1pm [1]. The displacement was found to range from 0.2-0.4 Å which was not detectable using HREM. An example on the study of a 5° tilt boundary in Bi/2212 using PARODI is presented in Fig.5. [3]

References

[1] Wu, Zhu and Tafto, *Phys.Rev.B,* **59,** 6035 (1999), *Phys. Rev. Lett.* **85** 5126 (2000).
[2] Y.Zhu, Q.Li, Y.N.Tsay, M.Suenaga, G.D.Gu, and N.Koshizuka, *Phys.Rev.B,* **57**, 8601 (1998).
[3] Work supported by US DOE under contract No. DE-AC02-98CH10886.

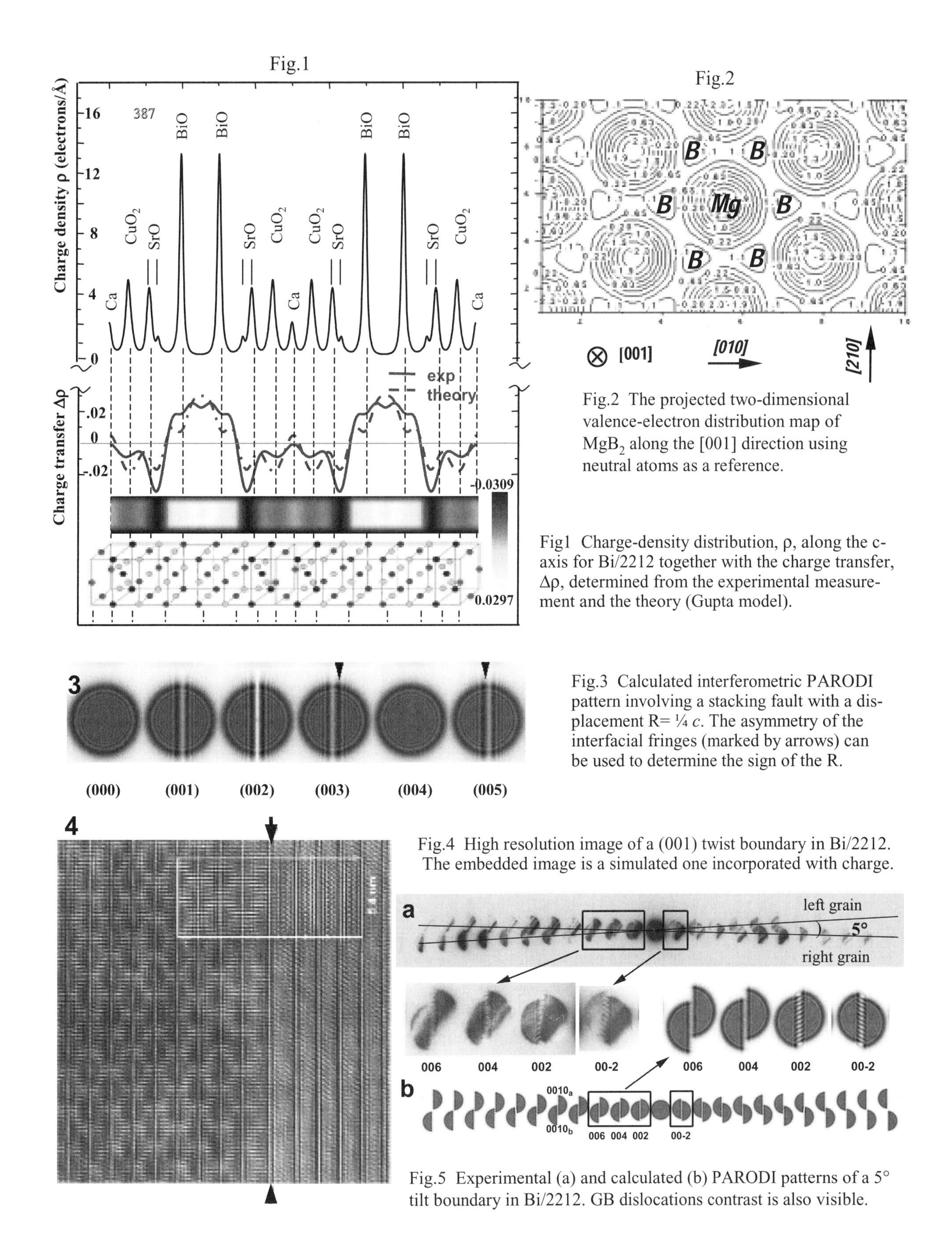

Fig.2 The projected two-dimensional valence-electron distribution map of MgB_2 along the [001] direction using neutral atoms as a reference.

Fig1 Charge-density distribution, ρ, along the c-axis for Bi/2212 together with the charge transfer, Δρ, determined from the experimental measurement and the theory (Gupta model).

Fig.3 Calculated interferometric PARODI pattern involving a stacking fault with a displacement R= ¼ *c*. The asymmetry of the interfacial fringes (marked by arrows) can be used to determine the sign of the R.

Fig.4 High resolution image of a (001) twist boundary in Bi/2212. The embedded image is a simulated one incorporated with charge.

Fig.5 Experimental (a) and calculated (b) PARODI patterns of a 5° tilt boundary in Bi/2212. GB dislocations contrast is also visible.

Microsc. Microanal. 8 (Suppl. 2), 2002
DOI 10.1017.S1431927602100705

HIGH-RESOLUTION AND LOW TEMPERATURE TEM STUDY OF SUPERCONDUCTING CUPRATES AND CMR-MANGANITES

YOSHIO MATSUI
National Institute for Materials Science, Tsukuba, Ibaraki 305-0044, Japan.

After the discovery of high -Tc cuprate in 1986 [1] as well as CMR -manganite in 1993 [2], we have developed the following different types of TEMs to examine "strongly -correlated transition metal oxides"; (1) 1.3MV high -voltage TEM (H -1500) with 1Å resolution [3], (2) 300kV analytical TEM (HF-3000S) with FEG and GIF, and (3) 300kV Lor entz TEM with FEG and external magnet up to 300G at the specimen. In the field of high -Tc superconductors, we have mainly examined the structures of new compound by HRTEM [4,5]. In the field of magnetic materials, we examined the charge-ordered superstruc tures and ferromagnetic domain structures formed at low temperature (typically below 150K) in various manganites [6,7]. Here, we present two examples of recent application of our three TEMs to superconductors and magnetic materials.

1. Order/Disorder of CO_3 (partly NO_3) Groups in Oxycarbonitrate Superconductors[5]

A series of new oxycarbonitrate superconductors, $(Cu,C,N)Sr_2Ca_{n-1}Cu_nO_y$ (n=1 to 6), prepared under high-pressure of 5 to 6 GPa, are examined by high-resolution TEM. Ordered arrangements of Cu and C (N) are observed in the compounds with n=1 to 4, while almost random arrangements for those with n=5 and 6. In the first two members, (Cu,C,N)-1201 (n=1) and 1212 (n=2), ordering scheme of -Cu-C(N)-C(N)-C(N)-Cu- with four-times periodicity is observed in the charge-reservoir blocks (CRB), as shown in Fig.1(a) to (d) for n=2 compound. For (Cu,C,N)-1234 (n=4), with highest T_c of 113K in the series, the ordering scheme of Cu-C(N)-Cu-C(N)-Cu- with twice periodicity is observed in the CRB. In the last two members, (Cu,C,N)-1245 (n=5) and 1256 (n=6), no evidence of ordering was observed, suggesting that Cu, C and N are distributed almost randomly in the CRB. Such relations between the order/disorder scheme and n-parameters, are also preserved locally in the crystals which contain plenty of intergrowth defects.

2. Direct Observations of Ferromagnetic Domains in Manganites [8]

Ferromagnetic domains in $Nd_{1/2}Sr_{1/2}MnO_3$, which undergoes the ferromagnetic to charge -order (antiferromagnetic) transitions at around T_{CO}=150K, are examined by Lorentz TEM using mainly the Fresnel-mode. On cooling from paramagnetic state of room temperature, magnetic domain walls started to appear, as black and white lines, below T_C=250 K. The direction of magnetization in each domain is along the long sides of domain, suggesting that the compound has a magnetocrystalline anisotropy. With a further decrease of temperature, the volume of magnetic domains increased with discontinuous domain-wall jumps, and then gradually disappeared below T_{CO}=150K. Clear satellite reflections due to lattice distortion, are observed in the hk0 electron diffraction pattern. On heating process, a characteristic granular contrast with 30-40 nm in size was observed at around 140 K, close to the T_{CO}, as sho wn in the Lorentz TEM image in Fig. 3(a) and (b), at underfocus and overfocus, respectively. Such a granular contrast was not observed in the cooling process . We consider that the origin of this contrast is the formations of ferromagnetic microclusters i n antiferromagnetic matrix.

References:

1.) J.G. Bednorz and K.A. Muller, Z. Phys. B64, 189, (1986).
2.) K. Chabara et al., Appl. Phys. Lett., 63, 1990 (1993).
3.) Y. Matsui et al., Ultramicroscopy, 39, 8, (1991).
4.) Y. Matsui and J. Akimitsu, Microsc. Res. and Tech., 30, 155, (1995).
5.) Y. Anan et al., Phil. Mag.B 81, 1847, (2001).
6.) T. Asaka et al., Phys. Rev. Lett., in press.
7.) T. Nagai et al., Phys. Rev. B., in press.
8.) T. Asaka et al., unpublished.

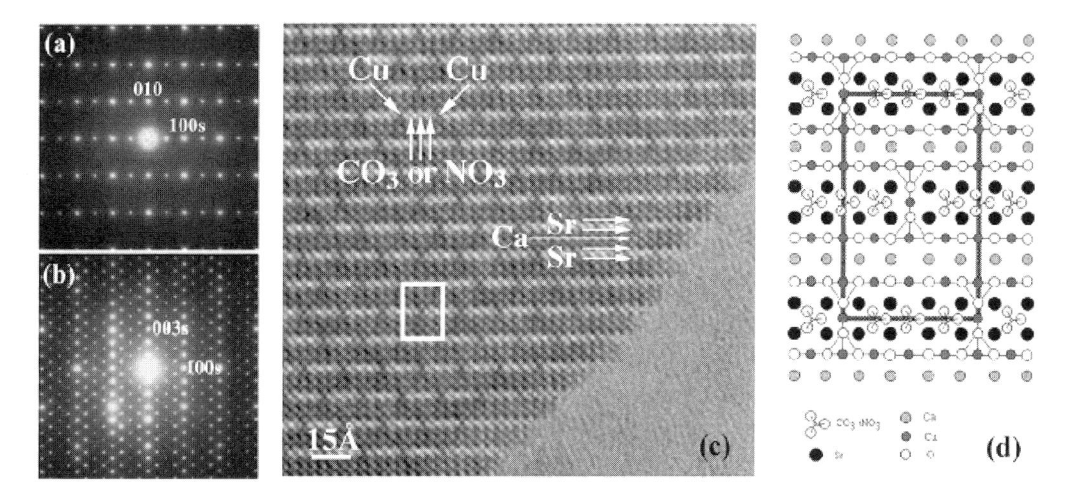

Fig. 1. The (a) *hk0* and (b) *h0l* electron diffraction patterns, (c) HRTEM image projected along *b*-direction and (d) the superstructure model of (Cu,C,N) -1212 type of oxycarbonitrate superconductor. Ordered arrangements of Cu and three CO_3 (partly NO_3 ones) are clearly observed. HRTEM image was taken at 800kV, by H-1500, to prevent electron-beam damage.

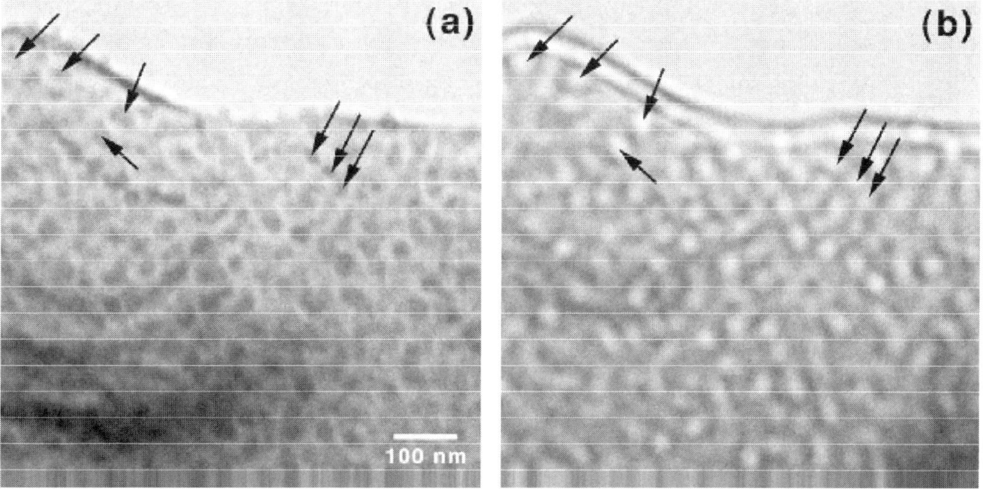

Fig. 2. Lorentz TEM images (300kV) of $Nd_{1/2}Sr_{1/2}MnO_3$ in the (a)underfocused and (b)overfocused conditions at 140K, near T_{CO}. Arrows indicate the reverse in contrast.

Microsc. Microanal. 8 (Suppl. 2), 2002
DOI 10.1017.S1431927602100717

In-Situ Lift-Out FIB Specimen Preparation for TEM of Magnetic Materials

B.W. Kempshall and L.A. Giannuzzi

Mechanical Materials and Aerospace Engineering, University of Central Florida, 4000 Central Florida Blvd, Orlando, FL 32816-2450

The focused ion beam (FIB) (ex-situ) lift-out technique has been used to prepare specimens for transmission electron microscopy (TEM) from numerous materials applications [1]. In this report, we use the in-situ lift-out technique [2] to prepare specimens from magnetic materials for TEM analysis. An FEI 200TEM FIB workstation equipped with an Omniprobe in-situ W probe was used for the specimen preparation. A Philips/FEI Tecnai F30 operating at 300 keV was used for the TEM analyses.

FIG. 1 shows a cross-section bright field (BF) TEM micrograph of a $Nd_2Fe_{14}B_1$ sample containing Nd, Dy, and Co. The sample was heated at 600°C for 30 min. Many of the grains are elongated and consist of dimensions that are 50-100 nm in width and 100-300 nm in length. Preliminary results indicate that the smaller equiaxed grains that are on the order of ~ 20 nm in dimension may consist of the αFe phase. FIG. 2 shows a lattice image from a region in FIG. 1 which indicates that the FIB/TEM technique for these materials are suitable for further analysis.

FIG. 3 shows a BF TEM image from a single particle from a $Sm(Co_{7.05-x}Cu_xTi_{0.25})$,x=0.8 powder sample. We have previously shown that individual particles may be prepared directly by FIB lift-out techniques [3]. The specimen in FIG. 3 consisted of particles embedded in an epoxy matrix. Note that the single particle shown in FIG. 3 is polycrystalline in nature. A convergent beam electron diffraction pattern from the dark grain diffracting in the Laue condition shown in FIG. 3 is shown in FIG. 4. This CBED pattern most closely resembles a [0001] pattern that would be expected from either of the hexagonal structures (e.g., Sm_2Co_{17}, $SmCo_5$) that should exist in this system.

High temperature superconducting $Bi_2Sr_2Ca_2Cu_3O_{10}$ (BSCCO) was obtained from Colorado Superconductor, Inc. FIG. 5 shows a BF TEM micrograph from the BSCCO sample. The polycrystalline nature of the specimen in FIG. 5 is evident by the selected area diffraction pattern of the orthorhombic structure [4] shown in FIG. 6. These images show that FIB specimen preparation techniques are viable for TEM studies of magnetic materials. TEM studies may now be performed to elucidate the microstructural details of these materials [5].

References

[1] L.A. Giannuzzi, et al., Mat. Res. Soc. Symp. Proc., 480 (1997) 19.
[2] T. Kamino et al., Microsc. Microanal., 6 (Suppl. 2) (2000) 510.
[3] J. K. Lomness et al., Microscopy and Microanalysis, 7 (2001) 418.
[4] D.R. Mishra et al., Pramana J. Physics, 54 2 (2000) 317.
[5] The support of NSF DMR #9703281 and Omniprobe is gratefully appreciated. Some samples were prepared by W. Liu and provided courtesy of Y. Liu.

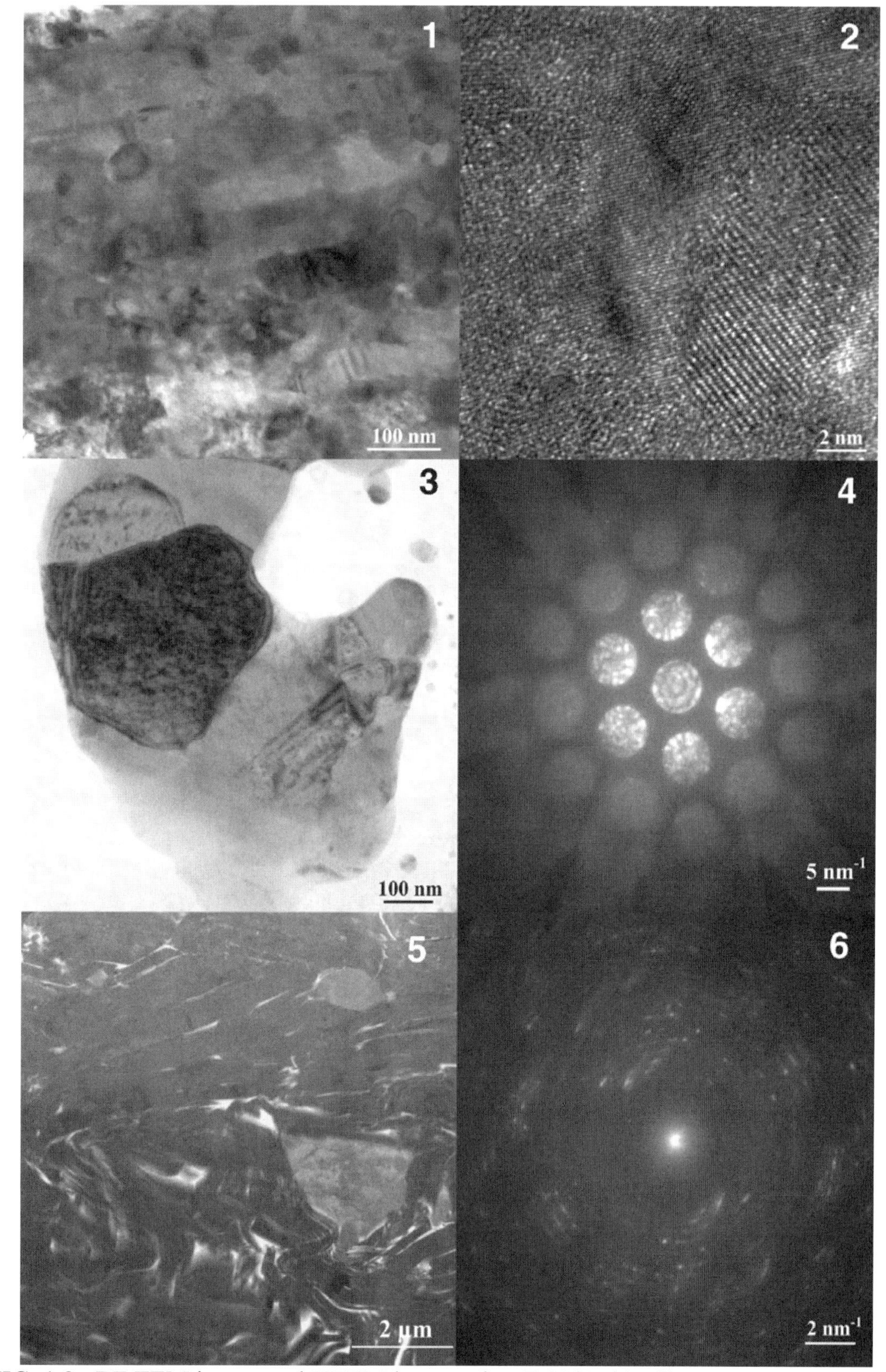

FIG. 1,2. BF TEM image and HREM image of an FeNdB thin film.
FIG. 3,4. BF TEM image and [111] CBED from a NDFCNB powder sample.
FIG. 5,6. BF TEM image and SADP of a BSCCO sample.

Microsc. Microanal. 8 (Suppl. 2), 2002
DOI 10.1017.S1431927602100729

Crystal Structure Determination of Superconductors and Related Compounds by Combining High-Resolution Electron Microscopy and Electron Diffraction

F.H. Li, B.H. Wang and H. Jiang

Institute of Physics & Center for Condensed Matter Physics, Chinese Academy of Sciences, Beijing 100080, P. R. China

High-temperature superconducting compounds generally consist of very small crystalline grains. The two-stage image processing technique [1] based on the combination of high-resolution electron microscopy and electron diffraction is effective to determine the crystal structure for materials of such kind. In the first stage the high-resolution electron microscope image taken at an arbitrary defocus condition is transformed into the structure image by image deconvolution [2]. In general not all atoms can be seen in such obtained structure image, because the image resolution is limited by the resolution of the electron microscope. In the second stage the electron diffraction data is introduced to enhance the image resolution to the diffraction resolution limit by means of the phase extension technique using the direct method developed in X-ray crystallography [3]. Usually all atoms are resolved in the final projected potential map (PPM).

A series of crystal structures of high-temperature superconductors and related compounds were determined by this method [4-9]. Recently, the crystal structure of $(Y_{0.6}Ca_{0.4})(SrBa)(Cu_{2.5}B_{0.5})O_{7-\delta}$ has been studied. The electron diffraction patterns and images were taken with JEM-200CX and JEM-2010 electron microscopes, respectively. The diffraction intensities were collected with a slow scan CCD camera in a Tacnai F20 electron microscope.

Fig. 1 shows the electron diffraction patterns of [001] and [010] zone axes. It was determined that the crystal belongs to the orthorhombic system and lattice parameters are a = 3.85 Å, b = 3.86 Å and c = 11.5 Å. It is reasonable to recognize that the crystal structure is isomorphous to that of $YBa_2Cu_3O_{7-\delta}$. The present work is aiming at confirming the assumption that boron atoms substitute for those copper atoms, which are located at the Cu-O chain positions. Fig. 2 shows the [010] image corresponding to Fig. 1b. Fig. 3a shows the Fourier filtered image obtained from a thin area in Fig. 2. The rectangle indicates the projected unit cell. The symmetry average image obtained from Fig. 3a is given in Fig. 3b. Fig. 3c is the structure image obtained from Fig. 3b by maximum entropy image deconvolution with the defocus value –650 Å [10], where dark dots represent atoms, but not all atoms can be seen. For enhancing the image resolution, the phase extension in combination with diffraction intensity correction [11] was performed. In order to determine the position of boron atoms, two structure models were set up in the second cycle of diffraction intensity correction. In one of models boron atoms distribute randomly in all Cu positions, while in the other they are located at Cu-O chain positions. It turns out that the later model is correct. The determined PPM is given in Fig. 3d.

References

[1] F.H. Li, Proc. International Congress on Electron Microscopy, Paris (1994) Vol. 1, 481.
[2] F.S. Han et al., Acta Cryst. A, 42 (1986) 353.
[3] H.F. Fan et al., Ultramicroscopy, 36 (1991) 361.
[4] Z.Q. Fu et al., Ultramicroscopy, 54 (1994) 229.
[5] B. Lu et al., Ultramicroscopy, 70, (1997) 13.
[6] J. Liu et al., Materials Transaction, JIM, 39 (1998) 920.
[7] H. Jiang and F.H. Li, Micron, 30 (1999) 417.
[8] H. Jiang and F.H. Li, Proc. 8th Beijing International Conference on Instrumental Analysis (1999) A97.
[9] J. Liu et al., Acta Cryst. A, 57 (2001) 540.
[10] J.J. Hu, and F.H. Li, Ultramicroscopy, 35 (1991) 339.
[11] D.X. Huang et al., Acta Cryst. A, 52 (1996) 152.
[12] This research is supported by the National Natural Science Foundation of China and the Ministry of Science and Technology of China (NKBRSF-G1999064603).

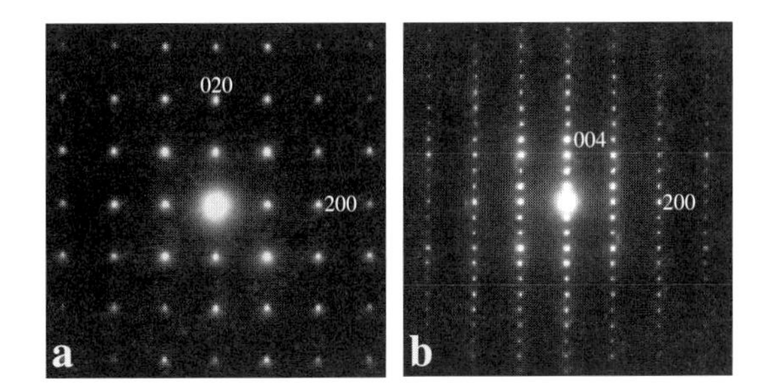

Fig. 1. Diffraction patterns of $(Y_{0.6}Ca_{0.4})(SrBa)(Cu_{2.5}B_{0.5})O_{7-}$

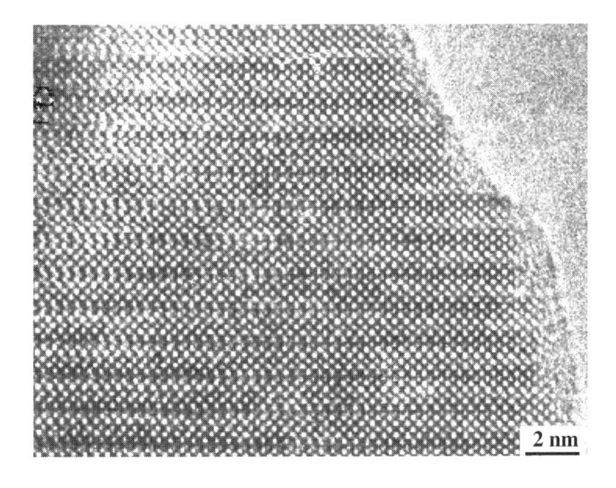

Fig. 2. [010] image corresponding to Fig 1b.

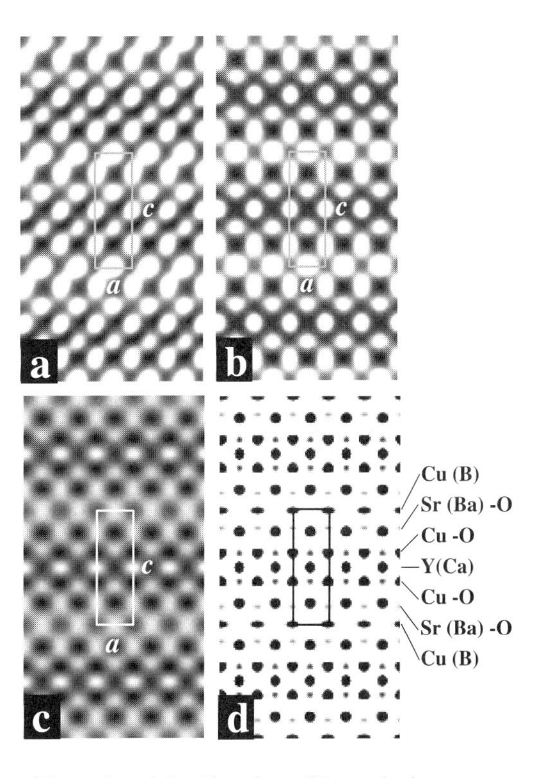

Fig. 3. (a) Fourier filtered image obtained from Fig. 2, (b) symmetry average image obtained from (a), (c) deconvoluted image obtained from (b) and (d) PPM obtained from (c).

Microsc. Microanal. 8 (Suppl. 2), 2002
DOI 10.1017.S1431927602100730

Exploring the Surface of Pb-doped Manganese Perovskite

C.N. Borca, and P.A. Dowben

Physics Department, University of Nebraska at Lincoln, Lincoln, NE 68588

Recently, there have been a surge of interest in transition metal oxides due to their unique physical properties, e.g. colossal magnetoresistance (CMR) [1] and extremely high degree of spin polarization [2]. Systems with 100% spin-polarization (or half-metals) are desirable in spin-valves, spin-tunnel junctions and other spin dependent devices and structures. In the present study, we investigated the importance of the surface and near-surface composition on the value of spin-asymmetry in $La_{0.65}Pb_{0.35}MnO_3$ thin films.

The $La_{0.65}Pb_{0.35}MnO_3$ thin films (1000 Å in thickness) were grown on (100) $LaAlO_3$ substrates by RF sputtering [3]. The films are polycrystalline and, highly oriented along the substrate normal direction and single phase. For the detailed studies of the surface composition, the films were installed into an ultra high vacuum system (base pressure $2x10^{-10}$ torr). To ensure surface cleanliness, the "as-deposited" samples were gently annealed to 250°C for 10-14 hours, in oxygen atmosphere of $1x10^{-6}$ torr. After careful analysis of this initial surface, the samples were further annealed in vacuum to 520°C for 10-14 hours. This second thermal treatment produced a heavily annealed surface with completely different composition and structure, as discussed below.

Changes in surface topography were obtained using a scanning probe microscope (STM), as seen in Figure 1. Panel 1(a) corresponds to the "as-deposited" sample, with a roughness of 5 nm, which indicates a smooth surface, with atomically flat terraces. The heavily annealed surface shown in Figure 1(b) suggests the existence of mixed phases and an increased surface roughness of 50 nm. The changes induced into the surface composition and structure due to heavy annealing have been confirmed by low energy electron diffraction (LEED) and X-ray photoemission spectroscopy (XPS). LEED shows the appearance of sharp superlattice spots in the case of the heavily annealed surfaces, indicating a possible four-fold modulation in the Mn-O planes. XPS of the La, Pb and Mn core levels, acquired as a function of photoelectron emission angles, show that the heavily annealed surface goes through a restructuring transition characterized by the formation of a Ruddlesden-Popper phase $(La_{1-x}Pb_x)MnO_4$ with a possibly embedded MnO columnar phase.

The effect of surface composition on the spin-resolved band structure of this material can be seen using inverse photoemission spectroscopy. Figure 2 shows the spin-resolved spectra of the unoccupied band structure of the as-deposited surface (a) and of the heavily annealed surface (b). The value of the surface spin-asymmetry is obtained by taking the ratio between the difference and the sum of the spin-up and spin-down spectra for each surface, and it is shown at the bottom of panels (a) and (b) in Figure 2. The spin-asymmetry at room temperature for the as-deposited surface reaches 80% above background, while the heavily annealed surface shows a maximum of 40% above background at an energy of 0.5 eV above the Fermi level. These values are far from the theoretically predicted 100% spin-asymmetry at the Fermi level, and we believe that the reason for this discrepancy is the existence of a surface region with a different composition than the bulk [4].

The results shown here are of great importance, not only in explaining lower than expected polarization values found in literature, but also because they underscore the importance of characterizing the interface in any effort to understand the properties of spin-electronic junctions. We present clear evidence that changes in surface composition induced by thermal annealing greatly affect the electronic structure of the surface region in Pb-doped manganese perovskite thin films.

[1] H. Kuwahara et al., *Science* 270 (1995) 961.
[2] R.J. Soulen Jr. et al., *Science* 282 (1998) 85.
[3] Q.L. Xu et al., *Matter. Res. Soc. Symp. Proc.* 602 (2000) 75.
[4] C.N. Borca et al., *Europhys. Lett.* 56 (2001) 722.

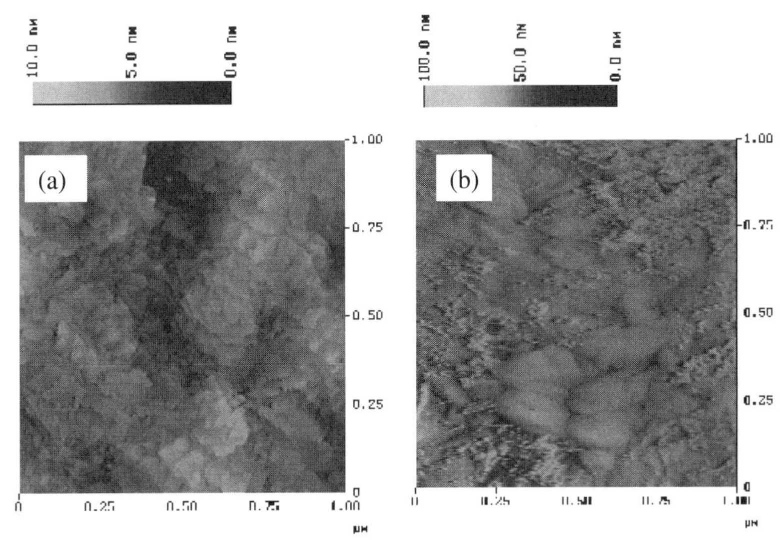

FIG. 1. STM images, 1x1 μm^2 in lateral dimensions, of the $La_{0.65}Pb_{0.35}MnO_3$ thin film 'as-deposited' surface (a) and after heavy annealing treatment (b).

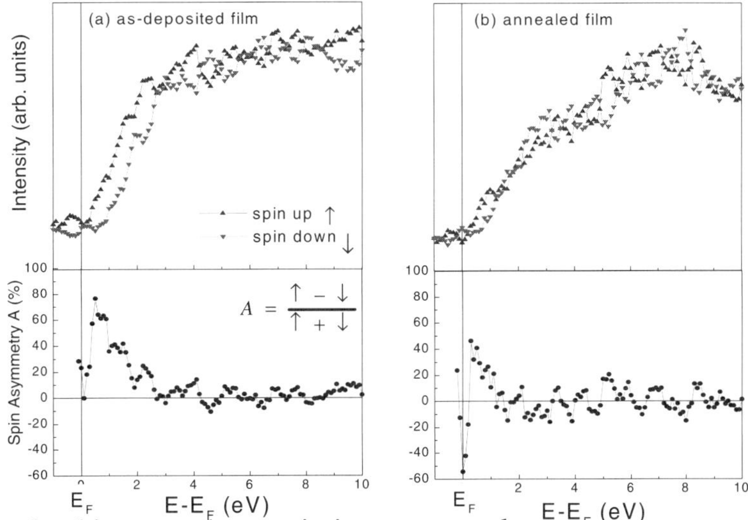

FIG. 2. Spin-resolved inverse photoemission spectra taken at room temperature and normal incidence for the 'as-deposited' $La_{0.65}Pb_{0.35}MnO_3$ surface (a) and for heavily annealed surface (b). Spectra at the bottom of each panel represent the spin asymmetry distributionas a function of incident electron energies.

Microsc. Microanal. 8 (Suppl. 2), 2002
DOI 10.1017.S1431927602100742

Mössbauer Studies of Fe-doped La-Ca-Mn-O Colossal Magnetoresistive Perovskites

Zhao-hua Cheng[*,**], Zhi-hong Wang*, Nai-li Di[*], Rui-wei Li*, R.A. Dunlap[**], and Bao-gen Shen[*]

[*]State Key Laboratory of Magnetism, Institute of Physics, Chinese Academy of Sciences, Beijing 100080, P.R. China
[**]Department of Physics, Dalhousie University, Halifax, Nova Scotia, Canada, B3H 3J5

The discovery of colossal magnetoresistance (CMR) in the manganite perovskites $La_{1-x}Ca_xMnO_3$ has recently attracted significant scientific attention[1,2]. A local structural distortion of MnO_6 octahedron resulting from the Mn^{3+} Jahn-Teller effect was assumed to play an important role in determining the resistivity behavior and the magnetic transition temperature[3]. Since the long-range structural distortion decreases rapidly in $La_{1-x}Ca_xMnO_3$ compounds with increasing Ca concentration, a local structural probe is required to analyze the distortion of the MnO_6 octahedron. Mössbauer spectroscopy is a powerful technique for investigating the hyperfine interactions. One of the hyperfine parameters, the quadrupole splitting (Δ) is very sensitive to the distribution of surrounding electrons of resonant nuclei. A few percent of Fe, which substitutes for Mn in $La_{1-x}Ca_xMnO_3$ compounds, can be used as a micro-probe to detect the symmetry of the nearest-neighbor O^{2-} ions in the $Mn(Fe)O_6$ octahedron. Since Mn^{3+} ions are mainly replaced by Fe^{3+} ions in this Fe-doping range and both ions have identical ionic radii in six-fold octahedral coordination [4], the substitution of Fe^{3+} for Mn^{3+} does not change the tolerance factor, and consequently, the Jahn-Teller effect can be investigated separately. In this work, we summarized the application of Mössbauer spectroscopy in La-Ca-Mn-O CMR pervoskites. The local structural distortion and Jahn-Teller coupling have been determined by Mössbauer spectroscopy on the basis of quadrupole splitting. Furthermore, the connection between Jahn-Teller coupling and CMR effect has been discussed.

Rietveld profile refinement of x-ray diffraction (XRD) patterns reveals that all samples are single-phase with an orthorhombically distorted perovskite structure (*Pbnm* space group). The difference of Mn(Fe)-O bond lengths in Ca-doped perovskites cannot be detected within the experimental error, implying that the long-range structural distortion rapidly decreases and becomes indistinguishable by XRD technique. However, a theoretical model has predicted that strong local distortions of MnO_6 octahedra would persist over a wide range of compositions in the $T>T_c$ insulating phase[5].

In order to verify this assumption, Mössbauer spectroscopy utilizes ^{57}Fe nuclei as a local micro-probe to determine the structural information about the $Mn(Fe)O_6$ octahedra. ^{57}Fe Mössbauer spectra not only clearly indicate the presence of quadrupole splitting but also show the occurrence of a second quadrupole split doublet. These two doublets are related to the contributions of the high-spin Fe^{3+} and low-spin Fe^{4+}, respectively. The quadrupole splitting for both Fe^{3+} and Fe^{4+} ions confirms the local distortion of the $Mn(Fe)O_6$ octahedron. These results demonstrate that Mössbauer effect is very sensitive to the local structural information.

According to the relationship between the second-order crystalline electrical field coefficient, which obtained from Mössbauer results, as well as the Jahn-Teller coupling, the value of Jahn-Teller coupling were determined. With increasing Ca concentration, the Jahn-Teller coupling energy decreases from about 1.4 eV for x=0 to about 0.83 eV for x=0.27~0.39. The transport properties of these perovskites are governed by the interplay of the double-exchange coupling between Mn^{3+} and Mn^{4+} ions as well as the Jahn-Teller coupling. For the samples with x=0.27~0.39, the relatively weak Jahn-Teller coupling delocalizes the electrons and makes the hoping process become easier via double-exchange coupling below T_c. Therefore, these samples show a ferromagnetic metals state at $T<T_c$. At $T>T_c$, disorders in the spins result in a decrease of double-exchange coupling, and the Jahn-Teller coupling plays an important role in determining electrical resistance. Therefore, a transition from ferromagnetic metal to paramagnetic insulator can be found near Curie temperature. An applied magnetic field aligns the spins, which increases the kinetic energy, and hence decreases the effective electron-phonon coupling, leading to a large change in resistance. With further increasing Ca concentration, Mössbauer results indicate that the Jahn-Teller coupling becomes stronger again. The larger energy separation between d_{z^2} and $d_{x^2-y^2}$ tends to trap electrons in the low-lying energy level orbitals, and hence localize the conduction electrons. Therefore, for the samples with x>0.5, the ground state becomes insulating and antiferromagnetic again.

Our work not only confirms the local structural distortion of MnO_6 octahedron, but also reveals that Mössbauer spectroscopy can be effectively employed to investigate the Jahn-Teller effect in coupling in $La_{1-x}Ca_xMnO_3$ perovskites.

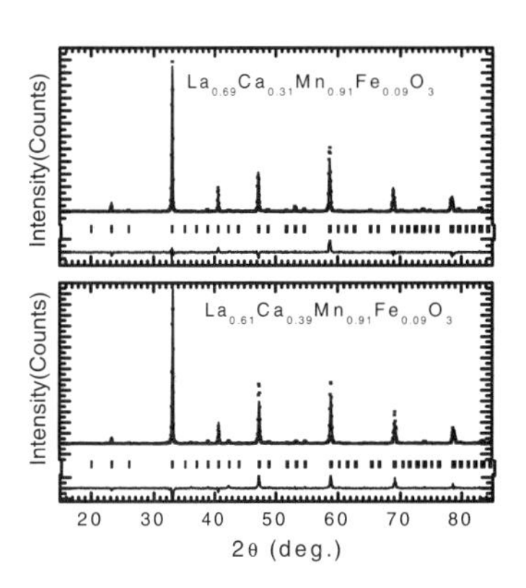

Fig.1. XRD patterns of $La_{1-x}Ca_xMn_{0.91}Fe_{0.09}O_3$.

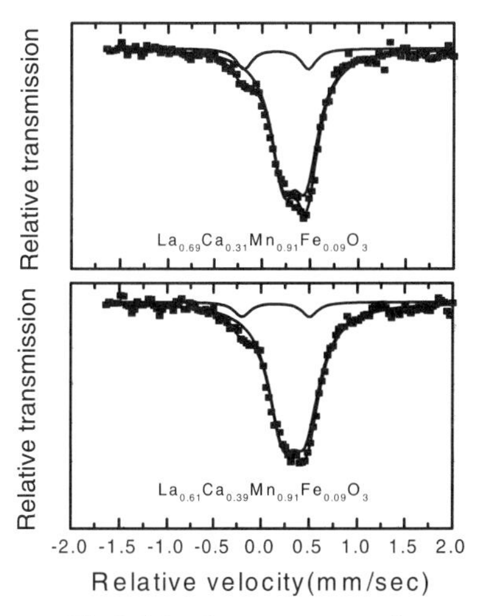

Fig.2. Mössbauer spectra of $La_{1-x}Ca_xMn_{0.91}Fe_{0.09}O_3$.

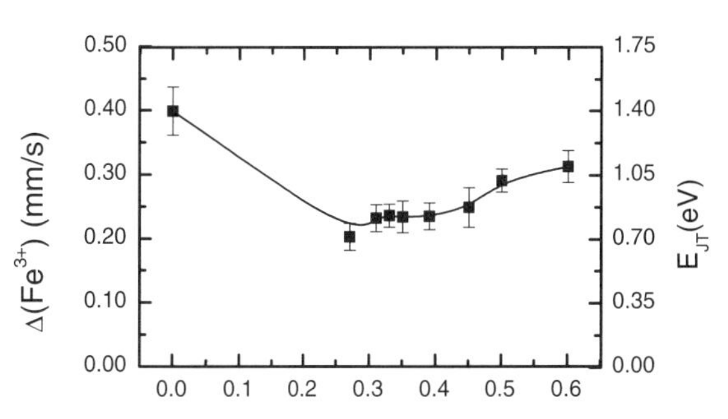

Fig. 3. Quadrupole splitting at Fe^{3+} ion of $La_{1-x}Ca_xMn_{0.91}Fe_{0.09}O_3$ perovskites at room temperature.

References

[1]S. Jinet al., *Science*, **264**(1994) 413.
[2]P.Schiffer et al., *Phys. Rev. Lett.* **75**(1995) 3336.
[3]A.J. Millis et al, *Phys. Rev. Lett.* **74**(1995)5144.
[4]J.B.A.A. Elemans et al., *J. Solid State. Chem.* **3**(1971) 238.
[5]A.J. Millis, *Nature*, **392**(1998) 147.

Microsc. Microanal. 8 (Suppl. 2), 2002
DOI 10.1017.S1431927602100754

Domain Structure of Magnetic Nanocrystalline Materials Studied by Electron Holography

D. Shindo
Institute of Multidisciplinary Research for Advanced Materials, Tohoku University, Sendai 980-8577, Japan

Recently, various advanced magnetic materials have been developed by fabricating fine grains in nanometer scale. Fe-based alloys $Fe_{73.5}Cu_1Nb_3Si_{13.5}B_9$ prepared by rapid quenching and annealing at appropriate temperatures show good properties for soft magnets [1]. It was reported that the magnetic properties drastically changed with the change of the annealing temperature or the change of the microstructure [2, 3]. In order to understand the magnetic properties of these materials, it is important to investigate not only the microstructure but also the magnetic domain structure. In this paper, by using the electron holography method, the magnetization distribution of $Fe_{73.5}Cu_1Nb_3Si_{13.5}B_9$ is investigated. Electron holography experiment was carried out with a JEM-3000F transmission electron microscope installed with a field emission gun and a biprism. The magnetic field around the specimen was reduced to 0.2 mT by switching off and then degaussing the objective lens.

Figures 1 (a) and 1 (b) show the coercive force (H_c) and the permeability (μ_r) of $Fe_{73.5}Cu_1Nb_3Si_{13.5}B_9$ with various heat treatments. The lowest coercive force is obtained at the annealing temperature of 823 K. It is also noted that the permeability takes large values at this temperature. Figure 2 (a) - (d) show TEM images of $Fe_{73.5}Cu_1Nb_3Si_{13.5}B_9$ with various heat treatments. While an as-quenched specimen (Fig. 2(a)) is amorphous state, fine grains of about 10 nm and enlarged grains of 50 . 200 nm are observed in Figs. 2 (b) and 2 (d), respectively. Figure 3 (a) – (c) show the reconstructed phase images of $Fe_{73.5}Cu_1Nb_3Si_{13.5}B_9$ films with the increase of tilting angle. In the top of Fig. 3 (a) of the as-quenched specimen with no tilt, monotonous closure domains are clearly seen. It is noted that the magnetic lines of force are parallel to the specimen edges, thereby eliminating the surface magnetic charge. It is interesting to note that the shape of the closure domain starts to change at the magnetic field of 8.3A/m which well corresponds to the coercive force of the bulk specimen of 6.9 A/m (Fig. 1 (a)). On the other hand, it is seen that the magnetic domain of the specimen annealed at 823 K is more sensitive to the magnetic field as seen in Fig. 3 (b), well corresponding to low coercivity and high permeability. In the case of the specimen annealed at 973 K, the size of magnetic domains becomes smaller, and the magnetic lines of force deviate significantly from the monotonous line shape. The irregularity in the shape of the magnetic lines of force is considered to result from the inhomogeneous magnetization distribution due to the bcc Fe and the Fe-B compounds. Being different from Figs. 3(a) and 3 (b), the magnetic lines of force do not change so much for the tilt. The difference directly indicates the strong pinning of magnetic domain walls due to the precipitates, resulting in the drastic increase of the coercive force and the decrease of the permeability [4].

References

[1] Y. Yoshizawa et al., J. Appl. Phys. 64 (1988) 6044.
[2] G. Herzer, IEEE Trans. Magn. 25 (1989) 3327.
[3] R. Schäfer et al., J. Appl. Phys. 69 (1991) 5325.
[4] The author wishes to thank Dr. Y. Yoshizawa and Mr. Y.-G. Park for providing the specimens and for the cooperation.

Fig. 1. Coercive force (a), permeability (b) of $Fe_{73.5}Cu_1Nb_3Si_{13.5}B_9$ as a function of annealing temperature.

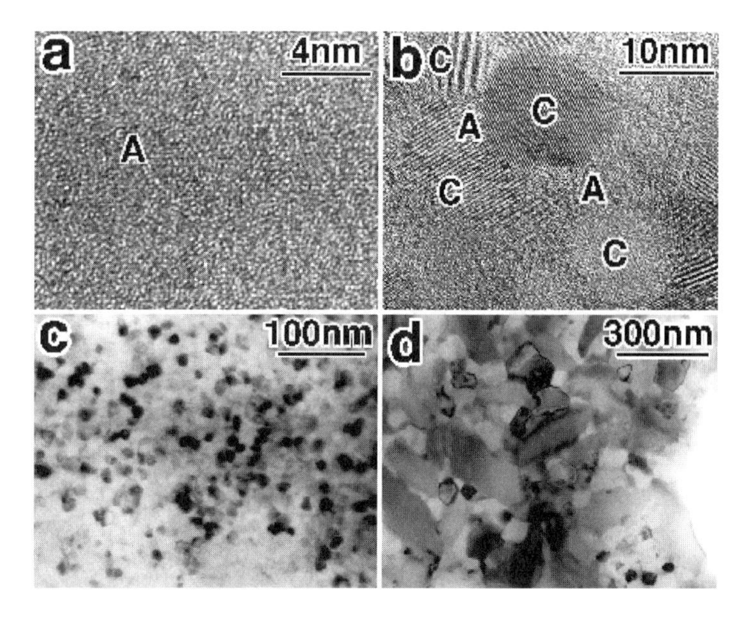

Fig. 2. TEM images of $Fe_{73.5}Cu_1Nb_3Si_{13.5}B_9$ as-quenched (a), annealed at 823 K (b), 923K (c) and 973 K (d).

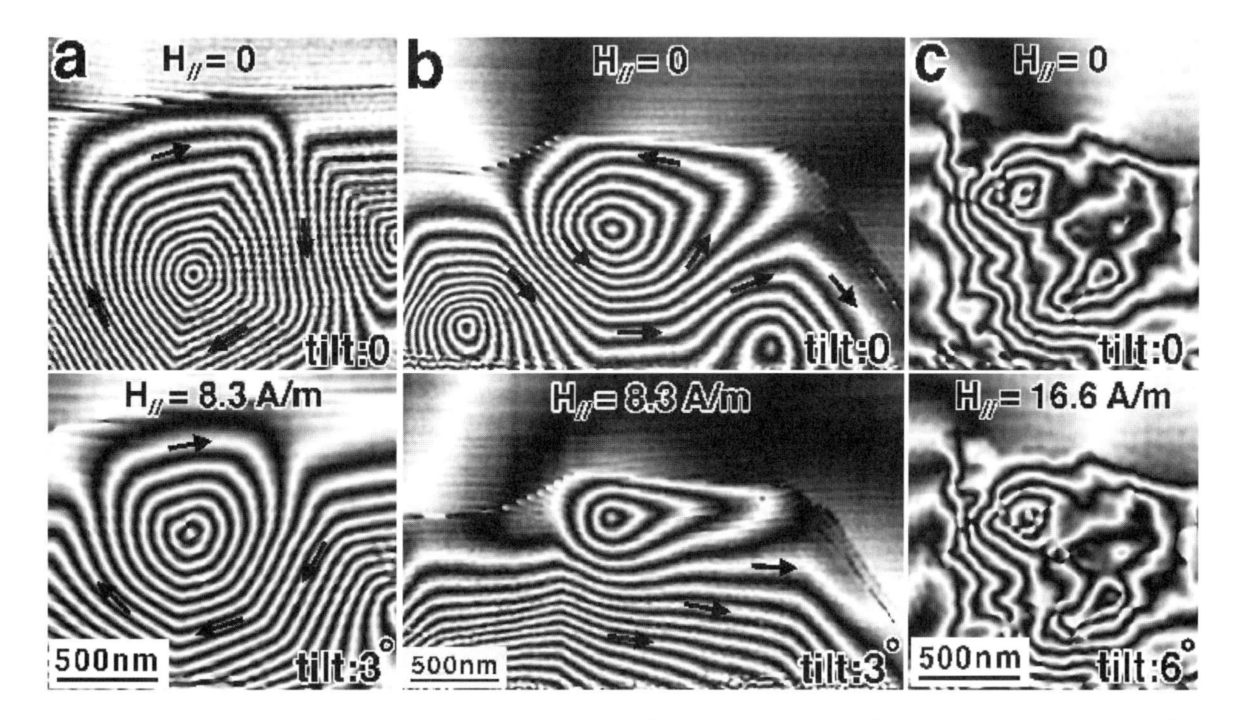

Fig. 3. Reconstructed phase images of $Fe_{73.5}Cu_1Nb_3Si_{13.5}B_9$ as-quenched (a), annealed at 823 K (b), and 973 K (d) with the increase of tilting angle. The direction of the magnetic lines of force is indicated by arrows. $H_{//}$ indicates the magnetic field in the film plane due to the tilt.

Microsc. Microanal. 8 (Suppl. 2), 2002
DOI 10.1017.S1431927602100766

Quantum Well Interference and Exchange Coupling in Double Quantum Well Thin Films

Zhang Zhi-dong

Shenyang National Laboratory for Materials Science and International Centre for Materials Physics, Institute of Metal Research, Academia Sinica, Wenhua Road 72, Shenyang 110016, Peoples' Republic of China

A comprehensive review is given of recent advances in understanding quantum well interference (QWI) in double quantum wells (DQW). The quantum well interference in double quantum wells is represented, based on the one – electron effective mass model. The probability of the electrons to be found in the several double quantum wells is derived analytically, providing with more detailed information on QWI. The calculation of the probability of the electron in the wells, based on effective mass theory, is in consistent with the density of states in metal thin films revealed by photoemission at the Fermi edge, which were performed at the third generation Advanced Light Source (ALS) at the Lawrence Berkeley National Laboratory. The beamline 7.0.1.2 at the ALS can focus the photon beam down to a 50-100 μm spot size with a photon flux (> 10^{12} photons per second at resolving power of 10000) to do the photoemission experiments. Thus, for a wedge of ~ 10 ML/mm, a scan of a 50 μm photon beam across the sample provides a systematic thickness-dependent measurement with ~ 0.5 ML thickness resolution.

According to our calculations, the exchange coupling is found to correlate directly to the quantum interference. The quantum well interference in the double quantum well depends sensitively on the relative magnitudes of the wavevectors of metals as well as the amplitudes of the density of states induced by different wavevectors. The phase accumulation method is still appropriate for interpreting the QWI in DQW systems. The symmetry of the quantum wells on quantization condition, and the special feature of the density of states are discussed. The phase accumulation relation could be linear for each well, but non-linear for the whole system in the symmetric DWQ. In the asymmetric DQW system, the linear quantization condition is valid only for part of layers whereas the quantization condition for others may be non-linear due to the quantum well interference. The non-linear quantization condition is a character of the DQW, which distinguishes with that of the SQW.

Neglecting the spin – dependent effects, we have illustrated successfully the oscillation of the probability of the electrons in the DQW systems. The oscillation periodicity of the probability due to the QWI is in good agreement with the experimental data for that of the exchange coupling and the giant magnetoresistance. We argue that the pure quantum interference is a possible mechanism for the oscillation of the exchange coupling and the giant magnetoresistance, besides the spin-dependent scattering. Because this mechanism is only the response of the electrons at the Fermi surface, the probability of electrons in the double quantum wells constructed all by ferromagnetic layers oscillates as the thickness of the spacer layer increases. Thus the oscillation of the exchange coupling between ferromagnetic layers separated by ferromagnetic layer is predicted.

All the results suggest that each layer in an entire multilayer stack is relevant to the magnetic coupling, thus favoring the QW coupling, in good agreement with the photoemission experiments. This work strongly supports the QW picture of the magnetic coupling, confirming that the oscillations of the interlayer exchange coupling correspond to the effective Fermi vectors k^{eff} of the layers.

The advantage and the limitation of the present model are discussed, in comparison with those of the first principles approach. The present work suggests that the effective mass theory could be used to fit and to interpret the experimental data of the multi-layers, even in the mono-layer limit. This model is an easy to use and rather effective tool for such systems. In some cases, this simple model could be used to predict new phenomena in the ultrathin films.

This work has been supported by the National Natural Sciences Foundation of China under Grant No. 59725103. The author thanks Prof. Dr. Z. Q. Qiu of the Department of Physics, University of California at Berkeley for helpful discussions.

Microsc. Microanal. 8 (Suppl. 2), 2002
DOI 10.1017.S1431927602100778

Solving the Structural Mysteries of Magnetic Materials in TEM

Y. Liu,
CLAIM, Department of Mechanical Engineering
University of Michigan
2350 Hayward St.
Ann Arbor, MI 48109-2125

We are entering the era of optimum nanostructure design of magnetic materials for applications such as recording media, magnetic memory, magnetic sensors, exchange-coupled permanent magnets and spin electronics etc. Transmission electron microscopy (TEM) is the single powerful tool that can provide all the necessary information. This paper will review my recent experience in TEM of various magnetic materials. The first step is to use proper preparation method for different samples. Methods for bulk sample, thin films, self-assembled nanowires [1], and mechanically milled powders will be reviewed. The next step is to choose a proper technique to answer a particular question. Examples using selected area electron diffraction, convergent beam electron diffraction, nanodiffraction, high resolution TEM [2], super-resolution TEM, selected reflection imaging (SRI) [3], compositional mapping by electron energy loss spectroscopy (EELS) [4] and by energy dispersive X-ray spectroscopy, and Lorentz microscopy will be shown [5].

Figure 1 is nanodiffraction patterns used to measure the volume fraction of amorphous phase and to identify the crystal structure of the crystallite phase in Co-Sm film. It is found that higher Ar pressure during deposition and higher concentration of Sm promote the formation of amorphous phase. This method is most accurate without limitation of the resolution of the TEM. Using nanodiffraction several stacking modes such as AB, ABC, and other modes were identified in Co-Sm films. Figure 2 (a) is diffraction pattern, (b) is (002) SRI image showing the grains with the preferred direction and (c) is (001) SRI image of FePt:Fe composite film. The bright grains in Figure 2 (c) is ordered $L1_0$ phase while those that are white in Figure 2 (b) but dark in Figure 13 (c) are Fe solid solution. This method is most sensitive to the ordering regardless of the chemical composition of the phases. The identification of the Fe phase and FePt phase is critical for explanation of the exchange coupling in this composite system. Figure 3 is the HRTEM image of $Sm_2(CoFeCuZr)_{17}$ system high temperature permanent magnet. Zr exists in the precipitate with a coherent lattice with the matrix. Such structure is stable at high temperature and the Zr-rich precipitates enhance the coercivity by pinning the domain wall movement.

References

[1] Y. Liu, M. Zhen, H.Zeng and D. J. Sellmyer, Nanophase and Nanostructured Materials, Eds. Z. L. Wang, Y. Liu and Z. Zhang, Tsinghua Univresity Press and Kluwer Accademic/ Plenum Publishers, Vol. 3, (2001) 210.
[2] Y. Liu, D.J. Sellmyer, B. W. Rob ertson, Z. S. Shan, and S. H. Liou, IEEE Trans. on Magn., Vol. 31, (1995) 2740.
[3] Y. Liu and D. J. Sellmyer. Proc. of Microscopy and Microanalysis 1998, Atlanta, Georgia, 752.
[4] Y. Liu, C. Nelson etc. Proc. of Microscopy and Microanalysis 2002.
[5] Y. Liu, Z. S. Shan, and D. J. Sellmyer, IEEE Trans. on Magn., Vol. 32,(1996) 3614.
[6] Most of this work was performed at CMRA, University of Nebraska at Lincoln and supported by CMRA, NSIC, DOE and NSF.

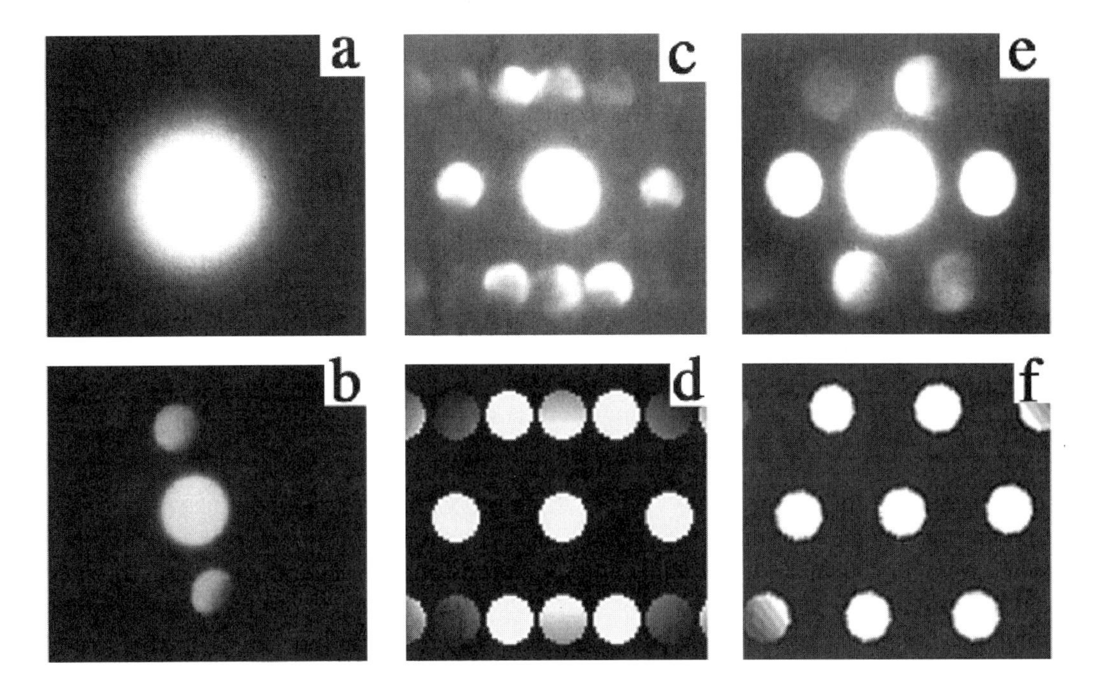

FIG. 1. Nano-diffraction patterns: (a) from amorphous phase, (b) from a crystallite phase, (c) is TEM pattern, (d) is simulated [11.0] zone axis pattern of HCP structure, (e) is TEM pattern, and (f) is simulated [110] zone axis pattern of FCC structure. Several stacking modes have been identified in the as-deposited Co-Sm films.

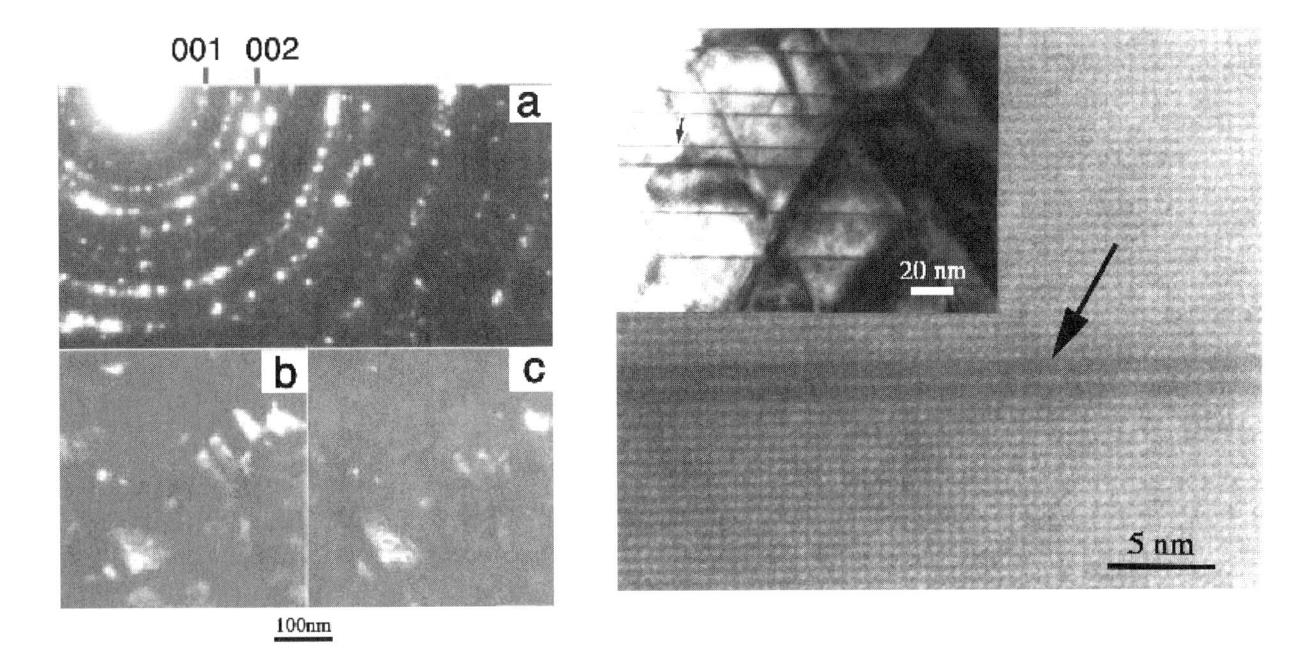

FIG. 2. (a) diffraction pattern
(b) is the (002) SRI image
(c) is (001) reflection image of
FePt:Fe composite film.

FIG. 3. Bright field and HRTEM image of a $Sm_2(CoFeCuZr)_{17}$ magnet.

Microsc. Microanal. 8 (Suppl. 2), 2002
DOI 10.1017.S1431927602100468

Progress Towards More Realistic In-Situ Microscopy Observations

A. Howie

Cavendish Laboratory, University of Cambridge, Madingley Road, Cambridge CB3 0HE, UK

As progress indicators in electron microscopy, advances in spatial resolution and in spectroscopy probably attract most frequent attention. Improved user-friendliness has also been significant even when judged in comparison with scanned probe microscopy. Evidence for developments in in-situ microscopy at least equally impressive can be found by comparing the relevant sections of the book by Hirsch et al. [1] and a more recent compilation [2].

Peter Hirsch's research group swiftly discovered the power and frustrations of in-situ microscopy. The cine film of dislocation motion observed in the earliest diffraction contrast studies [3,4], and attributed to the thermal stresses generated by the electron beam, was extremely effective in convincing the wider community that it was indeed dislocations that were being observed. In a remarkably successful in-situ hot stage experiment applied to a simple problem, the shrinkage of dislocation loops due to migration of vacancies to the foil surfaces was followed and fitted in detail to a quantitative model [5]. Any consequent feelings of euphoria were probably moderated when, in trying to employ the same equipment more ambitiously to study alloy phase transitions, the dominating effect of diffusion and reactions at the foil surfaces was observed [6]. Another hot stage project that was tackled with mixed success was to follow the recrystallisation process of grain boundary motion in heavily worked silver foils [7]. Successful cine photography was defeated by the tendency for recrystallisation to take place abruptly outside the field of view. Observations of dislocation motion under more controlled conditions of plastic deformation as well as in fracture have continued but the focus below is on surface science and catalysis work.

Despite the poor vacuum conditions, the power of TEM in studies of metal epitaxy was apparent [8] and encouraged Hirsch and Pashley to enlist the expertise of Ugo Valdre in the construction of a separately pumped UHV stage unit [9]. This system was used for several years in studies of both island and psuedomorphic growth as well as in metal oxidation [10] where much of the work has been confirmed in more modern equipment [11] capable of revealing the surface reconstructions associated with the oxide nuclei. For clean surface studies, these TEM methods are increasingly challenged by other techniques, including reflection electron microscopy (which first revealed the nucleation of surface reconstructions at surface steps [12]), photoemission electron microscopy (PEEM)[13] and, most particularly, scanned probe microscopy (SPM), whose popularity with the wider surface science community vividly illustrates the importance of user-friendliness. These other approaches have been compared [14] and offer varying levels of spatial resolution, nanoanalysis and imaging speed. They generally employ bulk sample surfaces rather than thin film surfaces with advantages for specimen preparation and surface cleaning but, in some cases, with disadvantages for ionisation damage and beam charging.

The drive towards in-situ microscopy of catalytic processes received early stimulation from some spectacular TEM observations [15] but places extreme demands if realistic pressures and temperatures are to be achieved. Peter Hirsch's long lasting influence here can be traced to the Oxford project for high voltage in-situ microscopy[16]. In some sense this is the ancestor of the much more ambitious medium voltage, high resolution, environmental TEM equipment designed and developed by E. Boyes and P.L. Gai [17] after their move from Oxford to Dupont and used with remarkable success in many catalyst investigations [2,18]. A later version of this equipment installed at Haldor Topsoe A/S recently yielded intriguing results on the distribution of Ba promoter in Ru ammonia synthesis catalysts [19].

Environmental SEM and PEEM both offer some competition to TEM methods for work under gas pressure but cannot compete in terms of spatial resolution. The most serious potential competitor here is undoubtedly SPM but for some reason the challenge has been slow to develop. Recent STM work in the few bar range [20,21] could surely be extended to higher pressures or even liquid operation conditions on catalysts (provided any oxide support is in the form of a thin coating on metal [22]). Atomic force microscopy and some other forms of SPM using optical methods, such as sum frequency generation [23] could also be potentially powerful methods for imaging catalytic reactions under gas pressure.

A general and incompletely resolved problem for all in-situ work, other than in SPM, is the possible influence of the electron beam. Atomic displacements are often manifest, for instance in HREM profile imaging of surfaces, but in other cases may still affect the course of any dynamic process even when they are not directly visible. In this respect we have not greatly advanced beyond the early observations of beam-induced dislocation motion.

References

[1] P.B. Hirsch et al., Electron Microscopy of Thin Crystals,Butterworth, London, 1965.
[2] P.L. Gai (ed), In-situ Microscopy in Materials Research, Kluwer, Boston, 1997.
[3] P.B. Hirsch et al., Phil. Mag. 1 (1956) 677.
[4] M.J. Whelan et al., Proc. Roy. Soc. A240 (1957) 524.
[5] J. Silcox and M.J. Whelan, Phil. Mag. 5 (1960) 1.
[6] G. Thomas and M.J. Whelan, Phil. Mag. 6 (1961) 1103.
[7] J.E. Bailey, Phil. Mag. 5 (1960) 833.
[8] D.W. Pashley et al., Phil. Mag. 10 (1964) 127.
[9] U. Valdre et al., J. Phys. E 3 (1970) 501.
[10] D.A. Goulden, Phil. Mag. 33 (1976) 393.
[11] J.C. Yang et al., Microscopy Microanal. 7 (2001) 486.
[12] K. Yagi, Surf. Sci. Rep. 17 (1993) 305.
[13] E. Bauer, J. Electron Spectrosc. & Rel. Phenom. 114 (2001) 975.
[14] E.D. Williams and L.D. Marks, Crit. Rev. in Surf. Chem. 5 (1995) 275.
[15] R.T.K. Baker and R.J. Waite, J. Catal. 37 (1975) 101.
[16] P.L.Gai and M.J. Goringe, Proc 39th EMSA Conf., San Francisco Press, 1981, 68.
[17] E.D. Boyes and P.L. Gai, Ultramicrosc. 67 (1997) 219.
[18] P.L. Gai, Advanced Mater. 10 (1998) 1259.
[19] T.W. Hansen et al., Science 294 (2001) 1508.
[20] B.L. Weeks et al., Rev. Sci. Instrum. 71 (2000) 3777.
[21] L. Osterland et al., Phys. Rev. Lett. 86 (2001) 460.
[22] H. Over et al., Science 287 (2000) 1474.
[23] T. Dellwig et al., Phys. Rev. Lett. 85, (2000) 776.

Microsc. Microanal. 8 (Suppl. 2), 2002
DOI 10.1017.S143192760210047X

In-situ Observation of Alloy Phase Formation in Isolated Nanometer-sized Particles

H. Mori*, J.G. Lee* and H. Yasuda**

*Research Center for Ultra-High Voltage Electron Microscopy, Osaka University, Yamadaoka, Suita, Osaka 565-0871, Japan
**Department of Mechanical Engineering, Kobe University, Rokkodai, Nada, Kobe 657-8501, Japan

Understanding the structure (or phase) stability of nm-sized condensed matter is one of the key issues in advancing materials reliability in a broad spectrum of nanotechnologies including microelectronics technologies. To the authors' knowledge, however, studies on the phase stability in nm-sized particles, particularly in alloy particles, are limited. Based upon this premise, in the present work, alloy phase formation in nm-sized particles has been studied by in-situ transmission electron microscopy, using particles in the Au-Sn and Sn-In systems [1-3].

Preparation of gold particles and subsequent vapor-deposition of tin onto gold particles was carried out using a miniature double-source evaporator that was set at the tip of a side-entry specimen holder. The evaporator essentially consisted of two spiral-shaped tungsten filaments. An amorphous carbon film was used as a supporting film and was mounted on a copper grid. Using this evaporator, gold was first evaporated onto an amorphous carbon film, and nm-sized gold particles were produced. Tin was then evaporated from the second source onto gold particles on the film kept at ambient temperature. Changes in the structure and chemical composition of particles associated with the tin deposition were studied. The chemical composition of individual particles on the film was analyzed by energy dispersive X-ray spectroscopy (EDS). The microscope used was a Hitachi HF-2000 TEM equipped with a field emission gun, operating at an accelerating voltage of 200 kV. The base pressure in the specimen chamber was below 5×10^{-7} Pa. Alloy formation in the Sn-In system was examined in a similar way.

When tin atoms were vapor-deposited onto nm-sized gold particles, rapid dissolution of tin atoms into gold particles took place and as a result of this, particles of an Au-rich fcc solid solution alloy, of a topologically-disordered amorphous alloy, and of the AuSn compound were formed with increasing concentration of tin. Figure 1 shows examples of high-resolution images (HRIs) taken from these alloy particles, with the corresponding EDS spectra. Each EDS spectrum was obtained from the region encircled in each figure.

Figure 1(a) shows an HRI of an Au-18at%Sn alloy particle and the corresponding EDS spectrum. The diameter of the particle is approximately 5 nm. In Fig. 1(a), there appear lattice fringes with a spacing of 0.24 nm. The spacing is very close to the (111) lattice spacing (i.e. 0.23_9 nm) of fcc Au-18at%Sn solid solution. This fact suggests that the particle is an Au-Sn solid solution and that the solid solubility of tin in nm-sized gold particles amounts to at least 18 at%Sn, which is much higher than that in bulk gold (i.e. ca. 3 at%Sn at room temperature). HRIs taken from Au-32at%Sn and Au-40at%Sn alloy particles are shown in Fig. 1(b) and 1(c), respectively. The diameter of these particles is approximately 6 nm. Both particles exhibit a contrast similar to the salt and pepper contrast characteristic of topologically-disordered materials. This observation indicates that a topologically-disordered, amorphous phase is produced in 6 nm-sized alloy particles in the composition range from 32 to 40 at%Sn. Figures 1(d) and 1(e) show HRIs of approximately 8 nm-sized AuSn compound particles with different chemical compositions. Both compound particles are single crystalline. The 0.31 nm- and 0.37 nm-spaced fringes in the images are the (011) and (100) lattice fringes of AuSn, respectively. EDS spectra in Figs. 1(d) and 1(e) indicate that the tin concentrations of particles are 46 at%Sn (which is by ca. 4 at% lower than the stoichiometric composition of AuSn) and 59 at%Sn (which is by ca. 9 at% higher than the stoichiometric composition of AuSn), respectively. From this result, it is evident that remarkable enhancement of solubility has been induced in AuSn compound particles, as compared with that in the corresponding bulk AuSn which is a line compound.

As mentioned above, in the Au-Sn system, a topologically-disordered, amorphous phase is produced in particles in the composition range from 32 to 40 at%Sn, when the size of particles is smaller than about 6 nm. A separate annealing experiment has revealed that the topologically-disordered, amorphous phase goes to melt with no preceding crystallization upon heating and the melt solidifies into the topologically-disordered, amorphous phase without any traces of crystallization upon cooling. This fact indicates that the phase is not a non-equilibrium phase. In this context, the phase is different from such non-equilibrium amorphous phases as those produced in bulk materials. It is interesting to note here that the composition range is very close to the bottom of the deep valley of the liquidus in the middle of the Au_5Sn-AuSn two-phase region in the phase diagram for the bulk material. On the other hand, in the Sn-In system, liquid alloy particles were formed over a composition range near the eutectic composition, when the size of particles is smaller than about 10 nm in diameter. When the size of particles is larger than these critical values, an essentially similar phase equilibrium was observed in nm-sized particles and bulk materials in the both systems. The formation of the amorphous and liquid phase in nm-sized alloy particles can be explained in terms of the large suppression of the eutectic temperature associated with the size reduction.

This work was supported by the Ministry of Education, Science and Culture under a Grant-in-Aid for Scientific Research (#13450260).

References

[1] H. Mori and H. Yasuda, Mat. Sci. Eng. A312 (2001), 99.
[2] H. Yasuda et al., Phys. Rev. B 64 (2001), 094101.
[3] J.G. Lee et al., Phys. Rev. B (2002) in press.

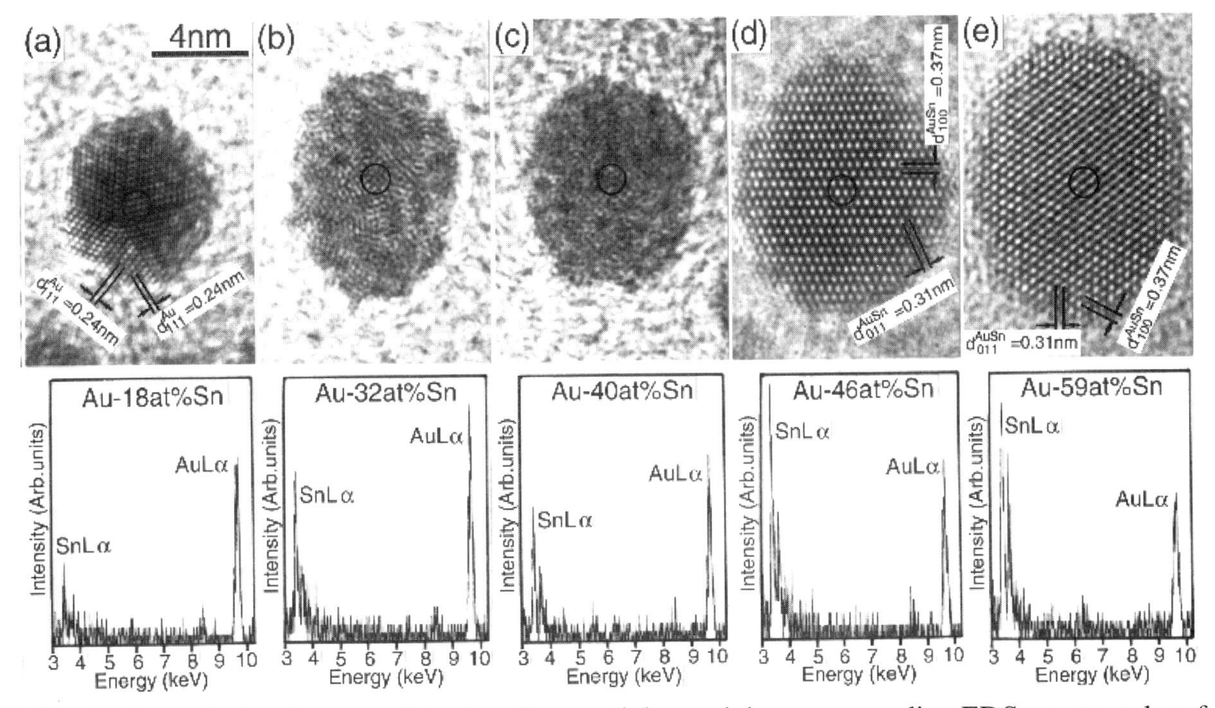

FIG.1. High resolution images of Au-Sn alloy particles and the corresponding EDS spectra taken from the region encircled.(a) Au -18at%Sn alloy particle, (b) Au-32at%Sn alloy particle, (c) Au-40at%Sn alloy particle, (d) Au-46at%Sn alloy particle, (e) Au-59at%Sn alloy particle.

Microsc. Microanal. 8 (Suppl. 2), 2002
DOI 10.1017.S1431927602100481

High Resolution In-situ SEM of Competitive Particle Sintering and Other Surface Processes

E D Boyes

DuPont Company, CR&D, PO Box 80356-383, Wilmington, DE 19880-0356, USA.

The SEM is a powerful platform for in-situ studies at very high, at more modest and even at low resolution; and in high vacuum and using controlled environments of gas, vapor or actual liquids. The primary advantage of the SEM is the use of robust, inherently stable and representative wide area (millimeters to inches) bulk samples as the primary subject of in-situ study or as substrates. The combination of lateral spatial resolution and chemical and crystallographic analyses is matched by few other surface science methods. Recent advances in imaging methods combine resolutions of almost 1nm at 1kV for imaging with sub-micron, and in some cases near-nm, sensitivities for nanoanalysis. The use of low voltages and digitally processed TV rate scanning separately, or preferably together, generally avoids the previous prerequisite for conductive coating of electrically insulating materials. This is a key development for surface specific in-situ sintering but it is less necessary for modest resolution mechanical deformation and fracture studies in which bulk processes are sampled at the surface which over large areas this may remain unmodified. Ashby et al [1] is a classical example of the latter and we will present data from sintering studies after Jagota et al [2] and Boyes [3] in which the substrate can be either a strong player or just a convenient 2D platform with minimal interactions.

Our current instrumentation utilizes ultra-high resolution FESEMs [3] and in some cases (S)TEMs operated in the SEM mode [2]. Samples are of modest size, which reduces outgassing and contamination issues, and they are mounted in regular TEM hot stages modified to accommodate disks 3mm in diameter and up to >0.5mm thick. These are effectively bulk samples. The in-lens FESEMs conveniently use similar TEM-style stages. In essence we use thicker versions of the disks from which TEM thin sections (the hard part) are typically prepared. This approach allows the high resolution study of much larger components and wider population distributions, and the interfaces between them, than could be accommodated in a TEM thin section. It also allows substrate surfaces to be selected and prepared in ways which are rarely compatible with the preservation of a thin sample. Moreover, on heating or other treatment gross artefact changes in the sample substrate are much less likely to be bothersome, or at least they are more reproducible. We are also developing a reciprocal space shuttle with air outside for the protected transfer of samples into the vacuum or other controlled atmosphere of the microscope [4]. This is designed to protect samples from contamination by transfer through air, including after preparation and evaluation in a UHV, dry box, high pressure liquid reactor or other process environment; and in other applications to protect the environment from people unfriendly samples with poisonous or pyrophoric etc tendencies. A hot stage for polymer sample transformations can be quite simple.

Overall bulk sintering data raise a number of issues which in-situ SEM is used to quantify and to resolve. Our application examples include sintering of soda lime silicate glass spherical beads sieved to <10um size with in some cases metal inclusions of 60Co.28Cr.7Mo of similar or smaller size. We are interested in the effect of constrained sintering between the beads and whether the early mechanisms involve a reduction in the viscosity of the glass by wide metal dissolution or there is more wetting of the metal by the glass than was initially assumed. Glass wetting of the substrate contributed to the results on Pt but partially graphitized carbon disk [2, 5] substrates supported the various beads without influencing their interactions significantly. It is as though the beads simply float freely and are supported and conveniently constrained by the substrate to a 2D pattern of interactions to observe wetting behavior and measure contact angles between the constituents. The fact that the sintering rate and contact angles are increased with oxidized metal particles and the influence of them in small amounts, together with more direct EDX analysis data, strongly suggest the dissolution model is the cause of the initial increase in sintering rate with metal addition. Low glass metal wetting was observed. At higher levels it is shown the there are stong metal-metal and glass-glass contacts. Unexpected reversible and angle changing glass contacts were also observed in very competitive processes. These mechanisms were studied effectively in the SEM with valuable additional information and a very high productivity for basic parameters compared to ex-situ methods. End point data were comparable by both approaches with an identical onset temperature of 704 (+/-10)^{0}C ex-situ in an air atmosphere DSC furnace and in-situ in the high vacuum SEM. The zero effect of the vacuum in this specific case is attributed to the high oxygen fugacity in the glass beads and to the pre-oxidation of the metal in the most interesting experiments. No influence of the TV rate beam was detected.

Low voltage ultra-high resolution SEM of wide areas containing infrequent substrate attributes are opening up in-situ studies of metal on complex ceramic systems of practical importance. At elevated temperatures conductive coatings are no longer required. The value and productivity of the SEM approach with robust bulk samples is extended when more intense ex-situ reaction conditions are required, e.g. high temperature oxidation or pressurized liquids, although with great difficulty and skill some of this data have also been recorded in the TEM by a number of authors [6]. Finally, the SEM allows direct plan view imaging of surface reactions uncomplicated by underlying bulk microstructures except as they interact with surface data and are therefore relevant. The value of combined in-situ, ex-situ, and other studies cannot be overemphasized.

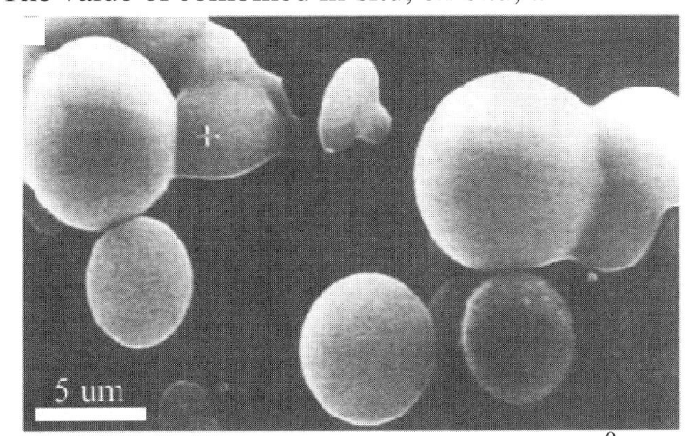

Fig.1 : Impeded glass sintering in-situ at 730^{0}C. Digitally integrated TV rate scanning.

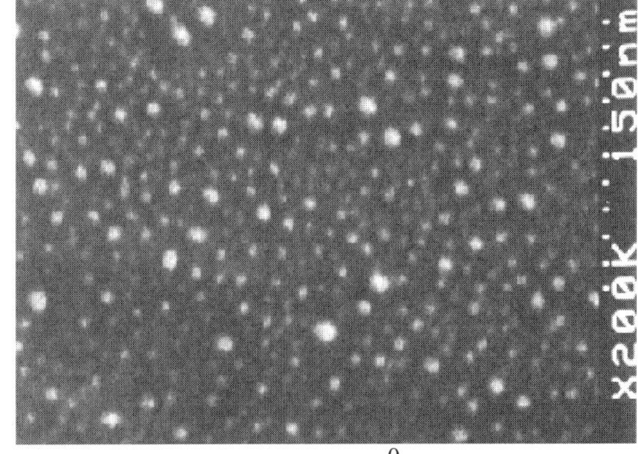

Fig. 2 : Ag on Al_2O_3 at 250^{0}C. 1kV, 200kX In-situ migration and sintering study.

References

1. M F Ashby et al, Proc Roy Soc A398 (1985) 261
2. A Jagota, E D Boyes and R K Bordia, Mat Res Soc Symp Proc 249 (1992) 475
3. E D Boyes, Mat Res Soc Symp Proc 404 (1996) 123
4. After G Parkinson and L Allard et al
5. Ted Pella Inc special order
6. P L Gai, P J F Harris et al, L Schmidt et al and L Allard et al, private and other communications

Microsc. Microanal. 8 (Suppl. 2), 2002
DOI 10.1017.S1431927602100493

REAL TIME UHV-HRTEM OBSERVATION OF Si(111)v3xv3-Pd SURFACE AND DYNAMIC MOTION OF Pd CLUSTERS

M. Takeguchi, K. Mitsuishi, M. Tanaka and K. Furuya

National Institute for Materials Science, Tsukuba, Ibaraki 305-0003, Japan

Recently, the development of advanced quantum devices requires the understanding of self-organized growth of metal or semiconductor nano-islands on semiconductor surfaces. For this purpose, the characterization and visualization tools for the surface growth with atomic resolution have been increasingly desired. Ultrahigh vacuum high-resolution transmission electron microscopy (UHV-HRTEM) has been a powerful tool to visualize atomic structure not only in a bulk but also on a surface. Moreover, well-established electron-diffraction and imaging theory can support the modeling of atomic structure. Individual atoms and clusters on a surface have often been imaged by HRTEM so far [1-2]. It is reported that Si(111) 3x 3-Bi was observed with HRTEM but image of surface atoms were overlaid with that of a substrate atoms [3]. A combination of HRTEM with computational process has provided clear atomic images of Si(111)5x2-Au and Si(111)7x7 surface without substrate image [4-5]. However, the atomic motions of atoms and/or clusters on the surface cannot be analyzed with such off-line processes.

This paper reports an atomic structure of a Si(111) 3x 3-Pd surface and the motion of Pd clusters on the surface with UHV-TEM in real time. Figure 1 shows a schematic drawing of Pd deposition. In-situ TEM observation was performed using a 200 kV UHV-TEM equipped with a field emission gun. Pd less than 1 ML was deposited onto a Si (111)7x7 surface using an electron beam evaporator attached to the microscope column at about 550 K. Figure 2A shows a conventional HRTEM image of the Si(111) surface, on which Pd exhibits the 3x 3 structure on the thin specimen (10 nm or less). Since the image was taken at near the Sherzer defocus, Si(111)1x1 lattice fringes (0.19nm) are dominantly observed and the surface structure is unclearly overlaid. From the image simulation, we found that the Si 1x1 lattice image disappears and the 3x 3 structure is enhanced when the convergent angle is larger than 1 mrad at the condition of over-focus of 20-50 nm and the thickness of 7.5-10.3 nm, and that the contrast enhancement of Pd clusters against the 3x 3 structure is maximized at the over-focus of around 34nm and the convergent angle of around 3 mrad. Furthermore, we introduced an objective aperture so that electron diffraction of 0.20 nm or less are filtered off. Figure 2B shows a HRTEM image of the same area, indicating that atomic arrangement of the 3x 3 structure is clearly seen without any Si(111)1x1 lattice fringes. At this condition, Pd clusters are visualized as a set of three bright dots as shown in Figure 3. We proposed two types of structure models for the cluster, e.g., Pd-trimer and silicide models, and the simulated image calculated using these models were compared with the experimental one. Figure 4 shows the each structure model and simulated images at 2 and 3 mrad convergent angles, indicating that the Pd-trimer model is consistent with our results rather than the silicide one.

References

[1] S. Iijima, Optik 48 (1977) 193.
[2] N. Tanaka, H. Kimata, and T. Kizuka, Surf. Sci. 386 (1997) 216.
[3] Y. Haga and K. Takayanagi, Ultramicrosc. 45 (1992) 95.
[4] L. D. Marks and R. Plass, Phys. Rev. Lett. 75 (1995) 2172.
[5] E. Bengu et. al., Phys. Rev. Lett. 77 (1996) 4226.

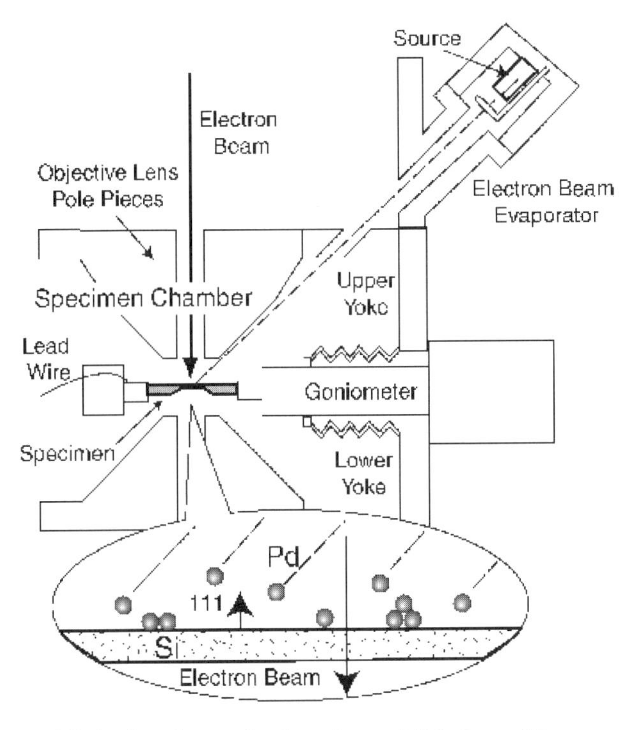

FIG 1. A schematic drawing of Pd deposition on Si (111) in the UHV-TEM

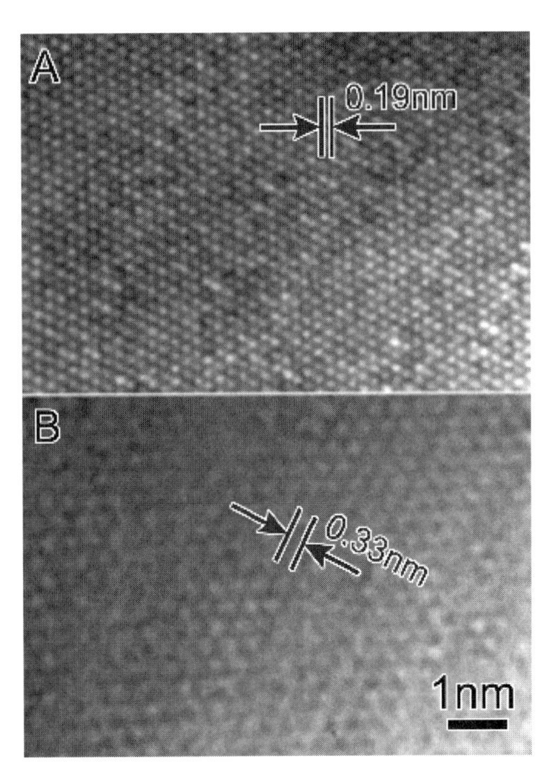

FIG 2. HRTEM images of Si(111) 3x 3-Pd taken by a conventional manner(A) and taken on the present experimental condition(B).

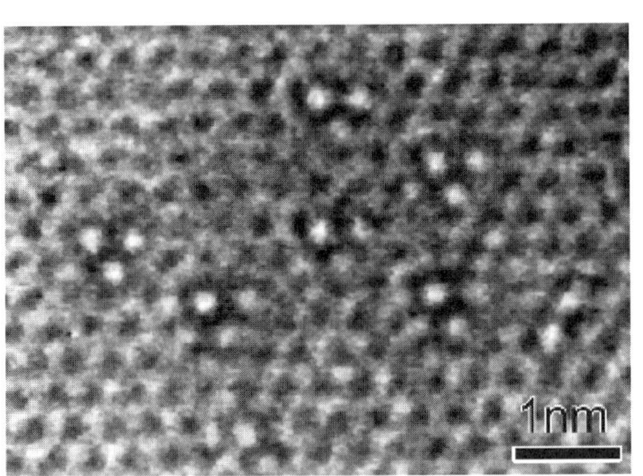

FIG 3. HRTEM image of Pd clusters moving around on the 3x 3 layer.

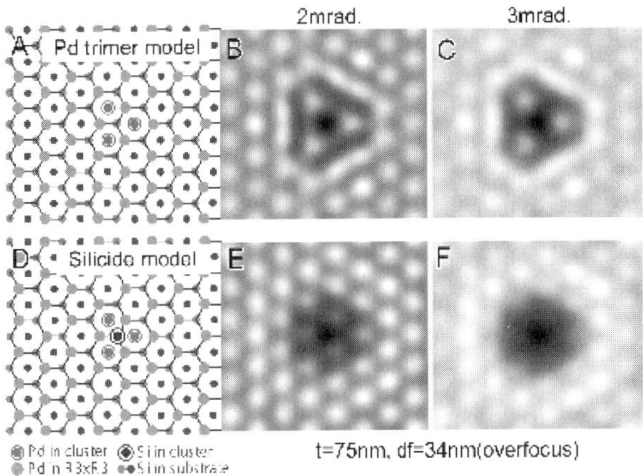

FIG 4. Structure models for a Pd cluster and their simulation images.

Microsc. Microanal. 8 (Suppl. 2), 2002
DOI 10.1017.S143192760210050X

In situ Molecular Imaging of Heterogeneous Catalytic Processes in Liquid Environments

Pratibha L. Gai

DuPont Central Research and Development, Wilmington, DE 19880-0356, U.S.A. and
Department of Materials Science, University of Delaware, U.S.A.

Understanding reacting materials at the atomic level is important for the design and development of new generation catalysts, new routes for polymers and green technological processes. Many commercial polymerization reactions and nanomaterials are derived from solutions and wet chemical methods. Probing reactions in solutions at the molecular level is therefore of great importance in the development of advanced catalytic process technologies. To probe gas molecule-solid catalyst reactions directly at the atomic level, we have pioneered the development of in-situ atomic resolution environmental transmission electron microscopy method (referred to as atomic resolution-ETEM, or as environmental-HRTEM) [1,2]. We have recently advanced the in situ technique to probe dynamic catalysis in liquid environments, and this method is referred to as wet-ETEM [3]. Our wet-ETEM capability allows high resolution imaging in liquids, with capabilities for heating samples to reaction temperatures. Here we demonstrate molecular imaging in liquid environments. We illustrate this development with examples of the catalytic hydrogenation of nitriles in the liquid phase and in the manufacture of nanomaterials.

Using wet-ETEM, we have investigated the hydrogenation of adiponitrile ($NC(CH_2)_4CN$) in liquid phase over cobalt modified ruthenium nanocatalysts supported on rutile titania. For the studies, the catalysts were immersed in ADN liquid (in methanol and 0.15wt% NaOH solvent) in a liquid injection sample stage with heating capabilities, which we have developed [3]. Upto microliters of the liquid were injected over the catalysts. Flowing hydrogen was passed over the samples simultaneously and the sample was heated to ~ 100 °C. Dynamic reactions were recorded in situ under the operating conditions. During the studies, we have observed that, bis-hexamethylene triamene (BHMT) is formed, along with the main product of hexamethylene diamene ($H_2N(CH_2)_6NH_2$) or HMD. Molecular imaging and electron diffraction under liquid environments have revealed the structure of BHMT. Figure 1 shows the nanocatalyst at room temperature (RT). Fig 2(a) and (b) show the catalyst in ADN liquid at RT and the formation of BHMT at the catalyst surface at 100 °C, respectively. Figure 2(c) shows the enlarged area from Figure 2(b), revealing the lattice layers in BHMT organic molecules. The dynamic images show that it is crystalline with the lattice spacing of about 5.6 angstroms. Figure 3 shows the formation of HMD at 100 °C. Wet-ETEM is also used to study the growth of gold nanorods [Figure 4]. The wet-ETEM method is opening up striking new opportunities in the direct molecular studies of dynamic polymerization reactions and in biomolecular nanotechnological applications.

References:

1. P.L. Gai et al., Science. 267 (1995) 666.
2. E,D. Boyes and P.L. Gai, Ultramicroscopy. 67 (1997) 219.
3. P.L. Gai, Microscopy and Microanalysis. 8 (2002).
4. L.G. Hanna, F.G. Gooding, K. Kourtakis and S. Ziemecki are thanked.

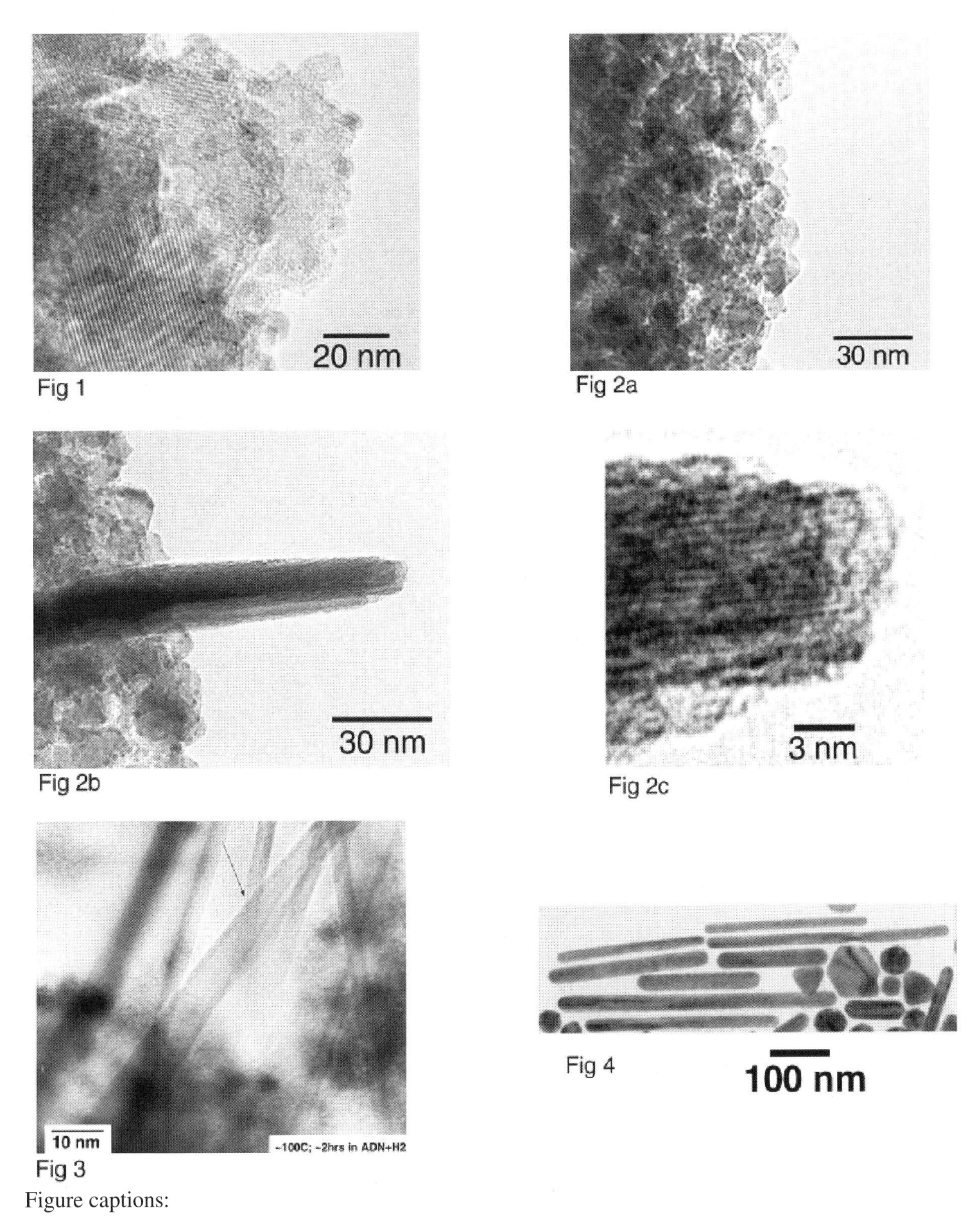

Figure captions:

FIG.1. Heterogeneous cobalt modified Ru/titania nanocatalyst.
FIG.2(a). Wet-ETEM of the nanocatalyst in ADN liquid and H_2 gas at RT;
FIG.2(b). Wet-ETEM of Fig 2(a) at ~ 100 °C, showing BHMT molecular structure;
FIG.2(c). Enlargement of FIG. 2(b) showing the molecular structure.
FIG.3. HMD formation;   FIG 4: high aspect ratio nanorods.

Microsc. Microanal. 8 (Suppl. 2), 2002
DOI 10.1017.S1431927602100511

In-situ UHV-Electron Microscopy with Scanning Tunneling Microscope

K.Takayanagi*, Y.Ohshima**, K.Mohri**, Y.Naitoh**, H.Hirayama**, Y.Tanishiro*, and Y.Kondo***

* Tokyo Institute of Technology, Physics Department, Tokyo, 152-8551, Japan
** Tokyo Institute of Technology, Material Science and Eng. Dept., Yokohama, 226-8502, Japan
*** JEOL LTD, EMG Electron Optics Division, Tokyo, 196-8558, Japan

Ultra-high-vacuum (UHV) electron microscopy has been used for analyses of thin film growths, surface structures, and monolayer adsorbates, in the last two decades. In recent years, a miniaturized scanning tunneling microscope (STM) has been built at the specimen stage of an UHV electron microscope, which allows us to study not only the structures but also electronic states of nanomaterials simultaneously. We present here the construction of an newly developed STM-combined-UHV TEM, and the observations on quantized conductance of the gold nanowire that is formed between the gold electrodes.

Figure 1(a) and (b) shows a new STM holder developed specially for the FE-UHV-TEM (JEOL-2000FV). The STM holder is transferred to the gap of the objective pole piece of the FE-UHV-TEM which was developed for the "particle surface" project of ERATO program at JST[1]. The STM holder head has two electrodes (see Fig.1(b)), one of which can be moved in three directions by a piezo driver. The electrode can further be moved by an inchworm as long as 5mm to make contact between the two electrodes. Once the two gold electrodes have a contact, they are slowly retracted by piezo drives so that the contact extends to form a gold nanowire. The electronic current that passes through the gold nanowire is measured in correlation with the TEM images, in order to obtain one-to-one correspondence between the structure and the conductance of the nanowire[2].

The new STM combined FE-UHV TEM was confirmed the previous experiments [3,4,5] : As shown in Figs 2 and 3, the gold nanowires change their structure from the six-prism[3] to the helical multi-shell (HMS) tubular structure[4]. The prism structure changes into the HMS tubular structure for diameters smaller than 2nm. The HMS structure in Fig.3 is the finest one found in the previous studies, and has the diameter of 0.56nm. The HMS gold tubes are designated by the number of gold atomic rows which composes the outer tube and that of the inner tube as n-n'-n" HMS. For the 7-1 HMS in Fig.3, the outer tube has 7 atomic rows and the inner tube (it is not a tube, in this case) has 1 atomic row. In addition the new STM system has succeeded to measure the conductance of the 11-4, 14-7-1, and 15-8-1 gold HMS tubes[6]. In he previous conductance measurements of the gold point contact [5], the gold nanowires formed at the contact were too short to have the HMS tubular structure

UHV TEM thus has an advantage to study nanowires, and thus, nanostructures without any attack from reactive residual gases. If any attack on nanostructures might cause unexpected change in electronic properties. UHV environment is the way to analyze their properties reliable. UHV condition is also reduces specimen contamination, and inevitably needed for nano-probe experiments. In this connection "super nano-probe UHV electron microscopy" which is the most powerful instrument for NANO-analysis [7] should only work at UHV condition.

References

[1] http://wwwsurf.phys.titech.ac.jp/tylab/index.html
[2] K.Mohri et al., Thesis at Tokyo Institute of Technology, 2002.
[3] Y.Kondo et al., Phys.Rev.Lett. 79 (1997) 3455.
[4] Y.Kondo et al., Science 289 (2000) 606.
[5] H.Ohnishi et al., Nature 395 (1998) 780.
[6] Y.Ohshima et al., in proceeding of ECOSS, 2002
[7] This work was supported by the grant-in-aid from the ministry of education, culture, art, science and technology.

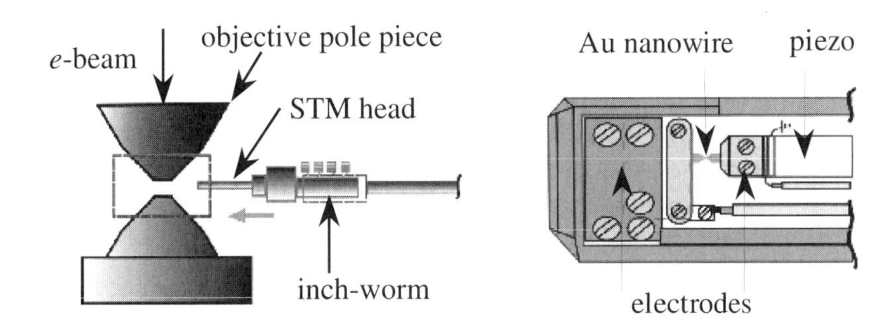

FIG. 1 Design of STM holder at the specimen stage of the UHV electron microscope.
(a) Side view of the STM holder and objective pole piece. (b) onstruction of the STM head.

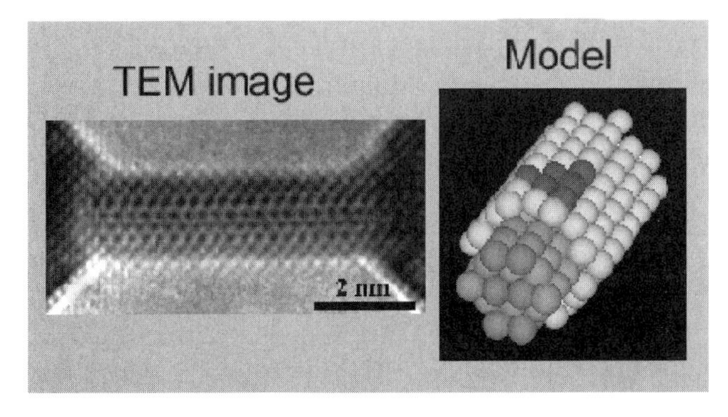

FIG 2 six-prism structure of gold nanowire. The surface has been reconstructed to (111) like 23x1 structure. The axis of the nanowire has the [110] atomic row of the gold face centered cubic crystal.

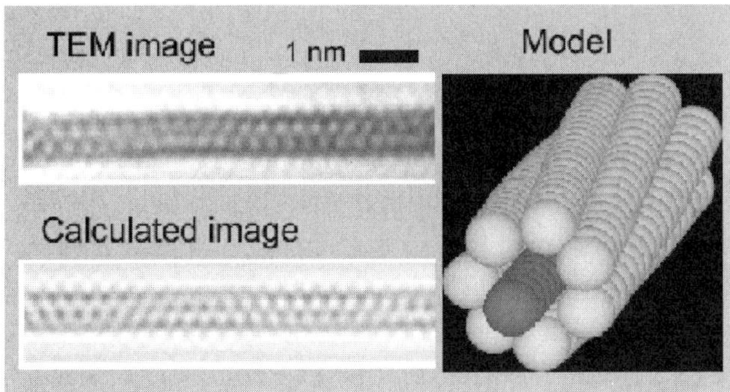

FIG 3 The 7 -1 helical multi-shell (HMS) gold nanotube. The outer tube has seven atomic rows which coil the axis of the tube axis. The observed and calculated images (multi-slice method) agree well with each other.

Microsc. Microanal. 8 (Suppl. 2), 2002
DOI 10.1017.S1431927602100523

In Situ HREM of Crystallization Reactions

R. Sinclair and K.-H. Min

Department of Materials Science and Engineering, Stanford University, Stanford, CA 94305-2205

In-situ high-resolution electron microscopy (HREM) is ideally suited for the investigation of solid-state crystallization behavior. The well-known image contrast of an amorphous material is quite clearly, and often dramatically, replaced by that of the crystalline phase. However compared to other techniques, the observations are now being made at the atomic or nano-scale level so the detail provided is quite different in scope. This article reviews work in this field.

Parker was the first researcher to utilize a heating holder to achieve controlled elevated temperatures while maintaining high-resolution imaging conditions (e.g. [1]). His work showed that kinetic data for the solid-phase epitaxial regrowth of silicon were the same by in situ HREM as those obtained by bulk Rutherford backscattering and high-voltage electron microscopy studies [2]. Morgiel et al. (e.g. [3]) extended this to the direct crystallization of amorphous silicon deposited on oxidized silicon substrates. Of particular note was the existence of pre-existing crystal embryos, the majority of which did not form viable nuclei. In both cases, there was no clear atomic ledge mechanism for the silicon crystal growth, even though this is an often-accepted mechanism [4]. Furthermore the growth often occurred in "spurts" with many atoms assuming the crystalline structure between individual video frames, an observation that was confirmed by Guillemet et al. in through-foil HREM investigation [5].

Konno utilized the in situ method combined with differential scanning calorimetry to study the metal-mediated crystallization of amorphous silicon and germanium [6-8]. In situ HREM clearly showed that the much-reduced crystallization temperatures arose from crystal nucleation within the metal, followed by more rapid substitutional diffusion of Si or Ge atoms through the metallic matrix than can occur at the amorphous-crystalline interface. This mechanism was also realized in the silicide-mediated crystallization of silicon, as described by Hayzelden and Batstone [9]. Kinetic and in situ studies demonstrated that an equivalent mechanism occurs during the crystallization of amorphous co-sputtered Al-Si alloys [10]. Progression of this work to carbon crystallization is described elsewhere [11]. A ledge mechanism for the Ag-Si system is also very clear when the metal-semiconductor interface is parallel to {111} type planes [7].

Room temperature thin film deposition of oxides also often results in an amorphous material. Because the melting points are generally higher than those of other materials (and so lower diffusion rates), crystallization may require higher annealing temperatures, which makes the in situ experiment more difficult. However, Fig. 1 shows in situ HREM taken during the crystallization of amorphous Ta_2O_5, a candidate material for high dielectric constant applications in integrated circuits. Pictures in this sequence show that the growth of the (001) planes take place via a ledge mechanism though it is not a perfect single plane one. The growth activation energy was found to be 4.8 eV, comparable to that found for Ta_2O_5 powder particle coarsening (5.6 eV) [12].

In summary, crystallization mechanisms are clearly elucidated by in situ HREM studies. One drawback is that because the atomic positions are not evident in the amorphous material, the exact attachments are not always clear, although a ledge mechanism normally is very distinctive when it is present.

References

[1] R. Sinclair and M.A. Parker, *Nature* 322 (1986) 531.
[2] M.A. Parker and R. Sinclair, *Proc. ICEM-XI 2* (1986) 991.
[3] R. Sinclair et al., *Ultramicroscopy* 51 (1993) 41.
[4] R.M. Drosd and J. Washburn, *J. Appl. Phys.* 53 (1982) 397.
[5] J.P. Guillemet et al., *Proc. ICEM-XIII* (1994) 449.
[6] T.J. Konno and R. Sinclair, *Philos. Mag. B*, 66 (1992) 749.
[7] T.J. Konno and R. Sinclair, *Philos. Mag. B*, 71 (1995) 163.
[8] T.J. Konno and R. Sinclair, *Philos. Mag. B*, 71 (1995) 179.
[9] C. Hayzelden and J.L. Batstone, *J. Appl. Phys.* 73 (1993) 8279.
[10] T.J. Konno and R. Sinclair, *Mater. Chem. Phys.* 35 (1993) 99.
[11] R. Sinclair et al., *Proc. ICEM-XV* (2002) in press.
[12] J.M. Heints et al., *Ceram. Int.* 18 (1992) 263.

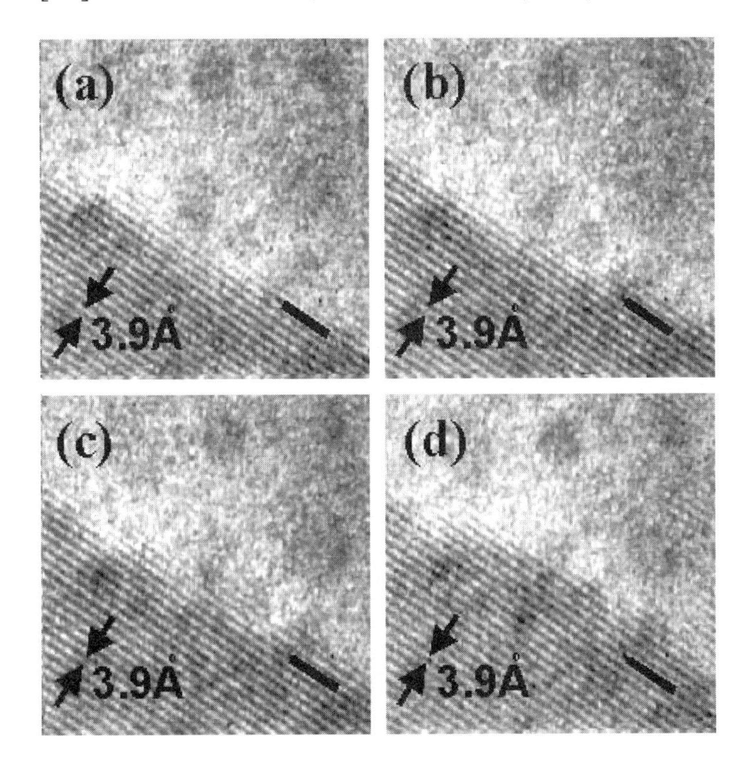

FIG. 1 A sequence of HREM images reproduced from videotape recorded during in-situ heating of Ta_2O_5 film with an interval of 2 seconds

Microsc. Microanal. 8 (Suppl. 2), 2002
DOI 10.1017.S1431927602100535

Aberration Correction for Analytical In Situ TEM – the NTEAM Concept.

B. Kabius, C.W. Allen, D.J. Miller
Materials Science Division, Argonne National Laboratory, Argonne, IL 60439

Future aberration corrected transmission electron microscopes (TEM) will have a strong impact in materials science, since such microscopes yield information on chemical bonding and structure of interfaces, grain boundaries and lattice defects at an atomic level. Beyond this, aberration correction offers new possibilities for in situ experiments performed under controlled temperature, magnetic field, strain etc. at atomic resolution. Such investigations are necessary for solving problems arising from electronic component miniaturization, for example. Significant progress can be expected by means of analytical aberration corrected TEM. These next generation microscopes will be equipped with an aberration corrected imaging system, a monochromator and aberration corrected energy filters. These novel elements have already been designed and partially realized [1,2,3]. Their incorporation alleviates three major obstacles that limit the capabilities of present TEMs:

a) Insufficient space for many in situ experiments.

High-resolution microscopy has been realized for medium acceleration voltages (200 and 300kV) by decreasing the coefficient of spherical aberration C_3, which has been achieved only by a small gap (3 and 5 mm) between the pole pieces of the objective lens. For most in situ experiments this is not sufficient space. High-voltage microscopes combine high resolution with a large gap. Unfortunately radiation damage reduces their usefulness especially for the study of semiconductors. Aberration correction and monochromators offer several solutions. The curves 1 and 2 in Fig. 1 represent contrast-transfer functions for an uncorrected and an aberration corrected TEM, respectively, for a 25 mm gap. Curve 2 applies to a corrector which eliminates both chromatic and spherical aberration. Correction of chromatic aberration has not been demonstrated for a commercial TEM yet. Nevertheless this calculation shows that state-of-the-art power supply stabilities are sufficient for a resolution limit smaller than 0.1 nm even if the high stability requirements for correction of chromatic aberration are taken into account.

b) Insufficient resolution and imaging artifacts.

A resolution limit of about 0.05nm is necessary for atomic resolution in non-periodic structures and crystalline materials along a sufficiently large number of zone axes to enable tomography. Present microscopes are capable of high-resolution only under favorable circumstances, such as appropriate materials in special orientations. For example, in metals only a few low index planes can be resolved. This allows the analysis of special types of grain boundaries, but the great majority of grain boundaries in polycrystalline materials is not accessible. According to calculation (curve 3 in Fig 1) a TEM with monochromator and C_3-corrector has the required optical properties to attain a resolution of 0.05 nm. Furthermore, a C_3-corrector eliminates contrast delocalization which is the origin of many artifacts in high-resolution imaging. Additional correction of chromatic aberration would enable a gap width of about 10 mm which is necessary for a large tilting range.

c) The energy width of the incident electron beam is too large.

Electron energy loss spectroscopy (EELS) is a powerful tool for qualitative and quantitative chemical analysis especially for light elements on very small areas. Furthermore, EELS is capable of analyzing the chemical bonding and the coordination of atoms in compounds requiring an energy resolution of about 0.1 eV. In this case it is possible to answer questions about the coordination and the bonding state in catalysts and to measure locally the band gap in semiconductors, for instance. In addition, it is often necessary to obtain such information on an area of atomic dimensions. Aberration corrected energy filters in combination with a monochromator meet these requirements. The plots of the envelope functions of temporal coherence in Fig. 2 reveal that chemical information can be obtained on an atomic scale using energy filtering techniques, if the chromatic aberration is also corrected. The parameters chosen for this calculation are valid for an in situ TEM with a gap width of 25 mm and chemical mapping using the Si-K absorption edge.

References
[1] J. Zach and M. Haider, Nucl. Instr. and Meth. in Phys. Res. A363 (1995) 316.
[2] M. Haider, H. Rose, S. Uhlemann, E. Schwan, B. Kabius and K.Urban, Nature 392 (1998) 768.
[3] O. Krivanek, N. Delby and A.R. Lupini, Ultramicroscopy 78 (1999) 1.
[4] This work was supported by the U.S. Department of Energy, Office of Science, under contract #W-31-109-ENG-38.

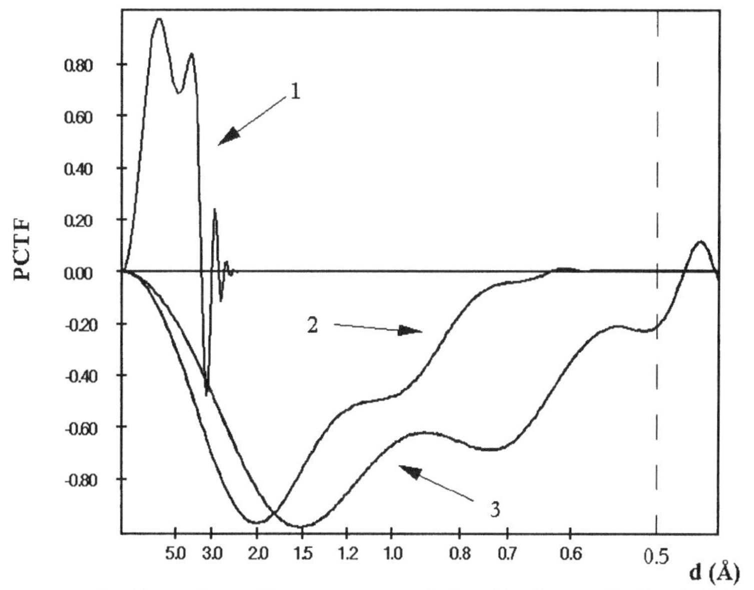

Fig. 1: Phase contrast transfer functions 1) uncorrected $C_c=C_3=5$mm 2) $C_c=0.01$mm, $C_s=-0.03$mm for an in situ TEM with a 25mm wide gap. 3) C_3 corrected with monochromator E=0.15eV.

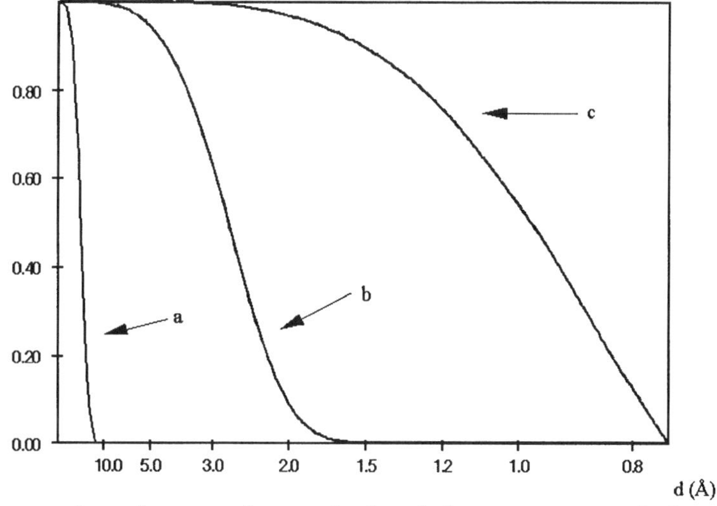

Fig. 2 : Temporal coherence damping envelops calculated for an energy window of 50 eV for a) $C_c=5$mm, b) $C_c=0.1$mm, c) $C_c=0.01$mm.

Microsc. Microanal. 8 (Suppl. 2), 2002
DOI 10.1017.S1431927602100547

In situ Transmission Electron Microscopy of Copper Electrodeposition

F. M. Ross[+], M. J. Williamson[*], R. M. Tromp[+], R. Hull[*] and P. M. Vereecken[+]

[+]IBM T. J. Watson Research Center, Yorktown Heights, NY 10598
[*]School of Engineering and Applied Science, University of Virginia, Charlottesville, VA 22903

Real time, *in situ* microscopy of surface and interface processes at low pressure has been very successful: it has allowed us to determine reaction mechanisms, observe transient structures and measure the parameters which control growth. Processes which take place in liquid environments are equally interesting scientifically and technologically but have not been as straightforward to observe *in situ* [1]. One important liquid process is the electrochemical deposition of copper, used in fabricating interconnects in advanced integrated circuits. To optimize the final structure it is useful to observe the nucleation of Cu clusters and to follow the growth of clusters to coalescence. In this paper we describe a simple cell in which we can control and observe this process in the TEM.

The cell is shown in figure 1. We etch two SiN-covered Si wafers to form windows and then glue them together with the windows aligned. An oxide layer maintains a separation of 0.5-1μm between the wafers. On one wafer we deposit an electron transparent Au working electrode partly over the viewing window and connect it electrically through the wafer to an external contact. The counter electrode is a Au wire extending into one of the reservoirs. The reservoirs, capped with glass spacers and sealed with sapphire lids [2], are designed to make it easier to introduce the electrolyte with a syringe. A $CuSO_4/H_2SO_4$ electrolyte is used. To improve the image quality, images are energy filtered and recorded digitally at 12 frames per second [3].

Figure 2 shows stills from a video recorded during Cu deposition and figure 3 illustrates the kinetics of cluster growth. Nucleation is clearly progressive rather than instantaneous, but the process is rapid with all clusters nucleating within 3 seconds. Progressive nucleation followed by diffusion-limited growth has been inferred indirectly from current transient measurements [4, 5]. Our real time data allows us to measure the initial nucleation rate (2.5×10^8 cm^{-2} sec^{-1}, within the range cited in literature [5]), and a linearly increasing nucleus density as expected theoretically [5]. After nucleation, we find that the growth of individual clusters can be understood with a simple electrochemical model [6].

It is important to consider the limitations arising from the finite volume of electrolyte available above the electrode. Voltammograms recorded in the cell are similar to those made in a standard three-electrode electrochemical cell, suggesting that the same process is taking place. However, we find that at higher current densities deposition becomes limited by diffusion of Cu ions from distant regions of the electrolyte. Calculations suggest that we can reproduce the bulk process if we choose appropriate length and time scales, and that the early stages of nucleation and growth are not too sensitive to the limited volume provided a low growth rate is used [6]. We can even extend the range of the experiments by applying the voltage in pulses. The liquid cell we have described may also be adapted to study phenomena such as corrosion, liquid crystal switching and biological systems, where the ability to analyze the solid/liquid interface in real-time and with reasonable resolution will help us gain a deeper understanding of the important processes involved [7].

[1] P. L. Gai, Microsc. Microanal. 8 (2002) 21.
[2] We used a heat curing epoxy (Measurements Group Inc., Raleigh, NC) to glue the wafers together and a UV-cured epoxy (Summers Optical, Fort Washington, PA) to glue the sapphire lids after introducing the electrolyte.
[3] Imaging energy filter and video rate CCD camera supplied by Gatan, Inc., Pleasanton, CA
[4] M. H. Hölzle et al., Electrochim. Acta 40 (1995), 1237.
[5] G. Gunawardena, G. Hills, I. Montenegro and B. Scharifker, J. Electroanal. Chem. 138 (1982) 225; P. M. Vereecken, K. Strubbe and W. P. Gomes, J. Electroanal. Chem. 433 (1997) 19; A. Radisic et al., J. Electrochem. Soc. 148 (2001) C41.
[6] M. J. Williamson, R. M. Tromp, R. Hull, P. M. Vereecken and F. M. Ross, in preparation.
[7] We acknowledge S. J. Chey, M. C. Reuter, A. Ellis P. C. Searson, J. Horkens, R. G. Kelly and J. M. Harper for their contributions.

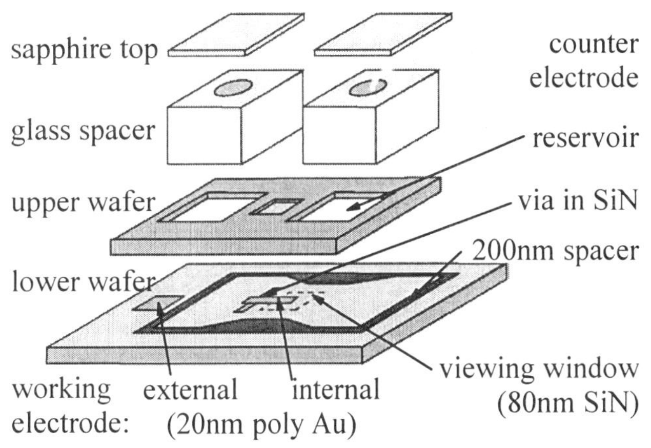

FIG 1: Components of the electrochemical cell. The viewing window is enlarged for clarity.

FIG 2: Images extracted from a video recorded during deposition of copper at a constant current density of $5mAcm^{-2}$. The scale bar is 1 micron. Images are shown after 0, 0.7, 1.3, 2.7 and 4.7 seconds.

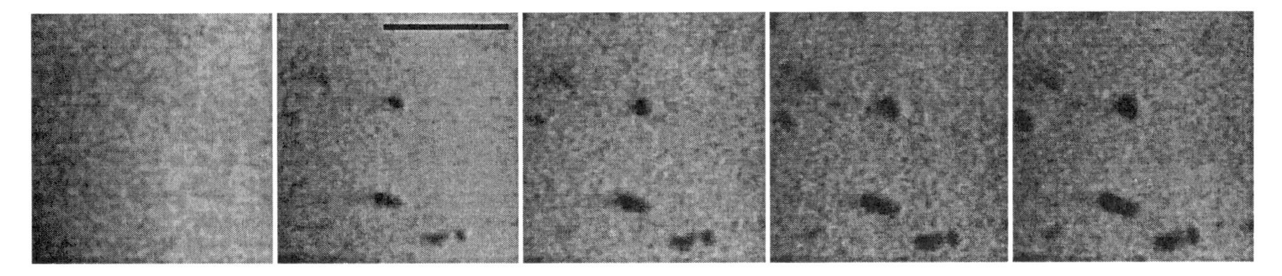

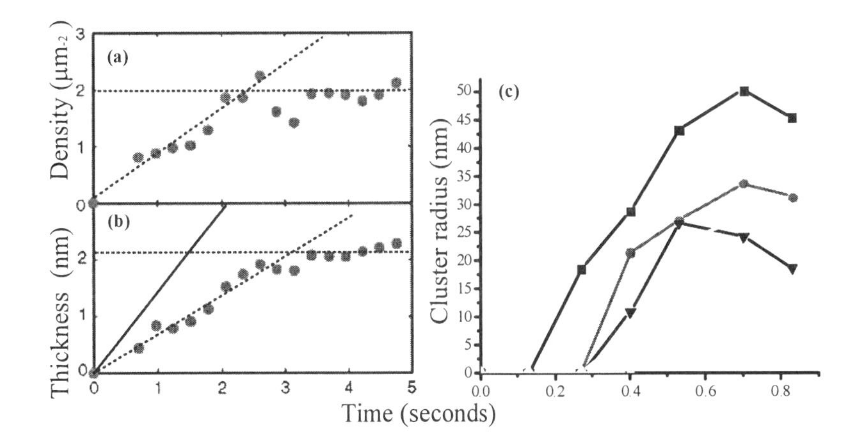

FIG 3: (a) Number density of nuclei obtained from Fig. 2. The density initially increases linearly as expected. (b) Average Cu thickness, calculated assuming nuclei are hemispherical (as seen in *ex situ* SEM). The straight line shows the expected deposition rate. The volume flattens out as ions are depleted from the liquid above the electrode. (c) Growth of individual clusters during deposition at $50mAcm^{-2}$.

Microsc. Microanal. 8 (Suppl. 2), 2002
DOI 10.1017.S1431927602100559

Local Measurement of Reaction Kinetics Using *in situ* Transmission Electron Microscopy

Renu Sharma and Peter Crozier

Center for Solid State Science, Arizona State University, Tempe, AZ 85287-1704

In solids, reaction rates and corresponding reaction mechanisms may be strongly influenced by nanostructural features such as defects, second phase precipitates and surface and interface properties. In order to correlate local changes in reaction rates to the microstructure, data must be obtained from the same sample regions. *In situ* microscopy is an excellent technique for correlating reaction kinetics with microstructure. In the TEM, the microstructure and chemical changes can be characterized using imaging, diffraction and spectroscopy signals. The kinetics can be determined by monitoring changes in the same signals. The time resolution of these kinetic measurements is limited only by signal-to-noise and detector technologies. We have developed several controlled atmosphere microscopes for studying local kinetics in gas-solid reactions. In such instruments, we are able to reveal fluctuations in local kinetics and correlate them with inhomogeneities in the solid. Here we illustrate this approach by application of the technique to a range of different problems in materials and solid-state chemistry.

The first example (Figure 1) demonstrates nucleation and growth of Cu particles during the nitridation of Cu-Ti thin films when heated up to 420°C in 3 Torr of NH_3. Cu is depleted from the matrix during nitridation and grows into large particles. Most of the Cu particles (marked A, C and E) grow as circular particles but two other particles (marked B and D) have different growth rates in two directions. This is due to variations in the growth rate of different crystallographic planes. Temperature resolved data was used to derive the kinetics of the reaction. The nitridation temperature for these samples was determined by electron diffraction and was found to reduce with increasing concentration of Cu in the thin films (Figure 2). Similar measurements were made for Cu-Cr thin films. The nucleation was dependent on the chemical reaction rate while the growth was controlled by Cu diffusion rates.

Growth rates for Au particles have also been measured from time resolved video sequences recorded during chemical vapor deposition (CVD) on clean Si {100} surfaces.[1] The ion milled Si samples were heated at different temperatures in 10^{-3} Torr of ethyl (trimethylphosphine) gold (I)(EtAu(PMe_3))and deposition was performed both with and without the electron irradiation. While a simple linear relationship could be observed between the growth rate and temperature (Figure 3) for depositions without the electron irradiation, data recorded during electron irradiation could not be fitted to a linear equation indicating a more complex reaction mechanism.

We have developed a method to obtain the local oxidation state of Ce in complex oxides during redox reactions[2]. The oxide compositions and surface areas were found to play important roles in the redox activity of Ce containing catalysts. The Ce oxidation is obtained by monitoring the change in white-line ratios (M_{45}) in the electron energy-loss spectrum during redox reactions. This technique can be used to obtain local kinetics during redox reactions in many transition metal systems.

In situ microscopy can provide direct information on the kinetics and dynamic mechanisms of polymerization processes. We have studied the initial stages of nanoscale polypropylene (PP) growth during gas phase Ze igler-Natta polymerization. Polymerizations were run *in situ* under C_3H_6 pressures ~ 1 Torr and revealed growth of non-uniform PP layers over areas ~100 nm in size with diffuse interfaces between adjacent globules. Local kinetics demonstrated a linear behav ior with a growth rate of 0.16 -0.23 nm/s implying that there is no significant diffusion barrier for C_3H_6 through PP up to 140 nm. Active sites are probably located at the interface between the PP layer and the catalyst. The measured growth rate is 6 order s of magnitude lower than a surface collision model would predict suggesting that there is a slow rate -determining step in the polymerization process.

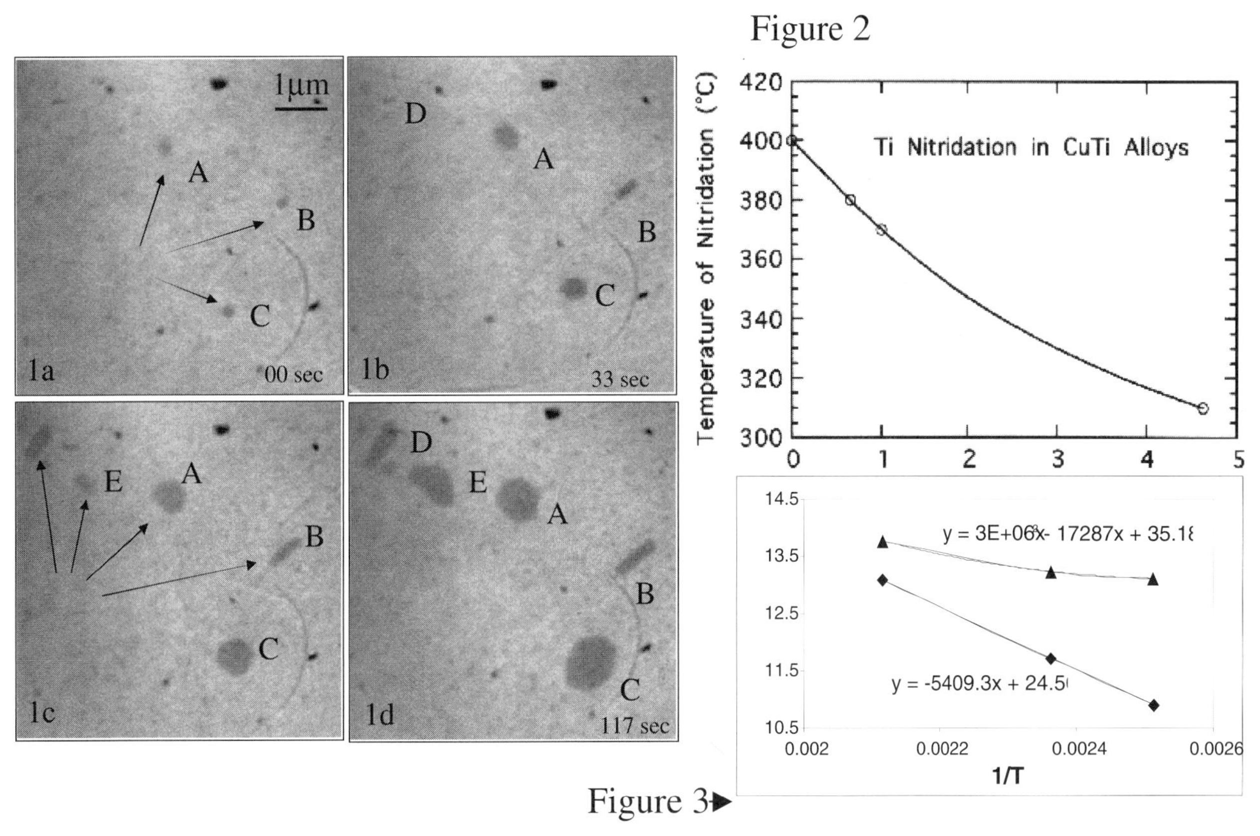

Figure 1a-d. Time resolved low-magnification images recorded at 420°C in 3 Torr of NH_3 showing the nucleation and growth of Cu particles during the nitridation of Ti in Cu/Ti thin films.
Figure 2. The nitridation temperature reduced with increasing Cu content of the matrix.
Figure 3. Arrhenius plots of $\log_{10}$ (growth rates (atoms/cm^2/sec)) of Au particles with electron irradiation (triangles) and without irradiation (diamonds).

References

1. Drucker et al, J. Appl. Phys. 77 (1995) 2846.
2. Sharma and Crozier, Microscopy and Microanalysis 2002 (these proceedings)
3. Oleshko et al, J. of Electron Microscopy, 2002 (in press)

Microsc. Microanal. 8 (Suppl. 2), 2002
DOI 10.1017.S1431927602100948

X-ray Microanalysis of Light Elements

G.F. Bastin and H.J.M. Heijligers

Laboratory of Solid State and Materials Chemistry, University of Technology, P.O. Box 513, NL-5600 MB Eindhoven, The Netherlands.

Quantitative x-ray microanalysis of the ultra-light elements boron, carbon, nitrogen, and oxygen by wavelength-dispersive methods requires a lot of dedication and rather complicated experimental procedures compared to the analysis of heavier (Z > 11) elements. The most difficult practical problems in the analysis are those related to the differences in the x-ray emission profiles from one ultra-light element compound to another. These peak shape variations are the result of the fact that in exciting ultra-light element x-rays electronic transitions of the bonding electrons are involved. As a consequence large peak shifts and peak shape alterations in WD analysis of ultra-light elements may occur [1]. Peak shifts can easily be accounted for by simply retuning the spectrometer when moving from standard to specimen. Peak shape alterations, however, are much more difficult to deal with because they force the operator to perform the intensity measurements in an integral fashion. The alterations in peak shape are most pronounced for the lightest element studied so far (boron) and they gradually decrease with increasing atomic number. They are also strongly dependent on the type of analyzer crystal used with its typical spectral resolution: synthetic multilayer crystals with their poorer spectral resolution exhibit less pronounced shape alteration effects than their conventional counterpart lead-stearate [2]. The analysis of boron presents additional problems because the shape of the B-K_α peak is found to be dependent also on the crystallographic orientation of the specimen with respect to the electron beam and the spectrometer. These peculiar effects must be attributed to the presence of polarized components in the B-K_α emission peak [3], which can partially be filtered out by the analyzer crystal. It is evident that all these effects together can make the analysis of an element such as boron decidedly tricky and the operator should at least be fully aware of all the problems associated with the analysis of the particular light element at hand.

Thanks to realistic $\varphi(\rho z)$ approaches modern matrix correction procedures are no longer a limiting factor in the analysis. An often underestimated, but major, physical problem in the quantification of ultra-light elements, however, is the uncertainty in the mass absorption coefficients. It can easily be shown that if final quantitative results are required with a precision of 1 % then mass absorption coefficients with a similar precision are necessary [1]. Unfortunately, the scatter in published mac data in literature is one to two orders of magnitude higher. It is out of the question to work with such a large scatter in vital physical input data in the matrix correction program. This is especially true, of course, for the absorption correction scheme. The latter has to be of outstanding reliability anyway in order to be able to cope with the extreme demands encountered in EPMA of ultra-light elements. We have pointed out [1] that the performance of a particular matrix correction procedure can only be judged in conjunction with consistent sets of mass absorption coefficients for each of the ultra-light elements. This assessment, in turn, can only be carried out when large databases of high-quality measurements are available over the widest possible range in experimental conditions.

Apart from the fundamental problems briefly mentioned so far there are also a number of practical problems during the actual analysis, some of which are not always easy to deal with.

One of the items of major importance is the specimen preparation, in which topics such as flatness and cleanliness of the specimen surface play a crucial role. Due to the usually strong absorption the x-ray emission volume in the specimen will be extremely shallow. Hence, anything that interferes with this shallow volume rapidly leads to a deterioration of the results.

For the same reason contamination phenomena can have disastrous effects on the analysis. Not only the (well-known) effects of carbon contamination have to be considered but also those of the much less well-known contamination with oxygen [4]. Carbon contamination is not only very bad for the analysis of carbon but also for the analysis of nitrogen because N-K_α x-rays are heavily absorbed in carbon, due to the presence of the carbon K-edge. Especially during the relatively long time period required for an integral WD intensity measurement on the same location the build-up of carbon can have deleterious effects on the analysis. Several devices, such as a liquid nitrogen cooling trap or the even more effective air-jet, can be used to reduce the build-up of carbon considerably. One has to be cautious, though, that while using an air-jet the specimen is not oxidized under the electron beam. There are cases, notably with some nitrides [4], where even without air-jet a process of oxidation sets in immediately after the electron beam has been positioned on the specimen. These examples demonstrate again how important it is that the operator is fully aware of all the problems that might disturb the analysis of ultra-light elements and that he constantly monitors the signals of all the elements involved.

Another major problem is usually the correct determination of the background, which is all the more important in case of relatively low intensities. Typical problems are strong curvatures or kinks in the background or the presence of multitudes of higher-order metal lines, which interfere with the light element peak. These problems are most pronounced with a conventional lead-stearate crystal, which is very effective in transmitting higher-orders of reflections. In combination with its low peak intensities such a crystal usually produces very low peak-to-background ratios, especially in the case of nitrogen. Fortunately, the new synthetic multilayer crystals have proved to be a great help [2] in two ways: They can supply considerably higher peak count rates than the conventional stearate crystal (more than an order of magnitude is not uncommon) and they can suppress higher orders of reflection quite effectively. The latter effect has to be checked for each particular light element because it is dependent on the wavelength range and the 2d-spacing of the crystal. The poorer spectral resolution of these crystals appears to be only a small price to pay for all these benefits. In some cases though, e.g., in Nb-borides, their use can lead to a complete overlap of B-K_α and Nb-M_ζ peaks, which can normally be resolved fairly well with a lead stearate crystal.

References

[1] G.F. Bastin and H.J.M. Heijligers, in K.F.J. Heinrich and D.E. Newbury, Eds., *Electron Probe Quantitation*, Plenum, New York (1991), 145.
[2] G.F. Bastin and H.J.M. Heijligers, in D. Williams, J. Goldstein, and D. Newbury, Eds., *X-Ray Spectrometry in Electron Beam Instruments*, Plenum, New York (1995), 239.
[3] G. Wiech, "X-ray emission spectroscopy", NATO Adv. Study Inst., P. Day, Ed., *Emission and Scattering Techniques Ser. C.*, (1981), 103.
[4] G.F. Bastin and H.J.M. Heijligers, *Microbeam Analysis*, D. E. Newbury, Ed., San Francisco Press, (1988), 325.

Microsc. Microanal. 8 (Suppl. 2), 2002
DOI 10.1017.S143192760210095X

Charging at the Steady State in EPMA, SEM and ESEM

J. Cazaux

DTI, CNRS UMR 6107,UFR Sciences, BP 1039, 51687 Reims Cedex 2, France.

Most of the specific effects observed in the e^--irradiated insulators result from charges trapped below the surface. At the very early beginning of this irradiation, $t \sim 0_+$, the charge distribution is very similar to that deduced from standard (non-charging) calculations but, via the rearrangement of electrons and holes generated by the beam, the system evolves towards a steady state rapidly attained under the standard conditions of EPMA, SEM and ESEM. To simplify, only homogeneous and thick (h~1mm) specimens widely illuminated by the beam are considered here.

In EPMA of ground coated specimens, δ is restricted to the conductive coating and a negative charge density, Q_-(C/cm^2)~-$J_0(1-\eta)t(0_+)$ (1), starts to be established via electrons trapped in the bulk down the range of the primaries, R(Fig.1a). The induced electric field and potential functions, F_Z and V(z), may be evaluated from models of negative charge distribution by solving a one dimensional Poisson equation [1]. Independently from the chosen distribution, the field is always maximum at the coating/dielectric interface where de-trapping processes start when a critical field value, F_C is reached while the de-trapped electrons are evacuated to the ground via the coating. The increase of the electric slowing down reduces the penetration depth of the primaries from R to R_C (subscript C for charging). Shown in fig.1, the steady state is characterized by a depleted region, ($0<z<R_C$), submitted to a uniform critical field F_C due remaining charges trapped between R_C and R in a low field region: $|Q(\infty)|$ (~ εF_C with ε :dielectric constant). The value of F_C depends upon the energy of the trapping sites. Similarly to R, R_C may be expressed in a form of a power law where E_0^n is changed into $(E_0-qF_CR_C)^n$ (2). The main consequence is a compression of the $\phi(\rho z)$ function similar to that previously obtained [1;2] but with a less field strength (see [3] for details).

For bare insulators investigated in SEM at large E_0 values, there is now a double layer system, +&-, at the early beginning of the irradiation: $Q_- \sim -J_0(1-\eta)t(0_+)$ of thickness R and $Q_+ \sim J_0\delta\, t(0_+)$ (3) of thickness s as a result of the SEE emission δ (Fig 2 left). The electric field is maximum near to z~s (between + and - charges) where recombination processes take place while the surface potential becomes more and more negative inducing a progressive external slowing down of the primaries. The steady state corresponds to $\delta_C + \eta_C = 1$; it is mainly characterized by a landing energy $E(R_C \sim s)$ (~E_{max}) in the 1-3 keV range[4]. $E(R_C \sim s)$ is independent from E_0 but it may change from place to place with the local change of s or of δ. The surface potential $V(0)=V_S$ is of the form:$-qV_S=E_0-E(R_C \sim s)$ and it results from some remaining electrons trapped between s and R in a low field region.

The initial situation of ESEM is similar to that of SEM but the deposition of ions+ leads rapidly to a 3-layer system with a 2^d positive (and very thin) layer on the vacuum side of the surface, Q(ion+). The steady state nearly corresponds to the neutrality, $|Q_-| \sim Q_+ + Q(ion+)$, when is neglected the contribution of the space charge in the gap. The potential function, V(z), takes an approximate S-shape form similar to that expected in SEM at an energy E_0 close to E_{02} [4] . Its surface value, V(0), is a function of the density of negative charges trapped below the surface: more electrons are trapped in the bulk and larger is the V(0) value, reducing then δ in consequence (Fig.2 right). This analysis is consistent with Griffin's images where the electron trapping sites appear darker [5].

References

[1] J. Cazaux , X-Ray Spectrometry 25 (1996) 265
[2] O.Jbara et al, X-Ray Spectrometry 26 (1997) 291
[3] J. Cazaux, Microscopy & Microanalysis, Special Issue: "Characterization of non-conductive …"
[4] J. Cazaux, Eur. Phys.; J.A.P. 15 (2001) 167
[5] B.J. Griffin, Scanning 22 (2001)234

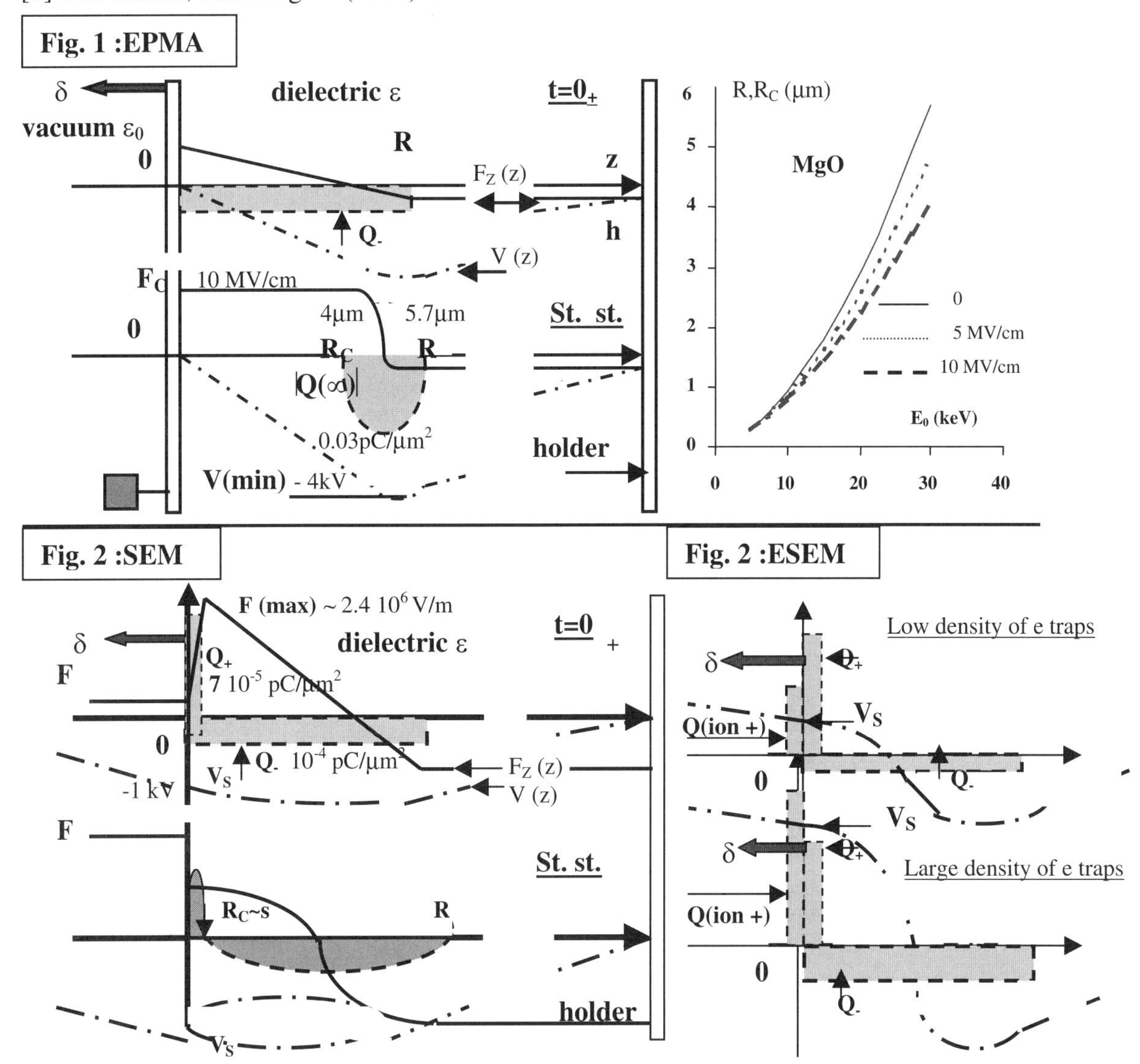

FIG.1. EPMA: V(z) and F_Z (z) functions at the early beginning of the irradiation; $t=0_+$ (top left) and at the steady state (bottom left). For MgO (as an example) with n=5/3 in eq.(2), influence of the field build up on the range R_C of the primaries (right).

FIG. 2.Left, SEM: V(z) and F_Z (z) functions at $t=0_+$ (top) and at the state state (bottom).Like for fig.1, note the very low charge densities, Q (that may be deduced from eqs (1)&(3)) giving very large electric field strengths, F, and potential values, V. Right, ESEM: Expected influence of the density (top: low; bottom: large) of trapped electrons on the surface potential value, V_S and on δ.

Microsc. Microanal. 8 (Suppl. 2), 2002
DOI 10.1017.S1431927602100961

Capability and Uncertainty in Multilayer Quantitative Procedure with Electron Probe Microanalysis

C. Merlet

ISTEEM, CNRS, Université de Montpellier II, Pl. E. Bataillon, 34095 Montpellier cedex 5, France

Electron probe microanalysis (EPMA) has become a well established technique for determining compositions and thicknesses (in the range from 1nm to 2000 nm) of multilayer samples. For the simple case of a thin film deposited on a substrate, this technique is efficient, and the accuracy for the concentration is similar to a bulk specimen. Depending of the instrument, the operator experience, the nature of the film and of the substrate, the uncertainty in the thickness determination can be expected to be less than 10% even when the difference in the atomic number of the film and of the substrate is very large (Fig. 1, Fig. 2). For a multilayer sample, quantitative results require hypothesis concerning the layer description and consequently, the operator experience is crucial in the quality of the results. In addition each defined layer must be homogeneous in depth, and in the majority of cases, when the same element is present in more than one layer, the solution is undetermined without an hypothesis on the concentration of this element. For a buried layer in a multilayer sample, the technique is less and less efficient when the layer thickness decreases and is deeply buried (Fig 3). In some cases, the X-ray lines are completely absorbed by the upper layers and are not detected.

To get reliable results, this technique requires an accurate description of the X-ray depth distribution from which the emitted X-ray intensities are calculated. Intensities can be estimated by two different methods: Monte Carlo simulation [1] and analytical approximations [2,3,4]. The first, although more accurate, is very time consuming, even with the fastest computers available. Moreover, quantitative results are obtained with the help of automatic iterative numerical procedures or with a manual trial and error approach. Consequently, for on line quantification of electron probe measurement only analytical models are used in practice. However, the lack of knowledge about the X-ray depth distribution for stratified samples limits the accuracy attainable with these analytical procedures, mainly when the atomic number between the different layers are largely different. In addition, analytical approximations as well as Monte Carlo simulation require the knowledge of many atomic parameters which describe the electron interaction and the X-ray emission, such as the ionization cross section, mass absorption coefficient, fluorescent yields, and others. Many of these atomic parameters are canceled or do not need to be known accurately by using standard in quantitative microanalysis on bulk sample. However, as shown by the figure 4 as example for the ionization cross section[5], multilayer quantitative analysis require more accurate knowledge of atomic parameters. Similarly, the uncertainty of mass absorption coefficient is less counterbalanced in multilayer than in bulk sample by using standards.

References

[1] J. Baro et al., Nucl. Instr. And Meth. B, 100 (1995) 31
[2] G.F. Bastin et al., Proc. 12th ICEM Meeting, San francisco Press, (1990) 216

[3] J.L. Pouchou and F. Pichoir. In Electron Probe Quantitation, Edited by K.F.J. Heinrich and D. Newbury, plenum Press, New York, (1991), 31
[4] C. Merlet, Proc. of Microbeam Analysis, Edited by E. S. Etz, VCH Publishers, (1995) 203
[5] X. Llovet et al., J. Phys. B: Atom. Mol. Opt. Phys., 33, (2000) 3761

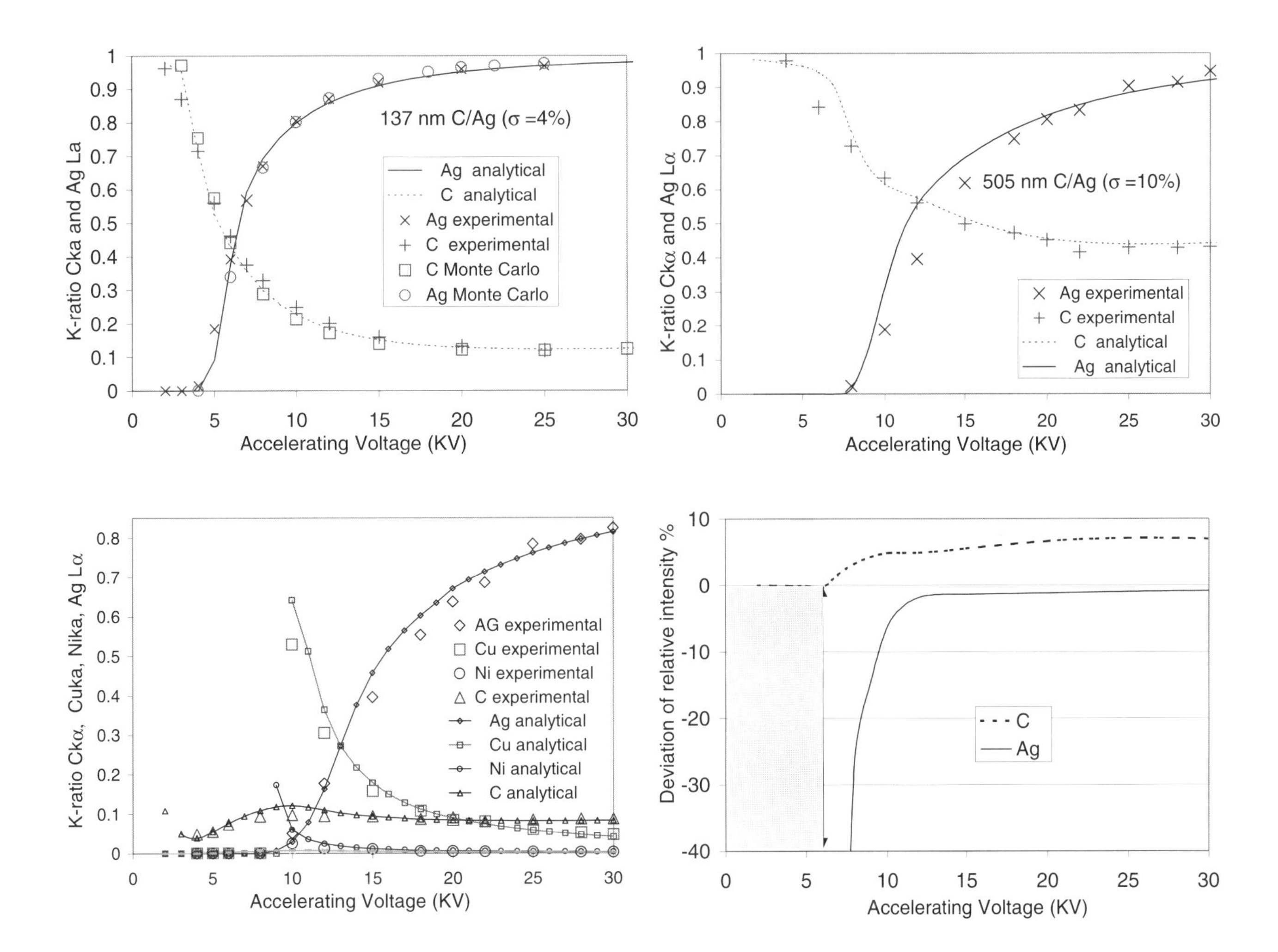

Fig. 1. Relative x-ray intensities for C/Ag specimen (C=137nm) as a function of the electron incident energy, (x, +) measurements, (o,) Monte Carlo calculation[1], (---) analytical procedure[4]
Fig. 2. Relative x-ray intensities for C/Ag specimen (C=505nm) as a function of the electron incident energy, (x, +) measurements, (---) analytical procedure[4]
Fig. 3. Relative x-ray intensities for Ni/Cu/C/Ag multilayer specimen (Ni=5.6nm, Cu=66nm, C=505nm), as a function of the electron incident energy, symbols represent the measurements and the lines represent the result of the analytical procedure[4]
Fig. 4. Deviation of the computation of the relative intensity obtained by using two different ionization cross section model in the analytical procedure, for the 505nm C film deposited onto the Ag substrates.

Microsc. Microanal. 8 (Suppl. 2), 2002
DOI 10.1017.S1431927602100973

On the Simulation of True EDS X-Ray Spectra

Raynald Gauvin[1] and Eric Lifshin[2]

1. Department of Mining, Metals and Materials Engineering, McGill University, Montréal, Québec, Canada, H3A 2B2
2. Albany Institute for Materials, CESTM, 251 Fuller Road, Albany, NY, 12301.

The Monte Carlo method has been used recently to simulate complete EDS X-Ray spectra of planar and of rough surfaces[1]. However, these simulations only include the statistics of the generation and emission processes that are shown by varying the number of simulated electrons. Above a critical number of incidents electrons (5000 to 20000 simulated trajectories), the effects on the number of electrons on the shape of the X-Ray spectrum becomes negligible. Since the X-Ray yields are very low (10^{-4} to 10^{-6} typically), the statistics of the EDS spectra are dictated by the number of detected photons, i.e. by the physics of the detection process. In order to simulate true X-Ray spectrum including the effect of the statistic of the detection process, a new Monte Carlo program was developed.

This new Monte Carlo program simulates the EDS detection process. From the simulated emitted spectrum, the ratio of the intensity of the characteristic lines to that of the total bremstrallung intensity is computed. A random number is generated and if it is smaller that the ratio, a characteristic lines is generated. From the weight of the characteristic lines, a random number is used to choose the appropriate line. If the first random number is greater than the ratio, a bremstrallung photon is generated and its energy is picked with another random number and the computed energy distribution. The photon with specific energy is then send through the detector and its absorption location computed. Then a photoelectron is generated and Monte Carlo simulations compute its diffusion through the crystal. Its energy dissipated in the crystal is used to compute the number of electron – holes pairs generated including the statistical fluctuations with the Fano factor. The effect of incomplete charge collection with the Si dead layer is included in this program[2]. Figure [1] shows the emitted X-Ray spectrum used to generate the photons that are sent through the EDS detector. It is a Fe- 50 (wt%) B allow at incident electron energy of 8.3 keV. Figure [2] shows the simulated EDS X-Ray spectrum with a total of 100 000 photons. Clearly, the effect of the statistical noise coming from the detection process on the bremstrallung is visible. Also, in this simulation, very bad parameters for incomplete charge collection were used, giving the disappearance of the B K_α line in the bremstrallung.

References

1. R. Gauvin and E. Lifshin (2000), "Simulation of X-Ray Emission from Rough Surfaces", Mikrochimica Acta, Vol. 132, pp. 201-204.
2. D. C. Joy (1995), "Modeling the Energy Dispersive X-Ray Detector", in: X-Ray Spectrometry in electron Beam Instruments, edited by D. B. Williams, J. I. Goldstein and D. E. Newbury, Plenum Press, New-York.

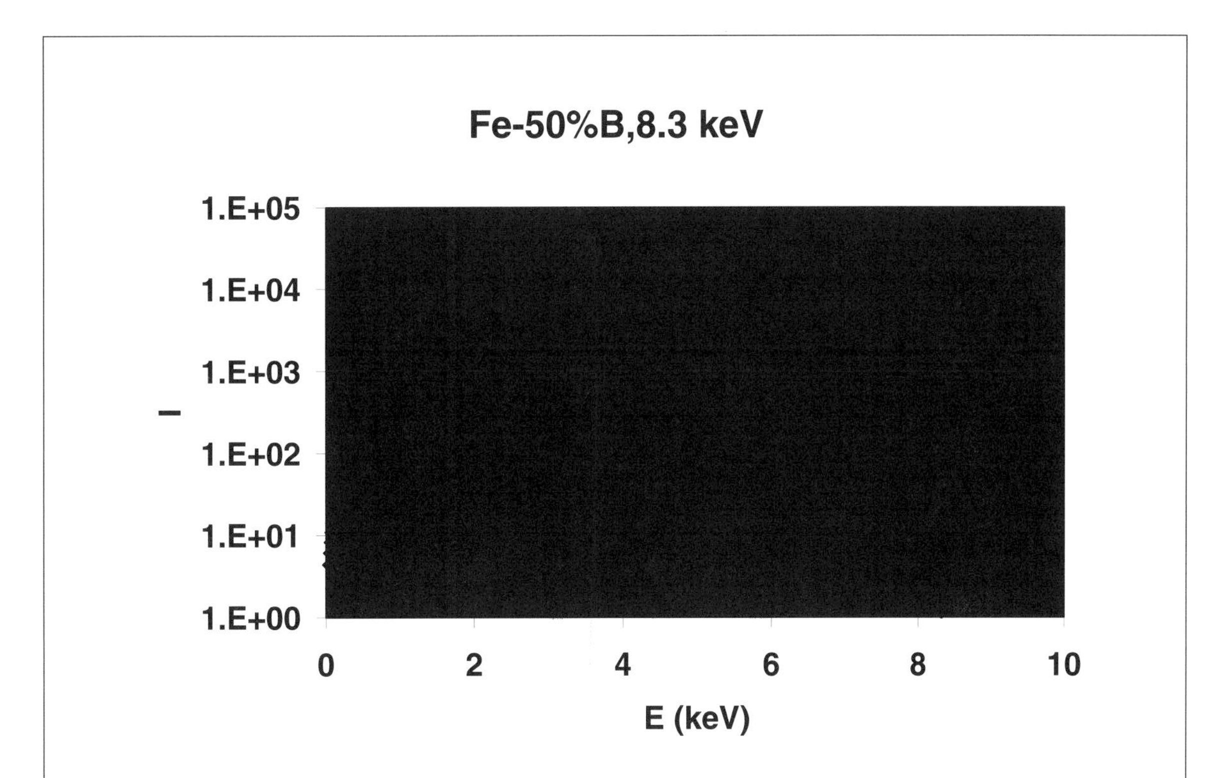

Figure 1 Emitted Fe- 50 (wt%) B X-Ray spectrum used to generate the photons that are sent through the EDS detector.

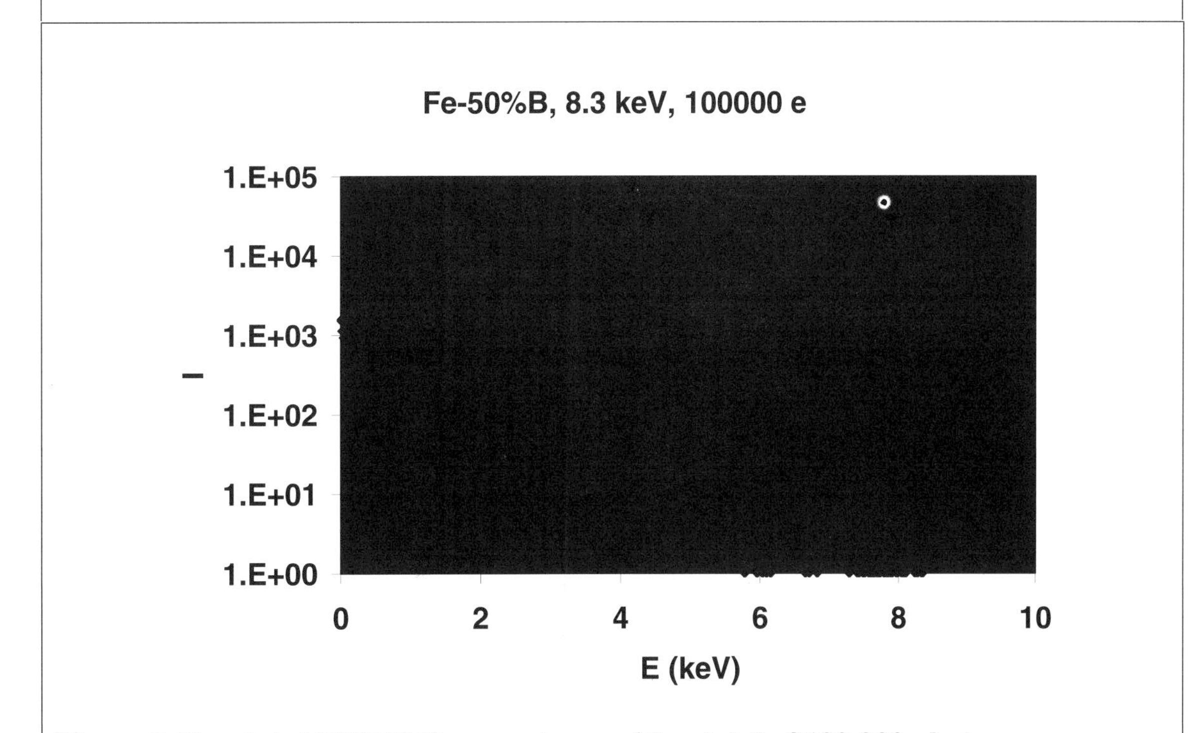

Figure 2 Simulated EDS X-Ray spectrum with a total of 100 000 photons.

Microsc. Microanal. 8 (Suppl. 2), 2002
DOI 10.1017.S1431927602100985

The Influence of X-ray Counting Statistics on Trace Analysis and Spatial Resolution

Eric Lifshin* and Raynald Gauvin**

* University at Albany, School of Nanosciences and Nanoengineering, Albany, NY, 12203
** McGill University, Department of Mining, Metals and Materials, Montreal, Quebec, Canada H3A2B2

It is well recognized that under proper circumstances the relative accuracy of quantitative electron microprobe analysis can be better than 2% [1]. Critical requirements include the use of well prepared polished samples, good standards, stable electron beam and x-ray detector operation, a small excitation volume relative to the phase of interest, generally more than 10% of the element of interest, minimal spectral interferences, and a good quantitative model containing accurate parameters. Precision, on the other hand, is determined by the variability of repeated measurements under what should be constant operating conditions. The factors affecting precision include stage and spectrometer reproducibility, noise in the electron beam, sample inhomogeneity and the inherent randomness of the x-ray emission process itself. The latter is described by x-ray counting statistics. Under the conditions stated above for the high accuracy case it is usually not difficult to acquire sufficient counts from samples and standards in reasonable times (100 seconds or less) to have the contribution to the precision from x-ray counting statistics be less than 2%. Here precision is given by the 95% confidence interval defined by + or - 1.96 s_c where s_c is the standard deviation in the composition [2,3]. This level is generally possible with either energy dispersive spectrometry (EDS) or wavelength dispersive spectrometery (WDS) since measurable count rates of at least several thousand counts per second are often possible for many pure elements. When sufficient counting times are available then precision for the high accuracy case may often be limited by factors other than counting statistics and s_c should be determined experimentally.

There are, however, frequently encountered circumstances where the precision due to counting statistics can be much larger than 2% and counting statistics may be the main factor limiting precision. These include the analysis of thin films and fine phases as well as trace element determination. In these cases the count rates may be so low and close to background levels that it is not always practical or even possible to count long enough to accumulate the needed numbers of counts because of drift, contamination or the lack of long term instrument availability. Monte Carlo modeling [4] optimized to give accurate characteristic line and continuum intensities can be very effectively used in conjunction with detector collection efficiency data serves as an excellent way to predict precision as a function of a variety of experimental parameters. These include SEM (or electron microprobe) operating conditions, x-ray detector operating conditions, sample composition and sample morphology including thickness for thin films or phase size for structures with lateral dimensions smaller than the electron beam excited volume. Fig. 1 is an example of a 95% confidence plot as a function of electron beam voltage for a WDS system measuring aluminum K_α in an AlGaAs sample. The different curves show how beam current and counting time influence precision.

Beam voltage was chosen as an independent variable because it must be reduced to decrease the excitation volume as shown in Fig. 2 illustrating the Anderson-Hasler range [5] for the production of Al Kα x-rays in both Al and AlGaAs. A more detailed three dimensional picture of the excitation volume can be obtained with Monte Carlo calculations which also show the effect of electron beam diameter. X-ray intensity is directly proportional to probe current, and probe current decreases rapidly with probe diameter for all the different types of electron sources used. The probe current also decreases with beam voltage for a given probe size. Obtaining a small probe (30nm<) with stable current of 1 nA at low voltages (5KV) can be a real challenge, but can be done with Schottky sources. In addition, the excitation efficiency of x-ray lines drops as beam voltage approaches the line excitation energy. All of these factors can combine to limit x-ray emission under conditions required for high spatial resolution, however expected performance and optimum conditions for each experimental situation can be defined by modeling. A similar approach has been developed for thin film and trace analysis. The authors wish to acknowledge the support of the MARCO and DARPA sponsored Interconnect Focus Center.

References
[1] J. Pouchou and J Pichoir, Electron Probe Quantitation, Heinrich and Newbury eds., Plenum Press, 31-75 (1991)
[2] T. O. Ziebold, Anal. Chem., 39, 858-861 (1967)
[3] E. Lifshin, N. Doganaksoy, J. Sirois and R. Gauvin, Microsc. Microanal, 4 598-604 (1999)
[4] E. Lifshin and R. Gauvin, Microsc and Microanal 4,232-233 (1998)
[5] C. A. Anderson and M. F. Hasler, Proc. 4th Intl. Conf. X-ray Optics and Microanalysis, Hermann Paris, p 310 (1966)

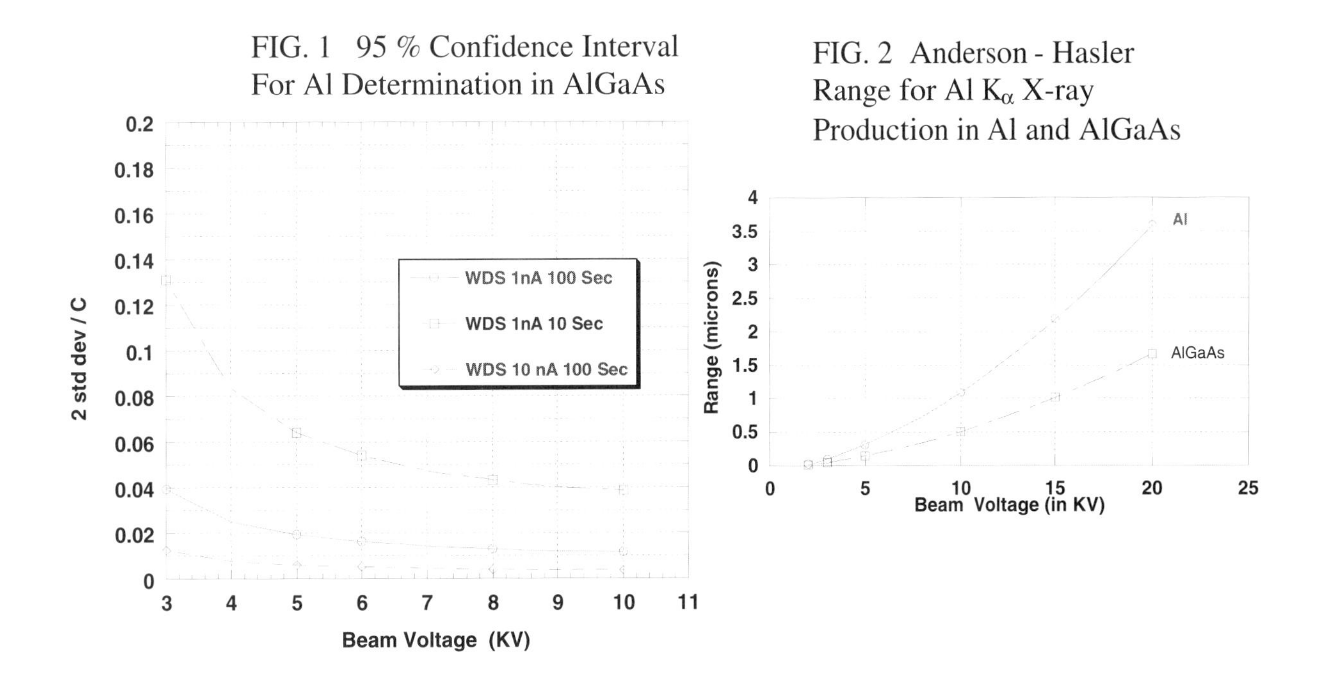

FIG. 1 95 % Confidence Interval For Al Determination in AlGaAs

FIG. 2 Anderson - Hasler Range for Al K_{α} X-ray Production in Al and AlGaAs

Microsc. Microanal. 8 (Suppl. 2), 2002
DOI 10.1017.S1431927602100997

Low-Overvoltage Microanalysis: an Alternative High Resolution Strategy to Low-Voltage Microanalysis

Dale E. Newbury

National Institute of Standards and Technology, Gaithersburg, MD 20899-8371

When improved spatial resolution for microanalysis is a critical requirement, the two leading strategies are (1) the thin specimen/high beam voltage option of analytical electron microscopy (AEM) and (2) the thick specimen/low beam voltage option of low-voltage scanning electron microscopy (LV-SEM).[1] If the specimen must be studied in bulk form, only option 2 is viable. However, with currently available wavelength dispersive (WDS) and semiconductor energy dispersive (Si-EDS) x-ray spectrometry techniques, the analyst frequently encounters severe constraints upon the suite of elements that can be effectively measured.[2] The choice of low voltage (defined as $V_0 \leq 5$ kV) greatly restricts the atomic shells that can be excited, which frequently forces the analyst to choose x-ray peaks with low fluorescence yields, e.g., Ti-L rather than Ti-K and Ba-M rather than Ba-L. WDS has sufficient spectral resolution to resolve most interferences encountered in low-voltage x-ray spectrometry and to produce a sufficiently high peak-to-background (P/B) for effective detection of most elements, even those with low fluorescence yield. However, the overall efficiency of the conventional WDS is so poor that it is not practical to operate with the low beam currents (typically $\leq$ nA) available in high resolution LV-SEM instruments. Si-EDS has adequate efficiency, but spectral resolution is so poor that low yield peaks are often not discernible above background, especially with the added complication of interfering carbon and oxygen K-peaks, a situation that is frequently encountered in practical materials.[2]

Low-overvoltage microanalysis offers an alternative route to achieving improved spatial resolution, but with the added advantage of utilizing the conventional choices for analytical peaks. This method is based upon the observation that the x-ray production range collapses very rapidly as the overvoltage, $U = E_0/E_c$, where E_0 (keV) is the incident beam energy and E_c (keV) is the critical excitation energy, decreases toward unity:

$$R\ (\mathrm{nm}) = [(27.6\ A)/(\rho\ Z^{0.89})]\ [E_0^{1.68} - E_c^{1.68}] \qquad (1)$$

A and Z are atomic weight and number, and ρ is the density.[1] Thus, the strategy consists of choosing an x-ray peak with high yield, e.g. K rather than L, L rather than M, and then selecting a beam energy that provides a low overvoltage, $U \leq 1.25$, for that peak. As an example, consider the range of production for copper K-shell x-rays ($E_c = 8.98$ keV) in copper. For $U = 1.15$, the range for Cu K production is about 100 nm, and with $U = 1.25$ ($E_0 = 11.2$ keV), the production range is 175 nm. By comparison, utilizing a low-voltage microanalysis strategy with a beam energy of $E_0 = 2.5$ keV, which would provide an overvoltage of 2.7 for the Cu L shell, gives an x-ray production range of 37 nm, while $E_0 = 5$ keV ($U = 5.4$) gives an x-ray range of 138 nm.

Are low-overvoltage EDS spectra useful? Figure 1 shows an example of a spectrum for a 70Cu – 30Zn brass, for which the K-edges of Cu (8.98 keV) and Zn (9.66 keV) are sufficiently close to simultaneously satisfy the low overvoltage condition for both the Cu K and Zn K x-ray peaks. The Si-EDS spectrum was recorded with a resolution of 129 eV (Mn Kα) at a beam energy of 11 keV, which gives an overvoltage of U = 1.22 for Cu and U = 1.14 for Zn. This level of excitation provides a peak-to-background (at 1.2 FWHM) of 6.15 for Cu and 1.86 for Zn. These P/B values and peak count rates (for a deadtime of 30%) result in limits of detection of 0.0098 mass fraction for Cu and 0.017 for Zn in this alloy with a single 100 s spectrum accumulation. Thus, major constituents (C > 0.1 mass fraction) are readily measurable under these conditions, while the full range of minor constituents ($0.01 \leq C \leq 0.1$) would require increased measurement time. Trace elements (C < 0.01) would only be accessible with high dose measurement conditions.

While there are other examples of technologically interesting combinations of elements with similar excitation edges, e.g., Fe-Cr-Ni alloys, GaAs, and InSb, that would be directly amenable to this low-overvoltage microanalysis strategy, the analyst will be confronted most often with mixtures of elements having excitation edges that span a broad range of energy. The low-overvoltage strategy can still be applied in this case through the use of two or more beam energies chosen to give similar overvoltage values for all peaks of interest. While procedurally more difficult, such a measurement strategy is entirely feasible with a modern computer-controlled instrument where even the beam position is well maintained while changing voltage. Moreover, there exist matrix correction procedures, such as NIST COR 2, that can calculate corrections when the suite of k-values is determined with different energies.[3] Moreover, low-overvoltage conditions result in matrix correction factors that tend strongly toward unity since absorption and fluorescence are negligible and backscattering effects are minimized.

References

[1] J.I. Goldstein et al., Scanning Electron Microcopy and X-ray Microanalysis, Plenum, New York, 1992, 417.
[2] D.E. Newbury, Micros .Microanal. 3(Suppl 2) (1997) 881.
[3] J. Henoc,, K.F.J. Heinrich, and R.L. Myklebust, A Rigorous Correction Procedure for EPMA (COR 2), NIST (NBS) Technical Note 769, (Gaithersburg, MD, 1973).

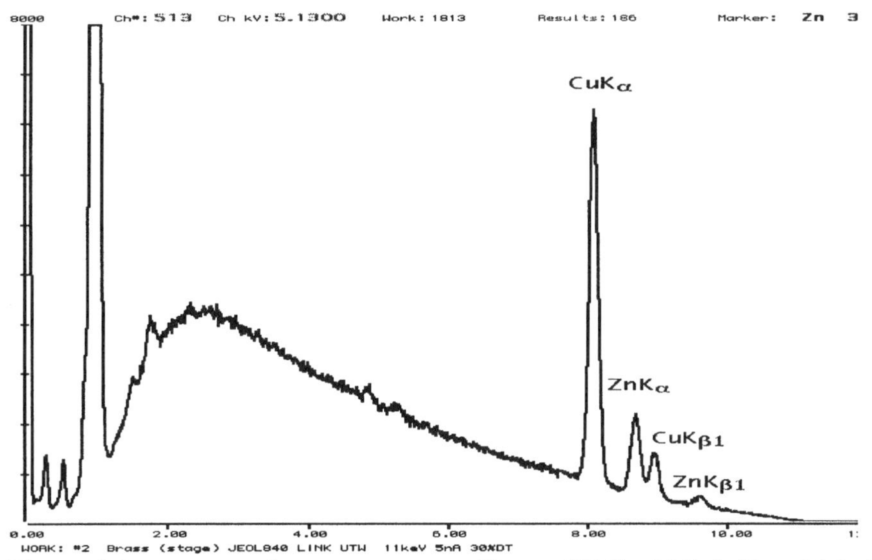

Figure 1 Low-overvoltage spectrum of brass (70Cu-30Zn), E_0 = 11 keV, 30% deadtime

Microsc. Microanal. 8 (Suppl. 2), 2002
DOI 10.1017.S1431927602101000

X-ray emission induced by low energy electrons

C. Bonnelle and P. Jonnard

Laboratoire de Chimie Physique Matière et Rayonnement, UMR 7614
Université Pierre et Marie Curie, 11 rue Pierre et Marie Curie,
75231 Paris cedex 05, FRANCE

The advantage of the x-ray emission induced by electrons, with respect to the x-ray fluorescence, or emission induced by photons, is that the interaction of the probe particles with the sample is selective in depth. Indeed, the electrons are slowing down along their path in the material. Then, by choosing an incident energy of the electrons near the threshold of the considered x-ray emission, only the superficial zone of the sample emits. The probed depth increases with a progressive increase of the energy of the incident electrons from the threshold.

From a simulation model [1,2], appropriate for describing the interaction of the low energy electrons with the matter, we calculate the in-depth distributions of the ionizations and consequently of the characteristic x-ray production. By using this model and the x-ray intensities measured at different incident electron energies, we deduce the depth distribution of the emitting atoms. The precision depends on the variation of the incident energy. When sufficiently small energy increments are used, we have shown that the analysis can be performed with a precision of the order of one nanometer [3].

The analysis of the buried interfaces is possible by this technique, by using a line emitted from an element present in the buried material. The incident electron energy must be such that the electrons having traversed the first material arrive in the second material with an energy near the appearance threshold of the chosen x-ray line. In these studies, it is important to determine the chemical interactions between the atoms present at the interface. Consequently, emissions of the soft x-ray range must be analyzed with a resolution sufficiently high for the changes due to the chemical environnement to be observed. For the emissions of this energy range, the fluorescence yield is generally small, making the emitted intensity weak. However, two factors partially compensate. The intensity of the Bremsstrahlung is weak, leading to a high peak/background ratio. The reabsorption of the radiation in the target is small.

Two choices in the experimental conditions can be made, either the analysis of a large surface, of the order of the cm^2, with a low electronic density or the analysis of a μm^2 surface with a higher electronic density. We dispose an apparatus [4] equipped of a Pierce electron gun having a spot diameter between 0.5 and 2 cm, and of an high-resolution curved-crystal spectrometer. The electronic density varies between 0.1 and 2 mA/cm^2. The target is cooled either by a water circuit or by a closed-cycle He circuit [5]. In these conditions, no modification of the sample is observed.

Interfaces between a film and a substrate have been studied. The thickness of the film is of some hundred nm, that is to say sufficiently thick to have the properties of the bulk[6]. Semiconductor/semiconductor, metal/metal, metal/semiconductor, metal/polymer and metal/ceramic have been analysed.

We present below the study of the Mo (10 nm) / SiO2 (70 nm) system prepared by magnetron sputtering and PECVD. The Si Kß emission (3p -1s transition) has been analyzed with electrons of 2.2, 2.5 and 3 keV (cf. figure). At high energy the spectrum is that of the silica. At low energy, the spectrum can be fitted by a sum of silica and molybdenum silicides spectra showing the presence of compounds at the interface [7]. Formation of compounds occurs at this interface which is diffuse and extended on some nm. Examples of abrupt interfaces will be presented.

[1] - P.F.Staub, X-ray Spectrom. **27**, 43, 1998.
[2] - P.F.Staub, P.Jonnard, F.Vergand, J.Thirion, C.Bonnelle, X -ray Spectrom. **27**, 58, 1998.
[3] - C.Hombourger, P.Jonnard, C.Bonnelle, E.Beauprez, M.Spirckel, B.Feltz, D.Boutard, J.-P.Gallien, Microscopy Microanalysis Microstructures **8**, 287, 1997.
[4] - C.Bonnelle, F.Vergand, P.Jonnard, J.M.André, P.F.Staub, P.Avila, P.Chargelègue, M.F.Fontaine, D.Laporte, P.Paquier, A.Ringuenet, B.Rodriguez, Rev. Sci. Instrum. **65**, 3466, 1994.
[5] - P. Jonnard, P. Chargelègue, C. Hombo urger, J. Thirion, F. Vergand, Rev. Sci. Instrum. **67**, 2417, (1996).
[6] - C. Bonnelle, Annual Report C, p. 201- 272, 1987.
[7] - P.Jonnard, C.Bonnelle, A.Bossebœuf, K.Danaie, E.Beauprez, Surf. Interface Anal. **29**, 255, 2000.

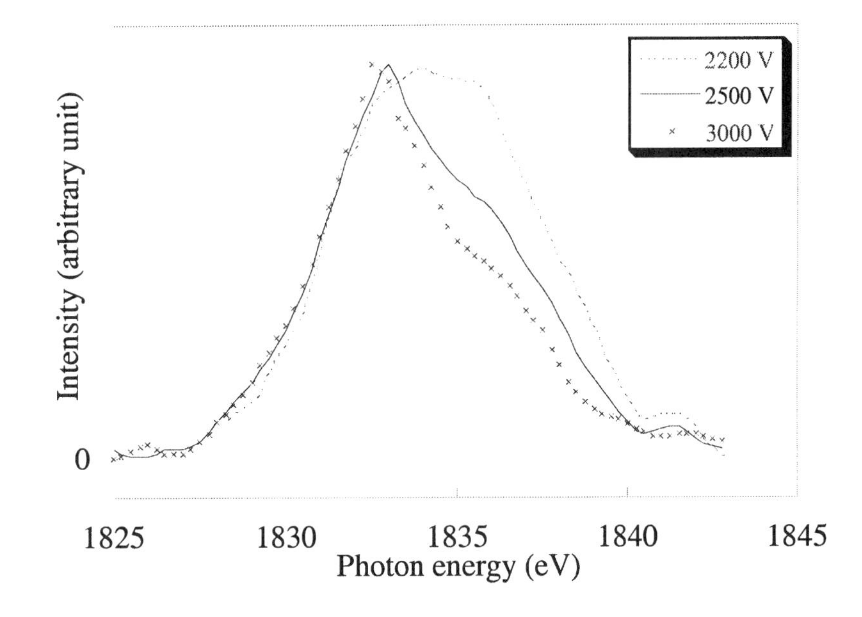

Microsc. Microanal. 8 (Suppl. 2), 2002
DOI 10.1017.S1431927602101012

A SIMPLE METHOD FOR DETERMINING OPTIMUM CORRECTIONS FOR HIGH-ACCURACY EPMA IN DIFFICULT CHEMICAL SYSTEMS

J. T. Armstrong*, R. B. Marinenko*, and J. M. Davis*,**

*Surface and Microanalysis Science Division, National Institute of Standards and Technology, Gaithersburg, MD 20899-8371
**Undergraduate School of Engineering, Clemson University, Clemson, SC 296340911

Increasingly, analysts are called to perform high accuracy quantitative electron microprobe analyses using x-ray lines for which the available correction procedures have significant uncertainties. Conventional evaluation of microprobe corrections typically involves analyzing a series of standards as 'unknowns' (often at several different beam energies), correcting the data through the algorithms and parameters being tested, and plotting the results as error histograms[1]. The problem with this method is that often there are simply not enough multi-element, homogeneous materials available to provide the necessary suite of secondary standards. Even ifthere are, independent determination of the exact compositions and degree of homogeneity of the candidate standards requires so much time, effort, and money as to make it impractical.

We propose a simplified version of the multi-keV evaluation procedure for binary systems that does not require a series of homogeneous secondary standards. In a specimen A_xB_{1-x} (where A and B are single elements or compounds of fixed atom proportions), as long as A+B = 100%-- a criterion easily checked by qualitative analysis, there should be a simple monotonic relation between the measured k-factor (relative intensity of the specimen to an end-member standard) and the relative concentration at any single set of analytical conditions. If one measures k-values at two different electron beam energies for a series of samples having different A/B concentration ratios (or a series different points on a inhomogeneous specimen of A:B), one should obtain monotonic curves (plotting the k-values of A or B at one E_0 vs. those at the other E_0) having end points of 0 and 1 (e.g., Fig. 1). The degree of curvature of the hyperbolic-like curve stretching from 0 to 1 (or the slope of a near-linear portion of the curve over a limited range of concentration) will depend on the difference in the magnitude of the correction factors for the two sets of analytical conditions. Bad analyses should appear as points lying off the curve formed by the majority of the data[2].

This $k(E_{0,1})$ vs $k(E_{0,2})$ plot can be used to evaluate the agreement to a series of correction algorithms without knowing, *a priori,* the compositions of any of the individual points. One simply calculates the k-values for a range of compositions in the binary system at the two beam energies for each of the correction procedures being considered (e.g., Fig. 2) and plots them along with the experimentally measured points. The expressions can be individually tested for each x-ray line analyzed and for each pair of electron beam energies employed. If a particular correction procedure well fits the analytical data, one can make the initial assumption that it would be the correction method that would most accurately correct the experimental data (e.g., Fig. 3). One can then process the measured data at the two beam energies using this 'best fit' method and evaluate the quality of the results on the basis of any independent knowledge of the samples being analyzed. The only absolute requirements of this method are that the sample is homogeneous over large enough areas and that the instrument is stable enough so that the microvolumes analyzed are of the same composition for the given point measured at the two different beam energies.

We are using this method to evaluate a series of materials involved in standard reference material and interactive reference material development at NIST, including the Cu_xAu_{1-x}, Si_xGe_{1-x}, Ni_xAl_{1-x}, and $Al_xGa_{1-x}As$ systems.

References

1. J.T. Armstrong, in *Electron Probe Quantitation*, K.F.J. Heinrich and D.E. Newbury, Eds., New York, Plenum Press (1991) 261; G.F. Bastin and H.J.M. Heijligers, *ibid.*, 145; J-L. Pouchou and F. Pichoir, *ibid.*, 31.
2. J.T. Armstrong, *Microscopy and Microanalysis* 7 (suppl. 2) (2001), 670.

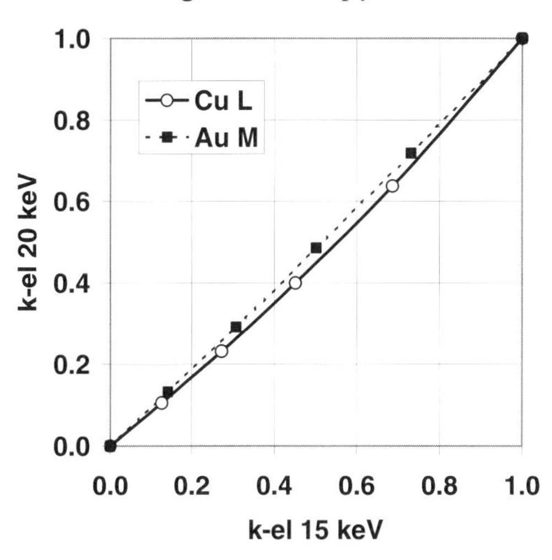

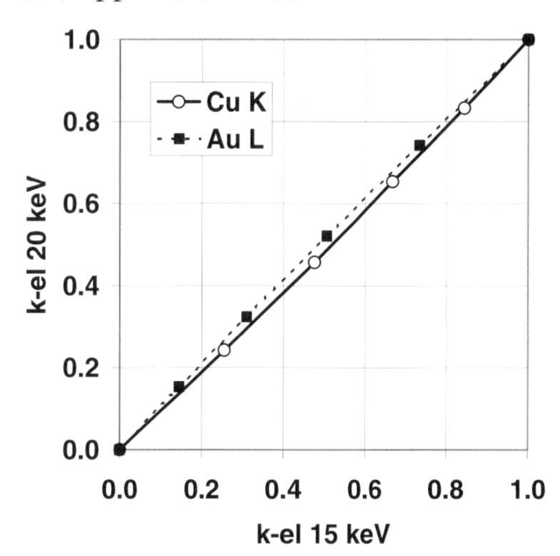

FIG. 1. Measured k-ratios (relative to the pure elements) for the Cu L, Cu K, Au L and Au M x-ray lines in NIST SRM 482 (100%, 80:20, 60:40, 40:60, 20:80, and 0 wt % Cu-Au alloys) at 15 and 20 keV. The measurement uncertainties are smaller than the size of the plotting symbols. The deviation of the data from a 1:1 line is indicative of the differences in matrix corrections for the two different beam energies. Note that the concentrations of the samples are not needed for this plot.

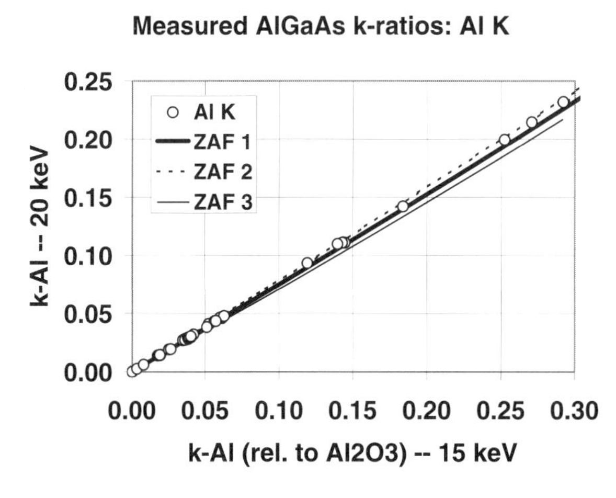

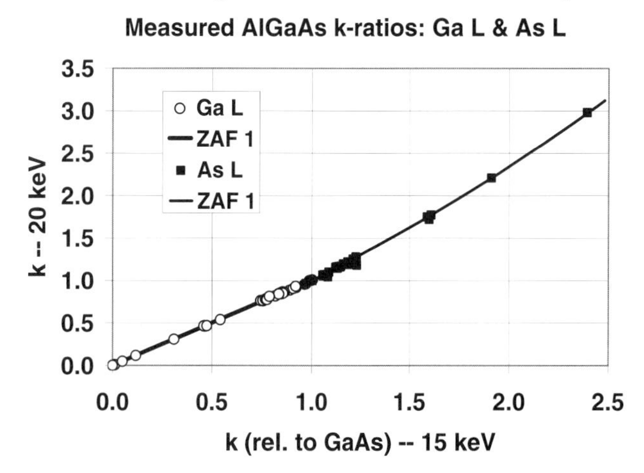

FIG. 2. Measured k-ratios of Al in a series of AlGaAs samples being evaluated for a NIST SRM and k-ratios calculated by various commonly used correction procedures (whose values range by over 20% relative). 'ZAF 1' fits data to better than 3%.

FIG. 3. Measured k-ratios of Ga L and As L in the same series of samples as shown in Fig. 2 (k-Ga < 1, k-As > 1). The same correction procedure, 'ZAF 1', that best fits the measured k-Al, also best fits the Ga L and As L data.

Microsc. Microanal. 8 (Suppl. 2), 2002
DOI 10.1017.S1431927602101024

Spectral imaging: Towards Quantitative X-ray Microanalysis

P.G. Kotula and M.R. Keenan

Sandia National Laboratories, PO Box 5800, Albuquerque, NM 87185-0886

Spectral imaging, where a series of complete x-ray spectra are collected from a 2D area, holds great promise for comprehensive microanalysis. The challenge remains, however, to acquire enough data and robustly analyze it all. Spectral images come in a range of sizes but 'small' ones typically comprise 64 by 64 pixels (4096 spectra) or 128 by 128 pixels (16384 spectra). In terms of typical SEM images, these are decidedly low-resolution. A prime goal of utilizing spectral images in routine and comprehensive microanalysis is therefore to surpass 1000 by 1000-pixel (10^6 spectra) spectral images. Automated multivariate statistical analysis (MSA) methods have been applied successfully to spectral images up to 128 by 128 pixels as described previously [1]. In the present paper we describe the extension of that work to much larger data sets.

As spectral images increase in size, manual analysis methods quickly become cumbersome. Additionally they are typically subjective and may require that the individual spectra in the spectral image contain a large number of counts [2]. Optimized MSA methods have been shown to be robust and capable of handling very noisy data [1] because statistically similar pixels are grouped together to produce a 'super spectrum' with the collective statistics. MSA techniques, where the energy channels (one thousand or more) are considered the variables, have been considered viable so long as the entire analysis is not bigger than the amount of free memory (RAM) on the computer. For 128, 256, 512 and 1024 pixel^2 (by 1024 energy channels) calculations a minimum of 135, 537, 2147, and 8590 Mbytes of free memory, respectively, would be required (assuming that the calculation does not require appreciably more space). We have developed an efficient method for automating the MSA algorithm in cases where the data set is too large to fit in memory. The method is based on spectral unmixing of pixel neighborhoods and retains full spectral and spatial resolution in the final spectral and component map estimates. To demonstrate this, a 512 by 512 pixel by 1024 energy-channel spectral image was acquired from a geologic material (described in more detail elsewhere [3]). 64x64 arrays of 8x8 pixel neighborhoods were formed from the original data set and then optimized MSA calculations were performed on this reduced-size data set. The results of this calculation (i.e., the 'pure' components) were then fit to the original full-resolution raw data. The data was acquired on a JEOL 5900LV SEM operating at 20kV equipped with a NORAN Vantage (Digital Imaging with Spectral Imaging), 2 μm/pixel (1mm field of view), a 9 μsec pulse-processor shaping time, 9 msec total dwell time per pixel, and 40% dead time resulting in an average of 100 counts per spectrum. The entire calculation took 3 minutes, resulting in 15 components. The results for 128 by 128 arrays of 4-pixel neighborhoods on the same data set were virtually indistinguishable from the 64 by 64 arrays of 8-pixel neighborhoods case, differing only by a few minutes in the amount of time to calculate the solution. Figure 1 shows component images and spectra from S, Pb, Fe and Cu delineating the different sulfides present in the specimen. Because S is present in three different phases (Fe-S, Cu-S and Pb-S) and Pb is present in two additional phases (Pb-Te and Pb-Te-O [3]), there will not be distinct 'phase' images and spectra and instead almost pure elemental images and spectra are observed [1]. Figure 2 is an overlay of the reconstructed spectrum from Pb-S, and Pb- and S-components for one pixel of Pb-S. The reconstructed spectrum represents the sum of the products of all the component images and spectra, not just Pb and S. Only Pb and S components contribute appreciably to the Pb-S reconstructed spectrum though.

References
[1] P.G. Kotula et al., *Microsc. Microanal.* In press (2002).
[2] P.G. Kotula and M.R. Keenan, *These proceedings* (2002).
[3] P. G. Kotula, P.F. Hlava, and M.R. Keenan, *These proceedings* (2002).
[4] Sandia is a multiprogram laboratory operated by Sandia Corporation, a Lockheed Martin Company, for the United Stated Department of Energy (DOE) under contract DE-AC04-94AL85000.

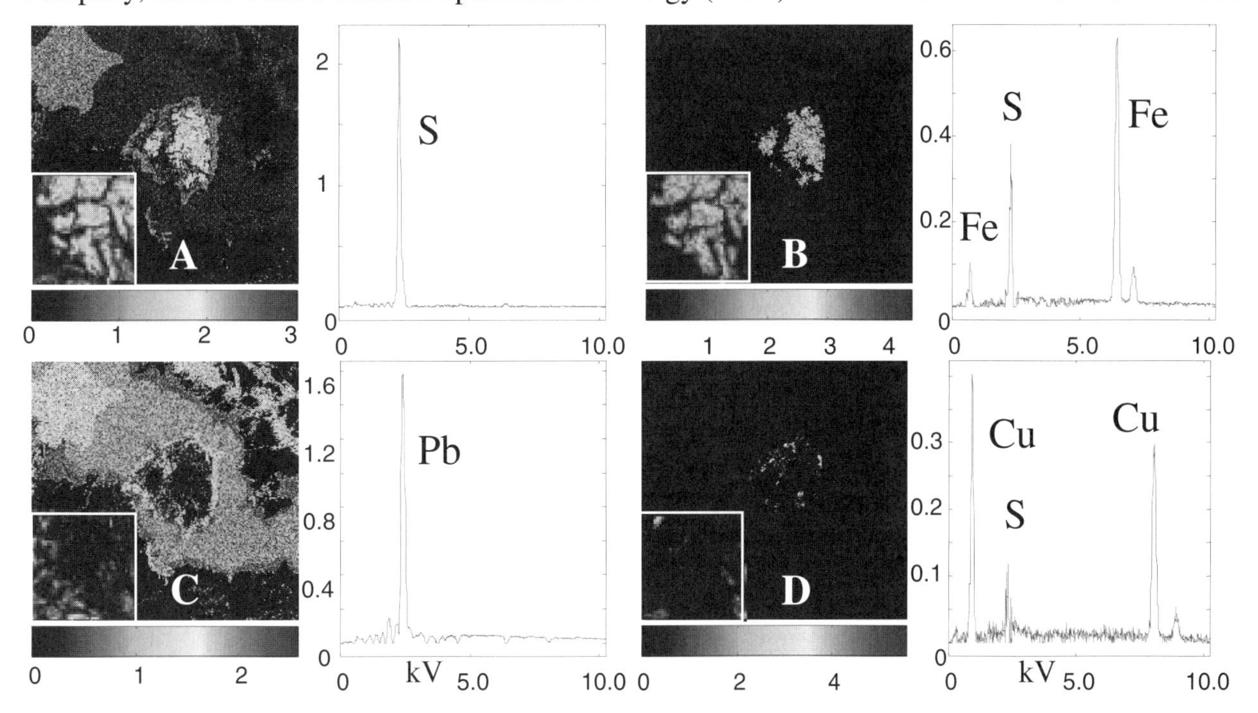

FIG. 1. Four 'pure' component images and spectra corresponding to: A. S; B. Fe-S; C. Pb and D. Cu-S. Spectral intensities are in normalized units. Inset are enlarged regions from the center.

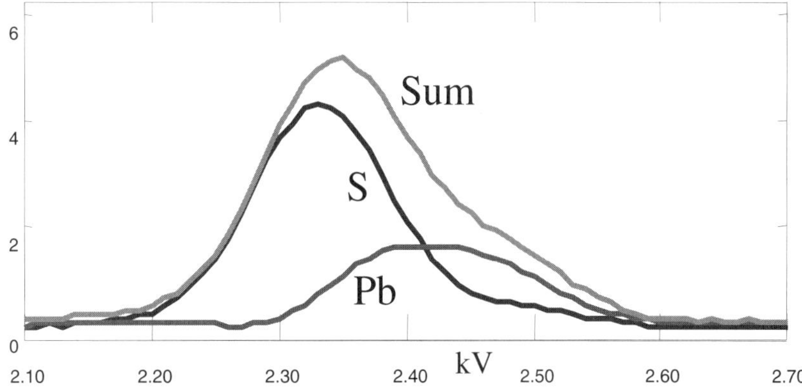

FIG. 2. Pure component spectra from Pb and S and the corresponding superposition of all pure components for one pixel from the Pb-S.

Microsc. Microanal. 8 (Suppl. 2), 2002
DOI 10.1017.S1431927602101036

Infrared microscopic analysis of tissues: a comparison of methodologies.

M Jackson[1], Janie Dubois[1], Richard Baydak[2] and Timothy Booth[2]

[1]Institute for Biodiagnostics, National Research Council Canada, 435 Ellice Ave., Winnipeg, Canada, R3B 1Y6 and [2]National Microbiology Laboratory, Canadian Science Center for Human and Animal Health, 1015 Arlington Street, Winnipeg, Manitoba,Canada R3E 3M4

Given that the infrared spectrum of most materials is one of the most characteristic properties of that material, it seems reasonable to assume that infrared spectra of human tissues will be equally characteristic. This has in fact been shown to be the case, with many reports appearing in the literature over the last 10 years that describe the spectral characteristics of both normal and abnormal tissues.

A consensus is now appearing that IR spectroscopic analysis of tissues is most reliably performed using an infrared microscope. This consensus is based upon the fact that IR microscopy allows tissues to be probed non-destructively with relatively high spatial resolution and direct correlation may be made between spectra and sample histology. However, there is by no means consensus with regards to sample preparation and data acquisition methods.

A number of important questions remain to be addressed before such a consensus can be reached. For example, is it realistic to analyse tissues that have been stained? What is the most appropriate instrumental approach (standard mapping microscopy, synchrotron based microscopy, microscopic imaging with a focal plane array system)? These questions will be addressed. Results from studies on a variety of tissues will be presented that demonstrate that significant spectral changes may be observed in tissues that have undergone the various manipulations involved in staining. For example the first stage in staining, exposure to alcohol/water, results in a decreased lipid content and removal of glycogen from tissues sections (Figure 1). These changes may be directly attributed to simple phenomena such as dissolution of materials in the staining media.

The potential advantages and disadvantages of infrared microscopy performed using three distinct approaches will be discussed. Data acquired by conventional infrared microscopy, synchrotron-based microscopy and microscopic imaging with a focal plane array detection system will be presented. Conventional microscopy has the advantage of simplicity, but is limited to a spatial resolution of around 25 μm. The use of a synchrotron source can significantly improve the achievable spatial resolution (5 μm or better, see Figure 2) but access may be problematic. Microscopic imaging using focal plane array detection holds the promise of improved spatial resolution and decreased measurement times but signal to noise may be compromised.

Choosing the appropriate measurement technique clearly involves compromise and requires an understanding of the strengths and weaknesses of each technique. The three techniques will be compared and contrasted using data obtained from skin tissue. Issues such as signal to noise, spatial resolution, speed and volume of data will be addressed.

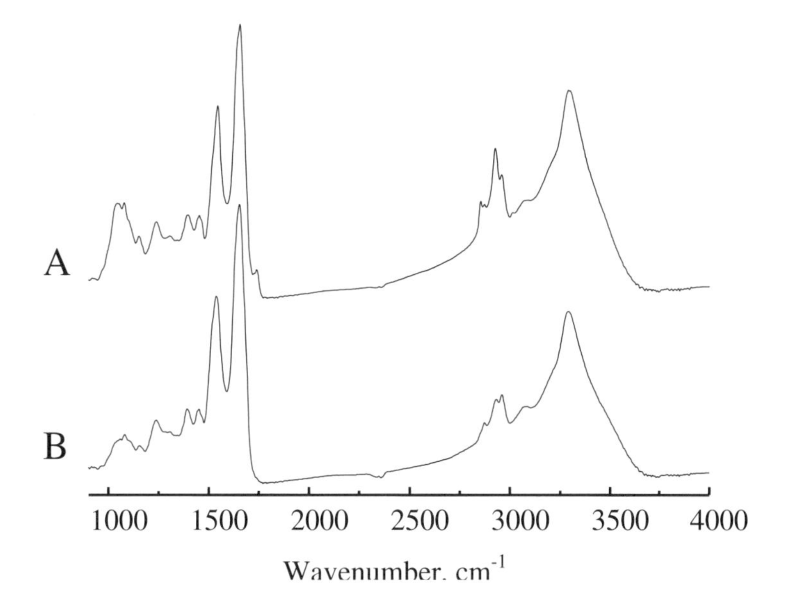

Figure 1. Infrared spectra of mouse liver before (A) and after (B) the first step in H& E staining (immersion in 100% alcohol followed by 95% alcohol followed by water).

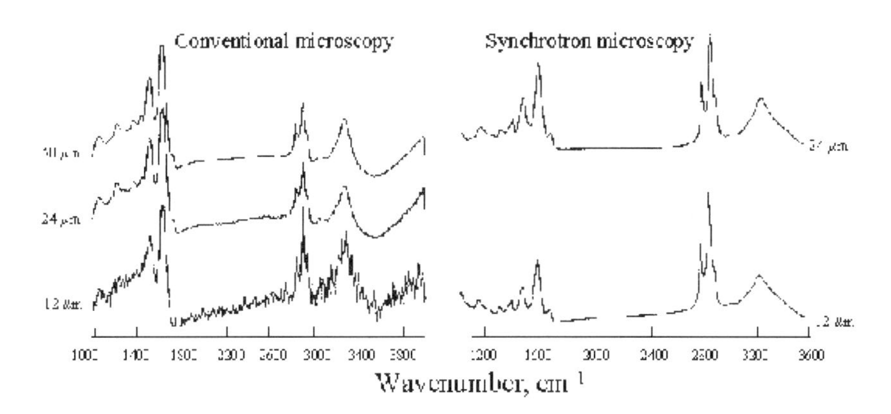

Figure 2. A comparison of spectra of human CNS tissue obtained at various aperture sizes using conventional and synchrotron based infrared microscopy.

References:

"Vibrational Spectroscopy and Pathology" M. Jackson and H. H. Manstch in Handbook of Vibrational Spectroscopy, in press.

"Medical Science Applications of Infrared Spectroscopy" Jackson, M. and Mantsch, H.H. in Encyclopedia of Spectroscopy and Spectrometry, Vol. 2, 1271-1281, 2000.

"Analysis and interpretation of infrared microscopic maps: visualisation and classification of skin components by digital staining and multivariate analysis" Mcintosh, L M., Mansfield, J.R., Crowson, A. N,. Mantsch, H.H. and Jackson, M. Biospectroscopy, 5, 265-275, 1999

"A LDA-guided search engine for the non-subjective analysis of infrared microscopic maps" Mansfield, J.R., McIntosh, L M., Crowson, A. N., Mantsch, H.H. and Jackson, M. Appl. Spectrosc. 53, 1323-1330, 1999

"*In-Situ* characterization of β-amyloid in Alzheimer's diseased tissue by synchrotron FTIR microspectroscopy" Choo, L.-P., Wetzel, D.L., Halliday, W.C., Jackson, M., LeVine, S.M. and Mantsch, H.H. Biophys J. 71, 1672-1679, 1996.

Microsc. Microanal. 8 (Suppl. 2), 2002
DOI 10.1017.S1431927602101048

Discriminating vital tumor from necrotic tissue in human glioblastoma samples by Raman microspectroscopy

S. Koljenovic*, L.-P. Choo-Smith*¶, T.C. Bakker Schut*, J.M. Kros‡, H.J. van den Berge† and G.J. Puppels*

* Dept. General Surgery, 10M, Laboratory for Intensive Care Research and Optical Spectroscopy, Erasmus University Rotterdam and University Hospital Rotterdam "Dijkzigt", Dr. Molewaterplein 40, 3015 GD, Rotterdam, The Netherlands
‡ Dept. Pathology, University Hospital Rotterdam "Dijkzigt", Dr. Molewaterplein 40, 3015 GD, Rotterdam, The Netherlands
† Dept. Neurosurgery, University Hospital Rotterdam "Dijkzigt", Dr. Molewaterplein 40, 3015 GD, Rotterdam, The Netherlands
¶ Present address: Institute for Biodiagnostics, National Research Council of Canada, 435 Ellice Avenue, Winnipeg, MB, R3B 1Y6, Canada

Brain tumour malignancy designation is performed according to the presence of histological parameters such as endothelial proliferation and necrosis, which are often not evenly distributed throughout the sample [1]. Furthermore, the grading of tumours as glioblastoma (grade IV, the highest malignancy classification of gliomas) is often performed on samples obtained from stereotactic surgery. This procedure, while suitable for sampling biopsies from regions of the brain which are otherwise not accessible, suffers from the disadvantage that the samples are small and subject to sampling error; resulting in the under-estimation of the malignancy grade. Although necrotic tissue is important for grading, a diagnosis cannot be made when only necrotic tissue is present in the sample [2]. In recent years, significant progress has been made in the application of Raman spectroscopy for ex vivo and in vivo tissue characterization [3]. Vital and necrotic glioblastoma tissues were studied by Raman microspectroscopy in order to identify possibilities for the development of an in vivo Raman method for real-time intra-operative brain biopsy guidance.

Raman microspectroscopic mapping studies were performed on unfixed thin cryo-sections of human glioblastoma samples obtained from 20 patients. Adjacent thin sections stained with hematoxilin and eosin (H&E) served to guide the localization of regions of interest on the tissue samples mapped. Raman spectroscopic mapping was performed using a near-infrared multichannel Raman microspectrometer built in-house and consisted of a Leica DM-RXE microscope coupled to a Renishaw System 100 Raman spectrometer with laser light at 847 nm focused on the samples. Using a computer-controlled xyz-stage that allowed scanning during measurements, consecutive Raman spectra were collected from the tissue. The calibrated spectra were used to construct maps using multivariate statistical techniques. Following measurements, the sections were stained with H&E for histological confirmation.

K-means cluster analysis of the spectra resulted in groups of similar spectra that could be assigned colours to generate pseudo-colour Raman maps of the tissue sections (Fig. 1). Comparing these maps with the H&E staining revealed that Raman spectra of vital tumour tissue differ significantly from spectra of necrotic tissue (Fig. 2). Taking a difference of the averaged spectra from each of these two regions revealed that necrotic tissue was found to consistently contain higher levels of cholesterol and cholesterol-esters. Further studies involving the development of a classification model for non-

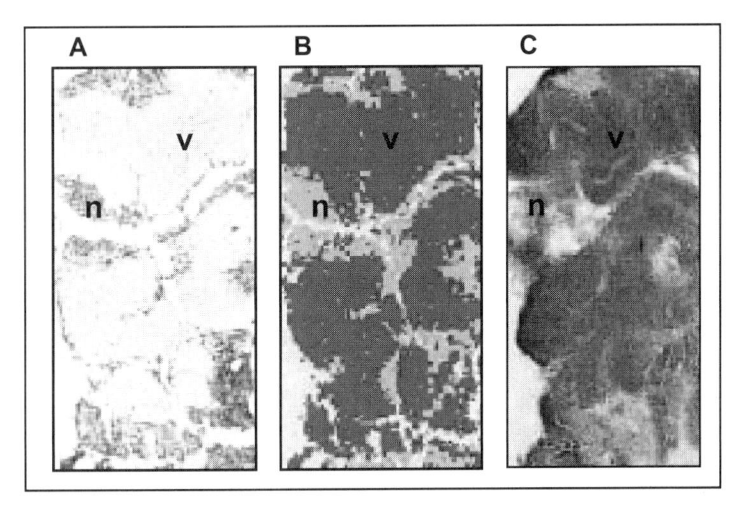

FIG. 1. A: Photomicrograph of unstained (25 μm thick) human glioblastoma cryo-section used in Raman mapping experiment. B: Corresponding psuedo-colour Raman map based on K-means cluster analysis and LDA-model prediction of Raman spectra. Dark-grey pixels denote vital tumour, light-grey pixels denote necrotic tissue. C: Photomicrograph of the same tissue section after H&E staining following Raman measurements. (vital tissue designated with **v** and necrotic tissue with **n**)

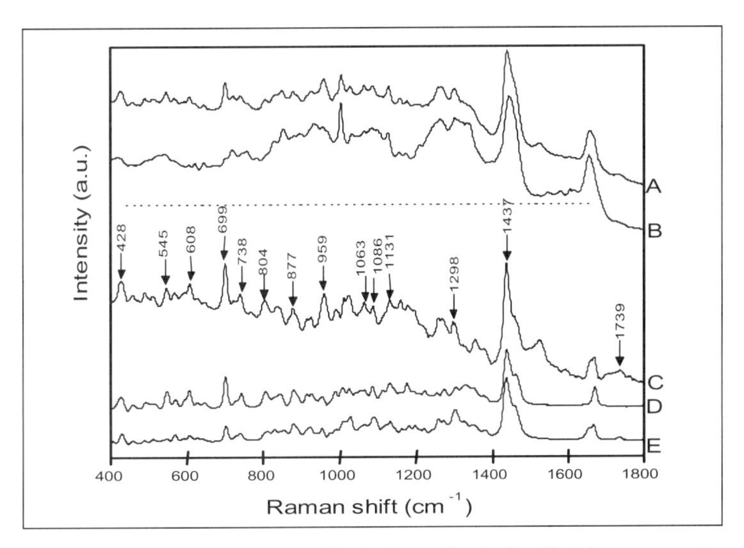

FIG. 2. Averaged Raman spectra from necrotic (A) and vital glioblastoma tissue (B). The difference spectrum (C is A minus B; intensity multiplied by factor of 2) is compared with the Raman spectra of pure cholesterol (D) and cholesterol oleate (E). (a.u., arbitrary units)

subjective discrimination between vital and necrotic tissue based on linear discriminant analysis (LDA) were also performed. Testing the model on 9 independent tissue sections yielded 100 % accuracy rates. This in vitro result indicates that Raman spectra contain biochemical information that can be used to distinguish vital glioblastoma from necrosis suggesting that Raman spectroscopy is a powerful candidate for guidance of stereotactic brain biopsy.

References

[1] J. Hildebrand et al., *Eur. Neurol.* 34 (1997) 238.
[2] M.J. Glantz et al., *Neurol.* 41 (1991) 1741.
[3] E.B. Hanlon et al., *Phys. Med. Biol.* 45 (2000) R1.

Microsc. Microanal. 8 (Suppl. 2), 2002
DOI 10.1017.S1431927602101103

Time Resolved Analysis of the Positive Ion Dynamics in the Variable Pressure Scanning Electron Microscope

M.R. Phillips and S.W. Morgan

Microstructural Analysis Unit, University of Technology Sydney, Broadway, NSW 2007 Australia

In a variable pressure scanning electron microscope (VPSEM) the secondary electron (SE) emission signal is amplified via a gas ionization cascade, which is produced by the introduction of a positively biased electrode into the specimen chamber. Current flow induced in the biased (or ground) electrode through the movement of charge within the cascade is used to form an image rich in SE contrast [1]. A consequence of this type SE detection process is the generation of a significant concentration of positive ions within the specimen chamber. The presence of these positive ions enables imaging and analysis of uncoated non-conductive specimens at all electron beam energies without charging artifacts. Recent studies, however, have revealed that the positive ions can have a significant affect on SE contrast by (i) suppressing SE emission [2], (ii) reducing the ionization cascade amplification of the SE emission signal [3] and (iii) increasing the landing energy of the primary beam [4]. A detailed knowledge of the dynamic behavior of positive ions will therefore enable optimization of the SE image quality and correct interpretation of SE contrast in the VPSEM.

In this work, the time dependent behavior of the positive ion current was used to investigate the ion dynamics in the VPSEM. Time resolved ion current profiles were measured as a function of specimen stage geometry, biased electrode voltage (+30V to +550V), sample conductivity and atomic number, type of chamber gas (water vapor, nitrogen and argon) and gas pressure using four different types of grounded electrode arrangements; a grounded straight copper wire inserted into the gap between the specimen and the biased electrode, copper wire rings with a range of diameters positioned at various heights above the sample, aluminum cylinders centered over the specimen with an assortment of diameter and height configurations and the specimen stage itself. The electric field distribution for each of the above electrode configurations was calculated using commercial software (Quickfield™). In each experiment, the electron beam was positioned on the specimen in spot mode with a zero electrode bias. The ion current was allowed to stabilize before the bias voltage was rapidly switched to a pre-determined positive voltage. Once a steady state current was observed the bias was then re-set to zero volts, and the ion current was collected until it reached the initial zero bias signal level. The ionization cascade amplification was measured in all experiments. A Keithley 617 electrometer connected to an Eagle Technology 330kHz, 12 bit A/D board was used to measure the ion current as a function of time.

A typical time resolved ion current profile (shown in figure 1) exhibits three distinct regions over a ~ 40 second time interval; region 1 where the ion current (I_{ion}) increases since the ion generation rate (dG) is greater the ion de-ionization rate (dI) reaching a maximum at time t_{max}, region 2 where the I_{ion} decays as dI < dG and region 3 where I_{ion} is constant where dG = dI. The results show that as the ion concentration grows the position of t_{max} decreases with an associated increase in the I_{ion} decay rate in region 2. Time resolved ion current profiles collected under identical experimental conditions and electrode arrangements varied significantly for water vapor, nitrogen and argon, reflecting the differences in the ionization / recombination efficiency, lifetime and mobility of each gas ion species. These profiles provide a measure of the capacity of each gas to form a positive ion space charge. A significant shift in the ionization gas gain curve as a function of pressure was observed for different grounded electrode configurations, particularly the VPSEM conditions for maximum amplification when compared with the conductive stage as the ground electr ode. The significance of these results in terms of SE image quality and interpretation of SE contrast in the VPSEM will be presented. A series of SE images were collected from a copper TEM grid under equivalent VPSEM conditions and grounded electrode configurations to illustrate these effects.

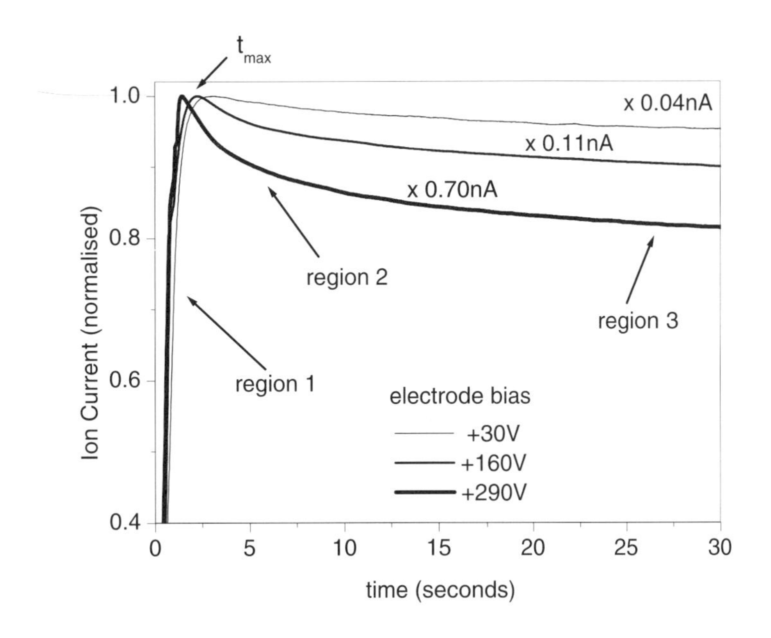

Figure 1: Typical ion current profile versus time for an insulating specimen measured from the grounded conductive stage. Pressure = 3.0 torr, Gas = Nitrogen, E_o=20keV, I_b=1nA, Working Distance =10mm

References:

[1] Danalatos G.D., Adv. Elec. Electron Phys, 71 (1988) 109
[2] Toth M. et al., *Applied Physics Letters* 77 9 (2000) 1342
[3] Toth M. and Phillips M.R, *Scanning* 22 (2000) 319
[4] Phillips M.R., Toth M. and Griffin B.J., *Microscopy and Microanalysis* 6 Suppl. 2 (2000) 786

Microsc. Microanal. 8 (Suppl. 2), 2002
DOI 10.1017.S1431927602101115

Considerations for Secondary Electron Imaging of Dielectric Materials in Low-Vacuum and Environmental SEM

B.L. Thiel and M. Toth

Polymers and Colloids Group, Cavendish Laboratory, University of Cambridge, Cambridge CB3 0HE, UK

Many secondary electron (SE) contrast effects observed during the imaging of dielectric substances in low vacuum scanning electron microscopes can be attributed to the electric fields that exist above and below the specimen surface. This paper will review the origins of these fields and examine their consequences for imaging and microanalysis.

If we consider a sample oriented such that the z direction is normal to the surface (*i.e.*, the x,y plane at z=0), the net field $\vec{E}_{net}(x,y,z)$ at any given point is the sum of three contributions:

$$\vec{E}_{net}(x,y,z) = \vec{E}_{bias}(x,y,z) + \vec{E}_q(x,y,z) + \vec{E}_{ion}(x,y,z). \qquad (1)$$

where E_{bias} is the contribution due to the detector bias V_0, E_q arises from trapped charge in the specimen, and E_{ion} is due to the presence of gaseous ions above the specimen surface. The significance of each contribution will be discussed.

Many of the SE detectors developed for poor vacuum environments use a positively biased detector located either above or near the specimen. The specimen is usually situated on a grounded support, thus creating an electric field from the detector, through the dielectric specimen, and terminating at the support. (Alternatively, the support may have a negative bias, establishing a field of the same sign, but terminating on the grounded pole-piece.) The intensity of the field above and below the specimen surface will be determined by the capacitance of the specimen, which in turn is a function of the specimen geometry and dielectric constant. It can be shown that the fields above and below the specimen surface are given respectively by

$$\vec{E}_{bias}(x,y,z>0) = V_0 \frac{\varepsilon_s}{d_g \varepsilon_s + d_s \varepsilon_g}; \vec{E}_{bias}(x,y,z<0) = V_0 \frac{\varepsilon_g}{d_g \varepsilon_s + d_s \varepsilon_g}. \qquad (2a,b)$$

The dielectric constants of the gas and specimen are given by ε_g and ε_s respectively, while the thicknesses of the gas region and specimen are given by d_g and d_s. The surface potential V_s of the specimen can be found similarly:

$$V_s = V_0 \frac{\varepsilon_g d_s}{\varepsilon_g d_s + \varepsilon_s d_g}. \qquad (3)$$

Thus, as experience supports, the change in surface potential due to the detector bias is negligible for most specimens, but could become an issue if bulk dielectrics are examined (*i.e.*, when $d_s > d_g$). In most cases, the detector field induced in the specimen is too weak to affect SE emission. Significantly, the field strength between the surface and detector determines the amplification efficiency of the gas cascade, and thus the SE signal intensity.

The field effects on SE emission due to implanted charge in the specimen have been treated rigorously by Cazaux.[1] In this case, the field in the region between the centre of mass of the implanted charge and the specimen surface [$\vec{E}_q(z<0)$] is of interest, as is the field that extends into the region above the surface[$\vec{E}_q(z>0)$]. In high vacuum, the (negative) surface potential can easily reach thousands of volts, resulting in deflection of the primary beam and other classic charging effects. Similarly, the associated internal fields are strong enough to result in enhancedSE emission. The associated field strength will depend on the amount of trapped charge, which in turn depends on the nature and density of available charge traps.

With the addition of a low pressure of gas and an ionization cascade to amplify SE signals, positive ions become an important factor. Ions created in the cascade accumulate above the surface of insulators. The field $\vec{E}_{ion}$ radiates outward from the centre of mass of the ion cloud and terminates on the detector and the grounded specimen support. The field intensity in the gas and specimen regions and the surface potential are determined by similar considerations that went into driving equations (2) and (3). The ions give rise to several effects[3,4]: The surface potential of the specimen can actually be shifted positive by a few hundred volts, which reduces the primary electron landing energy and reduces the cascade amplification field. Also, ions that accumulate at the surface can capture secondary electrons and diminish the SE signal. Conversely, a strong dipole field is set up between the positive ions at the surface and the implanted negative charge. Because the charge density can be fairly high and the charge separation (*i.e.*, the penetration depth of the deposited primary electrons) is small, the local field can be very great. This reduces the SE escape barrier at the surface, resulting in enhanced emission.

Inhomogeneities in trap/defect distribution, and therefore stored charge and resulting field $\vec{E}_q(x,y,z<0)$, can give rise to contrast via this mechanism. It should also be noted that not all defects behave in the same way. Depending on trap depth and density, some regions may be able to store more charge than the bulk material while others may in fact be more conductive, therefore storing less charge. Similarly, anything that perturbs the net electric field at the surface $\vec{E}_{net}(x,y,z>0)$ results in inhomogeneous ion flux, also allowing the possibility of differential contrast.[5]

In the final analysis, any intrinsic differences in SE emission from the specimen will be modulated by all of these effects. Image contrast will then be a net result of all the contributing factors. Different widows in the operating parameter space where each effect dominates make it possible to diagnose the probable origin of a particular contrast feature.

1) J. Cazaux, J. Electron Spect. and Rel. Phenom. **105** 155-185 (1999).
2) J.P. Craven, F.S. Baker, B.L. Thiel and A.M. Donald, J. Microsc. **205** Pt.1, 96-105 (2002).
3) M. Toth, M.R. Phillips, B.L. Thiel and A.M. Donald, J. Appl. Phys, (in press).
4) M. Toth, B.L. Thiel and A.M. Donald, J. Microsc. **205** Pt.1, 86-95 (2002).
5) This work was funded by: EPSRC (GR/M90139), FEI, and the Isaac Newton Trust.

Microsc. Microanal. 8 (Suppl. 2), 2002
DOI 10.1017.S1431927602101127

CHARGE CONTRAST IMAGING IN VARIABLE PRESSURE SEM: CORRELATION, OPTIMISATION AND APPLICATION

Brendan J Griffin and Alexandra S. Suvorova

Centre for Microscopy and Microanalysis, The University of Western Australia, 35 Stirling Highway, Crawley, WA 6009

Charge contrast imaging (CCI), as initially reported[1], revealed growth structure in hydrated aluminium hydroxide (gibbsite); data not attainable by any other electron, laser or light optical imaging technique. This new contrast result was achieved with a specific set of operating conditions of the variable pressure or environmental SEM. CCI have subsequently been reported from a range of poor and non-conductors. The detail of CCI has also been shown to correlate with cathodoluminescence in many natural minerals and synthesised materials[2,3]. The operating conditions for optimal contrast in CCI have been characterised for a range of previously reported responsive materials. CCI is found to be observable in a multi-dimensional envelope of VPSEM operating conditions. The more recently described techniques that demonstrate the benefits of reduction of positive ion fluxes above the sample have been found to extend to the 'CCI envelop', a confirmation of earlier studies (Doehne, pers. comm.).

New studies of the retardation of the primary electron landing energies, by measurement of Duane-Hunt limits in EDS X-ray spectra, have been undertaken in parallel with the characterisation of CCI conditions. Specific CCI detail are found not to be observed when the landing energy of the primary electron beam is significantly retarded, ie when the sample is negatively charged (figure 1). This result is reproducible and consistent within a range of studied materials. The compositional control on charge balance, as observed in Duane-Hunt values across a range of materials (figure 2), matches the variations in conditions required for optimal CCI in these materials.

A parallel study has confirmed that the detail observed in CCI are not visible when samples are examined under very low accelerating voltage conditions in field emission SEM (FESEM), even when uncoated, under conventional high vacuum conditions. These data confirm the view that CCI is separate from the long-recognised voltage contrast imagery. Significantly, the surface of our 'CCI responsive' test samples have been observed to have a thin surface contaminant layer, in these FESEM images. These observations are consistent with the recently reported observation that CCI can be used to images defects below surficial deposits[4].

The new data support the view that CCI is affected strongly by variations in sample conductivity at the micro or even 'nano' scale. CCI visibility through surface contaminant layers supports a view that induced field fluctuations from variations in trapped charge, and its decay, may carry the contrast data rather than direct or specific actual emissions.

References

1. B.J. Griffin *Microsc. Microanal.,* **3** (s2) (1997), 1197.
2. E. Doehne, *Scanning,* **19**(1997)75.
3. B.J. Griffin et al., *Microsc Microanal* **5**(1999)278.
4. S.P.Galvin et al., *J. Materials Science,* **36**(2001).

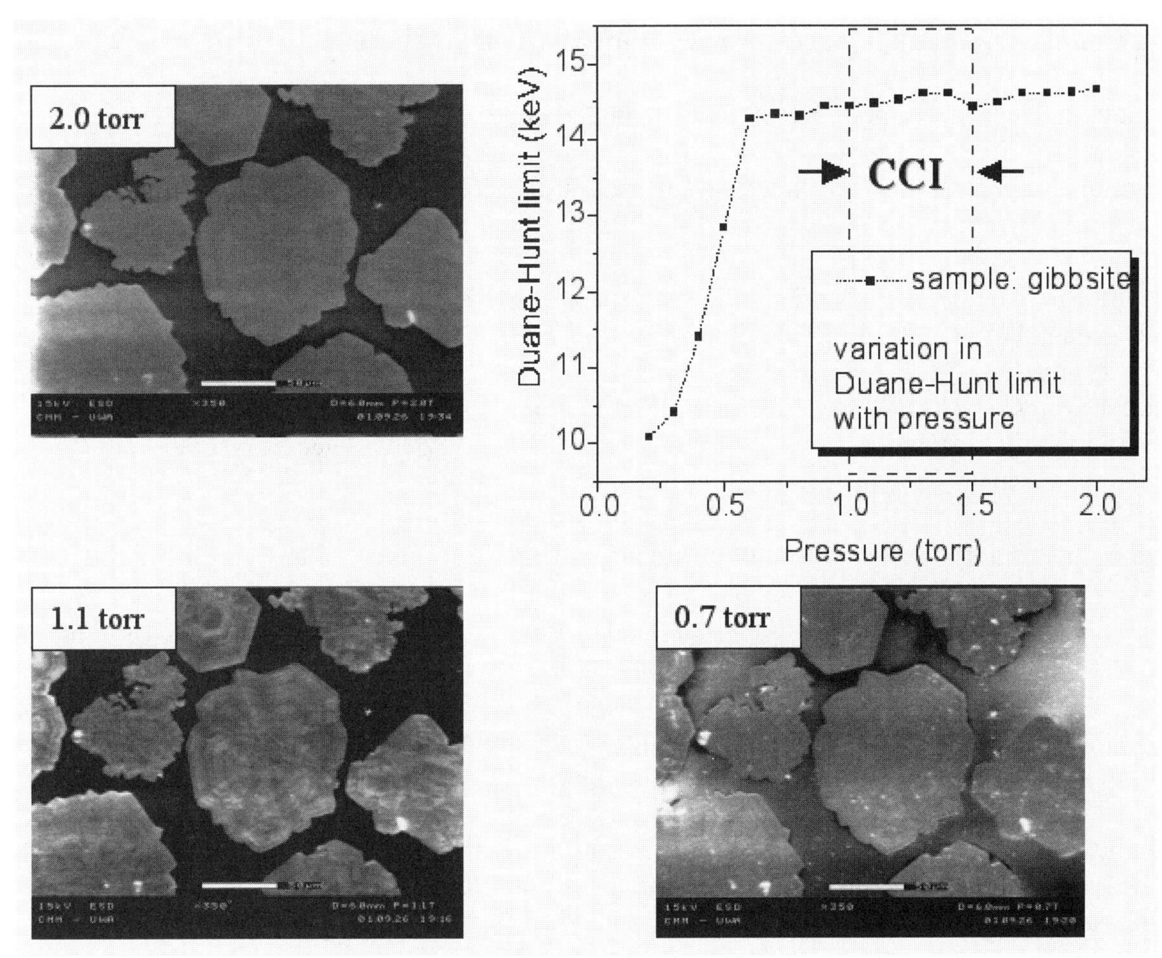

Figure 1: Comparison of the 'CCI envelope' and sample charging effects on the primary electron landing energy.

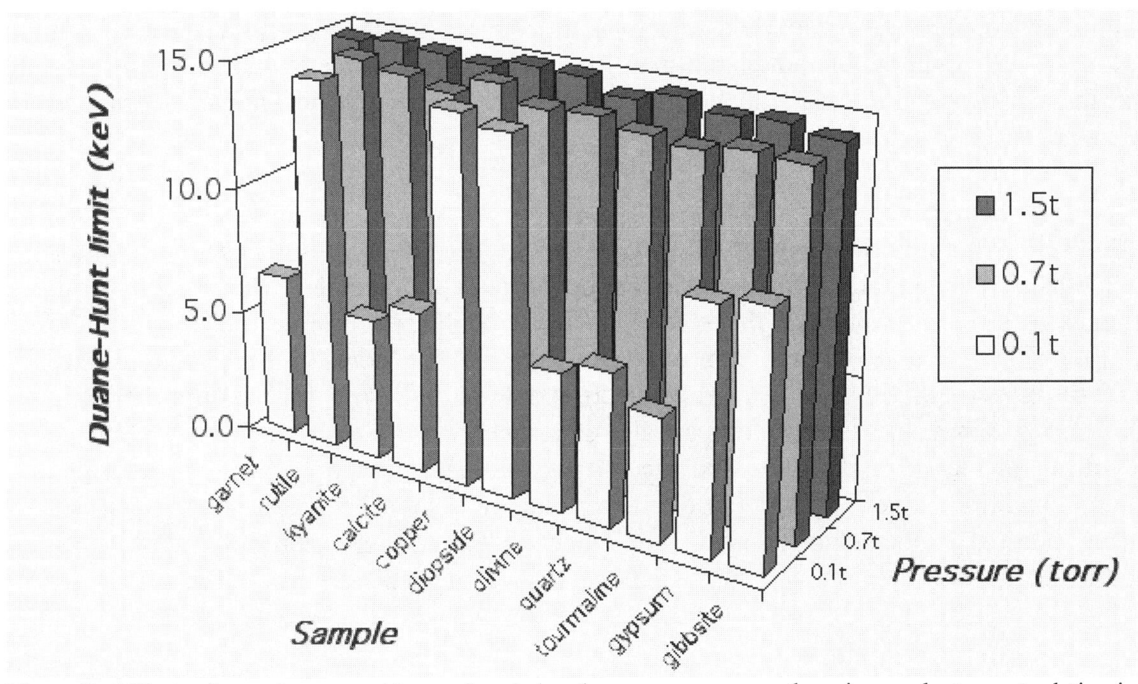

Figure 2: Effects of sample composition and and chamber gas pressure on the primary electron retardation in VPSEM.

Microsc. Microanal. 8 (Suppl. 2), 2002
DOI 10.1017.S1431927602101139

Modeling Contrasts in Variable Pressure Scanning Electron Microscopes

Raynald Gauvin, Hendrix Demers, Kevin Robertson and James Finch

Department of Mining, Metals and Materials Engineering, McGill University, Montréal, Québec, Canada, H3A 2B2

Charge Contrast Imaging (CCI) is a new imaging mode that has been discovered recently using the GSED Secondary Electron (SE) detector in the ESEM[1, 2.] CCI allows to image non-conductive specimens with details about the structures of these materials that are not seen with conventional SE and Backscattering Electron (BSE) imaging modes. CCI was also observed in a Hitachi S-3500N VP-SEM with the Shah detector[3]. The Hitachi ESED detector is a variation of the Shah detector which allow the observation of CCI[4].

Despite the incredible amount of new information that CCI gives, there is still a huge controversy concerning the mechanisms of CCI. It is believed that CCI are obtained when there is an optimal charge compensation allowing to map the surface potential differences of the materials, giving an enhanced sensitivity in SE emission. However, this explanation still remain speculative and dedicate and elaborate research must be performed in order to understand the mechanisms of CCI. It is clear that CCI is related with the charging of non-conductive materials that accumulate charges when irradiated by incident electrons.

The modeling of charging is a very difficult task because it is very difficult, if not impossible, to predict the trap charge density because of the drift mobility of electrons and ions, their recombination rate and their trapping by detects like vacancies, dislocation and grain boundaries. This is very unfortunate because once the trapped charge density is known, it is easy to solve numerically the Poisson equation to obtain the electrical field inside and outside the materials. With the knowledge of the electric field, electron trajectories inside the materials and ions trajectories in the gas could be computed and their effect on CCI could be estimated. In the case of the gas, the ion production rate and the drift velocity of ions must be known in order to compute accurately the electric field in the gas.

Since the trapped charge density is difficult to compute in a solid, it should be measured experimentally. It is possible to compute an X-Ray spectrum by assuming the strength of the electric field inside the material[5]. By comparison between experimental X-Ray Spectra with simulated ones at different values of the electric field, it could be possible to determine the electric field inside the material and hence the trapped charge density. Figure [1] shows simulated X-Ray spectra of Gibbsite at incident electron energy of 30 keV for maximum electric field of 0, 3 and 5 V/nm. In order to compare the experiment to these simulations, a Gibbsite specimen with a conductive coating must be used. Also, the measurement should be performed under vacuum conditions. Even if this it not the representative case of ESEM or VP-SEM, it is a good starting point to understand charging and hence, CCI. We are currently working to generalize this analysis to the case of non coated specimen under vacuum as well as under gas pressure.

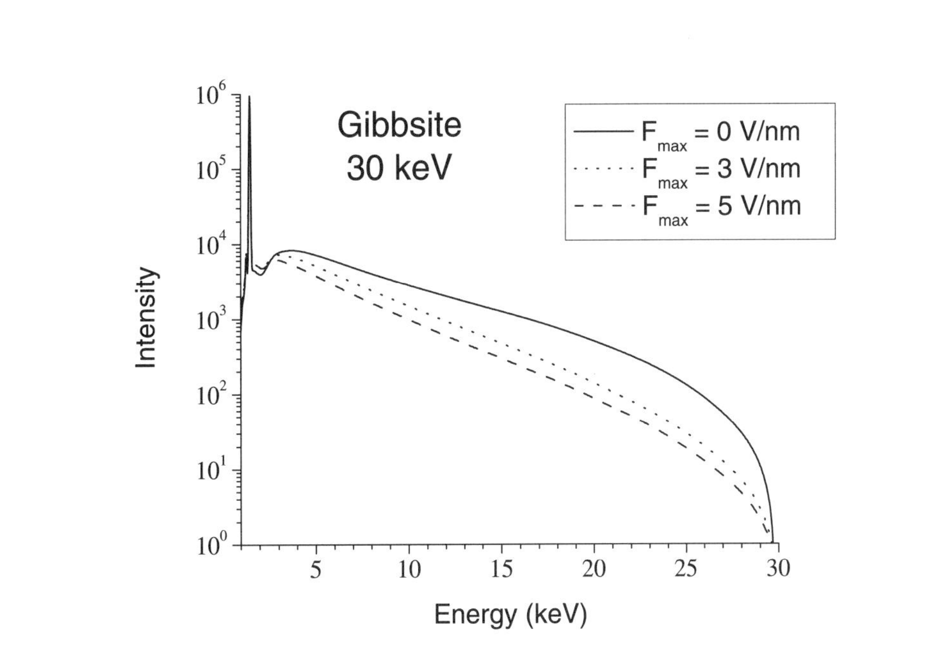

Figure 1 Simulated X-Ray spectra for Gibsite at 30 keV. The maximum electric field is ranging between 0 to 5 V/nm.

References

1. B. J. Griffin (1997), Microscopy & Microanalysis, Vol. 3, Supp. 2, pp. 1197 - 1198.
2. B. J. Griffin (2000), Scanning, Vol. 22, pp. 234– 242.
3. A. Mohan, N. Khanna, J. Hwu and D. C. Joy (1998), Scanning, Vol. 20, pp. 436– 441.
4. K. Robertson, R. Gauvin and J. Finch (2002), This volume.
5. H. Demers and R. Gauvin (2002), This volume.

Microsc. Microanal. 8 (Suppl. 2), 2002
DOI 10.1017.S1431927602101140

The WebSEM in K-12 Classrooms: Lessons Learned

L.S. Chumbley, D.J. Eisenmann*, A.E. Chumbley, T. Frizell†, ,T. Andre† and C.P. Hargrave†

Iowa State University, Materials Science and Engineering Department, Ames, IA 50011
† Iowa State University, Department of Curriculum and Instruction, Ames, IA 50011
* Now with the Center for Nondestructive Evaluation, Ames, IA 50011

The WebSEM is a scanning electron microscope (SEM) that provides teachers and students in K-12 classrooms with the ability to conduct on-line investigations. Jointly developed by RJ Lee Group and Iowa State University, the WebSEM is based on the "Personal SEM" (PSEM) marketed by ASPEX (formerly RJ Lee) Instruments [1]. A web-based interface allows students to move samples, change magnification, adjust contrast and brightness, take pictures, and conduct chemical studies using the energy dispersive spectrometer (EDS) which is integrated into the system.

In operation, a computer server accepts java commands from the web interface, translates them into instrument commands and passes them to the PSEM for execution, then receives the updated image from the PSEM for transmission over the internet, Fig 1. Initial lessons with the system were very disappointing. Although the system functioned well it was found to be too confusing for many teachers and students, and few lessons were conducted. Improvements were identified to increase system reliability while also making it easier to operate. Commands and functions deemed unnecessary for operators encountered in education were eliminated in a "second generation" server and web-interface developed specifically for K-12 instruction, Fig 2 [2]. This interface is reliable, easy to use, and teachers and students require little instruction to become proficient.

Recent efforts have been aimed at increasing the number of teachers using the WebSEM. These efforts have included 1) introducing the WebSEM to pre-service teachers at ISU through their science methods classes, 2) conducting live demonstrations of the WebSEM at conferences held for science teachers, and 3) conducting on-campus workshops for science teachers and education agency science coordinators, Fig. 3. A number of sample lesson plans have also been developed as an aid to teachers and are available for use. Finally, a short videotape showing the actually WebSEM and explaining the operation of the instrument is also available for teachers to show in their classrooms.

As a result of these actions the use of WebSEM has increased dramatically, such that in Fall 2001 approximately 2-3 sessions were conducted each week. This indicates that getting teachers to accept and use the SEM in their classroom is more dependent upon having good support for the teacher than it is on having the latest technology for delivering the image. [3]

References

[1] L.S. Chumbley, et al., "Project ExCEL - Web-based SEM for K-12 Education," to be published, J Eng Ed, April, 2002.
[2] L. S. Chumbley, et al., "Development of a Web-based SEM Specifically for K-12 Education," Micro Res. and Tech., 56, #6, March, 2002.
[3] This work is being conducted in cooperation with G. Casuccio and D. Kritikos of RJ Lee Instruments and is partially support by Qwest Inc. and the National Science Foundation.

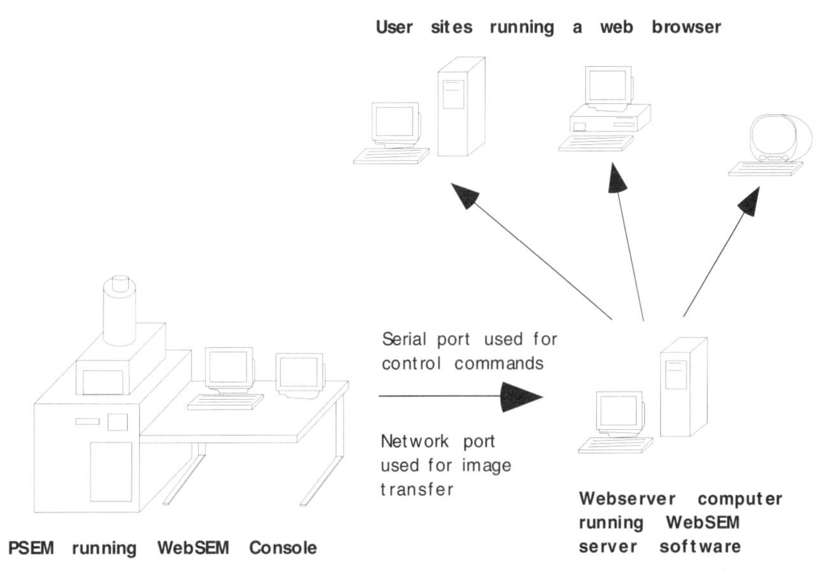

Figure 1: Schematic of the system.

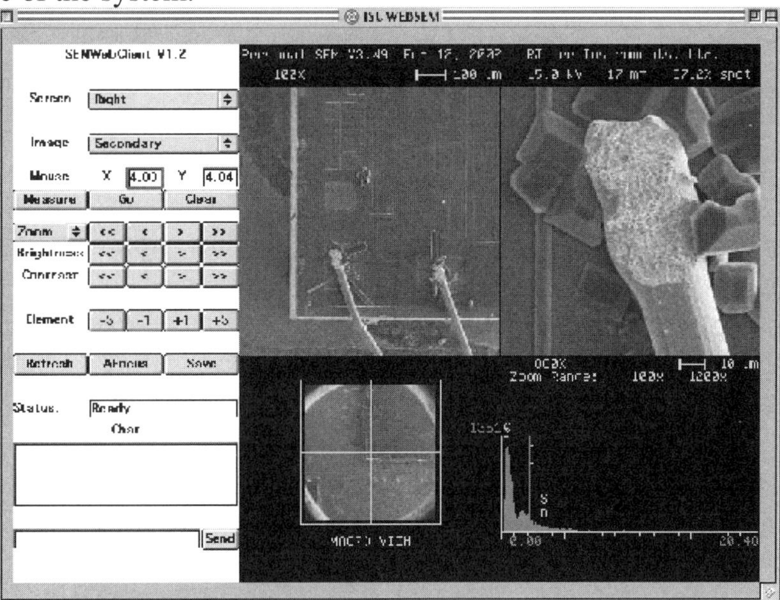

Figure 2: WebSEM console showing imaging, EDS, and measuring features.

Figure 3: Teachers attending workshop at ISU on WebSEM.

Microsc. Microanal. 8 (Suppl. 2), 2002
DOI 10.1017.S1431927602101152

Microscopic Digital Imaging in Introductory Biology

James Ekstrom*
*Phillips Exeter Academy, Exeter, NH 03833

K-12 instruction in biology has traditionally taken a very descriptive approach. This is in marked contrast to a quantitative as well as qualitative way of looking at things in physics and chemistry. This qualitative/descriptive approach even extends into the laboratory portion of the biological course. One way to introduce a more quantitative approach can take place is in the microscopy portion of the biology curriculum.

This area of biology ordinarily occurs first in the syllabus for several reasons. Because cellular structure is primarily a microscopic province it makes sense to introduce students to the different microscopic tools such as TEM and SEM as well as the light microscope that are used to investigate cell structure. Also, the light microscope is the principle, if sometimes only instrument, found in biology classrooms.

A typical introduction to the microscope can involve a measurement of the "field of view" as well as getting use to the various controls found on the instrument. If the lowest power student objective is 4X and the ocular 10X this measurement can occur with a fair degree of accuracy using a 6" mass-produced plastic ruler that also has a metric edge to it. Using a higher power objective would involve mathematically calculating what the field would be or using an inexpensive $15.00) micrometer. Once the student makes these calculations for 40X (4X x 10X), 100X and 400X they can record this and keep this with the microscope. At some future point if a "wee beastie" or some such thing should occupy one-half of their field under 100X then they would have an approximate size of the object.

The advent of inexpensive digital photography allows the instructor to carry out a more sophisticated approach to this exercise. Digital images of the three fields of magnification can be stored as calibrations for that microscope. A subsequent microscopic image can be digitized and the reference scale for that magnification can be cut and pasted on the image. Figure 1 on the second page shows a cheek cell that has been photographed and the relevant scale pasted on the image. This calibrated image can be printed and passed out to students as an introduction to measurement. The students would simply be given the sheets, told to work in small groups and allowed to have string and a millimeter ruler. Their goal is to determine the length of the cell. A discussion follows bearing on the accuracy of their results.

The same calibrated image can then be brought up under one of the following freeware programs. NIH Image (Macintosh), Scion Image (PC) or the new internet applet know as ImageJ. Students carrying out the above exercise in either of these applications can compare their results to what they received on the paper exercise. After calibrating their image, they can measure the area and perimeter of various structures and then compare these measurements to other ways of looking at the cheek cell like the SEM image shown in Figure 2[1]. A teacher using ImageJ can put class results on the internet and allow the students to interpret their results and write-up their conclusions after the laboratory period.

The biggest hurdle to the use of quantitative work in biology, with specific reference to digital imaging, is training teachers and the use of this technology and seeing that the exercises "fit-in" to the curricular standards.

[1] J. Ekstrom http://science.exeter.edu/jekstrom/WEB/CELLS/Epith/Epith.html

[2] J.Ekstrom *Cell Structure Study*, The Science Teacher Vol 67, No. 7, October 2000 http://science.exeter.edu/jekstrom/nsta/elodea.html

[3] J. Ekstrom Slicing for Biology, The Science Teacher Vol 68, No. 2, February 2001

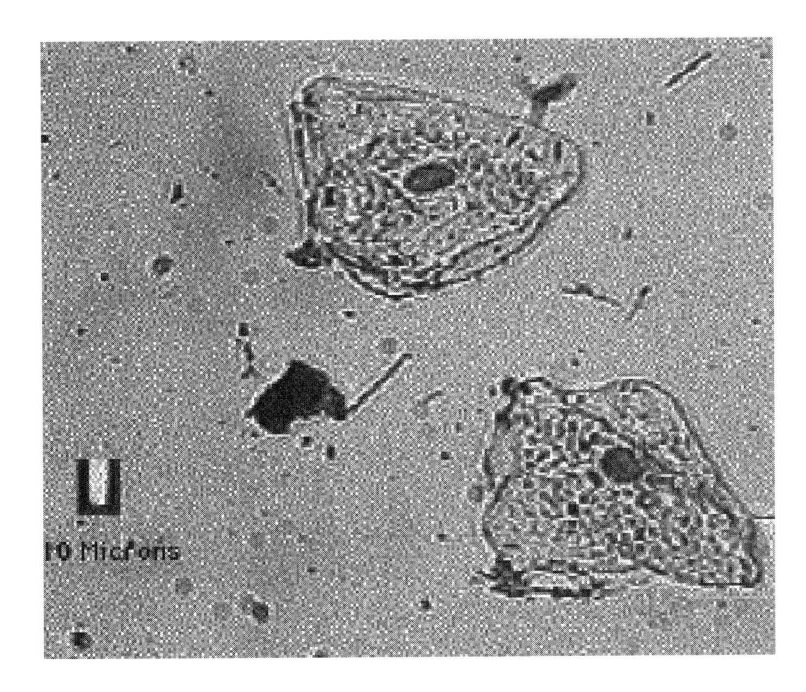

Figure 1: Cheek cells stained with methylene blue.

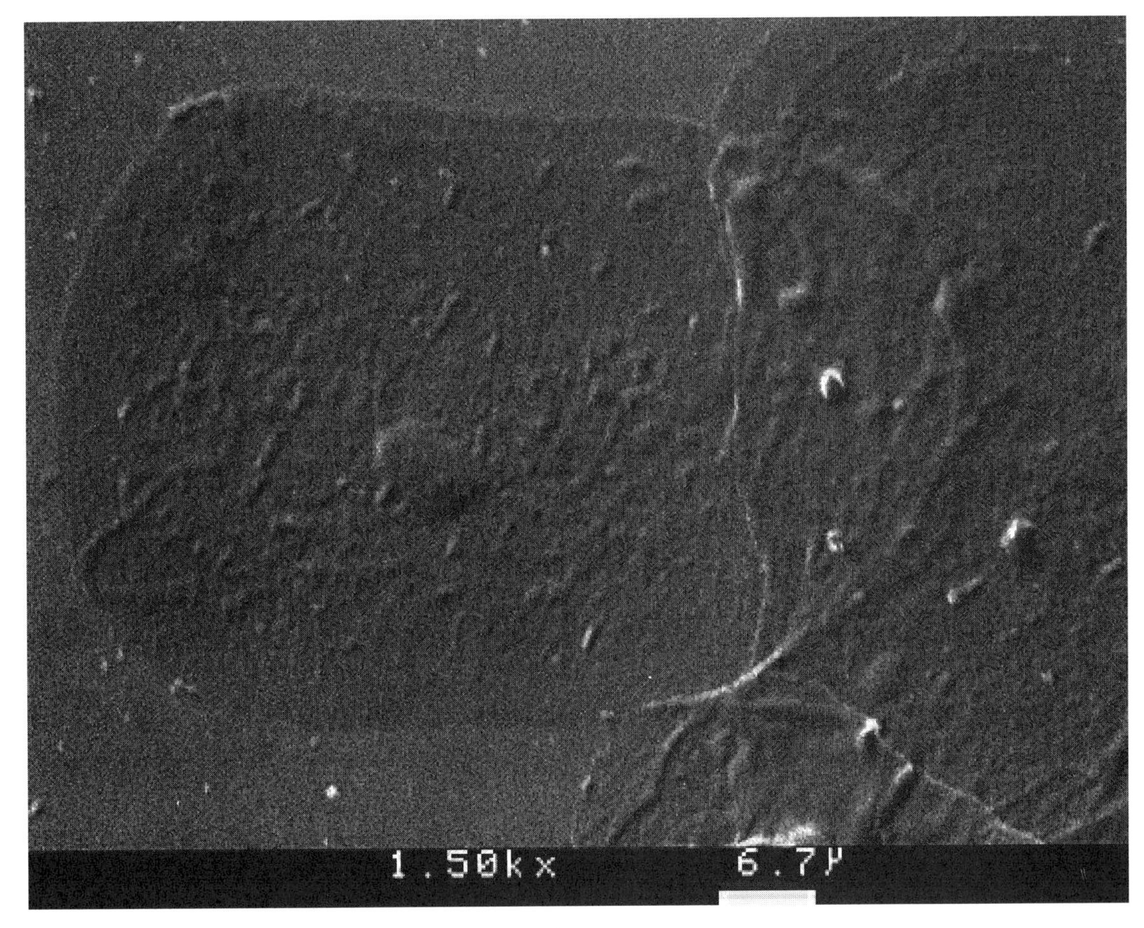

Figure 2: Cheek cells dried and sputter coated with gold.

Microsc. Microanal. 8 (Suppl. 2), 2002
DOI 10.1017.S1431927602101164

Developing and Implementing a Unified Imaging and Lab Information Management WorkFlow System – Lessons Learned and Insights Gained.

Christopher M. Yip*

*Departments of Chemical Engineering and Applied Chemistry, and Biochemistry, Institute of Biomaterials and Biomedical Engineering. University of Toronto 407 – 4 Taddle Creek Rd Toronto, Ontario, Canada M5S 3G9

The Centre for Studies in Molecular Imaging (CSMI) is a Canada Foundation for Innovation funded research enterprise located at the University of Toronto (http://bigten.ibme.utoronto.ca/CSMI). A small part of a large research enterprise in cellular and biomolecular research, the CSMI's research mandate is the active promotion of the application of functional molecular-scale imaging tools and techniques, such as confocal and scanning probe microscopy, to soft materials, and in particular, the development of new integrated systems. The unique instrumentation and expertise held within the CSMI, and its functional linkages with other University of Toronto and University Health Network research groups has led to extensive private and non-academic research collaborations. The CSMI provides both support and training facilities for teaching and research and is a showcase facility for its private sector sponsors.

Since its inception, the diverse nature of the data collected by CSMI users, which can range from epifluorescence and confocal images to scanning probe microscopy and force spectroscopy files, computational simulations, and spectroscopic data, has made it difficult to provide a systematic unified file format. Users are often faced with having to migrate between individual workstations in order to analyse their data while off-site users are hampered by not having access to the actual acquisition and analysis software. The large number of instruments, varied skill level of the user base (high school, undergraduate, graduate, professional), and a desire to accommodate the different needs of the users made it necessary to develop a unified imaging and lab information management system. The ideal system would be one that is platform-agnostic, networked, accessible from remote user locations, and secure. A workflow approach was felt to be the most appropriate with all data stored centrally but accessible from all workstations and through secure servers, over the World Wide Web. For our purposes, the ideal system would be one that incorporates a common software package that would allow the user to simultaneous compare data acquired from different instrument in an open architecture.

We will discuss our initial approaches to addressing these needs, which ranged from the implementation of secure FTP servers and dedicated back-up systems, to our most recent projects which bring together Java-enabling technologies, automated user booking and accounting systems, and custom data analysis routines.

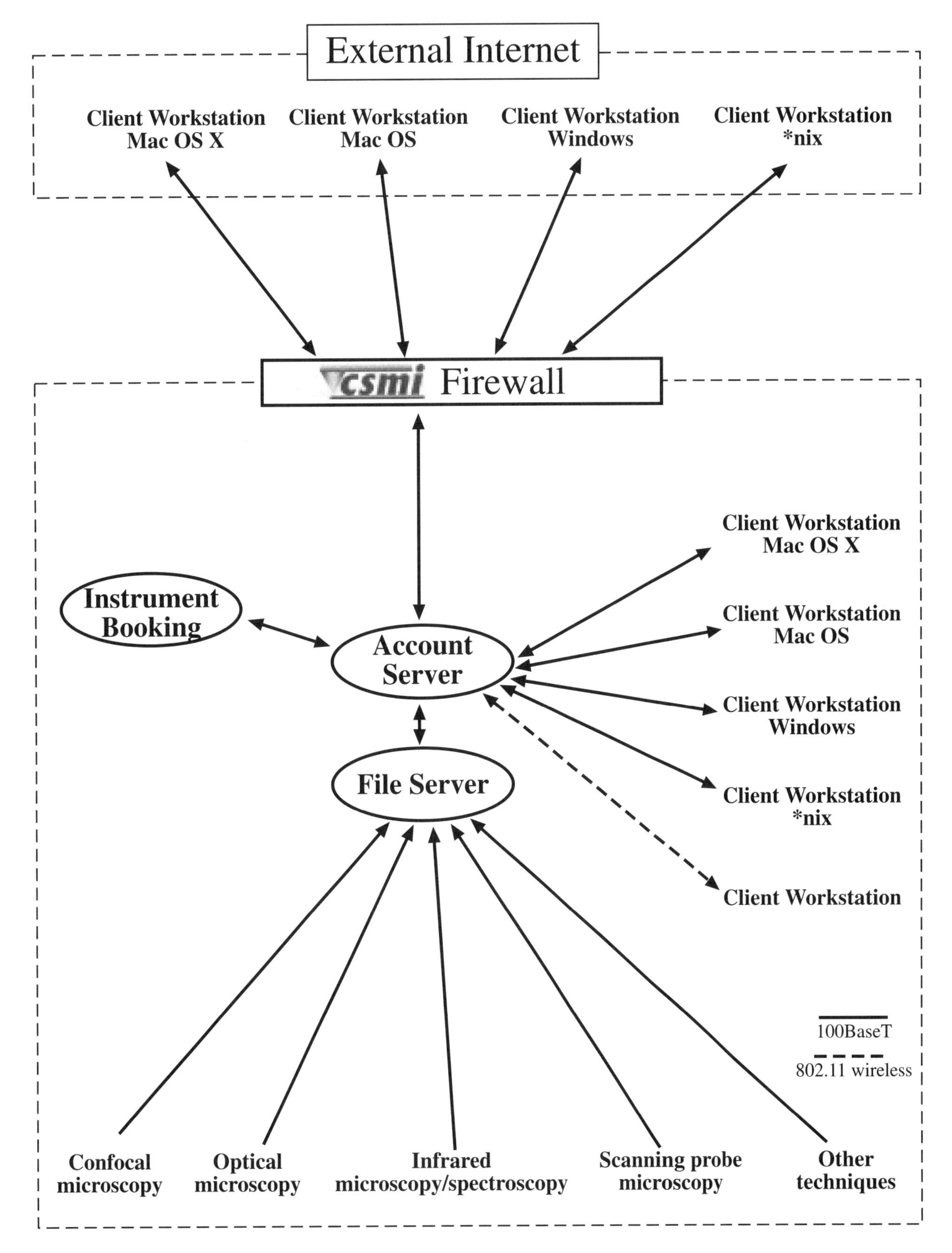

Figure 1. Schematic of CSMI instrumentation and image analysis workflow

Microsc. Microanal. 8 (Suppl. 2), 2002
DOI 10.1017.S1431927602101176

Valence Excitations in Electron Microscopy: Pursuing Zeitlerian Initiatives

A. Howie

Cavendish Laboratory, University of Cambridge, Madingley Road, Cambridge CB3 0HE, UK

Current exploitation of electron energy loss spectroscopy (EELS) in electron micro scopy mainly focuses on the core state excitation region for losses $\Delta E > 200$eV. Characteristic edge signals suitable for microanalysis are readily available, the spatial resolution is usually limited only by electron optics and the theoretical tools are now well developed to extract useful additional electronic information from the structure of the excited state. Valence excitations on the other hand involve more spatially extended initial states, as indicated by the Bohr impact parameter $\eta v/\Delta E$, and the final states are potentially of collective (plasmon) as well as sing le electron type. Though relatively much more intense, these valence loss signals depend on the dielectric response and geometry of the nearby nanostructure and are therefore less simple to interpret. Nevertheless valence excitations are of increasing interest, not only to the burgeoning company of electron microscopists with GIF equipment, but also to wider communities in surface science, ceramics and biology. Fortunately the dielectric response theory of valence excitation by a fast electron has now been extended and tested for inhomogeneous nanostructures even in the relativistic regime. The way is open for many more quantitative studies of some of the salient problems as outlined below where, in addition to their seminal work on core excitations note d elsewhere in this symposium, Elmar Zeitler and his colleagues made several early and valuable contributions [1-6].

Surface plasmons and other valence excitations on small particles have recently attracted great attention in the context of near field optical microscopy [7] but can be generated and detected with much higher spatial resolution in electron microscopes. The influence of particle size, shape and surface coating have all been studied both in the context of the hydrodynamical model [1] as wel l as in the more general dielectric formalism [8]. Interactions of neighboring particles have also been observed [9]. A problem of great topical interest in apertureless near field microscopy is the local field enhancement effect experienced by a light wave incident obliquely on the region between the sample and the scanning tip[10]. Resonance with the tip-sample surface excitations can result in very large enhancement factors useful for highly localized spectroscopy, particularly when non-linear processes such as two-photon fluorescence spectroscopy or Raman spectroscopy are used [11,12]. The ultimate usefulness of this form of highly localized spectroscopy will depend on more complete study of the dependence of the field enhancement factor on the tip s hape and dielectric coupling between tip and sample. Robust operating conditions that can routinely be achieved will be essential.

The dielectric response of a body also can be used to evaluate the image force experienced by a passing fast electron [2]. Similarly the Casimir or van der Waals forces between two or more dielectric bodies depend on their dielectric response functions and can be computed (including retardation effects if necessary) for any given geometry [13]. These forces are generally, though not always attractive, and are for instance involved in determining the thickness of intergranular layers in ceramics [14]. Transmission electron microscopy is obviously in a unique position to tackle such problems since, at the same time as the structure is imaged, the relevant dielectric functions of the various regions can be determined from the localized valence loss spectra. The variations in the valence loss spectra near interfaces in materials and from point to point in biological samples [15] may in any case be large enough to yield analytical information which can usefully supplement the core loss spectral data, particularly in samples susceptible to beam damage.

Ionisation damage places severe limitations on the more widespread application of high resolution electron microscopy despite the successful development of sophisticated cryomicroscopy and minimal dose techniques [3]. The final stage of electronic excitation leading to irreversible atomic displacement is usually some form of valence excitation, either through a direct valence loss event leaving a single hole in the valence band or through an initial core loss event followed by an Auger decay leaving two holes in the valence band. In aliphatic organic molecular crystals such as paraffin where high resolution imaging is achieved only with great difficulty [4], the former process appears to operate. Core excitation leading to Auger decay has been identified as the key process in many oxides [16] and possibly also in aromatic organic crystals [17]. The very low electron energies, such as those employed for PEEM imaging of surface processes [5,18], may sometimes lie below the damage threshold energy for all the above processes so that a damage-free energy window may exist. Exciton capture sometimes provides yet another damage process at these low energies [19] but is a resonant process with a sharply defined energy and may also be avoidable. In clarifying our picture of these phenomena new experiments, particularly comparison of photon and electron desorption cross sections and energy thresholds in different materials would be very valuable. In the valence excitation region, the two cross sections can differ not only because of momentum transfer effects [6] but also because of valence electron screening [20].

References

[1] D.B.T. Thoai and E. Zeitler, Phys. Stat. Sol. (a) 107 (1988) 791; ibid (b) 149 (1988) 169.
[2] D.B.T. Thoai and E. Zeitler, Phys. Stat. Sol. (b) 146 (1988) 137; ibid 158 (1990) 557.
[3] E. Zeitler and F. Zemlin, Ann. N.Y. Acad. Sci. 483 (1986) 5.
[4] J.R. Fryer et al., Proc. Roy. Soc. A453 (1997) 1929.
[5] M. Mundschau et al., Surf. Sci. 227 (1990) 246.
[6] D.S. Su et al., Ultramicrosc. 53 (1994) 97.
[7] T. Klar et al. Phys. Rev. Lett. 80 (1998) 4249.
[8] A. Rivacoba et al. Progr in Surface Science 65 (2000) 1.
[9] P. E. Batson, Phys. Rev. Lett. 49 (1982) 936.
[10] Y.C. Martin et al., J. Appl. Phys. 89 (2001) 5774.
[11] E.J. Sanchez et al., Phys. Rev. Lett. 82 (1999) 4014.
[12] M.S. Anderson, Appl. Phys. Lett. 76 (2000) 3130.
[13] E.M. Lifshitz, Sov. Phys. JETP 2 (1956) 73.
[14] R.H. French, J. Am. Ceram. Soc. 83 (2000) 2117.
[15] S.O. Sun et al., J. Microsc. 177 (1995) 18.
[16] M.L. Knotek and M.L. Feibelman, Phys. Rev. Lett. 40 (1978) 964.
[17] A. Howie et al., Phil. Mag. B52 (1985) 75.
[18] E. Bauer, J. Electron Spectrosc. & Rel. Phenom. 114 (2001) 975.
[19] P. Rowntree et al., J. Phys. Chem. 100 (1996) 4546.
[20] O. Kidum and J. Berakdar, Phys. Rev. Lett. 87 (2001) 263401-1.

Microsc. Microanal. 8 (Suppl. 2), 2002
DOI 10.1017.S1431927602101188

A Comparison of Microcompositional Imaging Methods

Michael S. Isaacson

Cornell University, College of Engineering, School of Applied and Engineering Physics,
W.M. Keck Foundation Program in Nanobiotechnology
241 Carpenter Hall, Ithaca, New York, 14853

More than a half century ago, in a classic paper, Zeitler and Bahr [1] outlined a method for obtaining quantitative information from electron micrographs, thereby extending classical optical methods of microanalysis [2] to even smaller spatial localization scales. In this last decade we have seen an explosion in the quantitative analysis of micro-images. In particular, there has been an almost exponential growth in the coupling of microscopic imaging with spectroscopic methods. This has occurred in all forms of microscopy, from optical, to electron, ion and X-ray microscopy, and more recently to scanned tip imaging microscopies.

Because these various microcharacterization methods (fashionable now to call them nanocharacterization methods) have evolved from many different disciplines using many different tools and techniques, the "local" language used to describe and evaluate the quantitation is not often translatable with ease from one technique to the next. In addition, because there are so many techniques available today which can couple an image with local compositional information, it becomes difficult for the user of a particular method (as opposed to the practitioner of the method) to understand the physical basis of the technique and thus to correlate information from various analytical methods. Moreover, with different "languages" or "currencies", the similarities and complementarities of various methods are not always transparent. In this presentation, I will try to develop a common "language" for microcharacterization so that similarities of the methods become obvious and one can easily look at the advantages and limitations of the various methods.

Of course, in any such discussion, for simplicity, one must make approximations (and a certain number of disclaimers and caveats) so as to not obscure the basic features of the methods. The aim in this paper is not so much to be all-inclusive, but rather to present a framework for looking at the different techniques.

Therefore, in this presentation we will make the following assumptions: 1) we will concentrate only on electromagnetic radiation "beams" (i.e., photons, electron, ions); 2) we will assume an incoherent beam interrogating the sample so that no beam particle influences the other; 3) we will assume a beam of circular cross-section and uniform flux rate (particles/sec/unit area). With these caveats, it can be shown [3,4] that the detected signal rate (counts per second), S_A, emanating from N_A atoms (or molecules) of type A in the irradiated volume using a beam of flux rate, J(particles/sec/unit area) is given by:

$$S_A = N_A J \sigma_A Y_A F_A \qquad (1)$$

Where σ_A is the cross-section (area) for the primary interaction process, Y_A is the yield of the secondary interaction process (e.g. if one is detecting the primary process, then $Y_A = 1$). F_A is the efficiency of the detection process which includes the efficiency of the detected interaction product reaching the detector, the production of the wanted interaction product by other primary and secondary products and the detection efficiency of the detector itself.

This basic microcharacterization equation is general so it is useful in comparing different techniques. This expression, when properly applied, can be used in "standardless analysis" methods found in the literature. For example, the ratio of the number of atoms of type A to those of type B can be obtained from equation (1) as:

$$N_A / N_B = [S_A / S_B] x [\sigma_B Y_B F_B / \sigma_A Y_A F_A] \qquad (2)$$

if the signals were acquired simultaneously. For example, for electron beam X-ray analysis, the quantity in the brackets is just related to the Cliff-Lorimer k factors [5]. For electron beam induced Auger analysis, the quantity in the brackets is just the relative sensitivity[6].

Of course, more exact calculations need to take into account that the signal rate is not simply a product of multiplicative factors. In this presentation we will present examples of the basic microanalysis equation (1) from EDX, EELS, RBS, AES, PIXE, etc. It is the purpose here to tie the various techniques together and provide a correlative framework between them.

References:

[1] E. Zeitler and G.F. Bahr, *Exp. Cell Res.* 12 (1957) 44.
[2] E.M. Chamot, *Elementary Chemical Microscopy*, John Wiley and Sons, New York, 1915.
[3] M.S. Isaacson, *Micro/Nanocharacterization: The Physics and Methodology of Materials Characterization from Volumes Less Than a Cubic Micron"*, Kluwer/Plenum Press, New York (in press).
[4[M.S. Isaacson, *Ultramicroscopy*, 49 (1993) 171.
[5] G. Cliff and G.W. Lorimer, *J. Microscopy*, 103 (1975)203.
[6] L.E. Davis et.al., *Handbook of Auger Electron Spectroscopy*, Perkin-Elmer Corp., Physical Electronics Industries, Inc., (1976).
[7] The author would like to thank Dangsheng Su and Gianluigi Botton, who organized this symposium in honor of Elmar Zeitler, a man who has always tried to instill in microscopy researchers the need to be quantitative and not mere"stamp collectors". In addition, I would like to thank Prof. Dr. Zeitler for the many years as a colleague, collaborator, friend and aficionado of jazz.

Microsc. Microanal. 8 (Suppl. 2), 2002
DOI 10.1017.S143192760210119X

The Future of EELS

R.F. Egerton

Physics Department, University of Alberta, Edmonton, Canada T6G 2J1

The combination of an electron energy-loss spectrometer and a TEM provides us with physical and chemical information in the form of energy-loss spectra and energy-filtered images. The field-emission electron source has made an energy resolution below 1 eV possible and, combined with improvements in the design of electron spectrometers and detectors, it is now feasible to obtain useful data with a spatialresolution close to atomic dimensions, given the right kind of instrumentation and specimen [1,2].

The use of an electron-source monochromator promises a further improvement in instrumental energy resolution [3], perhaps down to 0.1 eV. How can we explot this capability? The resolution of core-loss data is limited by the natural width of the inner-shell energy level, which is below 0.5 eV for K-edges below 2000eV or L- edges below 1000eV (see Fig. 1). A more common use of a monochromated EELS/TEM systemmay be for valence-electron spectroscopy, where a reduced incident-beam intensity is less of a problem and there is often a need to resolve fine structure, when measuring the water content of tissue [4] or distinguishing peaks due to chromophore dyes [5]for example. Structure below 2 eV loss has been particularly hard to measure because it lies within the tail of the zero-loss peak. High energy resolution would therefore benefit EELS studies of the local electronic structure at defects and interfaces in semiconductors [6,7] or of the electronic structure of carbon nanotubes or small particles [8,9]. Below 0.1 eV, there are vibrational modes of energy loss, as studied by IR spectroscopy and by Geiger et al. [8] using a monochromated source and transmitted-electron analyser, which achieved 3meV resolution at 30 kV. This concept has not yet applied to the TEM or applied to practical problems, but might become a useful additional tool for characterizing nanostructures.

The Heisenberg uncertainty principle suggests that high energy and spatial resolution are mutually incompatible. In fact, the wave nature of the electron imposes a spatial localization limit that approaches 1 nm at energy loss below 100 eV (see Fig. 1). In this energy region, a spherical-aberration corrector (which can reduce the point resolution for TEM imaging to below 0.1 nm) will not improve the spatial resolution of the inelastic signal. However, the corrector should increase the current available in a small probe, which will benefit the core-loss signal in fine-probe microanalysis or STEM imaging.

Although EELS has excellent spatial resolution in relation to other techniques, it has not been so successful for measuring low atomic concentrations. A large part of the reason is the high background underlying core-loss edges, plus the problem of edge overlap. One solution has been to employ multiple least squares (MLS) fitting to spectral quantification [11] and it is to be hoped that this procedure can be simplified and become more widely adopted in the future. An alternative approach to low-concentration analysis is to modify the background-extrapolation procedure, putting constraints on the background fit [12].

As always, the ultimate limit to spatial resolution and detection limits is set byradiation damage to the specimen, more severe for EELS than for x-ray absorption microscopy, even if less than for x-ray emission (EDX) spectroscopy. Does the damage depend only on the accumulated dose, or is the dose rate important? This question is relevant to the choice of TEM or STEM imaging, particularly for beam-sensitive specimens.

[1] O. Krivanek et al., Microsc. Microanal. Microstruct. 2 (1991) 257.
[2] N. Browning et al. Micron 30 (1999) 425.
[3] F. Kahl and E. Voekl, Microsc. Microanal 7 (Suppl. 2) (2001) 922.
[4] S. Sun et al., Ultramicroscopy 50 (1993) 127.
[5] X.G. Jiang and F.P. Ottensmeyer, Proc. ICEM-13 (Paris, 1994) 781.
[6] P.E. Batson et al., Phys. Rev. Lett. 57 (1986) 1729.
[7] U. Bangert et al. J. Microscopy 188 (1997) 237.
[8] B.W. Reed et al., Appl. Phys. Lett 78 (2001) 3358.
[9] M. Kociak et al., Phys. Rev. B 61 (2000) 13936
[10] J. Geiger, 39th Ann. Proc. Electron Microsc. Soc. Amer. (1981) 182.
[11] R.D. Leapman and C.R. Swyt, Ultramicroscopy 26 (1988) 393.
[12] R.F. Egerton and M. Malac, Ultramicroscopy 90 (2002), in press.
[13] M.O. Krause and J.H. Oliver, J. Phys. Chem. Ref. Data 8 (1979) 329.
[14] R.F. Egerton, J. Electron Microscopy 48 (1999) 711.

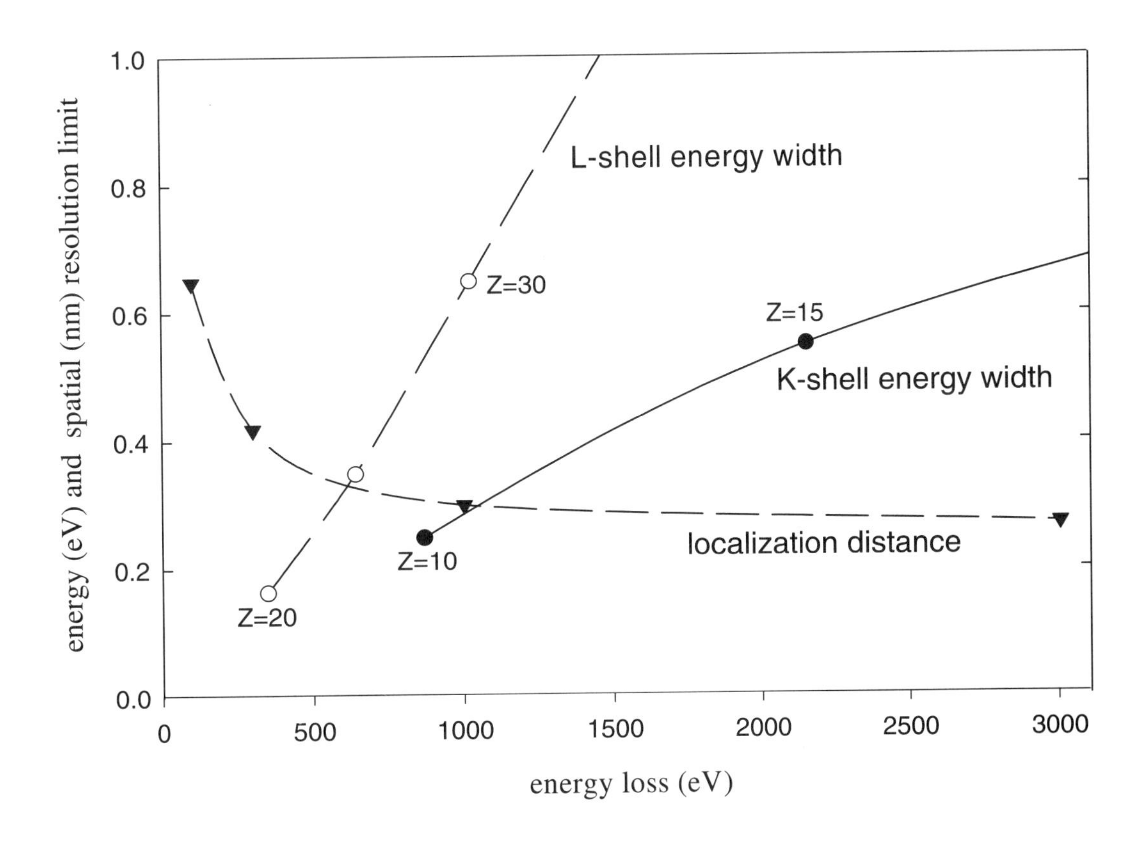

FIG. 1. K- and L-shell core-level widths [13] and localization distance d_{50} containing 50% of the inelastic scattering [14], as a function of energy loss and atomic number Z .

IMPORTANT

Pages 466CD through 1620CD appear only on the CD-ROM.

Author Index

***Note:** Invited papers appear in both the printed Proceedings and in the CD-ROM. The suffix "CD" following a page number indicates a submitted paper; these appear only on the CD-ROM.*

A

A-Hassan, E, 172
Abdollahi, K, 1088CD
Abe, E, 1218CD, 1220CD
Ackerley, C A, 992CD
Ackland, D W, 1598CD
Adachi, H, 606CD
Addadi, L, 164
Adem, E, 1234CD
Adler, M, 940CD
Agbessi, S, 968CD
Agemura, T, 716CD
Agnihotri, A, 260
Agrawal, R, 206
Aguiar, L E, 1570CD
Aharinejad, S, 222
Ahmad, I, 1012CD
Aitouchen, A, 284
Akase, Z, 652CD
Akashi, T, 514CD, 524CD
Aksenov, Y, 1386CD
Al-Bagdadi, F, 954CD, 956CD
Al-Jassim, M H, 1196CD
Al-Kindi, A, 888CD
Al-Kofahi, O, 1042CD
Al-Sharab, J, 1336CD
Alamgir, F M, 608CD
Alani, R, 372
Albrecht, R, 124, 132, 194
Alekseeva, L A, 1372CD
Alexander, C M O, 1550CD
Allameh, S, 56
Allard, L, 478CD
Alldredge, E S, 570CD
Allen, C W, 418
Allen, L, 1258CD
Allman, B E, 534CD
Amigó, V, 500CD, 502CD
Ancer-Rodríguez, J, 928CD
Anciso, A, 626CD
Anderhalt, R A, 1468CD
Andersen, S J, 1444CD
Anderson, I M, 1606CD
Anderson, M L, 1240CD
Anderson, R M, 44
Anderson, S, 1192CD
Anderson, S D, 556CD
Andre, T, 454
Andrei, C M, 1436CD
Angeliu, T M, 684CD, 698CD
Angert, I, 820CD, 856CD
Antler, C E, 834CD
Antonell, M, 556CD
Apkarian, R P, 830CD, 962CD, 1032CD
Aral, B, 1528CD
Araujo-Jorge, T C, 1570CD
Arevalo, S, 756CD
Armbruster, B L, 842CD
Armstrong, J T, 438, 1490CD
Arslan, I, 572CD
Asamura, S, 1024CD
Asayama, K, 48
Ashcroft, N W, 570CD
Asher, L S, 920CD
Asher, S A, 320
Asturias, F J, 202
Asurmendi, S, 966CD
Atar, M, 886CD, 932CD
Atar-Zwillenberg, D R, 886CD, 932CD
Attas, E M, 286
Atwater, H A, 1114CD
Aubin, A S, 1538CD
Aung, L H, 1076CD
Austin, R, 154
Avalos Borja, M, 1234CD
Aviram, M, 944CD
Ayars, E J, 1518CD

B

Ba-Omar, T, 888CD
Baayen, R P, 254
Bae, H J, 1084CD
Bagle, H, 1118CD
Bahlmann, K, 280
Bailey, J L, 924CD
Baker Schut, T C, 444
Baker, I, 1264CD, 1398CD, 1544CD, 1546CD
Baker, T S, 208, 822CD, 846CD, 852CD
Ball, T B, 922CD
Bandyopadhay, S, 1368CD
Barbosa, H L, 1570CD
Barbour, J C, 1156CD
Barfels, M, 614CD
Bargar, T, 1012CD
Baril, C, 1306CD
Barkan, S, 82
Barlow, S, 1578CD
Barnard, J A, 1120CD
Barnes, J A, 946CD
Barone-Nugent, E D, 534CD
Barr, B, 150
Bartuskova, I, 228
Basgall, E J, 1132CD
Bastin, G F, 424
Batcheler, R, 350
Batchelor, A D, 1208CD
Bates, I R, 834CD
Bator, C M, 822CD
Batson, P, 14, 472CD
Bauhammer, G, 546CD
Baumeister, W, 870CD, 1610CD
Bavarian, B, 344
Baydack, R, 1506CD
Baydak, R, 442
Bazett-Jones, D P, 74
Beachy, R N, 966CD
Bearat, H, 796CD
Beaulieu, C, 968CD, 996CD
Beckedorf, M, 250
Becker, L E, 992CD
Bednarova, L, 1150CD
Beerens, J, 712CD
Beers, K, 322
Bejar Gomez, L, 624CD
Beleggia, M, 512CD, 528CD, 530CD
Belhaj, M, 1460CD
Bell, D C, 330
Beniac, D, 200, 848CD
Benner, G, 484CD, 486CD, 586CD, 594CD
Bennett, J C, 312
Bentley, J, 1334CD, 1336CD
Berg, R H, 966CD
Berry, J, 950CD
Berta, Y, 1608CD
Bertenshaw, G P, 824CD
Bewlay, B P, 696CD, 1454CD
Bhargava, R, 1514CD
Biggin, M D, 1044CD
Bishop, G, 174
Blackburn, D H, 1494CD
Blackford, M, 1420CD
Blochel, A, 470CD
Blom, D A, 1148CD, 478CD
Blum, A S, 1116CD
Bo, X Z, 1210CD
Böhm, H, 312
Boissy, Y, 786CD
Boissy, Y L, 1066CD
Bolduc, N, 248, 970CD
Bond, J S, 824CD
Bonelle, C, 436
Booth, T, 442
Borca, C, 394
Bording, J, 658CD
Bosco, L, 1568CD
Boswell, F W, 312
Botton, G, 188, 338, 358, 580CD
Bouchard, R, 548CD
Bouwstra, J, 278
Bovin, J O, 1394CD
Bowdoin, S, 720CD
Bowman, V D, 852CD
Boyd, D, 352, 356
Boyes, E D, 110, 122, 408
Boylston, E, 192
Braithwaite, J, 1156CD
Brand, K, 536CD, 538CD
Braun, P V, 316
Breton, B C, 1566CD
Brewer, L N, 684CD, 696CD, 698CD
Bright, D S, 1574CD, 496CD, 498CD
Brink, F, 306
Brink, H, 614CD
Brink, H A, 372, 588CD, 630CD
Brinks, H, 1436CD
Brisson, L, 248, 970CD
Brito, M E, 1246CD, 1250CD
Brochu, M, 708CD
Brown, J, 354, 844CD
Brown, L M, 470CD
Browning, N D, 18, 572CD, 578CD, 814CD, 1114CD, 1380CD, 1396CD
Bruining, H A, 288
Bruley, J, 596CD
Brust, M, 1124CD
Brydson, R, 470CD
Buban, J, 578CD
Buchanan, J, 160
Bucher, E, 618CD
Bückins, M, 1488CD
Buckner, J, 1014CD
Buechele, A C, 1312CD
Buehler, C, 280
Bunker, K L, 54, 1208CD
Bunton, J H, 1094CD
Burch, R V, 1526CD
Burgner, P, 614CD
Burke, A S, 910CD
Burke, M G, 292
Burrell, M, 772CD
Bursill, L, 1182CD
Busby, J, 1422CD
Bushby, A J, 554CD
Busquets, D, 500CD, 502CD
Butler, M, 1028CD
Butnor, K J, 918CD
Buttcher, B, 212
Buttle, K, 1118CD
Butts, D A, 1322CD
Butts, M, 772CD

C

Caldwell, N H M, 1566CD
Caliman, N C, 906CD
Callahan, J W, 898CD
Cambridge, V, 214
Campagna, C, 924CD
Campanati, L, 506CD
Campbell, H, 1470CD
Campin, M J, 1156CD
Can, A, 1042CD
Capaldi, R A, 1006CD
Caragianis-Broadbridge, C, 774CD
Cardona, T S, 1570CD
Carlton, R A, 1298CD
Carpenter, G, 188
Carpenter, M A, 1482CD
Carpenter, R W, 796CD
Carpick, R W, 768CD
Carr, G L, 1524CD
Carragher, B, 218, 866CD, 872CD, 1580CD
Carroll, J W, 1440CD

Carter, C B, 1144CD, 562CD, 662CD
Casey Jr., J D, 52
Casey, J, 362
Caspar, D L, 860CD
Caspers, P J, 288
Castejon, H V, 1032CD
Castejon, O J, 1032CD
Casuccio, G, 1564CD, 1568CD
Catalfam, M, 1040CD
Cayer, M L, 946CD
Cazaux, J, 112, 426
Ceh, M, 492CD, 494CD
Chakrabart, A, 1056CD
Chalifour, R J, 952CD
Chamberland, H, 254
Chamberland, H L N, 1084CD
Chambers, S A, 1160CD
Chandra, A, 910CD
Chang, C R, 1370CD
Chao, W T, 942CD
Chaplin, T D, 1532CD
Chappell, S C, 274
Charest, M, 358
Charest, P M, 254, 256, 968CD
Chase, D B, 1530CD
Chase, E S, 852CD
Chein, K, 914CD
Chen, A, 176
Chen, F R, 98
Chen, H, 658CD, 1122CD
Chen, I S, 268
Chen, J H, 468CD
Chen, L C, 564CD
Chen, T T, 800CD
Chen, X, 304, 882CD
Chen, Y, 1222CD
Chen, Z, 66
Cheng, N, 844CD
Cheng, P C, 268, 1038CD, 1048CD, 1068CD, 1072CD, 1080CD
Cheng, T Z, 308
Cheng, V, 1068CD
Cheng, W Y, 1048CD, 1068CD, 1072CD, 1080CD
Cheng, Z H C, 396
Cheresh, D, 1046CD
Cheung, C S, 1124CD
Chevet, E, 166
Chew, D, 630CD
Chiang, H C, 988CD
Chin, Y H, 1318CD
Ching, W Y, 606CD
Chipman, M, 544CD
Chipman, P, 822CD
Chipman, P R, 852CD
Chiriac, H, 370
Chiu, W, 216, 1576CD
Chizmeshya, A V G, 796CD
Cho, E, 1040CD
Choo-Smith, LP, 444
Chopard, R P, 232
Chou, C T, 686CD, 688CD
Christensen, D C, 768CD
Chu, S W, 268
Chubinskaya, S, 778CD
Chue, B, 898CD
Chumbley, A E, 454
Chumbley, L S, 454
Chumbley, S, 1564CD
Chun, C H, 610CD
Chun, S H, 1620CD
Citrin, P H, 1614CD
Clancy, R, 1022CD
Clark, R J H, 1532CD
Clayton, F B, 656CD
Clegg, W J, 554CD
Clemmens, J, 1092CD
Cockayne, D J H, 10
Cohen, A H, 914CD
Cohen, D, 1166CD
Cohen, G, 1008CD
Colella, M, 1420CD
Coleman, R, 944CD
Colijn, H O, 58
Colliex, C, 574CD, 1384CD
Collins, S R, 1316CD
Conrad, M N, 1052CD
Constantinou, P, 908CD
Conway, C, 866CD
Cook, R E, 582CD, 1214CD
Cooper, S J, 1226CD
Costa, F M, 1352CD
Costello, M J, 1004CD
Costes, S, 1040CD
Cota Araiza, L, 1234CD
Coutinho, C M L M, 1570CD
Couture, ARA, 702CD
Cowin, P, 1060CD
Cox, B, 894CD
Cox, R A, 910CD
Craig, A, 822CD
Craig, R, 828CD
Crang, R F E, 1078CD
Craven, A, 470CD
Craven, J P, 1478CD
Crawford, P, 954CD
Cremer, R, 1488CD
Crewe, A V, 4
Cronin-Golomb, M, 1020CD
Crookham, H, 1528CD
Crooks, R, 638CD
Crozier, P, 422, 600CD, 1108CD, 1392CD
Crugnola, A, 758CD
Cullen, D, 1398CD, 1544CD, 1546CD
Cummings, J, 174
Czarnota, G, 1028CD

D

da Rocha, R C G, 232
da Silva, M, 1206CD
Dabrowski, B M, 582CD
Dahmen, U, 98, 1404CD, 1426CD
Dai, Z, 364
Dailey, M, 1032CD
Damon, I K, 904CD
Danino, D, 850CD, 868CD
Darling, C, 1192CD
Darveau, A, 730CD
Darwish, F A, 1280CD
Davidson, K G V, 130
Davis, C C, 1476CD
Davis, J, 634CD
Davis, J M, 438
Davis, K, 1326CD
Day, J, 156
De Andrade Lima, L R P, 1560CD
de Bruijn, J D, 1386CD
De Castro, S L, 1570CD
de Jong, A, 70
De Kloe, R, 680CD
De Koninck, Y, 1000CD
Deaton, J, 840CD
Decker, M, 542CD, 1192CD
Defoort, B, 1232CD
Degenhardt, L J, 478CD
Deiker, S, 82
Delaney, P, 902CD
Della Negra, S, 504CD
Dellby, N, 14, 18, 20, 476CD
Demaree, R S, 146, 156
Demers, H, 452, 1462CD, 1498CD
Dempere, L A, 1270CD
den Dekker, A, 94
DeRege, P, 322
Deshparde, S S, 940CD
Detry, J, 114
Di, N L, 396
Diaz-Avalos, R, 860CD
Dickey, E C, 1146CD, 1170CD, 1190CD, 1620CD
Dicks, K G, 670CD, 686CD, 688CD
Dijkstra, J M, 1468CD
Dikin, D, 304
Dimiduk, D M, 1432CD
Ding, W, 304
Dingley, D J, 678CD
Dionne, S, 188, 358
Disko, M M, 18
Ditrich, H, 226
Divry, E, 1272CD
Djachkov, A L, 1372CD
Dohnalkova, A C, 750CD
Dominguez, J E, 1154CD, 1168CD
Donald, A M, 958CD, 960CD, 1228CD, 1478CD
Dong, C Y, 280
Dong, J, 830CD
Dong, J W, 302
Dong, L, 1142CD
Dong, Z, 1548CD
Donlon, W, 176
Doole, R C, 10
Dorignac, D, 186
Dorin, C, 1102CD, 1222CD
Dovey-Hartman, B J, 140
Dowben, P, 394, 1374CD
Downer, R, 954CD
Drazba, J, 198
Dremailova, O, 354
Dresser, M E, 1052CD
Drew, R, 674CD, 708CD
Drouin, D, 702CD, 704CD, 712CD, 1538CD
Drury, M, 680CD
Drzal, L, 1232CD
Duarte Moller, J A, 616CD, 1234CD
Duarte, A, 756CD
Dubois, J, 442, 1506CD, 1510CD
Dunin-Borkowski, R, 42, 168, 518CD
Dunlap, R A, 396
Dunstan, D J, 554CD
Dupuis, E, 1306CD, 1388CD
Dupuis, R D, 1214CD
Duscher, G, 384, 1604CD
Dusevich, V M, 760CD, 1480CD
Dutartre, D, 1174CD
Dutrizac, J E, 800CD
Dye, K D, 1186CD

E

Eades, A, 690CD, 692CD
Ebadi, M, 1012CD
Echlin, P, 120
Eckels, K, 920CD
Edwards, J, 808CD
Edwards, M, 1390CD
Edwards, R, 614CD
Edwards, V, 806CD, 936CD
Eger, B, 848CD
Egerton, R F, 464
Ehlers, G, 374
Eick, J D, 760CD, 1480CD
Eilts, B, 954CD, 956CD
Eisen, R N, 916CD
Eisenmann, D J, 454
Ekstrom, J, 456, 1582CD
El-Refaey, H, 1012CD
Elboujdaini, M, 1282CD, 1284CD
Ellis, E A, 884CD, 1070CD
Elsaesser, C, 64
Elswick, D S, 1616CD
Elwazri, A M, 710CD
Emadi, D, 352, 354
Endo, J, 522CD
Endo, N, 1176CD
Engels, M, 986CD
Eom, C B, 1364CD
Eppell, S, 262, 1018CD
Erbe, E F, 722CD
Erbil, W K, 962CD
Erie, D A, 762CD
Erlandsen, S L, 114
Erlund, E, 1562CD
Errington, R J, 274
Esmacher, M J, 810CD
Espinosa Magana, F, 624CD
Espinoza, D, 1552CD
Essadiqi, E, 354, 1324CD
Essers, E, 484CD, 612CD
Esterman, M, 134
Etz, E S, 1520CD
Evans, N D, 638CD, 1336CD
Evans, W H, 274
Ewing, R C, 1136CD, 1424CD, 1558CD

F

Fagerland, J A, 740CD
Fajardo-Bermudez, A, 274
Fakhfakh, S, 1460CD
Fambrough, K, 972CD
Fan, H F, 308
Farber, L, 784CD
Farrer, J, 544CD, 680CD
Fasolka, M J, 322
Feldmann, Y, 1128CD
Fellmann, D, 218, 866CD, 1580CD
Fernandes, A, 200, 632CD, 634CD
Fernandez, D C, 1514CD
Fernández-Salas, E, 1002CD
Fialin, G M, 1496CD
Finch, J, 452, 1536CD
Fink, V, 1200CD
Finol, H J, 926CD, 934CD
Firbas, U, 222
Fisher, J, 174
Fletterick, R J, 210
Fliss, I, 730CD
Flores-Gutierrez, J P, 928CD
Foiles, S, 1166CD
Föll, H, 1242CD
Follstaedt, D M, 1188CD
Fong, H, 746CD
Fontano, E, 860CD
Forbes, G W, 152
Fortin, I, 248
Foss, S, 1586CD
Foster, J, 722CD
Fox, G, 1070CD
Foxley, S, 1142CD
Francis, L F, 330
Frank, J, 206, 854CD, 856CD, 878CD
Franz, AW, 852CD
Fredrickson, J W, 750CD

Freel, C D, 1004CD
Freeman, T, 1014CD
Freitag, B H, 70, 1592CD, 1612CD
Frethem, C, 114
Fried, G, 1036CD, 1274CD, 764CD
Friedlander, S K, 1130CD
Friel, J, 350
Friis, J, 90, 654CD
Frizell, T, 454
Frost, B, 28
Frost, B G, 516CD, 520CD
Fu, Z Q, 308
Fujii, T, 1434CD
Fujimori, H, 1348CD
Fukuda, Y, 332
Fuller, S, 1016CD
Fung, J, 632CD
Furman, C S, 130
Furmanova, T A, 1372CD
Furukawa, H, 842CD
Furutsu, T, 524CD
Furuya, K, 332, 410, 1418CD
FuruyaSong, M, 332

G

Gagne, G D, 740CD
Gahleitner, M, 1522CD
Gai, P, 412, 1412CD, 1414CD
Gajdardziska-Josifovska, M, 752CD, 754CD, 1108CD, 1158CD, 1278CD
Gale, W F, 1322CD
Gales, T, 984CD
Galindo, D, 340
Gallegos, L, 972CD
Gammon, L M, 1236CD
Gan, Z H, 788CD
Ganguli, A K, 1414CD
Gao, F, 1160CD
Garcia, J R, 768CD
Garcia, P E L, 1328CD
Garsha, K, 1036CD
Gauvin, R, 116, 430, 432, 452, 674CD, 700CD, 704CD, 708CD, 710CD, 1462CD, 1470CD, 1474CD, 1482CD, 1484CD, 1498CD, 1536CD
Gerits, W, 1612CD
Gerrity, R G, 152
Gerstl, S S A, 1096CD
Gervais, F, 952CD
Geuens, P, 662CD
Ghali, E, 1282CD
Ghebrehiwet, B, 726CD
Ghiabi, P, 166
Ghoshroy, S, 972CD, 1074CD
Gianetto, J A, 338
Giannuzzi, L, 50, 390, 656CD
Giberson, R, 156, 162
Giberson, R T, 148
Giersig, M, 1362CD
Gignac, L, 672CD
Gilkerson, R W, 1006CD
Gillen, G, 1298CD, 1520CD
Gilliland, K O, 1004CD
Gillis, T, 954CD
Gilliss, S R, 562CD
Gjønnes, J, 92, 100
Glaros, A G, 760CD
Glick, M, 718CD
Gloess, D, 552CD
Gloter, A, 574CD
Gnade, B, 1116CD
Gnägi, H, 324
Gnauck, P, 46, 546CD
Godleski, J J, 1500CD
Goff, D, 246
Goff, H D, 238
Goh, C, 982CD
Goh, M C, 1026CD, 776CD
Goldberg, J J, 976CD
Goldner, L S, 322
Goldsmith, C S, 904CD
Golla-Schindler, U, 594CD
Gonzalez, J, 1206CD
Gonzalez, J C, 54, 1208CD
Gonzalez, L, 176
Gonzalez, R, 914CD
Gonzalez, S, 282
González-Valenzuela, C, 616CD
Goodhew, P, 470CD
Goodhue, W, 1258CD
Goodwin, F, 358
Gossmann, HJ L, 1614CD
Gouma, P, 666CD
Goynes, W, 806CD
Grace, M J, 140
Graham, G W, 1112CD, 1154CD
Grams, Y, 278
Grande, T, 1428CD
Grassucci, R, 854CD
Grazul, J, 576CD, 1614CD
Greedan, J E, 580CD
Gregor-Svetec, D, 1522CD
Greve, J, 1386CD
Gribb, T T, 1094CD
Gribelyuk, M, 672CD
Griess, G A, 974CD
Griffin, B J, 450
Griffiths, M, 340
Griffs, D P, 54
Griggs, W, 922CD
Grogger, W, 72, 618CD
Grondin, G, 732CD, 996CD
Grudberg, P, 76
Grütter, P H, 1000CD, 1338CD
Guchard, A R, 54
Guijon, F B, 1504CD
Gum, R J, 740CD
Guo, R Q, 1278CD
Gupper, A, 1522CD

H

Hadfield, M G, 930CD
Hadjipanayis, G, 1360CD, 1376CD, 1378CD
Haeni, J H, 1162CD
Hagan-Brown, J, 978CD
Haider, M, 10, 12, 486CD, 584CD
Hainfeld, J, 832CD, 916CD, 1030CD
Hajime, M, 1446CD
Hall, D C, 1214CD
Hallen, H D, 1518CD
Hallett, F R, 834CD
Hamilton, T, 736CD, 900CD, 940CD
Han, J P, 774CD
Hanaguri, T, 514CD, 526CD
Hancewicz, T M, 286, 1064CD
Hangas, J, 176, 1276CD
Hanko, J A, 136
Hansen, V, 100
Hanson, J, 134
Hao, Z, 1368CD
Haque, A, 1274CD
Harada, K, 514CD, 522CD, 524CD, 526CD
Harauz, G, 834CD, 836CD
Haren, H, 676CD
Hargrave, C, 454, 1564CD
Harmon, R, 372, 592CD
Harris, J, 76
Harrison, R, 890CD
Hart, T K, 138
Hartel, P, 10
Hasebe, S, 30
Hasenkopf, A, 1202CD
Hashimoto, H, 1418CD, 1446CD, 1448CD
Hashimoto, S, 336
Hashimoto, T, 48, 1152CD
Hashimoto, T M, 1328CD, 1330CD
Hatzistergos, M S, 1484CD
Hauback, B, 1436CD
Hauffe, W, 552CD
Hävecker, M, 602CD
Hawkins, H K, 910CD
Hayashi, J, 1010CD
Hayek, T, 944CD
Hayes, B S, 1236CD
He, A, 1292CD
He, J, 1542CD
He, L L, 1294CD, 1342CD
He, W, 1060CD
He, X, 1088CD
Heard, E, 728CD
Hearn, S, 728CD
Heathershaw, M L, 914CD
Hébert, C C, 602CD, 636CD, 1594CD
Heckman, C A, 946CD
Hedley, D, 908CD
Hedrick, R, 150
Heijligers, H J M, 424
Hein Lehman, A, 774CD
Heindel, M A, 740CD
Heller, R D, 1214CD
Henderson, J, 166
Heng, Y M, 1028CD
Henkart, P, 1040CD
Henriques-Pons, A, 1570CD
Hernandez Carreon, C A, 624CD
Herring, R A, 508CD
Hess, H, 1092CD
Hess, W M, 922CD
Hetherington, C J D, 10, 622CD
Heusser, R C, 914CD
Hewko, M D, 1510CD
Heymann, J, 844CD
Heyne, T, 1054CD
Hibbs, A, 276
Hibbs, A R, 1048CD
Hibino, M, 24
Hidalgo, C, 934CD
Hill, C M, 834CD, 836CD
Hill, F, 984CD
Hillmann, D, 954CD, 956CD
Hillyer, J, 1024CD
Hinshaw, J, 850CD
Hiraga, K, 1224CD
Hiraga, T, 1606CD
Hirata, G A, 1234CD
Hirayama, H, 414
Hirayama, T, 24, 38
Hirokawa, N, 210
Hirschberg, J, 950CD
Hlava, P, 1554CD
Ho, K L, 938CD
Hockey, B J, 1244CD
Hodouin, D, 1560CD
Hofer, F, 72, 618CD, 1592CD
Hoffrogge, P, 546CD
Hogue, F, 1296CD
Hoh, J, 172
Höhne, J, 80
Høier, R, 1428CD
Holburn, D M, 1566CD
Holden, P, 990CD
Hollerith, C, 80
Holmen, A, 1150CD
Holmes, K C, 820CD, 856CD
Holmestad, R, 90, 654CD, 1436CD
Holzenburg, A, 840CD, 1254CD
Honda, S, 1366CD
Honda, T, 68, 1106CD
Hong, C, 634CD
Hoppe, B, 250
Hopson, T, 1184CD
Horny, P, 1474CD, 1498CD
Hosokawa, F, 10
Hossler, F E, 224
Hovington, P, 704CD, 1306CD, 1388CD, 1472CD
Hovmöller, S, 96
Howard, J, 1092CD
Howe, J M, 1430CD
Howie, A, 404, 460
Hren, J J, 1110CD, 1140CD, 1616CD
Hsieh, C E, 878CD
Hsu, Y C, 994CD
Huang, J Y, 564CD
Huang, S, 974CD
Huang, Y, 1376CD
Huber, B, 612CD
Hubert, D, 1104CD, 792CD
Huffman, S W, 1514CD
Hug, G, 608CD
Hull, R, 420
Humphrey, C, 980CD
Humphreys, C, 718CD
Hunt, J A, 588CD, 592CD, 620CD, 630CD
Hutchings, G, 1390CD
Hutchison, J L, 10
Hwang, H Y, 576CD
Hwang, J, 322, 936CD
Hyatt, C V, 338
Hyde, J M, 292
Hyman, A, 880CD
Hyttel, P, 924CD

I

Iadarola, S, 758CD
Iapicca, D, 1552CD
Ichihashi, Y, 904CD
Idrobo, J C, 1380CD
Ihnat, P M, 140
Iliescu, B, 1264CD
Iliescu, D, 1398CD, 1544CD
Imhoff, D, 574CD
Ingram, P, 918CD
Irwin, K, 78, 82
Irwin, R, 1192CD
Isaacson, M, 462, 1118CD
Isakozawa, S, 590CD
Isheim, D, 1098CD
Ishida, Y, 1172CD
Ishikawa, T, 204
Ishitani, T, 48
Ishizaki, K, 334
Isogai, N, 1024CD
Ito, Y, 582CD
Ivey, D G, 1252CD, 1292CD
Iwanczyk, J, 82
Iwasaki, T, 848CD

J

Jabaji-Hare, S H, 256
Jackson, M, 442, 1506CD, 1510CD

Jackson, M R, 1454CD
Jacobs, D, 786CD, 1532CD
Jacquet, R, 1024CD
Jaeger, D L, 1140CD
Jahn, W, 820CD
Jahncke, C L, 1518CD
Jain, H, 608CD
Jamieson, D, 1182CD
Jamison, M, 1192CD
Janney, D E, 1310CD
Jansen, J, 102, 468CD, 1444CD
Jaouen, C, 504CD
Jasinski, J, 1198CD
Javaid, A, 1326CD
Jbara, O, 1460CD
Jenkins, M L, 622CD
Jensen, K, 1532CD
Jensen, T E, 976CD, 978CD
Jeong, B Y, 674CD, 700CD
Jerome, W G, 818CD, 894CD, 1034CD
Jiang, B, 650CD, 654CD, 980CD
Jiang, H, 392
Jiao, J, 1142CD
Jin-Phillipp, N Y, 1164CD, 1204CD
Jinschek, JR, 466CD
Jobin, G, 996CD
Jodan, K, 334
Johnson, B, 1318CD
Johnson, E, 1426CD
Johnson, I, 270
Johnson, J, 1568CD
Jolly, R A, 740CD
Jones, G T, 234
Jones, K M, 1196CD
Jonnard, P, 436
Joshi, R, 758CD
Joshi, V, 832CD
Jouffrey, B, 636CD
Joy, D C, 28, 118, 516CD, 520CD, 716CD, 1486CD, 1540CD, 1542CD
Jug, N, 312
Jullian, S, 1174CD

K

Kabius, B, 418, 468CD
Kai, J J, 98
Kaji, K, 590CD
Kalceff, W, 1182CD
Kallender, H, 984CD
Kamimura, O, 514CD, 526CD
Kamino, T, 48, 1152CD
Kan, R K, 736CD, 900CD
Kaneyama, T, 68
Kangasniemi, K, 766CD
Kao, R L, 224
Kao, V, 1324CD
Kaplan, D L, 1020CD
Kaplan, P, 286, 1064CD
Kappel, R, 484CD
Karpova, T, 1040CD
Kasai, H, 514CD, 524CD, 526CD
Kasuya, A, 1152CD
Kato, M, 1434CD
Kato, N, 38
Kato, T, 38
Kaufman, M J, 1270CD
Kavanagh, K L, 1200CD
Kawabata, K, 1366CD
Kawana, M, 644CD
Kawasaki, K, 550CD
Kawasaki, M, 492CD, 494CD
Kawasaki, T, 474CD, 482CD, 524CD
Kean, W, 752CD
Kearns, S, 1562CD
Keenan, M, 1554CD
Keenan, M R, 440, 648CD
Keider, S, 944CD
Keim, E, 1346CD, 1438CD
Keller, K, 174
Kelly, T F, 1094CD
Kelsch, M, 1204CD
Kempshall, B, 50, 390, 656CD, 790CD
Kenik, E A, 1422CD, 638CD, 1126CD
Kennedy, D W, 750CD
Kennedy, S, 1568CD
Keranen, S, 1044CD, 1564CD
Kersker, M, 68, 1106CD
Kerstiing, A B, 1558CD
Ketchum, R J, 1010CD
Keyes, W J, 962CD
Kharas, B G, 666CD
Kidd, P, 554CD
Kidder, L H, 1516CD
Kido, T, 1008CD
Kiely, C J, 470CD, 1124CD, 1390CD
Kikkawa, M, 210
Killius, J, 1024CD
Kim, B, 1134CD
Kim, G H, 610CD
Kim, H S, 1416CD
Kim, J K, 1278CD
Kim, J S, 610CD
Kim, K H, 280
Kim, M J, 1116CD
Kim, Y W, 1096CD
Kim, Y S, 1084CD
Kimura, Y, 482CD
King, L, 82
Kingon, A I, 66
Kiritani, M, 1448CD
Kirk, M A, 622CD
Kirk, T B, 902CD
Kishio, K, 514CD, 526CD
Kisielowski, C, 466CD
Kitazawa, K, 514CD, 522CD, 524CD, 526CD
Klamut, P W, 582CD
Klansky, J, 348
Klaus, A V, 182
Klein, K, 774CD
Klein, M, 220
Kleinschmidt, J R, 212
Klenov, D, 66
Klie, R, 18, 1396CD, 578CD, 1380CD
Kliewer, C E, 1408CD
Klose, F, 374
Kneissl, A C, 1320CD
Knop-Gericke, A, 602CD
Knowles, D W, 1044CD
Knuutila, D, 708CD
Kocsis, E, 882CD
Koerten, H, 278
Koga, D, 652CD
Kohen, E, 950CD
Kohlstedt, D L, 1606CD
Kohn, J, 1022CD
Koinuma, H, 524CD
Kolb, U, 104
Kolios, M, 1028CD
Koljenovic, S, 444
Kolodziejczyk, E, 236
Kommichau, G, 84
Kondo, Y, 414, 1106CD, 1176CD
Kong, X, 952CD
Koo, H H, 610CD
Kosel, T H, 1214CD
Kotera, M, 1172CD
Kothleitner, G, 1522CD, 1592CD
Kotula, P G, 440, 562CD, 648CD, 1144CD, 1554CD, 1616CD
Koutrakis, P, 1500CD
Krahl, D, 612CD
Krautgartner, W D, 220
Krishnan, K M, 72
Kritikos, D G, 802CD
Krivanek, O L, 14, 18, 20, 476CD
Krivda, S, 920CD
Kronenberg, S, 212
Kros, J M, 444
Krsko, P, 1022CD
Kuchar, M, 922CD
Kuebel, C, 792CD, 1104CD
Kuhn, R J, 208, 822CD
Kukavica-Ibrulj, I, 730CD
Kular, R, 148
Kumao, A, 1450CD
Kumar, D, 384
Kundmann, M, 592CD
Kung, E, 56
Kunicki, T C, 1094CD
Kunisu, M, 606CD
Kuo, K H, 96
Kuo, M X, 268
Kuroda, Y, 48
Kurokawa, H, 550CD
Kutschej, K, 1320CD
Kuyucak, S, 1290CD
Kvit, A V, 1110CD, 1140CD
Kwon, U, 368
Kyrsta, S, 1488CD

L

Laabs, F C, 694CD
Labat, K B, 824CD
Laberge, S, 1084CD
Lagacé, M, 704CD, 1304CD, 1306CD, 1388CD, 1472CD
Lahkak, N, 712CD
Laiho, L H, 280, 1064CD
Lakis, R E, 1476CD
Lalik, M, 360
Lametschwandtner, A, 220
Lamond, A, 266
Landis, W J, 778CD, 1024CD
Lang, G, 486CD, 586CD, 594CD
Lang, J, 1564CD
Langa, S, 1242CD
Langford, R, 46
Lapek, L, 1118CD
Larabell, C, 826CD
Larsen, M, 1454CD
Lartey, R, 972CD
Lasek, S, 1042CD
Lau, J, 1344CD
Laurin, D, 508CD
Lavoie, L, 952CD
Lawrence, J, 1500CD
Lazarov, V, 1158CD
Leapman, R D, 882CD
Lee, E, 176, 1516CD
Lee, J G, 406
Lee, R J, 802CD
Lee, S F, 1370CD
Lee, S P, 268
Lee, S R, 1188CD
Lee, T C, 564CD
Lee, W E, 336
Leeson, D T, 1062CD
Lefebvre, J, 712CD
LeGros, M A, 826CD
Lehmann, M, 34, 536CD, 538CD
Lei, Y, 1114CD
Lentz, H, 1564CD
Lentz, H P, 802CD
Lentzen, M, 8, 466CD, 468CD
Lenz, D R, 1094CD
Leonardi, L, 1510CD
Leopold, R, 1014CD
Lersch, T, 1568CD
Letarte, M, 890CD
Letofsky-Papst, I, 618CD
Leunissen, J, 126
Levi, D H, 1196CD
Levi-Kalisman, Y, 164
Levin, I, 1178CD
Levin, I W, 1514CD
LeVine, S M, 1502CD
Levinsen, M T, 1426CD
Lewis, E N, 1516CD
Li, B Q, 658CD, 1122CD
Li, C, 1598CD
Li, D X, 1342CD, 1410CD, 1596CD
Li, F H, 308, 392
Li, H, 1420CD
Li, J, 52, 548CD, 664CD, 1264CD, 1598CD
Li, L, 318, 320, 1120CD
Li, P, 1430CD
Li, R W, 396
Li, X, 100, 1258CD
Li, X Z, 96
Li, Y, 308, 946CD
Lian, G D, 1146CD, 1170CD, 1190CD, 1620CD
Lian, J, 1136CD, 1424CD
Liang, L, 184
Libera, M, 284, 1022CD
Lichte, H, 22, 34, 536CD, 538CD
Lifshin, E, 430, 432, 1474CD, 1482CD, 1484CD, 1498CD
Liliental-Weber, Z, 1198CD
Lin, A C, 1026CD, 776CD
Lin, B L, 268
Lin, B Y, 912CD
Lin, D J, 268
Lin, P C, 1048CD
Lin, Z, 1378CD
Lindberg, S, 786CD
Liou, S H, 378, 1374CD
Lisowski, W, 1438CD
Litwinenko, J W, 244
Liu, D, 640CD
Liu, H L, 268
Liu, J, 18, 706CD, 1600CD
Liu, J P, 1354CD
Liu, J Q, 1180CD
Liu, K Z, 1510CD
Liu, L, 814CD
Liu, R J, 1392CD
Liu, S, 1452CD
Liu, T M, 268
Liu, W, 1030CD, 832CD
Liu, Y, 306, 402, 1354CD, 1356CD, 1358CD, 1368CD, 1374CD, 1440CD, 1452CD
Liu, Z, 854CD, 1348CD, 1418CD
Lloyd, D J, 794CD
Lloyd, S J, 554CD
Lockbridge, O, 940CD
Lockett, S, 1040CD
Lockley, A, 1288CD, 1308CD
Lodder, C, 1346CD
Loepfe, E, 986CD
Long, C, 860CD

Long, J W, 1240CD
Loong, C A, 1324CD
Lopatin, S, 1604CD
Lopes, A B, 1352CD
Lowe, S, 728CD
Lowther, S, 1556CD
Lubag Jr, A J M, 766CD
Lucadamo, G, 1166CD, 1402CD
Lucas, E M, 1240CD
Lucas, G, 1302CD
Lucas, J R, 1090CD
Lucassen, G W, 288
Ludtke, S, 216, 1576CD
Lumpkin, G R, 1420CD
Luo, R, 200
Luo, Z P, 1254CD
Lupini, A R, 16, 476CD
Lupu, N, 370
Luther, E, 744CD
Lyman, C E, 1150CD
Lynn, D, 830CD, 962CD
Lyons, A, 920CD
Lyubchenko, Y, 170

M

Ma, C, 1608CD
Ma, J, 1336CD
Ma, T, 774CD
Ma, Y R, 1370CD, 988CD
MacDonald, G, 1058CD
Magny, P, 712CD
Mahmoud, I, 888CD
Mai, T, 720CD
Maier, J, 1164CD
Majorovits, E, 540CD, 864CD
Makita, M A, 756CD
Maklakov, S A, 1372CD
Malac, M, 1344CD
Malecki, M, 128
Maleeff, B E, 138, 984CD
Malek, A, 1408CD
Malis, T, 188, 1618CD
Malm, J O, 1394CD
Mammen, M, 920CD
Mancilla Tolama, J E, 624CD
Mancini, D C, 298
Mannella, C, 878CD
Manov, V, 1300CD
Mantsch, H H, 1504CD, 1510CD
Marangoni, A G, 242, 244
Margarine, F, 1038CD
Margolin, A, 1128CD
Margosan, D A, 1076CD
Maria, J-P, 66
Marinenko, R B, 438, 1490CD
Marioara, C D, 1444CD
Marko, M, 878CD
Marks, L D, 660CD
Marquis, E A, 1100CD
Martens, R L, 1094CD
Martin, P E M, 274
Martin, R, 1020CD
Martin, S P, 622CD
Martinez, C, 964CD
Martinez, N, 500CD, 502CD
Maser, J, 298
Massover, W H, 862CD
Masuko, J, 530CD
Matsuda, T, 514CD, 524CD, 526CD
Matsuhata, H, 92
Matsui, Y, 388
Matsumaru, K, 334
Matsushita, M, 1106CD
Matthews, G, 174
Matveeva, G, 104
Mäurer, G, 84
Maurice, J L, 574CD
Maurizi, M R, 204
Maxwell, M, 582CD
Maxwell, R J, 136
Mayer, D, 566CD
Mayer, R, 602CD
Mayo, J A, 478CD
Mayville, R, 1308CD
Mazumder, J, 1440CD, 1452CD
Mazzulla, M, 984CD
McAlduff, D, 874CD
McCabe, R J, 1382CD
McCallum, J, 1182CD
McCammon, D, 640CD
McCartney, M, 40
McCartney, M R, 168
McClean, R, 752CD
McClure, H, 980CD
McComb, D W, 62
McCrae, K C, 1504CD
McCready, D E, 1160CD
McDonald, K, 144, 880CD
McGaw, I, 230
McKee, M D, 166
McKelvy, M J, 796CD
McKernan, S, 302
McKinley, J, 790CD, 1192CD
McKinley, J M, 542CD
McLaughlin, R W, 952CD
McMahon, G, 1212CD
McMahon, P J, 534CD
McNally, J, 1040CD
Meckenstock, R, 1340CD
Medlin, D L, 1166C, 1400CD D, 1402CD
Meek, W D, 1010CD
Meirelles, M N L, 1570CD
Meirelles, R M S, 1570CD
Meléndez, W, 1552CD
Mendes, C L S, 1570CD
Menon, N, 620CD
Menon, S K, 1432CD
Merlet, C, 428
Mershon, W, 808CD
Merzbacher, C I, 1240CD
Messing, G L, 1244CD
Meyer, D, 124
Meyer, S, 870CD
Michael, J R, 1144CD, 562CD, 724CD
Michaels, C A, 1530CD
Michaels, J N, 784CD
Michel, A, 504CD
Michel, M, 236
Micheva, K, 160
Midgley, P, 42, 518CD
Mikula, R, 804CD
Miller, A F, 1226CD, 1228CD
Miller, D J, 418
Miller, K, 902CD
Miller, L, 1512CD
Miller, L M, 1508CD
Miller, M K, 1126CD
Milligan, R, 218
Mills, R M, 556CD
Millunchick, J M, 1102CD
Milne, J L S, 214
Mims, C W, 252
Min, K H, 416
Minamino, Y, 676CD
Minnich, B, 220
Mirabile, R C, 138
Mirecki Millunchick, J, 1222CD
Mirkin, C A, 1020CD
Misra, A, 1382CD
Misra, M, 284, 286
Mitchell, T E, 1382CD
Mitra, R, 1412CD
Mitro, R J, 552CD
Mitsuishi, K, 410
Mizoguchi, T, 606CD
Mkhoyan, A, 570CD, 628CD
Mo, Y D, 308
Mo, Z M, 96
Möck, P, 1114CD
Moebus, G, 10
Mohri, K, 414
Moller, PC, 910CD
Mondy, W, 1086CD
Monteiro-Leal, L H, 506CD
Monti, M, 950CD
Monticello, T, 738CD
Mooers, C T, 1312CD
Moonen, A, 588CD
Moonen, D, 630CD
Mooney, P E, 630CD
Moore, M, 60
Morgan, C, 1584CD
Morgan, S W, 446
Mori, H, 406, 1418CD
Mori, N, 488CD
Moriya, N, 524CD
Mosher, C L, 1142CD
Moss, B, 882CD
Motomiya, K, 1152CD
Mount, R J, 890CD
Mueller, F M, 1380CD
Mueller, R L, 142
Mueller-Reichert, T, 880CD
Mukai, M, 68
Mukhopadhyay, M, 208
Müller, D, 892CD
Muller, D A, 576CD, 1614CD
Müller, H, 12
Müller, M, 986CD
Mullins, J, 1054CD
Munoz, T E, 156
Munoz, V, 804CD
Murfitt, M, 20
Murray, C E, 672CD
Muscat, C, 1544CD
Musselman, I H, 766CD
Myers, A, 1194CD

N

Nag, N, 814CD
Nagayama, K, 864CD
Nagy, J I, 130
Naidoo, G, 1082CD
Naidoo, Y, 1082CD
Naitoh, Y, 414
Nakanishi, N, 492CD
Nakayama, Y, 514CD, 526CD
Nam, S W, 82
Nangia, S, 1414CD
Nanko, M, 334
Narayan, J, 1604CD
Naruse, M, 68, 1106CD
Nason, L, 1576CD
Nast, C C, 914CD
Neff, M, 346
Nellist, P D, 16, 18, 20, 476CD
Nelms, K, 640CD
Nelson, B P, 1196CD
Nelson, C, 1356CD, 1360CD, 1376CD
Nelson, D, 1014CD
Nesper, J, 836CD
Neudeck, P G, 1180CD
Newbury, D E, 82, 434, 1464CD
Newcomb, W, 844CD
Newsholme, S J, 138
Nguyen, L, 1346CD
Nibert, M L, 846CD
Nicklee, T, 908CD
Nicoletti, E S M, 1280CD
Nicolosi, J, 80, 642CD, 714CD
Nielsen, C, 488CD
Ning, G, 734CD
Ning, Z H, 1088CD
Nishio, K, 1450CD
Nishioka, H, 842CD
Nittler, L R, 1550CD
Niu, D, 66
Nockholds, C E, 1496CD
Noebe, R D, 1098CD
Noel, P, 1306CD, 1388CD
Noheda, B, 664CD
Nolan, T, 478CD, 1564CD
Norcum, M T, 824CD
Nordhausen, R, 150
Norén, L, 306
Norton, A S, 920CD
Noseworthy, M D, 992CD
Nowell, M M, 682CD
Nugent, K A, 534CD
Nxumalo, J, 1212CD

O

O'Connell, M, 1074CD
O'Donnell, M E, 948CD
O'Grady, M, 1550CD
O'Keefe, M, 480CD, 1198CD
O'Toole, E, 880CD
Obenauer-Kutner, L J, 140
Ochoa, O, 1254CD
Ogata, Y, 88
Oh, A, 200
Ohnishi, T, 48
Ohnuma, S, 1348CD
Ohsaki, M, 1106CD
Ohshima, Y, 414
Ohtomo, A, 576CD
Ohyagi, M, 334
Oikawa, T, 842CD
Okada, Y, 210
Okayasu, S, 526CD
Okerstrom, S, 782CD
Okunishi, E, 1176CD
Olsen, A, 1586CD
Olson, J D, 1094CD
Oltman, E X, 1094CD
Onaka, S, 1434CD
Orchowski, A, 484CD, 486CD, 586CD, 612CD
Ornek, C, 950CD
Orr, B G, 1222CD
Orrantia, E, 756CD
Orue, E M, 1330CD
Orwa, J O, 1182CD
Osakabe, N, 522CD, 530CD
Oshel, P, 132
Otero-Díaz, L C, 310
Othon, M A, 684CD, 698CD
Otten, M, 792CD, 1104CD
Ottensmeyer, P, 200, 632CD, 634CD, 848CD, 874CD
Ottenwaelter, C, 114
Otto, C, 1386CD
Ouellet, M, 248
Ouellette, G B, 254
Owen, H, 754CD
Owen, N, 296

P

P'ng, K M Y, 554CD
Pacaud, J, 504CD
Pai, E, 848CD
Paige, M F, 776CD
Pailloux, F, 504CD

Pailloux, F D R, 574CD
Palatini, D J, 286, 1528CD
Palmstrøm, C J, 302
Pan, X Q, 1112CD, 1154CD, 1162CD, 1168CD, 1364CD
Pan, Z, 1288CD
Panessa-Warren, B, 726CD
Panglre, S, 1194CD
Pantel, R, 1174CD
Papworth, G C, 290
Paquet, M, 968CD
Paransky, E, 1272CD
Parikh, D V, 806CD
Park, C, 638CD
Parsons, G N, 66
Patchett, B M, 1252CD
Paterson, A, 596CD
Paterson, A D, 890CD
Patterson, R J, 566CD, 1266CD
Pawley, J, 1050CD
Pearson, C A, 1222CD
Pechkis, D, 774CD
Peck, M, 176
Peden, C H F, 1112CD, 1160CD
Pegg, I L, 1312CD
Peijper, R, 126
Pelzl, J, 1340CD
Pendleton, M W, 1070CD
Peng, J, 1182CD
Penninger, J, 634CD
Pennycook, S J, 16, 384, 476CD, 1218CD, 1220CD
Pereira, M D S, 1328CD, 1330CD
Perez-Berenguer, J, 930CD
Perham, N, 214
Perov, N S, 1372CD
Perovic, A, 1442CD
Perovic, D D, 1442CD
Perrey, C R, 1144CD
Perry, K L, 852CD
Petford-Long, A K, 46, 296
Petrali, J P, 736CD, 900CD
Petrov, I, 658CD
Pettersson, N, 1394CD
Pham, K, 812CD
Pham, N A, 898CD
Phaneuf, M W, 52, 188, 338, 566CD, 568CD, 1212CD, 1266CD
Phillipp, F, 1202CD, 1204CD
Phillips, M N, 234
Phillips, M R, 446, 1478CD, 1496CD, 1538CD
Phillips, R, 922CD
Pierce, S K, 1086CD
Pietron, J J, 1240CD
Pilgram, G, 278
Pint, B A, 1492CD
Pitre, F, 248
Pitts, O J, 1200CD
Platani, M, 266
Pletnev, S V, 208
Pleva, C M, 736CD, 900CD
Plitzko, J M, 1610CD, 870CD
Ponce-Camacho, M A, 928CD
Pooley, C, 722CD
Poorhaydari-A, K, 1252CD
Popovitz-Biro, R, 1128CD
Portella, P D, 1280CD
Potaman, V, 170
Potter, C, 218
Potter, C S, 866CD, 872CD, 1580CD
Powell, J A, 1180CD
Powell, R D, 1030CD, 832CD, 916CD
Pozzi, G, 32, 528CD, 530CD
Pralle, M U, 318
Prasad, M S, 1486CD
Prawer, S, 1182CD
Prenitzer, B, 790CD
Price, R, 1034CD
Prikhodko, S V, 1130CD
Prodan, A, 312
Prokofjef, S, 1426CD
Pugh, M D, 708CD
Pulido-Mendez, M, 926CD
Pulokas, J, 218, 866CD, 872CD
Puppels, G J, 288, 444
Putatunda, S K, 1314CD

Q

Qiang, Y, 1354CD
Querido, E, 728CD

R

Ra, H S, 1078CD
Radetic, T, 1404CD, 1426CD
Ragan, R, 1114CD
Raible, D, 1058CD
Rainey, J K, 776CD
Rajadhyaksha, M, 282
Rajsiri, S, 50
Ramírez-Bon, E, 928CD
Rammohan, J, 1018CD
Ranaware, Y, 1332CD
Randle, V, 108
Rango, A, 722CD
Rash, J E, 130
Rasmussen, B, 1256CD
Ratna, B R, 1116CD
Rau, W-D, 36, 486CD, 586CD
Ravel-Chapuis, R, 670CD
Ravishankar, N, 562CD
Ray, D A, 588CD
Raz, S, 164
Recnik, A, 492CD, 494CD
Redfern, D, 80, 642CD, 714CD
Reed, B, 598CD
Reese, T S, 882CD
Reeves, J L, 1484CD
Reffner, J A, 1526CD
Regand, A, 246
Reibold, M, 538CD
Rémond, G, 1496CD
Reno, J L, 1188CD
Renzaglia, K S, 1090CD
Retterer, S, 1118CD
Revie, W, 1284CD
Rez, P, 746CD
Rhodes, C, 1390CD
Richards, T, 1118CD
Richardson, E A, 252
Richardson, G, 956CD
Richardson, T, 898CD
Richter, S, 1488CD
Ridner, C W, 224
Rigaud, M, 1272CD
Ringer, S P, 294
Ringnalda, J, 626CD
Ris, H, 838CD
Risner, J, 368
Robbins, J L, 1408CD
Roberson, S V, 1520CD
Roberston, K, 1536CD
Robertson, D, 752CD
Robertson, D P, 1278CD
Robertson, K, 452
Robertson, R P, 1566CD
Robinson, S J, 1274CD
Roche, A D, 1276CD
Rocks, L, 640CD
Rodbell, K P, 672CD
Rodrigue, L, 1304CD, 1472CD
Rodriguez-Acosta, A, 926CD
Rodriguez-Sierra, J, 1012CD
Rodriguez-Uribe, L, 1074CD
Roggli, Y L, 918CD
Rohatgi, P K, 1278CD
Rolison, D R, 1240CD
Rolland, P, 670CD, 686CD, 688CD
Rom, I, 618CD
Root, J H, 342
Roques-Carmes, C, 1496CD
Rosa-Molinar, E, 132
Rose, H, 6
Roseman, M, 1338CD
Rosenberg, H, 936CD
Rosentsveig, R, 1128CD
Ross, F M, 420
Rossie, B, 542CD, 556CD, 1192CD
Rossignol, R, 1006CD
Rossmann, M G, 208, 822CD
Rothbard, D R, 178
Roysam, B, 1042CD
Rubakin, S, 764CD
Rubel, E, 1058CD
Rueger, D, 778CD
Rühle, M, 1164CD, 1204CD
Ruiz, P, 1500CD
Rulis, P, 606CD
Ruoff, R S, 304
Russell, A M, 694CD
Russell, P E, 54, 558CD, 770CD, 780CD, 1206CD, 1208CD
Rytter, E, 1150CD
Ryzhikov, I A, 1372CD

S

Sablin, E, 210
Sacchettini, J, 840CD
Safa-Sefat, A, 580CD
Sage, D, 340
Saitoh, K, 68
Saka, H, 38
Sakamoto, I, 1366CD
Sakata, T, 1418CD
Salanga, M, 998CD
Saleta, J L, 990CD
Salmon, M E, 770CD
Salvador, M D, 500CD, 502CD
Salzer, R, 1504CD
Samarth, N, 1620CD
Samet, L, 574CD
Sandborg, A, 1468CD, 1612CD
Sanders, M, 158
Sanders, W, 640CD
Sands, S S, 1010CD
Santa-Rita, R, 1570CD
Santamaria, J, 384
Santeufemio, C, 1138CD
Sanwald, R S, 1140CD
Sarazin, P, 952CD
Sarikaya, M, 598CD, 746CD
Sasaki, K, 38
Sasase, M, 526CD
Sata, N, 1164CD
Sato, T, 1152CD
Satomi, J, 890CD
Sattin, B, 982CD
Savva, C, 840CD
Sayar, M, 318
Schaaff, T G, 1148CD
Schaerble, M D, 1514CD
Schalek, R, 1232CD
Schaper, J, 1184CD
Schatten, H, 838CD, 1056CD
Schattka, B, 1510CD
Schattschneider, P, 636CD, 1594CD
Scheu, C, 1164CD
Schlaegle, S, 1568CD
Schlögl, R1592CD, 602CD, 1590CD,
Schlom, D G, 1162CD
Schmalstieg, F C, 910CD
Schmid, S, 904CD
Schneider, D, 896CD
Schneider, T, 948CD
Schofield, M, 752CD, 1344CD
Schofield, M A, 512CD, 532CD
Scholz, F, 1202CD
Schönjahn, C, 718CD
Schouten, A, 250
Schraner, E M, 986CD
Schreiber, K C, 1526CD
Schroeder, R R, 540CD, 820CD, 856CD, 858CD, 864CD
Schülein, T, 84
Schultz, C P, 1512CD
Schumann, M, 546CD
Schwappach, C, 180
Schwartz, L W, 138
Schwarz, S, 50, 656CD
Scott, J H J, 82, 646CD
Sears, S K, 166
Sedova, M V, 1372CD
Seidman, D N, 1096CD, 1098CD, 1100CD
Selker, J M L, 1006CD
Sellar, J R, 1216CD
Sellmyer, D, 366, 1354CD, 1356CD, 1358CD, 1368CD
Senger, C, 936CD
Sengupta, J, 206
Serquis, A C, 1380CD
Serrano-Vélez, J L, 132
Serwer, P, 974CD
Shaffer, K, 772CD
Shain, W, 1118CD
Shang, P, 296
Sharma, A C, 320
Sharma, R, 422, 600CD, 796CD
Sharma, S, 840CD
Shaw, R A, 1504CD
Shawon, J, 1230CD
Shechtman, D, 314
Shehata, M, 1284CD
Shehata, M T, 1282CD, 1324CD
Shen, B G, 396
Shen, R B, 1468CD
Shepard, J D, 1094CD
Shepelev, L, 1300CD
Sheridan, R E, 940CD
Sheybany, S, 1296CD
Shi, D, 214
Shi, S, 284
Shi, Y F, 658CD, 1122CD
Shield, J E, 1334CD
Shimizu, M, 842CD
Shimizu, T, 1366CD
Shimoda, K, 910CD
Shimoyama, J, 514CD, 526CD
Shindo, D, 398, 1348CD
Shiojiri, M, 492CD, 494CD
Shirakashi, J, 1350CD
Shirolski, H, 914CD
Shlyakhtenko, L, 170
Shutthanandan, V, 1160CD
Sides, W H, 478CD
Sidorov, M, 480CD, 560CD, 1572CD
Siedlecki, C, 260
Siew, S, 906CD
Silcox, J, 2, 472CD, 570CD, 604CD, 628CD

Silva, R F, 1352CD
Simard, M, 254
Simensen, C, 1586CD
Simmnacher, B, 80
Simmons, J P, 1432CD
Simon, M N, 912CD
Simoneau, M, 1388CD
Sinclair, R, 368, 416
Sinden, R, 170
Sinkler, W, 660CD
Siochi, E J, 638CD
Siperko, L, 778CD
Sirard, M-A, 924CD
Sitte, W, 618CD
Skomski, R, 1358CD
Skowronski, M, 1180CD
Skumryev, V, 1360CD
Small, J, 82
Smith, B A, 1000CD
Smith, C, 936CD
Smith, D J, 382
Smith, G D, 1524CD, 1532CD
Smith, K L, 1420CD
Smith, P J, 274
Smith, R, 190, 326
Smith, S J, 160
Smith, T J, 852CD
Smolinski, D, 902CD
Snugovsky, P, 1286CD
Snyder, K, 1006CD
So, P T, 280, 1064CD
Sobel, B, 896CD
Soboyejo, W, 56
Solazzi, M, 642CD
Solf, M, 250
Solórzano, I G, 1280CD
Sommer, C, 752CD
Song, M, 1418CD
Sosa, L, 934CD
Sosinsky, G, 876CD
Sovak, G, 748CD
Sowa, M G, 1510CD
Spanos, J, 1302CD
Spector, D, 728CD
Spence, A J, 1118CD
Spence, J C H, 650CD
Spencer, K, 1046CD
Spencer, M, 1118CD
Speransky, V V, 1002CD
Spiegel, C N, 1570CD
Spoddig, D, 1340CD
Sporn, T A, 918CD
Spornitz, U M, 228, 886CD, 932CD
Spring, H, 506CD
St. Romain, E, 478CD
Stark, T J, 54
Statham, P J, 1466CD
Staun, C, 1564CD
Stéa, D, 952CD
Stearns, D, 998CD
Stearns, R, 1500CD
Steel, E B, 816CD, 1490CD, 1494CD
Steele, D C, 794CD
Steele, J, 1458CD
Steele, J H, 1248CD, 1260CD, 1262CD, 1268CD
Steen, N, 1562CD
Stein, A, 330
Steiner, G, 1504CD
Stemmer, S, 66
Stephan, O, 1384CD
Stephens, T C, 1070CD
Stephenson, G B, 298
Steven, A C, 204, 844CD, 1002CD
Stevens-Kalceff, M A, 1182CD, 1242CD, 1534CD
Stevie, F A, 542CD, 556CD, 558CD, 1192CD, 1490CD
Still, D, 174
Stinebaugh, W, 790CD
Stoessel, S, 772CD
Stoker, T, 272
Stokes, D J, 958CD, 960CD
Stokes, D L, 1060CD
Stone, K J, 1066CD
Stöttinger, B, 220
Stout, R, 954CD
Strait, D R, 1094CD
Stranick, S J, 1530CD
Strauss, M, 874CD
Street, S C, 1120CD
Strennen, E M, 1094CD
Stroud, R M, 1240CD, 1550CD
Stupp, S I, 318
Sturm, J, 1210CD
Su, D H I, 564CD
Su, D S, 602CD, 1592CD
Subramaniam, S, 214
Subramanian, P R, 1454CD
Sudar, D, 1044CD
Sudbrack, C K, 1098CD
Sue, J, 1254CD
Suh, K S, 1002CD
Suh, Y J, 1130CD
Sui, M L, 1410CD, 1596CD
Sukedai, E, 1446CD, 1448CD
Sullivan, N, 720CD
Suloway, C, 218, 872CD
Sun, C K, 268
Sun, H P, 1112CD, 1162CD, 1168CD
Sun, J, 840CD
Sun, K, 18, 814CD
Sun, L, 1584CD
Sun, S, 364
Sun, W, 920CD, 1224CD
Sung, C, 758CD, 1230CD, 1258CD
Sunkara, M K, 1146CD
Superfine, R, 174
Suvorova, A S, 450
Suzuki, T, 30, 1176CD
Suzuki, Y, 842CD
Swatland, H J, 240
Swedlow, J, 266
Sweedler, J, 764CD
Swider Lyons, K E, 1240CD
Szarowski, D, 1042CD
Szarowski, D H, 1118CD
Szilagyi, Z, 20

T

Taatjes, D, 896CD
Tafto, J, 86, 386, 1586CD
Tague, T J, 1512CD
Takahashi, H, 488CD
Takai, Y, 474CD, 482CD
Takakura, M, 488CD
Takayangi, K, 414
Takeguchi, M, 410
Takemura, Y, 1350CD
Tamura, T, 1434CD
Tan, C M, 788CD
Tan, P, 640CD
Tanaka, I, 606CD
Tanaka, M, 68, 88, 410
Tanaka, Y, 652CD
Tang, H, 1356CD
Tang, X, 1540CD
Tangen, I L, 1428CD
Taniguchi, Y, 590CD
Tanimura, J, 550CD
Tanishiro, Y, 414
Tanji, T, 24, 30
Tao, J, 658CD
Tao, X, 690CD, 692CD
Taraschi, T F, 948CD
Tardos, G, 784CD
Taylor, G, 256
Taylor, J V, 962CD
Taylor, R M, 174
Teetsov, J, 772CD
Tenberge, K B, 250
Tenc, M, 574CD
Tenne, R, 1128CD
Terauchi, M, 68
Terauci, M, 644CD
terBrugge, K G, 890CD
Tessier, J, 758CD
Tew, G N, 318
Thesen, A, 28, 516CD, 520CD
Thevuthasan, S, 1112CD, 1160CD
Thiel, B L, 448, 960CD, 1478CD, 958CD
Thirumal, M, 1414CD
Thomas, E L, 322
Thomas, H C, 138
Thomas, P, 592CD
Thomas, P J, 372
Thomas, S, 920CD
Thorne, B B, 646CD
Thornton, J T, 742CD
Tian, W, 1162CD, 1364CD
Tiemeijer, P, 480CD
Tiemeijer, P C, 70, 1592CD
Tiginyanu, I M, 1242CD
Tilups, A, 992CD
Tiner, M J, 544CD, 678CD
Tinling, S P, 148
Titchmarsh, J M, 10
Tivol, W F, 1118CD
Todd, B A, 262
Tohji, K, 1152CD
Tomita, T, 68
Tomokiyo, Y, 652CD, 1588CD
Toms, A M, 802CD
Tonino, P, 934CD
Tonomura, A, 26, 380, 514CD, 522CD, 524CD, 526CD, 528CD, 530CD
Topuria, T, 1114CD
Tortora, G, 726CD
Toth, M, 448, 1478CD
Toyoda, T, 676CD
Traber, D L, 910CD
Traber, L D, 910CD
Traktman, P, 734CD
Trasobares, S, 1384CD
Tremblay, P, 952CD
Trendelenburg, M, 506CD
Trevor, C, 372, 588CD, 592CD, 630CD
Tripp, S L, 1134CD
Tromp, R M, 420
Tröster, H, 506CD
Troughton Jr., E B, 770CD
Trudeau, M, 1304CD
Trunz, M, 484CD
Trus, B, 844CD
Tsai, A P, 1218CD, 1220CD
Tsai, J L, 1370CD
Tsen, SC Y, 1108CD
Tsuda, K, 68, 88
Tsuji, N, 676CD
Tsung, L, 626CD
Tsuno, K, 68
Tubbs, R R, 916CD
Tuggle, D W, 1142CD
Turner, J, 1042CD, 1426CD
Turner, J N, 1118CD
Turner, S, 1490CD, 668CD
Tweddell, R, 964CD
Twesten, R, 622CD, 658CD
Twitchett, A, 42, 518CD
Tyler, T, 1110CD, 1140CD
Typke, D, 870CD
Tyson, W R, 548CD

U

Ueji, R, 676CD
Ueki, Y, 590CD
Ueno, M, 1146CD
Uhlemann, S, 12, 584CD
Ulfig, R M, 1094CD
Umemura, K, 48
Une, T, 1366CD
Urban, J M, 946CD
Urban, K, 8, 466CD, 468CD
Urbas, A M, 322
Urbina, C, 1552CD
Utsunomiya, S, 1558CD

V

Vaillancourt, J, 640CD
Valero, N, 500CD, 502CD
Vali, H, 166, 952CD
Valle, M, 206
van Aert, S, 94
van Apeldoorn, A A, 1386CD
van Balen, A, 792CD, 872CD
van Blitterswijk, C A, 1386CD
van Cappellen, E, 1612CD
van den Berge, H J, 444
van den Bos, A, 94
van Driel, R R, 534CD
van Dyck, D, 94, 662CD
van Lin, J H A, 70
van Midden, H J P, 312
van Rij, A M, 234
Vane, R, 720CD
Vangsness, M, 798CD
VanVianen, A, 1184CD
Varela, M, 384
Vartuli, C B, 1192CD
Vastenhout, J S, 324, 1238CD
Vaughan, J B, 894CD
Vaughn, D, 920CD
Vereecken, P M, 420
Verry, P, 932CD
Vicci, L, 174
Viera, J M, 1352CD
Vigo, T, 806CD
Vikas, S, 1584CD
Viterelli, J P, 54
Vogel, V, 1092CD
Vogen, W, 478CD
Volkov, V V, 510CD, 512CD, 1344CD
von den Driesch, M, 250
von Harrach, H S, 1612CD
Voorhout, W, 792CD
Voorhout, W F, 1104CD
Vossen, O, 858CD
Voyles, P M, 1614CD

W

Wadsworth, M, 896CD
Wagenknecht, T, 854CD
Wagner, G J, 304
Walden, D B, 1068CD, 1072CD, 1080CD
Walker, L R, 1492CD

Walker, S B, 846CD
Wall, F D, 1616CD
Wall, J, 490CD
Wall, J S, 912CD
Wallace, M, 890CD
Wallenberg, L R, 1394CD
Walmsley, J, 1436CD
Walther, P, 986CD
Wan, Z H, 308
Wang, C, 1160CD
Wang, H, 1378CD, 762CD
Wang, H B, 392
Wang, J, 54
Wang, L M, 1136CD, 1424CD, 1558CD
Wang, Q, 1170CD
Wang, Y G, 296
Wang, Z, 38
Wang, Z H, 396
Wang, Z L, 300, 328, 364, 1602CD, 1608CD
Warburton, W, 76
Ward, M S, 320
Warner, R R, 1066CD
Warren, J, 726CD
Was, G, 1422CD
Washburn, J, 1198CD
Watanabe, K, 492CD, 494CD
Watanabe, M, 292, 1588CD, 1598CD
Watkins, S, 1200CD
Watkins, S C, 290
Wauchope, C, 1102CD, 1222CD
Weaver, L, 566CD
Webber, R, 360
Weertman, J, 1412CD
Wei, A, 1134CD
Weigle, C, 174
Weiland, R, 80
Weiner, S, 164
Weisberg, A, 882CD
Weiss, I M, 164
Weller, D, 1376CD
Wells, O C, 106
Wen, C K, 776CD
Wergin, W P, 722CD
Westbrook, E, 930CD
Wetzel, D L, 1502CD
White, G F, 834CD
White, N S, 264
White, T, 1548CD
Whitfield, C, 836CD
Whittle, D, 1390CD
Wiener, S A, 1094CD
Wight, S A, 646CD, 1494CD, 1298CD
Wild, P, 986CD
Wilde, B, 174
Wildman, H, 596CD
Wilhelm, P, 1522CD
Wilks, D K, 1526CD
Williams, D B, 292, 608CD, 1588CD, 1598CD
Williamson, M J, 420
Williard, J N, 810CD
Wilson, D L, 1020CD
Wimmer, E, 822CD
Windsor, E S, 1298CD, 1494CD
Wisher, A, 1556CD
Withers, R L, 306
Wittig, J E, 1336CD
Wong, H, 920CD
Wong, S, 726CD
Woo, J, 908CD
Woodward, J H, 196
Wright, G A, 992CD
Wright, I G, 1492CD
Wright, S I, 678CD, 682CD
Wu, D, 1474CD
Wu, I, 974CD
Wu, J, 1190CD
Wu, J P, 902CD
Wu, L, 86, 386, 664CD
Wu, X, 952CD, 1320CD
Wylie, I, 790CD

X

Xiao, C, 822CD
Xu, F, 1120CD
Xu, S, 548CD
Xu, X, 320
Xu, Y, 180

Y

Yaguchi, T, 48, 1152CD
Yajima, H, 210
Yakubtsov, I, 352
Yamamoto, K, 24, 38
Yamazaki, T, 492CD, 494CD
Yan, H, 330
Yan, Z, 1360CD
Yang, J C, 320, 1120CD, 1406CD
Yang, L, 1352CD
Yang, T, 140
Yang, V C, 942CD, 994CD
Yang, Y, 762CD
Yanke, A, 1024CD
Yao, N, 56, 1210CD
Yao, Y D, 988CD, 1370CD
Yasuda, H, 406, 1418CD
Yasumura, T, 130
Ye, H Q, 1294CD
Yi, H, 126
Ying, S L, 1504CD
Ying, X, 738CD
Yip, C, 200, 258, 458
Yoda, R, 676CD
Yoshida, T, 514CD, 524CD
Yoshida, Y, 550CD
Yoshimitsu, D, 1448CD
Yoshiya, M, 606CD
You, L P, 1558CD
Young, L M, 684CD, 698CD
Young, R, 840CD
Youngblom, J H, 1054CD
Youngblom, J J, 1054CD
Youngman, R A, 1186CD
Yu, C, 1370CD
Yu, H, 1048CD
Yu, M J, 1354CD
Yu, MF, 304
Yu, Y, 332, 1436CD
Yu, Y D, 1428CD
Yu, Z, 472CD, 604CD
Yuan, L, 1374CD
Yue, S, 674CD, 700CD, 710CD
Yuspa, S H, 1002CD

Z

Zaki, S R, 904CD
Zaluzec, N J, 376
Zandbergen, H W, 468CD, 1444CD
Zavadil, R, 1290CD
Zbrzezny, A, 1456CD
Zeissler, C J, 1494CD
Zhan, Q, 1342CD
Zhang, H, 1294CD
Zhang, K, 330
Zhang, L, 1410CD
Zhang, L H, 1410CD
Zhang, M, 234
Zhang, S, 336
Zhang, S L, 286, 1528CD
Zhang, W, 208, 1596CD
Zhang, X, 846CD
Zhang, Y, 1360CD, 1376CD, 1378CD
Zhang, Z, 1052CD
Zhang, Z D, 400
Zhao, F Q, 828CD
Zhao, P, 356, 1558CD
Zhao, X, 1368CD
Zhen, M, 1368CD
Zheng, C Q, 1324CD
Zheng, M, 902CD
Zhirnov, V V, 1110CD, 1140CD
Zhou, G, 1406CD
Zhou, H, 666CD
Zhou, J, 1358CD
Zhou, T, 1322CD
Zhu, J, 1156CD
Zhu, Y, 86, 218, 386, 510CD, 512CD, 532CD, 664CD, 1344CD, 1396CD
Zipper, P, 1522CD
Zou, X D, 96
Zucker, R, 272
Zuo, J, 658CD
Zuo, J M, 1122CD

Subject Index

Note: *Invited papers appear in both the printed Proceedings and in the CD-ROM. The suffix "CD" following a page number indicates a submitted paper; these appear only on the CD-ROM.*

2-photon, 1036CD, 1064CD
2D crystals, 834CD, 836CD, 840CD
2D-dopant profiling, 536CD
3-D sample, 702CD
3-photon, 1036CD
3D reconstruction, 184, 204, 1040CD, 1042CD, 1072CD, 1104CD, 552CD, 844CD, 852CD, 876CD, 880CD
3D structure, 824CD
3D writing, 1048CD
3DAP, 1094CD
3DEM, 200

A

a-factor, 1476CD
Abalone, 656CD
Aberration correction, 2, 4, 6, 8, 10, 12, 14, 20, 22, 418, 468CD, 470CD, 474CD, 476CD, 478CD, 482CD, 612CD
Accumulative roll-bonding, 676CD
Accutom-50, 1256CD
Acetylcholinesterase, 940CD
Acid phosphatase, 1008CD
ACT, 1598CD
Actinides, 1558CD
Acto-Myosin, 820CD
Adeno-Associated Virus, 212
ADF, 1614CD, 472CD, 576CD
Adrenal, 926CD
Adsorbed, 772CD
AEM, 376, 418, 1150CD, 1334CD, 1336CD, 1590CD, 618CD
Aerogel, 1240CD
Aerosols, 1482CD
AFM, 170, 172, 174, 258, 260, 324, 1000CD, 1018CD, 1020CD, 1026CD, 1138CD, 1206CD, 1258CD, 1350CD, 726CD, 742CD, 746CD, 762CD, 764CD, 768CD, 772CD, 774CD, 778CD, 780CD, 782CD, 982CD, 988CD
AFM, 262
Ag-doped, 1342CD
Agglomeration, 784CD
Aging, 1054CD, 1294CD
Air Pollution, 1078CD
Al, 1278CD, 1444CD, 794CD
Al-alloy, 100, 344, 1282CD
ALCHEMI, 1586CD
Alexa stain, 132, 1030CD, 974CD
Alkane Crystalization, 798CD
Alloys, 406, 1300CD, 504CD
AlN, 1428CD, 650CD
Alumia, 1244CD
Amorphous, 50, 1300CD, 608CD
Amphibian, 886CD
Amyloid, 860CD
Angiogenesis, 934CD, 994CD
Angle-resolved Auger Spectroscopy, 596CD
Anisotropy, 1304CD
AOS, 250
Apatite, 1548CD
Apoptosis, 272, 1002CD, 1028CD, 1086CD, 900CD, 998CD
Aquaporin, 1004CD
Arabidopsis, 972CD
ArcView, 1556CD
Art history, 1532CD
Arthroscope, 902CD
Assembly, 1140CD
Atherosclerosis, 894CD, 896CD, 944CD
Atom clusters, 1224CD
Atom Probe, 292, 294, 1094CD, 1096CD, 1098CD, 1100CD, 1126CD
Atom-sized wire, 482CD
Atomic, 1218CD
Atomic imaging, 22, 1218CD, 1444CD, 468CD
Atomic scattering factors, 102
Atomistic modelling, 64
ATP synthase, 1058CD, 892CD, 984CD
Au, 52, 1108CD, 1120CD, 1134CD, 832CD
Au-Sn, 1292CD
Au/Pd, 812CD
Auger, 556CD
Austempering, 1314CD
Austenite, 1262CD, 1316CD
Autofluorescence, 280
Automation, 48, 218, 792CD, 802CD, 872CD
Automotive Materials, 176
Axial tomography, 856CD
AZ91D, 1324CD

B

Backscatter Electron Imaging, 116, 1262CD, 782CD
BaF2, 1164CD
Bainite, 1276CD
Ballistic conductance, 300
Band Structure, 436
Basement membrane, 934CD
BaTiO3, 1162CD
Battery Characterization, 1388CD
Bax inhibitor, 248, 970CD
Be, 918CD
Beam current, 670CD
Beam damage, 54, 118
Bee venom, 926CD
Berylliosis, 918CD
Beta Titanium Alloy, 1446CD
Beta-amyloid, 952CD
BGA, 1256CD
Bi supercondor, 1352CD
Bicubic, 488CD
Bio-photonic crystal, 268
Bioactive 3DOM glass, 330
Biological Control, 968CD
Biomaterials, 260, 330
Biomedical, 1504CD, 1508CD
Biomineralization, 164, 746CD
Biopolymer, 996CD
Bioremediation, 750CD
Birds, 226
Bladder, 224
Bloch waves, 16, 92
Block Copolymer, 322
Blood Vesels, 226, 934CD
Boersch phase plate, 540CD
Boiler Deposit, 810CD
Boilers, 1332CD
Bonding, 1322CD, 654CD
Bone, 1024CD, 1512CD, 778CD
Boron, 610CD
Botanical magnetite, 752CD, 754CD
Boundary, 496CD
Brain vascularization, 220
Brazing, 346
Brucite, 796CD
Bug, 1582CD
bZIP, 1074CD

C

c-Jun, 994CD
c. elegans, 880CD
CaF2, 1164CD
Calcium Oxalate, 746CD
Calcium tolerance, 1010CD
Calibration, 792CD
Cancer, 1056CD, 1504CD
Capsid, 844CD, 974CD
Carbide, 1326CD
Carbon fiber, 1254CD
Carbon nanotube, 1142CD, 1152CD, 726CD
Carbon sequestration, 796CD
Carbon-Carbon Composites, 1248CD
Carbonation, 796CD
Cardiovascular, 230
Cartilage, 1024CD
Casimir, 460
Casting, 362
Castings, 1300CD, 1326CD
Catalase, 862CD
Catalyst, 404, 1112CD, 1590CD, 1600CD
CAV21, 822CD
Caveolin, 942CD
Cavity imaging, 708CD
CBED, 88, 90, 92, 1586CD, 1608CD, 652CD, 658CD
CD SEM, 720CD
CdSe, 10
Cells, 144, 172, 1012CD, 1022CD, 1056CD, 876CD, 946CD
Cellulase, 1084CD
Cellulose, 772CD
CeO2, 600CD
Ceramics, 1428CD, 606CD
Cerium chloride, 962CD
Channeling, 1458CD, 636CD, 662CD
Channeling Contrast, 1186CD
Chaperonin, 840CD
Characterization, 358, 592CD
Charge Contrast Imaging, 450, 452, 1536CD
Charge density, 88, 654CD
Charge density waves, 312
Charge distribution, 386, 388
Charging, 118, 448, 1172CD, 1478CD, 1498CD, 1534CD
Chemical imaging, 648CD
Chemical mapping, 592CD
Chemometrics, 1510CD
Chitosan, 996CD
Chloride channel, 1002CD
Chloroplast, 1090CD
Cholera Toxin B-Subunit, 132
Cholesterol, 898CD, 942CD
Chromatic aberration, 12
Chromia, 1390CD
Chromium, 998CD
Chromophore mapping, 634CD
Chromosome, 1052CD
CL, 1182CD, 1242CD, 1272CD, 712CD
Classification, 946CD
Cleavage fracture, 548CD
Cluster, 294, 1126CD, 490CD, 504CD
CMOS, 1184CD
CMR Effect, 396
Co, 364
Co-Zr-O, 1348CD
Coatings, 1440CD, 674CD
Coductive Tissue, 1088CD
Coercivity, 398
Cold stage, 1398CD
Cold-rolling, 1264CD
ColdSpot, 154
Collagen, 1018CD, 1020CD, 1026CD, 1118CD, 776CD
Colloidal crystal, 316
Colloidal Gold, 132

Colloidal metals, 124, 194
Colloids, 320, 1558CD
Colocalization, 274, 1040CD
Column Stability, 484CD
Composite, 336, 1232CD, 1270CD, 1278CD, 1454CD
Composition modulation, 1222CD
Compositional analysis, 462, 1082CD, 1102CD, 1150CD, 1520CD, 556CD
Compositional imaging, 284, 372, 1554CD, 648CD, 966CD
compositional mapping, 440
Computational methods, 824CD, 842CD
Confocal microscopy, 130, 138, 182, 184, 198, 264, 266, 272, 276, 282, 334, 1034CD, 1042CD, 1044CD, 1046CD, 1048CD, 1050CD, 1062CD, 902CD, 966CD
Conformation, 170
Conical dark field, 656CD
Connexins, 130
Connexon, 892CD
Conservation, 1532CD
Contamination, 720CD
Contrast, 112, 282, 1416CD, 854CD, 992CD
Contrast Transfer Function, 1572CD
Controlled Growth, 1142CD
Cooling, 1456CD
CoppeRx, 52
Corpus luteum, 888CD
Corrected Imaging Filter, 586CD
Corrections, 438
Correlative microscopy, 132, 194, 1030CD, 1528CD, 580CD, 962CD
Corrosion, 338, 344, 1282CD, 1318CD, 500CD, 756CD
Corrosion casting, 220, 222, 224, 226, 228, 234, 890CD
Cortical granules, 272
CoSi2, 1184CD
Cotton, 192, 806CD
Covalent, 772CD
Cr/silica catalyst, 1392CD
Crack, 1202CD
Cristae, 1006CD
Critical voltage, 92
Criticality, 1216CD
Cross-section, 1486CD, 546CD
Crustacean, 230
Cryo EM, 204, 208, 210, 214, 216, 218, 1016CD, 1066CD, 1394CD, 1576CD, 1610CD, 786CD, 820CD, 822CD, 846CD, 848CD, 850CD, 852CD, 854CD, 864CD, 866CD, 868CD, 870CD, 986CD, 992CD
Cryptorchid testis, 954CD
Crystal structure, 102, 1372CD, 688CD
Crystallinity, 1254CD
Crystallization, 242, 244, 416, 1272CD
Crystallography, 664CD, 860CD
CSL boundaries, 670CD
CTF correction, 212, 216
Cu, 52, 1156CD, 1194CD, 1406CD, 654CD, 656CD, 708CD
Cu-Sb, 1598CD
Cu3Ga, 52
CuAlNi, 1320CD
CVD, 186, 422, 1196CD
Cytoskeleton, 1056CD, 892CD

D

Dark Field, 1512CD, 790CD
Data analysis, 598CD
Data storage, 296, 1488CD
Database, 1576CD, 802CD
Decalcification, 148
Deconvolution, 266, 272, 1038CD, 1052CD
Defects, 186, 1164CD, 1166CD, 1180CD, 1596CD, 622CD
Deformation, 670CD, 694CD
Degradation, 550CD
Dehydration, 760CD
Dehydroxylation, 796CD
Delayed Hydride Cracking, 1288CD
Delocalisation, 16
Dendrimer, 1120CD
Dengue Mast cells, 920CD
Density functional theory, 64
Dentin, 760CD
Deposition, 56, 568CD
Depth profiling, 1212CD
Depth resolution, 542CD
Dermatology, 282
Detection, 462, 916CD
Dewetting, 562CD
Diabetes, 932CD
Diamond, 186, 1146CD
DIC, 534CD
Dielectric, 448, 460, 1478CD
Diffraction, 98, 142, 376, 662CD
Diffraction Contrast, 690CD, 790CD
Diffraction Theory, 1416CD
Diffuse Scattering, 658CD
Diffusion, 974CD
Digital Imaging, 456, 738CD, 850CD, 868CD, 870CD
Direct Methods, 308, 660CD
Directional Recrystallization, 1264CD
Dislocations, 340, 404, 1160CD, 1162CD, 1166CD, 1198CD, 1382CD, 1402CD, 1416CD, 1422CD, 1430CD, 554CD
Disproportionation, 1436CD
Distance education, 454
Distribution, 368
DMMC, 694CD
DMRB, 750CD
DMSO, 924CD
DNA, 170, 982CD, 988CD
Dopant, 22, 36, 40, 1168CD, 516CD, 718CD
Drift, 866CD
Drosophila, 1044CD
Duane-Hunt limit, 1466CD, 1540CD
Ductile, 1326CD
Dyes, 270
Dynamic, 482CD, 982CD
Dynamical Scattering, 858CD
Dynamin, 850CD

E

EBIC, 1208CD, 1538CD
EBSD, 106, 108, 1264CD, 1454CD, 1468CD, 544CD, 670CD, 672CD, 674CD, 676CD, 678CD, 680CD, 682CD, 684CD, 686CD, 688CD, 690CD, 692CD, 694CD, 696CD, 698CD, 700CD, 724CD
EDS, 76, 80, 110, 122, 142, 338, 368, 424, 1102CD, 1168CD, 1170CD, 1192CD, 1334CD, 1446CD, 1466CD, 1468CD, 1470CD, 1478CD, 1480CD, 1598CD, 1606CD, 1612CD, 614CD, 642CD, 714CD, 918CD, 992CD
Education, 454, 458, 1566CD, 1568CD, 1570CD, 1578CD
EELS, 16, 18, 62, 66, 70, 74, 128, 284, 332, 338, 372, 384, 418, 422, 464, 1114CD, 1146CD, 1152CD, 1170CD, 1178CD, 1234CD, 1334CD, 1356CD, 1360CD, 1384CD, 1396CD, 1420CD, 1590CD, 1600CD, 570CD, 572CD, 574CD, 576CD, 578CD, 580CD, 582CD, 590CD, 592CD, 594CD, 596CD, 598CD, 604CD, 608CD, 610CD, 614CD, 616CD, 620CD, 624CD, 628CD, 636CD
EFM, 1206CD
EFTEM, 72, 128, 1174CD, 1194CD, 1610CD, 586CD, 592CD, 594CD, 614CD, 626CD, 638CD
EGF, 1012CD
Elasticity, 1000CD, 1130CD
Elctroless Ni, 1286CD
Electric fields, 538CD
Electrochemical deposition, 420
Electromagnetic Field, 38
Electron beam curing, 1232CD
Electron Beam Heating, 788CD
Electron crystallography, 94, 100, 102, 104, 308, 660CD, 840CD, 858CD, 862CD
Electron diffraction, 90, 96, 278, 386, 392, 1158CD, 1176CD, 1216CD, 660CD, 684CD, 696CD, 698CD, 752CD, 754CD, 860CD, 862CD
Electron Holography, 22, 24, 28, 30, 32, 34, 36, 38, 42, 1348CD, 512CD, 514CD, 516CD, 518CD, 520CD, 522CD, 524CD, 526CD, 532CD, 536CD, 538CD
Electron irradiation, 1418CD
Electron Nanodiffraction, 658CD
Electron optics, 6, 484CD, 486CD, 586CD
Electron range, 704CD
Electron spectrometry, 1234CD, 616CD
Electron Tomography, 1104CD, 1610CD, 792CD, 882CD
Electronic devices, 574CD
Electronic Structure, 62, 64, 1374CD
Electropolishing, 1332CD
Electropulsing, 1596CD
Electrospinning, 1230CD
Electrostatics, 1140CD
Elemental mapping, 1356CD, 590CD, 594CD, 626CD, 632CD
ELNES, 1592CD, 1594CD, 594CD, 602CD, 606CD
Elongation, 206
EM history, 4
Embedding, 144, 192, 734CD, 914CD
Emulsion, 786CD
Enamel, 932CD
Endometrium, 228
Endothelial, 1046CD
Energy Contrast, 626CD
Energy filtering, 68, 72, 74, 372, 1610CD, 612CD, 718CD, 882CD
Energy resolution, 602CD
Entomology, 1582CD
Entrapment, 996CD
Envelopment, 986CD
Environment, 478CD
Environmental TEM, 422
Epilayer, 302
Epitaxy, 1364CD, 1374CD
EPMA, 82, 426, 428, 438, 1270CD, 1460CD, 1488CD, 1490CD, 1562CD, 488CD
Epoxy, 1232CD, 884CD
Escherichia coli, 976CD
ESEM, 426, 446, 448, 1226CD, 1228CD, 1478CD, 1480CD, 1494CD, 1534CD, 1538CD, 760CD, 958CD, 960CD, 990CD
ESI, 74, 128, 1610CD, 594CD, 634CD
Etching, 362, 760CD
ETEM, 600CD
Ets-1, 994CD
Euler Space, 686CD
EXAFS, 608CD
Exchange Coupling, 366, 400
Excitations, 460
EXEELFS, 1234CD, 616CD
EXELFS, 608CD
Exit wave reconstruction, 466CD, 468CD
Exochitinase, 256
Explosives, 1520CD
Extraction, 804CD

F

Faceting, 1400CD, 1404CD
Facilities management, 1580CD
Failure analysis, 326, 546CD
Far-infrared, 1524CD
Fe, 1326CD
Fe-Nd-B, 1334CD
Fe-SiO2, 1366CD
FE-TEM, 1106CD
Fe-Zr, 1310CD
Fe3O4(111), 1158CD
FeCrAlY, 1318CD
FEG, 486CD, 584CD, 586CD
FEG-SEM, 122, 140, 182, 1132CD, 670CD, 676CD, 708CD, 710CD, 726CD
FEG-STEM, 292
FePt, 364
Ferrite, 356, 1326CD
Ferroelectric, 22, 1162CD, 538CD, 664CD, 774CD
Ferromagnetic, 302, 388, 1340CD, 1350CD
FFM, 1206CD
FIB, 44, 46, 48, 50, 52, 54, 56, 58, 60, 180, 188, 390, 1142CD, 1144CD, 1212CD, 1266CD, 1550CD, 542CD, 544CD, 546CD, 548CD, 550CD, 554CD, 556CD, 558CD, 560CD, 562CD, 564CD, 566CD, 568CD, 672CD

Fiber, 178, 192, 196, 240, 1352CD
FIBM, 1186CD
Fibrinogen, 260
Fibroblasts, 898CD
Fibronectin, 1022CD
Field Compensator, 588CD
Field Emission, 114, 300, 1110CD
File Format, 646CD
Filipin, 898CD
Filter, 334
Fine structure, 576CD
Fish-eye, 716CD
Flexible phases, 306
Fluorescence microscopy, 128, 198, 266, 270, 1022CD, 1030CD, 1486CD, 1508CD, 1528CD, 740CD, 896CD, 908CD, 950CD, 962CD, 974CD
Fluorite-related structures, 1424CD
Fluxons, 530CD
Fly-ash, 1278CD
Focal pair, 216
Focal-series reconstruction, 34
Food, 236, 238, 786CD, 958CD
Force spectroscopy, 172
Formalin, 736CD
Formation product, 1552CD
Formulations, 1528CD
Fourier methods, 32
Fractal, 242, 244, 1574CD, 496CD
Fracture, 548CD
Fracture Toughness, 1314CD
Freeze fracture, 130, 278
Freeze-substitution, 880CD
Fresnel, 24
Friction, 768CD
Frozen-hydrated, 490CD
FTIR, 1506CD, 1514CD, 1526CD
Fuse, 1184CD

G

GaAs, 1190CD, 1200CD, 1204CD
Gallium, 52
Galvannealing, 188, 358
Gamma', 1432CD
GaN, 1198CD, 1202CD, 1206CD, 1208CD, 572CD
Ganglioglioma, 938CD
Gap Junctions, 130, 892CD
Garnet, 378
Gas cluster ion beam, 1138CD
GaSb, 1200CD, 1258CD
Gastroenteritis, 934CD, 980CD
Ge-Si, 1490CD
Gene expression, 202
Generalized cylinder model, 1018CD
Genetic analysis, 1054CD
Genotoxicity, 998CD
Gentamicin, 956CD
Gerchberg-Saxton Algorithm, 98
GFP, 970CD
Glass, 370, 1312CD, 1494CD, 604CD, 808CD
Glial, 1012CD
Glioblastoma, 444
Glycogen, 886CD
Glycogenosis, 928CD, 930CD
GMR, 382, 1372CD
Gold labeling, 254
Gold labelling, 1030CD, 506CD
Gold ore, 1560CD
Golgi complex, 986CD
Grain Boundary, 64, 108, 1166CD, 1170CD, 1266CD, 1356CD, 1400CD, 1402CD, 1404CD, 1442CD, 1452CD, 1544CD, 1598CD, 578CD, 656CD
Grain growth, 1244CD, 1400CD
Grain size, 348, 368, 1308CD, 656CD, 672CD
Granular Film, 30, 1366CD
Graphite, 610CD
Grinding wheels, 334
Guard cells, 1090CD
Guinea Pig, 736CD

H

HAADF, 14, 1164CD, 1176CD, 1250CD, 1558CD, 492CD, 494CD
Hair, 1008CD, 1058CD, 922CD
Halloysite, 1552CD
Halophyte, 1082CD
Hardness, 1120CD, 1444CD
HBLO, 566CD
HDDR, 1436CD
HDL, 942CD
Heat Exchanger, 346
Heavy Metal Sequestration, 976CD
Helical reconstruction, 856CD
Helladic II Period, 1070CD
Hematite, 800CD
Hepatitis A, 730CD
Hepatocytes, 740CD
Herpes virus, 844CD, 986CD
Heterogeneity, 1490CD
Heterophase interface, 574CD
HFTEM, 1198CD
HHT, 890CD
High Current, 1236CD
High energy, 584CD
High Pressure Freezing, 252
High Speed Pulse Processor, 642CD
High Temperature Alloys, 1492CD
High-pressure freezing, 880CD
High-Tc superconductors, 530CD
HIS-TAG, 832CD
Holography, 40, 60, 168, 398, 508CD
Homogeneity, 1490CD
Hormonal regulation, 886CD
Host cell invasion, 838CD
Host-pathogen interaction, 250
Hot ductility, 354
Hot Mill, 360
Hot Stage, 408, 1152CD
Hot Wire, 1196CD
HREM, 72, 95, 122, 186, 368, 392, 416, 418, 434, 1108CD, 1152CD, 1164CD, 1204CD, 1224CD, 1360CD, 1418CD, 466CD, 492CD, 494CD, 504CD, 584CD, 618CD
HREM, 10, 332, 402, 1158CD, 1160CD, 1166CD, 1178CD, 1246CD, 1348CD, 1376CD, 1394CD, 1428CD, 1548CD, 474CD, 480CD, 540CD
HSS roll, 360
Humidity, 1156CD
HVEM, 1118CD, 630CD
Hydration, 1066CD, 958CD
Hydrides, 342
Hydrocarbon, 720CD
Hydrogenation, 1436CD
Hypertension, 222

I

IASCC, 1422CD
ICAM-1, 822CD
Ice, 238, 246, 1398CD, 1544CD, 1546CD
Ice cream, 246
III-V compound, 1242CD
Image analaysis, 134, 236, 348, 350, 462, 1044CD, 1304CD, 1306CD, 1326CD, 1570CD, 1574CD, 500CD, 502CD, 740CD, 766CD, 808CD, 908CD, 922CD
Image analysis, 210
Image processing, 492CD, 504CD, 506CD, 856CD, 946CD
Image Quality, 692CD
Image reconstruction, 98, 208, 212
Image Simulation, 32, 1108CD, 1140CD, 1394CD, 490CD, 620CD
Imaging filter, 630CD
Immunocytochemistry, 156, 270, 734CD
Immunoelectron microscopy, 146, 160, 162, 1002CD, 1030CD, 1084CD, 726CD, 730CD, 934CD
Immunofluorescence, 130, 730CD
Immunogold, 126, 1074CD, 732CD, 734CD
Immunohistochemistry, 736CD, 818CD, 906CD, 916CD, 934CD
Impact Compression, 1448CD
Implants, 748CD
Impurity location, 1546CD
In situ, 412, 414, 416, 418, 420, 1332CD, 1382CD, 1384CD, 1388CD, 1396CD, 1398CD, 1400CD, 1406CD, 1410CD, 1424CD, 600CD, 760CD, 772CD, 796CD, 916CD
In vivo, 276, 282, 286, 902CD
InAs, 1190CD
Incommensurate, 306, 310
Incunabula, 1532CD
Induction mapping, 1344CD, 510CD
Industrial, 786CD
Inelastic scattering, 622CD
Infectious diseases, 964CD, 980CD
Information extraction, 440, 1554CD, 648CD
Information limit, 12, 480CD
Infrared microscopy, 442, 1504CD, 1508CD, 1510CD, 1512CD, 1526CD, 1530CD
InGaN, 1206CD, 1208CD
Inhibition layer, 358
InN, 628CD
Inner ear, 1008CD
Insulating materials, 1460CD
Insulators, 426, 448
Insulin, 200
Interaction, 822CD
Intercellular junctions, 1004CD
Interconnects, 626CD
Interdiffusion, 1404CD
Interfaces, 18, 98, 386, 1100CD, 1160CD, 1164CD, 1434CD, 502CD, 578CD
Intergranular fracture, 354
Intergranular glassy phase, 1250CD
Intergranular segregation, 1606CD
Intermetallic, 1286CD, 1322CD, 1376CD
Intracellular communication, 892CD
Inverse photoemission, 394
Inverse pole figure, 686CD, 688CD
Ion Etching, 552CD
Ion irradiation, 1418CD
Ionic conductor, 1164CD
IQ map, 700CD
IR microscopy, 1524CD
Iron biominerals, 752CD, 754CD
Iron nitride, 1418CD
Iron precipitation, 800CD
Irradiation, 292, 1420CD
ISO, 816CD

J

Jet Fuel, 798CD
Jump-ratio, 590CD
Junction, 1402CD, 940CD

K

K-12 Teaching, 454, 1582CD
Kelvin Probe Microscopy, 1534CD
Keratinocyte, 1002CD
Kidney, 936CD
Kidney, 226, 746CD
Kinesin, 210
kinetics, 244, 422
Knee cartilage, 902CD
Knockout, 940CD
Koehler Illumination, 486CD

L

Labeling, 124, 194, 832CD
Laboratory design, 478CD
Laminin, 934CD
Langmuir Film, 1226CD
Larvae, 164
Laser Diodes, 1202CD, 550CD
Laser floating zone technique, 1352CD
Laser scanning cytometry, 744CD
Laser scanning microscopy, 264
Lattice spacing, 1204CD
LCMO films, 1342CD
Lead-Free, 1456CD
Lens, 1004CD
LFM, 772CD
Lichens, 1078CD
Lift out, 1266CD, 566CD
Light microscopy, 1272CD, 1322CD, 1494CD, 930CD
Lightning Strikes, 1236CD
Line broadening, 340
Line pipe, 362, 1284CD
Line profiles, 340
Lipid, 874CD, 898CD

Lipoproteins, 836CD
Liquid Crystal, 1228CD
Liquid environments, 412
Liquid Helium, 854CD
Liquid metal embrittlement, 344
Lithium Polymer Battery, 1306CD
Lithography, 1022CD, 1132CD
Live tissue, 276
Live-cell, 1046CD
Liver, 886CD
LMIS, 1212CD
Local atomic order, 1220CD
Local Electrode, 1094CD
Localization, 158
Lorentz microscopy, 26, 296, 376, 380, 1344CD, 1348CD, 510CD, 512CD, 514CD, 522CD, 524CD, 526CD, 528CD, 530CD
Low Energy Microanalysis, 80, 1466CD, 1472CD, 714CD
Low loss, 598CD
Low temperature, 120, 388, 722CD, 786CD
Low vacuum, 448, 1478CD
Low voltage microscopy, 110, 118, 120, 1474CD, 1540CD, 704CD, 706CD
Low-dimensional systems, 312
Low-energy ion-polishing, 1204CD
Low-k dielectric, 1194CD
LR White, 734CD
LVSEM, 112, 122
Lysosome, 894CD, 898CD

M

Machining, 568CD
Macromolecular complexes, 214, 876CD
Macrophage, 894CD
Magetotactic, 168
Magnesia, 1268CD
Magnet, 1334CD, 1358CD, 510CD, 716CD
Magnetic anisotropy, 1376CD
Magnetic domains, 378, 512CD
Magnetic films, 1344CD
Magnetic Fine Structure, 30
Magnetic force microscopy, 378, 1338CD, 1370CD
Magnetic Materials, 370, 390, 396, 1336CD
Magnetic nanoparticles, 1362CD
Magnetic Nanowires, 1340CD, 1368CD
Magnetic particle, 364
Magnetic thin films, 368, 382
Magnetism, 40, 174, 370, 374, 376, 398
Magnetite, 1390CD
Magnetization, 398
Magnetoresistance, 1370CD
Magnetostatics, 1372CD
Magnets, 366
Maize, 1068CD, 1072CD, 1080CD
Manganese perovskites, 394
Manganite, 384, 388, 574CD
MAP-2, 736CD
Mare, 956CD
Martensite, 1276CD, 700CD
Matrix Correction, 424
MBE, 1222CD
McArdle Disease, 930CD
Measurements, 272
Meat quality, 240
Mechanical Properties, 304, 1314CD, 1330CD
Meiosis, 1052CD
Membrane, 316
Membrane proteins, 840CD, 874CD
Membrane separation, 1118CD
Membranes parasitophorous vacuole, 838CD
MEMS, 1274CD
Meprin, 824CD
Mercury embrittlement, 344
Mesopore, 328
Metal, 320, 1120CD, 916CD
Metal injection molding, 1316CD
Metal matrix composites, 188
Metal oxides, 1590CD
Metal-matrix composites, 342
Metal/ceramic interfaces, 64
Metallic Clusters, 1122CD
Metallic glass, 608CD
Metallic sulphides, 310
Metallography, 350, 1256CD, 1278CD, 1290CD, 1296CD, 1298CD, 1302CD, 916CD
Meteorites, 1296CD, 1550CD
MeV ion-implantation, 1182CD
Mg Alloys, 1324CD
Mg-Ta-oxide, 1414CD
MgB2, 1364CD
MgO(111), 1158CD
Microanalysis, 338, 440, 464, 1298CD, 1474CD, 1486CD, 1554CD
Microbe, 966CD
Microcalorimeter, 78, 80, 82, 714CD
Microcirculation, 282
Microdiffraction, 1438CD
Microelectronics, 640CD
Microinhomogeneity, 1300CD
Microsatellites, 1054CD
Microscope control, 872CD
Microspectroscopy, 1526CD
Microstructure, 242, 352, 1156CD, 1196CD, 1210CD, 1260CD, 1268CD, 1270CD, 1286CD, 1294CD, 1296CD, 1316CD, 1328CD, 1332CD, 1342CD, 1364CD, 1400CD, 1456CD, 674CD
Microtomy, 878CD
Microtubule, 210
Microvasculature, 224, 232
Microvoid coalescence, 354
Microwave, 144, 146, 148, 150, 152, 156, 158, 160, 162, 736CD
Milkfat, 244
Minerals, 1558CD
Misfit dislocations, 1200CD
Mitochondria, 1002CD, 1006CD, 1056CD, 878CD
MLLS, 592CD
Modeling, 1166CD
Modulated Structures, 306, 308
Mold fluxes, 1272CD
Molecular analysis, 1526CD
Molecular imaging, 412
Molecular motor, 820CD
Molecular Shuttles, 1092CD
Mollusk, 164
Molten salt method, 1414CD
Monochromator, 68, 70, 464, 1592CD, 584CD
Monolayer, 770CD
Monomeric cyanine, 1058CD
Monte Carlo, 116, 430, 1172CD, 1462CD, 1474CD, 1498CD, 702CD, 704CD
Morphological Instability, 1410CD
Morphology, 324, 1238CD, 776CD, 946CD
Morphometry, 1078CD
Mossbauer Spectroscopy, 396, 1366CD
Mouse, 932CD, 940CD
Movement, 972CD
MRI, 128, 992CD
Mucus glands, 910CD
Multi-photon, 264, 268, 280, 1036CD, 1048CD
Multi-pole corrector, 6
Multi-slice, 490CD
Multilayer, 30, 428, 1164CD, 1376CD, 1488CD, 858CD
Multimedia, 1566CD
Multiphase material, 184
Multiple Scattering, 572CD
Multiple sclerosis, 834CD
Multiresolutional Analysis, 1584CD
Multislice theory, 1602CD
Multivariate statistical analysis, 440, 1554CD, 648CD
Murataite, 1424CD
Muscle fibers, 240
Mustard gas, 900CD
Mutant, 1072CD
Mycoparasite, 256
Myophosphorylase deficiency, 930CD

N

Nanobacteria, 166
Nanobelts, 328, 1608CD
Nanocomposite, 1354CD, 1378CD
Nanocrystalline, 1120CD, 1146CD, 1148CD, 1412CD
Nanodiamond, 1110CD, 1182CD
Nanodiffraction, 402
Nanofabrication, 54, 1118CD, 1132CD, 1142CD
Nanofiber, 770CD
Nanoforms, 166
Nanogold, 160, 832CD
nanoindentation, 554CD
Nanomaterials, 1412CD
Nanoparticle, 406, 1116CD, 1130CD, 1134CD, 1140CD, 1144CD, 1240CD, 1348CD, 1360CD, 1394CD, 1600CD, 786CD
Nanoparticles, 1124CD, 1150CD
Nanopattern, 1020CD
Nanopores, 1372CD
Nanoprobe, 298
Nanoscience, 294
Nanostructure, 318, 1122CD, 1126CD, 1350CD, 1354CD, 1358CD, 1362CD, 1372CD, 1452CD, 1596CD, 658CD
Nanotechnology, 1092CD, 1104CD, 1136CD
Nanotip, 118, 520CD
Nanotubes, 300, 1128CD, 1384CD
Nanowire, 304, 1240CD
NaOH, 332
Natural history, 182
Nb-silicide, 1454CD
NdFeB, 716CD
Near edge structures, 580CD
Near infrared, 1516CD
Near-Field Microscopy, 1518CD, 1530CD
Neck formation, 1246CD
Necrosis, 444
Negative Stain, 1026CD, 980CD
Neuromuscular, 940CD
Neuron, 1000CD, 1012CD, 938CD
Neutron diffraction, 342
Neutron scattering, 374
NEXAFS, 602CD
NH3, 332
Ni, 52, 1412CD, 608CD, 832CD
Nickel Superalloys, 1098CD, 1280CD
Nicrobraz 51, 346
Nile red, 898CD
NIR, 286
Nodule count, 1326CD
Non-Conductive Specimen, 1462CD
Non-uniformities, 604CD
Nondestructive Technique, 788CD
Nonlinear microscopy, 268
Nonlinear optical activity, 104
Nonwovens, 806CD
NSOM, 322, 1020CD, 1518CD, 1530CD
NTA, 832CD
Nuclear Body, 728CD
Nuclear Microprobe, 1082CD
Nuclear pores, 986CD
Nuclear ultrastructure, 728CD
Nuclear Waste, 1310CD, 1312CD
nucleation, 1448CD
Nucleus, 74, 728CD

O

Oblique illumination, 1068CD
ODF, 686CD
Oil sands, 804CD
OIM, 678CD, 694CD
Olivine, 1606CD
Olivine-Orthopyroxene, 680CD
One-way effect, 1320CD
ONO, 1178CD
Oocyte, 924CD
Opisthobranch, 1086CD
Optical gratings, 1132CD
Optical Microscopy, 282, 1328CD, 1330CD
Order-disorder transition, 1424CD
Ordered superlattices, 1124CD
Organelle, 274
Organic adsorbates, 766CD
Organochlorine, 924CD
Orientation, 1522CD
Oscillating diamond knife, 324
Osmolytes, 834CD
Osmotic regulation, 1090CD
OsO4, 1238CD
Osteogenic Protein, 778CD
Osteointegration, 748CD
Osteosarcoma, 950CD
Overexpression, 984CD
Ovulation, 888CD
Oxidation, 338, 1318CD, 1406CD, 600CD
Oxides, 62, 66, 580CD
Oxynitride, 596CD

P

P, 608CD
p-n junctions, 42
P/B ratio, 1612CD
p53, 140, 1002CD
Packaging, 1292CD
Paper, 178, 196
Paraffin, 736CD
Parasites, 838CD
Particle analysis, 1270CD, 498CD, 802CD
Particles, 1500CD
Pathogenic calcification, 166
Pathology, 152, 738CD, 900CD, 904CD, 910CD, 914CD, 944CD
Patterned arrays, 1344CD
Pb Film, 1410CD
PCB, 1256CD, 924CD
Pd, 410, 1112CD, 1154CD, 608CD
Pearlite, 1326CD
PEELS, 1608CD, 624CD
Pepsin, 1018CD
Perikaryon, 938CD
Permeability, 398, 1046CD
Perovskite, 1374CD, 582CD, 618CD, 664CD
Petrography, 1556CD
pH study, 776CD
Pharmaceutical Science, 738CD, 742CD
Phase contrast, 10, 540CD, 682CD, 864CD
Phase images, 466CD, 538CD
Phase retrieval, 1344CD, 474CD, 510CD
Phase transformation, 52, 294, 1136CD, 1446CD, 1448CD
Phases, 782CD
Phason disorders, 1218CD
Phloem, 1088CD
Phonon scattering, 1602CD
Phosphodiesterase, 138
Phospholipidosis, 740CD
Phosphorus, 978CD
Photonic, 316, 320, 322
Photopolymerization, 1036CD
Physiology, 230
Phytoferritin, 752CD, 754CD
Piezoelectric, 318
Pigments, 1532CD
Pine, 1088CD
Pitting, 1282CD
PL, 332
Plant Pathology, 252, 254, 256, 1090CD, 962CD, 966CD, 968CD
Plasmid DNA, 988CD
Plasmodium falciparum, 948CD
Plasmon, 636CD
Plastic sections, 882CD
Plasticity, 1000CD
Plutonium, 1476CD
PM Aluminum Composites, 500CD, 502CD
PML, 728CD
Point projection microscope, 520CD
Poisson equation, 1172CD
Polar oxide interfaces, 1158CD
Polarimetry, 322
Polarization Microscopy, 538CD, 798CD
Polarizing light microscopy, 240, 242, 1020CD, 896CD
Pole Figures, 686CD
Pollen dispersion, 1080CD
Poly(DTE carbonate), 1022CD
Polymer, 190, 324, 326, 1236CD, 1238CD, 758CD
Polymerization, 1392CD
Polyphosphate Bodies, 978CD
Polypropylene, 1522CD
Polysaccharide, 238, 836CD
Polysilicon film, 1210CD, 666CD
Polytype, 1428CD
Pores, 336, 1248CD
Porin, 892CD
Porosity, 1242CD, 784CD
Porous alumina, 1246CD
Porous Silicon, 332
Porous silicon nitride, 1250CD
Position-tagged Spectrometry, 350
Post-embedding, 732CD
Post-irradiation annealing, 1422CD
Powders, 1260CD
Precipitates, 1444CD, 624CD
Precipitation, 294, 1098CD, 1432CD
Precision, 94, 432
Primary structure, 362
Prion, 1506CD
Probe current measrument, 1538CD
Process control, 236
Processing, 142, 786CD, 804CD
Programmed Cell Death, 248, 970CD
Prostate, 1056CD
protease, 204
Protein, 258, 262, 274, 1040CD, 1092CD, 728CD, 762CD, 776CD, 834CD, 840CD, 850CD, 862CD, 876CD, 892CD
Pt, 812CD
Pt-Sn catalysts, 1150CD
PtO2, 1600CD
Pulp, 178
Pulse Dispersion, 1036CD
PVLO, 1266CD, 566CD
Pyrochlores, 1136CD

Q

QCBED, 90, 650CD, 654CD
Quadrupole, 1212CD
Quantification, 272, 428, 1482CD, 1516CD
Quantitative analysis, 432, 438, 1472CD, 1484CD
Quantitative EDX Mapping, 292
Quantitative Electron Diffraction, 652CD
Quantitative electron microscopy, 94, 662CD
Quantitative Phase, 534CD
Quantum dots, 10, 1200CD, 712CD
Quantum Well, 1190CD
Quantum Well Interference, 400
Quantum Wires, 1190CD
Quasicrystal, 314, 1218CD, 1220CD, 1224CD

R

R-values, 102
Radiation-induced segregation, 1422CD
Raman microscopy, 288, 444, 1386CD, 1512CD, 1518CD, 1520CD, 1522CD, 1528CD, 1532CD
Random orientation probability, 688CD
Rapid fixation, 828CD
Rat, 228, 748CD
Ray Path, 28
Real Time Microscope, 898CD
Real time processing, 482CD
RecA, 982CD
Receptor, 822CD
Reciprocal space, 668CD
Recombinant adenovirus, 140
Reconstruction, 200
Recording media, 366, 1356CD, 1358CD, 1368CD
Recrystallization, 356
Recursive Pixel Allocation, 350
Redeposition, 50
Reduction, 600CD
Reflection, 240
Refractive index, 1050CD
Refractories, 336
Refractory Brick, 1268CD
Remaining Life Assessment, 1332CD
Remote Microscopy, 454, 1584CD
Reovirus, 846CD
Replica, 360
Reproductive toxicology, 924CD
Resolution, 1524CD
Resonance, 304
Respiratory chain, 1006CD
Retained Austenite, 1330CD
Rethin, 1266CD, 566CD
Review, 818CD
RHEED, 106
Rheology, 242, 244
Ribosome, 206
Rietveld, 1548CD
Rodcoil polymer, 318
Roll microstructure, 360
Ronchigram, 1176CD
Root, 1074CD
Rose criterion, 1588CD
Rough surfaces, 1472CD
RTM, 898CD
Ru-doped, 1146CD
Rubber, 326
RuO4, 1238CD
Rust Fungi, 252

S

SAED, 618CD, 1278CD
Salicylic Acid, 248, 970CD
Salmon, 1054CD
Sample holder, 1026CD, 668CD
Sample preparation, 56, 560CD, 564CD
Sample thickness, 1608CD
SAMs, 768CD
Scales, 1054CD
SCC, 344, 1282CD, 1284CD
Scheduling, 1580CD
Schlieren optics, 1080CD
Scorpion venom, 926CD
SDD, 76, 84
Segregation, 368, 1100CD, 1126CD, 1170CD, 1334CD
Selected reflection imaging, 402, 1354CD
Self assembly, 316, 318, 320, 328, 366, 1020CD, 1122CD, 1124CD, 1134CD, 1368CD, 770CD, 868CD
Semiconductor, 60, 180, 1102CD, 1104CD, 1106CD, 1174CD, 1212CD, 1214CD, 1242CD, 1292CD, 516CD, 518CD, 536CD, 564CD, 626CD
Sensors, 320
Shape memory alloy, 1320CD
Sharpness, 488CD
Sheath, 964CD
Shell formation, 164
SHG, 268
Short range ordering, 1430CD
Si, 50, 1196CD, 790CD
Si/GaAs interface, 1604CD
SiC, 1180CD, 1550CD
SiGe, 1174CD
Sigma 3 regeneration model, 108
Signal Attenuation Correction, 1042CD
Signal trafficking, 892CD
Silicon Drift Detector, 76, 82, 642CD
Silicone, 772CD
SiLK, 1194CD
Silver enhancement, 1030CD
Silver scurf, 964CD
SIMS, 1192CD, 1212CD, 542CD
Single particle analysis, 216, 1576CD, 840CD, 870CD
Single-atom detection, 1614CD
Sintering, 408, 1428CD
Site specific, 1116CD, 1266CD
Skeletal Muscle, 930CD
Skin, 278, 280, 282, 286, 288, 1062CD, 1064CD, 1066CD, 900CD
Slide, 162
Slip bands, 782CD
Smoothing, 1138CD
Sn segregation, 354
Sn-Ag-Cu, 1456CD
Sn-Si Quantum Dots, 1114CD
Snow, 722CD
SNR, 1588CD
Soft condensed matter, 868CD
Soil, 990CD
Solder, 1292CD, 1456CD
Solidification Patterns, 1260CD
Spatial Feature Analysis, 1556CD
Spatial resolution, 670CD
Specimen holder, 866CD
Specimen preparation, 44, 46, 48, 60, 158, 180, 1204CD, 1346CD, 566CD, 1256CD, 546CD
Spectrometer, 76
Spectroscopy, 1064CD, 1386CD, 1518CD, 630CD
Spectrum Image, 350
Spectrum-line, 596CD
Spherical aberration, 1048CD, 1050CD
SPM, 174, 1000CD, 1206CD, 742CD, 772CD, 780CD
Sporobolus, 1082CD
Sputter Coater, 812CD
Sputtering, 1376CD
SSM, 1324CD
Stabilizers, 246
Stacking faults, 1166CD
Staining, 190, 192, 916CD
Standard Reference Material, 1298CD
Standardless analysis, 1464CD
Standards, 1476CD

Steel, 188, 352, 354, 356, 358, 360, 362, 1252CD, 1276CD, 1290CD, 1328CD, 1330CD, 1458CD, 548CD, 624CD, 700CD
STEM, 4, 12, 14, 16, 18, 20, 200, 376, 384, 1106CD, 1148CD, 1174CD, 1176CD, 1192CD, 1208CD, 1222CD, 1246CD, 1250CD, 1396CD, 1606CD, 470CD, 476CD, 576CD, 578CD, 590CD, 912CD, 976CD
STEM, 1598CD
Stem cells, 1012CD
Stereomicroscopy, 1382CD
Sterlet, 220
STM, 394, 1222CD, 766CD, 780CD
Stobbs-factor, 34
Storage diseases., 928CD
Strain, 670CD, 794CD
Strain energy, 1434CD
Strain relaxation, 1200CD
Strained boundaries, 670CD
Straining, 676CD
Stratum Corneum, 1066CD
Stray radiation, 1466CD
Streptomycetes, 968CD, 996CD
Structure, 236, 1254CD, 1300CD, 664CD, 822CD
Structure factor, 650CD
Structure refinement, 88
Sub-Angstrom, 480CD
Substrates, 1494CD
Substructure, 1458CD
Sulfide, 1156CD
Sulfide ore, 1560CD
Sulphate, 800CD
Superalloys, 1432CD
Superconductivity, 26, 86, 380, 384, 388, 392, 1338CD, 1484CD, 514CD, 522CD, 524CD, 526CD, 528CD, 652CD
Superficial veins, 234
Superlattice, 384, 554CD
Superparamagnetic, 320
Superplastic deformation, 1450CD
Supersphere, 1434CD
SuperSTEM, 470CD
Supramolecular, 318
Surface alloying, 1440CD
Surface structures, 312, 666CD
Surfaces, 408, 1258CD, 706CD, 776CD
Surfactants, 868CD
SWNT, 638CD
Symbiosis, 1086CD
Symmetry, 1220CD
Symphysis Pubis, 232
Synchrotron, 1508CD, 1524CD
Synechococcus leopoliensis, 978CD

T

Tantalum, 1500CD
Target, 1256CD
TbNiAl, 1436CD
TEC, 1010CD
Tensile testing, 1274CD
Tepary bean, 1074CD
Terbium, 1360CD
Tescan, 808CD
Test images, 1574CD
Textiles, 192, 806CD
Texture, 1244CD, 1354CD, 1378CD, 676CD, 686CD, 688CD, 696CD
TFT, 1210CD
Theoretical calculation, 606CD
Thermo-elastic response, 1340CD
THG, 268
Thin Film, 56, 296, 1120CD, 1160CD, 1274CD, 1320CD, 1404CD, 1438CD, 1484CD
Thin Wall, 1326CD
Thiobacillus ferroxidans, 756CD
Thixocasting, 1324CD
Thrombosis, 260
Thymic epithelium, 1010CD
Ti-based alloy, 1450CD
Ti-Mo alloy, 1448CD
TiAl, 1322CD
TiAl alloy, 1096CD, 1294CD
Time-lapse imaging, 170
TiN, 674CD
Tin Dioxide, 1154CD, 1168CD
Tire, 326
Tissue, 1024CD, 1056CD, 1502CD
Tissue, 878CD
Titanate, 1420CD
Titania, 1130CD
Titanium, 782CD
Titanium carbide, 342
Titanium Dioxide, 1170CD
Titanium Nitride, 748CD
To-Pro-3, 1058CD
Tomography, 168, 1126CD, 842CD, 880CD
Tracheid, 1088CD
Training aids, 1578CD
Transcription regulation, 202
Transformation, 1262CD
Transgene, 128, 1084CD
Transient structure, 828CD
Transistors, 40
Transition Metal Oxides, 1592CD, 582CD
Translation, 206
Transmembrane signalling, 200
Transmissivity, 612CD
Trehalose, 862CD
Tribology, 768CD
Tripod method, 774CD
Tumor, 906CD, 936CD
Turtle, 888CD
TVCV, 972CD
Tween, 244
Twinning, 108, 302, 554CD, 666CD
Two-way effect, 1320CD

U

UHV, 410, 414
Ultra-fine grain, 676CD
Ultramicrotomy, 190, 324, 1238CD
Ultrasound Imaging, 1028CD
Ultrastructure, 1006CD, 1028CD, 1090CD, 776CD, 926CD, 932CD, 938CD, 984CD
Upper detector, 708CD
Uranium, 750CD
Urban Tree, 1088CD
Urinary stone, 746CD
Uterus, 956CD

V

V2O5, 602CD
Valence, 460
Vapor hydration test, 1312CD
Variable cooler, 154
Vascular, 138, 890CD
Vasculature, 1024CD
Vega, 808CD
VEGF, 1046CD
Venous valves, 234
Vesicle, 892CD
Via, 790CD
Virology, 904CD
Virus, 208, 1016CD, 1116CD, 822CD, 852CD, 912CD, 966CD, 972CD
Voids, 1198CD, 788CD
Vortex, 26, 380, 1338CD
VPSEM, 450, 452, 1536CD, 808CD

W

W, 790CD
W-SEM, 670CD
Water, 1066CD, 1390CD
Wattage control mechanism, 154
Wavelets, 1584CD
WDS, 1490CD, 714CD
Wear, 1440CD, 802CD
Weathering product, 1552CD
Web, 1576CD
Weld, 1252CD, 1452CD
wet-ETEM, 412
Wide-field, 266
Wien-filter, 68
Wilm's Tumor, 936CD
Wire Bonds, 1186CD

X

X-Ray, 432, 1468CD, 1498CD
X-ray analysis, 1480CD
X-ray detector, 82, 84
X-ray microanalysis, 120, 430, 434, 1462CD, 1464CD, 1470CD, 1496CD, 640CD
x-ray microscopy, 298, 488CD, 826CD
X-ray spectroscopy, 436, 1150CD, 644CD
X-Ray Tomography, 784CD, 826CD
Xenopus laevis, 886CD
XPS, 1366CD
XRD, 278, 340, 694CD
XY model, 1216CD
Xylem, 1088CD

Y

YBa2Cu3O7-x, 652CD
Yo-Pro-1, 1058CD

Z

Z-contrast, 16, 18, 66, 384, 1114CD, 1190CD, 1218CD, 1220CD, 1396CD, 1604CD, 1614CD, 578CD
Zebrafish, 1058CD
Zernike phase plate, 864CD
Zinc process, 800CD
zirconia, 1216CD
Zn, 800CD
ZnO, 494CD
Zr alloys, 342, 1288CD
Zr-2.5Nb, 1308CD
Zwitterionic detergents, 952CD

Proceedings -- Microscopy and Microanalysis 2002

Technical Documentation (Macintosh)

Macintosh System Requirements

* Macintosh System 7 or better Operating System
* Minimum of 65Mb of RAM
* CD-ROM Drive
* Internet Browser (Microsoft Internet Explorer 4.0 or higher, or Netscape Communicator 4.0 or higher)
* Adobe Acrobat Reader 4.0 or higher

Usage Instructions

There is no installation of the Proceedings -- Microscopy and Microanalysis 2002 necessary. If you have an Internet Browser and Adobe Acrobat Reader, you should have no difficulty using the Proceedings CD-ROM if you follow these instructions.

Netscape Communicator 4.0 or higher
1. Insert the Proceedings -- Microscopy and Microanalysis 2002 CD-ROM into your CD-ROM drive.
2. Launch Netscape Communicator 4.0 or higher.
3. From Netscape Communicator, select the "File" menu, and select the "Open >" option from the "File" menu.
4. From the File: Open menu, select "Page in Navigator".
5. In the "Open Page" dialog box that appears, use the drop-down menu at the top left of the dialog box to find and select "Desktop".
5. Find the MAM_PROCEEDINGS CD-ROM, select it and click "Open".
6. Find the file "INDEX.HTM", select it and click "Open".

This will open the Proceedings -- Microscopy and Microanalysis 2002 home page in Netscape Communicator.

Microsoft Internet Explorer
1. Insert the Proceedings -- Microscopy and Microanalysis 2002 CD-ROM into your CD-ROM drive.
2. Launch Microsoft Internet Explorer 4.0 or higher.
3. From Internet Explorer, select the "File" menu, and select the "Open File" option from the "File" menu.
4. In the "Open File" dialog box that appears, use the drop-down menu at the top left of the dialog box to find and select "Desktop".
5. Find the MAM_PROCEEDINGS CD-ROM, select it and click "Open".
6. Find the file "INDEX.HTM", select it and click "Open".

This will open the Proceedings -- Microscopy and Microanalysis 2002 home page in Internet Explorer.

****

Proceedings -- Microscopy and Microanalysis 2002

Technical Documentation (Windows)

Windows System Requirements

* Windows 95, 98, 2000, NT, ME, or XP Operating System
* Minimum of 65Mb of RAM
* CD-ROM Drive
* Internet Browser (Microsoft Internet Explorer 4.0 or higher, or Netscape Communicator 4.0 or higher)
* Adobe Acrobat Reader 4.0 or higher

Usage Instructions

There is no installation of the Proceedings -- Microscopy and Microanalysis 2002 necessary. If you have an Internet Browser and Adobe Acrobat Reader, you should have no difficulty using the Proceedings CD-ROM if you follow these instructions.

Microsoft Internet Explorer
1. Insert the Proceedings -- Microscopy and Microanalysis 2002 CD-ROM into your CD-ROM drive.
2. Launch Microsoft Internet Explorer 4.0 or higher.
3. From Internet Explorer, select the "File" menu, and select the "Open" option from the "File" menu.
4. In the "Open" dialog box that appears, type: "D:\INDEX.HTM" (where "D:" is the letter name assigned to your CD-ROM drive; if your CD-ROM drive is assigned to a letter other than "D:", such as "E:", substitute that letter assignment for "D:").
5. Click "OK".

This will open the Proceedings -- Microscopy and Microanalysis 2002 home page in Internet Explorer.

Netscape Communicator 4.0 or higher
1. Insert the Proceedings -- Microscopy and Microanalysis 2002 CD-ROM into your CD-ROM drive.
2. Launch Netscape Communicator 4.0 or higher.
3. From Netscape Communicator, select the "File" menu, and select the "Open Page" option from the "File" menu.
4. In the "Open Page" dialog box that appears, type: "D:\INDEX.HTM" (where "D:" is the letter name assigned to your CD-ROM drive; if your CD-ROM drive is assigned to a letter other than "D:", such as "E:", substitute that letter assignment for "D:").
5. Click "OK".

This will open the Proceedings -- Microscopy and Microanalysis 2002 home page in Netscape Communicator